Springer Series in Statistics

Springer Series in Statistics

Andersen/Borgan/Gill/Keiding: Statistical Models Based on Counting Processes.
Atkinson/Riani: Robust Diagnostic Regression Analysis.
Atkinson/Riani/Cerioli: Exploring Multivariate Data with the Forward Search.
Berger: Statistical Decision Theory and Bayesian Analysis, 2nd edition.
Borg/Groenen: Modern Multidimensional Scaling: Theory and Applications, 2nd edition.
Brockwell/Davis: Time Series: Theory and Methods, 2nd edition.
Bucklew: Introduction to Rare Event Simulation.
Cappé/Moulines/Rydén: Inference in Hidden Markov Models
Chan/Tong: Chaos: A Statistical Perspective.
Chen/Shao/Ibrahim: Monte Carlo Methods in Bayesian Computation.
Coles: An Introduction to Statistical Modeling of Extreme Values.
David/Edwards: Annotated Readings in the History of Statistics.
Devroye/Lugosi: Combinatorial Methods in Density Estimation.
Efromovich: Nonparametric Curve Estimation: Methods, Theory, and Applications.
Eggermont/LaRiccia: Maximum Penalized Likelihood Estimation, Volume I: Density Estimation.
Fahrmeir/Tutz: Multivariate Statistical Modelling Based on Generalized Linear Models, 2nd edition.
Fan/Yao: Nonlinear Time Series: Nonparametric and Parametric Methods.
Farebrother: Fitting Linear Relationships: A History of the Calculus of Observations 1750-1900.
Federer: Statistical Design and Analysis for Intercropping Experiments, Volume I: Two Crops.
Federer: Statistical Design and Analysis for Intercropping Experiments, Volume II: Three or More Crops.
Ghosh/Ramamoorthi: Bayesian Nonparametrics.
Glaz/Naus/Wallenstein: Scan Statistics.
Good: Permutation Tests: A Practical Guide to Resampling Methods for Testing Hypotheses, 2nd edition.
Good: Permutation Tests: Parametric and Bootstrap Tests of Hypotheses, 3rd edition.
Gouriéroux: ARCH Models and Financial Applications.
Gu: Smoothing Spline ANOVA Models.
Györfi/Kohler/Krzyżak/ Walk: A Distribution-Free Theory of Nonparametric Regression.
Haberman: Advanced Statistics, Volume I: Description of Populations.
Hall: The Bootstrap and Edgeworth Expansion.
Härdle: Smoothing Techniques: With Implementation in S.
Harrell: Regression Modeling Strategies: With Applications to Linear Models, Logistic Regression, and Survival Analysis.
Hart: Nonparametric Smoothing and Lack-of-Fit Tests.
Hastie/Tibshirani/Friedman: The Elements of Statistical Learning: Data Mining, Inference, and Prediction.
Hedayat/Sloane/Stufken: Orthogonal Arrays: Theory and Applications.
Heyde: Quasi-Likelihood and its Application: A General Approach to Optimal Parameter Estimation.

(continued after index)

Olivier Cappé
Eric Moulines
Tobias Rydén

Inference in Hidden Markov Models

With 78 Illustrations

Springer

Olivier Cappé
CNRS LTCI
GET / Télécom Paris
46 rue Barrault
75634 Paris cedex 13
France
cappe@tsi.enst.fr

Eric Moulines
CNRS LTCI
GET / Télécom Paris
46 rue Barrault
75634 Paris cedex 13
France
moulines@tsi.enst.fr

Tobias Rydén
Centre for Mathematical Sciences
Lund University
Box 118
221 00 Lund
Sweden
tobias@maths.lth.se

Library of Congress Control Number: 2005923551

ISBN-10: 0-387-40264-0
ISBN-13: 978-0387-40264-2

Printed on acid-free paper.

Printed in the United States of America. (MVY)

9 8 7 6 5 4 3 2

springeronline.com

Preface

Hidden Markov models—most often abbreviated to the acronym "HMMs"—are one of the most successful statistical modelling ideas that have came up in the last forty years: the use of hidden (or unobservable) states makes the model generic enough to handle a variety of complex real-world time series, while the relatively simple prior dependence structure (the "Markov" bit) still allows for the use of efficient computational procedures. Our goal with this book is to present a reasonably complete picture of statistical inference for HMMs, from the simplest finite-valued models, which were already studied in the 1960's, to recent topics like computational aspects of models with continuous state space, asymptotics of maximum likelihood, Bayesian computation and model selection, and all this illustrated with relevant running examples. We want to stress at this point that by using the term *hidden Markov model* we do not limit ourselves to models with finite state space (for the hidden Markov chain), but also include models with continuous state space; such models are often referred to as *state-space models* in the literature.

We build on the considerable developments that have taken place during the past ten years, both at the foundational level (asymptotics of maximum likelihood estimates, order estimation, etc.) and at the computational level (variable dimension simulation, simulation-based optimization, etc.), to present an up-to-date picture of the field that is self-contained from a theoretical point of view and self-sufficient from a methodological point of view. We therefore expect that the book will appeal to academic researchers in the field of HMMs, in particular PhD students working on related topics, by summing up the results obtained so far and presenting some new ideas. We hope that it will similarly interest practitioners and researchers from other fields by leading them through the computational steps required for making inference in HMMs and/or providing them with the relevant underlying statistical theory.

The book starts with an introductory chapter which explains, in simple terms, what an HMM is, and it contains many examples of the use of HMMs in fields ranging from biology to telecommunications and finance. This chapter also describes various extension of HMMs, like models with autoregression or hierarchical HMMs. Chapter 2 defines some basic concepts like transition kernels and Markov chains. The remainder of the book is divided into three parts: *State Inference*, *Parameter Inference* and *Background and Complements*; there are also three appendices.

Part I of the book covers inference for the unobserved state process. We start in Chapter 3 by defining smoothing, filtering and predictive distributions and describe the forward-backward decomposition and the corresponding recursions. We do this in a general framework with no assumption on finiteness of the hidden state space. The special cases of HMMs with finite state space and Gaussian linear state-space models are detailed in Chapter 5. Chapter 3 also introduces the idea that the conditional distribution of the hidden Markov chain, given the observations, is Markov too, although non-homogeneous, for both ordinary and time-reversed index orderings. As a result, two alternative algorithms for smoothing are obtained. A major theme of Part I is simulation-based methods for state inference; Chapter 6 is a brief introduction to Monte Carlo simulation, and to Markov chain Monte Carlo and its applications to HMMs in particular, while Chapters 7 and 8 describe, starting from scratch, so-called sequential Monte Carlo (SMC) methods for approximating filtering and smoothing distributions in HMMs with continuous state space. Chapter 9 is devoted to asymptotic analysis of SMC algorithms. More specialized topics of Part I include recursive computation of expectations of functions with respect to smoothed distributions of the hidden chain (Section 4.1), SMC approximations of such expectations (Section 8.3) and mixing properties of the conditional distribution of the hidden chain (Section 4.3). Variants of the basic HMM structure like models with autoregression and hierarchical HMMs are considered in Sections 4.2, 6.3.2 and 8.2.

Part II of the book deals with inference for model parameters, mostly from the maximum likelihood and Bayesian points of views. Chapter 10 describes the expectation-maximization (EM) algorithm in detail, as well as its implementation for HMMs with finite state space and Gaussian linear state-space models. This chapter also discusses likelihood maximization using gradient-based optimization routines. HMMs with continuous state space do not generally admit exact implementation of EM, but require simulation-based methods. Chapter 11 covers various Monte Carlo algorithms like Monte Carlo EM, stochastic gradient algorithms and stochastic approximation EM. In addition to providing the algorithms and illustrative examples, it also contains an in-depth analysis of their convergence properties. Chapter 12 gives an overview of the framework for asymptotic analysis of the maximum likelihood estimator, with some applications like asymptotics of likelihood-based tests. Chapter 13 is about Bayesian inference for HMMs, with the focus being on models with finite state space. It covers so-called reversible jump MCMC algorithms for choosing between models of different dimensionality, and contains detailed examples illustrating these as well as simpler algorithms. It also contains a section on multiple imputation algorithms for global maximization of the posterior density.

Part III of the book contains a chapter on discrete and general Markov chains, summarizing some of the most important concepts and results and applying them to HMMs. The other chapter of this part focuses on order estimation for HMMs with both finite state space and finite output alphabet; in particular it describes how concepts from information theory are useful for elaborating on this subject.

Various parts of the book require different amounts of, and also different kinds of, prior knowledge from the reader. Generally we assume familiarity with probability and statistical estimation at the levels of Feller (1971) and Bickel and Doksum (1977), respectively. Some prior knowledge of Markov chains (discrete and/or general) is very helpful, although Part III does contain a primer on the topic; this chapter should however be considered more a brush-up than a comprehensive treatise of the subject. A reader with that knowledge will be able to understand most parts of the book. Chapter 13 on Bayesian estimation features a brief introduction to the subject in general but, again, some previous experience with Bayesian statistics will undoubtedly be of great help. The more theoretical parts of the book (Section 4.3, Chapter 9, Sections 11.2–11.3, Chapter 12, Sections 14.2–14.3 and Chapter 15) require knowledge of probability theory at the measure-theoretic level for a full understanding, even though most of the results as such can be understood without it.

There is no need to read the book in linear order, from cover to cover. Indeed, this is probably the wrong way to read it! Rather we encourage the reader to first go through the more algorithmic parts of the book, to get an overall view of the subject, and then, if desired, later return to the theoretical parts for a fuller understanding. Readers with particular topics in mind may of course be even more selective. A reader interested in the EM algorithm, for instance, could start with Chapter 1, have a look at Chapter 2, and then proceed to Chapter 3 before reading about the EM algorithm in Chapter 10. Similarly a reader interested in simulation-based techniques could go to Chapter 6 directly, perhaps after reading some of the introductory parts, or even directly to Section 6.3 if he/she is already familiar with MCMC methods. Each of the two chapters entitled "Advanced Topics in..." (Chapters 4 and 8) is really composed of three disconnected complements to Chapters 3 and 7, respectively. As such, the sections that compose Chapters 4 and 8 may be read independently of one another. Most chapters end with a section entitled "Complements" whose reading is not required for understanding other parts of the book—most often, this section mostly contains bibliographical notes—although in some chapters (9 and 11 in particular) it also features elements needed to prove the results stated in the main text.

Even in a book of this size, it is impossible to include all aspects of hidden Markov models. We have focused on the use of HMMs to model long, potentially stationary, time series; we call such models *ergodic HMMs*. In other applications, for instance speech recognition or protein alignment, HMMs are used to represent short variable-length sequences; such models are often called *left-to-right HMMs* and are hardly mentioned in this book. Having said that we stress that the computational tools for both classes of HMMs are virtually the same. There are also a number of generalizations of HMMs which we do not consider. In Markov random fields, as used in image processing applications, the Markov chain is replaced by a graph of dependency which may be represented as a two-dimensional regular lattice. The numerical techniques that can be used for inference in hidden Markov random fields are similar to some of the methods studied in this book but the statistical side is very different. Bayesian networks are even more general since the dependency structure is allowed to take any form represented by a (directed or undirected) graph. We do not consider Bayesian networks in their generality although some of the concepts developed in the Bayesian networks literature (the graph representation, the sum-product algorithm) are used. Continuous-time HMMs may also be seen as a further generalization of the models considered in this book. Some of these "continuous-time HMMs", and in particular partially observed diffusion models used in mathematical finance, have recently received considerable attention. We however decided this topic to be outside the range of the book; furthermore, the stochastic calculus tools needed for studying these continuous-time models are not appropriate for our purpose.

We acknowledge the help of Stéphane Boucheron, Randal Douc, Gersende Fort, Elisabeth Gassiat, Christian P. Robert, and Philippe Soulier, who participated in the writing of the text and contributed the two chapters that compose Part III (see next page for details of the contributions). We are also indebted to them for suggesting various forms of improvement in the notations, layout, etc., as well as helping us track typos and errors. We thank François Le Gland and Catherine Matias for participating in the early stages of this book project. We are grateful to Christophe Andrieu, Søren Asmussen, Arnaud Doucet, Hans Künsch, Steve Levinson, Ya'acov Ritov and Mike Titterington, who provided various helpful inputs and comments. Finally, we thank John Kimmel of Springer for his support and enduring patience.

Paris, France
& Lund, Sweden
March 2005

Olivier Cappé
Eric Moulines
Tobias Rydén

Contributors

We are grateful to

Randal Douc
Ecole Polytechnique
Christian P. Robert
CREST INSEE & Université Paris-Dauphine

for their contributions to Chapters 9 (Randal) and 6, 7, and 13 (Christian) as well as for their help in proofreading these and other parts of the book

Chapter 14 was written by

Gersende Fort
CNRS & LMC-IMAG
Philippe Soulier
Université Paris-Nanterre

with Eric Moulines

Chapter 15 was written by

Stéphane Boucheron
Université Paris VII-Denis Diderot
Elisabeth Gassiat
Université d'Orsay, Paris-Sud

Contents

1

Introduction

1.1 What Is a Hidden Markov Model?

A *hidden Markov model* (abbreviated HMM) is, loosely speaking, a Markov chain observed in noise. Indeed, the model comprises a Markov chain, which we will denote by $\{X_k\}_{k\geq 0}$, where k is an integer index. This Markov chain is often assumed to take values in a finite set, but we will not make this restriction in general, thus allowing for a quite arbitrary state space. Now, the Markov chain is *hidden*, that is, it is not observable. What is available to the observer is another stochastic process $\{Y_k\}_{k\geq 0}$, linked to the Markov chain in that X_k governs the distribution of the corresponding Y_k. For instance, Y_k may have a normal distribution, the mean and variance of which is determined by X_k, or Y_k may have a Poisson distribution whose mean is determined by X_k. The underlying Markov chain $\{X_k\}$ is sometimes called the *regime*, or *state*. All statistical inference, even on the Markov chain itself, has to be done in terms of $\{Y_k\}$ only, as $\{X_k\}$ is not observed. There is also a further assumption on the relation between the Markov chain and the observable process, saying that X_k must be the only variable of the Markov chain that affects the distribution of Y_k. This is expressed more precisely in the following formal definition.

A hidden Markov model is a bivariate discrete time process $\{X_k, Y_k\}_{k\geq 0}$, where $\{X_k\}$ is a Markov chain and, conditional on $\{X_k\}$, $\{Y_k\}$ is a sequence of independent random variables such that the conditional distribution of Y_k only depends on X_k. We will denote the state space of the Markov chain $\{X_k\}$ by X and the set in which $\{Y_k\}$ takes its values by Y.

The dependence structure of an HMM can be represented by a *graphical model* as in Figure 1.1. Representations of this sort use a directed graph without loops to describe dependence structures among random variables. The nodes (circles) in the graph correspond to the random variables, and the edges (arrows) represent the structure of the joint probability distribution, with the interpretation that the latter may be factored as a product of the conditional distributions of each node given its "parent" nodes (those that are directly

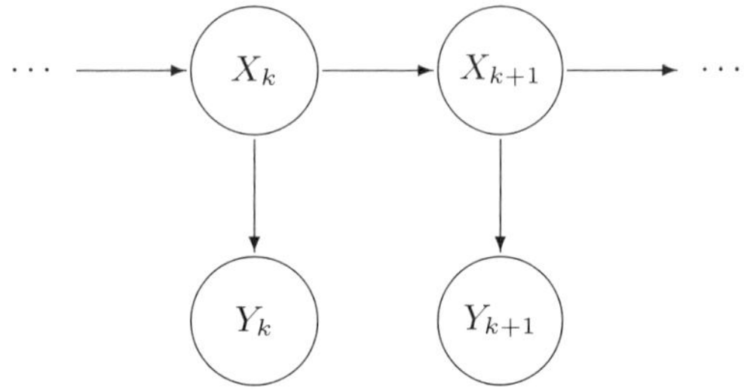

Fig. 1.1. Graphical representation of the dependence structure of a hidden Markov model, where $\{Y_k\}$ is the observable process and $\{X_k\}$ is the hidden chain.

connected to it by an arrow). Figure 1.1 thus implies that the distribution of a variable X_{k+1} conditional on the history of the process, $X_0, \ldots, X_k$, is determined by the value taken by the preceding one, X_k; this is called the *Markov property*. Likewise, the distribution of Y_k conditionally on the past observations $Y_0, \ldots, Y_{k-1}$ and the past values of the state, $X_0, \ldots, X_k$, is determined by X_k only (this is exactly the definition we made above). We shall not go into details about graphical models, but just sometimes use them as an intuitive means of illustrating various kinds of dependence. The interested reader is referred to, for example, Jensen (1996) or Jordan (2004) for introductory texts and to Lauritzen (1996), Cowell *et al.* (1999), or Jordan (1999) for in-depth coverage. Throughout the book, we will assume that each HMM is *homogeneous*, by which we mean that the Markov chain $\{X_k\}$ is homogeneous (its transition kernel does not depend on the time index k), and that the conditional law of Y_k given X_k does not depend on k either. In order to keep this introductory discussion simple, we do not embark into precise mathematical definitions of Markov chain concepts such as transition kernels for instance. The formalization of several of the ideas that are first reviewed on intuitive grounds here will be the topic of the first part of the book (Section 2.1).

As mentioned above, of the two processes $\{X_k\}$ and $\{Y_k\}$, only $\{Y_k\}$ is actually observed, whence inference on the parameters of the model must be achieved using $\{Y_k\}$ only. The other topic of interest is of course inference on the unobserved $\{X_k\}$: given a model and some observations, can we estimate the unobservable sequence of states? As we shall see later in the book, these two major statistical objectives are indeed strongly connected. Models that comprise unobserved random variables, as HMMs do, are called *latent variable models*, *missing data models*, or also *models with incomplete data*, where the latent variable refers to the unobservable random quantities.

Let us already at this point give a simple and illustrative example of an HMM. Suppose that $\{X_k\}$ is a Markov chain with state space $\{0, 1\}$ and that Y_k, conditional on $X_k = i$, has a Gaussian $\mathrm{N}(\mu_i, \sigma_i^2)$ distribution. In other words, the value of the regime governs the mean and variance of the Gaussian distribution from which we then draw the output. This model illustrates a common feature of HMMs considered in this book, namely that the conditional distributions of Y_k given X_k all belong to a single parametric family, with parameters indexed by X_k. In this case, it is the Gaussian family of distributions, but one may of course also consider the Gamma family, the Poisson family, etc. A meaningful observation, in the current example, is that the marginal distribution of $\{Y_k\}$ is that of a mixture of two Gaussian distributions. Hence we may also view HMMs as an extension of independent mixture models, including some degree of dependence between observations.

Indeed, even though the Y-variables are conditionally independent given $\{X_k\}$, $\{Y_k\}$ is not an independent sequence because of the dependence in $\{X_k\}$. In fact, $\{Y_k\}$ is not a Markov chain either: the joint process $\{X_k, Y_k\}$ is of course a Markov chain, but the observable process $\{Y_k\}$ does not have the loss of memory property of Markov chains, in the sense that the conditional distribution of Y_k given $Y_0, \ldots, Y_{k-1}$ generally depends on all the conditioning variables. As we shall see in Chapter 2, however, the dependence in the sequence $\{Y_k\}$ (defined in a suitable sense) is not stronger than that in $\{X_k\}$. This is a general observation that is valid not only for the current example.

Another view is to consider HMMs as an extension of Markov chains, in which the observation $\{Y_k\}$ of the state $\{X_k\}$ is distorted or blurred in some manner that includes some additional, independent randomness. In the previous example, the distortion is simply caused by additive Gaussian noise, as we may write this model as $Y_k = \mu_{X_k} + \sigma_{X_k} V_k$, where $\{V_k\}_{k \geq 0}$ is an i.i.d. (independent and identically distributed) sequence of standard Gaussian random variables. We could even proceed one step further by deriving a similar functional representation for the unobservable sequence of states. More precisely, if $\{U_k\}_{k \geq 0}$ denotes an i.i.d. sequence of of uniform random variables on the interval $[0, 1]$, we can define recursively $X_1, X_2, \ldots$ by the equation

$$X_{k+1} = \mathbb{1}(U_k \leq p_{X_k})$$

where p_0 and p_1 are defined respectively by $p_i = \mathrm{P}(X_{k+1} = 1 \,|\, X_k = i)$ (for $i = 0$ and 1). Such a representation of a Markov chain is usually referred to as a *stochastically recursive sequence* (and sometimes abbreviated to SRS) (Borovkov, 1998). An alternative view consists in regarding $\mathbb{1}(U_k \leq p.)$ as a random function (here on $\{0, 1\}$), hence the name *iterated random functions* also used to refer to the above representation of a Markov chain (Diaconis and Freedman, 1999). Our simple example is by no means a singular case and, in great generality, any HMM may be equivalently defined through a functional representation known as a (general) *state-space model*,

$$X_{k+1} = a(X_k, U_k) , \tag{1.1}$$

$$Y_k = b(X_k, V_k) , \tag{1.2}$$

where $\{U_k\}_{k \geq 0}$ and $\{V_k\}_{k \geq 0}$ are mutually independent i.i.d. sequences of random variables that are independent of X_0, and a and b are measurable functions. The first equation is known as the state or dynamic equation, whereas the second one is the observation equation. These two equations correspond to a recursive, generative form of the model, as opposed to our initial exposition, which focused on the specification of the joint probability distribution of the variables. Which view is most natural and fruitful typically depends on what the HMM is intended to model and for what purpose it is used (see the examples section below).

In the times series literature, the term "state-space model" is usually reserved for models in which a and b are linear functions and the sequences $\{U_k\}$, $\{V_k\}$, and X_0 are jointly Gaussian (Anderson and Moore, 1979; Brockwell and Davis, 1991; Kailath *et al.*, 2000). In this book, we reverse the perspective and refer to the family of models defined by (1.1) as (general) state-space models. The linear Gaussian sub-family of models will be covered in some detail, notably in Chapter 5, but is clearly not the main focus of this book. Similarly, in the classical HMM literature like the tutorial by Rabiner (1989) or the books by Elliott *et al.* (1995) and MacDonald and Zucchini (1997), it is tacitly assumed that the denomination "hidden Markov model" implies a finite state space X. This is a very important case indeed, but in this book we will treat more general state spaces as well. In our view, the terms "hidden Markov model" and "state-space model" refer to the same type of objects, although we will reserve the latter for describing the functional representation of the model given by (1.1).

1.2 Beyond Hidden Markov Models

The original works on (finite state space) hidden Markov models, as well as most of the theory regarding Gaussian linear state-space models, date back to the 1960s. Since then, the practical success of these models in several distinct application domains has generated an ever-increasing interest in HMMs and a similarly increasing number of new models based on HMMs. Several of these extensions of the basic HMM structure are, to some extent, also covered in this book.

A first simple extension is when the hidden state sequence $\{X_k\}_{k\geq 0}$ is a dth order Markov process, that is, when the conditional distribution of X_k given past values X_ℓ (with $0 \leq \ell < k$) depends on the d-tuple $X_{k-d}, X_{k-d+1}, \ldots, X_{k-1}$. At least conceptually this is not a very significant step, as we can fall back to the standard HMM setup by redefining the state to be the vector $(X_{k-d+1}, \ldots, X_k)$, which has Markovian evolution. Another variation consists in allowing for non-homogeneous transitions of the hidden chain or for non-homogeneous observation distributions. By this we mean that the distribution of X_k given X_{k-1}, or that of Y_k given X_k, can be allowed to depend on the index k. As we shall see in the second part of this book, non-homogeneous models lead to identical methods as far as state inference, i.e., inference about the hidden chain $\{X_k\}$, is concerned (except for the need to index conditional distributions with k).

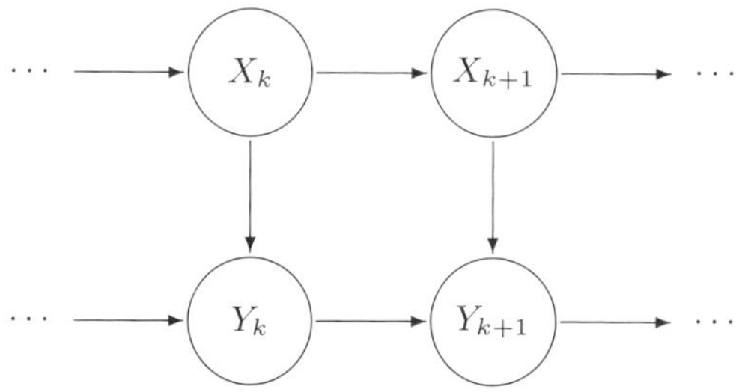

Fig. 1.2. Graphical representation of the dependence structure of a Markov-switching model, where $\{Y_k\}$ is the observable process and $\{X_k\}$ is the hidden chain.

Markov-switching models perhaps constitute the most significant generalization of HMMs. In such models, the conditional distribution of Y_{k+1}, given all past variables, depends not only on X_{k+1} but also on Y_k (and possibly more lagged Y-variables). Thus, conditional on the state sequence $\{X_k\}_{k\geq 0}$, $\{Y_k\}_{k\geq 0}$ forms a (non-homogeneous) Markov chain. Graphically, this is represented as in Figure 1.2. In state-space form, a Markov-switching model may be written as

$$X_{k+1} = a(X_k, U_k) , \tag{1.3}$$
$$Y_{k+1} = b(X_{k+1}, Y_k, V_{k+1}) . \tag{1.4}$$

The terminology regarding these models is not fully standardized and the term *Markov jump systems* is also used, at least in cases where the (hidden) state space is finite.

Markov-switching models have much in common with basic HMMs. In particular, virtually identical computational machinery may be used for both models. The statistical analysis of Markov-switching models is however much more intricate than for HMMs due to the fact that the properties of the observed process $\{Y_k\}$ are not directly controlled by those of the unobservable chain $\{X_k\}$ (as is the case in HMMs; see the details in Chapter 4). In particular, $\{Y_k\}$ is an infinite memory process whose dependence may be stronger than that of $\{X_k\}$ and it may even be the case that no stationary solution $\{Y_k\}_{k\geq 0}$ to (1.3)–(1.4) exists.

A final observation is that the computational tools pertaining to posterior inference, and in particular the smoothing equations of Chapter 3, hold in even greater generality. One could for example simply assume that $\{X_k, Y_k\}_{k\geq 0}$ jointly forms a Markov process, only a part $\{Y_k\}_{k\geq 0}$ of which is actually observed. We shall see however in the third part of the book that all statistical statements that we can currently make about the properties of estimators of the parameters of HMMs heavily rely on the fact that $\{X_k\}_{k\geq 0}$ is a Markov chain, and even more crucially, a uniformly ergodic Markov chain (see Chapter 4). For more general models such as partially observed Markov processes, it is not yet clear what type of (not overly restrictive and reasonably general) conditions are required to guarantee that reasonable estimators (such as the maximum likelihood estimator for instance) are well behaved.

1.3 Examples

HMMs and their generalizations are nowadays used in many different areas. The (partial) bibliography by Cappé (2001b) (which contains more than 360 references for the period 1990–2000) gives an idea of the reach of the domain. Several specialized books are available that largely cover applications of HMMs to some specific areas such as speech recognition (Rabiner and Juang, 1993; Jelinek, 1997), econometrics (Hamilton, 1989; Kim and Nelson, 1999), computational biology (Durbin *et al.*, 1998; Koski, 2001), or computer vision (Bunke and Caelli, 2001). We shall of course not try to compete with these in fully describing real-world applications of HMMs. We will however consider throughout the book a number of prototype HMMs (used in some of these applications) in order illustrate the variety of situations: finite-valued state space (DNA or protein sequencing), binary Markov chain observed in Gaussian noise (ion channel), non-linear Gaussian state-space model (stochastic volatility), conditionally Gaussian state-space model (deconvolution), etc.

It should be stressed that the idea one has about the nature of the hidden Markov chain $\{X_k\}$ may be quite different from one case to another. In some cases it does have a well-defined physical meaning, whereas in other cases it is conceptually more diffuse, and in yet other cases the Markov chain may be completely fictitious and the probabilistic structure of the HMM is then used only as a tool for modeling dependence in data. These differences are illustrated in the examples below.

1.3.1 Finite Hidden Markov Models

In a *finite hidden Markov model*, both the state space X of the hidden Markov chain and the set Y in which the output lies are finite. We will generally assume that these sets are $\{1, 2, \ldots, r\}$ and $\{1, 2, \ldots, s\}$, respectively. The HMM is then characterized by the transition probabilities $q_{ij} = \mathrm{P}(X_{k+1} = j \mid X_k = i)$ of the Markov chain and the conditional probabilities $g_{ij} = \mathrm{P}(Y_k = j \mid X_k = i)$.

Example 1.3.1 (Gilbert-Elliott Channel Model). The Gilbert-Elliott channel model, after Gilbert (1960) and Elliott (1963), is used in information theory to model the occurrence of transmission errors in some digital communication channels. Interestingly, this is a pre-HMM hidden Markov model, as it predates the seminal papers by Baum and his colleagues who introduced the term *hidden Markov model*.

In digital communications, all signals to be transmitted are first digitized and then transformed, a step known as *source coding*. After this preprocessing, one can safely assume that the bits that represent the signal to be transmitted form an i.i.d. sequence of fair Bernoulli draws (Cover and Thomas, 1991). We will denote by $\{B_k\}_{k\geq 0}$ the sequence of bits at the input of the transmission system.

Abstracted high-level models of how this sequence of bits may get distorted during the transmission are useful for devising efficient reception schemes and deriving performance bounds. The simplest model is the *(memoryless) binary symmetric channel* in which it is assumed that each bit may be randomly flipped by an independent error sequence,

$$Y_k = B_k \oplus V_k \ , \tag{1.5}$$

where $\{Y_k\}_{k\geq 0}$ are the observations and $\{V_k\}_{k\geq 0}$ is an i.i.d. Bernoulli sequence with $\mathrm{P}(V_k = 1) = q$, and $\oplus$ denotes modulo-two addition. Hence, the received bit is equal to the input bit B_k if $V_k = 0$; otherwise $Y_k \neq B_k$ and an error occurs.

The more realistic Gilbert-Elliott channel model postulates that errors tend to be more bursty than predicted by the memoryless channel. In this model, the channel regime is modeled as a two-state Markov chain $\{S_k\}_{k\geq 0}$, which represents low and high error conditions, respectively. The transition matrix of this chain is determined by the switching probabilities $p_0 = \mathrm{P}(S_{k+1} = 1 \mid S_k = 0)$ (transition into the high error regime) and $p_1 = \mathrm{P}(S_{k+1} = 0 \mid S_k = 1)$ (transition into the low error regime). In each regime, the model acts like the memoryless symmetric channel with error probabilities $q_0 = \mathrm{P}(Y_k \neq B_k \mid S_k = 0)$ and $q_1 = \mathrm{P}(Y_k \neq B_k \mid S_k = 1)$, where $q_0 < q_1$.

To recover the HMM framework, define the hidden state sequence as the joint process that collates the emitted bits and the sequence of regimes, $X_k = (B_k, S_k)$. This is a four-state Markov chain with transition matrix

	$(0,0)$	$(0,1)$	$(1,0)$	$(1,1)$
$(0,0)$	$(1-p_0)/2$	$p_0/2$	$(1-p_0)/2$	$p_0/2$
$(0,1)$	$p_1/2$	$(1-p_1)/2$	$p_1/2$	$(1-p_1)/2$
$(1,0)$	$(1-p_0)/2$	$p_0/2$	$(1-p_0)/2$	$p_0/2$
$(1,1)$	$p_1/2$	$(1-p_1)/2$	$p_1/2$	$(1-p_1)/2$

Neither the emitted bit B_k nor the channel regime S_k is observed directly, but the model asserts that conditionally on $\{X_k\}_{k\geq 0}$, the observations are independent Bernoulli draws with

$$\mathrm{P}(Y_k = b \mid B_k = b, S_k = s) = 1 - q_s \ .$$

■

Example 1.3.2 (Channel Coding and Transmission Over Memoryless Discrete Channel). We will consider in this example another elementary example of the use of HMMs, also drawn from the digital communication world. Assume we are willing to transmit a message encoded as a sequence $\{b_0, \ldots, b_m\}$ of bits, where $b_i \in \{0, 1\}$ are the bits and m is the length of the message. We wish to transmit this message over a channel, which will typically affect the transmitted message by introducing (at random) errors.

To go further, we need to have an abstract model for the channel. In this example, we will consider *discrete* channels, that is, the channel's inputs and outputs are assumed to belong to finite alphabets: $\{i_1, \ldots, i_q\}$ for the inputs and $\{o_1, \ldots, o_l\}$ for the outputs. In this book, we will most often consider binary channels only; then the inputs and the outputs of the transmission channel are bits, $q = l = 2$ and $\{i_1, i_2\} = \{o_1, o_2\} = \{0, 1\}$. A transmission channel is said to be *memoryless* if the probability of the channel's output $Y_{0:n} = y_{0:n}$ conditional on its input sequence $S_{0:n} = s_{0:n}$ factorizes as

$$\mathrm{P}(Y_{0:n} \mid S_{0:n}) = \prod_{i=0}^{n} \mathrm{P}(Y_i \mid S_i) .$$

In words, conditional on the input sequence $S_{0:n}$, the channel outputs are conditionally independent. The transition probabilities of the discrete memoryless channel are defined by a transition kernel $R : \{i_1, \ldots, i_q\} \times \{o_1, \ldots, o_l\} \to [0, 1]$, where for $i = 1, \ldots, q$ and $j = 1, \ldots, l$,

$$R(i_i, o_j) = \mathrm{P}(Y_0 = o_j \mid S_0 = i_i) . \tag{1.6}$$

The most classical example of a discrete memoryless channel is the *binary symmetric channel* (BSC) with binary input and binary output, for which $R(0, 1) = R(1, 0) = \varepsilon$ with $\varepsilon \in [0, 1]$. In words, every time a bit $S_k = 0$ or $S_k = 1$ is sent across the BSC, the output is also a bit $Y_k = \{0, 1\}$, which differs from the input bit with probability ε; that is, the error probability is $\mathrm{P}(Y_k \neq O_k) = \varepsilon$. As described in Example 1.3.1, the output of a binary symmetric channel can be modeled as a noisy version of the input sequence, $Y_k = S_k \oplus V_k$, where $\oplus$ is the modulo-two addition and $\{V_k\}_{k \geq 0}$ is an independent and identically distributed sequence of bits, independent of the input sequence $\{X_k\}_{k \geq 0}$ and with $\mathrm{P}\{V_k = 0\} = 1 - \varepsilon$. If we wish to transmit a message $S_{0:m} = b_{0:m}$ over a BSC without coding, the probability of getting an error will be

$$\mathrm{P}(Y_{0:m} \neq b_{0:m} \mid S_{0:m} = b_{0:m}) = \\ 1 - \mathrm{P}(Y_{0:m} = b_{0:m} \mid S_{0:m} = b_{0:m}) = 1 - (1 - \varepsilon)^m .$$

Therefore, as m becomes large, with probability close to 1, at least one bit of the message will be incorrectly received, which calls for practical solution. Channel coding is a viable method to increase reliability, but at the expense of reduced information rate. Increased reliability is achieved by adding redundancy to the information symbol vector, resulting in a longer coded vector of symbols that are distinguishable at the output of the channel. There are many ways to construct codes, and we consider in this example only a very elementary example of a rate 1/2 convolutional coder with memory length 2.

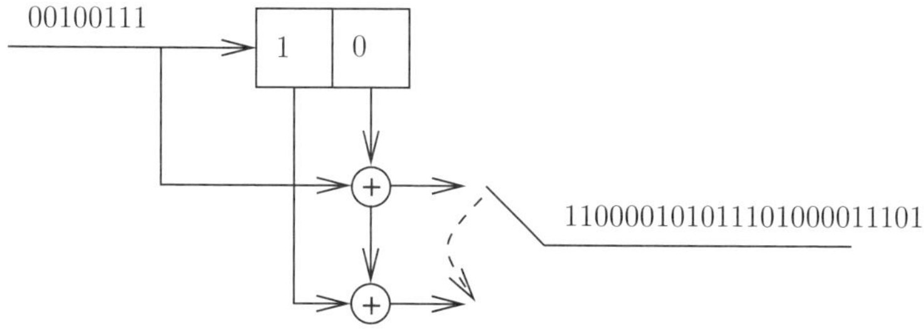

Fig. 1.3. Rate 1/2 convolutional code with memory length 2.

The rate 1/2 means that a message of length m will be transformed into a message of length $2m$, that is, we will send $2m$ bits over the transmission channel in order to introduce some kind of redundancy to increase our chance of getting an error-free message. The principle of this convolutional coder is depicted in Figure 1.3.

Because the memory length is 2, there are 4 different states and the behavior of this convolutional encoder can be captured as 4-state machine, where the state alphabet is $\mathsf{X} = \{(0,0),(0,1),(1,0),(1,1)\}$. Denote by X_k the value of the state at time k, $X_k = (X_{k,1}, X_{k,2}) \in \mathsf{X}$. Upon the arrival of the bit B_{k+1}, the state is transformed to

$$X_{k+1} = (X_{k+1,1}, X_{k+1,2}) = (B_{k+1}, X_{k,1}) .$$

In the engineering literature, X_k is said to be a shift register. If the sequence $\{B_k\}_{k\geq 0}$ of input bits is i.i.d. with probability $\mathrm{P}(B_k = 1) = p$, then $\{X_k\}_{k\geq 0}$ is a Markov chain with transition probabilities

$$\begin{aligned}
\mathrm{P}[X_{k+1} = (1,1) \mid X_k = (1,0)] &= \mathrm{P}[X_{k+1} = (1,1) \mid X_k = (1,1)] = p , \\
\mathrm{P}[X_{k+1} = (1,0) \mid X_k = (0,1)] &= \mathrm{P}[X_{k+1} = (1,0) \mid X_k = (0,0)] = p , \\
\mathrm{P}[X_{k+1} = (0,1) \mid X_k = (1,0)] &= \mathrm{P}[X_{k+1} = (0,1) \mid X_k = (1,1)] = 1-p , \\
\mathrm{P}[X_{k+1} = (0,0) \mid X_k = (0,1)] &= \mathrm{P}[X_{k+1} = (0,0) \mid X_k = (0,0)] = 1-p ,
\end{aligned}$$

all other transition probabilities being zero. To each input bit, the convolutional encoder generates two outputs according to

$$S_k = (S_{k,1}, S_{k,2}) = (B_k \oplus X_{k,2}, B_k \oplus X_{k,2} \oplus X_{k,1}) .$$

These encoded bits, referred to as *symbols*, are then sent on the transmission channel. A graphical interpretation of the problem is quite useful. A convolutional encoder (or, more generally, a finite state Markovian machine) can be represented by a state transition diagram of the type in Figure 1.4. The nodes are the states and the branches represent transitions having non-zero probability. If we index the states with both the time index k and state index m, we get the *trellis diagram* of Figure 1.4. The trellis diagram shows the time

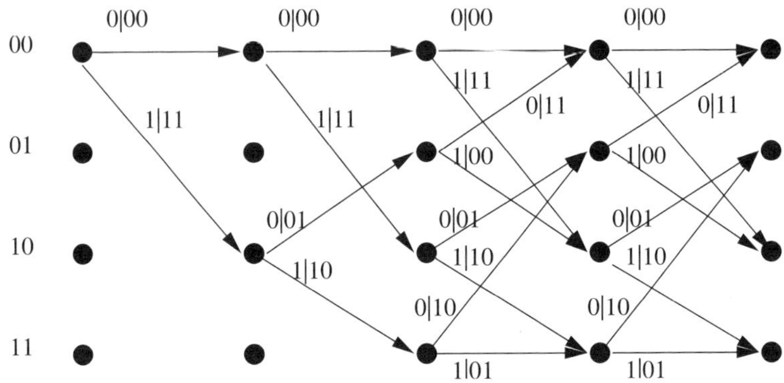

Fig. 1.4. Trellis representation of rate 1/2 convolutional code with memory length 2.

progression of the state sequences. For every state sequence, there is a unique path through the trellis diagram and *vice versa*.

More generally, the channel encoder is a finite state machine that transforms a message encoded as a finite stream of bits into an output sequence whose length is increased by a multiplicative factor that is the inverse of the rate of the encoder. If the input bits are i.i.d., the state sequence of this finite state machine is a finite state Markov chain. The m distinct states of the Markov source are $\{t_1, \ldots, t_m\}$. The outputs of this finite state machine is a sequence S_k with values in a finite alphabet $\{o_1, \ldots, o_q\}$. The state transitions of the Markov source are governed by the transition probabilities $p(i, j) = \mathrm{P}(X_n = t_j \mid X_{n-1} = t_i)$ and the output of the finite-state machine by the probabilities $q(i; j, k) = \mathrm{P}(S_n = o_i \mid X_n = t_j, X_{n-1} = t_k)$.

The Markov source always starts from the same initial state, $X_0 = t_1$ say, and produces an output sequence $S_{0:n} = (S_0, S_1, \ldots, S_n)$ ending in the terminal state $X_n = t_1$. $S_{0:n}$ is the input to a noisy *discrete memoryless channel* whose output is the sequence $Y_{0:n} = (Y_0, \ldots, Y_n)$. This discrete memoryless channel is also governed by transition probabilities (1.6). It is easy to recognize the general set-up of hidden Markov models, which are an extremely useful and popular tool in the digital communication community.

The objective of the decoder is to examine $Y_{0:n}$ and estimate the *a posteriori probability* of the states and transitions of the Markov source, i.e., the conditional probabilities $\mathrm{P}(X_k = t_i \mid Y_{0:n})$ and $\mathrm{P}(X_k = t_i, X_{k+1} = t_j \mid Y_{0:n})$. ■

Example 1.3.3 (HMM in Biology). Another example featuring finite HMMs is stochastic modeling of biological sequences. This is certainly one of the most successful examples of applications of HMM methodology in recent years. There are several different uses of HMMs in this context (see Churchill, 1992; Durbin *et al.*, 1998; Koski, 2001; Baldi and Brunak, 2001, for further references and details), and we only briefly describe the application of HMMs to gene finding in DNA, or more generally, functional annotation of sequenced genomes.

In their genetic material, all living organisms carry a blueprint of the molecules they need for the complex task of living. This genetic material is (usually) stored in the form of DNA—short for deoxyribonucleic acid—sequences. The DNA is not actually a sequence, but a long, chain-like molecule that can be specified uniquely by listing the sequence of amine bases from which it is composed. This process is known as *sequencing* and is a challenge on its own, although the number of complete sequenced genomes is growing at an impressive rate since the early 1990s. This motivates the abstract view of DNA as a sequence over a four-letter alphabet A, C, G, and T (for adenine, cytosine, guanine, and thymine—the four possible instantiations of the amine base).

The role of DNA is as a storage medium for information about the individual molecules needed in the biochemical processes of the organism. A region of the DNA that encodes a single functional molecule is referred to as a gene. Unfortunately, there is no easy way to discriminate coding regions (those that correspond to genes) from non-coding ones. In addition, the dimension of the problem is enormous as typical bacterial genomes can be millions of bases long with the number of genes to be located ranging from a few hundreds to a few thousands.

The simplistic approach to this problem (Churchill, 1992) consists in modeling the observed sequence of bases $\{Y_k\}_{k\geq 0} \in \{A, C, G, T\}$ by a two-state hidden Markov model such that the non-observable state is binary-valued with one state corresponding to non-coding regions and the other one to coding regions. In the simplest form of the model, the conditional distribution of Y_k given X_k is simply parameterized by the vector of probabilities of observing A, C, G, or T when in the coding and non-coding states, respectively. Despite its deceptive simplicity, the results obtained by estimating the parameters of this basic two-state finite HMM on actual genome sequences and then determining the smoothed estimate of the state sequence X_k (using techniques to be discussed in Chapter 3) were sufficiently promising to generate an important research effort in this direction.

The basic strategy described above has been improved during the years to incorporate more and more of the knowledge accumulated about the behavior of actual genome sequences—see Krogh *et al.* (1994), Burges and Karlin (1997), Kukashin and Borodovsky (1998), Jarner *et al.* (2001) and references therein. A very important fact, for instance, is that in coding regions the DNA is structured into *codons*, which are composed of three successive symbols in our A, C, G, T alphabet. This property can be accommodated by using higher order HMMs in which the distribution of Y_k does not only depend on the current state X_k but also on the previous two observations Y_{k-1} and Y_{k-2}. Another option consists in using non-homogeneous models such that the distribution of Y_k does not only depend on the current state X_k but also on the value of the index k modulo 3. In addition, some particular sub-sequences have a specific function, at least when they occur in a coding region (there are start and end codons for instance). Needless to say, enlarging the state space X to add specific states corresponding to those well identified functional sub-sequences is essential. Finally and most importantly, the functional description of the DNA sequence is certainly not restricted to just the coding/non-coding dichotomy, and most models use many more hidden states to differentiate between several distinct functional regions in the genome sequence. ■

Example 1.3.4 (Capture-Recapture). Capture-recapture models are often used in the study of populations with unknown sizes as in surveys, census undercount, animal abundance evaluation, and software debugging to name a few of their numerous applications. To set up the model in its original framework, we consider here the setting examined in Dupuis (1995) of a population of lizards (*Lacerta vivipara*) that move between three spatially connected zones, denoted 1, 2, and 3, the focus being on modeling these moves. For a given lizard, the sequence of the zones where it stays can be modeled as a Markov chain with transition matrix Q. This model still pertains to HMMs as, at a given time, most lizards are not observed: this is therefore a partly hidden Markov model. To draw inference on the matrix Q, the capture-recapture experiment is run as follows. At time $k = 0$, a (random) number of lizards are captured, marked, and released. This operation is repeated at times $k = 1, \ldots, n$ by tagging the newly captured animals and by recording at each capture the position (zone) of the recaptured animals. Therefore, the model consists of a series of capture events and positions (conditional on a capture) of $n+1$ cohorts of animals marked at times $k = 0, \ldots, n$. To account for open populations (as lizards can either die or leave the region of observation for good), a fourth state is usually added to the three spatial zones. It is denoted † (*dagger*) and, from the point of view of the underlying Markov chain, it is an absorbing state while, from the point of view of the HMM, it is always hidden.[1]

The observations may thus be summarized by the series $\{Y_{km}\}_{0 \le k \le n}$ of capture histories that indicate, for each lizard at least captured once (m being the lizard index), in which zone it was at each of the times it was captured. We may for instance record

$$\{y_{km}\}_{0 \le k \le n} = (0, \ldots, 0, 1, 1, 2, 0, 2, 0, 0, 3, 0, 0, 0, 1, 0, \ldots, 0) ,$$

where 0 means that the lizard was not captured at that particular time index. To each such observed sequence, there corresponds a (partially) hidden sequence $\{X_{km}\}_{0 \le k \le n}$ of lizard locations, for instance

$$\{x_{km}\}_{0 \le k \le n} = (1, \ldots, 2, \mathbf{1}, \mathbf{1}, \mathbf{2}, 2, \mathbf{2}, 3, 2, \mathbf{3}, 3, 2, 2, \mathbf{1}, \dagger, \ldots, \dagger)$$

[1]One could argue that lizards may also enter the population, either by migration or by birth. The latter reason is easily accounted for, as the age of the lizard can be assessed at the first capture. The former reason is real but will be ignored.

which indicates that the animal disappeared right after the last capture (where the values that are deterministically known from the observations have been stressed in bold).

The purposes in running capture-recapture experiments are often twofold: first, inference can be drawn on the size of the whole population based on the recapture history as in the Darroch model (Castledine, 1981; Seber, 1983), and, second, features of the population can be estimated from the captured animals, like capture and movement probabilities. ■

1.3.2 Normal Hidden Markov Models

By a *normal hidden Markov model* we mean an HMM in which the conditional distribution of Y_k given X_k is Gaussian. In many applications, the state space is finite, and we will then assume it is $\{1, 2, \ldots, r\}$. In this case, given $X_k = i$, $Y_k \sim \mathrm{N}(\mu_i, \sigma_i^2)$, so that the marginal distribution of Y_k is a finite mixture of normals.

Example 1.3.5 (Ion Channel Modeling). A cell, for example in the human body, needs to exchange various kinds of ions (sodium, potassium, etc.) with its surrounding for its metabolism and for purposes of chemical communication. The cell membrane itself is impermeable to such ions but contains so-called ion channels, each tailored for a particular kind of ion, to let ions pass through. Such a channel is really a large molecule, a protein, that may assume different configurations, or states. In some states, the channel allows ions to flow through—the channel is open—whereas in other states ions cannot pass—the channel is closed. A flow of ions is a transportation of electrical charge, hence an electric current (of the order of picoamperes). In other words, each state of the channel is characterized by a certain conductance level. These levels may correspond to a fully open channel, a closed channel, or something in between. The current through the channel can be measured using special probes (this is by no means trivial!), with the result being a time series that switches between different levels as the channel reconfigures. In this context, the main motivation is to study the characteristics of the dynamic of these ion channels, which is only partly understood, based on sampled measurements.

In the basic model, the channel current is simply assumed to be corrupted by additive white (i.i.d.) Gaussian measurement noise. If the state of the ion channel is modeled as a Markov chain, the measured time series becomes an HMM with conditionally Gaussian output and with the variances σ_i^2 not depending on i. A limitation of this basic model is that if each physical configuration of the channel (say closed) corresponds to a single state of the underlying Markov chain, we are implicitly assuming that each visit to this state has a duration drawn from a geometric distribution. A work-around that makes it possible to keep the HMM framework consists in modeling each physical configuration by a compound of distinct states of the underlying Markov chain, which are constrained to have a common conditional Gaussian output distribution. Depending on the exact transition matrix of the hidden chain, the durations spent in a given physical configuration can be modeled by negative binomial, mixtures of geometric or more complicated discrete distributions.

Further reading on ion-channel modeling can be found, for example, in Ball and Rice (1992) for basic references and Ball *et al.* (1999) and Hodgson (1998) for more advanced statistical approaches. ■

Example 1.3.6 (Speech Recognition). As yet another example of normal HMMs, we consider applications to speech recognition, which was the first area where HMMs were used extensively, starting in the early 1980s. The basic task is to, from a recording of a person's voice (or in real time, on-line), automatically determine what he or she said.

To do that, the recorded and sampled speech signal is slotted into short sections (also called frames), typically representing about 20 milliseconds of the original signal. Each section is then analyzed separately to produce a set of coefficients that represent the estimated power spectral density of the signal in the frame. This preprocessing results in a discrete-time multivariate time series of spectral coefficients. For a given word to be recognized (imagine, for simplicity, that speakers only pronounce single words), the length of the series of vectors resulting from this preprocessing is not determined beforehand but depends on the time taken for the speaker to utter the word. A primary requirement on the model is thus to cope with the time alignment problem so as to be able to compare multivariate sequences of unequal lengths.

In this application, the hidden Markov chain corresponds to sub-elements of the utterance that are expected to have comparable spectral characteristics. In particular, we may view each word as a sequence of phonemes (for instance, red: [r-e-d]; class: [k-l-a:-s]). The state of the Markov chain is then the hypothetical phoneme that is currently being uttered at a given time slot. Thus, for a word with three phonemes, like "red" for example, the state of the Markov chain may evolve according to Figure 1.5. Note that as opposed to Figures 1.1 and 1.2, Figure 1.5 is an automaton description of the Markov chain that indicates where the chain may jump to given its current state. Each arrow thus represents a possible transition that is associated with a non-zero transition probability. In this book, we shall use double circles for the nodes of such automata, as in Figure 1.5, to distinguish them from graphical models. We see that each state corresponding to a phoneme has a transition back

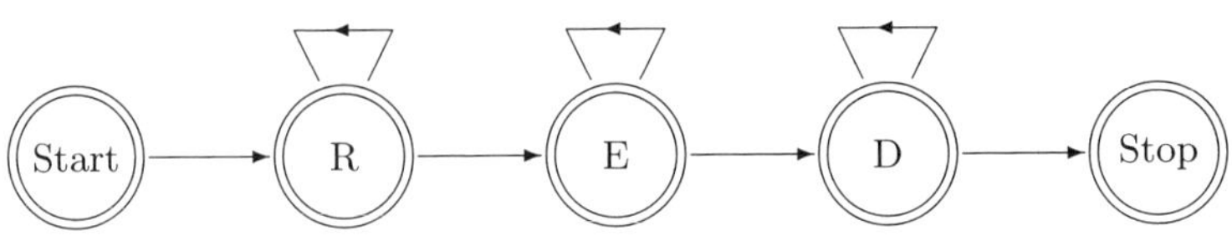

Fig. 1.5. Automaton representation of the Markov chain structure of an HMM for recognizing the word "red".

to itself, that is, a loop; this is to allow the phoneme to last for as long as the recording of it does. The purposes of the initial state *Start* and terminal state *Stop* is simply to have well-defined starts and terminations of the Markov chain; the stop state may be thought of as an absorbing state with no associated observation.

The observation vectors associated with a particular (unobservable) state are assumed to be independent and are assigned a multivariate distribution, most often a mixture of Gaussian distributions. The variability induced by this distribution is used to model spectral variability within and between speakers. The actual speech recognition is realized by running the recorded word as input to several different HMMs, each representing a particular word, and selecting the one that assigns the largest likelihood to the observed sequence. In a prior training phase, the parameters of each word model have been estimated using a large number of recorded utterances of the word. Note that the association of the states of the hidden chain with the phonemes in Figure 1.5 is more a conceptual view than an actual description of what the model does. In practice, the recognition performance of HMM-based speech recognition engines is far better than their efficiency at segmenting words into phonemes.

Further reading on speech recognition using HMMs can be found in the books by Rabiner and Juang (1993) and Jelinek (1997). The famous tutorial by Rabiner (1989) gives a more condensed description of the basic model, and Young (1996) provides an overview of current large-scale speech recognition systems. ■

1.3.3 Gaussian Linear State-Space Models

The standard state-space model that we shall most often employ in this book takes the form

$$X_{k+1} = AX_k + RU_k \,, \tag{1.7}$$
$$Y_k = BX_k + SV_k \,, \tag{1.8}$$

where

- $\{U_k\}_{k\geq 0}$, called the *state* or *process noise*, and $\{V_k\}_{k\geq 0}$, called the *measurement noise*, are independent standard (multivariate) Gaussian white noise (sequences of i.i.d. multidimensional Gaussian random variables with zero mean and identity covariance matrices);
- The initial condition X_0 is Gaussian with mean μ_ν and covariance Γ_ν and is uncorrelated with the processes $\{U_k\}$ and $\{V_k\}$;
- The *state transition matrix* A, the *measurement transition matrix* B, the square-root of the state noise covariance R, and the square-root of the measurement noise covariance S are known matrices with appropriate dimensions.

Ever since the pioneering work by Kalman and Bucy (1961), the study of the above model has been a favorite both in the engineering (automatic control, signal processing) and time series literature. Recommended readings on the state-space model include the books by Anderson and Moore (1979), Caines (1988), and Kailath *et al.* (2000). In addition to its practical importance, the Gaussian linear state-space model is interesting because it corresponds to one of the very few cases for which exact and reasonably efficient numerical procedures are available to compute the distributions of the X-variables given Y-variables (see Chapters 3 and 5).

Remark 1.3.7. The form adopted for the model (1.7)–(1.8) is rather standard (except for the symbols chosen for the various matrices, which vary widely in the literature), but the role of the matrices R and S deserve some comments. We assume in the following that both noise sequences $\{U_k\}$ and $\{V_k\}$ are i.i.d. with identity covariance matrices. Hence R and S serve as square roots of the noise covariances, as

$$\mathrm{Cov}(RU_k) = RR^t \quad \text{and} \quad \mathrm{Cov}(SV_k) = SS^t ,$$

where the superscript t denotes matrix transposition. In some cases, and in particular when either the X- or Y-variables are scalar, it would probably be simpler to use $U'_k = RU_k$ and $V'_k = SV_k$ as noise variables, adopting their respective covariance matrices as parameters of the model. In many situations, however, the covariance matrices have a special structure that is most naturally represented by using R and S as parameters. In Example 1.3.8 below for instance, the dynamic noise vector U_k has a dimension much smaller than that of the state vector X_k. Hence R is a tall matrix (with more rows than columns) and the covariance matrix of $U'_k = RU_k$ is rank deficient. It is then much more natural to work only with the low-dimensional unit covariance disturbance vector U_k rather than with $U'_k = RU_k$. In the following, we will assume that SS^t is a full rank covariance matrix (for reasons discussed in Section 5.2), but RR^t will often be rank deficient as in Example 1.3.8.

In many respects, the case in which the state and measurement noises $\{U_k\}$ and $\{V_k\}$ are correlated is not much more complicated. It however departs from our usual assumptions in that $\{X_k, Y_k\}$ then forms a Markov chain but $\{X_k\}$ itself is no longer Markov. We will thus restrict ourselves to the case in which $\{U_k\}$ and $\{V_k\}$ are independent and refer, for instance, to Kailath *et al.* (2000) for further details on this issue. ■

Example 1.3.8 (Noisy Autoregressive Process). We shall define a pth order scalar autoregressive (AR) process $\{Z_k\}_{k\geq 0}$ as one that satisfies the stochastic difference equation

$$Z_{k+1} = \phi_1 Z_k + \cdots + \phi_p Z_{k-p+1} + U_k , \tag{1.9}$$

where $\{U_k\}_{k\geq 0}$ is white noise. Define the lag-vector

$$X_k = (Z_k, \ldots, Z_{k-p+1})^t \,, \tag{1.10}$$

and let A be the so-called *companion matrix*

$$A = \begin{pmatrix} \phi_1 & \phi_2 & \ldots & & \phi_p \\ 1 & 0 & \ldots & & 0 \\ 0 & 1 & \ldots & & 0 \\ \vdots & \vdots & \ddots & & \vdots \\ 0 & 0 & \ldots & 1 & 0 \end{pmatrix} . \tag{1.11}$$

Using these notations, (1.9) can be equivalently rewritten in state-space form:

$$X_k = AX_{k-1} + \begin{pmatrix}1 \; 0 \ldots 0\end{pmatrix}^t U_{k-1} \,, \tag{1.12}$$

$$Y_k = \begin{pmatrix}1 \; 0 \ldots 0\end{pmatrix} X_k \,. \tag{1.13}$$

If the autoregressive process is not directly observable but only a noisy version of it is available, the measurement equation (1.13) is replaced by

$$Y_k = \begin{pmatrix}1 \; 0 \ldots 0\end{pmatrix} X_k + V_k \,, \tag{1.14}$$

where $\{V_k\}_{k \geq 0}$ is the measurement noise. When there is no feedback between the measurement noise and the autoregressive process, it is sensible to assume that the state and measurement noises $\{U_k\}$ and $\{V_k\}$ are independent. ■

1.3.4 Conditionally Gaussian Linear State-Space Models

We gradually move toward more complicated models for which the state space X of the hidden chain is no more finite. The previous example is, as we shall see in Chapter 5, a singular case because of the unique properties of the multivariate Gaussian distribution with respect to linear transformations. We now describe a related, although more complicated, situation in which the state X_k is composed of two components C_k and W_k where the former is finite-valued whereas the latter is a continuous, possibly vector-valued, variable. The term "conditionally Gaussian linear state-space models", or CGLSSMs in short, corresponds to structures by which the model, *when conditioned on the finite-valued process* $\{C_k\}_{k \geq 0}$, reduces to the form studied in the previous section.

Conditionally Gaussian linear state-space models belong to a class of models that we will refer to as *hierarchical hidden Markov models*, whose dependence structure is depicted in Figure 1.6. In such models the variable C_k, which is the highest in the hierarchy, influences both the transition from W_{k-1} to W_k as well as the observation Y_k. When $\{C_k\}$ takes its values in a finite set, it is also common to refer to such models as *jump Markov models*, where the jumps correspond to the instants k at which the value of C_k differs from that of C_{k-1}. Of course, Figure 1.6 also corresponds to a standard HMM structure by

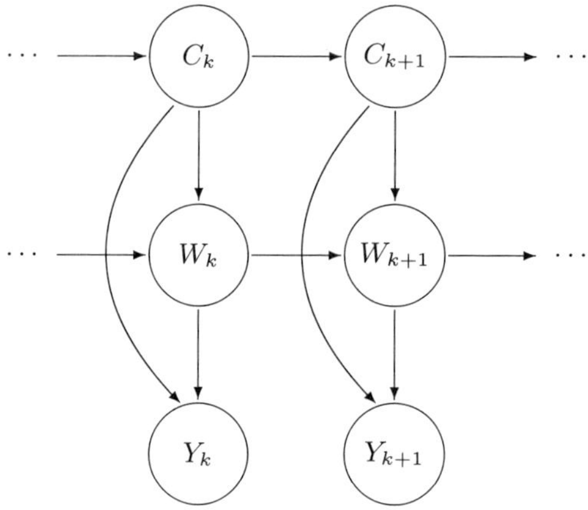

Fig. 1.6. Graphical representation of the dependence structure of a hierarchical HMM.

considering the composite state $X_k = (C_k, W_k)$. But for hierarchical HMMs in general and CGLSSMs in particular, it is often advantageous to consider the intermediate state sequence $\{W_k\}_{k \geq 0}$ as a *nuisance parameter* to focus on the $\{C_k\}$ component that stands at the top of the hierarchy in Figure 1.6. To do so, one needs to integrate out the influence of $\{W_k\}$, conditioning on $\{C_k\}$ only. This principle can only be made effective in situations where the model belongs to a simple class (such as Gaussian linear state-space models) once conditioned on $\{C_k\}$. Below we give several simple examples that illustrate the potential of this important class of models.

Example 1.3.9 (Rayleigh-fading Channel). We will now follow up on Example 1.3.1 and again consider a model of interest in digital communication. The point is that for wireless transmissions it is possible, and desirable, to model more explicitly (than in Example 1.3.1) the physical processes that cause errors during transmissions. As in Example 1.3.1, we shall assume that the signal to be transmitted forms an i.i.d. sequence of fair Bernoulli draws. Here the sequence is denoted by $\{C_k\}_{k \geq 0}$ and we assume that it takes its values in the set $\{-1, 1\}$ rather than in $\{0, 1\}$. This sequence is transmitted through a suitable modulation (Proakis, 1995) that is not of direct interest to us.

At the receiving side, the signal is first demodulated and the simplest model, known as the *additive white Gaussian noise* (AWGN) channel, postulates that the demodulated signal $\{Y_k\}_{k \geq 0}$ may be written

$$Y_k = hC_k + V_k \ , \tag{1.15}$$

where h is a (real) channel gain, also known as a fading coefficient, and $\{V_k\}_{k \geq 0}$ is an i.i.d. sequence of Gaussian observation noise with zero mean and variance σ^2. For reasons that are inessential for the discussion that follows, the actual model features complex channel gain and noise (Proakis, 1995), a fact that we will ignore in the following.

The AWGN channel model ignores inter-symbol interference in the sense that under (1.15) the observations $\{Y_k\}$ are i.i.d. In many practical situations, it is necessary to account for channel memory to obtain a reasonable model of the received signal. Another issue is that, in particular in wireless communication, the physical characteristics of the propagation path or channel are continuously changing over time. As a result, the fading coefficient h will typically not stay constant but vary with time. A very simple model consists in assuming that the fading coefficient follows a (complex) autoregressive model of order 1, giving the model

$$
\begin{aligned}
W_{k+1} &= \rho W_k + U_k \ , \\
Y_k &= W_k C_k + V_k \ ,
\end{aligned}
$$

where the time-varying h is denoted by W_k, and $\{U_k\}_{k \geq 0}$ is white Gaussian noise (an i.i.d. sequence of zero mean Gaussian random variables). With this model, it is easily checked that if we assume that W_0 is a Gaussian random variable independent of both the observation noise $\{V_k\}$ and the state noise $\{U_k\}$, $\{Y_k\}$ is the observation sequence corresponding to an HMM with hidden state $X_k = (C_k, W_k)$ (the emitted bit and the fading coefficient). This is a general state-space HMM, as W_k is a real random variable. In this application, the aim is to estimate the sequence $\{C_k\}$ of bits, which is thus a component of the unobservable state sequence, given the observations $\{Y_k\}$. The fading coefficients $\{W_k\}$ are of no direct interest and constitute nuisance variables.

This model however has a unique feature among general state-space HMMs in that conditionally on the sequence $\{C_k\}$ of bits, it reduces to a Gaussian linear state-space model with state variables $\{W_k\}$. The only difference to Section 1.3.3 is that the observation equation becomes non-homogeneous in time,

$$
Y_k = W_k c_k + V_k \ ,
$$

where $\{C_k = c_k\}$ is the event on which we are conditioning. As a striking consequence, we shall see in Chapters 4 and 5 that the distribution of W_k given the observations $Y_0, Y_1, \ldots, Y_k$ is a mixture of 2^{k+1} Gaussian distributions. Because this is clearly not a tractable form when k is a two-digit number, the challenge consists in finding practical approaches to approximate the exact distributions. ■

Conditionally Gaussian models related to the previous example are also commonly used to approximate non-Gaussian state-space models. Imagine that we are interested in the linear model given by Eqs. (1.7)–(1.8) with both noise sequences still being i.i.d. but at least one of them with a non-Gaussian distribution. Assuming a very general form of the noise distribution would directly lead us into the world of (general) continuous state-space HMMs. As a middle ground, we may however assume that the distribution of the noise is a finite mixture of Gaussian distributions.

Let $\{C_k\}_{k\geq 0}$ denote an i.i.d. sequence of random variables taking values in a set C, which can be finite or infinite. We refer to these variables as the *indicator variables* when C is finite and *latent variables* otherwise. To model non-Gaussian system dynamics we will typically replace the evolution equation (1.7) by

$$W_{k+1} = \mu_W(C_{k+1}) + A(C_{k+1})W_k + R(C_{k+1})U_k \;, \quad U_k \sim \mathrm{N}(0, I) \;,$$

where, μ_W, A and R are respectively vector-valued and matrix-valued functions of suitable dimensions on C. When $\mathsf{C} = \{1, \ldots, r\}$ is finite, the distribution of the noise $\mu_W(C_{k+1}) + R(C_{k+1})U_k$ driving the state equation is a finite mixture of multivariate Gaussian distributions,

$$\sum_{i=1}^{r} m_i \, \mathrm{N}\left(\mu_W(i), R(i)R^t(i)\right) \qquad \text{with } m_i = \mathrm{P}(C_0 = i) \;.$$

Another option consists in using the same modeling to represent non-Gaussian observation noise by replacing the observation equation (1.8) by

$$Y_k = \mu_Y(C_k) + B(C_k)W_k + S(C_k)V_k \;, \quad V_k \sim \mathrm{N}(0, I) \;,$$

where μ_Y, B and S are respectively vector-valued and matrix-valued functions of suitable dimensions on C. Of course, by doing this the state of the HMM has to be extended to the joint process $\{X_k\}_{k\geq 0}$, where $X_k = (W_k, C_k)$, taking values in the product set $\mathsf{X} \times \mathsf{C}$. At first sight, it is not obvious that anything has been gained at all by introducing additional mixture indices with respect to our basic objective, which is to allow for linear state-space models with non-Gaussian noises. We shall see however in Chapter 8 that the availability of computational procedures that evaluate quantities such as $\mathrm{E}[W_k \mid Y_0, \ldots Y_k, C_0, \ldots, C_k]$ is a distinct advantage of conditionally linear state-space models over more general (unstructured) continuous state-space HMMs. Conditionally Gaussian linear state-space models (CGLSSM) have found an exceptionally broad range of applications.

Example 1.3.10 (Change Point Detection). A simple yet useful example of CGLSSMs appears in change point detection problems (Shumway and Stoffer, 1991; Fearnhead, 1998). In a Gaussian linear state-space model, the dynamics of the state depends on the state transition matrix and on the state noise covariance. These quantities may change over time, and if the changes, when they occur, do so unannounced and at unknown time points, then the associated inferential problem is referred to as a change point problem. Various important application areas of statistics involve change detection in a central way (for instance, environmental monitoring, quality assurance, biology). In the simplest change point problem, the state variable is the level

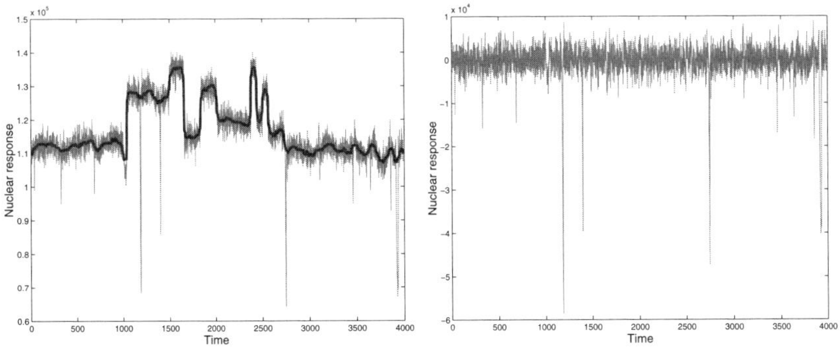

Fig. 1.7. Left: well-log data waveform with a median smoothing estimate of the state. Right: median smoothing residual.

of a quantity of interest, which is modeled as a step function; the time instants at which the step function jumps are the change points. An example of this situation is provided by the well-log data considered in Chapter 5 of the book by Ó Ruanaidh and Fitzgerald (1996) and analyzed, among others, by Fearnhead (1998) and Fearnhead and Clifford (2003).

In this example, the data, which is plotted in Figure 1.7, consists of measurements of the nuclear magnetic response of underground rocks that are obtained whilst drilling for oil. The data contains information about the rock structure that is being drilled through. In particular, it contains information about boundaries between rock strata; jumps in the step function relate to the rock strata boundaries. As can be seen from the data, the underlying state is a step function, which is corrupted by a fairly large amount of noise. It is the position of these jumps that one needs to estimate. To model this situation, we put $\mathsf{C} = \{0, 1\}$, where $C_k = 0$ means that there is no change point at time index k, whereas $C_k = 1$ means that a change point has occurred. The state-space model is

$$\begin{aligned} W_{k+1} &= A(C_{k+1})W_k + R(C_{k+1})U_k \;, \\ Y_k &= W_k + V_k \;, \end{aligned}$$

where $A(0) = I$, $R(0) = 0$ and $A(1) = 0$ and $R(1) = R$. The simplest model consists in taking for $\{C_k\}_{k \geq 0}$ an i.i.d. sequence of Bernoulli random variables with probability of success p. The time between two change points (period of time during which the state variable is constant) is then distributed as a geometric random variable with mean $1/p$;

$$W_{k+1} = \begin{cases} W_k & \text{with probability } p \;, \\ U_k & \text{otherwise} \;. \end{cases} \tag{1.16}$$

It is possible to allow a more general form for the prior distribution of the durations of the periods by introducing dependence among the indicator variables.

Note that it is also possible to consider such multiple change point models under the different, although strictly equivalent, perspective of a Bayesian model with an unknown number of parameters. In this alternative representation, the hidden state trajectory is parameterized by the succession of its levels (between two change points), which thus form a variable dimension set of parameters (Green, 1995; Lavielle and Lebarbier, 2001). Bayesian inference about such parameters, equipped with a suitable prior distribution, is then carried out using simulation-based techniques to be discussed further in Chapter 13. ■

Example 1.3.11 (Linear State-Space Model with Observational Outliers and Heavy-Tailed Noise). Another interesting application of conditional Gaussian linear state-space models pertains to the field of robust statistics (Schick and Mitter, 1994). In the course of model building and validation, statisticians are often confronted with the problem of dealing with outliers. Routinely ignoring unusual observations is neither wise nor statistically sound, as such observations may contain valuable information about unmodeled system characteristics, model degradation and breakdown, measurement errors and so forth.

The well-log data considered in the previous example illustrates this situation. A visual inspection of the nuclear response reveals the presence of outliers, which tend to clump together in bursts (or clusters). This is confirmed when plotting the quantile-quantile regression plot (see Figure 1.8) of the residuals of the well-log data obtained from a crude moving median estimate of the state variable (the median filter applies a sliding window to a sequence and outputs the median value of all points in the window as a smoothed estimate at the window center). It can be seen that the normal distribution does not fit the measurement noise well in the tails. Following Fearnhead and Clifford (2003), we model the measurement noise as a mixture of two Gaussian distributions. The model can be written

$$\begin{aligned} W_{k+1} &= A(C_{k+1,1})W_k + R(C_{k+1,1})U_k \;, \qquad & U_k &\sim \mathrm{N}(0,1) \;, \\ Y_k &= \mu(C_{k,2}) + B(C_{k,2})W_k + S(C_{k,2})V_k \;, \qquad & V_k &\sim \mathrm{N}(0,1) \;, \end{aligned}$$

where $C_{k,1} \in \{0,1\}$ and $C_{k,2} \in \{0,1\}$ are indicators of a change point and of the presence of an outlier, respectively. As above, the level is assumed to be constant between two change points. Therefore we put $A(0) = 1$, $R(0) = 0$, $A(1) = 0$, and $R(1) = \sigma_U$. When there is no outlier, that is, $C_{k,2} = 0$, we assume that the level is observed in additive Gaussian noise. Therefore $\{\mu(0), B(0), S(0)\} = (0, 1, \sigma_{V,0})$. In the presence of an outlier, the measurement does no longer carry information about the current value of the level, that is, $B(1) = 0$, and the measurement noise is assumed to follow a Gaussian distribution with negative mean μ and (large) variance $\sigma_{V,1}$. Therefore

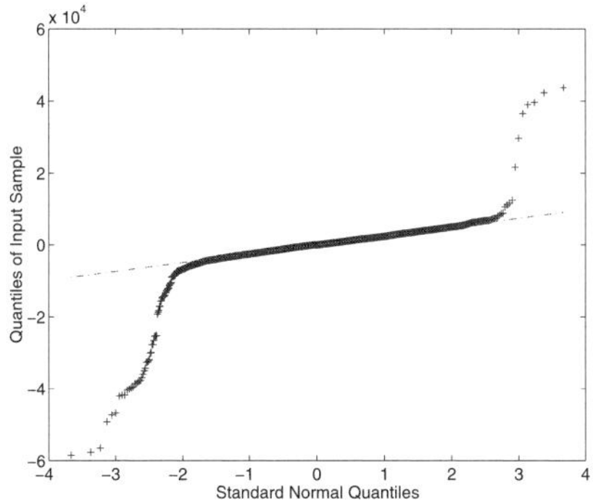

Fig. 1.8. Quantile-quantile regression of empirical quantiles of the well-log data residuals with respect to quantiles of the standard normal distribution.

$\{\mu(1), B(1), S(1)\} = (\mu, 0, \sigma_{V,1})$. One possible model for $\{C_{k,2}\}$ would be a Bernoulli model in which we could include information about the ratio of outliers/non-outliers in the success probability. However, this does not incorporate any information about the way samples of outliers cluster together, as samples are assumed independent in such a model. A better model might be a two-state Markov chain in which the state transition probabilities allow a preference for "cohesion" within outlier bursts and non-outlier sections. Similar models have been used for audio signal restoration, where an outlier is a local degradation of the signal (click, scratch, etc.).

There are of course, in the framework of CGLSSMs, many additional degrees of freedom. For example, Ó Ruanaidh and Fitzgerald (1996) claimed that the distribution of the measurement noise in the "clean" segments (segments free from outliers) of the nuclear response measurement have tails heavier than those of the Gaussian distribution, and they advocated a Laplacian additive noise model. The use of heavy-tailed distributions to model either the observation noise or the measurement noise, which finds its roots in the field of robust statistics, is very popular and has been worked out in many different fields. One can of course consider to use Laplace, Weibull, or Student t-distributions, depending on the expected "size" of the tails, but if one is willing to exploit the full strength of conditionally Gaussian linear systems, it is wiser to consider using Gaussian scale mixtures. A random vector V is a Gaussian scale mixture if it can be expressed as the product of a Gaussian vector W with zero mean and identity covariance matrix and an independent positive scalar random variable $\sqrt{C}$: $V = \sqrt{C}W$ (Andrews and Mallows, 1974). The variable C is the *multiplier* or the *scale*. If C has finite support, then V is a finite mixture of Gaussian vectors, whereas if C has a density with respect to Lebesgue measure on $\mathbb{R}$, then V is a continuous mixture of Gaussian vectors. Gaussian scale mixtures are symmetric, zero mean, and have leptokurtic marginal densities (tails heavier than those of a Gaussian distribution). ■

1.3.5 General (Continuous) State-Space HMMs

Example 1.3.12 (Bearings-only Tracking). Bearings-only tracking concerns online estimation of a target trajectory when the observations consist solely of the direction of arrivals (*bearings*) of a plane wavefront radiated by a target as seen from a known observer position (which can be fixed, but is, in most applications, moving). The measurements are blurred by noise, which accounts for the errors occurring when estimating the bearings. In this context, the range information (the distance between the object and the sensor) is not available. The target is usually assumed to be traveling in a two-dimensional space, the state of the target being its position and its velocity. Although the observations occur at regularly spaced instants, we describe the movement of the object in continuous time to be able to define the derivatives of the motion. The system model that we describe here is similar to that used in Gordon *et al.* (1993) and Chapter 6 of Ristic *et al.* (2004)—see also (Pitt and Shephard, 1999; Carpenter *et al.*, 1999).

The state vector at time kT is $X_k = (P_{x,k}, \dot{P}_{x,k}, P_{y,k}, \dot{P}_{y,k})^t$, representing the target's position at time kT and its velocity, where T denotes the sampling period. One possible discretization of this model, based on a second order Taylor expansion, is given by (Gordon *et al.*, 1993)

$$X_{k+1} = AX_k + RU_k \; , \tag{1.17}$$

where

$$A = \begin{pmatrix} 1 & T & 0 & 0 \\ 0 & 1 & 0 & 0 \\ 0 & 0 & 1 & T \\ 0 & 0 & 0 & 1 \end{pmatrix} , \qquad R = \sigma_U \begin{pmatrix} T^2/2 & 0 \\ T & 0 \\ 0 & T^2/2 \\ 0 & T \end{pmatrix}$$

and $\{U_k\}_{k\geq 0}$ is bivariate standard white Gaussian noise, $U_k \sim \mathrm{N}(0, I_2)$. The scale σ_U characterizes the magnitude of the random fluctuations of the acceleration between two sampling points. The initial position X_0 is multivariate Gaussian with mean $(\mu_x, \dot{\mu}_x, \; \mu_y, \dot{\mu}_y)$ and covariance matrix $\mathrm{diag}(\sigma_x^2, \dot{\sigma}_x^2, \sigma_y^2, \dot{\sigma}_y^2)$. The measurements $\{Y_k\}_{k\geq 0}$ are modeled as

$$Y_k = \tan^{-1}\left(\frac{P_{y,k} - R_{y,k}}{P_{x,k} - R_{x,k}}\right) + \sigma_V V_k \; , \tag{1.18}$$

where $\{V_k\}_{k\geq 0}$ is white Gaussian noise with zero mean and unit variance, and $(R_{x,k}, R_{y,k})$ is the (known) observer position. It is moreover assumed that $\{U_k\}$ and $\{V_k\}$ are independent. One important feature of this model is that the amount of information about the range of the target that is present in the measurements is, in general, small. The only range information in the observations arise due to the knowledge of the state equations, which are informative about the maneuvers that the target is likely to perform. Therefore, the majority of range information contained in the model is that which is included in the prior model of the target motion. ■

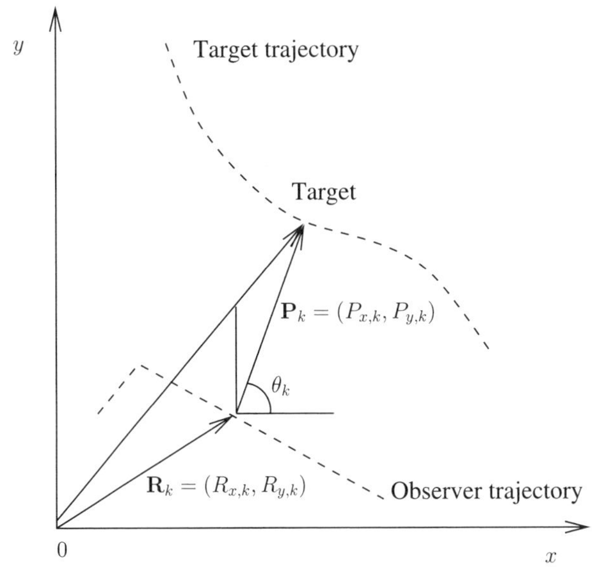

Fig. 1.9. Two-dimensional bearings-only target tracking geometry.

Example 1.3.13 (Stochastic Volatility). The distributional properties of speculative prices have important implications for several financial models. Let S_k be the price of a financial asset—such as a share price, stock index, or foreign exchange rate—at time k. Instead of the prices, it is more customary to consider the *relative returns* $(S_k - S_{k-1})/S_{k-1}$ or the *log-returns* $\log(S_k/S_{k-1})$, which both describe the relative change over time of the price process. In what follows we often refer, for short, to *returns* instead of relative or log-returns (see Figure 1.10). The unit of the discrete time index k may be for example an hour, a day, or a month. The famous Black-Scholes model, which is a continuous-time model and postulates a geometric Brownian motion for the price process, corresponds to log-returns that are i.i.d. and with a Gaussian $\mathrm{N}(\mu, \sigma^2)$ distribution, where σ is the *volatility* (the word *volatility* is the word used in econometrics for standard deviation). The Black and Scholes option pricing model provides the foundation for the modern theory of option valuation.

In actual applications, however, this model has certain well-documented deficiencies. Data from financial markets clearly indicate that the distribution of returns usually have tails that are heavier than those of the normal distribution (see Figure 1.11). In addition, even though the returns are approximately uncorrelated over times (as predicted by the Black and Scholes model), they are not independent. This can be readily verified by the fact that the sample autocorrelations of the absolute values (or squares) of the returns are non-zero for a large number of lags (see Figure 1.12). Whereas the former

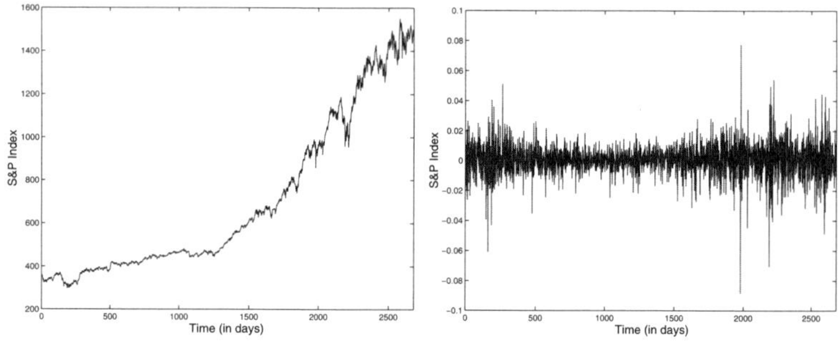

Fig. 1.10. Left: opening values of the Standard and Poors index 500 (S&P 500) over the period January 2, 1990–August 25, 2000. Right: log-returns of the opening values of the S&P 500, same period.

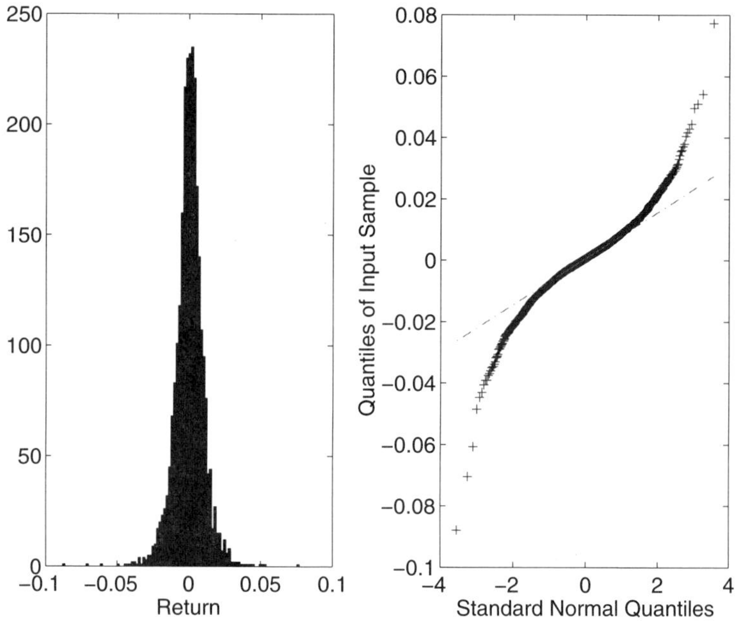

Fig. 1.11. Left: histogram of S&P 500 log-returns. Right: quantile-quantile regression plot of empirical quantiles of S&P 500 log-returns against quantiles of the standard normal distribution.

property indicates that the returns can be modeled by a white noise sequence (a stationary process with zero autocorrelation at all positive lags), the latter property indicates that the returns are dependent and that the dependence may even span a rather long period of time.

The variance of returns tends to change over time: the large and small values in the sample occur in clusters. Large changes tend to be followed by

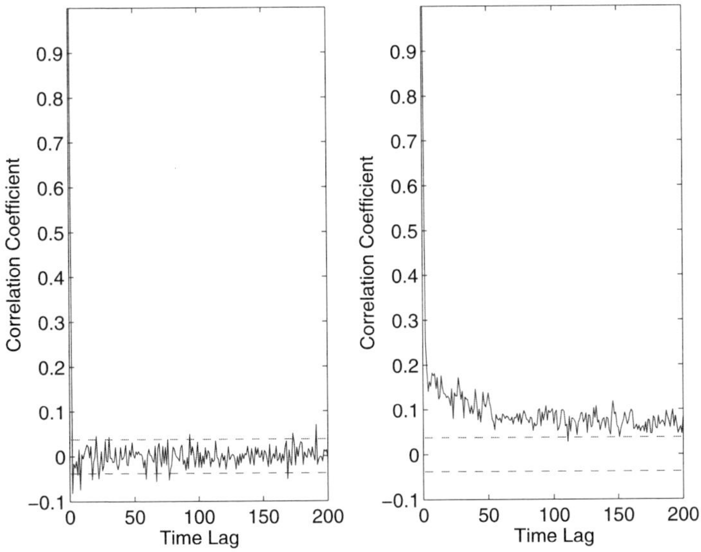

Fig. 1.12. Left: correlation coefficients of S&P 500 log-returns over the period January 2, 1990–August 25, 2000. The dashed lines are 95% confidence bands ($\pm 1.96/\sqrt{n}$) corresponding to the autocorrelation function of i.i.d. white Gaussian noise. Right: correlation coefficients of absolute values of log-returns, same period.

large changes—of either sign—and small changes tend to be followed by small changes, a phenomenon often referred to as *volatility clustering*.

Most models for return data that are used in practice are of a multiplicative form,

$$Y_k = \sigma_k V_k \ , \tag{1.19}$$

where $\{V_k\}_{k \geq 0}$ is an i.i.d. sequence and the *volatility process* $\{\sigma_k\}_{k \geq 0}$ is a non-negative stochastic process such that σ_k and V_k are independent for all k. Mostly, $\{\sigma_k\}$ is assumed to be strict sense stationary. It is often assumed that V_k is symmetric or, at least, has zero mean. The rationale for using these models is quite simple. First of all, the direction of the price changes is modeled by the sign of V_k only, independently of the order of magnitude of this change, which is directed by the volatility. Because σ_k and V_k are independent and V_k is assumed to have unit variance, σ_k^2 is then the conditional variance of X_k given σ_k. Most models assume that σ_k is a function of past values. The simplest model assumes that σ_k is a function of the squares of the previous observations. This leads to the celebrated autoregressive conditional heteroscedasticity (ARCH) model developed by Engle (1982),

$$\begin{aligned} Y_k &= \sqrt{X_k} V_k \ , \\ X_k &= \alpha_0 + \sum_{i=1}^{p} \alpha_i Y_{k-i}^2 \ , \end{aligned} \tag{1.20}$$

where $\alpha_0, \ldots, \alpha_p$ are non-negative constants. In the Engle (1982) model, $\{V_k\}$ is normal; hence the *conditional* error distribution is normal, but with *conditional* variance equal to a linear function of the p past squared observations. ARCH models are thus able to reproduce the tendency for extreme values to be followed by other extreme values, but of unpredictable sign. The autoregressive structure can be seen by the following argument. Writing $\nu_k = Y_k^2 - X_k = X_k(V_k^2 - 1)$, one obtains

$$Y_k^2 - \sum_{i=1}^{p} \alpha_i Y_{k-i}^2 = \alpha_0 + \nu_k \,. \tag{1.21}$$

Because $\{V_k\}$ is an i.i.d. sequence with zero mean and unit variance, $\{\nu_k\}_{k \geq 0}$ is an uncorrelated sequence. Because ARCH(p) processes do not fit log-returns very well unless the order p is quite large, various people have thought about improvements. As (1.21) bears some resemblance to an AR structure, a possible generalization is to introduce an ARMA structure. This construction leads to the so-called GARCH(p, q) process (Bollerslev *et al.*, 1994). This model displays some striking similarities to autoregressive models with Markov regime; this will be discussed in more detail below.

An alternative to the ARCH/GARCH framework is a model in which the variance is specified to follow some latent stochastic process. Such models, referred to as stochastic volatility (SV) models, appear in the theoretical literature on option pricing and exchange rate modeling. In contrast to GARCH-type processes, there is no direct feedback from past returns to the volatility process, which has been questioned as unnatural by some authors. Empirical versions of the SV model are typically formulated in discrete time, which makes inference problems easier to deal with. The canonical model in SV for discrete-time data is (Hull and White, 1987; Jacquier *et al.*, 1994),

$$\begin{aligned} X_{k+1} &= \phi X_k + \sigma U_k \,, & U_k &\sim \mathrm{N}(0,1) \,, \\ Y_k &= \beta \exp(X_k/2) V_k \,, & V_k &\sim \mathrm{N}(0,1) \,, \end{aligned} \tag{1.22}$$

where the observations $\{Y_k\}_{k \geq 0}$ are the log-returns, $\{X_k\}_{k \geq 0}$ is the log-volatility, which is assumed to follow a stationary autoregression of order 1, and $\{U_k\}_{k \geq 0}$ and $\{V_k\}_{k \geq 0}$ are independent i.i.d. sequences. The parameter β plays the role of the constant scaling factor, ϕ is the persistence (memory) in the volatility, and σ is the volatility of the log-volatility. Despite a very parsimonious representation, this model is capable of exhibiting a wide range of behaviors. Like ARCH/GARCH models, the model can give rise to a high persistence in volatility ("volatility clustering"). Even with $\phi = 0$, the model is a Gaussian scale mixture that will give rise to excess kurtosis in the marginal distribution of the data. In ARCH/GARCH models with normal errors, the degree of kurtosis is tied to the roots of the volatility equation; as the volatility becomes more correlated, the degree of kurtosis also increases. In the stochastic volatility model, the parameter σ governs the degree of mixing independently of the degree of smoothness in the variance evolution.

It is interesting to note that stochastic volatility models are related to conditionally Gaussian linear state-space models. By taking logarithms of the squared returns, one obtains,

$$\begin{aligned} X_k &= \phi X_{k-1} + \sigma U_{k-1} , \\ \log Y_k^2 &= \log \beta^2 + X_k + Z_k , \qquad \text{where } Z_k = \log V_k^2 . \end{aligned}$$

If V_k is standard normal, Z_k follows the $\log \chi_1^2$ distribution. This distribution may be approximated with arbitrary accuracy by a finite mixture of Gaussian distributions, and then the SV model becomes a conditionally Gaussian linear state-space model (Sandmann and Koopman, 1998; Durbin and Koopman, 2000). This time, the latent variable C_k is the mixture component and the model writes

$$\begin{aligned} W_{k+1} &= \phi W_k + U_k , & U_k &\sim \mathrm{N}(0,1) , \\ Y_k &= W_k + (\mu(C_k) + \sigma_V(C_k) V_k) , & V_k &\sim \mathrm{N}(0,1) . \end{aligned}$$

This representation of the stochastic volatility model may prove useful when deriving numerical algorithms to filter the hidden state or estimate the model parameters. ■

1.3.6 Switching Processes with Markov Regime

We now consider several examples that are not HMMs but belong to the class of Markov-switching models already mentioned in Section 1.2. Perhaps the most famous example of Markov-switching processes is the switching autoregressive process that was introduced by Hamilton (1989) to model econometric data.

1.3.6.1 Switching Linear Models

A switching linear autoregression is a model of the form

$$Y_k = \mu(C_k) + \sum_{i=1}^{d} a_i(C_k)(Y_{k-i} - \mu(C_{k-i}) + \sigma(C_k) V_k , \qquad k \geq 1 , \tag{1.23}$$

where $\{C_k\}_{k\geq 0}$, called the *regime*, is a Markov chain on a finite state space $\mathsf{C} = \{1, 2, \ldots, r\}$, and $\{V_k\}_{k\geq 0}$ is white noise independent of the regime; the functions $\mu : \mathsf{C} \to \mathbb{R}$, $a_i : \mathsf{C} \to \mathbb{R}$, $i = 1, \ldots, r$, and $\sigma : \mathsf{C} \to \mathbb{R}$ describe the dependence of the parameters on the realized regime. In this model, we change only the scale of the innovation as a function of the regime, but we can of course more drastically change the innovation distribution conditional on each state.

Remark 1.3.14. A model closely related to (1.23) is

$$Y_k = \mu(C_k) + \sum_{i=1}^{d} a_i(C_k)Y_{k-i} + \sigma(C_k)V_k \,, \qquad k \geq 1 \,. \tag{1.24}$$

In (1.23), $\mu(C_k)$ is the mean of Y_k conditional on the sequence of states $C_1, \ldots, C_k$, whereas in (1.24) the shift is on the intercept of the autoregressive process. ■

A model like this is not an HMM because, given $\{C_k\}$, the Y_k are not conditionally independent but rather form a non-homogeneous autoregression. Hence it is a Markov-switching model. Obviously, the conditional distribution of Y_k does not only depend on C_k and Y_{k-1} but also on other lagged Cs and Ys back to C_{k-d} and Y_{k-d}. By vectorizing the Ys and Cs, that is, stacking them in groups of d elements, we can obtain a process whose conditional distribution depends on one lagged variable only, as in Figure 1.2.

This model can be rewritten in state-space form. Let

$$\begin{aligned}
\mathbf{Y}_k &= [Y_k, Y_{k-1}, \ldots, Y_{k-d+1}]^t \,, \\
\mathbf{C}_k &= [C_k, C_{k-1}, \ldots, C_{k-d+1}]^t \,, \\
\boldsymbol{\mu}(\mathbf{C}_k) &= [\mu(C_k), \ldots, \mu(C_{k-d+1})]^t \,, \\
\mathbf{V}_k &= [V_k, 0, \ldots, 0]^t \,,
\end{aligned}$$

and denote by $C(c)$ the $d \times d$ companion matrix associated with the autoregressive coefficients of the state c,

$$A(c) = \begin{bmatrix} a_1(c) & a_2(c) & \ldots & \ldots & a_d(c) \\ 1 & 0 & & & 0 \\ 0 & 1 & 0 & & \vdots \\ \vdots & \ddots & \ddots & \ddots & \vdots \\ 0 & \ldots & 0 & 1 & 0 \end{bmatrix} . \tag{1.25}$$

The stacked observation vector $\mathbf{Y}_k$ then satisfies

$$\mathbf{Y}_k = \boldsymbol{\mu}(C_k) + A(C_k)\left(\mathbf{Y}_{k-1} - \boldsymbol{\mu}(\mathbf{C}_{k-1})\right) + \sigma(C_k)\mathbf{V}_k \,. \tag{1.26}$$

Interestingly enough, switching autoregressive processes have a rather rich probabilistic structure and have proven to be useful in many different contexts. We focus here on applications in econometrics and finance, but the scope of potential applications of these models span many different areas.

Example 1.3.15 (Regime Switches in Econometrics). The Hamilton (1989) model for the U.S. business cycle fostered a great deal of interest in Markov-switching autoregressive models as an empirical vehicle for characterizing macro-economic fluctuations. This model provides a formal statistical representation of the old idea that expansion and contraction constitute two distinct economic phases: Hamilton's model assumes that a macro-economic aggregate (real output growth, country's gross national product measured per quarter, annum, etc.) follows one of two different autoregressions depending on whether the economy is expanding or contracting, with the shift between regimes governed by the outcome of an unobserved Markov chain. The simple business cycle model advocated by Hamilton takes the form

$$Y_k = \mu(C_k) + \sum_{i=1}^{d} a_i(Y_{k-i} - \mu(C_{k-i})) + \sigma V_k \,, \tag{1.27}$$

where $\{V_k\}_{k\geq 0}$ is white Gaussian noise with zero mean and unit variance, and $\{C_k\}_{k\geq 0}$ is the unobserved latent variable that reflects the state of the business cycle (the autoregressive coefficients do not change; only the mean of the process is effectively modulated). In the simplest model, $\{C_k\}$ takes only two values; for example, $C_k = 0$ could indicate that the economy is in recession and $C_k = 1$ that it is in expansion. When $C_k = 0$, the average growth rate is given by $\mu(0)$, whereas when $C_k = 1$ the average growth rate is $\mu(1)$. This simple model can be made more sophisticated by making the variance a function of the state C_k as well,

$$Y_k = \mu(C_k) + \sum_{i=1}^{d} a_i(Y_{k-i} - \mu(C_{k-i})) + \sigma(C_k) V_k \,.$$

The Markov assumption on the hidden states basically says that if the economy was, say, in expansion the last period, the probability of going into recession is a fixed constant that does not depend on how long the economy has been in expansion or other measures of the strength of the expansion. This assumption, though rather naive, does not appear to be a bad representation of historical experience, though several researchers have suggested that more complicated specifications of the transition matrix ought to be considered.

Further reading on applications of switching linear Gaussian autoregressions in economics and finance can be found in, for instance, Krolzig (1997), Kim and Nelson (1999), Raj (2002), and Hamilton and Raj (2003). ■

It is possible to include an additional degree of sophistication by considering instead of a linear autoregression, linear state-space models (see for instance Tugnait, 1984; West and Harrison, 1989; Kim and Nelson, 1999; Doucet *et al.*, 2000a; Chen and Liu, 2000):

$$\begin{aligned} W_{k+1} &= \mu_W(C_{k+1}) + A(C_{k+1}) W_k + R(C_{k+1}) U_k \,, \\ Y_k &= \mu_Y(C_k) + B(C_k) W_k + S(C_k) V_k \,, \end{aligned} \tag{1.28}$$

where $\{C_k\}_{k\geq 0}$ is a Markov chain on a discrete state space, $\{U_k\}_{k\geq 0}$ and $\{V_k\}_{k\geq 0}$ are mutually independent i.i.d. sequences independent of $\{C_k\}_{k\geq 0}$, and μ_W, μ_Y, A, B, R, and S are vector- and matrix-valued functions of appropriate dimensions. Each state of the underlying Markov chain is then associated with a particular *regime* of the dynamic system, specified by particular values of $(\mu_W, \mu_Y, A, B, R, S)$ governing the behavior the state and observations. Switching linear state-space models approximate complex non-linear dynamics with a dynamic mixture of linear processes. This type of model has found a broad range of applications in econometrics (Kim and Nelson, 1999), in engineering including control (hybrid system, target tracking), signal processing (blind channel equalization) and communications (interference suppression) (Doucet *et al.*, 2000b, 2001b).

Example 1.3.16 (Maneuvering Target). Recall that in Example 1.3.12, we considered the motion of a single target that evolves in 2-D space with (almost) constant velocity. To represent changes in the velocity (either speed or direction or both), we redefine the model that describes the evolution of the state $W_k = (P_{x,k}, \dot{P}_{x,k}, P_{y,k}, \dot{P}_{y,k})$ by making it conditional upon a maneuver indicator $C_k = c_k \in \{1, \dots, r\}$ that is assumed to take only a finite number of values corresponding to various predefined maneuver scenarios. The state now evolves according to the following conditionally Gaussian linear equation

$$W_k = A(C_{k+1})W_k + R(C_{k+1})U_k \;, \quad U_k \sim \mathrm{N}(0, I) \;,$$

where $A(c)$ and $R(c)$ describe the parameters of the dynamic system characterizing the motion of the target for the maneuver labeled by c. Assuming that the observations are linear, $Y_k = BW_k + V_k$, the system is a switching Gaussian linear state-space model. ∎

1.3.6.2 Switching Non-linear Models

Switching autoregressive processes with Markov regime can be generalized by allowing non-linear autoregressions. Such models were considered in particular by Francq and Roussignol (1997) and take the form

$$Y_k = \phi(Y_{k-1}, \dots, Y_{k-d}, X_k) + \sigma(Y_{k-1}, \dots, Y_{k-d}, X_k)V_k \;, \tag{1.29}$$

where $\{X_k\}_{k \geq 0}$, called the *regime*, is a Markov chain on a discrete state space X, $\{V_k\}$ is an i.i.d. sequence, independent of the regime, with zero mean and unit variance, and $\phi : \mathbb{R}^d \times \mathsf{X} \to \mathbb{R}$ and $\sigma : \mathbb{R}^d \times \mathsf{X} \to \mathbb{R}^+$ are (measurable) functions. Of particular interest are the switching ARCH models (Francq *et al.*, 2001),

$$Y_k = \sqrt{\zeta_0(X_k) + \xi_1(X_k)Y_{t-1}^2 + \xi_d(X_k)Y_{t-d}^2}\, V_k \;.$$

Krishnamurthy and Rydén (1998) studied an even more general class of switching autoregressive processes that do not necessarily admit an additive decomposition; these are characterized by

$$Y_k = \phi(Y_{k-1}, \dots, Y_{k-d}, X_k, V_k) \;, \tag{1.30}$$

where $\{X_k\}_{k\geq 0}$, the regime, is a Markov chain on a discrete state space, $\{V_k\}_{k\geq 0}$ is an i.i.d. sequence independent of the regime, and $\phi : \mathbb{R}^d \times \mathsf{X} \to \mathbb{R}$ is a (measurable) function. Conditional on the regime, $\{Y_k\}$ is thus a dth order Markov chain on a general state space. Douc *et al.* (2004) studied the same kind of model but allowing the regime to evolve on a general state space.

Example 1.3.17 (Switching ARCH Models). Hamilton's (1989) switching autoregression (1.27) models a change in the business cycle phase as a shift in the average growth rate. By contrast, Hamilton and Susmel (1994) modeled changes in the volatility of the stock market as a shift in the overall scale of the ARCH process modeling stock returns. They suggested to model the monthly excess return of a financial asset (for example, the excess return of a financial index over the treasury bill yield) as

$$
\begin{aligned}
W_k &= \sqrt{\zeta_0 + \xi_1 W_{k-1}^2 + \cdots + \xi_m W_{k-m}^2}\, U_k \;, \\
Y_k &= \delta_0 + \delta_1 Y_{k-1} + \cdots + \delta_q Y_{k-q} + \sigma(C_k) W_k \;. \qquad (1.31)
\end{aligned}
$$

where $\{U_k\}_{k\geq 0}$ is Gaussian white noise with zero mean and $\{C_k\}_{k\geq 0}$ is an unobserved Markov chain on a discrete state space that represents the volatility phase of the stock market; $\{C_k\}$ and $\{U_k\}$ are independent. In the absence of such phases, the parameter $\sigma(C_k)$ would simply be constant over k, and (1.31) would describe stock returns by an autoregressive model whose innovations $\{U_k\}$ follow an mth order ARCH process.

More generally, when the function $\sigma : \mathsf{C} \to \mathbb{R}^+$ is not identically equal to unity, the latent ARCH process W_k is multiplied by a scale factor $\sigma(C_k)$ representing the current phase C_k that characterizes overall stock volatility. Assuming again that the market has two phases, $\mathsf{C} = \{0, 1\}$, and normalizing $\sigma(0) = 1$, $\sigma(1)$ has the interpretation as the ratio of the average variance of stock returns when $C_k = 1$ compared to that observed when $C_k = 0$. ■

1.4 Left-to-Right and Ergodic Hidden Markov Models

Most HMMs fall into one of two principally different classes of models: *left-to-right HMMs* and *ergodic HMMs*. By a left-to-right HMM is meant an HMM with a Markov chain that starts in a particular initial state, traverses a number of intermediate states, and finally terminates in a final state (this state may be considered as absorbing). When traversing the intermediate states the chain may not go backwards—toward the initial state—but only toward the final state. This progression is usually pictured from left to right; thus the term "left-to-right HMM". Speech recognition, discussed in Example 1.3.6 above, is typically a case where only left-to-right HMMs are used. A left-to-right HMM is not ergodic, but produces a sequence, typically of random length, of output. The number of states is also usually large.

In contrast, an ergodic HMM is one for which the underlying Markov chain is ergodic, or at least is irreducible and admits a unique stationary distribution (thus allowing for periodicity). Such a model can thus produce an infinitely long sequence of output, which is an ergodic sequence as well. The number of states, if the state space is finite, is typically small. Most of the examples mentioned in Section 1.3 correspond to ergodic HMMs.

Left-to-right HMMs and ergodic HMMs have much in common, in particular on the computational side. Indeed, computational algorithms like the EM algorithm, which is widely used for HMMs, may be implemented similarly whatever the structure of the Markov chain. Of course, because left-to-right HMMs often have many states, in such models it is often considerably more difficult to find the maximum likelihood estimator, say, among all local maxima of the likelihood function.

Having said that, when it comes to matters of theoretical statistics, there are noticeable differences between ergodic and left-to-right HMMs. Inference in left-to-right HMMs cannot be based on a single observed sequence of output, but is based on many, usually independent sequences. In contrast, inference in ergodic HMMs is usually based on a single long observed sequence, within which there is no independence. For this reason, issues regarding asymptotics of estimators and statistical tests are to be treated quite differently. For ergodic HMMs, one cannot rely on statistical theory for i.i.d. data but must develop specific methods. This development was initiated in the late 1960s by Baum and Petrie (1966) but was not continued until the 1990s. The case of left-to-right HMMs is simpler because it involves only independent observations, even though each observation is a sequence of random length.

It should however be stressed that, when dealing with left-to-right HMMs, finding the global maximum of the log-likelihood function, that is, the maximum likelihood estimator, or computing confidence intervals for parameters, etc., is not always a main goal, as for left-to-right HMMs the focus is often on how the model performs with respect to the particular application at hand: how good is the DNA sequence alignment, how large is the percentage of correctly recognized words, etc.? Indeed, even comparisons between models of different structure are often done by evaluating their performance on the actual application rather than applying statistical model selection procedures. For these reasons, one can argue that left-to-right HMMs are often applied in a "data fitting way" or "data mining way", rather than in a "statistical way".

Throughout this book, most examples given are based on ergodic HMMs, but the methodologies described are with few exceptions applicable to left-to-right HMMs either directly or after minor modifications.

2

Main Definitions and Notations

We now formally describe hidden Markov models, setting the notations that will be used throughout the book. We start by reviewing the basic definitions and concepts pertaining to Markov chains.

2.1 Markov Chains

2.1.1 Transition Kernels

Definition 2.1.1 (Transition Kernel). *Let* $(\mathsf{X}, \mathcal{X})$ *and* $(\mathsf{Y}, \mathcal{Y})$ *be two measurable spaces. An* unnormalized transition kernel *from* $(\mathsf{X}, \mathcal{X})$ *to* $(\mathsf{Y}, \mathcal{Y})$ *is a function* $Q : \mathsf{X} \times \mathcal{Y} \to [0, \infty]$ *that satisfies*

(i) for all $x \in \mathsf{X}$, $Q(x, \cdot)$ *is a positive measure on* $(\mathsf{Y}, \mathcal{Y})$*;*
(ii) for all $A \in \mathcal{Y}$*, the function* $x \mapsto Q(x, A)$ *is measurable.*

If $Q(x, \mathsf{Y}) = 1$ *for all* $x \in \mathsf{X}$*, then* Q *is called a* transition kernel*, or simply a* kernel. *If* $\mathsf{X} = \mathsf{Y}$ *and* $Q(x, \mathsf{X}) = 1$ *for all* $x \in \mathsf{X}$*, then* Q *will also be referred to as a* Markov transition kernel *on* $(\mathsf{X}, \mathcal{X})$.

An (unnormalized) transition kernel Q *is said to* admit a density *with respect to the positive measure* μ *on* Y *if there exists a non-negative function* $q : \mathsf{X} \times \mathsf{Y} \to [0, \infty]$*, measurable with respect to the product* σ*-field* $\mathcal{X} \otimes \mathcal{Y}$*, such that*

$$Q(x, A) = \int_A q(x, y)\, \mu(dy) \;, \qquad A \in \mathcal{Y} \;.$$

The function q *is then referred to as an (unnormalized)* transition density function.

When X *and* Y *are countable sets it is customary to write* $Q(x, y)$ *as a shorthand notation for* $Q(x, \{y\})$*, and* Q *is generally referred to as a* transition matrix *(whether or not* X *and* Y *are finite sets).*

We summarize below some key properties of transition kernels, introducing important pieces of notation that are used in the following.

- Let Q and R be unnormalized transition kernels from $(\mathsf{X}, \mathcal{X})$ to $(\mathsf{Y}, \mathcal{Y})$ and from $(\mathsf{Y}, \mathcal{Y})$ to $(\mathsf{Z}, \mathcal{Z})$, respectively. The product QR, defined by

$$QR(x, A) \stackrel{\text{def}}{=} \int Q(x, dy)\, R(y, A) , \qquad x \in \mathsf{X},\ A \in \mathcal{Z} ,$$

is then an unnormalized transition kernel from $(\mathsf{X}, \mathcal{X})$ to $(\mathsf{Z}, \mathcal{Z})$. If Q and R are transition kernels, then so is QR, that is, $QR(x, \mathsf{Z}) = 1$ for all $x \in \mathsf{X}$.
- If Q is an (unnormalized) Markov transition kernel on $(\mathsf{X}, \mathcal{X})$, its iterates are defined inductively by

$$Q^0(x, \cdot) = \delta_x \text{ for } x \in \mathsf{X} \text{ and } Q^k = QQ^{k-1} \text{ for } k \geq 1 .$$

These iterates satisfy the *Chapman-Kolmogorov* equation: $Q^{n+m} = Q^n Q^m$ for all $n, m \geq 0$. That is, for all $x \in \mathsf{X}$ and $A \in \mathcal{X}$,

$$Q^{n+m}(x, A) = \int Q^n(x, dy)\, Q^m(y, A) . \tag{2.1}$$

If Q admits a density q with respect to the measure μ on $(\mathsf{X}, \mathcal{X})$, then for all $n \geq 2$ the kernel Q^n is also absolutely continuous with respect to μ. The corresponding transition density is

$$q_n(x, y) = \int_{\mathsf{X}^{n-1}} q(x, x_1) \cdots q(x_{n-1}, y)\, \mu(dx_1) \cdots \mu(dx_{n-1}) . \tag{2.2}$$

- Positive measures operate on (unnormalized) transition kernels in two different ways. If μ is a positive measure on $(\mathsf{X}, \mathcal{X})$, the positive measure μQ on $(\mathsf{Y}, \mathcal{Y})$ is defined by

$$\mu Q(A) \stackrel{\text{def}}{=} \int \mu(dx)\, Q(x, A) , \qquad A \in \mathcal{Y} .$$

Moreover, the measure $\mu \otimes Q$ on the product space $(\mathsf{X} \times \mathsf{Y}, \mathcal{X} \otimes \mathcal{Y})$ is defined by

$$\mu \otimes Q(C) \stackrel{\text{def}}{=} \iint_C \mu(dx)\, Q(x, dy) , \qquad C \in \mathcal{X} \otimes \mathcal{Y} .$$

If μ is a probability measure and Q is a transition kernel, then μQ and $\mu \otimes Q$ are probability measures.
- (Unnormalized) transition kernels operate on functions. Let f be a real measurable function on Y. The real measurable function Qf on X is defined by

$$Qf(x) \stackrel{\text{def}}{=} \int Q(x, dy)\, f(y) , \qquad x \in \mathsf{X} ,$$

provided the integral is well-defined. It will sometimes be more convenient to use the alternative notation $Q(x, f)$ instead of $Qf(x)$. In particular, for $x \in \mathsf{X}$ and $A \in \mathcal{Y}$, $Q(x, A)$, $\delta_x Q(A)$, $Q\mathbb{1}_A(x)$, and $Q(x, \mathbb{1}_A)$, where $\mathbb{1}_A$ denotes the indicator function of the set A, are four equivalent ways of denoting the same quantity. In general, we prefer using the $Q(x, \mathbb{1}_A)$ and $Q(x, A)$ variants, which are less prone to confusion in complicated expressions.

- For any positive measure μ on $(\mathsf{X}, \mathcal{X})$ and any real measurable function f on $(\mathsf{Y}, \mathcal{Y})$,

$$(\mu Q)(f) = \mu(Qf) = \iint \mu(dx)\, Q(x, dy)\, f(y) ,$$

provided the integrals are well-defined. We may thus use the simplified notation $\nu Q f$ instead of $(\nu Q)(f)$ or $\nu(Qf)$.

Definition 2.1.2 (Reverse Kernel). *Let Q be a transition kernel from $(\mathsf{X}, \mathcal{X})$ to $(\mathsf{Y}, \mathcal{Y})$ and let ν be a probability measure on $(\mathsf{X}, \mathcal{X})$. The reverse kernel $\overleftarrow{Q}_\nu$ associated to ν and Q is a transition kernel from $(\mathsf{Y}, \mathcal{Y})$ to $(\mathsf{X}, \mathcal{X})$ such that for all bounded measurable functions f defined on $\mathsf{X} \times \mathsf{Y}$,*

$$\iint_{\mathsf{X}\times\mathsf{Y}} f(x, y)\, \nu(dx) Q(x, dy) = \iint_{\mathsf{X}\times\mathsf{Y}} f(x, y)\, \nu Q(dy)\, \overleftarrow{Q}_\nu(y, dx) . \tag{2.3}$$

The reverse kernel does not necessarily exist and is not uniquely defined. Nevertheless, if $\overleftarrow{Q}_{\nu,1}$ and $\overleftarrow{Q}_{\nu,2}$ satisfy (2.3), then for all $A \in \mathcal{X}$, $\overleftarrow{Q}_{\nu,1}(y, A) = \overleftarrow{Q}_{\nu,2}(y, A)$ for νQ-almost every y in Y. The reverse kernel does exist if X and Y are Polish spaces endowed with their Borel σ-fields (see Appendix A.1 for details). If Q admits a density q with respect to a measure μ on $(\mathsf{Y}, \mathcal{Y})$, then $\overleftarrow{Q}_\nu$ can be defined for all y such that $\int_{\mathsf{X}} q(z, y)\, \nu(dz) \neq 0$ by

$$\overleftarrow{Q}_\nu(y, dx) = \frac{q(x, y)\, \nu(dx)}{\int_{\mathsf{X}} q(z, y)\, \nu(dz)} . \tag{2.4}$$

The values of $\overleftarrow{Q}_\nu$ on the set $\{y \in \mathsf{Y} : \int_{\mathsf{X}} q(z, y)\, \nu(dz) = 0\}$ are irrelevant because this set is νQ-negligible. In particular, if X is discrete and μ is counting measure, then for all $(x, y) \in \mathsf{X} \times \mathsf{Y}$ such that $\nu Q(y) \neq 0$,

$$\overleftarrow{Q}_\nu(y, x) = \frac{\nu(x) Q(x, y)}{\nu Q(y)} . \tag{2.5}$$

2.1.2 Homogeneous Markov Chains

Let $(\Omega, \mathcal{F}, \mathrm{P})$ be a probability space and let $(\mathsf{X}, \mathcal{X})$ be a measurable space. An X-valued (discrete index) *stochastic process* $\{X_n\}_{n\geq 0}$ is a collection of X-valued random variables. A *filtration* of $(\Omega, \mathcal{F})$ is a non-decreasing sequence $\{\mathcal{F}_n\}_{n\geq 0}$ of sub-σ-fields of $\mathcal{F}$. A *filtered space* is a triple $(\Omega, \mathcal{F}, \mathbb{F})$, where $\mathbb{F}$ is a filtration; $(\Omega, \mathcal{F}, \mathbb{F}, \mathrm{P})$ is called a *filtered probability space*. For any filtration

$\mathbb{F} = \{\mathcal{F}_n\}_{n\geq 0}$, we denote by $\mathcal{F}_\infty = \vee_{n=0}^{\infty}\mathcal{F}_n$ the σ-field generated by $\mathbb{F}$ or, in other words, the minimal σ-field containing $\mathbb{F}$. A stochastic process $\{X_n\}_{n\geq 0}$ is *adapted* to $\mathbb{F} = \{\mathcal{F}_n\}_{n\geq 0}$, or simply $\mathbb{F}$-*adapted*, if X_n is $\mathcal{F}_n$-measurable for all $n \geq 0$ The *natural filtration* of a process $\{X_n\}_{n\geq 0}$, denoted by $\mathbb{F}^X = \{\mathcal{F}_n^X\}_{n\geq 0}$, is the smallest filtration with respect to which $\{X_n\}$ is adapted.

Definition 2.1.3 (Markov Chain). *Let $(\Omega, \mathcal{F}, \mathbb{F}, \mathrm{P})$ be a filtered probability space and let Q be a Markov transition kernel on a measurable space $(\mathsf{X}, \mathcal{X})$. An X-valued stochastic process $\{X_k\}_{k\geq 0}$ is said to be a* Markov chain *under* P, *with respect to the filtration $\mathbb{F}$ and with transition kernel Q, if it is $\mathbb{F}$-adapted and for all $k \geq 0$ and $A \in \mathcal{X}$,*

$$\mathrm{P}\left(X_{k+1} \in A \,|\, \mathcal{F}_k\right) = Q(X_k, A) \,. \tag{2.6}$$

The distribution of X_0 is called the initial distribution *of the chain, and X is called the* state space.

If $\{X_k\}_{k\geq 0}$ is $\mathbb{F}$-adapted, then for all $k \geq 0$ it holds that $\mathcal{F}_k^X \subseteq \mathcal{F}_k$; hence a Markov chain with respect to a filtration $\mathbb{F}$ is also a Markov chain with respect to its natural filtration. Hereafter, a Markov chain with respect to its natural filtration will simply be referred to as a Markov chain. When there is no risk of confusion, we will not mention the underlying probability measure P.

A fundamental property of a Markov chain is that its finite-dimensional distributions, and hence the distribution of the process $\{X_k\}_{k\geq 0}$, are entirely determined by the initial distribution and the transition kernel.

Proposition 2.1.4. *Let $\{X_k\}_{k\geq 0}$ be a Markov chain with initial distribution ν and transition kernel Q. For any $k \geq 0$ and any bounded $\mathcal{X}^{\otimes(k+1)}$-measurable function f on $\mathsf{X}^{(k+1)}$,*

$$\mathrm{E}\left[f(X_0, \ldots, X_k)\right] = \int f(x_0, \ldots, x_k)\, \nu(dx_0)\, Q(x_0, dx_1) \cdots Q(x_{k-1}, dx_k) \,.$$

In the following, we will use the generic notation $f \in \mathcal{F}_{\mathrm{b}}(\mathsf{Z})$ to denote the fact that f is a measurable bounded function on $(\mathsf{Z}, \mathcal{Z})$. In the case of Proposition 2.1.4 for instance, one considers functions f that are in $\mathcal{F}_{\mathrm{b}}\left(\mathsf{X}^{(k+1)}\right)$. More generally, we will usually describe measures and transition kernels on $(\mathsf{Z}, \mathcal{Z})$ by specifying the way they operate on the functions of $\mathcal{F}_{\mathrm{b}}(\mathsf{Z})$.

2.1.2.1 Canonical Version

Let $(\mathsf{X}, \mathcal{X})$ be a measurable space. The *canonical space* associated to $(\mathsf{X}, \mathcal{X})$ is the infinite-dimensional product space $(\mathsf{X}^{\mathbb{N}}, \mathcal{X}^{\otimes \mathbb{N}})$. The *coordinate process* is the X-valued stochastic process $\{X_k\}_{k\geq 0}$ defined on the canonical space by $X_n(\omega) = \omega(n)$. The canonical space will always be endowed with the natural filtration $\mathbb{F}^{\mathsf{X}}$ of the coordinate process.

Let $(\Omega, \mathcal{F}) = (\mathsf{X}^{\mathbb{N}}, \mathcal{X}^{\otimes \mathbb{N}})$ be the canonical space associated to the measurable space $(\mathsf{X}, \mathcal{X})$. The *shift operator* $\theta : \Omega \to \Omega$ is defined by

$$\theta(\omega)(n) = \omega(n+1) , \qquad n \geq 0 .$$

The iterates of the shift operator are defined inductively by $\theta^0 = \mathrm{Id}$ (the identity), $\theta^1 = \theta$ and $\theta^k = \theta \circ \theta^{k-1}$ for $k \geq 1$. If $\{X_k\}_{k \geq 0}$ is the coordinate process with associated natural filtration $\mathbb{F}^X$, then for all $k, n \geq 0$, $X_k \circ \theta^n = X_{k+n}$, and more generally for any $\mathcal{F}^X_k$-measurable random variable Y, $Y \circ \theta^n$ is $\mathcal{F}^X_{n+k}$-measurable.

The following theorem, which is a particular case of the Kolmogorov consistency theorem, states that it is always possible to define a Markov chain on the canonical space.

Theorem 2.1.5. *Let $(\mathsf{X}, \mathcal{X})$ be a measurable set, ν a probability measure on $(\mathsf{X}, \mathcal{X})$, and Q a transition kernel on $(\mathsf{X}, \mathcal{X})$. Then there exists an unique probability measure on $(\mathsf{X}^{\mathbb{N}}, \mathcal{X}^{\otimes \mathbb{N}})$, denoted by P_ν, such that the coordinate process $\{X_k\}_{k \geq 0}$ is a Markov chain (with respect to its natural filtration) with initial distribution ν and transition kernel Q.*

For $x \in \mathsf{X}$, let P_x be an alternative simplified notation for P_{δ_x}. Then for all $A \in \mathcal{X}^{\otimes \mathbb{N}}$, the mapping $x \to \mathrm{P}_x(A) = Q(x, A)$ is $\mathcal{X}$-measurable, and for any probability measure ν on $(\mathsf{X}, \mathcal{X})$,

$$\mathrm{P}_\nu(A) = \int \nu(dx)\, \mathrm{P}_x(A) . \tag{2.7}$$

The Markov chain defined in Theorem 2.1.5 is referred to as the *canonical version* of the Markov chain. The probability P_ν defined on $(\mathsf{X}^{\mathbb{N}}, \mathcal{X}^{\otimes \mathbb{N}})$ depends on ν and on the transition kernel Q. Nevertheless, the dependence with respect to Q is traditionally omitted in the notation. The relation (2.7) implies that $x \to \mathrm{P}_x$ is a regular version of the conditional probability $\mathrm{P}_\nu\,(\cdot \mid X_k = x)$ in the sense that one can rewrite (2.6) as

$$\mathrm{P}_\nu\left(X_{k+1} \in A \mid \mathcal{F}^X_k\right) = \mathrm{P}_\nu\left(X_1 \circ \theta^k \in A \,\middle|\, \mathcal{F}^X_k\right) = \mathrm{P}_{X_k}(X_1 \in A) \qquad \mathrm{P}_\nu\text{-a.s.}$$

2.1.2.2 Markov Properties

More generally, an induction argument easily yields the *Markov property*: for any $\mathcal{F}^X_\infty$-measurable random variable Y,

$$\mathrm{E}_\nu[Y \circ \theta^k \mid \mathcal{F}^X_k] = \mathrm{E}_{X_k}[Y] \qquad \mathrm{P}_\nu\text{-a.s.} \tag{2.8}$$

The Markov property can be extended to a specific class of random times known as *stopping times*. Let $\bar{\mathbb{N}} = \mathbb{N} \cup \{+\infty\}$ denote the extended integer set and let $(\Omega, \mathcal{F}, \mathbb{F})$ be a filtered space. Then, a mapping $\tau : \Omega \to \bar{\mathbb{N}}$ is said to be an $\mathbb{F}$-stopping time if $\{\tau = n\} \in \mathcal{F}_n$ for all $n \geq 0$. Intuitively, this means that at any time n one should be able to tell, based on the information $\mathcal{F}_n$

available at that time, if the stopping time occurs at this time n (or before then) or not. The class $\mathcal{F}_\tau$ defined by

$$\mathcal{F}_\tau = \{B \in \mathcal{F}_\infty : B \cap \{\tau = n\} \in \mathcal{F}_n \text{ for all } n \geq 0\} ,$$

is a σ-field, referred to as the σ-field of the events occurring before τ.

Theorem 2.1.6 (Strong Markov Property). *Let $\{X_k\}_{k\geq 0}$ be the canonical version of a Markov chain and let τ be an $\mathbb{F}^X$-stopping time. Then for any bounded $\mathcal{F}^X_\infty$-measurable function Ψ,*

$$\mathrm{E}_\nu[\mathbb{1}_{\{\tau<\infty\}}\Psi \circ \theta^\tau \mid \mathcal{F}^X_\tau] = \mathbb{1}_{\{\tau<\infty\}}\,\mathrm{E}_{X_\tau}[\Psi] \qquad \mathrm{P}_\nu\text{-a.s.} \tag{2.9}$$

We note that an $\mathcal{F}^X_\infty$-measurable function, or random variable, Ψ, is typically a function of potentially the whole trajectory of the Markov chain, although it may of course be a rather simple function like X_1 or $X_2 + X_3^2$.

2.1.3 Non-homogeneous Markov Chains

Definition 2.1.7 (Non-homogeneous Markov Chain). *Let $(\Omega, \mathcal{F}, \mathbb{F}, \mathrm{P})$ be a filtered probability space and let $\{Q_k\}_{k\geq 0}$ be a family of transition kernels on a measurable space $(\mathsf{X}, \mathcal{X})$. An X-valued stochastic process $\{X_k\}_{k\geq 0}$ is said to be a* non-homogeneous Markov chain *under* P, *with respect to the filtration $\mathbb{F}$ and with transition kernels $\{Q_k\}$, if it is $\mathbb{F}$-adapted and for all $k \geq 0$ and $A \in \mathcal{X}$,*

$$\mathrm{P}(X_{k+1} \in A \mid \mathcal{F}_k) = Q_k(X_k, A) .$$

For $i \leq j$ we define

$$Q_{i,j} = Q_i Q_{i+1} \cdots Q_j .$$

With this notation, if ν denotes the distribution of X_0 (which we refer to as the initial distribution as in the homogeneous case), the distribution of X_n is $\nu\, Q_{0,n-1}$. An important example of a non-homogeneous Markov chain is the so-called reverse chain. The construction of the reverse chain is based on the observation that if $\{X_k\}_{k\geq 0}$ is a Markov chain, then for any index $n \geq 1$ the time-reversed (or, index-reversed) process $\{X_{n-k}\}_{k=0}^n$ is a Markov chain too. The definition below provides its transition kernels.

Definition 2.1.8 (Reverse Chain). *Let Q be a Markov kernel on some space X, let ν be a probability measure on this space, and let $n \geq 1$ be an index. The reverse chain is the non-homogeneous Markov chain with initial distribution νQ^n, (time) index set $k = 0, 1, \ldots, n$ and transition kernels*

$$Q_k = \overleftarrow{Q}_{\nu Q^{n-k-1}} , \qquad k = 0, \ldots, n-1 ,$$

assuming that the reverse kernels are indeed well-defined.

If the transition kernel Q admits a transition density function q with respect to a measure μ on $(\mathsf{X}, \mathcal{X})$, then Q_k also admits a density with respect to the same measure μ, namely

$$h_k(y,x) = \frac{\int q_{n-k-1}(z,x)q(x,y)\,\nu(dz)}{\int q_{n-k}(z,y)\,\nu(dz)} . \tag{2.10}$$

Here, q_l is the transition density function of Q^l with respect to μ as defined in (2.2). If the state space is countable, then

$$Q_k(y,x) = \frac{\nu Q^{n-k-1}(x)Q(x,y)}{\nu Q^{n-k}(y)} . \tag{2.11}$$

An interesting question is in what cases the kernels Q_k do not depend on the index k and are in fact all equal to the forward kernel Q. A Markov chain with this property is said to be *reversible*. The following result gives a necessary and sufficient condition for reversibility.

Theorem 2.1.9. *Let* X *be a Polish space. A Markov kernel* Q *on* X *is reversible with respect to a probability measure* ν *if and only if for all bounded measurable functions* f *on* $\mathsf{X} \times \mathsf{X}$,

$$\iint f(x,x')\,\nu(dx)\,Q(x,dx') = \iint f(x,x')\,\nu(dx')\,Q(x',dx) . \tag{2.12}$$

The relation (2.12) is referred to as the *local balance equations* (or *detailed balance equations*). If the state space is countable, these equations hold if for all $x, x' \in \mathsf{X}$,

$$\nu(x)Q(x,x') = \nu(x')Q(x',x) . \tag{2.13}$$

Upon choosing a function f that only depends on the second variable in (2.12), it is easily seen that $\nu Q(f) = \nu(f)$ for all functions $f \in \mathcal{F}_b(\mathsf{X})$. We can also write this as $\nu = \nu Q$. This equation is referred to as the *global balance equations*. By induction, we find that $\nu Q^n = \nu$ for all $n \geq 0$. The left-hand side of this equation is the distribution of X_n, which thus does not depend on n when global balance holds. This is a form of stationarity, obviously implied by local balance. We shall tie this form of stationarity to the following customary definition.

Definition 2.1.10 (Stationary Process). *A stochastic process* $\{X_k\}$ *is said to be* stationary *(under* P*) if its finite-dimensional distributions are translation invariant, that is, if for all* $k, n \geq 1$ *and all* $n_1, \ldots, n_k$*, the distribution of the random vector* $(X_{n_1+n}, \ldots, X_{n_k+n})$ *does not depend on* n.

A stochastic process with index set $\mathbb{N}$, stationary but otherwise general, can always be extended to a process with index set $\mathbb{Z}$, having the same finite-dimensional distributions (and hence being stationary). This is a consequence of Kolmogorov's existence theorem for stochastic processes.

For a Markov chain, any multi-dimensional distribution can be expressed in terms of the initial distribution and the transition kernel—this is Proposition 2.1.4—and hence the characterization of stationarity becomes much simpler than above. Indeed, a Markov chain is stationary if and only if its initial distribution ν and transition kernel Q satisfy $\nu Q = \nu$, that is, satisfy global balance. Much more will be said about stationary distributions of Markov chains in Chapter 14.

2.2 Hidden Markov Models

A hidden Markov model is a doubly stochastic process with an underlying stochastic process that is not directly observable (it is "hidden") but can be observed only through another stochastic process that produces the sequence of observations. As shown in the introduction, the scope of HMMs is large and covers a variety of situations. To accommodate these conceptually different models, we now define formally a hidden Markov model.

2.2.1 Definitions and Notations

In simple cases such as fully discrete models, it is common to define hidden Markov models by using the concept of conditional independence. Indeed, this was the view taken in Chapter 1, where an HMM was defined as a bivariate process $\{(X_k, Y_k)\}_{k\geq 0}$ such that

- $\{X_k\}_{k\geq 0}$ is a Markov chain with transition kernel Q and initial distribution ν;
- Conditionally on the state process $\{X_k\}_{k\geq 0}$, the observations $\{Y_k\}_{k\geq 0}$ are independent, and for each n the conditional distribution of Y_n depends on X_n only.

It turns out that conditional independence is mathematically more difficult to define in general settings (in particular, when the state space X of the Markov chain is not countable), and we will adopt a different route to define general hidden Markov models. The HMM is defined as a bivariate Markov chain, only partially observed though, whose transition kernel has a special structure. Indeed, its transition kernel should be such that both the joint process $\{X_k, Y_k\}_{k\geq 0}$ *and* the marginal unobservable (or hidden) chain $\{X_k\}_{k\geq 0}$ are Markovian. From this definition, the usual conditional independence properties of HMMs will then follow (see Corollary 2.2.5 below).

Definition 2.2.1 (Hidden Markov Model). *Let* $(\mathsf{X}, \mathcal{X})$ *and* $(\mathsf{Y}, \mathcal{Y})$ *be two measurable spaces and let* Q *and* G *denote, respectively, a Markov transition kernel on* $(\mathsf{X}, \mathcal{X})$ *and a transition kernel from* $(\mathsf{X}, \mathcal{X})$ *to* $(\mathsf{Y}, \mathcal{Y})$. *Consider the Markov transition kernel defined on the product space* $(\mathsf{X} \times \mathsf{Y}, \mathcal{X} \otimes \mathcal{Y})$ *by*

$$T\left[(x,y),C\right] = \iint_C Q(x,dx')\,G(x',dy')\,, \qquad (x,y) \in \mathsf{X}\times\mathsf{Y},\, C \in \mathcal{X}\otimes\mathcal{Y}\,. \tag{2.14}$$

The Markov chain $\{X_k, Y_k\}_{k\geq 0}$ *with Markov transition kernel* T *and initial distribution* $\nu \otimes G$*, where* ν *is a probability measure on* $(\mathsf{X}, \mathcal{X})$*, is called a* hidden Markov model.

Although the definition above concerns the joint process $\{X_k, Y_k\}_{k\geq 0}$, the term *hidden* is only justified in cases where $\{X_k\}_{k\geq 0}$ is not observable. In this respect, $\{X_k\}_{k\geq 0}$ can also be seen as a fictitious intermediate process that is useful only in defining the distribution of the observed process $\{Y_k\}_{k\geq 0}$. We shall denote by P_ν and E_ν the probability measure and corresponding expectation associated with the process $\{X_k, Y_k\}_{k\geq 0}$ on the canonical space $\left((\mathsf{X}\times\mathsf{Y})^{\mathbb{N}}, (\mathcal{X}\otimes\mathcal{Y})^{\otimes\mathbb{N}}\right)$. Notice that this constitutes a slight departure from the Markov notations introduced previously, as ν is a probability measure on X only and not on the state space $\mathsf{X}\times\mathsf{Y}$ of the joint process. This slight abuse of notation is justified by the special structure of the model considered here. Equation (2.14) shows that whatever the distribution of the initial joint state (X_0, Y_0), even if it were not of the form $\nu \times G$, the law of $\{X_k, Y_k\}_{k\geq 1}$ only depends on the marginal distribution of X_0. Hence it makes sense to index probabilities and expectations by this marginal initial distribution only.

If both X and Y are countable, the hidden Markov model is said to be *discrete*, which is the case originally considered by Baum and Petrie (1966). Many of the examples given in the introduction (those of Section 1.3.2 for instance) correspond to cases where Y is uncountable and is a subset of $\mathbb{R}^d$ for some d. In such cases, we shall generally assume that the following holds true.

Definition 2.2.2 (Partially Dominated Hidden Markov Model). *The model of Definition 2.2.1 is said to be partially dominated if there exists a probability measure* μ *on* $(\mathsf{Y}, \mathcal{Y})$ *such that for all* $x \in \mathsf{X}$*,* $G(x,\cdot)$ *is absolutely continuous with respect to* μ*,* $G(x,\cdot) \ll \mu(\cdot)$*, with transition density function* $g(x,\cdot)$*. Then, for* $A \in \mathcal{Y}$*,* $G(x,A) = \int_A g(x,y)\,\mu(dy)$ *and the joint transition kernel* T *can be written as*

$$T\left[(x,y),C\right] = \iint_C Q(x,dx')g(x',y')\,\mu(dy')\; C \in \mathcal{X}\otimes\mathcal{Y}\,. \tag{2.15}$$

In the third part of the book (Chapter 10 and following) where we consider statistical estimation for HMMs with unknown parameters, we will require even stronger conditions and assume that the model is fully dominated in the following sense.

Definition 2.2.3 (Fully Dominated Hidden Markov Model). *If, in addition to the requirements of Definition 2.2.2, there exists a probability measure* λ *on* $(\mathsf{X}, \mathcal{X})$ *such that* $\nu \ll \lambda$ *and, for all* $x \in \mathsf{X}$*,* $Q(x,\cdot) \ll \lambda(\cdot)$ *with transition density function* $q(x,\cdot)$*. Then, for* $A \in \mathcal{X}$*,* $Q(x,A) = \int_A q(x,x')\,\lambda(dx')$

and the model is said to be fully dominated. *The joint Markov transition kernel T is then dominated by the product measure $\lambda \otimes \mu$ and admits the transition density function*

$$t\left[(x,y),(x',y')\right] \stackrel{\text{def}}{=} q(x,x')g(x',y') . \tag{2.16}$$

Note that for such models, we will generally re-use the notation ν to denote the *probability density function* of the initial state X_0 (with respect to λ) rather than the distribution itself.

2.2.2 Conditional Independence in Hidden Markov Models

In this section, we will show that the "intuitive" way of thinking about an HMM, in terms of conditional independence, is justified by Definition 2.2.1. Readers unfamiliar with conditioning in general settings may want to read more on this topic in Appendix A.4 before reading the rest of this section.

Proposition 2.2.4. *Let $\{X_k, Y_k\}_{k \geq 0}$ be a Markov chain over the product space $\mathsf{X} \times \mathsf{Y}$ with transition kernel T given by (2.14). Then, for any integer p, any ordered set $\{k_1 < \cdots < k_p\}$ of indices and all functions $f_1, \ldots, f_p \in \mathcal{F}_{\mathrm{b}}(\mathsf{Y})$,*

$$\mathrm{E}_\nu\left[\prod_{i=1}^{p} f_i(Y_{k_i}) \,\middle|\, X_{k_1}, \ldots, X_{k_p}\right] = \prod_{i=1}^{p} \int_{\mathsf{Y}} f_i(y)\, G(X_{k_i}, dy) . \tag{2.17}$$

Proof. For any $h \in \mathcal{F}_{\mathrm{b}}(\mathsf{X}^p)$, it holds that

$$\begin{aligned}
&\mathrm{E}_\nu\left[\prod_{i=1}^{p} f_i(Y_{k_i}) h(X_{k_1}, \ldots, X_{k_p})\right] \\
&= \int \cdots \int \nu(dx_0) G(x_0, dy_0) \left[\prod_{i=1}^{k_p} Q(x_{i-1}, dx_i)\, G(x_i, dy_i)\right] \\
&\qquad \times \left[\prod_{i=1}^{p} f_i(y_{k_i})\right] h(x_{k_1}, \ldots, x_{k_p}) \\
&= \int \cdots \int \nu(dx_0) \prod_{i=1}^{k_p} Q(x_{i-1}, dx_i) h(x_{k_1}, \ldots, x_{k_p}) \\
&\qquad \int \cdots \int \left[\prod_{i \notin \{k_1, \ldots, k_p\}} G(x_i, dy_i)\right] \left[\prod_{i \in \{k_1, \ldots, k_p\}} \int f_i(y_i)\, G(x_i, dy_i)\right] .
\end{aligned}$$

Because $\int G(x_i, dy_i) = 1$,

$$\mathrm{E}_\nu\left[\prod_{i=1}^{p} f_i(Y_{k_i}) h(X_{k_1},\dots,X_{k_p})\right] = \\ \mathrm{E}_\nu\left[h(X_{k_1},\dots,X_{k_p}) \prod_{i\in\{k_1,\dots,k_p\}} \int f_i(y_i)\, G(X_i, dy_i)\right] .$$

□

Corollary 2.2.5.

(i) For any integer p and any ordered set $\{k_1 < \cdots < k_p\}$ of indices, the random variables $Y_{k_1},\dots,Y_{k_p}$ are P_ν-conditionally independent given $(X_{k_1}, X_{k_2},\dots,X_{k_p})$.

(ii) For any integers k and p and any ordered set $\{k_1 < \cdots < k_p\}$ of indices such that $k \notin \{k_1,\dots,k_p\}$, the random variables Y_k and $(X_{k_1},\dots,X_{k_p})$ are P_ν-conditionally independent given X_k.

Proof. Part (i) is an immediate consequence of Proposition 2.2.4. To prove (ii), note that for any $f \in \mathcal{F}_\mathrm{b}(\mathsf{Y})$ and $h \in \mathcal{F}_\mathrm{b}(\mathsf{X}^p)$,

$$\begin{aligned} &\mathrm{E}_\nu\left[f(Y_k) h(X_{k_1},\dots,X_{k_p}) \mid X_k\right] \\ &\quad= \mathrm{E}_\nu\left[\mathrm{E}_\nu[f(Y_k) \mid X_{k_1},\dots,X_{k_p},X_k] h(X_{k_1},\dots,X_{k_p}) \,\middle|\, X_k\right] \\ &\quad= \mathrm{E}_\nu\left[f(Y_k) \mid X_k\right] \mathrm{E}_\nu[h(X_{k_1},\dots,X_{k_p}) \mid X_k\right] . \end{aligned}$$

□

As a direct application of Propositions A.4.2 and A.4.3, the conditional independence of the observations given the underlying sequence of states implies that for any integers p and p', any indices $k_1 < \cdots < k_p$ and $k'_1 < \cdots < k'_{p'}$ such that $\{k_1,\dots,k_p\} \cap \{k'_1,\dots,k'_{p'}\} = \emptyset$ and any function $f \in \mathcal{F}_\mathrm{b}(\mathsf{Y}^p)$,

$$\begin{aligned} &\mathrm{E}_\nu[f(Y_{k_1},\dots,Y_{k_p}) \mid X_{k_1},\dots,X_{k_p},X_{k'_1},\dots,X_{k'_p},Y_{k'_1},\dots Y_{k'_p}] \\ &\qquad\qquad= \mathrm{E}_\nu[f(Y_{k_1},\dots,Y_{k_p}) \mid X_{k_1},\dots,X_{k_p}] . \end{aligned} \tag{2.18}$$

Indeed, in terms of conditional independence of the variables,

$$(Y_{k_1},\dots,Y_{k_p}) \perp\!\!\!\perp (Y_{k'_1},\dots,Y_{k'_{p'}}) \mid (X_{k_1},\dots,X_{k_p},X_{k'_1},\dots,X_{k'_{p'}}) \ [\mathrm{P}_\nu]$$

and

$$(Y_{k_1},\dots,Y_{k_p}) \perp\!\!\!\perp (X_{k'_1},\dots,X_{k'_{p'}}) \mid (X_{k_1},\dots,X_{k_p}) \ [\mathrm{P}_\nu] .$$

Hence, by the contraction property of Proposition A.4.3,

$$(Y_{k_1},\dots,Y_{k_p}) \perp\!\!\!\perp (X_{k'_1},\dots,X_{k'_{p'}},Y_{k'_1},\dots,Y_{k'_p}) \mid (X_{k_1},\dots,X_{k_p}) \ [\mathrm{P}_\nu] ,$$

which implies (2.18).

2.2.3 Hierarchical Hidden Markov Models

In examples such as 1.3.16 and 1.3.15, we met hidden Markov models whose state variable naturally decomposes into two distinct sub-components. To accommodate such structures, we define a specific sub-class of HMMs for which the state X_k consists of two components, $X_k = (C_k, W_k)$. This additional structure will be used to introduce a level of hierarchy in the state variables. We call this class *hierarchical hidden Markov models*. In general, the hierarchical structure will be as follows.

- $\{C_k\}_{k\geq 0}$ is a Markov chain on a state space $(\mathsf{C}, \mathcal{C})$ with transition kernel Q_C and initial distribution ν_C. Thus, for any $f \in \mathcal{F}_\mathrm{b}(\mathsf{C})$ and any $k \geq 1$,

$$\mathrm{E}\left[f(C_k) \mid C_{0:k-1}\right] = Q_C(C_{k-1}, f) \quad \text{and} \quad \mathrm{E}_{\nu_C}[f(C_0)] = \nu_C(f) .$$

- Conditionally on $\{C_k\}_{k\geq 0}$, $\{W_k\}_{k\geq 0}$ is a Markov chain on $(\mathsf{W}, \mathcal{W})$. More precisely, there exists a transition kernel $Q_W : (\mathsf{X} \times \mathsf{C}) \times \mathcal{W} \to [0, 1]$ such that for any $k \geq 1$ and any function $f \in \mathcal{F}_\mathrm{b}(\mathsf{W})$,

$$\mathrm{E}\left[f(W_k) \mid W_{0:k-1}\, , C_{0:k}\right] = Q_W\left[(W_{k-1}, C_k), f\right] .$$

  In addition, there exists a transition kernel $\nu_W : \mathsf{C} \times \mathcal{W} \to [0, 1]$ such that for any $f \in \mathcal{F}_\mathrm{b}(\mathsf{W})$,

$$\mathrm{E}\left[f(W_0) \mid C_0\right] = \nu_W(C_0, f) .$$

We denote by $X_k = (C_k, W_k)$ the composite state variable. Then, $\{X_k\}_{k\geq 0}$ is a Markov chain on $\mathsf{X} = \mathsf{C} \times \mathsf{W}$ with transition kernel

$$Q\left[(c, w), A \times B\right] = \int_A \int_B Q_C(c, dc')\, Q_W\left[(w, c'), dw'\right], \quad A \in \mathcal{C},\ B \in \mathcal{W} ,$$

and initial distribution

$$\nu(A \times B) = \int_A \nu_C(dc)\, \nu_W(c, B) .$$

As before, we assume that $\{Y_k\}_{k\geq 0}$ is conditionally independent of $\{X_k\}_{k\geq 0}$ and such that the conditional distribution of Y_n depends on X_n only, meaning that (2.17) holds.

The distinctive feature of hierarchical HMMs is that it is often advantageous to consider that the state variables are $\{C_k\}_{k\geq 0}$ rather than $\{X_k\}_{k\geq 0}$. Of course, the model is then no longer an HMM because the observation Y_k depends on all partial states C_l for $l \leq k$ due to the marginalization of the intermediate component W_l (for $l = 0, \ldots, k$). Nonetheless, this point of view is often preferable, particularly in cases where the structure of $\{C_k\}_{k\geq 0}$ is very simple, such as when C is finite. The most common example of hierarchical HMM is the conditionally Gaussian linear state-space model (CGLSSM), which we already met in Examples 1.3.9, 1.3.11, and 1.3.16. We now formally define this model.

Definition 2.2.6 (Conditionally Gaussian Linear State-Space Model). *A CGLSSM is a model of the form*

$$\begin{aligned} W_{k+1} &= A(C_{k+1})W_k + R(C_{k+1})U_k \;, \quad W_0 \sim \mathrm{N}(\mu_\nu, \Sigma_\nu) \;, \\ Y_k &= B(C_k)W_k + S(C_k)V_k \;, \end{aligned} \tag{2.19}$$

subject to the following conditions.

- *The indicator process $\{C_k\}_{k\geq 0}$ is a Markov chain with transition kernel Q_C and initial distribution ν_C. Usually, C is finite and then identified with the set $\{1, \ldots, r\}$.*
- *The state (or process) noise $\{U_k\}_{k\geq 0}$ and the measurement noise $\{V_k\}_{k\geq 0}$ are independent multivariate Gaussian white noises with zero mean and identity covariance matrices. In addition, the indicator process $\{C_k\}_{k\geq 0}$ is independent of both the state noise and of the measurement noise.*
- *A, B, R, and S are known matrix-valued functions of appropriate dimensions.*

Part I

State Inference

3

Filtering and Smoothing Recursions

This chapter deals with a fundamental issue in hidden Markov modeling: given a fully specified model and some observations $Y_0, \ldots, Y_n$, what can be said about the corresponding unobserved state sequence $X_0, \ldots, X_n$? More specifically, we shall be concerned with the evaluation of the conditional distributions of the state at index k, X_k, given the observations $Y_0, \ldots, Y_n$, a task that is generally referred to as *smoothing*. There are of course several options available for tackling this problem (Anderson and Moore, 1979, Chapter 7) and we focus, in this chapter, on the *fixed-interval smoothing* paradigm in which n is held fixed and it is desired to evaluate the conditional distributions of X_k for all indices k between 0 and n. Note that only the general mechanics of the smoothing problem are dealt with in this chapter. In particular, most formulas will involve integrals over X. We shall not, for the moment, discuss ways in which these integrals can be effectively evaluated, or at least approximated, numerically. We postpone this issue to Chapter 5, which deals with some specific classes of hidden Markov models, and Chapters 6 and 7, in which generally applicable Markov chain Monte Carlo methods or sequential importance sampling techniques are reviewed.

The driving line of this chapter is the existence of a variety of smoothing approaches that involve a number of steps that only increase linearly with the number of observations. This is made possible by the fact (to be made precise in Section 3.3) that conditionally on the observations $Y_0, \ldots, Y_n$, the state sequence still is a Markov chain, albeit a non-homogeneous one.

Readers already familiar with the field could certainly object that as the probabilistic structure of any hidden Markov model may be represented by the generic probabilistic network drawn in Figure 1.1 (Chapter 1), the fixed interval smoothing problem under consideration may be solved by applying the general principle known as probability propagation or sum-product—see Cowell *et al.* (1999) or Frey (1998) for further details and references. As patent however from Figure 1.1, the graph corresponding to the HMM structure is so simple and systematic in its design that efficient instances of the probability propagation approach are all based on combining two systematic phases:

one in which the graph is scanned systematically from left to right (or *forward pass*), and one in which the graph is scanned in reverse order (*backward pass*). In this context, there are essentially only three different ways of implementing the above principle, which are presented below in Sections 3.2.2, 3.3.1, and 3.3.2.

From a historical perspective, it is interesting to recall that most of the early references on smoothing, which date back to the 1960s, focused on the specific case of Gaussian linear state-space models, following the pioneering work by Kalman and Bucy (1961). The classic book by Anderson and Moore (1979) on *optimal filtering*, for instance, is fully devoted to linear state-space models—see also Chapter 10 of the recent book by Kailath *et al.* (2000) for a more exhaustive set of early references on the smoothing problem. Although some authors such as (for instance) Ho and Lee (1964) considered more general state-space models, it is fair to say that the Gaussian linear state-space model was the dominant paradigm in the automatic control community[1]. In contrast, the work by Baum and his colleagues on hidden Markov models (Baum *et al.*, 1970) dealt with the case where the state space X of the hidden state is finite. These two streams of research (on Gaussian linear models and finite state space models) remained largely separated. Approximately at the same time, in the field of probability theory, the seminal work by Stratonovich (1960) stimulated a number of contributions that were to compose a body of work generally referred to as *filtering theory*. The object of filtering theory is to study inference about partially observable Markovian processes in *continuous time*. A number of early references in this domain indeed consider some specific form of discrete state space continuous-time equivalent of the HMM (Shiryaev, 1966; Wonham, 1965)—see also Lipster and Shiryaev (2001), Chapter 9. Working in continuous time, however, implies the use of mathematical tools that are definitely more complex than those needed to tackle the discrete-time model of Baum *et al.* (1970). As a matter of fact, filtering theory and hidden Markov models evolved as two mostly independent fields of research. A poorly acknowledged fact is that the pioneering paper by Stratonovich (1960) (translated from an earlier Russian publication) describes, in its first section, an equivalent to the forward-backward smoothing approach of Baum *et al.* (1970). It turns out, however, that the formalism of Baum *et al.* (1970) generalizes well to models where the state space is *not* discrete anymore, in contrast to that of Stratonovich (1960) (see Section 3.4 for the exact correspondence between both approaches).

[1]Interestingly, until the early 1980s, the works that *did not* focus on the linear state-space model were usually advertised by the use of the words "Bayes" or "Bayesian" in their title—see, e.g., Ho and Lee (1964) or Askar and Derin (1981).

3.1 Basic Notations and Definitions

In the rest of this chapter, the principles of smoothing as introduced by Baum *et al.* (1970) are exposed in a general setting that is suitable for all the examples introduced in Section 1.3.

3.1.1 Likelihood

The joint probability of the unobservable states and observations up to index n is such that for any function $f \in \mathcal{F}_b\left(\{\mathsf{X} \times \mathsf{Y}\}^{n+1}\right)$,

$$\mathrm{E}_\nu[f(X_0, Y_0, \ldots, X_n, Y_n)] = \int \cdots \int f(x_0, y_0, \ldots, x_n, y_n) \\ \times \nu(dx_0) g(x_0, y_0) \prod_{k=1}^{n} \{Q(x_{k-1}, dx_k) g(x_k, y_k)\} \mu_n(dy_0, \ldots, dy_n) , \quad (3.1)$$

where μ_n denotes the product distribution $\mu^{\otimes(n+1)}$ on $(\mathsf{X}^{n+1}, \mathcal{X}^{\otimes(n+1)})$. Marginalizing with respect to the unobservable variables $X_0, \ldots, X_n$, one obtains the marginal distribution of the observations only,

$$\mathrm{E}_\nu[f(Y_0, \ldots, Y_n)] = \int \cdots \int f(y_0, \ldots, y_n) \, \mathrm{L}_{\nu,n}(y_0, \ldots, y_n) \, \mu_n(dy_0, \ldots, dy_n) , \quad (3.2)$$

where $\mathrm{L}_{\nu,n}$ is an important quantity which we define below for future reference.

Definition 3.1.1 (Likelihood). *The* likelihood *of the observations is the probability density function of $Y_0, Y_1, \ldots, Y_n$ with respect to μ_n defined, for all $(y_0, \ldots, y_n) \in \mathsf{Y}^{n+1}$, by*

$$\mathrm{L}_{\nu,n}(y_0, \ldots, y_n) = \\ \int \cdots \int \nu(dx_0) g(x_0, y_0) Q(x_0, dx_1) g(x_1, y_1) \cdots Q(x_{n-1}, dx_n) g(x_n, y_n) . \quad (3.3)$$

In addition,

$$\ell_{\nu,n} \stackrel{\text{def}}{=} \log \mathrm{L}_{\nu,n} , \quad (3.4)$$

is referred to as the log-likelihood function.

Remark 3.1.2 (Concise Notation for Sub-sequences). For the sake of conciseness, we will use in the following the notation $Y_{l:m}$ to denote the collection of consecutively indexed variables $Y_l, \ldots, Y_m$ wherever possible (proceeding the same way for the unobservable sequence $\{X_k\}$). In quoting (3.3) for instance, we shall write $\mathrm{L}_{\nu,n}(y_{0:n})$ rather than $\mathrm{L}_{\nu,n}(y_0, \ldots, y_n)$. By transparent convention, $Y_{k:k}$ refers to the single variable Y_k, although the second notation (Y_k) is to be preferred in this particular case. In systematic expressions, however, it may be helpful to understand $Y_{k:k}$ as a valid replacement of Y_k. For similar reasons, we shall, when needed, accept $Y_{k+1:k}$ as a valid empty set. The latter convention should easily be recalled by programmers, as instructions of the form "for i equals k+1 to k, do...", which do nothing, constitute a well-accepted ingredient of most programming idioms. ■

3.1.2 Smoothing

We first define generically what is meant by the word *smoothing* before deriving the basic results that form the core of the techniques discussed in the rest of the chapter.

Definition 3.1.3 (Smoothing, Filtering, Prediction). *For positive indices k, l, and n with $l \geq k$, denote by $\phi_{\nu,k:l|n}$ the conditional distribution of $X_{k:l}$ given $Y_{0:n}$, that is*

(a) $\phi_{\nu,k:l|n}$ is a transition kernel from $\mathsf{Y}^{(n+1)}$ to $\mathsf{X}^{(l-k+1)}$:
- *for any given set $A \in \mathcal{X}^{\otimes(l-k+1)}$, $y_{0:n} \mapsto \phi_{\nu,k:l|n}(y_{0:n}, A)$ is a $\mathcal{Y}^{\otimes(n+1)}$-measurable function,*
- *for any given sub-sequence $y_{0:n}$, $A \mapsto \phi_{\nu,k:l|n}(y_{0:n}, A)$ is a probability distribution on $(\mathsf{X}^{l-k+1}, \mathcal{X}^{\otimes(l-k+1)})$.*

(b) $\phi_{\nu,k:l|n}$ satisfies, for any function $f \in \mathcal{F}_{\mathrm{b}}\left(\mathsf{X}^{l-k+1}\right)$,

$$\mathrm{E}_\nu\left[f(X_{k:l}) \mid Y_{0:n}\right] = \int \cdots \int f(x_{k:l})\, \phi_{\nu,k:l|n}(Y_{0:n}, dx_{k:l}) \;,$$

where the equality holds P_ν-almost surely. Specific choices of k and l give rise to several particular cases of interest:

Joint Smoothing: $\phi_{\nu,0:n|n}$, for $n \geq 0$;

(Marginal) Smoothing: $\phi_{\nu,k|n}$ for $n \geq k \geq 0$;

Prediction: $\phi_{\nu,n+1|n}$ for $n \geq 0$; In describing algorithms, it will be convenient to extend our notation to use $\phi_{\nu,0|-1}$ as a synonym for the initial distribution ν;

p-step Prediction: $\phi_{\nu,n+p|n}$ for $n, p \geq 0$.

Filtering: $\phi_{\nu,n|n}$ for $n \geq 0$; Because the use of filtering will be preeminent in the following, we shall most often abbreviate $\phi_{\nu,n|n}$ to $\phi_{\nu,n}$.

In more precise terms (see details in Section A.2 of Appendix A), $\phi_{\nu,k:l|n}$ is *a version* of the conditional distribution of $X_{k:l}$ given $Y_{0:n}$. It is however not obvious that such a quantity indeed exists in great generality. The proposition below complements Definition 3.1.3 by a constructive approach to defining the smoothing quantities from the elements of the hidden Markov model.

Proposition 3.1.4. *Consider a hidden Markov model compatible with Definition 2.2.2, let n be a positive integer and $y_{0:n} \in \mathsf{Y}^{n+1}$ a sub-sequence such that $\mathrm{L}_{\nu,n}(y_{0:n}) > 0$. The joint smoothing distribution $\phi_{\nu,0:n|n}$ then satisfies*

$$\phi_{\nu,0:n|n}(y_{0:n}, f) = \mathrm{L}_{\nu,n}(y_{0:n})^{-1} \int \cdots \int f(x_{0:n}) \times \nu(dx_0) g(x_0, y_0) \prod_{k=1}^{n} Q(x_{k-1}, dx_k) g(x_k, y_k) \quad (3.5)$$

for all functions $f \in \mathcal{F}_{\mathrm{b}}\left(\mathsf{X}^{n+1}\right)$. *Likewise, for indices* $p \geq 0$,

$$\phi_{\nu,0:n+p|n}(y_{0:n}, f) = \int \cdots \int f(x_{0:n+p}) \times \phi_{\nu,0:n|n}(y_{0:n}, dx_{0:n}) \prod_{k=n+1}^{n+p} Q(x_{k-1}, dx_k) \quad (3.6)$$

for all functions $f \in \mathcal{F}_{\mathrm{b}}\left(\mathsf{X}^{n+p+1}\right)$.

Proof. Equation (3.5) defines $\phi_{\nu,0:n|n}$ in a way that obviously satisfies part (a) of Definition 3.1.3. To prove the (b) part of the definition, recall the characterization of the conditional expectation given in Appendix A.2 and consider a function $h \in \mathcal{F}_{\mathrm{b}}\left(\mathsf{Y}^{n+1}\right)$. By (3.1),

$$\mathrm{E}_\nu[h(Y_{0:n}) f(X_{0:n})] = \int \cdots \int h(y_{0:n}) f(x_{0:n}) \times \nu(dx_0) g(x_0, y_0) \left[\prod_{k=1}^{n} Q(x_{k-1}, dx_k) g(x_k, y_k)\right] \mu_n(dy_{0:n}) .$$

Using Definition 3.1.1 of the likelihood $\mathrm{L}_{\nu,n}$ and (3.5) for $\phi_{\nu,0:n|n}$ yields

$$\begin{aligned} \mathrm{E}_\nu[h(Y_{0:n}) f(X_{0:n})] &= \int \cdots \int h(y_{0:n})\, \phi_{\nu,0:n|n}(y_{0:n}, f) \mathrm{L}_{\nu,n}(y_{0:n})\, \mu_n(dy_{0:n}) \\ &= \mathrm{E}_\nu[h(Y_{0:n}) \phi_{\nu,0:n|n}(Y_{0:n}, f)] . \end{aligned} \quad (3.7)$$

Hence $\mathrm{E}_\nu[f(X_{0:n}) \mid Y_{0:n}]$ equals $\phi_{\nu,0:n|n}(Y_{0:n}, f)$, P_ν-a.e., for any function $f \in \mathcal{F}_{\mathrm{b}}\left(\mathsf{X}^{n+1}\right)$.

For (3.6), proceed similarly and consider two functions $f \in \mathcal{F}_{\mathrm{b}}\left(\mathsf{X}^{n+p+1}\right)$ and $h \in \mathcal{F}_{\mathrm{b}}\left(\mathsf{Y}^{n+1}\right)$. First apply (3.1) to obtain

$$\mathrm{E}_\nu[h(Y_{0:n}) f(X_{0:n+p})] = \int \cdots \int f(x_{0:n+p}) \times \nu(dx_0) g(x_0, y_0) \left[\prod_{k=1}^{n} Q(x_{k-1}, dx_k) g(x_k, y_k)\right] h(y_{0:n}) \times \left[\prod_{l=n+1}^{n+p} Q(x_{l-1}, dx_l) g(x_l, y_l)\right] \mu_{n+p}(dy_{0:n+p}) .$$

When integrating with respect to the subsequence $y_{n+1:n+p}$, the third line of the previous equation reduces to $\prod_{l=n+1}^{n+p} Q(x_{l-1}, dx_l)\mu_n(dy_{0:n})$. Finally use (3.3) and (3.5) to obtain

$$\mathrm{E}_\nu[h(Y_{0:n})f(X_{0:n+p})] = \int \cdots \int h(y_{0:n})f(x_{0:n+p})$$
$$\phi_{\nu,0:n|n}(y_{0:n}, dx_{0:n}) \left[\prod_{k=n+1}^{n+p} Q(x_{k-1}, dx_k)\right] \mathrm{L}_{\nu,n}(y_{0:n})\mu_n(dy_{0:n}) , \quad (3.8)$$

which concludes the proof. □

Remark 3.1.5. The requirement that $\mathrm{L}_{\nu,n}(y_{0:n})$ be non-null is obviously required to guarantee that (3.5) makes sense and that (3.7) and (3.8) are correct. Note that if S is a set such that $\int_{\mathsf{S}} \mathrm{L}_{\nu,n}(y_{0:n})\mu_n(dy_{0:n}) = 0$, $\mathrm{P}_\nu(Y_{0:n} \in \mathsf{S}) = 0$ and the value of $\phi_{\nu,0:n|n}(y_{0:n}, \cdot)$ for $y_{0:n} \in \mathsf{S}$ is irrelevant (see discussion in Appendix A.3).

In the sequel, it is implicit that results similar to those in Proposition 3.1.4 hold for values of $y_{0:n} \in \mathsf{S}_{\nu,n} \subset \mathsf{Y}^{n+1}$, where the set $\mathsf{S}_{\nu,n}$ is such that $\mathrm{P}_\nu(Y_{0:n} \in \mathsf{S}_{\nu,n}) = 1$. In most models of practical interest, this nuance can be ignored as it is indeed possible to set $\mathsf{S}_{\nu,n} = \mathsf{Y}^{n+1}$. This is in particular the case when $g(x, y)$ is strictly positive for all values of $(x, y) \in \mathsf{X} \times \mathsf{Y}$. There are however more subtle cases where, for instance, the set $\mathsf{S}_{\nu,n}$ really depends upon the initial distribution ν (see Example 4.3.28). ■

Proposition 3.1.4 also implicitly defines all other particular cases of smoothing kernels mentioned in Definition 3.1.3, as these are obtained by marginalization. For instance, the marginal smoothing kernel $\phi_{\nu,k|n}$ for $0 \leq k \leq n$ is such that for any $y_{0:n} \in \mathsf{Y}^{n+1}$ and $f \in \mathcal{F}_\mathrm{b}(\mathsf{X})$,

$$\phi_{\nu,k|n}(y_{0:n}, f) \stackrel{\mathrm{def}}{=} \int \cdots \int f(x_k)\, \phi_{\nu,0:n|n}(y_{0:n}, dx_{0:k}) , \quad (3.9)$$

where $\phi_{\nu,0:n|n}$ is defined by (3.5).

Likewise, for any given $y_{0:n} \in \mathsf{Y}^{n+1}$, the p-step predictive distribution $\phi_{\nu,n+p|n}(y_{0:n}, \cdot)$ may be obtained by marginalization of the joint distribution $\phi_{\nu,0:n+p|n}(y_{0:n}, \cdot)$ with respect to all variables x_k except the last one (the one with index $k = n + p$). A closer examination of (3.6) together with the use of the Chapman-Kolmogorov equations introduced in (2.1) (cf. Chapter 14) directly shows that $\phi_{\nu,n+p|n}(y_{0:n}, \cdot) = \phi_{\nu,n}(y_{0:n}, \cdot)Q^p$, where $\phi_{\nu,n}$ refers to the filter (conditional distribution of X_n given $Y_{0:n}$).

3.1.3 The Forward-Backward Decomposition

As stated in the introduction, the rest of the chapter is devoted to techniques upon which the marginal smoothing kernels $\phi_{\nu,k|n}$ may be efficiently computed *for all values* of k in $\{0, \ldots, n\}$ for a given, pre-specified, value of n. This is the task that we referred to as *fixed interval smoothing*. In doing so, our main tool will be a simple representation of $\phi_{\nu,k|n}$, which we now introduce.

Replacing $\phi_{\nu,0:n|n}$ in (3.9) by its expression given in (3.5) shows that it is always possible to rewrite $\phi_{\nu,k|n}(y_{0:n}, f)$, for functions $f \in \mathcal{F}_b(\mathsf{X})$, as

$$\phi_{\nu,k|n}(y_{0:n}, f) = \mathrm{L}_{\nu,n}(y_{0:n})^{-1} \int f(x)\, \alpha_{\nu,k}(y_{0:k}, dx) \beta_{k|n}(y_{k+1:n}, x) , \tag{3.10}$$

where $\alpha_{\nu,k}$ and $\beta_{k|n}$ are defined below in (3.11) and (3.12), respectively. In simple terms, $\alpha_{\nu,k}$ correspond to the factors in the multiple integral that are to be integrated with respect to the state variables x_l with indices $l \leq k$ while $\beta_{k|n}$ gathers the remaining factors (which are to be integrated with respect to x_l for $l > k$). This simple splitting of the multiple integration in (3.9) constitutes the forward-backward decomposition.

Definition 3.1.6 (Forward-Backward "Variables"). *For* $k \in \{0, \ldots, n\}$, *define the following quantities.*
Forward Kernel $\alpha_{\nu,k}$ *is the non-negative finite kernel from* $(\mathsf{Y}^{k+1}, \mathcal{Y}^{\otimes(k+1)})$ *to* $(\mathsf{X}, \mathcal{X})$ *such that*

$$\alpha_{\nu,k}(y_{0:k}, f) = \int \cdots \int f(x_k)\, \nu(dx_0) g(x_0, y_0) \prod_{l=1}^{k} Q(x_{l-1}, dx_l) g(x_l, y_l) , \tag{3.11}$$

with the convention that the rightmost product term is empty for $k = 0$.
Backward Function $\beta_{k|n}$ *is the non-negative measurable function on* $\mathsf{Y}^{n-k} \times \mathsf{X}$ *defined by*

$$\beta_{k|n}(y_{k+1:n}, x) = \int \cdots \int Q(x, dx_{k+1}) g(x_{k+1}, y_{k+1}) \prod_{l=k+2}^{n} Q(x_{l-1}, dx_l) g(x_l, y_l) , \tag{3.12}$$

for $k \leq n-1$ *(with the same convention that the rightmost product is empty for* $k = n-1$*);* $\beta_{n|n}(\cdot)$ *is set to the constant function equal to 1 on* X.

The term "forward and backward variables" as well as the use of the symbols α and β is part of the HMM credo and dates back to the seminal work of Baum and his colleagues (Baum *et al.*, 1970, p. 168). It is clear however that for a general model as given in Definition 2.2.2, these quantities as defined in (3.11) and (3.12) are very different in nature, and indeed sufficiently so to prevent the use of the loosely defined term "variable". In the original framework studied by Baum and his coauthors where X is a finite set, both the forward measures $\alpha_{\nu,k}(y_{0:k}, \cdot)$ and the backward functions $\beta_{k|n}(y_{k+1:n}, \cdot)$ can be represented by vectors with non-negative entries. Indeed, in this case $\alpha_{\nu,k}(y_{0:k}, x)$ has the interpretation $\mathrm{P}_\nu(Y_0 = y_0, \ldots, Y_k = y_k, X_k = x)$ while

$\beta_{k|n}(y_{k+1:n}, x)$ has the interpretation $\mathrm{P}(Y_{k+1} = y_{k+1}, \ldots, Y_n = y_n \mid X_k = x)$. This way of thinking of $\alpha_{\nu,k}$ and $\beta_{k|n}$ may be extended to general state spaces: $\alpha_{\nu,k}(y_{0:k}, dx)$ is then the joint density (with respect to μ_{k+1}) of $Y_0, \ldots, Y_k$ and distribution of X_k, while $\beta_{k|n}(y_{k+1:n}, x)$ is the conditional joint density (with respect to μ_{n-k}) of $Y_{k+1}, \ldots, Y_n$ given $X_k = x$. Obviously, these entities may then not be represented as vectors of finite length, as when X is finite; this situation is the exception rather than the rule.

Let us simply remark at this point that while the forward kernel at index k is defined irrespectively of the length n of the observation sequence (as long as $n \geq k$), the same is not true for the backward functions. The sequence of backward functions clearly depends on the index where the observation sequence stops. In general, for instance, $\beta_{k|n-1}$ differs from $\beta_{k|n}$ even if we assume that the same sub-observation sequence $y_{0:n-1}$ is considered in both cases. This is the reason for adding the terminal index n to the notation used for the backward functions. This notation also constitute a departure from HMM traditions in which the backward functions are simply indexed by k. For $\alpha_{\nu,k}$, the situation is closer to standard practice and we simply add the subscript ν to recall that the forward kernel $\alpha_{\nu,k}$, in contrast with the backward measure, does depend on the distribution ν postulated for the initial state X_0.

3.1.4 Implicit Conditioning (Please Read This Section!)

We now pause to introduce a convention that will greatly simplify the exposition of the material contained in the first part of the book (from this chapter on, starting with the next section), both from terminological and notational points of view. This convention would however generate an acute confusion in the mind of a hypothetical reader who, having read Chapter 3 up to now, would decide to skip our friendly encouragement to read what follows carefully.

In the rest of Part I (with the notable exception of Section 4.3), we focus on the evaluation of quantities such as $\phi_{\nu,0:n|n}$ or $\phi_{\nu,k|n}$ for a given value of the observation sequence $y_{0:n}$. In this context, we *expunge from our notations the fact that all quantities depend on* $y_{0:n}$. In particular, we rewrite (3.5) for any $f \in \mathcal{F}_{\mathrm{b}}\left(\mathsf{X}^{n+1}\right)$ more concisely as

$$\phi_{\nu,0:n|n}(f) = \mathrm{L}_{\nu,n}^{-1} \int \cdots \int f(x_{0:n})\, \nu(dx_0) g_0(x_0) \prod_{i=1}^{n} Q(x_{i-1}, dx_i) g_i(x_i) \;, \tag{3.13}$$

where g_k are the data-dependent functions on X defined by $g_k(x) \stackrel{\text{def}}{=} g(x, y_k)$ for the particular sequence $y_{0:n}$ under consideration. The sequence of functions $\{g_k\}$ is about the only new notation that is needed as we simply re-use the previously defined quantities omitting their explicit dependence on the observations. For instance, in addition to writing $\mathrm{L}_{\nu,n}$ instead of $\mathrm{L}_{\nu,n}(y_{0:n})$, we will also use $\phi_n(\cdot)$ rather than $\phi_n(y_{0:n}, \cdot)$, $\beta_{k|n}(\cdot)$ rather than $\beta_{k|n}(y_{k+1:n}, \cdot)$, etc. This notational simplification implies a corresponding terminological adjustment. For instance, $\alpha_{\nu,k}$ will be referred to as the *forward measure* at index k and considered as a positive finite measure on $(\mathsf{X}, \mathcal{X})$. In all cases, the conversion should be easy to do mentally, as in the case of $\alpha_{\nu,k}$, for instance, what is meant is really "the measure $\alpha_{\nu,k}(y_{0:k}, \cdot)$, for a particular value of $y_{0:k} \in \mathsf{Y}^{k+1}$".

At first sight, omitting the observations may seem a weird thing to do in a statistically oriented book. However, for *posterior state inference* in HMMs, one indeed works conditionally on a given fixed sequence of observations. Omitting the observations from our notation will thus allow more concise expressions in most parts of the book. There are of course some properties of the hidden Markov model for which dependence with respect to the distribution of the observations does matter (hopefully!) This is in particular the case of Section 4.3 on forgetting and Chapter 12, which deals with statistical properties of the estimates for which we will make the dependence with respect to the observations explicit.

3.2 Forward-Backward

The forward-backward decomposition introduced in Section 3.1.3 is just a rewriting of the multiple integral in (3.9) such that for $f \in \mathcal{F}_{\mathrm{b}}(\mathsf{X})$,

$$\phi_{\nu,k|n}(f) = \mathrm{L}_{\nu,n}^{-1} \int f(x)\, \alpha_{\nu,k}(dx)\beta_{k|n}(x) \,, \tag{3.14}$$

where

$$\alpha_{\nu,k}(f) = \int \cdots \int f(x_k)\, \nu(dx_0) g_0(x_0) \prod_{l=1}^{k} Q(x_{l-1}, dx_l) g_l(x_l) \tag{3.15}$$

and

$$\beta_{k|n}(x) = \int \cdots \int Q(x, dx_{k+1}) g_{k+1}(x_{k+1}) \prod_{l=k+2}^{n} Q(x_{l-1}, dx_l) g_l(x_l) \,. \tag{3.16}$$

The last expression is, by convention, equal to 1 for the final index $k = n$. Note that we are now using the implicit conditioning convention discussed in the previous section.

3.2.1 The Forward-Backward Recursions

The point of using the forward-backward decomposition for the smoothing problem is that both the forward measures $\alpha_{\nu,k}$ and the backward functions

$\beta_{k|n}$ can be expressed *recursively* rather than by their integral representations (3.15) and (3.1.4). This is the essence of the *forward-backward algorithm* proposed by Baum *et al.* (1970, p. 168), which we now describe. Section 3.4 at the end of this chapter gives further comments on historical and terminological aspects of the forward-backward algorithm.

Proposition 3.2.1 (Forward-Backward Recursions). *The forward measures defined by* (3.15) *may be obtained, for all* $f \in \mathcal{F}_{\mathrm{b}}(\mathsf{X})$, *recursively for* $k = 1, \ldots, n$ *according to*

$$\alpha_{\nu,k}(f) = \int f(x') \int \alpha_{\nu,k-1}(dx) Q(x, dx') g_k(x') \tag{3.17}$$

with initial condition

$$\alpha_{\nu,0}(f) = \int f(x) g_0(x) \, \nu(dx) \; . \tag{3.18}$$

Similarly, the backward functions defined by (3.16) *may be obtained, for all* $x \in \mathsf{X}$, *by the recursion*

$$\beta_{k|n}(x) = \int Q(x, dx') g_{k+1}(x') \beta_{k+1|n}(x') \tag{3.19}$$

operating on decreasing indices $k = n - 1$ *down to* 0*; the initial condition is*

$$\beta_{n|n}(x) = 1 \; . \tag{3.20}$$

Proof. The proof of this result is straightforward and similar for both recursions. For $\alpha_{\nu,k}$ for instance, simply rewrite (3.15) as

$$\alpha_{\nu,k}(f) = \int_{x_k \in \mathsf{X}} f(x_k) \int_{x_{k-1} \in \mathsf{X}} \left[\int \cdots \int_{x_0 \in \mathsf{X}, \ldots, x_{k-2} \in \mathsf{X}} \nu(dx_0) g_0(x_0) \prod_{l=1}^{k-1} Q(x_{l-1}, dx_l) g_l(x_l) \right] Q(x_{k-1}, dx_k) g_k(x_k) \; ,$$

where the term in brackets is recognized as $\alpha_{\nu,k-1}(dx_{k-1})$. □

Remark 3.2.2 (Concise Markov Chain Notations). In the following, we shall often quote the above results using the concise Markov chain notations introduced in Chapter 2. For instance, instead of (3.17) and (3.19) one could write more simply $\alpha_{\nu,k}(f) = \alpha_{\nu,k-1} Q(f g_k)$ and $\beta_{k|n} = Q(g_{k+1} \beta_{k+1|n})$. Likewise, the decomposition (3.14) may be rewritten as

$$\phi_{\nu,k|n}(f) = \mathrm{L}_{\nu,n}^{-1} \alpha_{\nu,k} \left(f \beta_{k|n} \right) \; .$$

■

The main shortcoming of the forward-backward representation is that the quantities $\alpha_{\nu,k}$ and $\beta_{k|n}$ do not have an immediate probabilistic interpretation. Recall, in particular, that the first one is a finite (positive) measure but certainly not a probability measure, as $\alpha_{\nu,k}(1) \neq 1$ (in general). There is however an important solidarity result between the forward and backward quantities $\alpha_{\nu,k}$ and $\beta_{k|n}$, which is summarized by the following proposition.

Proposition 3.2.3. *For all indices* $k \in \{0, \dots, n\}$,

$$\alpha_{\nu,k}(\beta_{k|n}) = \mathrm{L}_{\nu,n}$$

and

$$\alpha_{\nu,k}(1) = \mathrm{L}_{\nu,k} \ ,$$

where $\mathrm{L}_{\nu,k}$ *refers to the likelihood of the observations up to index* k *(included) only, under* P_ν.

Proof. Because (3.14) must hold in particular for $f = 1$ and the marginal smoothing distribution $\phi_{\nu,k|n}$ is a probability measure,

$$\phi_{\nu,k|n}(1) \stackrel{\text{def}}{=} 1 = \mathrm{L}_{\nu,n}^{-1} \alpha_{\nu,k}\left(\beta_{k|n}\right) \ .$$

For the final index $k = n$, $\beta_{n|n}$ is the constant function equal to 1 and hence $\alpha_{\nu,n}(1) = \mathrm{L}_{\nu,n}$. This observation is however not specific to the final index n, as $\alpha_{\nu,k}$ only depends on the observations up to index k and thus any particular index may be selected as a potential final index (in contrast to what happens for the backward functions). □

3.2.2 Filtering and Normalized Recursion

The forward and backward quantities $\alpha_{\nu,k}$ and $\beta_{k|n}$, as defined in previous sections, are unnormalized in the sense that their scales are largely unknown. On the other hand, we know that $\alpha_{\nu,k}(\beta_{k|n})$ is equal to $\mathrm{L}_{\nu,n}$, the likelihood of the observations up to index n under P_ν.

The long-term behavior of the likelihood $\mathrm{L}_{\nu,n}$, or rather its logarithm, is a result known as the asymptotic equipartition property, or AEP (Cover and Thomas, 1991) in the information theoretic literature and as the Shannon-McMillan-Breiman theorem in the statistical literature. For HHMs, Proposition 12.3.3 (Chapter 12) shows that under suitable mixing conditions on the underlying unobservable chain $\{X_k\}_{k \geq 0}$, the AEP holds in that $n^{-1} \log \mathrm{L}_{\nu,n}$ converges P_ν-a.s. to a limit as n tends to infinity. The likelihood $\mathrm{L}_{\nu,n}$ will thus either grow to infinity or shrink to zero, depending on the sign of the limit, exponentially fast in n. This has the practical implication that in all cases where the recursions of Proposition 3.2.1 are effectively computable (like in the case of finite state space to be discussed in Chapter 5), the dynamics of the numerical values needed to represent $\alpha_{\nu,k}$ and $\beta_{k|n}$ is so large that it rapidly exceeds the available machine representation possibilities (even with high accuracy floating-point representations). The famous tutorial by Rabiner (1989) coined the term *scaling* to describe a practical solution to this problem. Interestingly, scaling also partly answers the question of the probabilistic interpretation of the forward and backward quantities.

Scaling as described by Rabiner (1989) amounts to normalizing $\alpha_{\nu,k}$ and $\beta_{k|n}$ by positive real numbers to keep the numeric values needed to represent $\alpha_{\nu,k}$ and $\beta_{k|n}$ within reasonable bounds. There are clearly a variety of options available, especially if one replaces (3.14) by the equivalent auto-normalized form

$$\phi_{\nu,k|n}(f) = [\alpha_{\nu,k}(\beta_{k|n})]^{-1} \int \alpha_{\nu,k}(f\beta_{k|n}) , \tag{3.21}$$

assuming that $\alpha_{\nu,k}(\beta_{k|n})$ is indeed finite and non-zero.

In our view, the most natural scaling scheme (developed below) consists in replacing the measure $\alpha_{\nu,k}$ and the function $\beta_{k|n}$ by scaled versions $\bar{\alpha}_{\nu,k}$ and $\bar{\beta}_{k|n}$ of these quantities, satisfying both

(i) $\bar{\alpha}_{\nu,k}(1) = 1$, and
(ii) $\bar{\alpha}_{\nu,k}(\bar{\beta}_{k|n}) = 1$.

Item (i) implies that the normalized forward measures $\bar{\alpha}_{\nu,k}$ are probability measures that have a probabilistic interpretation given below. Item (ii) implies that the normalized backward functions are such that $\phi_{\nu,k|n}(f) = \int f(x)\bar{\beta}_{k|n}(x)\,\bar{\alpha}_{\nu,k}(dx)$ for all $f \in \mathcal{F}_{\mathrm{b}}(\mathsf{X})$, without the need for a further renormalization. We note that this scaling scheme differs slightly from the one described by Rabiner (1989). The reason for this difference, which only affects the scaling of the backward functions, is non-essential and will be discussed in Section 3.4.

To derive the probabilistic interpretation of $\bar{\alpha}_{\nu,k}$, observe that (3.14) and Proposition 3.2.3, instantiated for the final index $k = n$, imply that the filtering distribution $\phi_{\nu,n}$ at index n (recall that $\phi_{\nu,n}$ is used as a simplified notation for $\phi_{\nu,n|n}$) may be written $[\alpha_{\nu,n}(1)]^{-1}\alpha_{\nu,n}$. This finding is of course not specific to the choice of the index n as already discussed when proving the second statement of Proposition 3.2.3. Thus, the normalized version $\bar{\alpha}_{\nu,k}$ of the forward measure $\alpha_{\nu,k}$ coincides with the filtering distribution $\phi_{\nu,k}$ introduced in Definition 3.1.3. This observation together with Proposition 3.2.3 implies that there is a unique choice of scaling scheme that satisfies the two requirements of the previous paragraph, as

$$\begin{aligned}\int f(x)\,\phi_{\nu,k|n}(dx) &= \mathrm{L}_{\nu,n}^{-1} \int f(x)\,\alpha_{\nu,k}(dx)\beta_{k|n}(x) \\ &= \int f(x)\,\underbrace{\mathrm{L}_{\nu,k}^{-1}\alpha_{\nu,k}(dx)}_{\bar{\alpha}_{\nu,k}(dx)}\,\underbrace{\mathrm{L}_{\nu,n}^{-1}\mathrm{L}_{\nu,k}\beta_{k|n}(x)}_{\bar{\beta}_{k|n}(x)}\end{aligned}$$

must hold for any $f \in \mathcal{F}_b(\mathsf{X})$. The following definition summarizes these conclusions, using the notation $\phi_{\nu,k}$ rather than $\bar{\alpha}_{\nu,k}$, as these two definitions refer to the same object—the filtering distribution at index k.

Definition 3.2.4 (Normalized Forward-Backward Variables). *For $k \in \{0,\ldots,n\}$, the normalized forward measure $\bar{\alpha}_{\nu,k}$ coincides with the filtering distribution $\phi_{\nu,k}$ and satisfies*

$$\phi_{\nu,k} = [\alpha_{\nu,k}(1)]^{-1}\alpha_{\nu,k} = \mathrm{L}_{\nu,k}^{-1}\alpha_{\nu,k} .$$

The normalized backward functions $\bar{\beta}_{k|n}$ are defined by

$$\bar{\beta}_{k|n} = \frac{\alpha_{\nu,k}(1)}{\alpha_{\nu,k}(\beta_{k|n})}\beta_{k|n} = \frac{\mathrm{L}_{\nu,k}}{\mathrm{L}_{\nu,n}}\beta_{k|n} .$$

The above definition would be pointless if computing $\alpha_{\nu,k}$ and $\beta_{k|n}$ was indeed necessary to obtain the normalized variables $\phi_{\nu,k}$ and $\bar{\beta}_{k|n}$. The following result shows that this is not the case.

Proposition 3.2.5 (Normalized Forward-Backward Recursions).

Forward Filtering Recursion *The filtering measures may be obtained, for all $f \in \mathcal{F}_b(\mathsf{X})$, recursively for $k = 1,\ldots,n$ according to*

$$c_{\nu,k} = \int\int \phi_{\nu,k-1}(dx)Q(x,dx')g_k(x') ,$$
$$\phi_{\nu,k}(f) = c_{\nu,k}^{-1}\int f(x)\int \phi_{\nu,k-1}(dx)Q(x,dx')g_k(x') , \tag{3.22}$$

with initial condition

$$c_{\nu,0} = \int g_0(x)\nu(dx) ,$$
$$\phi_{\nu,0}(f) = c_{\nu,0}^{-1}\int f(x)g_0(x)\,\nu(dx) .$$

Normalized Backward Recursion *The normalized backward functions may be obtained, for all $x \in \mathsf{X}$, by the recursion*

$$\bar{\beta}_{k|n}(x) = c_{\nu,k+1}^{-1}\int Q(x,dx')g_{k+1}(x')\bar{\beta}_{k+1|n}(x') \tag{3.23}$$

operating on decreasing indices $k = n-1$ down to 0; the initial condition is $\bar{\beta}_{n|n}(x) = 1$.

Once the two recursions above have been carried out, the smoothing distribution at any given index $k \in \{0,\ldots,n\}$ is available via

$$\phi_{\nu,k|n}(f) = \int f(x)\,\bar{\beta}_{k|n}(x)\phi_{\nu,k}(dx) \tag{3.24}$$

for all $f \in \mathcal{F}_b(\mathsf{X})$.

Proof. Proceeding by forward induction for $\phi_{\nu,k}$ and backward induction for $\beta_{k|n}$, it is easily checked from (3.22) and (3.23) that

$$\phi_{\nu,k} = \left(\prod_{l=0}^{k} c_{\nu,l}\right)^{-1} \alpha_{\nu,k} \quad \text{and} \quad \bar{\beta}_{k|n} = \left(\prod_{l=k+1}^{n} c_{\nu,l}\right)^{-1} \beta_{k|n} \; . \tag{3.25}$$

Because $\phi_{\nu,k}$ is normalized,

$$\phi_{\nu,k}(1) \stackrel{\text{def}}{=} 1 = \left(\prod_{l=0}^{k} c_{\nu,l}\right)^{-1} \alpha_{\nu,k}(1) \; .$$

Proposition 3.2.3 then implies that for any integer k,

$$\mathrm{L}_{\nu,k} = \prod_{l=0}^{k} c_{\nu,l} \; . \tag{3.26}$$

In other words, $c_{\nu,0} = \mathrm{L}_{\nu,0}$ and for subsequent indices $k \geq 1$, $c_{\nu,k} = \mathrm{L}_{\nu,k}/\mathrm{L}_{\nu,k-1}$. Hence (3.25) coincides with the normalized forward and backward variables as specified by Definition 3.2.4. □

We now pause to state a series of remarkable consequences of Proposition 3.2.5.

Remark 3.2.6. The forward recursion in (3.22) may also be rewritten to highlight a two-step procedure involving both the predictive and filtering measures. Recall our convention that $\phi_{\nu,0|-1}$ refers to the predictive distribution of X_0 when no observation is available and is thus an alias for ν, the distribution of X_0. For $k \in \{0, 1, \ldots, n\}$ and $f \in \mathcal{F}_{\mathrm{b}}(\mathsf{X})$, (3.22) may be decomposed as

$$\begin{aligned} c_{\nu,k} &= \phi_{\nu,k|k-1}(g_k) \; , \\ \phi_{\nu,k}(f) &= c_{\nu,k}^{-1}\, \phi_{\nu,k|k-1}(f g_k) \; , \\ \phi_{\nu,k+1|k} &= \phi_{\nu,k} Q \; . \end{aligned} \tag{3.27}$$

The equivalence of (3.27) with (3.22) is straightforward and is a direct consequence of the remark that $\phi_{k+1|k} = \phi_{\nu,k} Q$, which follows from Proposition 3.1.4 in Section 3.1.2. In addition, each of the two steps in (3.27) has a very transparent interpretation.

Predictor to Filter: The first two equations in (3.27) may be summarized as

$$\phi_{\nu,k}(f) \propto \int f(x)\, g(x, Y_k)\, \phi_{\nu,k|k-1}(dx) \; , \tag{3.28}$$

where the symbol $\propto$ means "up to a normalization constant" (such that $\phi_{\nu,k}(1) = 1$) and the full notation $g(x, Y_k)$ is used in place of $g_k(x)$ to highlight the dependence on the current observation Y_k. Equation (3.28) is recognized as Bayes' rule applied to a very simple equivalent Bayesian pseudo-model in which

- X_k is distributed *a priori* according to the predictive distribution $\phi_{\nu,k|k-1}$,
- g is the conditional probability density function of Y_k given X_k.

The filter $\phi_{\nu,k}$ is then interpreted as the posterior distribution of X_k given Y_k in this simple equivalent Bayesian pseudo-model.

Filter to Predictor: The last equation in (3.27) simply means that the updated predicting distribution $\phi_{\nu,k+1|k}$ is obtained by applying the transition kernel Q to the current filtering distribution $\phi_{\nu,k}$. We are thus left with the very basic problem of determining the one-step distribution of a Markov chain given its initial distribution.

■

Remark 3.2.7. In many situations, using (3.27) to determine $\phi_{\nu,k}$ is indeed the goal rather than simply a first step in computing smoothed distributions. In particular, for sequentially observed data, one may need to take actions based on the observations gathered so far. In such cases, filtering (or prediction) is the method of choice for inference about the unobserved states, a topic that will be developed further in Chapter 7. ■

Remark 3.2.8. Another remarkable fact about the filtering recursion is that (3.26) together with (3.27) provides a method for evaluating the likelihood $\mathrm{L}_{\nu,k}$ of the observations up to index k recursively in the index k. In addition, as $c_{\nu,k} = \mathrm{L}_{\nu,k}/\mathrm{L}_{\nu,k-1}$ from (3.26), $c_{\nu,k}$ may be interpreted as the conditional likelihood of Y_k given the previous observations $Y_{0:k-1}$. However, as discussed at the beginning of Section 3.2.2, using (3.26) directly is generally impracticable for numerical reasons. In order to avoid numerical under- or overflow, one can equivalently compute the log-likelihood $\ell_{\nu,k}$. Combining (3.26) and (3.27) gives the important formula

$$\ell_{\nu,k} \stackrel{\text{def}}{=} \log \mathrm{L}_{\nu,k} = \sum_{l=0}^{k} \log \phi_{\nu,l|l-1}(g_l) , \tag{3.29}$$

where $\phi_{\nu,l|l-1}$ is the one-step predictive distribution computed according to (3.27) (recalling that by convention, $\phi_{\nu,0|-1}$ is used as an alternative notation for ν). ■

Remark 3.2.9. The normalized backward function $\bar{\beta}_{k|n}$ does not have a simple probabilistic interpretation when isolated from the corresponding filtering measure. However, (3.24) shows that the marginal smoothing distribution, $\phi_{\nu,k|n}$, is dominated by the corresponding filtering distribution $\phi_{\nu,k}$ and that $\bar{\beta}_{k|n}$ is by definition the Radon-Nikodym derivative of $\phi_{\nu,k|n}$ with respect to $\phi_{\nu,k}$,

$$\bar{\beta}_{k|n} = \frac{d\phi_{\nu,k|n}}{d\phi_{\nu,k}}$$

As a consequence,

$$\inf\left\{M \in \mathbb{R} : \phi_{\nu,k}(\{\bar{\beta}_{k|n} \geq M\}) = 0\right\} \geq 1$$

and

$$\sup\left\{M \in \mathbb{R} : \phi_{\nu,k}(\{\bar{\beta}_{k|n} \leq M\}) = 0\right\} \leq 1 ,$$

with the conventions $\inf \emptyset = \infty$ and $\sup \emptyset = -\infty$. As a consequence, all values of $\bar{\beta}_{k|n}$ cannot get simultaneously large or close to zero as was the case for $\beta_{k|n}$, although one cannot exclude the possibility that $\bar{\beta}_{k|n}$ still has important dynamics without some further assumptions on the model.

The normalizing factor $\prod_{l=k+1}^{n} c_{\nu,l} = \mathrm{L}_{\nu,n}/\mathrm{L}_{\nu,k}$ by which $\bar{\beta}_{k|n}$ differs from the corresponding unnormalized backward function $\beta_{k|n}$ may be interpreted as the conditional likelihood of the future observations $Y_{k+1:n}$ given the observations up to index k, $Y_{0:k}$. ■

3.3 Markovian Decompositions

The forward-backward recursions (Proposition 3.2.1) and their normalized versions (Proposition 3.2.5) were probably already well-known to readers familiar with the hidden Markov model literature. A less widely observed fact is that the smoothing distributions may also be expressed using Markov transitions. In contrast to the forward-backward algorithm, this second approach will already be familiar to readers working with dynamic (or state-space) models (Kailath *et al.*, 2000, Chapter 10). Indeed, the method to be described in Section 3.3.2, when applied to the specific case of Gaussian linear state-space models, is known as Rauch-Tung-Striebel (sometimes, abbreviated to RTS) smoothing after Rauch *et al.* (1965). The important message here is that $\{X_k\}_{k\geq 0}$ (as well as the index-reversed version of $\{X_k\}_{k\geq 0}$, although greater care is needed to handle this second case) is a *non-homogeneous* Markov chain when conditioned on some observed values $\{Y_k\}_{0\leq k\leq n}$. The use of this approach for HMMs with finite state spaces as an alternative to the forward-backward recursions is due to Askar and Derin (1981)—see also (Ephraim and Merhav, 2002, Section V) for further references.

3.3.1 Forward Decomposition

Let n be a given positive index and consider the finite-dimensional distributions of $\{X_k\}_{k\geq 0}$ given $Y_{0:n}$. Our goal will be to show that the distribution of X_k given $X_{0:k-1}$ and $Y_{0:n}$ reduces to that of X_k given X_{k-1} only and $Y_{0:n}$, this for any positive index k. The following definition will be instrumental in decomposing the joint posterior distributions $\phi_{\nu,0:k|n}$.

Definition 3.3.1 (Forward Smoothing Kernels). *Given $n \geq 0$, define for indices $k \in \{0, \ldots, n-1\}$ the transition kernels*

$$\mathrm{F}_{k|n}(x, A) \stackrel{\text{def}}{=} \begin{cases} [\beta_{k|n}(x)]^{-1} \int_A Q(x, dx') g_{k+1}(x') \beta_{k+1|n}(x') & \textit{if } \beta_{k|n}(x) \neq 0 \\ 0 & \textit{otherwise} , \end{cases} \tag{3.30}$$

for any point $x \in \mathsf{X}$ *and set* $A \in \mathcal{X}$. *For indices* $k \geq n$, *simply set*

$$\mathrm{F}_{k|n} \stackrel{\text{def}}{=} Q , \tag{3.31}$$

where Q *is the transition kernel of the unobservable chain* $\{X_k\}_{k \geq 0}$.

Note that for indices $k \leq n-1$, $\mathrm{F}_{k|n}$ depends on the future observations $Y_{k+1:n}$ through the backward variables $\beta_{k|n}$ and $\beta_{k+1|n}$ only. The subscript n in the $\mathrm{F}_{k|n}$ notation is meant to underline the fact that, like the backward functions $\beta_{k|n}$, the forward smoothing kernels $\mathrm{F}_{k|n}$ depend on the final index n where the observation sequence ends. The backward recursion of Proposition 3.2.1 implies that $[\beta_{k|n}(x)]^{-1}$ is the correct normalizing constant. Thus, for any $x \in \mathsf{X}$, $A \mapsto \mathrm{F}_{k|n}(x, A)$ is a probability measure on $\mathcal{X}$. Because the functions $x \mapsto \beta_{k|n}(x)$ are measurable on $(\mathsf{X}, \mathcal{X})$, for any set $A \in \mathcal{X}$, $x \mapsto \mathrm{F}_{k|n}(x, A)$ is $\mathcal{X}/\mathcal{B}(\mathbb{R})$-measurable. Therefore, $\mathrm{F}_{k|n}$ is indeed a Markov transition kernel on $(\mathsf{X}, \mathcal{X})$. The next proposition provides a probabilistic interpretation of this definition in terms of the posterior distribution of the state at time $k+1$, given the observations up to time n and the state sequence up to time k.

Proposition 3.3.2. *Given* n, *for any index* $k \geq 0$ *and function* $f \in \mathcal{F}_{\mathrm{b}}(\mathsf{X})$,

$$\mathrm{E}_\nu[f(X_{k+1}) \,|\, X_{0:k}, Y_{0:n}] = \mathrm{F}_{k|n}(X_k, f) ,$$

where $\mathrm{F}_{k|n}$ *is the forward smoothing kernel defined by* (3.30) *for indices* $k \leq n-1$ *and* (3.31) *for indices* $k \geq n$.

Proof. First consider an index $0 \leq k \leq n$ and let f and h denote functions in $\mathcal{F}_{\mathrm{b}}(\mathsf{X})$ and $\mathcal{F}_{\mathrm{b}}(\mathsf{X}^{k+1})$, respectively. Then

$$\mathrm{E}_\nu[f(X_{k+1}) h(X_{0:k}) \,|\, Y_{0:n}] = \int \cdots \int f(x_{k+1}) h(x_{0:k}) \, \phi_{\nu,0:k+1|n}(dx_{0:k+1}) ,$$

which, using (3.13) and the definition (3.16) of the backward function, expands to

$$\begin{aligned} \mathrm{L}_{\nu,n}^{-1} \int \cdots \int h(x_{0:k}) \, \nu(dx_0) g_0(x_0) \prod_{i=1}^{k} Q(x_{i-1}, dx_i) g_i(x_i) \\ \times \int Q(x_k, dx_{k+1}) f(x_{k+1}) g_{k+1}(x_{k+1}) \\ \times \underbrace{\int \cdots \int \prod_{i=k+2}^{n} Q(x_{i-1}, dx_i) g_i(x_i)}_{\beta_{k+1|n}(x_{k+1})} . \end{aligned} \tag{3.32}$$

From Definition 3.3.1, $\int Q(x_k, dx_{k+1}) f(x_{k+1}) g_{k+1}(x_{k+1}) \beta_{k+1|n}(x_{k+1})$ is equal to $\mathrm{F}_{k|n}(x_k, f)\beta_{k|n}(x_k)$. Thus, (3.32) may be rewritten as

$$\mathrm{E}_\nu[f(X_{k+1})h(X_{0:k}) \,|\, Y_{0:n}] = \mathrm{L}_{\nu,n}^{-1} \int \cdots \int \mathrm{F}_{k|n}(x_k, f) h(x_{0:k}) \times \nu(dx_0) g_0(x_0) \left[\prod_{i=1}^{k} Q(x_{i-1}, dx_i) g_i(x_i) \right] \beta_{k|n}(x_k) \,. \quad (3.33)$$

Using the definition (3.16) of $\beta_{k|n}$ again, this latter integral is easily seen to be similar to (3.32) except for the fact that $f(x_{k+1})$ has been replaced by $\mathrm{F}_{k|n}(x_k, f)$. Hence

$$\mathrm{E}_\nu[f(X_{k+1})h(X_{0:k}) \,|\, Y_{0:n}] = \mathrm{E}_\nu[\mathrm{F}_{k|n}(X_k, f) h(X_{0:k}) \,|\, Y_{0:n}] \,,$$

for all functions $h \in \mathcal{F}_\mathrm{b}\left(\mathsf{X}^{k+1}\right)$ as requested.

For $k \geq n$, the situation is simpler because (3.6) implies that $\phi_{\nu,0:k+1|n} = \phi_{\nu,0:k|n} Q$. Hence,

$$\mathrm{E}_\nu[f(X_{k+1})h(X_{0:k}) \,|\, Y_{0:n}] = \int \cdots \int h(x_{0:k}) \, \phi_{\nu,0:k|n}(dx_{0:k}) \int Q(x_k, dx_{k+1}) f(x_{k+1}) \,,$$

and thus

$$\begin{aligned} \mathrm{E}_\nu[f(X_{k+1})h(X_{0:k}) \,|\, Y_{0:n}] &= \int \cdots \int h(x_{0:k}) \phi_{\nu,0:k|n}(dx_{0:k}) Q(x_k, f) \,, \\ &= \mathrm{E}_\nu[Q(X_k, f) h(X_{0:k}) \,|\, Y_{0:n}] \,. \end{aligned}$$

□

Remark 3.3.3. A key ingredient of the above proof is (3.32), which gives a representation of the joint smoothing distribution of the state variables $X_{0:k}$ given the observations up to index n, with $n \geq k$. This representation, which states that

$$\phi_{\nu,0:k|n}(f) = \mathrm{L}_{\nu,n}^{-1} \int \cdots \int f(x_{0:k}) \, \nu(dx_0) g_0(x_0) \left[\prod_{i=1}^{k} Q(x_{i-1}, dx_i) g_i(x_i) \right] \beta_{k|n}(x_k) \quad (3.34)$$

for all $f \in \mathcal{F}_\mathrm{b}\left(\mathsf{X}^{k+1}\right)$, is a generalization of the marginal forward-backward decomposition as stated in (3.14). ■

Proposition 3.3.2 implies that, *conditionally on* the observations $Y_{0:n}$, the state sequence $\{X_k\}_{k \geq 0}$ is a non-homogeneous Markov chain associated with the family of Markov transition kernels $\{\mathrm{F}_{k|n}\}_{k\geq 0}$ and initial distribution $\phi_{\nu,0|n}$. The fact that the Markov property of the state sequence is preserved when conditioning sounds surprising because the (marginal) smoothing distribution of the state X_k depends on both past and future observations. There is however nothing paradoxical here, as the Markov transition kernels $\mathrm{F}_{k|n}$ indeed depend (and depend only) on the future observations $Y_{k+1:n}$.

As a consequence of Proposition 3.3.2, the joint smoothing distributions may be rewritten in a form that involves the forward smoothing kernels using the Chapman-Kolmogorov equations (2.1).

Proposition 3.3.4. *For any integers* n *and* m*, function* $f \in \mathcal{F}_{\mathrm{b}}\left(\mathsf{X}^{m+1}\right)$ *and initial probability* ν *on* $(\mathsf{X}, \mathcal{X})$*,*

$$\mathrm{E}_\nu[f(X_{0:m}) \mid Y_{0:n}] = \int \cdots \int f(x_{0:m})\, \phi_{\nu,0|n}(dx_0) \prod_{i=1}^{m} \mathrm{F}_{i-1|n}(x_{i-1}, dx_i)\ , \tag{3.35}$$

where $\{\mathrm{F}_{k|n}\}_{k\geq 0}$ *are defined by* (3.30) *and* (3.31) *and* $\phi_{\nu,0|n}$ *is the marginal smoothing distribution defined, for any* $A \in \mathcal{X}$*, by*

$$\phi_{\nu,0|n}(A) = \left[\nu(g_0\beta_{0|n})\right]^{-1} \int_A \nu(dx) g_0(x) \beta_{0|n}(x)\ . \tag{3.36}$$

If one is only interested in computing the fixed point marginal smoothing distributions, (3.35) may also be used as the second phase of a smoothing approach which we recapitulate below.

Corollary 3.3.5 (Alternative Smoothing Algorithm).

Backward Recursion *Compute the backward variables* $\beta_{n|n}$ *down to* $\beta_{0|n}$ *by backward recursion according to* (3.19) *in Proposition 3.2.1.*

Forward Smoothing $\phi_{\nu,0|n}$ *is given by* (3.36) *and for* $k \geq 0$*,*

$$\phi_{\nu,k+1|n} = \phi_{\nu,k|n} \mathrm{F}_{k|n}\ ,$$

where $\mathrm{F}_{k|n}$ *are the forward kernels defined by* (3.30).

For numerical implementation, Corollary 3.3.5 is definitely less attractive than the normalized forward-backward approach of Proposition 3.2.5 because the backward pass cannot be carried out in normalized form without first determining the forward measures $\alpha_{\nu,k}$. We will discuss in Chapter 5 some specific models where these recursions can be implemented with some form of normalization, but generally speaking the backward decomposition to be described next is preferable for practical computation of the marginal smoothing distributions.

On the other hand, Proposition 3.3.4 provides a general decomposition of the joint smoothing distribution that will be instrumental in establishing some form of ergodicity of the Markov chain that corresponds to the unobservable states $\{X_k\}_{k\geq 0}$, conditional on some observations $Y_{0:n}$ (see Section 4.3).

3.3.2 Backward Decomposition

In the previous section it was shown that, conditionally on the observations up to index n, $Y_{0:n}$, the state sequence $\{X_k\}_{k\geq 0}$ is a Markov chain, with transition kernels $\mathrm{F}_{k|n}$. We now turn to the so-called *time-reversal* issue: is it true in general that the unobserved chain *with the indices in reverse order*, forms a non-homogeneous Markov chain, conditionally on some observations $Y_{0:n}$?

We already discussed time-reversal for Markov chains in Section 2.1 where it has been argued that the main technical difficulty consists in guaranteeing that the reverse kernel does exist. For this, we require somewhat stronger assumptions on the nature of X by assuming for the rest of this section that X is a Polish space and that $\mathcal{X}$ is the associated Borel σ-field. From the discussion in Section 2.1 (see Definition 2.1.2 and comment below), we then know that the reverse kernel does exist although we may not be able to provide a simple closed-form expression for it. The reverse kernel does have a simple expression, however, as soon as one assumes that the kernel to be reversed and the initial distribution admit densities with respect to some measure on X.

Let us now return to the smoothing problem. For positive indices k such that $k \leq n-1$, the posterior distribution of (X_k, X_{k+1}) given the observations up to time k satisfies

$$\mathrm{E}_\nu[f(X_k, X_{k+1}) \,|\, Y_{0:k}] = \iint f(x_k, x_{k+1})\, \phi_{\nu,k}(dx_k)\, Q(x_k, dx_{k+1}) \tag{3.37}$$

for all $f \in \mathcal{F}_{\mathrm{b}}(\mathsf{X} \times \mathsf{X})$. From the previous discussion, there exists a Markov transition kernel $\mathrm{B}_{\nu,k}$ which satisfies Definition 2.1.2, that is

$$\mathrm{B}_{\nu,k} \stackrel{\text{def}}{=} \{\mathrm{B}_{\nu,k}(x, A), x \in \mathsf{X}, A \in \mathcal{X}\}$$

such that for any function $f \in \mathcal{F}_{\mathrm{b}}(\mathsf{X} \times \mathsf{X})$,

$$\mathrm{E}_\nu[f(X_k, X_{k+1}) \,|\, Y_{0:k}] = \iint f(x_k, x_{k+1})\, \phi_{\nu,k+1|k}(dx_{k+1})\, \mathrm{B}_{\nu,k}(x_{k+1}, dx_k) \;, \tag{3.38}$$

where $\phi_{\nu,k+1|k} = \phi_{\nu,k} Q$ is the one-step predictive distribution.

Proposition 3.3.6. *Given a strictly positive index n, initial distribution ν, and index $k \in \{0, \ldots, n-1\}$,*

$$\mathrm{E}_\nu[f(X_k) \,|\, X_{k+1:n}, Y_{0:n}] = \mathrm{B}_{\nu,k}(X_{k+1}, f)$$

for any $f \in \mathcal{F}_{\mathrm{b}}(\mathsf{X})$. Here, $\mathrm{B}_{\nu,k}$ is the backward smoothing kernel defined in (3.38).

Before giving the proof of this result, we make a few remarks to provide some intuitive understanding of the backward smoothing kernels.

Remark 3.3.7. Contrary to the forward kernel, the backward transition kernel is only defined implicitly through the equality of the two representations (3.37) and (3.38). This limitation is fundamentally due to the fact that the backward kernel implies a non-trivial time-reversal operation.

Proposition 3.3.6 however allows a simple interpretation of the backward kernel: Because $\mathrm{E}_\nu[f(X_k) \mid X_{k+1:n}, Y_{0:n}]$ is equal to $\mathrm{B}_{\nu,k}(X_{k+1}, f)$ and thus depends neither on X_l for $l > k+1$ nor on Y_l for $l \geq k+1$, the tower property of conditional expectation (Proposition A.2.3) implies that not only is $\mathrm{B}_{\nu,k}(X_{k+1}, f)$ equal to $\mathrm{E}_\nu[f(X_k) \mid X_{k+1}, Y_{0:n}]$ but also coincides with $\mathrm{E}_\nu[f(X_k) \mid X_{k+1}, Y_{0:k}]$, for any $f \in \mathcal{F}_\mathrm{b}(\mathsf{X})$. In addition, the distribution of X_{k+1} given X_k and $Y_{0:k}$ reduces to $Q(X_k, \cdot)$ due to the particular form of the transition kernel associated with a hidden Markov model (see Definition 2.2.1). Recall also that the distribution of X_k given $Y_{0:k}$ is denoted by $\phi_{\nu,k}$. Thus, $\mathrm{B}_{\nu,k}$ can be interpreted as a Bayesian posterior in the equivalent pseudo-model where

- X_k is distributed *a priori* according to the filtering distribution $\phi_{\nu,k}$,
- The conditional distribution of X_{k+1} given X_k is $Q(X_k, \cdot)$.

$\mathrm{B}_{\nu,k}(X_{k+1}, \cdot)$ is then interpreted as the posterior distribution of X_k given X_{k+1} in this equivalent pseudo-model.

In particular, for HMMs that are "fully dominated" in the sense of Definition 2.2.3, Q has a transition probability density function q with respect to a measure λ on X. This is then also the case for $\phi_{\nu,k}$, which is a marginal of (3.13). In such cases, we shall use the slightly abusive but unambiguous notation $\phi_{\nu,k}(dx) = \phi_{\nu,k}(x)\,\lambda(dx)$ (that is, $\phi_{\nu,k}$ denotes the probability density function with respect to λ rather than the probability distribution). The backward kernel $\mathrm{B}_{\nu,k}(x_{k+1}, \cdot)$ then has a probability density function with respect to λ, which is given by Bayes' formula,

$$\mathrm{B}_{\nu,k}(x_{k+1}, x) = \frac{\phi_{\nu,k}(x) q(x, x_{k+1})}{\int_\mathsf{X} \phi_{\nu,k}(x) q(x, x_{k+1})\,\lambda(dx)} . \tag{3.39}$$

Thus, in many cases of interest, the backward transition kernel $\mathrm{B}_{\nu,k}$ can be written straightforwardly as a function of $\phi_{\nu,k}$ and Q. Several examples of such cases will be dealt with in some detail in Chapter 5. In these situations, Proposition 3.3.9 is the method of choice for smoothing, as it only involves normalized quantities, whereas Corollary 3.3.5 is not normalized and thus can generally not be implemented as it stands. ■

Proof (of Proposition 3.3.6). Let $k \in \{0, \ldots, n-1\}$ and $h \in \mathcal{F}_\mathrm{b}\left(\mathsf{X}^{n-k}\right)$. Then

$$\mathrm{E}_\nu[f(X_k) h(X_{k+1:n}) \mid Y_{0:n}] = \int \cdots \int f(x_k) h(x_{k+1:n})\, \phi_{\nu,k:n|n}(dx_{k:n}) . \tag{3.40}$$

Using the definition (3.13) of the joint smoothing distribution $\phi_{\nu,k:n|n}$ yields

$$
\begin{aligned}
&\mathrm{E}_\nu[f(X_k)h(X_{k+1:n}) \,|\, Y_{0:n}] \\
&= \mathrm{L}_{\nu,n}^{-1} \int \cdots \int \nu(dx_0) g_0(x_0) \prod_{i=1}^{k} Q(x_{i-1}, dx_i) g_i(x_i) f(x_k) \\
&\qquad \times \left[\prod_{i=k+1}^{n} Q(x_{i-1}, dx_i) g_i(x_i) \right] h(x_{k+1:n}) \;, \\
&= \frac{\mathrm{L}_{\nu,k}}{\mathrm{L}_{\nu,n}} \iint \phi_{\nu,k|n}(dx_k) Q(x_k, dx_{k+1}) f(x_k) g_{k+1}(x_{k+1}) \\
&\qquad \times \int \cdots \int \left[\prod_{i=k+2}^{n} Q(x_{i-1}, dx_i) g_i(x_i) \right] h(x_{k+1:n}) \;, \qquad (3.41)
\end{aligned}
$$

which implies, by the definition (3.38) of the backward kernel, that

$$
\begin{aligned}
&\mathrm{E}_\nu[f(X_k)h(X_{k+1:n}) \,|\, Y_{0:n}] \\
&= \frac{\mathrm{L}_{\nu,k}}{\mathrm{L}_{\nu,n}} \iint \mathrm{B}_{\nu,k}(x_{k+1}, dx_k) f(x_k) \phi_{\nu,k+1|k}(dx_{k+1}) g_{k+1}(x_{k+1}) \\
&\qquad \times \int \cdots \int \left[\prod_{i=k+2}^{n} Q(x_{i-1}, dx_i) g_i(x_i) \right] h(x_{k+1:n}) \,. \qquad (3.42)
\end{aligned}
$$

Taking $f \equiv 1$ shows that for any function $h' \in \mathcal{F}_{\mathrm{b}}\left(\mathsf{X}^{n-k}\right)$,

$$
\begin{aligned}
\mathrm{E}_\nu[h'(X_{k+1:n}) \,|\, Y_{0:n}] = \frac{\mathrm{L}_{\nu,k}}{\mathrm{L}_{\nu,n}} \int \cdots \int h'(x_{k+1:n}) \\
\times \phi_{\nu,k+1|k}(dx_{k+1}) g_{k+1}(x_{k+1}) \prod_{i=k+2}^{n} Q(x_{i-1}, dx_i) g_i(x_i) \,.
\end{aligned}
$$

Identifying h' with $h(x_{k+1:n}) \int f(x)\, \mathrm{B}_{\nu,k}(x_{k+1}, dx)$, we find that (3.42) may be rewritten as

$$
\begin{aligned}
&\mathrm{E}_\nu[f(X_k)h(X_{k+1:n}) \,|\, Y_{0:n}] \\
&\qquad = \mathrm{E}_\nu \left[h(X_{k+1:n}) \int \mathrm{B}_{\nu,k}(X_{k+1}, dx) f(x) \,\middle|\, Y_{0:n} \right] ,
\end{aligned}
$$

which concludes the proof. □

The next result is a straightforward consequence of Proposition 3.3.6, which reformulates the joint smoothing distribution $\phi_{\nu,0:n|n}$ in terms of the backward smoothing kernels.

Corollary 3.3.8. *For any integer $n > 0$ and initial probability ν,*

$$
\mathrm{E}_\nu[f(X_{0:n}) \,|\, Y_{0:n}] = \int \cdots \int f(x_{0:m})\, \phi_{\nu,n}(dx_n) \prod_{k=0}^{n-1} \mathrm{B}_{\nu,k}(x_{k+1}, dx_k) \qquad (3.43)
$$

for all $f \in \mathcal{F}_b(\mathsf{X}^{n+1})$. *Here,* $\{\mathrm{B}_{\nu,k}\}_{0 \le k \le n-1}$ *are the backward smoothing kernels defined in* (3.38) *and* $\phi_{\nu,n}$ *is the marginal filtering distribution corresponding to the final index* n.

It follows from Proposition 3.3.6 and Corollary 3.3.8 that, conditionally on $Y_{0:n}$, the joint distribution of the index-reversed sequence $\{\bar{X}_k\}_{0 \le k \le n}$, with $\bar{X}_k = X_{n-k}$, is that of a non-homogeneous Markov chain with initial distribution $\phi_{\nu,n}$ and transition kernels $\{\mathrm{B}_{\nu,n-k}\}_{1 \le k \le n}$. This is an exact analog of the forward decomposition where the ordering of indices has been reversed, starting from the end of the observation sequence and ending with the first observation. Three important differences versus the forward decomposition should however be kept in mind.

(i) The backward smoothing kernel $\mathrm{B}_{\nu,k}$ depends on the initial distribution ν and on the observations up to index k but it depends neither on the future observations nor on the index n where the observation sequence ends. As a consequence, the sequence of backward transition kernels $\{\mathrm{B}_{\nu,k}\}_{0 \le k \le n-1}$ may be computed by forward recurrence on k, irrespectively of the length of the observation sequence. In other terms, the backward smoothing kernel $\mathrm{B}_{\nu,k}$ depends only on the filtering distribution $\phi_{\nu,k}$, whereas the forward smoothing kernel $\mathrm{F}_{k|n}$ was to be computed from the backward function $\beta_{k|n}$.

(ii) Because $\mathrm{B}_{\nu,k}$ depends on $\phi_{\nu,k}$ rather than on the unnormalized forward measure $\alpha_{\nu,k}$, its computation involves only properly normalized quantities (Remark 3.3.7). The backward decomposition is thus more adapted to the actual computation of the smoothing probabilities than the forward decomposition. The necessary steps are summarized in the following result.

Proposition 3.3.9 (Forward Filtering/Backward Smoothing).

Forward Filtering *Compute, forward in time, the filtering distributions* $\phi_{\nu,0}$ *to* $\phi_{\nu,n}$ *using the recursion* (3.22). *At each index* k, *the backward transition kernel* $\mathrm{B}_{\nu,k}$ *may be computed according to* (3.38).

Backward Smoothing *From* $\phi_{\nu,n}$, *compute, for* $k = n-1, n-2, \ldots, 0$,

$$\phi_{\nu,k|n} = \phi_{\nu,k+1|n} \, \mathrm{B}_{\nu,k} \ ,$$

recalling that $\phi_{\nu,n|n} \stackrel{\text{def}}{=} \phi_{\nu,n}$.

(iii) A more subtle difference between the forward and backward Markovian decompositions is the observation that Definition 3.3.1 does provide an expression of the forward kernels $\mathrm{F}_{k|n}$ for any $k \ge 0$, that is, also for indices *after* the end of the observation sequence. Hence, the process $\{X_k\}_{k \ge 0}$, when conditioned on some observations $Y_{0:n}$, really forms a non-homogeneous Markov chain whose finite-dimensional distributions are defined by Proposition 3.3.4. In contrast, the backward kernels $\mathrm{B}_{\nu,k}$

are defined for indices $k \in \{0, \ldots, n-1\}$ only, and thus the index-reversed process $\{X_{n-k}\}$ is also defined, by Proposition 3.3.6, for indices k in the range $\{0, \ldots, n\}$ only. In order to define the index-reversed chain for negative indices, a minimal requirement is that the underlying chain $\{X_k\}$ also be well defined for $k < 0$. Defining Markov chains $\{X_k\}$ with indices $k \in \mathbb{Z}$ is only meaningful in the stationary case, that is when ν is the stationary distribution of Q. As both this stationarization issue and the forward and backward Markovian decompositions play a key role in the analysis of the statistical properties of the maximum likelihood estimator, we postpone further discussion of this point to Chapter 12.

3.4 Complements

The forward-backward algorithm is known to many, especially in the field of speech processing, as the *Baum-Welch algorithm*, although the first published description of the approach is due to Leonard E. Baum, Ted Petrie, George Soules, and Norman Weiss (1970, p. 168). The denomination refers to the collaboration between Baum and Lloyd R. Welch (Welch, 2003) who also worked out together an early version of the EM approach (to be discussed in Chapter 10). To the best of our knowledge however, the note entitled "A Statistical Estimation Procedure for Probabilistic Functions of Finite Markov Processes", co-authored by Baum and Welch and mentioned in the bibliography of Baum *et al.* (1970), has never been published.

The forward-backward algorithm was discovered several times in the early 1970s. A salient example is the paper by Bahl *et al.* (1974) on the computation of posterior probabilities for a finite-state Markov channel encoder for transmission over a discrete memoryless channel (see Example 1.3.2 in the introductory chapter). The algorithm described by Bahl *et al.* (1974) is fully equivalent to the forward-backward and is known in digital communication as the BCJR (for Bahl, Cocke, Jelinek, and Raviv) algorithm. Chang and Hancock (1966) is another less well-known reference, contemporary of the work of Baum and his colleagues, which also describes the forward-backward decomposition and its use for decoding in communication applications.

It is important to keep in mind that the early work on HMMs by Baum and his colleagues was conducted at the Institute for Defense Analyses (IDA) in Princeton under a contract from the U.S. National Security Agency. Although there are a few early publications of theoretical nature, most of the practical work that dealt with cryptography was kept secret and has never been published. It explains why some significant practical aspects (like the need for scaling to be discussed below) remained unpublished until HMMs became the *de facto* standard approach to speech recognition in the 1980s.

The famous tutorial by Rabiner (1989) is considered by many as the standard source of information for practical implementation of hidden Markov models. The impact of this publication has been very significant in speech processing but also in several other domains of application such as bioinformatics (Durbin *et al.*, 1998). It was Rabiner (1989) who coined the term *scaling* to describe the need for normalization when implementing the forward-backward recursions. There is indeed a subtle difference between the normalization scheme described in Section 3.2.2 and the solution advocated by Rabiner (1989), which was first published by Levinson *et al.* (1983). As was done in Section 3.2.2, Rabiner (1989) recommends normalizing the forward measures so that they integrate to one. However, the normalized backward functions are defined as $\breve{\beta}_{k|n} = (\prod_{l=k}^{n} c_{\nu,l})^{-1}\beta_{k|n}$ rather than $\bar{\beta}_{n|n} = (\prod_{l=k+1}^{n} c_{\nu,l})^{-1}\beta_{k|n}$. This difference is a consequence of the normalized backward recursion being carried out as

$$\breve{\beta}_{n|n}(x) = c_{\nu,n}^{-1} \quad \text{and}$$
$$\breve{\beta}_{k|n}(x) = c_{\nu,k}^{-1} \int_{\mathsf{X}} Q(x, dx') g_{k+1}(x') \breve{\beta}_{k+1|n}(x') \quad \text{for } k = n-1 \text{ down to } 0,$$

rather than as prescribed by (3.23). In contrast to our approach, Rabiner's scaling implies that normalization is still required for computing the marginal smoothing distributions as

$$\phi_{\nu,k|n}(dx) = [\phi_{\nu,k}(\breve{\beta}_{k|n})]^{-1} \breve{\beta}_{k|n}(x) \phi_{\nu,k}(dx) \; .$$

On the other hand, the joint smoothing distribution $\phi_{\nu,k:k+1|n}$ of X_k and X_{k+1} may be obtained directly, without normalization, as

$$\phi_{\nu,k:k+1|n}(dx, dx') = \phi_{\nu,k}(dx) Q(x, dx') g_{k+1}(x') \breve{\beta}_{k+1|n}(x') \; .$$

Indeed, $\phi_{\nu,k} = (\prod_{l=0}^{k} c_{\nu,l})^{-1} \alpha_{\nu,k}$ and thus

$$\phi_{\nu,k:k+1|n}(dx, dx') = \left(\prod_{l=0}^{n} c_{\nu,l} \right)^{-1} \alpha_{\nu,k}(dx) Q(x, dx') g_{k+1}(x') \beta_{k+1|n}(x') \; ,$$

as requested, as $\mathrm{L}_{\nu,n} = \prod_{l=0}^{n} c_{\nu,l}$ is the normalization factor common to all smoothing distributions from (3.13).

Easy computation of bivariate smoothing distributions does not, in our view, constitute a strong motivation for preferring a particular scaling scheme. The Markovian structure of the joint smoothing distribution exhibited in Section 3.3 in particular provides an easy means of evaluating bivariate smoothing distributions. For instance, with the scaling scheme described in Section 3.2.2, the forward Markovian decomposition of Section 3.3.1 implies that

$$\phi_{\nu,k:k+1|n}(dx, dx') = c_{\nu,k} \phi_{\nu,k|n}(dx) \frac{Q(x, dx') g_{k+1}(x') \bar{\beta}_{k+1|n}(x')}{\bar{\beta}_{k|n}(x)} \; .$$

As stated in the introduction, Stratonovich (1960) proposed a decomposition that is largely related to the forward-backward approach when the state space X is discrete. The forward measure, named w in the work of Stratonovich (1960), is defined as

$$w_k(x) = \mathrm{P}_\nu(X_k = x \mid Y_{0:k}) ,$$

which coincides with the definition of the filtering probability $\phi_{\nu,k}$ for a discrete X. Also recall that $\phi_{\nu,k}$ corresponds to the normalized forward variable $\bar{\alpha}_{\nu,k} = [\alpha_{\nu,k}(1)]^{-1}\alpha_{\nu,k}$. Instead of the backward function, Stratonovich (1960) defined

$$\bar{w}_k(x) = \mathrm{P}_\nu(X_k = x \mid Y_{k:n}) .$$

Forward and backward recursions for w_k and $\bar{w}_k$, respectively, as well as the relation for computing the marginal smoothing probability from w_k and $\bar{w}_k$, are given in the first section of Stratonovich (1960) on pages 160–162. Although $\bar{w}_k$ as defined by Stratonovich (1960) obviously has a probabilistic interpretation that the backward function lacks, the resulting recursion is more complicated because it requires the evaluation of the prior probabilities $\mathrm{P}_\nu(X_k = x)$ for $k \geq 0$. In addition, generalizing the definition of $\bar{w}_k$ to general state spaces X would require using the more restrictive index- (or time-) reversal concept discussed in Section 3.3.2. In contrast, the forward-backward decomposition of Baum *et al.* (1970) provides a very general framework for smoothing as discussed in this chapter.

The fact that, in some cases, a probabilistic interpretation may be given to the backward function $\beta_{k|n}$ (or to equivalent quantities) also explains why in the control and signal processing literatures, the forward-backward recursions are known under the generic term of *two-filter formulas* (Kitagawa, 1996; Kailath *et al.*, 2000, Section 10.4). This issue will be discussed in detail for Gaussian linear state-space models in Section 5.2.5.

4

Advanced Topics in Smoothing

This chapter covers three distinct complements to the basic smoothing relations developed in the previous chapter.

In the first section, we provide recursive smoothing relations for computing smoothed expectations of general functions of the hidden states. In many respects, this technique is reminiscent of the filtering recursion detailed in Section 3.2.2, but somewhat harder to grasp because the quantity that needs to be updated recursively is less directly interpretable.

In the second section, it is shown that the filtering and smoothing approaches discussed so far (including those of Section 4.1) may be applied, with minimal adaptations, to a family of models that is much broader than simply the hidden Markov models. We consider in some detail the case of hierarchical HMMs (introduced in Section 1.3.4) for which marginal filtering and smoothing formulas are still available, despite the fact that the hierarchic component of the state process is not *a posteriori* Markovian.

The third section is different in nature and is devoted to the so-called *forgetting* property of the filtering and smoothing recursions, which are instrumental in the statistical theory of HMMs (see Chapter 12). Forgetting refers to the fact that observations that are either far back in the past or in the remote future (relative to the current time index) have little impact on the posterior distribution of the current state. Although this section is written to be self-contained, its content is probably better understood after some exposure to the stability properties of Markov chains as can be found in Chapter 14.

4.1 Recursive Computation of Smoothed Functionals

Chapter 3 mostly dealt with *fixed-interval smoothing*, that is, computation of $\phi_{k|n}$[1] for a fixed value of the observation horizon n and for all indices

[1]Note that we omit the dependence with respect to the initial distribution ν, which is not important in this section.

$0 \le k \le n$. For Gaussian linear state-space models, it is well-known however that recursive (in n) evaluation of $\phi_{k|n}$ for a fixed value of k, also called *fixed-point smoothing*, is feasible (Anderson and Moore, 1979, Chapter 7). Gaussian linear state-space models certainly constitute a particular case, as the smoothing distributions $\phi_{k|n}$ are then entirely defined by their first and second moments (see Chapter 5). But fixed-point smoothing is by no means limited to some specific HMMs and (3.13) implies the existence of recursive update equations for evaluating $\phi_{k|n}$ with k fixed and increasing values of n. Remember that, as was the case in the previous chapter, we consider for the moment that evaluating integrals on X is a feasible operation.

The good news is that there also exist recursive formulas for computing a large class of smoothed quantities, which include in particular expressions like $\mathrm{E}[\sum_{k=0}^{n} s(X_n) \mid Y_{0:n}]$ and $\mathrm{E}[(\sum_{k=0}^{n} s(X_n))^2 \mid Y_{0:n}]$, where s is a real-valued measurable function on $(\mathsf{X}, \mathcal{X})$ such that both expectations are well-defined. Although one can of course consider arbitrary functions in this class, we will see in Chapter 10 that smoothed expectations of the state variables, for some specific choices of the function of interest, are instrumental in numerical approximations of the maximum likelihood estimate for parameter-dependent HMMs.

4.1.1 Fixed Point Smoothing

The fundamental equation here is (3.13), which upon comparing the expressions corresponding to n and $n+1$ gives the following update equation for the joint smoothing distribution:

$$\phi_{0:n+1|n+1}(f_{n+1}) = \left(\frac{\mathrm{L}_{n+1}}{\mathrm{L}_n}\right)^{-1} \int \cdots \int f_{n+1}(x_{0:n+1}) \\ \phi_{0:n|n}(dx_0, \ldots, dx_n)\, Q(x_n, dx_{n+1})\, g_{n+1}(x_{n+1}) \quad (4.1)$$

for functions $f_{n+1} \in \mathcal{F}_{\mathrm{b}}\left(\mathsf{X}^{n+2}\right)$. Recall that we used the notation c_{n+1} for the scaling factor $\mathrm{L}_{n+1}/\mathrm{L}_n$ that appears in (4.1), where, according to (3.27), c_{n+1} may also be evaluated as $\phi_{n+1|n}(g_{n+1})$.

Equation (4.1) corresponds to a simple, yet rich, structure in which the joint smoothing distribution is modified by applying an operator that only affects the last coordinate[2]. The probabilistic interpretation of this finding is that X_{n+1} and $X_{0:n-1}$ are conditionally independent given both $Y_{0:n+1}$ *and* X_n. This remark suggests that while the objective of updating $\phi_{k|n}$ recursively in n (for a fixed k) may not be achievable directly, $\phi_{k,n|n}$—the joint distribution of X_k and X_n given $Y_{0:n}$—does follow a simple recursion.

Proposition 4.1.1 (Fixed Point Smoothing). *For $k \ge 0$ and any $f \in \mathcal{F}_{\mathrm{b}}\left(\mathsf{X}^2\right)$,*

[2]This structure also has deep implications, which we do not comment on here, for sequential Monte Carlo approaches (to be discussed in Chapters 7 and 8).

$$\phi_{k,k+1|k+1}(f) = c_{k+1}^{-1} \iint f(x_k, x_{k+1})\, \phi_k(dx_k)\, Q(x_k, dx_{k+1})\, g_{k+1}(x_{k+1}) ,$$

where ϕ_k is the filtering distribution and $c_{k+1} = \phi_k Q g_{k+1}$. For $n \geq k+1$ and any $f \in \mathcal{F}_{\mathrm{b}}(\mathsf{X}^2)$,

$$\phi_{k,n+1|n+1}(f) = \\ c_{n+1}^{-1} \iint f(x_k, x_{n+1}) \int \phi_{k,n|n}(dx_k, dx_n)\, Q(x_n, dx_{n+1})\, g_{n+1}(x_{n+1}) .$$

Both relations are obtained by integrating (4.1) over all variables but those of relevant indices (k and $k+1$ for the first one, k, n, and $n+1$ for the second one). At any index n, the marginal smoothing distribution may be evaluated through $\phi_{k|n} = \phi_{k,n|n}(\cdot, \mathsf{X})$. Similarly the filtering distribution, which is required to evaluate c_{n+1}, is given by $\phi_n = \phi_{k,n|n}(\mathsf{X}, \cdot)$.

4.1.2 Recursive Smoothers for General Functionals

From Proposition 4.1.1, one can easily infer a smoothing scheme that applies to the specific situation where the only quantity of interest is $\mathrm{E}[s(X_k) \,|\, Y_{0:n}]$ *for a particular function* s, and not the full conditional distribution $\phi_{k,n|n}$. To this aim, define the finite signed measure τ_n on $(\mathsf{X}, \mathcal{X})$ by

$$\tau_n(f) = \int f(x_n)\, s(x_k)\, \phi_{k,n|n}(dx_k, dx_n) , \qquad f \in \mathcal{F}_{\mathrm{b}}(\mathsf{X}) ,$$

so that $\tau_n(\mathsf{X}) = \mathrm{E}[s(X_k) \,|\, Y_{0:n}]$. Proposition 4.1.1 then implies that

$$\tau_{k+1}(f) = c_{k+1}^{-1} \int f(x_{k+1}) \int s(x_k)\, \phi_k(dx_k)\, Q(x_k, dx_{k+1})\, g_{k+1}(x_{k+1}) ,$$

and

$$\tau_{n+1}(f) = c_{n+1}^{-1} \int f(x_{n+1}) \int \tau_n(dx_n)\, Q(x_n, dx_{n+1})\, g_{n+1}(x_{n+1}) \qquad (4.2)$$

for $n \geq k+1$ and $f \in \mathcal{F}_{\mathrm{b}}(\mathsf{X})$. Equation (4.2) is certainly less informative than Proposition 4.1.1, as one needs to fix the function s whose smoothed conditional expectation is to be updated recursively. On the other hand, this principle may be adapted to compute smoothed conditional expectations for a general class of functions that depend on the whole trajectory of the hidden states $X_{0:n}$ rather than on just a single particular hidden state X_k.

Before exposing the general framework, we first need to clarify a matter of terminology. In the literature on continuous time processes, and particularly in works that originate from the automatic control community, it is fairly common to refer to quantities similar to τ_n as *filters*—see for instance Elliott *et al.* (1995, Chapters 5 and 6) or Zeitouni and Dembo (1988). A filter is then defined as an object that may be evaluated recursively in n and is helpful in computing a quantity of interest that involves the observations up to index n. A more formal definition, which will also illustrate what is the precise meaning of the word *recursive*, is that a filter $\{\tau_n\}_{n\geq 0}$ is such that $\tau_0 = \mathfrak{R}_\nu(Y_0)$ and $\tau_{n+1} = \mathfrak{R}_n(\tau_n, Y_{n+1})$ where $\mathfrak{R}_\nu$ and $\{\mathfrak{R}_n\}_{n\geq 0}$ are some non-random operators. In the case discussed at the beginning of this section, $\mathfrak{R}_n$ is defined by (4.2) where Q is fixed (this is the transition kernel of the hidden chain) and Y_{n+1} enters through $g_{n+1}(x) = g(x, Y_{n+1})$. Note that because the normalizing constant c_{n+1}^{-1} in (4.2) depends on ϕ_n, Q and g_{n+1}, to be coherent with our definition we should say that $\{\phi_n, \tau_n\}_{n\geq 0}$ jointly forms a filter. In this book, we however prefer to reserve the use of the word *filter* to designate the state filter ϕ_n. We shall refer to quantities similar to $\{\tau_n\}_{n\geq 0}$ as the *recursive smoother associated with the functional* $\{t_n\}_{n\geq 0}$, where the previous example corresponds to $t_n(x_0, \ldots, x_n) = s(x_k)$. It is not generally possible to derive a recursive smoother without being more explicit about the family of functions $\{t_n\}_{n\geq 0}$. The device that we will use in the following consists in specifying $\{t_n\}_{n\geq 0}$ using a recursive formula that involves a set of fixed-dimensional functions.

Definition 4.1.2 (Smoothing Functional). *A* smoothing functional *is a sequence $\{t_n\}_{n\geq 0}$ of functions such that t_n is a function $\mathsf{X}^{n+1} \to \mathbb{R}$, and which may be defined recursively by*

$$t_{n+1}(x_{0:n+1}) = m_n(x_n, x_{n+1}) t_n(x_{0:n}) + s_n(x_n, x_{n+1}) \tag{4.3}$$

for all $x_{0:n+1} \in \mathsf{X}^{n+2}$ and $n \geq 0$, where $\{m_n\}_{n\geq 0}$ and $\{s_n\}_{n\geq 0}$ are two sequences of measurable functions $\mathsf{X} \times \mathsf{X} \to \mathbb{R}$ and t_0 is a function $\mathsf{X} \to \mathbb{R}$.

This definition can be extended to cases in which the functions t_n are d-dimensional vector-valued functions. In that case, $\{s_n\}_{n\geq 0}$ also are vector-valued functions $\mathsf{X} \times \mathsf{X} \to \mathbb{R}^d$ while $\{m_n\}_{n\geq 0}$ are matrix-valued functions $\mathsf{X} \times \mathsf{X} \to \mathbb{R}^d \times \mathbb{R}^d$.

In simpler terms, a smoothing functional is such that the value of t_{n+1} in $x_{0:n+1}$ differs from that of t_n, applied to the sub-vector $x_{0:n}$, only by a multiplicative and an additive factor that both only depend on the last two components x_n and x_{n+1}. The whole family is thus entirely specified by t_0 and the two sequences $\{m_n\}_{n\geq 0}$ and $\{s_n\}_{n\geq 0}$. This form has of course been chosen because it reflects the structure observed in (4.1) for the joint smoothing distributions. It does however encompass some important functionals of interest. The first and most obvious example is when t_n is a homogeneous additive functional, that is, when

$$t_n(x_{0:n}) = \sum_{k=0}^{n} s(x_k)$$

for a given measurable function s. In that case, $s_n(x, x')$ reduces to $s(x')$ and m_n is the constant function equal to 1.

The same strategy also applies for more complicated functions such as the squared sum $(\sum_{k=0}^{n} s(x_k))^2$. This time, we need to define two functions

$$t_{n,1}(x_{0:n}) = \sum_{k=0}^{n} s(x_k) \;,$$

$$t_{n,2}(x_{0:n}) = \left[\sum_{k=0}^{n} s(x_k)\right]^2 \;, \tag{4.4}$$

for which we have the joint update formula

$$t_{n+1,1}(x_{0:n+1}) = t_{n,1}(x_{0:n}) + s(x_{n+1}) \;,$$
$$t_{n+1,2}(x_{0:n+1}) = t_{n,2}(x_{0:n}) + s^2(x_{n+1}) + 2s(x_{n+1})t_{n,1}(x_{0:n}) \;.$$

Note that these equations can also be considered as an extension of Definition 4.1.2 for the vector valued function $t_n = (t_{n,1}, t_{n,2})^t$.

We now wish to compute $\mathrm{E}[t_n(X_{0:n}) \,|\, Y_{0:n}]$ recursively in n, assuming that the functions t_n are such that these expectations are indeed finite. We proceed as previously and define the family of finite signed measures $\{\tau_n\}$ on $(\mathsf{X}, \mathcal{X})$ such that

$$\tau_n(f) \stackrel{\text{def}}{=} \int \cdots \int f(x_n)\, t_n(x_{0:n})\, \phi_{0:n|n}(dx_0, \ldots, dx_n) \tag{4.5}$$

for all functions $f \in \mathcal{F}_{\mathrm{b}}(\mathsf{X})$. Thus, $\tau_n(\mathsf{X}) = \mathrm{E}[t_n(X_{0:n}) \,|\, Y_{0:n}]$. We then have the following direct consequence of (4.1).

Proposition 4.1.3. *Let $(t_n)_{n \geq 0}$ be a sequence of functions on $\mathsf{X}^{n+1} \to \mathbb{R}$ possessing the structure of Definition 4.1.2. The finite signed measures $\{\tau_n\}_{n \geq 0}$ on $(\mathsf{X}, \mathcal{X})$ defined by (4.5) may then be updated recursively according to*

$$\tau_0(f) = \{\nu(g_0)\}^{-1} \int f(x_0)\, \nu(dx_0)\, t_0(x_0)\, g_0(x_0)$$

and

$$\tau_{n+1}(f) = c_{n+1}^{-1} \iint f(x_{n+1}) \Big[\tau_n(dx_n)\, Q(x_n, dx_{n+1})\, g_{n+1}(x_{n+1}) m_n(x_n, x_{n+1})$$
$$+ \phi_n(dx_n)\, Q(x_n, dx_{n+1})\, g_{n+1}(x_{n+1}) s_n(x_n, x_{n+1}) \Big] \tag{4.6}$$

for $n \geq 0$, where f denotes a generic function in $\mathcal{F}_{\mathrm{b}}(\mathsf{X})$. At any index n, $\mathrm{E}[t_n(X_{0:n}) \,|\, Y_{0:n}]$ may be evaluated by computing $\tau_n(\mathsf{X})$.

In order to use (4.6), it is required that the standard filtering recursions (Proposition 3.2.5) be computed in parallel to (4.6). In particular, the normalizing constant c_{n+1} is given by (3.22) as

$$c_{n+1} = \phi_n Q g_{n+1} \;.$$

As was the case for Definition 4.1.2, Proposition 4.1.3 can obviously be extended to cases where the functional $(t_n)_{n\geq 0}$ is vector-valued, without any additional difficulty. Because the general form of the recursion defined by Proposition 4.1.3 is quite complex, we first examine the simple case of homogeneous additive functionals mentioned above.

Example 4.1.4 (First and Second Moment Functionals). Let s be a fixed function on X and assume that the functionals of interest are the sum and squared sum in (4.4). A typical example is when the base function s equals $\mathbb{1}_A$ for a some measurable set A. Then, $\mathrm{E}[t_{n,1}(X_{0:n}) \mid Y_{0:n}]$ is the conditional expected occupancy of the set A by the hidden chain $\{X_k\}_{k\geq 0}$ between indices 0 and n. Likewise, $\mathrm{E}[t_{n,2}(X_{0:n}) \mid Y_{0:n}] - (\mathrm{E}[t_{n,1}(X_{0:n}) \mid Y_{0:n}])^2$ is the conditional variance of the occupancy of the set A.

We define the signed measures $\tau_{n,1}$ and $\tau_{n,2}$ associated to $t_{n,1}$ and $t_{n,2}$ by (4.5). We now apply the general formula given by Proposition 4.1.3 to obtain a recursive update for $\tau_{n,1}$ and $\tau_{n,2}$:

$$\tau_{0,1}(f) = [\nu(g_0)]^{-1} \int f(x_0)\, \nu(dx_0)\, s(x_0) g_0(x_0) \;,$$
$$\tau_{0,2}(f) = [\nu(g_0)]^{-1} \int f(x_0)\, \nu(dx_0)\, s^2(x_0) g_0(x_0)$$

and, for $n \geq 0$,

$$\tau_{n+1,1}(f) = \int f(x_{n+1}) \left[\phi_{n+1}(dx_{n+1})\, s(x_{n+1}) + c_{n+1}^{-1} \int \tau_{n,1}(dx_n)\, Q(x_n, dx_{n+1})\, g_{n+1}(x_{n+1})\right] ,$$

$$\tau_{n+1,2}(f) = \int f(x_{n+1}) \left[\phi_{n+1}(dx_{n+1})\, s^2(x_{n+1}) + c_{n+1}^{-1} \int \tau_{n,2}(dx_n)\, Q(x_n, dx_{n+1})\, g_{n+1}(x_{n+1}) \right.$$
$$\left. + 2c_{n+1}^{-1} \int \tau_{n,1}(dx_n)\, Q(x_n, dx_{n+1})\, g_{n+1}(x_{n+1}) s(x_{n+1})\right] .$$

■

4.1.3 Comparison with Forward-Backward Smoothing

It is important to contrast the approach of Section 4.1.2 above with the techniques discussed previously in Chapter 3. What are exactly the differences between the recursive smoother of Proposition 4.1.3 and the various versions of forward-backward smoothing discussed in Sections 3.2 and 3.3? Is it always possible to apply either of the two approaches? If yes, is one of them preferable to the other? These are important issues that we review below. Note that for the moment we only compare these two approaches on principle grounds and we do not even try to discuss the computational burden associated with the effective implementation of either approach. This latter aspect is of course entirely dependent of the way in which we are to evaluate (or approximate) integrals, which is itself highly dependent on the specific model under consideration. Several concrete applications of this approach will be considered in Chapters 10 and 11.

4.1.3.1 Recursive Smoothing Is More General

Remember that in Chapter 3 our primary objective was to develop approaches for computing *marginal smoothing distributions* $\phi_{k|n} = \mathrm{P}(X_k \in \cdot \,|\, Y_{0:n})$. A closer inspection of the results indicate that both in the standard forward-backward approach (Section 3.2) or when using a Markovian (forward or backward) decomposition (Section 3.3), one may easily obtain the bivariate joint smoothing distribution $\phi_{k+1:k|n} = \mathrm{P}((X_{k+1}, X_k) \in \cdot \,|\, Y_{0:n})$ as a by-product of evaluating $\phi_{k|n}$, with essentially no additional calculation (see in particular Section 3.4).

If we consider however the second-order functional $t_{n,2}$ discussed in Example 4.1.4, we may write

$$\mathrm{E}[t_{n,2}(X_{0:n}) \,|\, Y_{0:n}] = \sum_{i=0}^{n} \sum_{j=0}^{n} \mathrm{E}[s(X_i)s(X_j) \,|\, Y_{0:n}] \,.$$

The conditional expectations on the right-hand side indeed only involve the bivariate joint smoothing distributions but *for indices that are not consecutive*: it is not sufficient to determine $\phi_{k+1:k|n}$ for $k = 0, \ldots, n-1$ to evaluate $\mathrm{E}[t_{n,2}(X_{0:n}) \,|\, Y_{0:n}]$ directly. One would require the complete set of distributions $\mathrm{P}[(X_i, X_j) \in \cdot \,|\, Y_{0:n}]$ for $0 \le i \le j \le n$.

From this example we may conclude that computing $\mathrm{E}[t_n(X_{0:n}) \,|\, Y_{0:n}]$ using forward-backward smoothing is not possible for the whole class of functionals defined in (4.3) but only for a subset of it. If we are to use only the bivariate joint smoothing distributions $\phi_{k+1:k|n}$, then t_n must be an *additive functional* for which the multipliers m_n are constant (say, equal to 1). In that case, t_n reduces to

$$t_n(x_{0:n}) = t_0(x_0) + \sum_{k=0}^{n-1} s_k(x_k, x_{k+1}) \,,$$

and the expected value of t_n may be directly evaluated as

$$\mathrm{E}[t_n(X_{0:n}) \,|\, Y_{0:n}] = \int t_0(x_0)\, \phi_{0|n}(dx_0) + \sum_{k=0}^{n-1} \int s_k(x_k, x_{k+1})\, \phi_{k:k+1|n}(dx_k, dx_{k+1}) \,. \quad (4.7)$$

Recursive smoothing is more general in the sense that it is not restricted to sum functionals but applies to the whole class of functions whose structure agrees with (4.3).

4.1.3.2 For Additive Functionals, Forward-Backward Is More General

A distinctive feature however of recursive smoothing is that it may only be applied once a particular function in the class has been selected. The recursive smoother τ_n is associated with a specific choice of the functional t_n. As an example, denote by $\tau_{n,A}$ the recursive smoother associated with the homogeneous sum functional

$$t_{n,A}(x_{0:n}) = \sum_{k=0}^{n} \mathbb{1}_A(x_k)$$

for a given set A. We may compute $\tau_{n,A}$, recursively in n using Proposition 4.1.3 and evaluate $\sum_{k=0}^{n} \mathrm{P}(X_k \in A \mid Y_{0:n})$ as $\tau_{n,A}(\mathsf{X})$. If we now consider a different set B, there is no way of evaluating $\sum_{k=0}^{n} \mathrm{P}(X_k \in B \mid Y_{0:n})$ from the previous recursive smoother $\tau_{n,A}$. It is thus required to *run a specific recursive smoother for each function that we are possibly interested in.*

In contrast, once we have evaluated $\phi_{k+1:k|n}$ for all indices k between 0 and $n-1$, we may apply (4.7) to obtain the expectation of any particular sum functional that we might be interested in.

4.1.3.3 Recursive Smoothing Is Recursive!

A final element of the comparison of the two approaches is the fact that forward-backward is fundamentally intended for a fixed amount of observations, a situation usually referred to as *block* or *batch* processing. Consider again, as an example, a simple sum functional of the form

$$t_n(x_{0:n}) = \sum_{k=0}^{n} s(x_k) \ ,$$

and suppose that we are given our n observations not as a whole but one by one, starting with X_0 and then X_1, X_2, etc.

If we use the normalized forward-backward recursions (Proposition 3.2.5) or the equivalent backward Markovian decomposition (Proposition 3.3.9), the only quantities that are available at an intermediate index k (with k less than n) are the filtering distributions ϕ_0 to ϕ_k. Although we could evaluate $\mathrm{E}[s(X_j) \mid Y_{0:j}]$ for $j \leq k$, it is not yet possible to evaluate $\mathrm{E}[t_k(X_{0:k}) \mid Y_{0:k}]$. To be able to compute smoothed quantities, one must decide on an endpoint, say $k = n$, from which the backward recursion is started. The backward recursion then provides us with the smoothed marginal distributions $\phi_{k|n}$ from which $\mathrm{E}[t_k(X_{0:n}) \mid Y_{0:n}]$ can be evaluated. This is even more obvious for the forward

Markovian decomposition (Corollary 3.3.5), which starts by the backward recursion initialized at the final index n.

In contrast, for the recursive smoother, the update equation (4.6) in Proposition 4.1.3 provides a means of computing $\mathrm{E}[t_k(X_{0:k}) \mid Y_{0:k}]$ for all indices $k = 1, 2, \ldots$, whether or not we have reached the final observation index. There need not even be a final observation index, and the method can be applied also when $n = \infty$ or when the final observation index is not specified. Note that in cases where n is finite but quite large, forward-backward (or the equivalent Markovian decompositions) requires that all the intermediate results be stored: before we can compute $\phi_{k|n}$ we first need to evaluate and keep track of all the filtering distributions ϕ_0 to ϕ_n (or, for the forward Markovian decomposition, the backward functions $\beta_{n|n}$ down to $\beta_{0|n}$). Thus for large values of n, recursive smoothing approaches are also preferable to those based on forward-backward ideas.

Remember however that the price to pay for deriving a recursive smoother is the need to particularize the function of interest. We will discuss in Chapter 10 the exact computational cost of both approaches in examples of HMMs for which the computation corresponding to Proposition 4.1.3 is actually feasible.

4.1.3.4 Bibliographic Notes

The recursive smoothing approach discussed in this section was first described by Zeitouni and Dembo (1988) and Elliott (1993) for continuous time discrete state Markov processes observed in (Gaussian) noise. The approach is also at the core of the book by Elliott *et al.* (1995). The application of the same principle to the specific case of Gaussian linear state-space models is considered, among others, by Elliott and Krishnamurthy (1999) (see also references therein). The common theme of these works is to use the EM algorithm (see Chapter 10), replacing forward-backward smoothing by recursive smoothing. For reasons to be explained in Section 10.2, the functionals of interest in this context are sums (that is, $m_n = 1$ in Definition 4.1.2). We will see in Section 10.2.4 that the same approach (always with sum functionals) also applies for computing the gradient of the log-likelihood with respect to the parameters in parameterized models. The fact that the same approach applies for more general functionals such as squared sums is, to the best of our knowledge, new (see also Section 10.3.4 for an example of this latter case).

4.2 Filtering and Smoothing in More General Models

Although our main interest is hidden Markov models as defined in Section 2.2, the smoothing decompositions and recursions derived so far turn out to be far more general. We briefly discuss below the case of several non-HMM models of practical interest before considering the specific case of hierarchical HMMs as defined in Section 2.2.3.

4.2.1 Smoothing in Markov-switching Models

In Markov-switching models (see Section 1.3.6), the distribution of Y_k given $X_{0:k}$ and $Y_{0:k-1}$ does not only depend on X_k but also on a number of past values of the observed sequence. Assume for ease of notation that the dependence with respect to previous observations is only on the last observation Y_{k-1}. It is easily checked that (3.1), which defines the joint distribution of a number of consecutive hidden states and observations, should then be replaced by

$$\mathrm{E}_\nu[f(X_0, Y_0, \ldots, X_n, Y_n)] = \int \cdots \int f(x_0, y_0, \ldots, x_n, y_n)$$
$$\times\, \nu(dx_0) h(x_0, y_0) \prod_{k=1}^{n} \{Q(x_{k-1}, dx_k)\, g\,[(x_k, y_{k-1}), y_k]\}\; \mu_n(dy_0, \ldots, dy_n) \tag{4.8}$$

for all $f \in \mathcal{F}_\mathrm{b}\left(\{\mathsf{X} \times \mathsf{Y}\}^{n+1}\right)$, where $g[(x_k, y_{k-1}), \cdot]$ is the transition density function of Y_k given X_k and Y_{k-1}. Note that for Markov-switching models, it is more natural to define the initial distribution as the joint distribution of X_0 and Y_0 and hence as a probability measure on $(\mathsf{X} \times \mathsf{Y}, \mathcal{X} \otimes \mathcal{Y})$. In (4.8), we have adopted a particular and equivalent way of representing this distribution as $\nu(dx_0)\, h(x_0, dy_0)\, \mu(dy_0)$ for some transition density function h.

Equation (4.8) is similar to (3.1) and will be even more so once we adopt the implicit conditioning convention introduced in Section 3.1.4. Indeed, upon defining

$$g_0(\cdot) \stackrel{\text{def}}{=} h(\cdot, Y_0)\ ,$$
$$g_k(\cdot) \stackrel{\text{def}}{=} g\,[(\cdot, Y_{k-1}), Y_k] \qquad \text{for } k \geq 1\ ,$$

the joint distribution $\phi_{\nu,0:n|n}$ of the hidden states $X_{0:n}$ given the observations $Y_{0:n}$ is still given by (3.13), and hence the mechanics of smoothing for switching autoregressive models are the same as for the standard HMM (see for instance Hamilton, 1994, Chapter 22).

4.2.2 Smoothing in Partially Observed Markov Chains

It should also be clear that the same remark holds, *mutatis mutandis*, for other variants of the model such as non-homogeneous ones—if Q depends on the index k for instance—or if the transition from X_k to X_{k+1} also depends on some function of the past observations $Y_{0:k-1}$. Moreover, a closer inspection of the smoothing relations obtained previously indicate that, except when one wishes to exhibit predicted quantities—as in (3.27)—only the *unnormalized product kernel* $R_{k-1}(x_{k-1}, dx_k) = Q(x_{k-1}, dx_k)\, g_k(x_k)$ does play a role[3]. In

[3] We will come back to this remark when examining sequential Monte Carlo approaches in Chapter 7.

particular, for the general class of models in which it is only assumed that $\{X_k, Y_k\}_{k\geq 0}$ jointly form a Markov chain, the joint distribution of Y_k and X_k given Y_{k-1} and X_{k-1} may be represented as

$$Q\left[(x_{k-1}, y_{k-1}), dx_k\right] g\left[(x_{k-1}, y_{k-1}, x_k), y_k\right] \mu(dy_k) ,$$

assuming that the second conditional distribution is dominated by μ. Hence in this case also, one may define

$$R_{k-1}(x_{k-1}, dx_k) \stackrel{\text{def}}{=} Q\left[(x_{k-1}, Y_{k-1}), dx_k\right] g\left[(x_{k-1}, Y_{k-1}, x_k), Y_k\right]$$

and use the same filtering and smoothing relations as before. With this notation, it is a simple matter of rewriting, replacing the product of Q and g_k by R_{k-1} to obtain, for instance, the filtering update from (3.22):

$$c_{\nu,k} = \iint \phi_{\nu,k-1}(dx)\, R_{k-1}(x, dx') ,$$

$$\phi_{\nu,k}(f) = c_{\nu,k}^{-1} \int f(x') \int \phi_{\nu,k-1}(dx)\, R_{k-1}(x, dx') , \qquad f \in \mathcal{F}_{\mathrm{b}}(\mathsf{X}) .$$

4.2.3 Marginal Smoothing in Hierarchical HMMs

An example that nicely illustrates the previous discussion on the generality of the filtering and smoothing recursions of Chapter 3 is the case of hierarchical HMMs. These models defined in Section 2.2.3 are hidden Markov models in which the unobservable chain $\{X_k\}_{k\geq 0}$ is split into two components $\{C_k\}_{k\geq 0}$ and $\{W_k\}_{k\geq 0}$ such that the component $\{C_k\}_{k\geq 0}$, which is the highest in the hierarchy, marginally forms a Markov chain. Of course, these models are HMMs and can be handled as such. In many cases, it is however advantageous to consider that the component of interest is $\{C_k\}_{k\geq 0}$ only, marginalizing with respect to the intermediate component $\{W_k\}_{k\geq 0}$. A typical example is the case of conditionally Gaussian linear state-space models (Definition 2.2.6), where the indicator component C_k takes values in a finite set, whereas the intermediate component W_k is a vector-valued, possibly high-dimensional, variable. It is clear however that the pair (C_k, Y_k) does not correspond to a hidden Markov model. In particular, the distribution of Y_n depends on all indicator variables C_0 up to C_n (rather than on C_n only), due to the marginalization of the intermediate variables $W_{0:n}$. Because of the generality of the smoothing relations obtained in Chapter 3, the implementation of marginal smoothing—that is, estimation of $\{C_k\}_{k\geq 0}$ only given $\{Y_k\}_{k\geq 0}$—however bears some similarity with the (simpler) case of HMMs.

For notational simplicity, we consider in the remainder of this section that the hierarchic component $\{C_k\}_{k\geq 0}$ takes values in the finite set $\{1, \ldots, r\}$. As usual in this context, we use the notations $Q_C(x, x')$ and $\nu_C(x)$ rather than $Q_C(x, \{x'\})$ and $\nu_C(\{x\})$. The other notations pertaining to hierarchical hidden Markov models can be found in Section 2.2.3. Let $\psi_{\nu,0:k|k}$ denote the posterior distribution of $C_{0:k}$ given $Y_{0:k}$,

$$\psi_{\nu,0:k|k}(c_{0:k}) \stackrel{\text{def}}{=} \mathrm{P}_\nu \left(C_{0:k} = c_{0:k} \,|\, Y_{0:k} \right) . \tag{4.9}$$

Using (3.13) for the hierarchical HMM and integrating with respect to the intermediate component $w_{0:n}$ readily gives

$$\psi_{\nu,0:n|n}\left(c_{0:n}\right) = \mathrm{L}_{\nu,n}^{-1}\, \nu_C(c_0) \prod_{k=1}^{n} Q_C(c_{k-1}, c_k) \\ \int \cdots \int \nu_W(c_0, dw_0) \prod_{k=1}^{n} Q_W\left[(w_{k-1}, c_k), dw_k\right] g_k(c_k, w_k) , \tag{4.10}$$

where $g_k(c_k, w_k) \stackrel{\text{def}}{=} g\left[(c_k, w_k), Y_k\right]$. Comparing the above expression for two successive indices, say n and $n+1$, yields

$$\psi_{\nu,0:n+1|n+1}\left(c_{0:n+1}\right) = \left(\frac{\mathrm{L}_{\nu,n+1}}{\mathrm{L}_{\nu,n}}\right)^{-1} \psi_{\nu,0:n|n}\left(c_{0:n}\right) Q_C(c_n, c_{n+1}) \\ \int \varphi_{\nu,n+1|n}(c_{0:n+1}, dw_{n+1})\, g_{n+1}(c_{n+1}, w_{n+1}) , \tag{4.11}$$

where

$$\varphi_{\nu,n+1|n}(c_{0:n+1}, f) \stackrel{\text{def}}{=} \\ \frac{\int_{\mathsf{W}^{n+1}} \nu_W(c_0, dw_0) \left\{ \prod_{k=1}^{n} Q_W[(w_{k-1}, c_k), dw_k] g_k(c_k, w_k) \right\} Q_W[(w_n, c_{n+1}), f]}{\int_{\mathsf{W}^{n+1}} \nu_W(c_0, dw_0) \prod_{k=1}^{n} Q_W[(w_{k-1}, c_k), dw_k] g_k(c_k, w_k)} \tag{4.12}$$

for $f \in \mathcal{F}_b(\mathsf{W})$, which is recognized as the predictive distribution of the intermediate component W_{n+1} given the observations $Y_{0:n}$ up to index n and the indicator variables $C_{0:n+1}$ up to index $n+1$.

In the example of conditionally Gaussian linear state-space models, the conditional predictive distribution $\varphi_{\nu,n+1|n}(c_{0:n+1}, \cdot)$ given in (4.12) is Gaussian and may indeed be evaluated recursively *for a given sequence of indicator variables* $c_{0:n+1}$ using the Kalman recursions (see Section 5.2). Moreover, in these models the integral featured on the second line of (4.11) may also be evaluated exactly. It is important however to understand that even in this (favorable) case, the existence of (4.11) does not provide an easy solution to updating the marginal filtering distribution $\psi_{\nu,n|n}$ as it does for HMMs. The fundamental problem is that (4.12) also directly indicates that the predictive distribution W_{n+1} given $Y_{0:n}$, but *without conditioning on the indicator variables* $C_{0:n+1}$, is a mixture distribution with a number of components equal to the number of possible configurations of $C_{0:n+1}$, that is, r^{n+2}. Hence in practice, even in cases such as the conditionally Gaussian linear state-space models for which evaluation of (4.12) is feasible, it is not possible to implement the exact marginal filtering relations for the sequence $\{C_k\}_{k\geq 0}$ because of the combinatorial explosion due to the need to enumerate all configurations of the indicator variables.

Thus, (4.1) will only be helpful in approaches where it is possible to impute values to (part of) the unknown sequence $\{C_k\}_{k\geq 0}$, making it possible to avoid exhaustive enumeration of all configurations of the indicator variables. This is precisely the aim of sequential Monte Carlo methods to be described in Chapters 7 and 8, where the specific case of hierarchical HMMs will be detailed in Section 8.2.

Note that while (4.1) obviously suggests a recursion in increasing values of n, it is also possible to write an analog to the forward-backward decomposition (see Section 3.2) starting from (4.10):

$$\psi_{\nu,0:n|n}(c_{0:n}) = \mathrm{L}_{\nu,n}^{-1} \int \alpha_{\nu,k}(c_{0:k}, dw_k)\, \beta_{k|n}(c_{k:n}, w_k) \;, \tag{4.13}$$

where

$$\alpha_{\nu,k}(c_{0:k}, f) \stackrel{\text{def}}{=} \int \cdots \int f(w_k)$$
$$\nu_C(c_0)\, \nu_W(c_0, dw_0) \prod_{l=1}^{k} Q_C(c_{l-1}, c_l)\, Q_W\left[(w_{l-1}, c_l), dw_l\right] g_l(c_l, w_l)$$

for $f \in \mathcal{F}_{\mathrm{b}}(\mathsf{W})$ and

$$\beta_{k|n}(c_{k:n}, w_k) \stackrel{\text{def}}{=} \int \cdots \int \prod_{l=k+1}^{n} Q_C(c_{l-1}, c_l)\, Q_W\left[(w_{l-1}, c_l), dw_l\right] g_l(c_l, w_l) \;.$$

The same comment as before applies regarding the fact that both the forward and backward variables do depend on complete sub-sequences of indicator variables; $c_{0:k}$ for $\alpha_{\nu,k}$ and $c_{k:n}$ for $\beta_{k|n}$. This property of hierarchical HMMs restricts the practical use of (4.13) to cases in which it is possible, for instance, to condition on all values of C_l in the sequence $C_{0:n}$ except C_k. The main application of this decomposition is to be found in Markov chain Monte Carlo methods (Chapter 6) and, more precisely, in the so-called Gibbs sampling approach (Section 6.2.5). The use of (4.13) in this context will be fully illustrated for conditionally Gaussian linear state space models in Sections 5.2.6 and 6.3.2.

4.3 Forgetting of the Initial Condition

Recall from previous chapters that in a partially dominated HMM model (see Definition 2.2.2), we denote by

- P_ν the probability associated to the Markov chain $\{X_k, Y_k\}_{k\geq 0}$ on the canonical space $\left((\mathsf{X}\times\mathsf{Y})^{\mathbb{N}}, (\mathcal{X}\otimes\mathcal{Y})^{\otimes\mathbb{N}}\right)$ with initial probability measure ν and transition kernel T defined by (2.15);
- $\phi_{\nu,k|n}$ the distribution of the hidden state X_k conditionally on the observations $Y_{0:n}$, under the probability measure P_ν.

Forgetting properties pertain to the dependence of $\phi_{\nu,k|n}$ with respect to the initial distribution ν. A typical question is to ask whether $\phi_{\nu,k|n}$ and $\phi_{\nu',k|n}$ are close (in some sense) for large values of k and arbitrary choices of ν and ν'. This issue will play a key role both when studying the convergence of sequential Monte Carlo methods (Chapter 9) and when analyzing the asymptotic behavior of the maximum likelihood estimator (Chapter 12).

In the following, it is shown more precisely that, under appropriate conditions on the kernel Q of the hidden chain and on the transition density function g, the total variation distance $\left\|\phi_{\nu,k|n} - \phi_{\nu',k|n}\right\|_{\mathrm{TV}}$ converges to zero as k tends to infinity. Remember that, following the implicit conditioning convention (Section 3.1.4), we usually omit to indicate explicitly that $\phi_{\nu,k|n}$ indeed depends on the observations $Y_{0:n}$. In this section however we cannot use this convention anymore, as we will meet both situations in which, say, $\|\phi_{\nu,n} - \phi_{\nu',n}\|_{\mathrm{TV}}$ converges to zero (as n tends to infinity) *for all possible values of* the sequence $\{y_n\}_{n\geq 0} \in \mathsf{Y}^{\mathbb{N}}$ (*uniform forgetting*) and cases where $\|\phi_{\nu,n} - \phi_{\nu',n}\|_{\mathrm{TV}}$ can be shown to converge to zero almost surely only when $\{Y_k\}_{k\geq 0}$ is assumed to be distributed under a specific distribution (typically $\mathrm{P}_{\nu_\star}$ for some initial distribution $\nu_\star$). In this section, we thus make dependence with respect to the observations explicit by indicating the relevant subset of observation between brackets, using, for instance, $\phi_{\nu,k|n}[y_{0:n}]$ rather than $\phi_{\nu,k|n}$.

We start by recalling some elementary facts and results about the total variation norm of a signed measure, providing in particular useful characterizations of the total variation as an operator norm over appropriately defined function spaces. We then discuss the contraction property of Markov kernels, using the measure-theoretic approach introduced in an early paper by Dobrushin (1956) and recently revisited and extended by Del Moral *et al.* (2003). We finally present the applications of these results to establish forgetting properties of the smoothing and filtering recursions and discuss the implications of the technical conditions required to obtain these results.

4.3.1 Total Variation

Let $(\mathsf{X}, \mathcal{X})$ be a measurable space and let ξ be a signed measure on $(\mathsf{X}, \mathcal{X})$. Then there exists a measurable set $H \in \mathcal{X}$, called a *Jordan set*, such that

(i) $\xi(A) \geq 0$ for each $A \in \mathcal{X}$ such that $A \subseteq H$;
(ii) $\xi(A) \leq 0$ for each $A \in \mathcal{X}$ such that $A \subseteq \mathsf{X} \setminus H$.

The set H is not unique, but any other such set $H' \in \mathcal{X}$ satisfies $\xi(H \cap H') = 1$. Hence two Jordan sets differ by at most a set of zero measure. If X is finite or countable and $\mathcal{X} = \mathcal{P}(\mathsf{X})$ is the collection of all subsets of X, then $H = \{x : \xi(x) \geq 0\}$ and $H' = \{x : \xi(x) > 0\}$ are two Jordan sets. As another example, if ξ is absolutely continuous with respect to a measure ν on $(\mathsf{X}, \mathcal{X})$ with Radon-Nikodym derivative f, then $\{f \geq 0\}$ and $\{f > 0\}$ are two Jordan sets. We define two measures on $(\mathsf{X}, \mathcal{X})$ by

$$\xi_+(A) = \xi(H \cap A) \quad \text{and} \quad \xi_-(A) = -\xi(H^c \cap A) , \qquad A \in \mathcal{X} .$$

The measures ξ_+ and ξ_- are referred to as the *positive* and *negative variations* of the signed measure ξ. By construction, $\xi = \xi_+ - \xi_-$. This decomposition of ξ into its positive and negative variations is called the *Hahn-Jordan decomposition* of ξ. The definition of the positive and negative variations above is easily shown to be independent of the particular Jordan set chosen.

Definition 4.3.1 (Total Variation of a Signed Measure). *Let $(\mathsf{X}, \mathcal{X})$ be a measurable space and let ξ be a signed measure on $(\mathsf{X}, \mathcal{X})$. The* total variation norm *of ξ is defined as*

$$\|\xi\|_{\mathrm{TV}} = \xi_+(\mathsf{X}) + \xi_-(\mathsf{X}) ,$$

where (ξ_+, ξ_-) is the Hahn-Jordan decomposition of ξ.

If X is finite or countable and ξ is a signed measure on $(\mathsf{X}, \mathcal{P}(\mathsf{X}))$, then $\|\xi\|_{\mathrm{TV}} = \sum_{x \in \mathsf{X}} |\xi(x)|$. If ξ has a density g with respect to a measure λ on $(\mathsf{X}, \mathcal{X})$, then $\|\xi\|_{\mathrm{TV}} = \int |f(x)| \, \lambda(dx)$.

Definition 4.3.2 (Total Variation Distance). *Let $(\mathsf{X}, \mathcal{X})$ be a measurable space and let ξ and ξ' be two measures on $(\mathsf{X}, \mathcal{X})$. The* total variation distance *between ξ and ξ' is the total variation norm of the signed measure $\xi - \xi'$.*

Denote by $\mathrm{M}(\mathsf{X}, \mathcal{X})$ the set of finite signed measures on the measurable space $(\mathsf{X}, \mathcal{X})$, by $\mathrm{M}_1(\mathsf{X}, \mathcal{X})$ the set of probability measures on $(\mathsf{X}, \mathcal{X})$ and by $\mathrm{M}_0(\mathsf{X}, \mathcal{X})$ the set of finite signed measures ξ on $(\mathsf{X}, \mathcal{X})$ satisfying $\xi(\mathsf{X}) = 0$. $\mathrm{M}(\mathsf{X}, \mathcal{X})$ is a Banach space with respect to the total variation norm. In this Banach space, the subset $\mathrm{M}_1(\mathsf{X}, \mathcal{X})$ is closed and convex.

Let $\mathcal{F}_{\mathrm{b}}(\mathsf{X})$ denote the set of bounded measurable real functions on X. This set embedded with the supremum norm $\|f\|_\infty = \sup\{f(x) : x \in \mathsf{X}\}$ also is a Banach space. For any $\xi \in \mathrm{M}(\mathsf{X}, \mathcal{X})$ and $f \in \mathcal{F}_{\mathrm{b}}(\mathsf{X})$, we may define $\xi(f) = \int f \, d\xi$. Therefore any finite signed measure ξ in $\mathrm{M}(\mathsf{X}, \mathcal{X})$ defines a linear functional on the Banach space $(\mathcal{F}_{\mathrm{b}}(\mathsf{X}), \|\cdot\|_\infty)$. We will use the same notation for the measure and for the functional. The following lemma shows that the total variation of the signed measure ξ agrees with the operator norm of ξ.

Lemma 4.3.3.

(i) For any $\xi \in \mathrm{M}(\mathsf{X}, \mathcal{X})$ and $f \in \mathcal{F}_{\mathrm{b}}(\mathsf{X})$,

$$\left| \int f \, d\xi \right| \leq \|\xi\|_{\mathrm{TV}} \; \|f\|_\infty \; .$$

(ii) For any $\xi \in \mathrm{M}(\mathsf{X}, \mathcal{X})$,

$$\|\xi\|_{\mathrm{TV}} = \sup\left\{\xi(f) : f \in \mathcal{F}_{\mathrm{b}}(\mathsf{X}, \mathcal{X}), \|f\|_\infty = 1\right\} .$$

(iii) For any $f \in \mathcal{F}_{\mathrm{b}}(\mathsf{X})$,

$$\|f\|_\infty = \sup\left\{\xi(f) : \xi \in \mathrm{M}(\mathsf{X}, \mathcal{X}), \|\xi\|_{\mathrm{TV}} = 1\right\} .$$

Proof. Let H be a Hahn-Jordan set of ξ. Then $\xi_+(H) = \xi(H)$ and $\xi_-(H^c) = -\xi(H^c)$. For $f \in \mathcal{F}_{\mathrm{b}}(\mathsf{X})$,

$$|\xi(f)| \le |\xi_+(f)| + |\xi_-(f)| \le \|f\|_\infty (\xi_+(\mathsf{X}) + \xi_-(\mathsf{X})) = \|f\|_\infty \|\xi\|_{\mathrm{TV}} ,$$

showing (i). It also shows that the suprema in (ii) and (iii) are no larger than $\|\xi\|_{\mathrm{TV}}$ and $\|f\|_\infty$, respectively. To establish equality in these relations, first note that $\|\mathbb{1}_H - \mathbb{1}_{H^c}\|_\infty = 1$ and $\xi(\mathbb{1}_H - \mathbb{1}_{H^c}) = \xi(H) - \xi(H^c) = \|\xi\|_{\mathrm{TV}}$. This proves (ii). Next pick f and let let $\{x_n\}$ be a sequence in X such that $\lim_{n\to\infty} |f(x_n)| = \|f\|_\infty$. Then $\|f\|_\infty = \lim_{n\to\infty} |\delta_{x_n}(f)|$, proving (iii). □

The set $\mathrm{M}_0(\mathsf{X}, \mathcal{X})$ possesses some interesting properties that will prove useful in the sequel. Let ξ be in this set. Because $\xi(\mathsf{X}) = 0$, for any $f \in \mathcal{F}_{\mathrm{b}}(\mathsf{X})$ and any real c it holds that $\xi(f) = \xi(f - c)$. Therefore by Lemma 4.3.3(i), $|\xi(f)| \le \|\xi\|_{\mathrm{TV}} \|f - c\|_\infty$, which implies that

$$|\xi(f)| \le \|\xi\|_{\mathrm{TV}} \inf_{c\in\mathbb{R}} \|f - c\|_\infty .$$

It is easily seen that for any $f \in \mathcal{F}_{\mathrm{b}}(\mathsf{X})$, $\inf_{c\in\mathbb{R}} \|f - c\|_\infty$ is related to the oscillation semi-norm of f, also called the global modulus of continuity,

$$\operatorname{osc}(f) \stackrel{\mathrm{def}}{=} \sup_{(x,x')\in\mathsf{X}\times\mathsf{X}} |f(x) - f(x')| = 2 \inf_{c\in\mathbb{R}} \|f - c\|_\infty . \tag{4.14}$$

The lemma below provides some additional insight into this result.

Lemma 4.3.4. *For any* $\xi \in \mathrm{M}(\mathsf{X}, \mathcal{X})$ *and* $f \in \mathcal{F}_{\mathrm{b}}(\mathsf{X})$,

$$|\xi(f)| \le \sup_{(x,x')\in\mathsf{X}\times\mathsf{X}} |\xi_+(\mathsf{X}) f(x) - \xi_-(\mathsf{X}) f(x')| , \tag{4.15}$$

where (ξ_+, ξ_-) *is the Hahn-Jordan decomposition of* ξ. *In particular, for any* $\xi \in \mathrm{M}_0(\mathsf{X}, \mathcal{X})$ *and* $f \in \mathcal{F}_{\mathrm{b}}(\mathsf{X})$,

$$|\xi(f)| \le \frac{1}{2} \|\xi\|_{\mathrm{TV}} \operatorname{osc}(f) , \tag{4.16}$$

where $\operatorname{osc}(f)$ *is given by* (4.14).

Proof. First note that

$$\begin{aligned}\xi(f) &= \int f(x)\,\xi_+(dx) - \int f(x)\,\xi_-(dx) \\ &= \frac{\iint f(x)\,\xi_+(dx)\,\xi_-(dx')}{\xi_-(\mathsf{X})} - \frac{\iint f(x')\,\xi_+(dx)\,\xi_-(dx')}{\xi_+(\mathsf{X})}\ .\end{aligned}$$

Therefore

$$\begin{aligned}|\xi(f)| &\le \iint |f(x)/\xi_-(\mathsf{X}) - f(x')/\xi_+(\mathsf{X})|\,\xi_+(dx)\,\xi_-(dx') \\ &\le \sup_{(x,x')\in\mathsf{X}\times\mathsf{X}} |f(x)/\xi_-(\mathsf{X}) - f(x')/\xi_+(\mathsf{X})|\,\xi_+(\mathsf{X})\xi_-(\mathsf{X})\ ,\end{aligned}$$

which shows (4.15). If $\xi(\mathsf{X}) = 0$, then $\xi_+(\mathsf{X}) = \xi_-(\mathsf{X}) = \frac{1}{2}\,\|\xi\|_{\mathrm{TV}}$, showing (4.16). □

Therefore, for $\xi \in \mathrm{M}_0(\mathsf{X},\mathcal{X})$, $\|\xi\|_{\mathrm{TV}}$ is the operator norm of ξ considered as an operator over the space $\mathcal{F}_{\mathrm{b}}(\mathsf{X})$ equipped with the oscillation semi-norm (4.14). As a direct application of this result, if ξ and ξ' are two probability measures on $(\mathsf{X},\mathcal{X})$, then $\xi - \xi' \in \mathrm{M}_0(\mathsf{X},\mathcal{X})$ which implies that for any $f \in \mathcal{F}_{\mathrm{b}}(\mathsf{X})$,

$$|\xi(f) - \xi'(f)| \le \frac{1}{2}\,\|\xi - \xi'\|_{\mathrm{TV}}\ \mathrm{osc}\,(f)\ . \tag{4.17}$$

This inequality is sharper than the bound $|\xi(f) - \xi'(f)| \le \|\xi - \xi'\|_{\mathrm{TV}}\ \|f\|_\infty$ provided by Lemma 4.3.3(i), because $\mathrm{osc}\,(f) \le 2\,\|f\|_\infty$.

We conclude this section by establishing some alternative expressions for the total variation distance between two probability measures.

Lemma 4.3.5. *For any ξ and ξ' in $\mathrm{M}_1(\mathsf{X},\mathcal{X})$,*

$$\frac{1}{2}\,\|\xi - \xi'\|_{\mathrm{TV}} = \sup_A |\xi(A) - \xi'(A)| \tag{4.18}$$

$$= 1 - \sup_{\nu\le\xi,\xi'} \nu(\mathsf{X}) \tag{4.19}$$

$$= 1 - \inf \sum_{p=1}^{n} \xi(A_i) \wedge \xi'(A_i)\ . \tag{4.20}$$

Here the supremum in (4.18) is taken over all measurable subsets of X, the supremum in (4.19) is taken over all finite signed measures ν on $(\mathsf{X},\mathcal{X})$ satisfying $\nu \le \xi$ and $\nu \le \xi'$, and the infimum in (4.20) is taken over all finite measurable partitions $A_1,\dots,A_n$ of X.

Proof. To prove (4.18), first write $\xi(A) - \xi'(A) = (\xi - \xi')\mathbb{1}_A$ and note that $\mathrm{osc}\,(\mathbb{1}_A) = 1$. Thus (4.17) shows that the supremum in (4.18) is no larger than $(1/2)\,\|\xi - \xi'\|_{\mathrm{TV}}$. Now let H be a Jordan set of the signed measure $\xi - \xi'$.

The supremum is bounded from below by $\xi(H) - \xi'(H) = (\xi - \xi')_+(\mathsf{X}) = (1/2)\,\|\xi - \xi'\|_{\mathrm{TV}}$. This establishes equality in (4.18).

We now turn to (4.19). For any $p, q \in \mathbb{R}$, $|p-q| = p+q-2(p \wedge q)$. Therefore for any $A \in \mathcal{X}$,

$$\frac{1}{2}|\xi(A) - \xi'(A)| = \frac{1}{2}(\xi(A) + \xi'(A)) - \xi(A) \wedge \xi'(A) .$$

Applying this relation to the sets H and H^c, where H is as above, shows that

$$\frac{1}{2}\,(\xi - \xi')\,(H) = \frac{1}{2}\,[\xi(H) + \xi'(H)] - \xi(H) \wedge \xi'(H) ,$$
$$\frac{1}{2}\,(\xi' - \xi)\,(H^c) = \frac{1}{2}\,[\xi(H^c) + \xi'(H^c)] - \xi(H^c) \wedge \xi'(H^c) .$$

For any measure ν such that $\nu \le \xi$ and $\nu \le \xi'$, it holds that $\nu(H) \le \xi(H) \wedge \xi'(H)$ and $\nu(H^c) \le \xi(H^c) \wedge \xi'(H^c)$, showing that

$$\frac{1}{2}\,(\xi - \xi')\,(H) + \frac{1}{2}\,(\xi' - \xi)\,(H^c) = \frac{1}{2}\,\|\xi - \xi'\|_{\mathrm{TV}} \le 1 - \nu(\mathsf{X}) .$$

Thus (4.19) is no smaller than the left-hand side. To show equality, let ν be the measure defined by

$$\nu(A) = \xi(A \cap H^c) + \xi'(A \cap H) . \tag{4.21}$$

By the definition of H, $\xi(A \cap H^c) \le \xi'(A \cap H^c)$ and $\xi'(A \cap H) \le \xi(A \cap H)$ for any $A \in \mathcal{X}$. Therefore $\nu(A) \le \xi(A)$ and $\nu(A) \le \xi'(A)$. In addition, $\nu(H) = \xi'(H) = \xi(H) \wedge \xi'(H)$ and $\nu(H^c) = \xi(H^c) = \xi(H^c) \wedge \xi'(H^c)$, showing that $\frac{1}{2}\,\|\xi - \xi'\|_{\mathrm{TV}} = 1 - \nu(\mathsf{X})$ and concluding the proof of (4.19).

Finally, because $\nu(\mathsf{X}) = \xi(H) \wedge \xi'(H) + \xi(H^c) \wedge \xi'(H^c)$ we have

$$\sup_{\nu \le \xi, \xi'} \nu(\mathsf{X}) \ge \inf \sum_{i=1}^{n} \xi(A_i) \wedge \xi'(A_i) .$$

Conversely, for any measure ν satisfying $\nu \le \xi$ and $\nu \le \xi'$, and any partition $A_1, \ldots, A_n$,

$$\nu(\mathsf{X}) = \sum_{i=1}^{n} \nu(A_i) \le \sum_{i=1}^{n} \xi(A_i) \wedge \xi'(A_i) ,$$

showing that

$$\sup_{\nu \le \xi, \xi'} \nu(\mathsf{X}) \le \inf \sum_{i=1}^{n} \xi(A_i) \wedge \xi'(A_i) .$$

The supremum and the infimum thus agree, and the proof of (4.20) follows from (4.19). □

4.3.2 Lipshitz Contraction for Transition Kernels

In this section, we study the contraction property of transition kernels with respect to the total variation distance. Such results have been discussed in a seminal paper by Dobrushin (1956) (see Del Moral, 2004, Chapter 4, for a modern presentation and extensions of these results to a general class of distance-like entropy criteria). Let $(\mathsf{X}, \mathcal{X})$ and $(\mathsf{Y}, \mathcal{Y})$ be two measurable spaces and let K be a transition kernel from $(\mathsf{X}, \mathcal{X})$ to $(\mathsf{Y}, \mathcal{Y})$ (see Definition 2.1.1). The kernel K is canonically associated to two linear mappings:

(i) a mapping $\mathrm{M}(\mathsf{X}, \mathcal{X}) \to \mathrm{M}(\mathsf{Y}, \mathcal{Y})$ that maps any ξ in $\mathrm{M}(\mathsf{X}, \mathcal{X})$ to a (possibly signed) measure ξK given by $\xi K(A) = \int_{\mathsf{X}} \xi(dx)\, K(x, A)$ for any $A \in \mathcal{Y}$;
(ii) a mapping $\mathcal{F}_{\mathrm{b}}(\mathsf{Y}) \to \mathcal{F}_{\mathrm{b}}(\mathsf{X})$ that maps any f in $\mathcal{F}_{\mathrm{b}}(\mathsf{Y})$ to the function Kf given by $Kf(x) = \int K(x, dy)\, f(y)$.

Here again, with a slight abuse in notation, we use the same notation K for these two mappings. If we equip the spaces $\mathrm{M}(\mathsf{X}, \mathcal{X})$ and $\mathrm{M}(\mathsf{Y}, \mathcal{Y})$ with the total variation norm and the spaces $\mathcal{F}_{\mathrm{b}}(\mathsf{X})$ and $\mathcal{F}_{\mathrm{b}}(\mathsf{Y})$ with the supremum norm, a first natural problem is to compute the operator norm(s) of the kernel K.

Lemma 4.3.6. *Let $(\mathsf{X}, \mathcal{X})$ and $(\mathsf{Y}, \mathcal{Y})$ be two measurable spaces and let K be a transition kernel from $(\mathsf{X}, \mathcal{X})$ to $(\mathsf{Y}, \mathcal{Y})$. Then*

$$\begin{aligned} 1 &= \sup\{\|\xi K\|_{\mathrm{TV}} : \xi \in \mathrm{M}(\mathsf{X}, \mathcal{X}), \|\xi\|_{\mathrm{TV}} = 1\} \\ &= \sup\{\|Kf\|_{\infty} : f \in \mathcal{F}_{\mathrm{b}}(\mathsf{Y}), \|f\|_{\infty} = 1\} . \end{aligned}$$

Proof. By Lemma 4.3.3,

$$\begin{aligned} &\sup\{\|\xi K\|_{\mathrm{TV}} : \xi \in \mathrm{M}(\mathsf{X}, \mathcal{X}), \|\xi\|_{\mathrm{TV}} = 1\} \\ &\quad = \sup\{\xi K f : \xi \in \mathrm{M}(\mathsf{X}, \mathcal{X}), f \in \mathcal{F}_{\mathrm{b}}(\mathsf{Y}), \|f\|_{\infty} = 1, \|\xi\|_{\mathrm{TV}} = 1\} \\ &\quad = \sup\{\|Kf\|_{\infty} : f \in \mathcal{F}_{\mathrm{b}}(\mathsf{Y}, \mathcal{Y}), \|f\|_{\infty} = 1\} \le 1 . \end{aligned}$$

If ξ is a probability measure then so is ξK. Because the total variation of any probability measure is one, we see that the left-hand side of this display is indeed equal to one. Thus all members equate to one, and the proof is complete. □

To get sharper results, we will have to consider K as an operator acting on a smaller set of finite measures than $\mathrm{M}(\mathsf{X}, \mathcal{X})$. Of particular interest is the subset $\mathrm{M}_0(\mathsf{X}, \mathcal{X})$ of signed measures with zero total mass. Note that if ξ lies in this subset, then ξK is in $\mathrm{M}_0(\mathsf{Y}, \mathcal{Y})$. Below we will bound the operator norm of the restriction of the operator K to $\mathrm{M}_0(\mathsf{X}, \mathcal{X})$.

Definition 4.3.7 (Dobrushin Coefficient). *Let K be a transition kernel from $(\mathsf{X}, \mathcal{X})$ to $(\mathsf{Y}, \mathcal{Y})$. Its* Dobrushin coefficient *$\delta(K)$ is given by*

$$\begin{aligned}\delta(K) &= \frac{1}{2} \sup_{(x,x')\in\mathsf{X}\times\mathsf{X}} \|K(x,\cdot) - K(x',\cdot)\|_{\mathrm{TV}} \\ &= \sup_{(x,x')\in\mathsf{X}\times\mathsf{X}, x\neq x'} \frac{\|K(x,\cdot) - K(x',\cdot)\|_{\mathrm{TV}}}{\|\delta_x - \delta_{x'}\|_{\mathrm{TV}}} .\end{aligned}$$

We remark that as $K(x,\cdot)$ and $K(x',\cdot)$ are probability measures, it holds that $\|K(x,\cdot)\|_{\mathrm{TV}} = \|K(x',\cdot)\|_{\mathrm{TV}} = 1$. Hence $\delta(K) \leq \frac{1}{2}(1+1) = 1$, so that the Dobrushin coefficient satisfies $0 \leq \delta(K) \leq 1$.

Lemma 4.3.8. *Let ξ be a finite signed measure on $(\mathsf{X}, \mathcal{X})$ and let K be a transition kernel from $(\mathsf{X}, \mathcal{X})$ to $(\mathsf{Y}, \mathcal{Y})$. Then*

$$\|\xi K\|_{\mathrm{TV}} \leq \delta(K)\, \|\xi\|_{\mathrm{TV}} + (1 - \delta(K))\, |\xi(\mathsf{X})| . \tag{4.22}$$

Proof. Pick $\xi \in \mathrm{M}(\mathsf{X}, \mathcal{X})$ and let, as usual, ξ_+ and ξ_- be its positive and negative part, respectively. If $\xi_-(\mathsf{X}) = 0$ (ξ is a measure), then $\|\xi\|_{\mathrm{TV}} = \xi(\mathsf{X})$ and (4.22) becomes $\|\xi K\|_{\mathrm{TV}} \leq \|\xi\|_{\mathrm{TV}}$; this follows from Lemma 4.3.6. If $\xi_+(\mathsf{X}) = 0$, an analogous argument applies.

Thus assume that both ξ_+ and ξ_- are non-zero. In view of Lemma 4.3.3(ii), it suffices to prove that for any $f \in \mathcal{F}_{\mathrm{b}}(\mathsf{Y})$ with $\|f\|_\infty = 1$,

$$|\xi K f| \leq \delta(K)(\xi_+(\mathsf{X}) + \xi_-(\mathsf{X})) + (1 - \delta(K))|\xi_+(\mathsf{X}) - \xi_-(\mathsf{X})| . \tag{4.23}$$

We shall suppose that $\xi_+(\mathsf{X}) \geq \xi_-(\mathsf{X})$, if not, replace ξ by $-\xi$ and (4.23) remains the same. Then as $|\xi_+(\mathsf{X}) - \xi_-(\mathsf{X})| = \xi_+(\mathsf{X}) - \xi_-(\mathsf{X})$, (4.23) becomes

$$|\xi K f| \leq 2\xi_-(\mathsf{X})\delta(K) + \xi_+(\mathsf{X}) - \xi_-(\mathsf{X}) . \tag{4.24}$$

Now, by Lemma 4.3.4, for any $f \in \mathcal{F}_{\mathrm{b}}(\mathsf{Y})$ it holds that

$$\begin{aligned}|\xi K f| &\leq \sup_{(x,x')\in\mathsf{X}\times\mathsf{X}} |\xi_+(\mathsf{X})Kf(x) - \xi_-(\mathsf{X})Kf(x')| \\ &\leq \sup_{(x,x')\in\mathsf{X}\times\mathsf{X}} \|\xi_+(\mathsf{X})K(x,\cdot) - \xi_-(\mathsf{X})K(x',\cdot)\|_{\mathrm{TV}}\, \|f\|_\infty .\end{aligned}$$

Finally (4.24) follows upon noting that

$$\begin{aligned}&\|\xi_+(\mathsf{X})K(x,\cdot) - \xi_-(\mathsf{X})K(x',\cdot)\|_{\mathrm{TV}} \\ &\qquad \leq \xi_-(\mathsf{X})\, \|K(x,\cdot) - K(x',\cdot)\|_{\mathrm{TV}} + [\xi_+(\mathsf{X}) - \xi_-(\mathsf{X})]\, \|K(x,\cdot)\|_{\mathrm{TV}} \\ &\qquad = 2\xi_-(\mathsf{X})\delta(K) + \xi_+(\mathsf{X}) - \xi_-(\mathsf{X}) .\end{aligned}$$

□

Corollary 4.3.9.

$$\delta(K) = \sup\left\{\|\xi K\|_{\mathrm{TV}} : \xi \in \mathrm{M}_0(\mathsf{X}, \mathcal{X}),\ \|\xi\|_{\mathrm{TV}} \leq 1\right\} . \tag{4.25}$$

Proof. If $\xi(\mathsf{X}) = 0$, then (4.22) becomes $\|\xi K\|_{\mathrm{TV}} \le \delta(K)\,\|\xi\|_{\mathrm{TV}}$, showing that

$$\sup\left\{\|\xi K\|_{\mathrm{TV}} : \xi \in \mathrm{M}_0(\mathsf{X}, \mathcal{X}),\ \|\xi\|_{\mathrm{TV}} \le 1\right\} \le \delta(K) .$$

The converse inequality is obvious, as

$$\delta(K) = \sup\left\{(x, x') \in \mathsf{X} \times \mathsf{X}, \left\|\frac{1}{2}(\delta_x - \delta_{x'})K\right\|_{\mathrm{TV}}\right\} \\ \le \sup\left\{\|\xi K\|_{\mathrm{TV}} : \xi \in \mathrm{M}_0(\mathsf{X}, \mathcal{X}),\ \|\xi\|_{\mathrm{TV}} = 1\right\} .$$

□

If ξ and ξ' are two probability measures on $(\mathsf{X}, \mathcal{X})$, Corollary 4.3.9 implies that

$$\|\xi K - \xi' K\|_{\mathrm{TV}} \le \delta(K)\,\|\xi - \xi'\|_{\mathrm{TV}} .$$

Thus the Dobrushin coefficient is the norm of K considered as a linear operator from $\mathrm{M}_0(\mathsf{X}, \mathcal{X})$ to $\mathrm{M}_0(\mathsf{Y}, \mathcal{Y})$.

Proposition 4.3.10. *The Dobrushin coefficient is sub-multiplicative. That is, if $K : (\mathsf{X}, \mathcal{X}) \to (\mathsf{Y}, \mathcal{Y})$ and $R : (\mathsf{Y}, \mathcal{Y}) \to (\mathsf{Z}, \mathcal{Z})$ are two transition kernels, then $\delta(KR) \le \delta(K)\delta(R)$.*

Proof. This is a direct consequence of the fact that the Dobrushin coefficient is an operator norm. By Corollary 4.3.9, if $\xi \in \mathrm{M}_0(\mathsf{X}, \mathcal{X})$, then $\xi K \in \mathrm{M}_0(\mathsf{Y}, \mathcal{Y})$ and $\|\xi K\|_{\mathrm{TV}} \le \delta(K)\,\|\xi\|_{\mathrm{TV}}$. Likewise, $\|\nu R\|_{\mathrm{TV}} \le \delta(R)\,\|\nu\|_{\mathrm{TV}}$ holds for any $\nu \in \mathrm{M}_0(\mathsf{Y}, \mathcal{Y})$. Thus

$$\|\xi K R\|_{\mathrm{TV}} = \|(\xi K)R\|_{\mathrm{TV}} \le \delta(R)\,\|\xi K\|_{\mathrm{TV}} \le \delta(K)\delta(R)\,\|\xi\|_{\mathrm{TV}}$$

□

4.3.3 The Doeblin Condition and Uniform Ergodicity

Anticipating results on general state-space Markov chains presented in Chapter 14, we will establish, using the contraction results developed in the previous section, some ergodicity results for a class of Markov chains $(\mathsf{X}, \mathcal{X})$ satisfying the so-called Doeblin condition.

Assumption 4.3.11 (Doeblin Condition). *There exist an integer $m \ge 1$, $\epsilon \in (0, 1)$, and a transition kernel $\nu = \{\nu_{x,x'}, (x, x') \in \mathsf{X} \times \mathsf{X}\}$ from $(\mathsf{X} \times \mathsf{X}, \mathcal{X} \otimes \mathcal{X})$ to $(\mathsf{X}, \mathcal{X})$ such that for all $(x, x') \in \mathsf{X} \times \mathsf{X}$ and $A \in \mathcal{X}$,*

$$Q^m(x, A) \wedge Q^m(x', A) \ge \epsilon \nu_{x,x'}(A) .$$

We will frequently consider a strengthened version of this assumption.

Assumption 4.3.12 (Doeblin Condition Reinforced). *There exist an integer $m \geq 1$, $\epsilon \in (0, 1)$, and a probability measure ν on $(\mathsf{X}, \mathcal{X})$ such that for any $x \in \mathsf{X}$ and $A \in \mathcal{X}$,*

$$Q^m(x, A) \geq \epsilon\nu(A) .$$

By Lemma 4.3.5, the Dobrushin coefficient of Q^m may be equivalently written as

$$\delta(Q^m) = 1 - \inf \sum_{i=1}^{n} Q^m(x, A_i) \wedge Q^m(x', A_i) , \tag{4.26}$$

where the infimum is taken over all $(x, x') \in \mathsf{X} \times \mathsf{X}$ and all finite measurable partitions $A_1, \ldots, A_n$ of X of X. Under the Doeblin condition, the sum in this display is bounded from below by $\epsilon \sum_{i=1}^{n} \nu_{x,x'}(A_i) = \epsilon$. Hence the following lemma is true.

Lemma 4.3.13. *Under Assumption 4.3.11, $\delta(Q^m) \leq 1 - \epsilon$.*

Stochastic processes that are such that for any k, the distribution of the random vector $(X_n, \ldots, X_{n+k})$ does not depend on n are called *stationary* (see Definition 2.1.10). It is clear that in general a Markov chain will not be stationary. Nevertheless, given a transition kernel Q, it is possible that with an appropriate choice of the initial distribution ν we may produce a stationary process. Assuming that such a distribution exists, the stationarity of the marginal distribution implies that $\mathrm{E}_\nu[\mathbb{1}_A(X_0)] = \mathrm{E}_\nu[\mathbb{1}_A(X_1)]$ for any $A \in \mathcal{X}$. This can equivalently be written as $\nu(A) = \nu Q(A)$, or $\nu = \nu Q$. In such a case, the Markov property implies that all finite-dimensional distributions of $\{X_k\}_{k \geq 0}$ are also invariant under translation in time. These considerations lead to the definition of *invariant measure.*

Definition 4.3.14 (Invariant Measure). *If Q is a Markov kernel on $(\mathsf{X}, \mathcal{X})$ and π is a σ-finite measure satisfying $\pi Q = \pi$, then π is called an* invariant measure.

If an invariant measure is finite, it may be normalized to an *invariant probability measure.* In practice, this is the main situation of interest. If an invariant measure has infinite total mass, its probabilistic interpretation is much more difficult. In general, there may exist more than one invariant measure, and if X is not finite, an invariant measure may not exist. As a trivial example, consider $\mathsf{X} = \mathbb{N}$ and $Q(x, x + 1) = 1$.

Invariant probability measures are important not merely because they define stationary processes. Invariant probability measures also define the long-term or *ergodic* behavior of a stationary Markov chain. Assume that for some initial measure ν, the sequence of probability measures $\{\nu Q^n\}_{n \geq 0}$ converges to a probability measure γ_ν in total variation norm. This implies that for any function $f \in \mathcal{F}_\mathrm{b}(\mathsf{X})$, $\lim_{n \to \infty} \nu Q^n(f) = \gamma_\nu(f)$. Therefore

$$\gamma_\nu(f) = \lim_{n\to\infty} \iint \nu(dx)\, Q^n(x, dx')\, f(x')$$
$$= \lim_{n\to\infty} \iint \nu(dx)\, Q^{n-1}(x, dx')\, Qf(x') = \gamma_\nu(Qf) \; .$$

Hence, if a limiting distribution exists, it is an invariant probability measure, and if there exists a unique invariant probability measure, then the limiting distribution γ_ν will be independent of ν, whenever it exists. These considerations lead to the following definitions.

Definition 4.3.15. *Let Q be a Markov kernel admitting a unique invariant probability measure π. The chain is said to be* ergodic *if for all x in a set $A \in \mathcal{X}$ such that $\pi(A) = 1$, $\lim_{n\to\infty} \|Q^n(x,\cdot) - \pi\|_{\mathrm{TV}} = 0$. It is said to be* uniformly ergodic *if $\lim_{n\to\infty} \sup_{x\in\mathsf{X}} \|Q^n(x,\cdot) - \pi\|_{\mathrm{TV}} = 0$.*

Note that when a chain is uniformly ergodic, it is indeed uniformly geometrically ergodic because $\lim_{n\to\infty} \sup_{x\in\mathsf{X}} \|Q^n(x,\cdot) - \pi\|_{\mathrm{TV}} = 0$ implies that there exists an integer m such that $\frac{1}{2}\sup_{(x,x')\in\mathsf{X}\times\mathsf{X}} \|Q^m(x,\cdot) - Q^m(x',\cdot)\|_{\mathrm{TV}} < 1$ by the triangle inequality. Hence the Dobrushin coefficient $\delta(Q^m)$ is strictly less than 1, and Q^m is contractive with respect to the total variation distance by Lemma 4.3.8. Thus there exist constants $C < \infty$ and $\rho \in [0,1)$ such that $\sup_{x\in\mathsf{X}} \|Q^n(x,\cdot) - \pi\|_{\mathrm{TV}} \le C\rho^n$ for all n.

The following result shows that if a power Q^m of the Markov kernel Q satisfies Doeblin's condition, then the chain admits a unique invariant probability and is uniformly ergodic.

Theorem 4.3.16. *Under Assumption 4.3.11, Q admits a unique invariant probability measure π. In addition, for any $\xi \in \mathrm{M}_1(\mathsf{X}, \mathcal{X})$,*

$$\|\xi Q^n - \pi\|_{\mathrm{TV}} \le (1-\epsilon)^{\lfloor n/m \rfloor} \; \|\xi - \pi\|_{\mathrm{TV}} \; ,$$

where $\lfloor u \rfloor$ is the integer part of u.

Proof. Let ξ and ξ' be two probability measures on $(\mathsf{X}, \mathcal{X})$. Corollary 4.3.9, Proposition 4.3.10, and Lemma 4.3.13 yield that for all $k \ge 1$,

$$\left\|\xi Q^{km} - \xi' Q^{km}\right\|_{\mathrm{TV}} \le \delta^k(Q^m)\, \|\xi - \xi'\|_{\mathrm{TV}} \le (1-\epsilon)^k \, \|\xi - \xi'\|_{\mathrm{TV}} \; . \quad (4.27)$$

Taking $\xi' = \xi Q^{pm}$, we find that

$$\left\|\xi Q^{km} - \xi Q^{(k+p)m}\right\|_{\mathrm{TV}} \le (1-\epsilon)^k \; ,$$

showing that $\{\xi Q^{km}\}$ is a Cauchy sequence in $\mathrm{M}_1(\mathsf{X}, \mathcal{X})$ endowed with the total variation norm. Because this metric space is complete, there exists a probability measure π such that $\xi Q^{km} \to \pi$. In view of the discussion above, π is invariant for Q^m. Moreover, by (4.27) this limit does not depend on ξ. Thus Q^m admits π as unique invariant probability measure. The Chapman-Kolmogorov equations imply that $(\pi Q)Q^m = (\pi Q^m)Q = \pi Q$, showing that πQ is also invariant for Q^m and hence that $\pi Q = \pi$ as claimed. □

Remark 4.3.17. Classical uniform convergence to equilibrium for Markov processes has been studied during the first half of the 20th century by Doeblin, Kolmogorov, and Doob under various conditions. Doob (1953) gave a unifying form to these conditions, which he named *Doeblin type conditions*. More recently, starting in the 1970s, an increasing interest in non-uniform convergence of Markov processes has arisen. An explanation for this interest is that many useful processes do not converge uniformly to equilibrium, while they do satisfy weaker properties such as a geometric convergence. It later became clear that non-uniform convergence relates to *local* Doeblin type condition and to hitting times for so-called *small sets*. These types of conditions are detailed in Chapter 14. ■

4.3.4 Forgetting Properties

Recall from Chapter 3 that the smoothing probability $\phi_{\nu,k|n}[Y_{0:n}]$ is defined by

$$\phi_{\nu,k|n}[Y_{0:n}](f) = \mathrm{E}_\nu[f(X_k) \,|\, Y_{0:n}] \;, \qquad f \in \mathcal{F}_\mathrm{b}(\mathsf{X}) \;.$$

Here, k and n are integers, and ν is the initial probability measure on $(\mathsf{X}, \mathcal{X})$. The filtering probability is defined by $\phi_{\nu,n}[Y_{0:n}] = \phi_{\nu,n|n}[Y_{0:n}]$. In this section, we will establish that under appropriate conditions on the transition kernel Q and on the function g, the sequence of filtering probabilities satisfies a property referred to in the literature as "forgetting of the initial condition". This property can be formulated as follows: given two probability measures ν and ν' on $(\mathsf{X}, \mathcal{X})$,

$$\lim_{n\to\infty} \|\phi_{\nu,n}[Y_{0:n}] - \phi_{\nu',n}[Y_{0:n}]\|_{\mathrm{TV}} = 0 \quad \mathrm{P}_{\nu_\star}\text{-a.s.} \tag{4.28}$$

where $\nu_\star$ is the initial probability measure that defines the law of the observations $\{Y_k\}$. Forgetting is also a concept that applies to the smoothing distributions, as it is often possible to extend the previous results showing that

$$\lim_{k\to\infty} \sup_{n\geq 0} \left\|\phi_{\nu,k|n}[Y_{0:n}] - \phi_{\nu',k|n}[Y_{0:n}]\right\|_{\mathrm{TV}} = 0 \quad \mathrm{P}_{\nu_\star}\text{-a.s.} \tag{4.29}$$

Equation (4.29) can also be strengthened by showing that, under additional conditions, the forgetting property is uniform with respect to the observed sequence $Y_{0:n}$ in the sense that there exists a deterministic sequence $\{\rho_k\}$ satisfying $\rho_k \to 0$ and

$$\sup_{y_{0:n}\in\mathsf{Y}^{n+1}} \sup_{n\geq 0} \left\|\phi_{\nu,k|n}[y_{0:n}] - \phi_{\nu',k|n}[y_{0:n}]\right\|_{\mathrm{TV}} \leq \rho_k \;.$$

Several of the results to be proven in the sequel are of this latter type (uniform forgetting).

As shown in (3.5), the smoothing distribution is defined as the ratio

$$\phi_{\nu,k|n}[y_{0:n}](f) = \frac{\int \cdots \int f(x_k)\, \nu(dx_0)\, g(x_0, y_0) \prod_{i=1}^n Q(x_{i-1}, dx_i)\, g(x_i, y_i)}{\int \cdots \int \nu(dx_0)\, g(x_0, y_0) \prod_{i=1}^n Q(x_{i-1}, dx_i)\, g(x_i, y_i)} \; .$$

Therefore, the mapping associating the probability measure $\nu \in \mathrm{M}_1(\mathsf{X}, \mathcal{X})$ to the probability measure $\phi_{\nu,k|n}[y_{0:n}]$ is non-linear. The theory developed above allows one to separately control the numerator and the denominator of this quantity but does not lend a direct proof of the forgetting properties (4.28) or (4.29). To achieve this, we use the alternative representation of the smoothing probability $\phi_{\nu,k|n}[y_{0:n}]$ introduced in Proposition 3.3.4, which states that

$$\begin{aligned}\phi_{\nu,k|n}[y_{0:n}](f) &= \int \cdots \int \phi_{\nu,0|n}[y_{0:n}](dx_0) \prod_{i=1}^k \mathrm{F}_{i-1|n}[y_{i:n}](x_{i-1}, dx_i)\, f(x_k) \\ &= \phi_{\nu,0|n}[y_{0:n}] \prod_{i=1}^k \mathrm{F}_{i-1|n}[y_{i:n}] f \; . \end{aligned} \tag{4.30}$$

Here we have used the following notations and definitions from Chapter 3.

(i) $\mathrm{F}_{i|n}[y_{i+1:n}]$ are the *forward smoothing kernels* (see Definition 3.3.1) given for $i = 0, \ldots, n-1$, $x \in \mathsf{X}$ and $A \in \mathcal{X}$, by

$$\begin{aligned}\mathrm{F}_{i|n}[y_{i+1:n}](x, A) &\stackrel{\text{def}}{=} \left(\beta_{i|n}[y_{i+1:n}](x)\right)^{-1} \\ &\quad \times \int_A Q(x, dx_{i+1})\, g(x_{i+1}, y_{i+1}) \beta_{i+1|n}[y_{i+2:n}](x_{i+1}) \; , \end{aligned} \tag{4.31}$$

where $\beta_{i|n}[y_{i+1:n}](x)$ are the backward functions (see Definition 3.1.6)

$$\beta_{i|n}[y_{i+1:n}](x) = \int Q(x, dx_{i+1})\, g(x_{i+1}, y_{i+1}) \beta_{i+1|n}[y_{i+2:n}](x_{i+1}) \; . \tag{4.32}$$

Recall that, by Proposition 3.3.2, $\{\mathrm{F}_{i|n}\}_{i \ge 0}$ are the transition kernels of the non-homogeneous Markov chain $\{X_k\}$ conditionally on $Y_{0:n}$,

$$\mathrm{E}_\nu[f(X_{i+1}) \mid X_{0:i}, Y_{0:n}] = \mathrm{F}_{i|n}[Y_{i+1:n}](X_i, f) \; .$$

(ii) $\phi_{\nu,0|n}[y_{0:n}]$ is the posterior distribution of the state X_0 conditionally on $Y_{0:n} = y_{0:n}$, defined for any $A \in \mathcal{X}$ by

$$\phi_{\nu,0|n}[y_{0:n}](A) = \frac{\int_A \nu(dx_0)\, g(x_0, y_0) \beta_{0|n}[y_{1:n}](x_0)}{\int \nu(dx_0)\, g(x_0, y_0) \beta_{0|n}[y_{1:n}](x_0)} \; . \tag{4.33}$$

We see that the non-linear mapping $\nu \mapsto \phi_{\nu,k|n}[y_{0:n}]$ is the composition of two mappings on $\mathcal{M}_1(\mathsf{X}, \mathcal{X})$.

(i) The mapping $\nu \mapsto \phi_{\nu,0|n}[y_{0:n}]$, which associates to the initial distribution ν the posterior distribution of the state X_0 given $Y_{0:n} = y_{0:n}$. This mapping consists in applying Bayes' formula, which we write as

$$\phi_{\nu,0|n}[y_{0:n}] = \mathrm{B}[g(\cdot, y_0)\beta_{0|n}[y_{1:n}](\cdot), \nu] .$$

Here

$$\mathrm{B}[\phi, \xi](f) = \frac{\int f(x)\phi(x)\,\xi(dx)}{\int \phi(x)\,\xi(dx)} , \qquad f \in \mathcal{F}_{\mathrm{b}}(\mathsf{X}) , \tag{4.34}$$

for any probability measure ξ on $(\mathsf{X}, \mathcal{X})$ and any non-negative measurable function ϕ on X. Note that $\mathrm{B}[\phi, \xi]$ is a probability measure on $(\mathsf{X}, \mathcal{X})$. Because of the normalization, this step is non-linear.

(ii) The mapping $\xi \mapsto \xi \prod_{i=1}^{k} \mathrm{F}_{i-1|n}[y_{i:n}]$, which is a linear mapping being defined as product of Markov transition kernels.

For two initial probability measures ν and ν' on $(\mathsf{X}, \mathcal{X})$, the difference of the associated smoothing distributions may thus be expressed as

$$\phi_{\nu,k|n}[y_{0:n}] - \phi_{\nu',k|n}[y_{0:n}] = \\ \left(\mathrm{B}[g(\cdot, y_0)\beta_{0|n}[y_{1:n}], \nu] - \mathrm{B}[g(\cdot, y_0)\beta_{0|n}[y_{1:n}], \nu']\right) \prod_{i=1}^{k} \mathrm{F}_{i-1|n}[y_{i:n}] . \tag{4.35}$$

Note that the function $g(x, y_0)\beta_{0|n}[y_{1:n}](x)$ defined for $x \in \mathsf{X}$ may also be interpreted as the likelihood of the observation $\mathrm{L}_{\delta_x,n}[y_{0:n}]$ when starting from the initial condition $X_0 = x$ (Proposition 3.2.3). In the sequel, we use the likelihood notation whenever possible, writing, in addition, $\mathrm{L}_{x,n}[y_{0:n}]$ rather than $\mathrm{L}_{\delta_x,n}[y_{0:n}]$ and $\mathrm{L}_{\bullet,n}[y_{0:n}]$ when referring to the whole function.

Using Corollary 4.3.9, (4.35) implies that

$$\left\|\phi_{\nu,k|n}[y_{0:n}] - \phi_{\nu,k|n}[y_{0:n}]\right\|_{\mathrm{TV}} \leq \\ \|\mathrm{B}[\mathrm{L}_{\bullet,n}[y_{0:n}], \nu] - \mathrm{B}[\mathrm{L}_{\bullet,n}[y_{0:n}], \nu']\|_{\mathrm{TV}}\, \delta\left(\prod_{i=1}^{k} \mathrm{F}_{i-1|n}[y_{i:n}]\right) , \tag{4.36}$$

where the final factor is a Dobrushin coefficient. Because Bayes operator B returns probability measures, the total variation distance in the right-hand side of this display is always bounded by 2. Although this bound may be sufficient, it is often interesting to relate the total variation distance between $\mathrm{B}[\phi, \xi]$ and $\mathrm{B}[\phi, \xi']$ to the total variation distance between ξ and ξ'. The following lemma is adapted from (Künsch, 2000)—see also (Del Moral, 2004, Theorem 4.3.1).

Lemma 4.3.18. *Let ξ and ξ' be two probability measures on $(\mathsf{X}, \mathcal{X})$ and let ϕ be a non-negative measurable function such that $\xi(\phi) > 0$ or $\xi'(\phi) > 0$. Then*

$$\|\mathrm{B}[\phi, \xi] - \mathrm{B}[\phi, \xi']\|_{\mathrm{TV}} \leq \frac{\|\phi\|_\infty}{\xi(\phi) \vee \xi'(\phi)} \|\xi - \xi'\|_{\mathrm{TV}} . \tag{4.37}$$

Proof. We may assume, without loss of generality, that $\xi(\phi) \geq \xi'(\phi)$. For any $f \in \mathcal{F}_b(\mathsf{X})$,

$$
\begin{aligned}
&\mathrm{B}[\phi,\xi](f) - \mathrm{B}[\phi,\xi'](f) \\
&= \frac{\int f(x)\phi(x)\,(\xi-\xi')(dx)}{\int \phi(x)\,\xi(dx)} + \frac{\int f(x)\phi(x)\,\xi'(dx)}{\int \phi(x)\,\xi'(dx)} \frac{\int \phi(x)\,(\xi'-\xi)(dx)}{\int \phi(x)\,\xi(dx)} \\
&= \frac{1}{\xi(\phi)} \int (\xi-\xi')(dx)\,\phi(x)(f(x) - \mathrm{B}[\phi,\xi'](f)) \, .
\end{aligned}
$$

By Lemma 4.3.5,

$$
\begin{aligned}
&\left| \int (\xi-\xi')(dx)\,\phi(x)(f(x) - \mathrm{B}[\phi,\xi'](f)) \right| \leq \|\xi-\xi'\|_{\mathrm{TV}} \times \\
&\qquad \frac{1}{2} \sup_{(x,x')\in\mathsf{X}\times\mathsf{X}} |\phi(x)(f(x)-\mathrm{B}[\phi,\xi'](f)) - \phi(x')(f(x')-\mathrm{B}[\phi,\xi'](f))| \, .
\end{aligned}
$$

Because $|\mathrm{B}[\phi,\xi'](f)| \leq \|f\|_\infty$ and $\phi \geq 0$, the supremum on the right-hand side of this display is bounded by $2\|\phi\|_\infty \|f\|_\infty$. This concludes the proof. □

As mentioned by Künsch (2000), the Bayes operator may be non-contractive: the numerical factor in the right-hand side of (4.37) is sometimes larger than one and the bound may be shown to be tight on particular examples. The intuition that the posteriors should at least be as close as the priors if the same likelihood (the same data) is applied is thus generally wrong.

Equation (4.30) also implies that for any integer j such that $j \leq k$,

$$
\begin{aligned}
\phi_{\nu,k|n}[y_{0:n}] &= \phi_{\nu,0|n}[y_{0:n}] \prod_{i=1}^{j} \mathrm{F}_{i-1|n}[y_{i:n}] \prod_{i=j+1}^{k} \mathrm{F}_{i-1|n}[y_{i:n}] \\
&= \phi_{\nu,j|n}[y_{0:n}] \prod_{i=j+1}^{k} \mathrm{F}_{i-1|n}[y_{i:n}] \, .
\end{aligned} \tag{4.38}
$$

This decomposition and Corollary 4.3.9 shows that for any $0 \leq j \leq k$, any initial distributions ν and ν' and any sequence $y_{0:n}$ such that $\mathrm{L}_{\nu,n}[y_{0:n}] > 0$ and $\mathrm{L}_{\nu',n}[y_{0:n}] > 0$,

$$
\begin{aligned}
&\left\| \phi_{\nu,k|n}[y_{0:n}] - \phi_{\nu',k|n}[y_{0:n}] \right\|_{\mathrm{TV}} \\
&\qquad \leq \delta\left(\prod_{i=j+1}^{k} \mathrm{F}_{i-1|n}[y_{i:n}] \right) \left\| \phi_{\nu,j|n}[y_{0:n}] - \phi_{\nu',j|n}[y_{0:n}] \right\|_{\mathrm{TV}} \, .
\end{aligned}
$$

Because the Dobrushin coefficient of a Markov kernel is bounded by one, this relation implies that the total variation distance between the smoothing distributions associated with two different initial distributions is non-expanding. To summarize this discussion, we have obtained the following result.

Proposition 4.3.19. *Let ν and ν' be two probability measures on $(\mathsf{X}, \mathcal{X})$. For any non-negative integers j, k, and n such that $j \le k$ and any sequence $y_{0:n} \in \mathsf{Y}^{n+1}$ such that $\mathrm{L}_{\nu,n}[y_{0:n}] > 0$ and $\mathrm{L}_{\nu',n}[y_{0:n}] > 0$,*

$$\begin{aligned}\left\|\phi_{\nu,k|n}[y_{0:n}] - \phi_{\nu',k|n}[y_{0:n}]\right\|_{\mathrm{TV}} \\ \le \delta\left(\prod_{i=j+1}^{k} \mathrm{F}_{i-1|n}[y_{i:n}]\right) \left\|\phi_{\nu,j|n}[y_{0:n}] - \phi_{\nu',j|n}[y_{0:n}]\right\|_{\mathrm{TV}} ,\end{aligned} \quad (4.39)$$

$$\begin{aligned}\left\|\phi_{\nu,k|n}[y_{0:n}] - \phi_{\nu',k|n}[y_{0:n}]\right\|_{\mathrm{TV}} \\ \le \frac{\|\mathrm{L}_{\bullet,n}[y_{0:n}]\|_\infty}{\mathrm{L}_{\nu,n}[y_{0:n}] \vee \mathrm{L}_{\nu',n}[y_{0:n}]} \delta\left(\prod_{i=1}^{k} \mathrm{F}_{i-1|n}[y_{i:n}]\right) \|\nu - \nu'\|_{\mathrm{TV}} .\end{aligned} \quad (4.40)$$

Along the same lines, we can compare the posterior distribution of the state X_k given observations $Y_{j:n}$ for different values of j. To avoid introducing new notations, we will simply denote these conditional distributions by $\mathrm{P}_\nu(X_k \in \cdot \mid Y_{j:n} = y_{j:n})$. As mentioned in the introduction of this chapter, it is sensible to expect that $\mathrm{P}_\nu(X_k \in \cdot \mid Y_{j:n})$ gets asymptotically close to $\mathrm{P}_\nu(X_k \in \cdot \mid Y_{0:n})$ as $k - j$ tends to infinity. Here again, to establish this alternative form of the forgetting property, we will use a representation of $\mathrm{P}_\nu(X_k \in \cdot \mid Y_{j:n})$ similar to (4.30).

Because $\{(X_k, Y_k)\}$ is a Markov chain, and assuming that $k \ge j$,

$$\mathrm{P}_\nu(X_k \in \cdot \mid X_j, Y_{j:n}) = \mathrm{P}_\nu(X_k \in \cdot \mid X_j, Y_{0:n}) .$$

Moreover, we know that conditionally on $Y_{0:n}$, $\{X_k\}$ is a non-homogeneous Markov chain with transition kernels $\mathrm{F}_{k|n}[Y_{k+1:n}]$ where $\mathrm{F}_{i|n} = Q$ for $i \ge n$ (Proposition 3.3.2). Therefore the Chapman-Kolmogorov equations show that for any function $f \in \mathcal{F}_\mathrm{b}(\mathsf{X})$,

$$\begin{aligned}&\mathrm{E}_\nu[f(X_k) \mid Y_{j:n}] = \mathrm{E}_\nu[\mathrm{E}_\nu[f(X_k) \mid X_j, Y_{j:n}] \mid Y_{j:n}] \\ &= \mathrm{E}_\nu\left[\prod_{i=j+1}^{k} \mathrm{F}_{i-1|n}[Y_{i:n}] f(X_j) \,\middle|\, Y_{j:n}\right] = \tilde{\phi}_{\nu,j|n}[Y_{j:n}] \prod_{i=j+1}^{k} \mathrm{F}_{i-1|n}[Y_{i:n}] f ,\end{aligned}$$

cf. (4.38), where the probability measure $\tilde{\phi}_{\nu,j|n}[Y_{j:n}(f)]$ is defined by

$$\tilde{\phi}_{\nu,j|n}[Y_{j:n}](f) = \mathrm{E}_\nu[f(X_j) \mid Y_{j:n}] , \qquad f \in \mathcal{F}_\mathrm{b}(\mathsf{X}) .$$

Using (4.38) as well, we thus find that the difference between $\mathrm{P}_\nu(X_k \in \cdot \mid Y_{j:n})$ and $\mathrm{P}_\nu(X_k \in \cdot \mid Y_{0:n})$ may be expressed by

$$\mathrm{E}_\nu[f(X_k) \mid Y_{j:n}] - \mathrm{E}_\nu[f(X_k) \mid Y_{0:n}] = (\tilde{\phi}_{\nu,j|n} - \phi_{\nu,j|n}) \prod_{i=j+1}^{k} \mathrm{F}_{i-1|n}[Y_{i:n}] f .$$

Proceeding like in Proposition 4.3.19, we may thus derive a bound on the total variation distance between these probability measures.

Proposition 4.3.20. *For any integers j, k, and n such that $0 \le j \le k$ and any probability measure ν on $(\mathsf{X}, \mathcal{X})$,*

$$\|\mathrm{P}_\nu(X_k \in \cdot \mid Y_{0:n}) - \mathrm{P}_\nu(X_k \in \cdot \mid Y_{j:n})\|_{\mathrm{TV}} \le 2\delta \left(\prod_{i=j+1}^{k} \mathrm{F}_{i-1|n}[Y_{i:n}] \right) . \tag{4.41}$$

4.3.5 Uniform Forgetting Under Strong Mixing Conditions

In light of the discussion above, establishing forgetting properties amounts to determining non-trivial bounds on the Dobrushin coefficient of products of forward transition kernels and, if required, on ratio of likelihoods $\mathrm{L}_{x,n}(y_{0:n})/(\mathrm{L}_{\nu,n}(y_{0:n}) \vee \mathrm{L}_{\nu',n}(y_{0:n}))$. To do so, we need to impose additional conditions on Q and g. We consider in this section the following assumption, which was introduced by Le Gland and Oudjane (2004, Section 2).

Assumption 4.3.21 (Strong Mixing Condition). *There exist a transition kernel $K : (\mathsf{Y}, \mathcal{Y}) \to (\mathsf{X}, \mathcal{X})$ and measurable functions ς^- and ς^+ from Y to $(0, \infty)$ such that for any $A \in \mathcal{X}$ and $y \in \mathsf{Y}$,*

$$\varsigma^-(y) K(y, A) \le \int_A Q(x, dx')\, g(x', y) \le \varsigma^+(y) K(y, A) . \tag{4.42}$$

We first show that under this condition, one may derive a non-trivial upper bound on the Dobrushin coefficient of the forward smoothing kernels.

Lemma 4.3.22. *Under Assumption 4.3.21, the following hold true.*

(i) For any non-negative integers k and n such that $k < n$ and $x \in \mathsf{X}$,

$$\prod_{j=k+1}^{n} \varsigma^-(y_j) \le \beta_{k|n}[y_{k+1:n}](x) \le \prod_{j=k+1}^{n} \varsigma^+(y_j) . \tag{4.43}$$

(ii) For any non-negative integers k and n such that $k < n$ and any probability measures ν and ν' on $(\mathsf{X}, \mathcal{X})$,

$$\frac{\varsigma^-(y_{k+1})}{\varsigma^+(y_{k+1})} \le \frac{\int_{\mathsf{X}} \nu(dx)\, \beta_{k|n}[y_{k+1:n}](x)}{\int_{\mathsf{X}} \nu'(dx)\, \beta_{k|n}[y_{k+1:n}](x)} \le \frac{\varsigma^+(y_{k+1})}{\varsigma^-(y_{k+1})} .$$

(iii) For any non-negative integers k and n such that $k < n$, there exists a transition kernel $\lambda_{k,n}$ from $(\mathsf{Y}^{n-k}, \mathcal{Y}^{\otimes(n-k)})$ to $(\mathsf{X}, \mathcal{X})$ such that for any $x \in \mathsf{X}$, $A \in \mathcal{X}$, and $y_{k+1:n} \in \mathsf{Y}^{n-k}$,

$$\frac{\varsigma^-(y_{k+1})}{\varsigma^+(y_{k+1})}\lambda_{k,n}(y_{k+1:n}, A) \leq \mathrm{F}_{k|n}[y_{k+1:n}](x, A) \leq \frac{\varsigma^+(y_{k+1})}{\varsigma^-(y_{k+1})}\lambda_{k,n}(y_{k+1:n}, A) . \quad (4.44)$$

(iv) For any non-negative integers k and n, the Dobrushin coefficient of the forward smoothing kernel $\mathrm{F}_{k|n}[y_{k+1:n}]$ satisfies

$$\delta(\mathrm{F}_{k|n}[y_{k+1:n}]) \leq \begin{cases} \rho_0(y_{k+1}) & k < n , \\ \rho_1 & k \geq n , \end{cases}$$

where for any $y \in \mathsf{Y}$,

$$\rho_0(y) \stackrel{\text{def}}{=} 1 - \frac{\varsigma^-(y)}{\varsigma^+(y)} \quad \text{and} \quad \rho_1 \stackrel{\text{def}}{=} 1 - \int \varsigma^-(y)\,\mu(dy) . \quad (4.45)$$

Proof. Take $A = \mathsf{X}$ in Assumption 4.3.21 to see that $\int_{\mathsf{X}} Q(x, dx')\, g(x', y)$ is bounded from above and below by $\varsigma^+(y)$ and $\varsigma^-(y)$, respectively. Part (i) then follows from (3.16).

Next, (3.19) shows that

$$\int \nu(dx)\, \beta_{k|n}[y_{k+1:n}](x) = \iint \nu(dx)\, Q(x, dx_{k+1})\, g(x_{k+1}, y_{k+1}) \beta_{k+1|n}[y_{k+2:n}](x_{k+1}) .$$

This expression is bounded from above by

$$\varsigma^+(y_{k+1}) \int K(y_{k+1}, dx_{k+1})\, \beta_{k+1|n}[y_{k+2:n}](x_{k+1}) ,$$

and similarly a lower bound, with $\varsigma^-(y_{k+1})$ rather than $\varsigma^+(y_{k+1})$, holds too. These bounds are independent of ν, and (ii) follows.

We turn to part (iii). Using the definition (3.30), the forward kernel $\mathrm{F}_{k|n}[y_{k+1:n}]$ may be expressed as

$$\mathrm{F}_{k|n}[y_{k+1:n}](x, A) = \frac{\int_A Q(x, dx_{k+1})\, g(x_{k+1}, y_{k+1}) \beta_{k+1|n}[y_{k+2:n}](x_{k+1})}{\int_{\mathsf{X}} Q(x, dx_{k+1})\, g(x_{k+1}, y_{k+1}) \beta_{k+1|n}[y_{k+2:n}](x_{k+1})} .$$

Using arguments as above, (4.44) holds with

$$\lambda_{k,n}(y_{k+1:n}, A) \stackrel{\text{def}}{=} \frac{\int_A K(y_{k+1}, dx_{k+1})\, \beta_{k+1|n}[y_{k+2:n}](x_{k+1})}{\int_{\mathsf{X}} K(y_{k+1}, dx_{k+1})\, \beta_{k+1|n}[y_{k+2:n}](x_{k+1})} .$$

Finally, part (iv) for $k < n$ follows from part (iii) and Lemma 4.3.13. In the opposite case, recall from (3.31) that $\mathrm{F}_{k|n} = Q$ for indices $k \geq n$. Integrating

(4.42) with respect to μ and using $\int g(x,y)\,\mu(dy) = 1$, we find that for any $A \in \mathcal{X}$ and any $x \in \mathsf{X}$,

$$Q(x, A) \geq \int \varsigma^-(y) K(y, A)\,\mu(dy) = \int \varsigma^-(y)\,\mu(dy) \times \frac{\int \varsigma^-(y) K(y, A)\,\mu(dy)}{\int \varsigma^-(y)\,\mu(dy)} ,$$

where the ratio on the right-hand side is a probability measure. The proof of part (iv) again follows from Lemma 4.3.13. □

The final part of the above lemma shows that under Assumption 4.3.21, the Dobrushin coefficient of the transition kernel Q satisfies $\delta(Q) \leq 1 - \epsilon$ for some $\epsilon > 0$. This is in fact a rather stringent assumption, which fails to be satisfied in many of the examples considered in Chapter 1. When X is finite, this condition is satisfied if $Q(x, x') \geq \epsilon$ for any $(x, x') \in \mathsf{X} \times \mathsf{X}$. When X is countable, $\delta(Q) < 1$ is satisfied under the Doeblin condition 4.3.11 with $n = 1$. When $\mathsf{X} \subseteq \mathbb{R}^d$ or more generally is a topological space, $\delta(Q) < 1$ typically requires that X is compact, which is, admittedly, a serious limitation.

Proposition 4.3.23. *Under 4.3.21 the following hold true.*

(i) For any non-negative integers k and n and any probability measures ν and ν' on $(\mathsf{X}, \mathcal{X})$,

$$\left\|\phi_{\nu,k|n}[y_{0:n}] - \phi_{\nu',k|n}[y_{0:n}]\right\|_{\mathrm{TV}} \leq \prod_{j=1}^{k\wedge n} \rho_0(y_j) \times \rho_1^{k-k\wedge n} \left\|\phi_{\nu,0|n}[y_{0:n}] - \phi_{\nu',0|n}[y_{0:n}]\right\|_{\mathrm{TV}} ,$$

where ρ_0 and ρ_1 are defined in (4.45).

(ii) For any non-negative integer n and any probability measures ν and ν' on $(\mathsf{X}, \mathcal{X})$ such that $\int \nu(dx_0)\, g(x_0, y_0) > 0$ and $\int \nu'(dx_0)\, g(x_0, y_0) > 0$,

$$\left\|\phi_{\nu,0|n}[y_{0:n}] - \phi_{\nu',0|n}[y_{0:n}]\right\|_{\mathrm{TV}} \leq \frac{\varsigma^+(y_1)}{\varsigma^-(y_1)} \frac{\|g\|_\infty}{\nu(g(\cdot, y_0)) \vee \nu'(g(\cdot, y_0))} \|\nu - \nu'\|_{\mathrm{TV}} .$$

(iii) For any non-negative integers j, k, and n such that $j \leq k$ and any probability measure ν on $(\mathsf{X}, \mathcal{X})$,

$$\|\mathrm{P}_\nu(X_k \in \cdot \,|\, Y_{0:n} = y_{0:n}) - \mathrm{P}_\nu(X_k \in \cdot \,|\, Y_{j:n} = y_{j:n})\|_{\mathrm{TV}} \leq 2 \prod_{i=j\wedge n+1}^{k\wedge n} \rho_0(y_i) \times \rho_1^{k-j-(k\wedge n - j\wedge n)} .$$

Proof. Using Lemma 4.3.22(iv) and Proposition 4.3.10, we find that for $j \leq k$,

$$\delta(\mathrm{F}_{j|n}[y_{j+1:n}] \cdots \mathrm{F}_{k|n}[y_{k+1:n}]) \leq \prod_{i=j\wedge n+1}^{k\wedge n} \rho_0(y_i) \times \rho_1^{k-j-(k\wedge n - j\wedge n)} .$$

Parts (i) and (iii) then follow from Propositions 4.3.19 and 4.3.20, respectively. Next we note that (4.33) shows that

$$\phi_{\nu,0|n}[y_{0:n}] = \mathrm{B}\left[\beta_{0|n}[y_{1:n}](\cdot), \mathrm{B}[g(\cdot, y_0), \nu]\right] .$$

Apply Lemma 4.3.18 twice to this form to arrive at a bound on the total variation norm of the difference $\phi_{\nu,0|n}[y_{0:n}] - \phi_{\nu',0|n}[y_{0:n}]$ given by

$$\frac{\left\|\beta_{0|n}[y_{1:n}]\right\|_\infty}{\mathrm{B}[g(\cdot, y_0), \nu](\beta_{0|n}[y_{1:n}])} \times \frac{\|g(\cdot, y_0)\|_\infty}{\nu(g(\cdot, y_0)) \vee \nu'(g(\cdot, y_0))} \|\nu - \nu'\|_{\mathrm{TV}} .$$

Finally, bound the first ratio of this display using Lemma 4.3.22(ii); the supremum norm is obtained by taking one of the initial measures as an atom at some point $x \in \mathsf{X}$. This completes the proof of part (ii). □

From the above it is clear that forgetting properties stem from properties of the product

$$\prod_{i=j\wedge n+1}^{k\wedge n} \rho_0(Y_i)\rho_1^{k-j-(k\wedge n - j\wedge n)} . \tag{4.46}$$

The situation is elementary when the factors of this product are (non-trivially) upper-bounded uniformly with respect to the observations $Y_{0:n}$. To obtain such bounds, we consider the following strengthening of the strong mixing condition, first introduced by Atar and Zeitouni (1997).

Assumption 4.3.24 (Strong Mixing Reinforced).

(i) There exist two positive real numbers σ^- and σ^+ and a probability measure κ on $(\mathsf{X}, \mathcal{X})$ such that for any $x \in \mathsf{X}$ and $A \in \mathcal{X}$,

$$\sigma^- \kappa(A) \leq Q(x, A) \leq \sigma^+ \kappa(A) .$$

(ii) For all $y \in \mathsf{Y}$, $0 < \int_{\mathsf{X}} \kappa(dx)\, g(x, y) < \infty$.

It is easily seen that this implies Assumption 4.3.21.

Lemma 4.3.25. *Assumption 4.3.24 implies Assumption 4.3.21 with $\varsigma^-(y) = \sigma^- \int_{\mathsf{X}} \kappa(dx)\, g(x, y)$, $\varsigma^+(y) = \sigma^+ \int_{\mathsf{X}} \kappa(dx)\, g(x, y)$, and*

$$K(y, A) = \frac{\int_A \kappa(dx)\, g(x, y)}{\int_{\mathsf{X}} \kappa(dx)\, g(x, y)} .$$

In particular, $\varsigma^-(y)/\varsigma^+(y) = \sigma^-/\sigma^+$ for any $y \in \mathsf{Y}$.

Proof. The proof follows immediately upon observing that

$$\sigma^- \int_A \kappa(dx')\, g(x', y) \leq \int_A Q(x, dx')\, g(x', y) \leq \sigma^+ \int_A \kappa(dx')\, g(x', y) .$$

□

Replacing Assumption 4.3.21 by Assumption 4.3.24, Proposition 4.3.23 may be strengthened as follows.

Proposition 4.3.26. *Under Assumption 4.3.24, the following hold true.*

(i) For any non-negative integers k and n and any probability measures ν and ν' on $(\mathsf{X}, \mathcal{X})$,

$$\left\|\phi_{\nu,k|n}[y_{0:n}] - \phi_{\nu',k|n}[y_{0:n}]\right\|_{\mathrm{TV}} \\ \leq \left(1 - \frac{\sigma^-}{\sigma^+}\right)^{k\wedge n} (1-\sigma^-)^{k-k\wedge n} \left\|\phi_{\nu,0|n}[y_{0:n}] - \phi_{\nu',0|n}[y_{0:n}]\right\|_{\mathrm{TV}} \cdot$$

(ii) For any non-negative integer n and any probability measures ν and ν' on $(\mathsf{X}, \mathcal{X})$ such that $\int \nu(dx_0)\, g(x_0, y_0) > 0$ and $\int \nu'(dx_0)\, g(x_0, y_0) > 0$,

$$\left\|\phi_{\nu,0|n}[y_{0:n}] - \phi_{\nu',0|n}[y_{0:n}]\right\|_{\mathrm{TV}} \\ \leq \frac{\sigma^+}{\sigma^-} \frac{\|g\|_\infty}{\nu[g(\cdot, y_0)] \vee \nu'[g(\cdot, y_0)]} \|\nu - \nu'\|_{\mathrm{TV}} \cdot$$

(iii) For any non-negative integers j, k, and n such that $j \leq k$ and any probability measure ν on $(\mathsf{X}, \mathcal{X})$,

$$\|\mathrm{P}_\nu\left(X_k \in \cdot \,|\, Y_{0:n} = y_{0:n}\right) - \mathrm{P}_\nu\left(X_k \in \cdot \,|\, Y_{j:n} = y_{j:n}\right)\|_{\mathrm{TV}} \\ \leq 2\left(1 - \frac{\sigma^-}{\sigma^+}\right)^{k\wedge n - j\wedge n} \left(1-\sigma^-\right)^{k-j-(k\wedge n - j\wedge n)} .$$

Thus, under Assumption 4.3.24 the filter and the smoother forget their initial conditions exponentially fast, uniformly with respect to the observations. This property, which holds under rather stringent assumptions, plays a key role in the sequel (see for instance Chapters 9 and 12).

Of course, the product (4.46) can be shown to vanish asymptotically under conditions that are less stringent than Assumption 4.3.24. A straightforward adaptation of Lemma 4.3.25 shows that the following result is true.

Lemma 4.3.27. *Assume 4.3.21 and that there exists a set $C \in \mathcal{Y}$ and constants $0 < \sigma^- \leq \sigma^+ < \infty$ satisfying $\mu(C) > 0$ and, for all $y \in C$, $\sigma^- \leq \varsigma^-(y) \leq \varsigma^+(y) \leq \sigma^+$. Then, $\rho_0(y) \leq 1 - \sigma^-/\sigma^+$, $\rho_1 \geq 1 - \sigma^-\mu(C)$ and*

$$\prod_{i=j\wedge n+1}^{k\wedge n} \rho_0(Y_i)\rho_1^{k-j-(k\wedge n - j\wedge n)} \\ \leq \left(1 - \sigma^-/\sigma^+\right)^{\sum_{i=j\wedge n+1}^{k\wedge n} \mathbb{1}_C(Y_i)} \left[1 - \sigma^-\mu(C)\right]^{k-j-(k\wedge n - j\wedge n)} . \quad (4.47)$$

In words, forgetting is guaranteed to occur when $\{Y_k\}$ visits a given set C infinitely often in the long run. Of course, such a property cannot hold true for all possible sequences of observations but it may hold with probability one under appropriate assumptions on the law of $\{Y_k\}$, assuming in particular that the observations are distributed under the model, perhaps with a different initial distribution $\nu_\star$. To answer whether this happens or not requires additional results from the general theory of Markov chains, and we postpone this discussion to Section 14.3 (see in particular Proposition 14.3.8 on the recurrence of the joint chain in HMMs).

4.3.6 Forgetting Under Alternative Conditions

Because Assumptions 4.3.21 and 4.3.24 are not satisfied in many contexts of interest, it is worthwhile to consider ways in which these assumptions can be weakened. This happens to raise difficult mathematical challenges that largely remain unsolved today. Perhaps surprisingly, despite many efforts in this direction, there is up to now no truly satisfactory assumption that covers a reasonable fraction of the situations of practical interest. The problem really is more complicated than appears at first sight. In particular, Example 4.3.28 below shows that the forgetting property does not necessarily hold under assumptions that imply that the underlying Markov chain is uniformly ergodic. This last section on forgetting is more technical and requires some knowledge of Markov chain theory as can be found in Chapter 14.

Example 4.3.28. This example was first discussed by Kaijser (1975) and recently worked out by Chigansky and Lipster (2004). Let $\{X_k\}$ be a Markov chain on $\mathsf{X} = \{0, 1, 2, 3\}$, defined by the recurrence equation $X_k = (X_{k-1} + U_k)$ mod 4, where $\{U_k\}$ is an i.i.d. binary sequence with $\mathrm{P}(B_k = 0) = p$ and $\mathrm{P}(B_k = 1) = 1 - p$ for some $0 < p < 1$. For any $(x, x') \in \mathsf{X} \times \mathsf{X}$, $Q^4(x, x') > 0$, which implies that $\delta(Q^4) < 1$ and, by Theorem 4.3.16, that the chain is uniformly geometrically ergodic. The observations $\{Y_k\}$ are a deterministic binary function of the chain, namely

$$Y_k = \mathbb{1}_{\{0,2\}}(X_k) .$$

The function mapping X_k to Y_k is not injective, but knowledge of Y_k indicates two possible values of X_k. The filtering distribution is given recursively by

$$\begin{aligned}
\phi_{\nu,k}[y_{0:k}](0) &= y_k \left\{\phi_{\nu,k-1}[y_{0:k-1}](0) + \phi_{\nu,k-1}[y_{0:k-1}](3)\right\} , \\
\phi_{\nu,k}[y_{0:k}](1) &= (1 - y_k) \left\{\phi_{\nu,k-1}[y_{0:k-1}](1) + \phi_{\nu,k-1}[y_{0:k-1}](0)\right\} , \\
\phi_{\nu,k}[y_{0:k}](2) &= y_k \left\{\phi_{\nu,k-1}[y_{0:k-1}](2) + \phi_{\nu,k-1}[y_{0:k-1}](1)\right\} , \\
\phi_{\nu,k}[y_{0:k}](3) &= (1 - y_k) \left\{\phi_{\nu,k-1}[y_{0:k-1}](3) + \phi_{\nu,k-1}[y_{0:k-1}](2)\right\} .
\end{aligned}$$

In particular, either one of the two sets $\{0, 2\}$ and $\{1, 3\}$ has null probability under $\phi_{\nu,k}[y_{0:k}]$, depending on the value of y_k, and irrespectively of the choice of ν. We also notice that

$$
\begin{aligned}
y_k\, \phi_{\nu,k}[y_{0:k}](j) &= \phi_{\nu,k}[y_{0:k}](j)\;, && \text{for } j = 0, 2,\\
(1 - y_k)\, \phi_{\nu,k}[y_{0:k}](j) &= \phi_{\nu,k}[y_{0:k}](j)\;, && \text{for } j = 1, 3.
\end{aligned}
\tag{4.48}
$$

In addition, it is easily checked that, except when $\nu(\{0,2\})$ or $\nu(\{1,3\})$ equals 1 (which rules out one of the two possible values for y_0), the likelihood $\mathrm{L}_{\nu,n}[y_{0:n}]$ is strictly positive for any integer n and any sequence $y_{0:n} \in \{0,1\}^{n+1}$.

Dropping the dependence on $y_{0:k}$ for notational simplicity and using (4.48) we obtain

$$
\begin{aligned}
&|\phi_{\nu,k}(0) - \phi_{\nu',k}(0)| \\
&= y_k |\phi_{\nu,k-1}(0) - \phi_{\nu',k-1}(0) + \phi_{\nu,k-1}(3) - \phi_{\nu',k-1}(3)| \\
&= y_k \left\{ y_{k-1} |\phi_{\nu,k-1}(0) - \phi_{\nu',k-1}(0)| + (1 - y_{k-1}) |\phi_{\nu,k-1}(3) - \phi_{\nu',k-1}(3)| \right\} .
\end{aligned}
$$

Proceeding similarly, we also find that

$$
\begin{aligned}
&|\phi_{\nu,k}(1) - \phi_{\nu',k}(1)| = \\
&(1 - y_k) \left\{ (1 - y_{k-1}) |\phi_{\nu,k-1}(1) - \phi_{\nu',k-1}(1)| + y_{k-1} |\phi_{\nu,k-1}(0) - \phi_{\nu',k-1}(0)| \right\} , \\
&|\phi_{\nu,k}(2) - \phi_{\nu',k}(2)| = \\
&y_k \left\{ y_{k-1} |\phi_{\nu,k-1}(2) - \phi_{\nu',k-1}(2)| + (1 - y_{k-1}) |\phi_{\nu,k-1}(1) - \phi_{\nu',k-1}(1)| \right\} , \\
&|\phi_{\nu,k}(3) - \phi_{\nu',k}(3)| = \\
&(1 - y_k) \left\{ (1 - y_{k-1}) |\phi_{\nu,k-1}(3) - \phi_{\nu',k-1}(3)| + y_{k-1} |\phi_{\nu,k-1}(2) - \phi_{\nu',k-1}(2)| \right\} .
\end{aligned}
$$

Adding the above equalities using (4.48) again shows that for any $k = 1, \ldots, n$,

$$
\begin{aligned}
\| \phi_{\nu,k}[y_{0:k}] - \phi_{\nu',k}[y_{0:k}] \|_{\mathrm{TV}} &= \| \phi_{\nu,k-1}[y_{0:k-1}] - \phi_{\nu',k-1}[y_{0:k-1}] \|_{\mathrm{TV}} \\
&= \| \phi_{\nu,0}[y_0] - \phi_{\nu',0}[y_0] \|_{\mathrm{TV}} .
\end{aligned}
$$

By construction, $\phi_{\nu,0}[y_0](j) = y_0\, \nu(j)/(\nu(0) + \nu(2))$ for $j = 0$ and 2, and $\phi_{\nu,0}[y_0](j) = (1 - y_0)\, \nu(j)/(\nu(1) + \nu(3))$ for $j = 1$ and 3. This implies that $\| \phi_{\nu,0}[y_0] - \phi_{\nu',0}[y_0] \|_{\mathrm{TV}} \neq 0$ if $\nu \neq \nu'$.

In this model, the hidden Markov chain $\{X_k\}$ is uniformly ergodic, but the filtering distributions $\phi_{\nu,k}[y_{0:k}]$ never forget the influence of the initial distribution ν, whatever the observed sequence. ■

In the above example, the kernel Q does not satisfy Assumption 4.3.24 with $m = 1$ (one-step minorization), but the condition is verified for a power Q^m (here for $m = 4$). This situation is the rule rather than the exception. In particular, a Markov chain on a finite state space has a unique invariant probability measure and is ergodic if and only if there exists an integer $m > 0$ such that $Q^m(x, x') > 0$ for all $(x, x') \in \mathsf{X} \times \mathsf{X}$ (but the condition may not hold for $m = 1$). This suggests considering the following assumption (see for instance Del Moral, 2004, Chapter 4).

Assumption 4.3.29.

(i) There exist an integer m, two positive real numbers σ^- and σ^+, and a probability measure κ on $(\mathsf{X}, \mathcal{X})$ such that for any $x \in \mathsf{X}$ and $A \in \mathcal{X}$,

$$\sigma^- \kappa(A) \le Q^m(x, A) \le \sigma^+ \kappa(A) .$$

(ii) There exist two measurable functions g^- and g^- from Y to $(0, \infty)$ such that for any $y \in \mathsf{Y}$,

$$g^-(y) \le \inf_{x \in \mathsf{X}} g(x, y) \le \sup_{x \in \mathsf{X}} g(x, y) \le g^+(y) .$$

Compared to Assumption 4.3.24, the condition on the transition kernel has been weakened, but at the expense of strengthening the assumption on the function g. Note in particular that part (ii) is *not* satisfied in Example 4.3.28.

Using (4.30) and writing $k = jm + r$ with $0 \le r < m$, we may express $\phi_{\nu,k|n}[y_{0:n}]$ as

$$\phi_{\nu,k|n}[y_{0:n}] = \phi_{\nu,0|n}[y_{0:n}] \prod_{u=0}^{j-1} \left(\prod_{i=um}^{(u+1)m-1} \mathrm{F}_{i|n}[y_{i+1:n}] \right) \prod_{i=jm}^{k-1} \mathrm{F}_{i|n}[y_{i+1:n}] .$$

This implies, using Corollary 4.3.9, that for any probability measures ν and ν' on $(\mathsf{X}, \mathcal{X})$ and any sequence $y_{0:n}$ satisfying $\mathrm{L}_{\nu,n}[y_{0:n}] > 0$ and $\mathrm{L}_{\nu',n}[y_{0:n}] > 0$,

$$\begin{aligned} &\left\| \phi_{\nu,k|n}[y_{0:n}] - \phi_{\nu',k|n}[y_{0:n}] \right\|_{\mathrm{TV}} \\ &\le \prod_{u=0}^{j-1} \delta \left(\prod_{i=um}^{(u+1)m-1} \mathrm{F}_{i|n}[y_{i+1:n}] \right) \left\| \phi_{\nu,0|n}[y_{0:n}] - \phi_{\nu',0|n}[y_{0:n}] \right\|_{\mathrm{TV}} . \end{aligned} \tag{4.49}$$

This expression suggest computing a bound on $\delta(\prod_{i=um}^{um+m-1} \mathrm{F}_{i|n}[y_{i+1:n}])$ rather than a bound on $\delta(\mathrm{F}_{i|n})$. The following result shows that such a bound can be derived under Assumption 4.3.29.

Lemma 4.3.30. *Under Assumption 4.3.29, the following hold true.*

(i) For any non-negative integers k and n such that $k < n$ and $x \in \mathsf{X}$,

$$\prod_{j=k+1}^{n} g^-(y_j) \le \beta_{k|n}[y_{k+1:n}](x) \le \prod_{j=k+1}^{n} g^+(y_j) , \tag{4.50}$$

where $\beta_{k|n}$ is the backward function (3.16).

(ii) For any non-negative integers u and n such that $0 \le u < \lfloor n/m \rfloor$ and any probability measures ν and ν' on $(\mathsf{X}, \mathcal{X})$,

$$\frac{\sigma^-}{\sigma^+} \prod_{i=um+1}^{(u+1)m} \frac{g^-(y_i)}{g^+(y_i)} \le \frac{\int_{\mathsf{X}} \nu(dx)\, \beta_{um|n}[y_{um+1:n}](x)}{\int_{\mathsf{X}} \nu'(dx)\, \beta_{um|n}[y_{um+1:n}](x)} \le \frac{\sigma^+}{\sigma^-} \prod_{i=um+1}^{(u+1)m} \frac{g^+(y_i)}{g^-(y_i)} .$$

(iii) For any non-negative integers u and n such that $0 \le u < \lfloor n/m \rfloor$, there exists a transition kernel $\lambda_{u,n}$ from $\left(\mathsf{Y}^{(n-(u+1)m)}, \mathcal{Y}^{\otimes(n-(u+1)m)}\right)$ to $(\mathsf{X}, \mathcal{X})$ such that for any $x \in \mathsf{X}$, $A \in \mathcal{X}$ and $y_{um+1:n} \in \mathsf{Y}^{(n-um)}$,

$$\frac{\sigma^-}{\sigma^+} \prod_{i=um+1}^{(u+1)m} \frac{g^-(y_i)}{g^+(y_i)} \lambda_{u,n}(y_{(u+1)m+1:n}, A) \le \prod_{i=um}^{(u+1)m-1} \mathrm{F}_{i|n}[y_{i+1:n}](x, A)$$
$$\le \frac{\sigma^+}{\sigma^-} \prod_{i=um+1}^{(u+1)m} \frac{g^+(y_i)}{g^-(y_i)} \lambda_{u,n}(y_{(u+1)m+1:n}, A) \; . \quad (4.51)$$

(iv) For any non-negative integers u and n,

$$\delta\left(\prod_{i=um}^{(u+1)m-1} \mathrm{F}_{i|n}[y_{i+1:n}]\right) \le \begin{cases} \rho_0(y_{um+1:(u+1)m}) & u < \lfloor n/m \rfloor \; , \\ \rho_1 & u \ge \lceil n/m \rceil \; , \end{cases}$$

where for any $y_{um+1:(u+1)m} \in \mathsf{Y}^m$,

$$\rho_0(y_{um+1:(u+1)m}) \stackrel{\text{def}}{=} 1 - \frac{\sigma^-}{\sigma^+} \prod_{i=um+1}^{(u+1)m} \frac{g^-(y_i)}{g^+(y_i)} \quad \text{and} \quad \rho_1 \stackrel{\text{def}}{=} 1 - \sigma^- \; . \quad (4.52)$$

Proof. Part (i) can be proved using an argument similar to the one used for Lemma 4.3.22(i).

Next notice that for $0 \le u < \lfloor n/m \rfloor$,

$$\beta_{um|n}[y_{um+1:n}](x_{um})$$
$$= \int \cdots \int \prod_{i=um+1}^{(u+1)m} Q(x_{i-1}, dx_i)\, g(x_i, y_i)\, \beta_{(u+1)m|n}[y_{(u+1)m+1:n}](x_{(u+1)m}) \; .$$

Under Assumption 4.3.29, dropping the dependence on the ys for notational simplicity, the right-hand side of this display is bounded from above by

$$\prod_{i=um+1}^{(u+1)m} g^+(y_i) \int \cdots \int \prod_{i=um+1}^{(u+1)m} Q(x_{i-1}, dx_i)\, \beta_{(u+1)m|n}(x_{(u+1)m})$$
$$\le \sigma^+ \prod_{i=um+1}^{(u+1)m} g^+(y_i) \int \beta_{(u+1)m|n}(x_{(u+1)m})\, \kappa(dx_{(u+1)m}) \; .$$

In a similar fashion, a lower bound may be obtained, containing σ^- and g^- rather than σ^+ and g^+. Thus part (ii) follows.

For part (iii), we use (3.30) to write

$$\prod_{i=um}^{(u+1)m-1} \mathrm{F}_{i|n}[y_{i+1:n}](x_{um}, A)$$
$$= \frac{\int \cdots \int \prod_{i=um+1}^{(u+1)m} Q(x_{i-1}, x_i)\, g(x_i, y_i) \mathbb{1}_A(x_{(u+1)m}) \beta_{(u+1)m|n}(x_{(u+1)m})}{\int \cdots \int \prod_{i=um+1}^{(u+1)m} Q(x_{i-1}, x_i)\, g(x_i, y_i) \beta_{(u+1)m|n}(x_{(u+1)m})} .$$

The right-hand side is bounded from above by

$$\frac{\sigma^+}{\sigma^-} \prod_{i=um+1}^{(u+1)m} \frac{g^+(y_i)}{g^-(y_i)} \times \frac{\int_A \kappa(dx)\, \beta_{(u+1)m|n}[y_{(u+1)m+1:n}](x)}{\int \kappa(dx)\, \beta_{(u+1)m|n}[y_{(u+1)m+1:n}](x)} .$$

We define $\lambda_{u,n}$ as the second ratio of this expression. Again a corresponding lower bound is obtained similarly, proving part (iii).

Part (iv) follows from part (iii) and Lemma 4.3.13. □

Using this result together with (4.49), we may obtain statements analogous to Proposition 4.3.23. In particular, if there exist positive real numbers γ^- and γ^+ such that for all $y \in \mathsf{Y}$,

$$\gamma^- \le g^-(y) \le g^+(y) \le \gamma^+ ,$$

then the smoothing and the filtering distributions both forget uniformly the initial distribution.

Assumptions 4.3.24 and 4.3.29 are still restrictive and fail to hold in many interesting situations. In both cases, we assume that either the one-step or the m-step transition kernel is uniformly bounded from above and below. The following weaker condition is a first step toward handling more general settings.

Assumption 4.3.31. *Let Q be dominated by a probability measure κ on $(\mathsf{X}, \mathcal{X})$ such that for any $x \in \mathsf{X}$ and $A \in \mathcal{X}$, $Q(x, A) = \int_A q_\kappa(x, x')\, \kappa(dx')$ for some transition density function q_κ. Assume in addition that*

(i) There exists a set $C \in \mathcal{X}$, two positive real numbers σ^- and σ^+ such that for all $x \in C$ and $x' \in \mathsf{X}$,

$$\sigma^- \le q_\kappa(x, x') \le \sigma^+ .$$

(ii) For all $y \in \mathsf{Y}$ and all $x \in \mathsf{X}$, $\int_C q_\kappa(x, x')\, g(x', y)\, \kappa(dx') > 0$;
(iii) There exists a (non-identically null) function $\alpha : \mathsf{Y} \to [0, 1]$ such that for any $(x, x') \in \mathsf{X} \times \mathsf{X}$ and $y \in \mathsf{Y}$,

$$\frac{\int_C \rho[x, x'\,; y](x'')\, \kappa(dx'')}{\int_{\mathsf{X}} \rho[x, x'\,; y](x'')\, \kappa(dx'')} \ge \alpha(y) ,$$

where for $(x, x', x'') \in \mathsf{X}^3$ and $y \in \mathsf{Y}$,

$$\rho[x, x'\,; y](x'') \stackrel{\text{def}}{=} q_\kappa(x, x'') g(x'', y) q_\kappa(x'', x') . \tag{4.53}$$

Part (i) of this assumption implies that the set C is 1-small for the kernel Q (see Definition 14.2.10). It it shown in Section 14.2.2.2 that such small sets do exist under conditions that are weak and generally simple to check. Assumption 4.3.31 is trivially satisfied under Assumption 4.3.24 using the whole state space X as the state C: in that case, their exists a transition density function $q_\kappa(x, x')$ that is bounded from above and below for all $(x, x') \in \mathsf{X}^2$. It is more interesting to consider cases in which the hidden chain is not uniformly ergodic. One such example, first addressed by Budhiraja and Ocone (1997), is a Markov chain observed in noise with bounded support.

Example 4.3.32 (Markov Chain in Additive Bounded Noise). We consider real states $\{X_k\}$ and observations $\{Y_k\}$, assuming that the states form a Markov chain with a transition density $q(x, x')$ with respect to Lebesgue measure. Furthermore we assume the following.

(i) $Y_k = X_k + V_k$, where $\{V_k\}$ is an i.i.d. sequence of satisfying $\mathrm{P}(|V| \geq M) = 0$ for some finite M (the essential supremum of the noise sequence is bounded). In addition, V_k has a probability density g with respect to Lebesgue measure.

(ii) The transition density satisfies $q(x, x') > 0$ for all (x, x') and there exists a positive constant A, a probability density h and positive constants σ^- and σ^+ such that for all $x \in C = [-A - M, A + M]$,

$$\sigma^- h(x') \leq q(x, x') \leq \sigma^+ h(x') \ .$$

The results below can readily be extended to cover the case $Y_k = \psi(X_k) + V_k$, provided that the level sets $\{x \in \mathbb{R} : |\psi(x)| \leq K\}$ of the function ψ are compact. This is equivalent to requiring $|\psi(x)| \to \infty$ as $|x| \to \infty$. Likewise extensions to multivariate states and/or observations are obvious.

Under (ii), Assumption 4.3.31(i) is satisfied with C as above and $\kappa(dx) = h(x)\, dx$. Denote by ϕ the probability density of the random variables V_k. Then $g(x, y) = \phi(y - x)$. The density ϕ may be chosen such that $\mathrm{supp}\phi \subseteq [-M, +M]$, so that $g(x, y) > 0$ if and only if $x \in [y - M, y + M]$. To verify Assumption 4.3.31(iii), put $\Gamma = [-A, A]$. For $y \in \Gamma$, we then have $g(x, y) = 0$ if $x \notin [-A - M, A + M]$, and thus

$$\int q(x, x'')g(x'', y)q(x'', x')\, dx'' = \int_{-A-M}^{A+M} q(x, x'')g(x'', y)q(x'', x')\, dx'' \ .$$

This implies that for all $(x, x') \in \mathsf{X} \times \mathsf{X}$,

$$\frac{\int_C q(x, x'')g(x'', y)q(x'', x')\, dx''}{\int_{\mathsf{X}} q(x, x'')g(x'', y)q(x'', x')\, dx''} = 1 \ .$$

The bounded noise case is of course very specific, because an observation Y_k allows locating the corresponding state X_k within a bounded set. ■

Under assumption 4.3.31, the lemma below establishes that the set C is a 1-small set for the forward transition kernels $\mathrm{F}_{k|n}[y_{k+1:n}]$ and that it is also uniformly accessible from the whole space X (for the same kernels).

Lemma 4.3.33. *Under Assumption 4.3.31, the following hold true.*

(i) For any initial probability measure ν on $(\mathsf{X}, \mathcal{X})$ and any sequence $y_{0:n} \in \mathsf{Y}^{n+1}$ satisfying $\int_C \nu(dx_0)\, g(x_0, y_0) > 0$,

$$\mathrm{L}_{\nu,n}(y_{0:n}) > 0\ .$$

(ii) For any non-negative integers k and n such that $k < n$ and any $y_{0:n} \in \mathsf{Y}^{n+1}$, the set C is a 1-small set for the transitions kernels $\mathrm{F}_{k|n}$. Indeed there exists a transition kernel $\lambda_{k,n}$ from $(\mathsf{Y}^{(n-k)}, \mathcal{Y}^{\otimes(n-k)})$ to $(\mathsf{X}, \mathcal{X})$ such that for all $x \in C$, $y_{k+1:n} \in \mathsf{Y}^{n-k}$ and $A \in \mathcal{X}$,

$$\mathrm{F}_{k|n}[y_{k+1:n}](x, A) \geq \frac{\sigma^-}{\sigma^+} \lambda_{k,n}[y_{k+1:n}](A)\ .$$

(iii) For any non-negative integers k and n such that $n \geq 2$ and $k < n-1$, and any $y_{k+1:n} \in \mathsf{Y}^{n-k}$,

$$\inf_{x \in \mathsf{X}} \mathrm{F}_{k|n}[y_{k+1:n}](x, C) \geq \alpha(y_{k+1})\ .$$

Proof. Write

$$\begin{aligned}
\mathrm{L}_{\nu,n}(y_{0:n}) &= \int \cdots \int \nu(dx_0)\, g(x_0, y_0) \prod_{i=1}^{n} Q(x_{i-1}, dx_i)\, g(x_i, y_i) \\
&\geq \int \cdots \int \nu(dx_0)\, g(x_0, y_0) \prod_{i=1}^{n} Q(x_{i-1}, dx_i)\, g(x_i, y_i) \mathbb{1}_C(x_{i-1}) \\
&\geq \int_C \nu(dx_0)\, g(x_0, y_0)\, (\sigma^-)^n \prod_{i=1}^{n} \int_C g(x_i, y_i)\, \kappa(dx_i)\ ,
\end{aligned}$$

showing part (i). The proof of (ii) is similar to that of Lemma 4.3.22(iii). For (iii), write

$$\begin{aligned}
&\mathrm{F}_{k|n}[y_{k+1:n}](x, C) \\
&= \frac{\iint \rho[x, x_{k+2}\,; y_{k+1}](x_{k+1}) \mathbb{1}_C(x_{k+1}) \varphi[y_{k+2:n}](x_{k+2})\, \kappa(dx_{k+1:k+2})}{\iint \rho[x, x_{k+2}\,; y_{k+1}](x_{k+1}) \varphi[y_{k+2:n}](x_{k+2})\, \kappa(dx_{k+1:k+2})} \\
&= \frac{\iint \Phi[y_{k+1}](x, x_{k+2}) \rho[x, x_{k+2}\,; y_{k+1}](x_{k+1}) \varphi[y_{k+2:n}](x_{k+2})\, \kappa(dx_{k+1:k+2})}{\iint \rho[x, x_{k+2}\,; y_{k+1}](x_{k+1}) \varphi[y_{k+2:n}](x_{k+2})\, \kappa(dx_{k+1:k+2})}\ .
\end{aligned}$$

where ρ is defined in (4.53) and

$$\begin{aligned}
\varphi[y_{k+2:n}](x_{k+2}) &= g(x_{k+2}, y_{k+2}) \beta_{k+2|n}[y_{k+3:n}](x_{k+2})\ , \\
\Phi[y_{k+1}](x, x_{k+2}) &= \frac{\int \rho[x, x_{k+2}\,; y_{k+1}](x_{k+1}) \mathbb{1}_C(x_{k+1})\, \kappa(dx_{k+1})}{\int \rho[x, x_{k+2}\,; y_{k+1}](x_{k+1})\, \kappa(dx_{k+1})}\ .
\end{aligned}$$

Under Assumption 4.3.31, $\Phi(x, x'; y) \geq \alpha(y)$ for all $(x, x') \in \mathsf{X} \times \mathsf{X}$ and $y \in \mathsf{Y}$, which concludes the proof. □

The corollary below then shows that the whole set X is a 1-small set for the composition $\mathrm{F}_{k|n}[y_{k+1:n}]\mathrm{F}_{k+1|n}[y_{k+2:n}]$. This generalizes a well-known result for homogeneous Markov chains (see Proposition 14.2.12).

Corollary 4.3.34. *Under Assumption 4.3.31, for positive indices $2 \leq k \leq n$,*

$$\left\|\phi_{\nu,k|n}[y_{0:n}] - \phi_{\nu',k|n}[y_{0:n}]\right\|_{\mathrm{TV}} \leq 2 \prod_{j=0}^{\lfloor k/2 \rfloor - 1} \left[1 - \frac{\sigma^-}{\sigma^+}\alpha(y_{2j+1})\right] .$$

Proof. Because of Lemma 4.3.33(i), we may use the decomposition in (4.39) with $j = 0$ bounding the total variation distance by 2 to obtain

$$\left\|\phi_{\nu,k|n}[y_{0:n}] - \phi_{\nu',k|n}[y_{0:n}]\right\|_{\mathrm{TV}} \leq 2 \prod_{j=0}^{k-1} \delta\left(\mathrm{F}_{j|n}[y_{j+1:n}]\right) .$$

Now, using assertions (ii) and (iii) of Lemma 4.3.33,

$$\begin{aligned} &\mathrm{F}_{j|n}[y_{j+1:n}]\mathrm{F}_{j+1|n}[y_{j+2:n}](x, A) \\ &\qquad \geq \int_C \mathrm{F}_{j|n}[y_{j+1:n}](x, dx')\mathrm{F}_{j+1|n}[y_{j+2:n}](x', A) \\ &\qquad\qquad \geq \alpha(y_{j+1})\frac{\sigma^-}{\sigma^+}\lambda_{j+1,n}[y_{j+2:n}](A) , \end{aligned}$$

for all $x \in \mathsf{X}$ and $A \in \mathcal{X}$. Hence the composition $\mathrm{F}_{j|n}[y_{j+1:n}]\mathrm{F}_{j+1|n}[y_{j+2:n}]$ satisfies Doeblin's condition (Assumption 4.3.12) and the proof follows by Application of Lemma 4.3.13. □

Corollary 4.3.34 is only useful in cases where the function α is such that the obtained bound indeed decreases as k and n grow. In Example 4.3.32, one could set $\alpha(y) = \mathbb{1}_\Gamma(y)$, for an interval Γ. In such a case, it suffices that the joint chain $\{X_k, Y_k\}_{k \geq 0}$ be recurrent under $\mathrm{P}_{\nu_\star}$—which was the case in Example 4.3.32—to guarantee that $\mathbb{1}_\Gamma(Y_k)$ equals one infinitely often and thus that $\left\|\phi_{\nu,k|n}[Y_{0:n}] - \phi_{\nu',k|n}[Y_{0:n}]\right\|_{\mathrm{TV}}$ tends to zero $\mathrm{P}_{\nu_\star}$-almost surely as $k, n \to \infty$. The following example illustrates a slightly more complicated situation in which Assumption 4.3.31 still holds.

Example 4.3.35 (Non-Gaussian Autoregressive Process in Gaussian Noise). In this example, we consider a first-order non-Gaussian autoregressive process, observed in Gaussian noise. This is a practically relevant example for which there is apparently no results on forgetting available in the literature. The model is thus

$$\begin{aligned} X_{k+1} &= \phi X_k + U_k , \qquad X_0 \sim \nu , \\ Y_k &= X_k + V_k , \end{aligned}$$

where

(i) $\{U_k\}_{k\geq 0}$ is an i.i.d. sequence of random variables with Laplace (double exponential) distribution with scale parameter λ;
(ii) $\{V_k\}_{k\geq 0}$ is an i.i.d. sequence of Gaussian random variable with zero mean and variance σ^2.

We will see below that the fact that the tails of the Xs are heavier than the tails of the observation noise is important for the derivations that follow. It is assumed that $|\phi| < 1$, which implies that the chain $\{X_k\}$ is positive recurrent, that is, admits a single invariant probability measure π. It may be shown (see Chapter 14) that although the Markov chain $\{X_k\}$ is geometrically ergodic, that is, $\|Q^n(x,\cdot) - \pi\|_{\mathrm{TV}} \to 0$ geometrically fast, it is not uniformly ergodic as $\liminf_{n\to\infty} \sup_{x\in\mathbb{R}} \|Q^n(x,\cdot) - \pi\|_{\mathrm{TV}} > 0$. We will nevertheless see that the forward smoothing kernel is uniformly geometrically ergodic.

Under the stated assumptions,

$$
\begin{aligned}
q(x,x') &= \frac{1}{2\lambda} \exp\left(-\lambda|x' - \phi x|\right) , \\
g(x,y) &= \frac{1}{\sqrt{2\pi}\sigma} \exp\left[-\frac{(y-x)^2}{2\sigma^2}\right] .
\end{aligned}
$$

Here we set, for some $M > 0$ to be specified later, $C = [-M - 1/2, M + 1/2]$, and we let $y \in [-1/2, +1/2]$. Note that

$$
\begin{aligned}
&\frac{\int_{-M-1/2}^{M+1/2} \exp(-\lambda|u - \phi x| - |y-u|^2/2\sigma^2 - \lambda|x' - \phi u|)\, du}{\int_{-\infty}^{\infty} \exp(-\lambda|u - \phi x| - |y-u|^2/2\sigma^2 - \lambda|x' - \phi u|)\, du} \\
&\qquad \geq \frac{\int_{-M}^{M} \exp(-\lambda|u - x| - u^2/2\sigma^2 - \phi\lambda|x' - u|)\, du}{\int_{-\infty}^{\infty} \exp(-\lambda|u - x| - u^2/2\sigma^2 - \phi\lambda|x' - u|)\, du} ,
\end{aligned}
$$

and to show Assumption 4.3.31(iii) it suffices to show that the right-hand side is bounded from below. This in turn is equivalent to showing that $\sup_{(x,x')\in\mathbb{R}\times\mathbb{R}} R(x,x') < 1$, where

$$
R(x,x') = \frac{\left(\int_{-\infty}^{-M} + \int_{M}^{\infty}\right) \exp(-\alpha|u-x| - \beta u^2 - \gamma|x'-u|)\, du}{\int_{-\infty}^{\infty} \exp(-\alpha|u-x| - \beta u^2 - \gamma|x'-u|)\, du} \tag{4.54}
$$

with $\alpha = \lambda$, $\beta = 1/2\sigma^2$ and $\gamma = \phi\lambda$.

To do this, first note that any $M > 0$ we have $\sup\{R(x,x') : |x| \leq M, |x'| \leq M\} < 1$, and we thus only need to study the behavior of this quantity when x and/or x' become large. We first show that

$$
\limsup_{M\to\infty} \sup_{x\geq M,\, |x'|\leq M} R(x,x') < 1 . \tag{4.55}
$$

For this we note that for $|x'| \leq M$ and $x \geq M$, it holds that

$$\left(\int_M^x + \int_x^\infty\right) \exp\left[-\alpha|x-u| - \beta u^2 - \gamma(u-x')\right] du$$
$$\leq \mathrm{e}^{-\alpha x}\mathrm{e}^{\gamma M} \frac{\exp[-\beta M^2 + (\alpha-\gamma)M]}{2\beta M - (\alpha-\gamma)} + \mathrm{e}^{\gamma M} \frac{\exp(-\beta x^2 - \gamma x)}{2\beta x + (\gamma+\alpha)} ,$$

where we used the bound

$$\int_y^\infty \exp(\lambda u - \beta u^2)\, du \leq (2\beta y - \lambda) \exp(-\beta y^2 + \lambda y) ,$$

which holds as soon as $2\beta y - \lambda \geq 0$. Similarly, we have

$$\int_{-\infty}^{-M} \exp\left[-\alpha(x-u) - \beta u^2 - \gamma(x'-u)\right] du$$
$$\leq \mathrm{e}^{-\alpha x}\mathrm{e}^{\gamma M} \frac{\exp[-\beta M^2 - (\gamma+\alpha)M]}{2\beta M + (\gamma+\alpha)} ,$$

$$\int_{-M}^{M} \exp\left[-\alpha(x-u) - \beta u^2 - \gamma|u-x'|\right] du$$
$$\geq \mathrm{e}^{-2\gamma M}\mathrm{e}^{-\alpha x} \int_{-M}^{M} \exp(-\beta u^2 + \alpha u)\, du .$$

Thus, (4.54) is bounded by

$$\mathrm{e}^{3\gamma M} \frac{\dfrac{2\exp[-\beta M^2 + (\alpha-\gamma)M]}{2\beta M + \gamma - \alpha} + \sup_{x \geq M} \dfrac{\exp[-\beta x^2 + (\alpha-\gamma)x]}{\beta x + (\gamma+\alpha)}}{\int_{-M}^{M} \exp(-\beta u^2 + \alpha u)\, du}$$

proving (4.55).

Next we show that

$$\limsup_{M\to\infty} \sup_{x \geq M,\, x' \geq M} R(x,x') < 1 . \tag{4.56}$$

We consider the case $M \leq x \leq x'$; the other case can be handled similarly. The denominator in (4.54) is then bounded by

$$\mathrm{e}^{-\alpha x - \gamma x'} \int_{-M}^{M} \exp(-\beta u^2 + (\alpha+\gamma)u)\, du .$$

The two terms in the numerator are bounded by, respectively,

$$\int_{-\infty}^{-M} \exp\left[-\alpha(x-u) - \beta u^2 - \gamma(x'-u)\right] du$$
$$\leq \mathrm{e}^{-\alpha x - \gamma x'} \frac{\exp[-\beta M^2 - (\alpha+\gamma)M]}{2\beta M + \alpha + \gamma}$$

and

$$\int_M^\infty \exp\left(-\alpha|x-u| - \beta u^2 - \gamma|x'-u|\right) du$$

$$\leq \mathrm{e}^{-\alpha x - \gamma x'} \frac{\exp[-\beta M^2 + (\alpha+\gamma)M]}{2\beta M - \alpha - \gamma}$$

$$+ \frac{\exp(-\beta x^2 + \gamma x - \gamma x')}{2\beta x - \gamma + \alpha} + \frac{\exp[-\beta (x')^2 + \alpha x - \alpha x']}{2\beta x' + \alpha + \gamma} ,$$

and (4.56) follows by combining the previous bounds.

We finally have to check that

$$\limsup_{M\to\infty} \sup_{x' \leq -M,\, x \geq M} R(x,x') < 1 .$$

This can be done along the same lines. ■

5

Applications of Smoothing

Remember that in the previous two chapters, we basically considered that integration over X was a feasible operation. This is of course not the case in general, and numerical evaluation of the integrals involved in the smoothing recursions turns out to be a difficult task. In Chapters 6 and 7, generally applicable methods for approximate smoothing, based on Monte Carlo simulations, will be considered. Before that, we first examine two very important particular cases in which an exact numerical evaluation is feasible: models with finite state space in Section 5.1 and Gaussian linear state-space models in Section 5.2. Most of the concepts to be used below have already been introduced in Chapters 3 and 4, and the current chapter mainly deals with computational aspects and algorithms. It also provides concrete examples of application of the methods studied in the previous chapters.

Note that we do not yet consider examples of application of the technique studied in Section 4.1, as the nature of functionals that can be computed recursively will only become more explicit when we discuss the EM framework in Chapter 10. Corresponding examples will be considered in Section 10.2.

5.1 Models with Finite State Space

We first consider models for which the state space X of the hidden variables is finite, that is, when the unobservable states may only take a finite number of distinct values. In this context, the smoothing recursions discussed in Chapter 3 take the familiar form described in the seminal paper by Baum *et al.* (1970) as well as Rabiner's (1989) tutorial (which also covers scaling issues). Section 5.1.2 discusses a technique that is of utmost importance in many applications, for instance digital communications and speech processing, by which one can determine the maximum *a posteriori* sequence of hidden states given the observations.

5.1.1 Smoothing

5.1.1.1 Filtering

Let X denote a finite set that we will, without loss of generality, identify with $\mathsf{X} = \{1, \ldots, r\}$. Probability distributions on X can be represented by vectors belonging to the simplex of $\mathbb{R}^r$, that is, the set

$$\left\{(p_1, \ldots, p_r) : p_i \geq 0 \text{ for every } 1 \leq i \leq r, \sum_{i=1}^{r} p_i = 1\right\} .$$

The components of the transition matrix Q and the initial distribution ν of the hidden chain are denoted by $(q_{ij})_{1 \leq i,j \leq r}$ and $(\nu_i)_{1 \leq i \leq r}$, respectively. Similarly, for the filtering and smoothing distributions, we will use the slightly abusive but unambiguous notation $\phi_k(i) = \mathrm{P}(X_k = i \,|\, Y_{0:k})$, for $1 \leq i \leq r$, instead of $\phi_k(\{i\})$. Finally, because we are mainly concerned with computational aspects given a particular model specification, we do not need to indicate dependence with respect to the initial distribution ν of X_0 and will simply denote the filter (and all associated quantities) by ϕ_k instead of $\phi_{\nu,k}$.

The first item below describes the specific form taken by the filtering recursions—or, in Rabiner's (1989) terminology, the normalized forward recursion—when the state space X is finite.

Algorithm 5.1.1 (Forward Filtering). Assume $\mathsf{X} = \{1, \ldots, r\}$.

Initialization: For $i = 1, \ldots, r$,

$$\phi_{0|-1}(i) = \nu(i) .$$

Forward Recursion: For $k = 0, \ldots, n$,

$$c_k = \sum_{i=1}^{r} \phi_{k|k-1}(i) g_k(i) , \tag{5.1}$$

$$\phi_k(j) = \phi_{k|k-1}(j) g_k(j) / c_k , \tag{5.2}$$

$$\phi_{k+1|k}(j) = \sum_{i=1}^{r} \phi_k(i) q_{ij} , \tag{5.3}$$

for each $j = 1, \ldots, r$.

The computational cost of filtering is thus proportional to n, the number of observations, and scales like r^2 (squared cardinality of the state space X) because of the r vector matrix products corresponding to (5.3). Note however that in models with many zero entries in the transition matrix, in particular for left-to-right models like speech processing HMMs (Example 1.3.6), the complexity of (5.3) is at most of order r times the maximal number of non-zero elements along the rows of Q, which can be significantly less. In addition, and this is also the case for speech processing HMMs, if the Y_k are high-dimensional multivariate observations, the main computational load indeed lies in (5.1)–(5.2) when computing the numerical values of the conditional densities of Y_k given $X_k = j$ for all r possible states j.

Recall from Section 3.2.2 that the likelihood of the observations $Y_{0:n}$ can be computed directly on the log scale according to

$$\ell_n \stackrel{\text{def}}{=} \log \mathrm{L}_n = \sum_{k=0}^{n} \log c_k \; . \tag{5.4}$$

This form is robust to numerical over- or underflow and should be systematically preferred to the product of the normalization constants c_k, which would evaluate the likelihood on a linear scale.

5.1.1.2 The Forward-Backward Algorithm

As discussed in Section 3.4, the standard forward-backward algorithm as exposed by Rabiner (1989) adopts the scaling scheme described by Levinson *et al.* (1983). The *forward pass* is given by Algorithm 5.1.1 as described above, where both the normalization constants c_k and the filter vectors ϕ_k have to be stored for $k = 0, \ldots, n$. Note that the tradition consists in denoting the forward variables by the letter α, but we reserved this notation for the unscaled forward variables (see Section 3.2). Here we actually only store the filter vectors ϕ_k, as their unnormalized versions would quickly under- or overflow the machine precision for any practical value of n.

Algorithm 5.1.2 (Backward Smoothing). Given stored values of $\phi_0, \ldots, \phi_n$ and $c_0, \ldots, c_n$, computed during the forward filtering pass (Algorithm 5.1.2), and starting from the end of the data record, do the following.

Initialization: For $j = 1, \ldots, r$,

$$\breve{\beta}_{n|n}(j) = c_n^{-1} .$$

Backward Recursion: For $k = n-1, \ldots, 0$,

$$\breve{\beta}_{k|n}(i) = c_k^{-1} \sum_{j=1}^{r} q_{ij} g_{k+1}(j) \breve{\beta}_{k+1|n}(j) \tag{5.5}$$

for each $i = 1, \ldots, r$.

For all indices $k < n$, the marginal smoothing probabilities may be evaluated as

$$\phi_{k|n}(i) \stackrel{\text{def}}{=} \mathrm{P}(X_k = i \mid Y_{0:n}) = \frac{\phi_k(i) \breve{\beta}_{k|n}(i)}{\sum_{j=1}^{r} \phi_k(j) \breve{\beta}_{k|n}(j)} \, , \tag{5.6}$$

and the bivariate smoothing probabilities as

$$\phi_{k:k+1|n}(i,j) \stackrel{\text{def}}{=} \mathrm{P}(X_k = i, X_{k+1} = j \mid Y_{0:n}) = \phi_k(i) q_{ij} g_{k+1}(j) \breve{\beta}_{k+1|n}(j) \, .$$

The correctness of the algorithm described above has already been discussed in Section 3.4. We recall that it differs from the line followed in Section 3.2.2 only by the choice of the normalization scheme. Algorithms 5.1.1 and 5.1.2 constitute the standard form of the two-pass algorithm known as forward-backward introduced by Baum *et al.* (1970), where the normalization scheme is first mentioned in Levinson *et al.* (1983) (although the necessity of scaling was certainly known before that date, as discussed in Section 3.4).

The complexity of the backward pass is comparable to that of the forward filtering, that is, it scales as $n \times r^2$. Note however that for high-dimensional observations Y_k, the computational cost of the backward pass is largely reduced, as it is not necessary to evaluate the $(n+1)r$ conditional densities $g_k(i)$ that have already been computed (given that these have been stored in addition to the filter vectors $\phi_0, \dots \phi_n$).

5.1.1.3 Markovian Backward Smoothing

The backward pass as described in Algorithm 5.1.2 can be replaced by the use of the backward Markovian decomposition introduced in Section 3.3.2. Although this second form of backward smoothing is equivalent to Algorithm 5.1.2 from a computational point of view, it is much more transparent on principle grounds. In particular, it shows that the smoothing distributions may be evaluated from the filtering ones using backward Markov transition matrices. In addition, these transition matrices only depend on the filtering distributions themselves and not on the data anymore. In this respect, the computation of the observation densities in (5.5) is thus inessential.

The algorithm, which has been described in full generality in Section 3.3.2, goes as follows,

Algorithm 5.1.3 (Markovian Backward Smoothing). Given stored values of $\phi_0, \dots, \phi_n$ and starting from the end of the data record, do the following.

Initialization: For $j = 1, \dots, r$,

$$\phi_{n|n}(j) = \phi_n(j) .$$

Backward Recursion: For $k = n-1, \dots, 0$,

- Compute the backward transition kernel according to

$$\mathrm{B}_k(j, i) = \frac{\phi_k(i) q_{ij}}{\sum_{m=1}^{r} \phi_k(m) q_{mj}} \tag{5.7}$$

  for $j, i = 1, \dots, r$ (if the denominator happens to be null for index j, then $\mathrm{B}_k(j, i)$ can be set to arbitrary values for $i = 1, \dots, r$).
- Compute

$$\phi_{k:k+1|n}(i, j) = \phi_{k+1|n}(j) \mathrm{B}_k(j, i)$$

  and

$$\phi_{k|n}(i) = \sum_{m=1}^{r} \phi_{k+1|n}(m) \mathrm{B}_k(m, i)$$

for $i, j = 1, \ldots, r$.

Compared to the general situation investigated in Section 3.3.2, the formulation of Algorithm 5.1.3 above takes profit of (3.39) in Remark 3.3.7, which provides an explicit form for the backward kernel B_k in cases where the hidden Markov model is fully dominated (which is always the case when the state space X is finite). Note also that the value of $\mathrm{B}_k(j, i)$ in cases where the denominator of (5.7) happens to be null is irrelevant. The condition $\sum_{m=1}^{r} \phi_k(m) q_{mj} = 0$ is equivalent to stating that $\phi_{k+1|k}(j) = 0$ by (5.3), which in turn implies that $\phi_{k+1}(j) = 0$ by (5.2) and finally that $\phi_{k+1|n}(j) = 0$ for $n \geq k+1$ by (5.6). Hence the value of $\mathrm{B}_k(j, i)$ is arbitrary and is (hopefully) never used in Algorithm 5.1.3, as it is multiplied by zero.

As noted in Section 3.3.2, the idea of using this form of smoothing for finite state space models is rarely ever mentioned except by Askar and Derin (1981) who illustrated it on a simple binary-valued example—see also discussion in Ephraim and Merhav (2002) about "stable" forms of the forward-backward recursions. Of course, one could also consider the forward Markovian decomposition, introduced in Section 3.3.1, which involves the kernels $\mathrm{F}_{k|n}$ that are computed from the backward variables $\beta_{k|n}$. We tend to prefer Algorithm 5.1.2, as it is more directly connected to the standard way of computing smoothed estimates in Gaussian linear state-space models to be discussed later in Section 5.2.

5.1.2 Maximum *a Posteriori* Sequence Estimation

When X is finite, it turns out that it is also possible to carry out a different type of inference concerning the unobservable sequence of states $X_0, \ldots X_n$. This second form is non-probabilistic in the sense that it does not provide a distributional statement concerning the unknown states. On the other hand, the result that is obtained is the jointly optimal, in terms of maximal conditional probability, sequence $X_0, \ldots X_n$ of unknown states given the corresponding observations, which is in some sense much stronger a result than just the marginally (or bivariate) optimal sequence of states. However, neither optimality property implies the other. To express this precisely, let x_k maximize the conditional probability $\mathrm{P}(X_k = x_k \mid Y_{0:n})$ for each $k = 0, 1, \ldots, n$, and let the sequence $x'_{0:n}$ maximize the joint conditional probability $\mathrm{P}(X_{0:n} = x'_{0:n} \mid Y_{0:n})$. Then, in general, the sequences $x_{0:n}$ and $x'_{0:n}$ do not agree. It may even be that a transition (x_k, x_{k+1}) of the marginally optimal sequence is disallowed in the sense that $q_{x_k, x_{k+1}} = 0$.

In the HMM literature, the algorithm that makes possible to compute efficiently the *a posteriori most likely sequence of states* is known as the *Viterbi algorithm*, after Viterbi (1967). It is based on the well-known *dynamic programming* principle. The key observation is indeed (4.1), which we rewrite in log form with notations appropriate for the finite state space case under consideration:

$$\log \phi_{0:k+1|k+1}(x_0, \ldots, x_{k+1}) = (\ell_k - \ell_{k+1}) \\ + \log \phi_{0:k|k}(x_0, \ldots, x_k) \\ + \log q_{x_k x_{k+1}} + \log g_{k+1}(x_{k+1}) , \quad (5.8)$$

where ℓ_k denotes the log-likelihood of the observations up to index k and $\phi_{0:k|k}$ is the joint distribution of the states $X_{0:k}$ given the observations $Y_{0:k}$. The salient feature of (5.8) is that, except for a constant term that does not depend on the state sequence (on the right-hand side of the first line), the *a posteriori* log-probability of the subsequence $x_{0:k+1}$ is equal to that of $x_{0:k}$ up to terms that only involve the pair (x_k, x_{k+1}).

Define

$$m_k(i) = \max_{\{x_0,\ldots,x_{k-1}\} \in \mathsf{X}^k} \log \phi_{0:k|k}(x_0, \ldots, x_{k-1}, i) + \ell_k , \quad (5.9)$$

that is, up to a number independent of the state sequence, the maximal conditional probability (on the log scale) of a sequence up to time k and ending with state i. Also define $b_k(i)$ to be that value in X of x_{k-1} for which the optimum is achieved in (5.9); in other words, $b_k(i)$ is the second final state in an optimal state sequence of length $k+1$ and ending with state i. Using (5.8), we then have the simple recursive relation

$$m_{k+1}(j) = \max_{i \in \{1,\ldots,r\}} [m_k(i) + \log q_{ij}] + \log g_{k+1}(j) , \quad (5.10)$$

and $b_{k+1}(j)$ equals the index i for which the maximum is achieved. This observation immediately leads us to formulate the Viterbi algorithm.

Algorithm 5.1.4 (Viterbi Algorithm).

Forward Recursion (for optimal conditional probabilities): Let

$$m_0(i) = \log(\nu(i) g_0(i)) .$$

Then for $k = 0, 1, \ldots, n-1$, compute $m_{k+1}(j)$ for all states j as in (5.10).

Backward Recursion (for optimal sequence): Let $\hat{x}_n$ be the state j for which $m_n(j)$ is maximal. Then for $k = n-1, n-2, \ldots, 0$, let $\hat{x}_k$ be the state i for which the maximum is attained in (5.10) for $m_{k+1}(j)$ with $j = \hat{x}_{k+1}$. That is, $\hat{x}_k = b_{k+1}(\hat{x}_{k+1})$.

The backward recursion first identifies the final state of the optimal state sequence. Then, once the final state is known, the next to final one can be determined as the state that gives the optimal probability for sequences ending with the now known final state. After that, the second next to final state can be determined in the same manner, and so on. Thus the algorithm requires storage of all the $m_k(j)$. Storage of the $b_k(j)$ is not necessary but makes the backward recursion run faster. In cases where there is no unique maximizing state i in (5.10), there may be no unique optimal state sequence either, and $b_{k+1}(j)$ can be taken arbitrarily within the set of maximizing indices i.

5.2 Gaussian Linear State-Space Models

Gaussian linear state-space models form another important class for which the tools introduced in Chapter 3 provide implementable algorithms. Sections 5.2.1 to 5.2.4 review two different variants of the general principle outlined in Proposition 3.3.9. The second form, exposed in Section 5.2.4, is definitely more involved, but also more efficient in several situations, and is best understood with the help of linear prediction tools that are reviewed in Sections 5.2.2 and 5.2.3. Finally, the exact counterpart of the forward-backward approach, examined in great generality in Section 3.2, is exposed in Section 5.2.5.

5.2.1 Filtering and Backward Markovian Smoothing

We here consider a slight generalization of the Gaussian linear state-space model defined in Section 1.3.3:

$$X_{k+1} = A_k X_k + R_k U_k, \tag{5.11}$$

$$Y_k = B_k X_k + S_k V_k, \tag{5.12}$$

where $\{U_k\}_{k \geq 0}$ and $\{V_k\}_{k \geq 0}$ are two independent vector-valued i.i.d. Gaussian sequences such that $U_k \sim \mathrm{N}(0, I)$ and $V_k \sim \mathrm{N}(0, I)$ where I is a generic notation for the identity matrices (of suitable dimensions). In addition, X_0 is assumed to be $\mathrm{N}(0, \Sigma_\nu)$ distributed and independent of $\{U_k\}$ and $\{V_k\}$. Recall from Chapter 1 that while we typically assume that $S_k S_k^t = \mathrm{Cov}(S_k V_k)$ is a full-rank covariance matrix, the dimension of the state noise vector (also referred to as the *excitation* or *disturbance*) U_k is in many situations smaller than that of the state vector X_k and hence $R_k R_k^t$ may be rank deficient.

Compared to the basic model introduced in Section 1.3.3, the difference lies in the fact that the parameters of the state-space model, A_k, B_k, R_k, and S_k, depend on the time index k. This generalization is motivated by conditionally Gaussian state-space models, as introduced in Section 1.3.4. For such models, neither is the state space finite nor is the complete model equivalent to a Gaussian linear state-space model. However, it is indeed possible, and often advantageous, to perform filtering *while conditioning on the state of the unobservable indicator variables*. In this situation, although the basic model is homogeneous in time, the conditional model features time-dependent parameters. There are also cases in which the means of $\{U_k\}$ and $\{V_k\}$ depend on time. To avoid notational blow-up, we consider only the zero-mean case:

the modifications needed to handle non-zero means are straightforward as explained in Remark 5.2.14 below.

A feature that is unique to the Gaussian linear state-space model defined by (5.11)–(5.12) is that because the states $X_{0:n}$ and the observations $Y_{0:n}$ are jointly multivariate Gaussian (for any n), all smoothing distributions are also Gaussian. Hence any smoothing distribution is fully determined by its mean vector and covariance matrix. We consider in particular below the predictive state estimator $\phi_{k|k-1}$ and filtered state estimator ϕ_k and denote by

$$\phi_{k|k-1} = \mathrm{N}\left(\hat{X}_{k|k-1}, \Sigma_{k|k-1}\right) , \tag{5.13}$$

$$\phi_k = \mathrm{N}\left(\hat{X}_{k|k}, \Sigma_{k|k}\right) , \tag{5.14}$$

their respective means and covariance matrices.

Remark 5.2.1. Note that up to now we have always used ϕ_k as a simplified notation for $\phi_{k|k}$, thereby expressing a default interest in the filtering distribution. To avoid all ambiguity, however, we will adopt the notations $\hat{X}_{k|k}$ and $\Sigma_{k|k}$ to denote the first two moments of the filtering distributions in Gaussian linear state-space models. The reason for this modification is that the conventions used in the literature on state-space models are rather variable, but with a marked general preference for using $\hat{X}_k$ and Σ_k to refer to the moments of predictive distribution $\phi_{k|k-1}$—see, e.g., Anderson and Moore (1979) or Kailath *et al.* (2000). In contrast, the more explicit notations $\hat{X}_{k|k}$ and $\Sigma_{k|k}$ are self-explaining and do not rely on an implicit knowledge of whether the focus is on the filtering or prediction task. ■

The following elementary lemma is instrumental in computing the predictive and the filtered state estimator.

Proposition 5.2.2 (Conditioning in the Gaussian Linear Model). *Let X and V be two independent Gaussian random vectors with $\mathrm{E}[X] = \mu_X$, $\mathrm{Cov}(X) = \Sigma_X$, and $\mathrm{Cov}(V) = \Sigma_V$, and assume $\mathrm{E}[V] = 0$. Consider the model*

$$Y = BX + V , \tag{5.15}$$

where B is a deterministic matrix of appropriate dimensions. Further assume that $B\Sigma_X B^t + \Sigma_V$ is a full rank matrix. Then

$$\begin{aligned}\mathrm{E}[X \mid Y] &= \mathrm{E}[X] + \mathrm{Cov}(X,Y)\left\{\mathrm{Cov}(Y)\right\}^{-1}(Y - \mathrm{E}[Y]) \\ &= \mu_X + \Sigma_X B^t \left\{B\Sigma_X B^t + \Sigma_V\right\}^{-1}(Y - B\mu_X)\end{aligned} \tag{5.16}$$

and

$$\begin{aligned}\mathrm{Cov}(X \mid Y) &= \mathrm{Cov}(X - \mathrm{E}[X|Y]) = \mathrm{E}\left[(X - \mathrm{E}[X|Y])X^t\right] \\ &= \Sigma_X - \Sigma_X B^t \left\{B\Sigma_X B^t + \Sigma_V\right\}^{-1} B\Sigma_X .\end{aligned} \tag{5.17}$$

Proof. Denote by $\hat{X}$ the right-hand side of (5.16). Then

$$X - \hat{X} = X - \mathrm{E}(X) - \mathrm{Cov}(X, Y)\{\mathrm{Cov}(Y)\}^{-1}(Y - \mathrm{E}[Y]) ,$$

which implies that

$$\mathrm{Cov}(X - \hat{X}, Y) = \mathrm{Cov}(X, Y) - \mathrm{Cov}(X, Y)\{\mathrm{Cov}(Y)\}^{-1}\,\mathrm{Cov}(Y) = 0 . \quad (5.18)$$

The random vectors Y and $X - \hat{X}$ thus are jointly Gaussian (as linear transformations of a Gaussian multivariate random vector) and uncorrelated. Hence, Y and $X - \hat{X}$ are also independent. Writing

$$X = \hat{X} + (X - \hat{X}) ,$$

where $\hat{X}$ is $\sigma(Y)$ measurable (as a linear combination of the components of Y) and $X - \hat{X}$ is independent of $\hat{X}$, it is then easily checked (see Appendix A.2) that $\hat{X} = \mathrm{E}(X \mid Y)$ and that, in addition,

$$\mathrm{Cov}\,(X \mid Y) \stackrel{\text{def}}{=} \mathrm{Cov}\left[(X - \hat{X})(X - \hat{X})' \,\middle|\, Y\right] = \mathrm{Cov}(X - \hat{X}) .$$

Finally, (5.17) is obtained upon noting that

$$\mathrm{Cov}(X - \hat{X}) = \mathrm{E}[(X - \hat{X})(X - \hat{X})^t] = \mathrm{E}[(X - \hat{X})X^t] ,$$

using (5.18) and the fact that $\hat{X}$ is a linear transform of Y. The second lines of (5.16) and (5.17) follow from the linear structure of (5.15). □

For Gaussian linear state-space models, Proposition 5.2.2 implies in particular that while the mean vectors $\hat{X}_{k|k-1}$ or $\hat{X}_{k|k}$ do depend on the observations, the covariance matrices $\Sigma_{k|k-1}$ and $\Sigma_{k|k}$ are completely determined by the model parameters. Our first result below simply consists in applying the formula derived in Proposition 5.2.2 for the Gaussian linear model to obtain an explicit equivalent of (3.27) in terms of the model parameters.

Proposition 5.2.3 (Filtering in Gaussian Linear State-Space Models). *The filtered and predictive mean and covariance matrices may be updated recursively as follows, for $k \geq 0$.*

Filtering:

$$\hat{X}_{k|k} = \hat{X}_{k|k-1} + \Sigma_{k|k-1} B_k^t (B_k \Sigma_{k|k-1} B_k^t + S_k S_k^t)^{-1} (Y_k - B_k \hat{X}_{k|k-1}) , \quad (5.19)$$

$$\Sigma_{k|k} = \Sigma_{k|k-1} - \Sigma_{k|k-1} B_k^t (B_k \Sigma_{k|k-1} B_k^t + S_k S_k^t)^{-1} B_k \Sigma_{k|k-1} , \quad (5.20)$$

with the conventions $\hat{X}_{0|-1} = 0$ and $\Sigma_{0|-1} = \Sigma_\nu$.

Prediction:

$$\hat{X}_{k+1|k} = A_k \hat{X}_{k|k} , \quad (5.21)$$

$$\Sigma_{k+1|k} = A_k \Sigma_{k|k} A_k^t + R_k R_k^t , \quad (5.22)$$

Proof. As mentioned in Remark 3.2.6, the predictor-to-filter update is obtained by computing the posterior distribution of X_k given Y_k in the equivalent pseudo-model $X_k \sim \mathrm{N}(\hat{X}_{k|k-1}, \Sigma_{k|k-1})$ and

$$Y_k = B_k X_k + V_k \ ,$$

where V_k is $\mathrm{N}(0, S_k S_k^t)$ distributed and independent of X_k. Equations (5.19) and (5.20) thus follow from Proposition 5.2.2. Equations (5.21) and (5.22) correspond to the moments of

$$X_{k+1} = A_k X_k + R_k U_k$$

when X_k and U_k are independent and, respectively, $\mathrm{N}(\hat{X}_{k|k}, \Sigma_{k|k})$ and $\mathrm{N}(0, I)$ distributed (see discussion in Remark 3.2.6). □

Next we consider using the backward Markovian decomposition of Section 3.3.2 to derive the smoothing recursion. We will denote by $\hat{X}_{k|n}$ and $\Sigma_{k|n}$ respectively the mean and covariance matrix of the smoothing distribution $\phi_{k|n}$. According to Remark 3.3.7, the backward kernel B_k corresponds to the distribution of X_k given X_{k+1} in the pseudo-model

$$X_{k+1} = A_k X_k + R_k U_k \ ,$$

when $X_k \sim \mathrm{N}(\hat{X}_{k|k}, \Sigma_{k|k})$ and $U_k \sim \mathrm{N}(0, I)$ independently of X_k. Using Proposition 5.2.2 once again, $\mathrm{B}_k(X_{k+1}, \cdot)$ is seen to be the Gaussian distribution with mean and covariance matrix given by, respectively,

$$\hat{X}_{k|k} + \Sigma_{k|k} A_k^t (A_k \Sigma_{k|k} A_k^t + R_k R_k^t)^{-1} (X_{k+1} - A_k \hat{X}_{k|k}) \ , \tag{5.23}$$

and covariance matrix

$$\Sigma_{k|k} - \Sigma_{k|k} A_k^t (A_k \Sigma_{k|k} A_k^t + R_k R_k^t)^{-1} A_k \Sigma_{k|k} \ . \tag{5.24}$$

Proposition 3.3.9 asserts that B_k is the transition kernel that maps $\phi_{k+1|n}$ to $\phi_{k|n}$. Hence, if we assume that $\phi_{k+1|n} = \mathrm{N}(\hat{X}_{k+1|n}, \Sigma_{k+1|n})$ is already known,

$$\hat{X}_{k|n} = \hat{X}_{k|k} + \Sigma_{k|k} A_k^t M_k (\hat{X}_{k+1|n} - A_k \hat{X}_{k|k}) \ , \tag{5.25}$$
$$\Sigma_{k|n} = \Sigma_{k|k} - \Sigma_{k|k} A_k^t M_k A_k \Sigma_{k|k} + \Sigma_{k|k} A_k^t M_k \Sigma_{k+1|n} M_k A_k \Sigma_{k|k} \ , \tag{5.26}$$

give the moments of $\phi_{k|n}$, where

$$M_k = (A_k \Sigma_{k|k} A_k^t + R_k R_k^t)^{-1} \ .$$

To derive these two latter equations, we must observe that (i) $\mathrm{B}_k(X_{k+1}, \cdot)$ may be interpreted as an affine transformation of X_{k+1} as in (5.23) followed by adding an independent zero mean Gaussian random vector with covariance matrix as in (5.24), and that (ii) mapping $\phi_{k+1|n}$ into $\phi_{k|n}$ amounts to replacing the fixed X_{k+1} by a random vector with distribution $\mathrm{N}(\hat{X}_{k+1|n}, \Sigma_{k+1|n})$.

The random vector obtained through this mapping is Gaussian with mean and covariance as in (5.25)–(5.26), the third term of (5.26) being the "extra term" arising because of (ii).

We summarize these observations in the form of an algorithm.

Algorithm 5.2.4 (Rauch-Tung-Striebel Smoothing). Assume that the filtering moments $\hat{X}_{k|k}$ and $\Sigma_{k|k}$ are available (for instance by application of Proposition 5.2.3) for $k = 0, \ldots, n$. The smoothing moments $\hat{X}_{k|n}$ and $\Sigma_{k|n}$ may be evaluated *backwards* by applying (5.25) and (5.26) from $k = n - 1$ down to $k = 0$.

This smoothing approach is generally known as *forward filtering, backward smoothing* or *RTS (Rauch-Tung-Striebel) smoothing* after Rauch *et al.* (1965). From the discussion above, it clearly corresponds to an application of the general idea that the backward posterior chain is a Markov chain as discussed in Section 3.3.2. Algorithm 5.2.4 is thus the exact counterpart of Algorithm 5.1.3 for Gaussian linear state-space models.

5.2.2 Linear Prediction Interpretation

The approach that we have followed so far to derive the filtering and smoothing recursions is simple and efficient and has the merit of being directly connected with the general framework investigated in Chapter 3. It however suffers from two shortcomings, the latter being susceptible of turning into a real hindrance in practical applications of the method.

The first concern has to do with the interpretability of the obtained recursions. Indeed, by repeated applications of Proposition 5.2.2, we rapidly obtain complicated expressions such as (5.26). Although such expressions are usable in practice granted that one identifies common terms that need only be computed once, they are hard to justify on intuitive grounds. This may sound like a vague or naive statement, but interpretability turns out to be a key issue when considering more involved algorithms such as the disturbance smoothing approach of Section 5.2.4 below.

The second remark is perhaps more troublesome because it concerns the numerical efficiency of the RTS smoothing approach described above. Several of the state-space models that we have considered so far share a common feature, which is dramatically exemplified in the noisy $AR(p)$ model (Example 1.3.8 in Chapter 1). In this model, the disturbance U_k is scalar, and there is a deterministic relationship between the state variables X_k and X_{k+1}, which is that the last $p-1$ components of X_{k+1} are just a copy of the first $p-1$ components of X_k. In such a situation, it is obvious that the same deterministic relation should be reflected in the values of $\hat{X}_{k|n}$ and $\hat{X}_{k+1|n}$, in the sense that the last $p-1$ components of $\hat{X}_{k+1|n}$ must coincide with the first $p-1$ components of $\hat{X}_{k|n}$. In contrast, Algorithm 5.2.4 implies a seemingly complex recursion, which involves a $p \times p$ matrix inversion, to determine $\hat{X}_{k|n}$ from $\hat{X}_{k+1|n}$ and $\hat{X}_{k|k}$.

In order to derive a smoothing algorithm that takes advantage of the model structure (5.11)–(5.12), we will need to proceed more cautiously. For models like the noisy $AR(p)$ model, it is in fact more appropriate to perform the smoothing on the disturbance (or dynamic noise) variables U_k rather than the states X_k themselves. This idea, which will be developed in Section 5.2.4 below, does not directly fit into the framework of Chapter 3 however because the pairs $\{U_k, Y_k\}_{k \geq 0}$ are not Markovian, in contrast to $\{X_k, Y_k\}_{k \geq 0}$.

The rest of this section thus follows a slightly different path by developing the theory of best linear prediction in mean squared error sense. The key point here is that linear prediction can be interpreted "geometrically" using (elementary) Hilbert space theory. In state-space models (and more generally, in time series analysis), this geometric intuition serves as a valuable guide in the development and construction of algorithms. As a by-product, this approach also constitutes a framework that is not limited to the Gaussian case considered up to now and applies to all linear state-space models with finite second moments. However, the fact that this approach also fully characterizes the marginal smoothing distributions is of course particular to Gaussian models.

5.2.2.1 Best Linear Prediction

This section and the following require basic familiarity with the key notions of L^2 projections, which are reviewed briefly in Appendix B. Let $Y_0, \ldots, Y_k$ and X be elements of $L^2(\Omega, \mathcal{F}, \mathrm{P})$. We will assume for the moment that $Y_0, \ldots, Y_k$ and X are scalar random variables. The *best linear predictor of X given $Y_0, \ldots, Y_k$* is the L^2 projection of X on the linear subspace

$$\operatorname{span}(1, Y_0, \ldots, Y_k) \stackrel{\text{def}}{=} \left\{ Y : Y = \mu + \sum_{i=0}^{k} \alpha_i Y_i, \quad \mu, \alpha_0, \ldots, \alpha_k \in \mathbb{R} \right\} .$$

The best linear predictor will be denoted by $\operatorname{proj}(X|1, Y_0, \ldots, Y_k)$, or simply by $\hat{X}$ in situations where there is no possible confusion regarding the subspace on which X is projected. The notation "1" refers to the constant (deterministic) random variable, whose role will be made clearer in Remark 5.2.5 below.

According to the projection theorem (Theorem B.2.4 in Appendix B), $\hat{X}$ is characterized by the equations

$$\mathrm{E}\{(X - \hat{X})Y\} = 0 \quad \text{for all} \quad Y \in \operatorname{span}(1, Y_0, \ldots, Y_k) .$$

Because $1, Y_0, \ldots, Y_k$ is a generating family of $\operatorname{span}(1, Y_0, \ldots, Y_k)$, this condition may be equivalently rewritten as

$$\mathrm{E}[(X - \hat{X})1] = 0 \quad \text{and} \quad \mathrm{E}[(X - \hat{X})Y_i] = 0, \quad \text{for all } i = 0, \ldots, k .$$

The notations $X - \hat{X} \perp \mathrm{span}(1, Y_0, \ldots, Y_k)$ and $X - \hat{X} \perp Y_i$ will also be used to denote concisely these orthogonality relations, where orthogonality is to be understood in the $L^2(\Omega, \mathcal{F}, \mathrm{P})$ sense. Because $\hat{X} \in \mathrm{span}(1, Y_0, \ldots, Y_k)$, the projection may be represented as

$$\hat{X} = \mu + \phi_0(Y_0 - \mathrm{E}[Y_0]) + \ldots + \phi_k(Y_k - \mathrm{E}[Y_k]) \tag{5.27}$$

for some scalars $\mu, \varphi_0, \ldots, \varphi_k$. Denoting by Γ_k the matrix $[\mathrm{Cov}(Y_i, Y_j)]_{0 \le i,j \le k}$ and γ_k the vector $[\mathrm{Cov}(X, Y_0), \ldots, \mathrm{Cov}(X, Y_k)]^t$, the prediction equations may be summarized as

$$\mu = \mathrm{E}[X] \quad \text{and} \quad \Gamma_n \varphi = \gamma_k, \quad \text{where} \quad \varphi = (\varphi_1, \ldots, \varphi_k)^t \;. \tag{5.28}$$

The projection theorem guarantees that there is at least one solution φ. If the covariance matrix Γ_k is singular, there are infinitely many solutions, but all of them correspond to the same (uniquely defined) optimal linear predictor. An immediate consequence of Proposition B.2.6(iii) is that the *covariance of the prediction error* may be written in two equivalent, and often useful, ways,

$$\mathrm{Cov}(X - \hat{X}) = \mathrm{E}[X(X - \hat{X})] = \mathrm{Cov}(X) - \mathrm{Cov}(\hat{X}) \;. \tag{5.29}$$

Remark 5.2.5. The inclusion of the deterministic constant in the generating family of the prediction subspace is simply meant to capture the prediction capacity of $\mathrm{E}[X]$. Indeed, because

$$\mathrm{E}[(X-\mu)^2] = \mathrm{E}\{[X - \mathrm{E}(X)]^2\} + [\mu - \mathrm{E}(X)]^2 \le \mathrm{E}(X^2) + [\mu - \mathrm{E}(X)]^2 \;,$$

predicting X by $\mathrm{E}(X)$ is the optimal guess that always reduces the mean squared error in the absence of observations.

In (5.27), we used a technique that will be recurrent in the following and consists in replacing some variables by orthogonalized ones. Because $\mathrm{E}[(Y_i - \mathrm{E}(Y_i))1] = 0$ for $i = 0, \ldots, k$, the projection on $\mathrm{span}(1, Y_0, \ldots, Y_k)$ may be decomposed as the projection on $\mathrm{span}(1)$, that is, $\mathrm{E}(X)$, plus the projection on $\mathrm{span}(Y_0 - \mathrm{E}[Y_0], \ldots, Y_k - \mathrm{E}[Y_k])$. Following (5.28), projecting a non-zero mean variable X is then achieved by first considering the projection on the centered observations $Y_i - \mathrm{E}(Y_i)$ and then adding the expectation of X to the obtained prediction. For this reason, considering means is not crucial, and we assume in the sequel that all variables under consideration have zero mean. Hence, $\hat{X}$ is directly defined as the projection on $\mathrm{span}(Y_0, \ldots, Y_k)$ only and the covariances $\mathrm{Cov}(Y_i, Y_j)$ and $\mathrm{Cov}(X, Y_i)$ can be replaced by $\mathrm{E}(Y_i Y_j)$ and $\mathrm{E}(XY_i)$, respectively. ■

We now extend these definitions to the case of vector-valued random variables.

Definition 5.2.6 (Best Linear Predictor). *Let $X = [X(1), \ldots, X(d_x)]^t$ be a d_x-dimensional random vector and $Y_0, \ldots, Y_k$ a family of d_y-dimensional* *random vectors, all elements of* $L^2(\Omega, \mathcal{F}, \mathrm{P})$. *It is further assumed that* $\mathrm{E}(X) = 0$ *and* $\mathrm{E}(Y_i) = 0$ *for* $i = 0, \ldots, k$. *The minimum mean square error prediction of* X *given* $Y_0, \ldots, Y_k$ *is defined as the vector* $[\hat{X}(1), \ldots, \hat{X}(d_x)]^t$ *such that every component* $\hat{X}(j)$, $j = 1, \ldots, d_x$, *is the* L^2*-projection of* $X(j)$ *on*

$$\mathrm{span}\left(\{Y_i(j)\}_{0 \leq i \leq k, 1 \leq j \leq d_y}\right) .$$

As a convention, we will also use the notations

$$\hat{X} = \mathrm{proj}(X | Y_0, \ldots, Y_k) = \mathrm{proj}(X | \, \mathrm{span}(Y_0, \ldots, Y_k)) ,$$

in this context.

Definition 5.2.6 asserts that each component $X(j)$ of X is to be projected on the linear subspace spanned by linear combinations of the components of the vectors Y_i,

$$\left\{ Y : Y = \sum_{i=0}^{k} \sum_{j=1}^{d_y} \alpha_{i,j} Y_i(j) \, , \quad \alpha_{i,j} \in \mathbb{R} \right\} .$$

Proceeding as in the case of scalar variables, the projection $\hat{X}$ may be written

$$\hat{X} = \sum_{i=0}^{k} \Phi_i Y_i \, ,$$

where $\Phi_0, \ldots, \Phi_k$ are $d_x \times d_y$ matrices. The orthogonality relations that characterize the projection of $\hat{X}$ may the be summarized as

$$\sum_{i=0}^{k} \Phi_i \, \mathrm{E}(Y_i Y_j^t) = \mathrm{E}(X Y_j^t) \quad \text{for } j = 0, \ldots, k \, , \tag{5.30}$$

where $\mathrm{E}(Y_i Y_j^t)$ and $\mathrm{E}(X Y_j^t)$ are respectively $d_y \times d_y$ and $d_x \times d_y$ matrices such that

$$\left[\mathrm{E}(Y_i Y_j^t)\right]_{l_1 l_2} = \mathrm{E}[Y_i(l_1) Y_j(l_2)] \, ,$$
$$\left[\mathrm{E}(X Y_j^t)\right]_{l_1 l_2} = \mathrm{E}[X(l_1) Y_j(l_2)] \, .$$

The projection theorem guarantees that there is at least one solution to this system of linear equations. The solution is unique if the $d_y(k+1) \times d_y(k+1)$ block matrix

$$\Gamma_k = \begin{pmatrix} \mathrm{E}(Y_0 Y_0^t) & \cdots & \mathrm{E}(Y_0 Y_k^t) \\ \vdots & & \vdots \\ \mathrm{E}(Y_n Y_0^t) & \cdots & \mathrm{E}(Y_n Y_n^t) \end{pmatrix}$$

is invertible. As in the scalar case, the covariance matrix of the prediction error may be written in any of the two forms

$$\operatorname{Cov}(X - \hat{X}) = \mathrm{E}[X(X - \hat{X})^t] = \mathrm{E}(XX^t) - \mathrm{E}(\hat{X}\hat{X}^t) . \tag{5.31}$$

An important remark, which can be easily checked from (5.30), is that

$$\operatorname{proj}(AX|Y_0, \ldots, Y_k) = A \operatorname{proj}(X|Y_0, \ldots, Y_k) , \tag{5.32}$$

whenever A is a deterministic matrix of suitable dimensions. This simply says that the projection operator is linear.

Clearly, solving for (5.30) directly is only possible in cases where the dimension of Γ_k is modest. In all other cases, an incremental way of computing the predictor would be preferable. This is exactly what the innovation approach to be described next is all about.

5.2.2.2 The Innovation Approach

Let us start by noting that when $k = 0$, and when the covariance matrix $\mathrm{E}(YY^t)$ is invertible, then the best linear predictor of the vector X in terms of Y only satisfies

$$\begin{aligned} \hat{X} &= \mathrm{E}(XY^t) \left[\mathrm{E}(YY^t)\right]^{-1} Y , \\ \operatorname{Cov}(X - \hat{X}) &= \mathrm{E}[X(X - \hat{X})^t] = \mathrm{E}(XX^t) - \mathrm{E}(XY^t) \left[\mathrm{E}(YY^t)\right]^{-1} \mathrm{E}(XY^t) . \end{aligned} \tag{5.33}$$

Interestingly, (5.33) is an expression that we already met in Proposition 5.2.2. Equation (5.33) is equivalent to the first expressions given in (5.16) and (5.17), assuming that X is a zero mean variable. This is not surprising, as the proof of Proposition 5.2.2 was based on the fact that $\hat{X}$, as defined by (5.33), is such that $X - \hat{X}$ is uncorrelated with Y. The only difference is that in the (multivariate) Gaussian case, the best linear predictor and the covariance of the prediction error also correspond to the first two moments of the conditional distribution of X given Y, which is Gaussian, and hence entirely define this distribution.

Another case of interest is when the random variables $Y_0, \ldots, Y_k$ are uncorrelated in the sense that $\mathrm{E}(Y_i Y_j^t) = 0$ for any $i, j = 0, \ldots, k$ such that $i \neq j$. In this case, provided that the covariance matrices $\mathrm{E}(Y_i Y_i^t)$ are positive definite for every $i = 0, \ldots, k$, the best linear predictor of X in terms of $\{Y_0, \ldots, Y_k\}$ is given by

$$\hat{X} = \sum_{i=0}^{k} \mathrm{E}(XY_i^t) \left[\mathrm{E}(Y_i Y_i^t)\right]^{-1} Y_i . \tag{5.34}$$

The best linear predictor of X in terms of $Y_0, \ldots, Y_k$ thus reduces to the sum of the best linear predictors of X in terms of each individual vector Y_i, $i = 0, \ldots, k$.

Of course, in most problems the vectors $Y_0, \ldots, Y_k$ are correlated, but there is a generic procedure by which we may fall back to this simple case, irrespectively of the correlation structure of the Y_k. This approach is the analog of the Gram-Schmidt orthogonalization procedure used to obtain a basis of orthogonal vectors from a set of linearly independent vectors.

Consider the linear subspace $\mathrm{span}(Y_0, \ldots, Y_j)$ spanned by the observations up to index j. By analogy with the Gram-Schmidt procedure, one may replace the set $\{Y_0, \ldots, Y_j\}$ of random vectors by an equivalent set $\{\epsilon_0, \ldots, \epsilon_j\}$ of uncorrelated random vectors spanning the same linear subspace,

$$\mathrm{span}(Y_0, \ldots, Y_j) = \mathrm{span}(\epsilon_0, \ldots, \epsilon_j) \quad \text{for all } j = 0, \ldots, k \; . \tag{5.35}$$

This can be achieved by defining recursively the sequence of ϵ_j by $\epsilon_0 = Y_0$ and

$$\epsilon_{j+1} = Y_{j+1} - \mathrm{proj}(Y_{j+1} | \, \mathrm{span}(Y_0, \ldots, Y_j)) \tag{5.36}$$

for $j \geq 0$. The projection of Y_{j+1} on $\mathrm{span}(Y_0, \ldots, Y_j) = \mathrm{span}(\epsilon_0, \ldots, \epsilon_j)$ has an explicit form, as $\epsilon_0, \ldots, \epsilon_j$ are uncorrelated. According to (5.34),

$$\mathrm{proj}(Y_{j+1} | \, \mathrm{span}(\epsilon_0, \ldots, \epsilon_j)) = \sum_{i=0}^{j} \mathrm{E}(Y_{j+1} \epsilon_i^t) \left[\mathrm{E}(\epsilon_i \epsilon_i^t) \right]^{-1} \epsilon_i \; , \tag{5.37}$$

which leads to the recursive expression

$$\epsilon_{j+1} = Y_{j+1} - \sum_{i=0}^{j} \mathrm{E}(Y_{j+1} \epsilon_i^t) \left[\mathrm{E}(\epsilon_i \epsilon_i^t) \right]^{-1} \epsilon_i \; . \tag{5.38}$$

For any $j = 0, \ldots, k$, ϵ_j may be interpreted as the part of the random variable Y_j that cannot be linearly predicted from the history $Y_0, \ldots, Y_{j-1}$. For this reason, ϵ_j is called the *innovation*. The innovation sequence $\{\epsilon_j\}_{j \geq 0}$ constructed recursively from (5.38) is uncorrelated but is also in a causal relationship with $\{Y_j\}_{j \geq 0}$ in the sense that for every $j \geq 0$,

$$\epsilon_j \in \mathrm{span}(Y_0, \ldots, Y_j) \quad \text{and} \quad Y_j \in \mathrm{span}(\epsilon_0, \ldots, \epsilon_j) \; . \tag{5.39}$$

In other words, the sequences $\{Y_j\}_{j \geq 0}$ and $\{\epsilon_j\}_{j \geq 0}$ are related by a causal and causally invertible linear transformation.

To avoid degeneracy in (5.37) and (5.38), one needs to assume that the covariance matrix $\mathrm{E}(\epsilon_j \epsilon_j^t)$ is positive definite. Hence we make the following definition, which guarantees that none of the components of the random vector Y_{j+1} can be predicted without error by some linear combination of past variables $Y_0, \ldots, Y_j$.

Definition 5.2.7 (Non-deterministic Process). *The process* $\{Y_k\}_{k \geq 0}$ *is said to be* non-deterministic *if for any* $j \geq 0$ *the matrix*

$$\mathrm{Cov}\left[Y_{j+1} - \mathrm{proj}(Y_{j+1} | Y_0, \ldots, Y_j) \right]$$

is positive definite.

The innovation sequence $\{\epsilon_k\}_{k\geq 0}$ is useful for deriving recursive prediction formulas for variables of interest. Let $Z \in L^2(\Omega, \mathcal{F}, \mathrm{P})$ be a random vector and denote by $\hat{Z}_{|k}$ the best linear prediction of Z given observations up to index k. Using (5.34), $\hat{Z}_{|k}$ satisfies the recursive relation

$$\begin{aligned} \hat{Z}_{|k} &= \sum_{i=0}^{k} \mathrm{E}(Z\epsilon_i^t) \left[\mathrm{E}(\epsilon_i\epsilon_i^t)\right]^{-1} \epsilon_i \\ &= \hat{Z}_{|k-1} + \mathrm{E}(Z\epsilon_k^t) \left[\mathrm{E}(\epsilon_k\epsilon_k^t)\right]^{-1} \epsilon_k \; . \end{aligned} \tag{5.40}$$

The covariance of the prediction error is given by

$$\begin{aligned} \mathrm{Cov}(Z - \hat{Z}_{|k}) &= \mathrm{Cov}(Z) - \mathrm{Cov}(\hat{Z}_{|k}) \\ &= \mathrm{Cov}(Z) - \sum_{i=0}^{k} \mathrm{E}(Z\epsilon_i^t) \left[\mathrm{E}(\epsilon_i\epsilon_i^t)\right]^{-1} \mathrm{E}(\epsilon_i Z^t) \\ &= \mathrm{Cov}(Z) - \mathrm{Cov}(\hat{Z}_{|k-1}) - \mathrm{E}(Z\epsilon_k^t) \left[\mathrm{E}(\epsilon_k\epsilon_k^t)\right]^{-1} \mathrm{E}(\epsilon_k Z^t) \; . \end{aligned} \tag{5.41}$$

5.2.3 The Prediction and Filtering Recursions Revisited

5.2.3.1 Kalman Prediction

We now consider again the state-space model

$$X_{k+1} = A_k X_k + R_k U_k, \tag{5.42}$$

$$Y_k = B_k X_k + S_k V_k, \tag{5.43}$$

where $\{U_k\}_{k\geq 0}$ and $\{V_k\}_{k\geq 0}$ are now only assumed to be uncorrelated second-order white noise sequences with zero mean and identity covariance matrices. The initial state variable X_0 is assumed to be uncorrelated with $\{U_k\}$ and $\{V_k\}$ and such that $\mathrm{E}(X_0) = 0$ and $\mathrm{Cov}(X_0) = \Sigma_\nu$. It is also assumed that $\{Y_k\}_{k\geq 0}$ is non-deterministic in the sense of Definition 5.2.7. The form of (5.43) shows that a simple sufficient (but not necessary) condition that guarantees this requirement is that $S_k S_k^t$ be positive definite for all $k \geq 0$.

As a notational convention, for any (scalar or vector-valued) process $\{Z_k\}_{k\geq 0}$, the projection of Z_k onto the linear space spanned by the random vectors $Y_0, \ldots, Y_n$ will be denoted by $\hat{Z}_{k|n}$. Particular cases of interest are $\hat{X}_{k|k-1}$, which corresponds to the (one-step) state prediction as well as $\hat{Y}_{k|k-1}$ for the observation prediction. The innovation ϵ_k discussed in the previous section is by definition equal to the observation prediction error $Y_k - \hat{Y}_{k|k-1}$. We finally introduce two additional notations,

$$\Gamma_k \stackrel{\mathrm{def}}{=} \mathrm{Cov}(\epsilon_k) \quad \text{and} \quad \Sigma_{k|n} \stackrel{\mathrm{def}}{=} \mathrm{Cov}(X_k - \hat{X}_{k|n}) \; .$$

Remark 5.2.8. The careful reader will have noticed that we overloaded the notations $\hat{X}_{k|k-1}$ and $\Sigma_{k|k-1}$, which correspond, in Proposition 5.2.3, to the mean and covariance matrix of $\phi_{k|k-1}$ and, in Algorithm 5.2.9, to the best mean square linear predictor of X_k in terms of $Y_0, \ldots, Y_{k-1}$ and the covariance of the linear prediction error $X_k - \hat{X}_{k|k-1}$. This abuse of notation is justified by Proposition 5.2.2, which states that these concepts are equivalent *in the Gaussian case.* In the general non-Gaussian model, only the second interpretation (linear prediction) is correct. ■

We first consider determining the innovation sequence from the observations. Projecting (5.43) onto $\text{span}(Y_0, \ldots, Y_{k-1})$ yields

$$\hat{Y}_{k|k-1} = B_k \hat{X}_{k|k-1} + S_k \hat{V}_{k|k-1} . \tag{5.44}$$

Our assumptions on the state-space model imply that $\mathrm{E}(V_k Y_j^t) = 0$ for $j = 0, \ldots, k-1$, so that $\hat{V}_{k|k-1} = 0$. Hence

$$\epsilon_k = Y_k - \hat{Y}_{k|k-1} = Y_k - B_k \hat{X}_{k|k-1} . \tag{5.45}$$

We next apply the general decomposition obtained (5.40) to the variable X_{k+1} to obtain the state prediction update. Equation (5.40) applied with $Z = X_{k+1}$ yields

$$\hat{X}_{k+1|k} = \hat{X}_{k+1|k-1} + \mathrm{E}(X_{k+1}\epsilon_k^t) \left[\mathrm{E}(\epsilon_k \epsilon_k^t)\right]^{-1} \epsilon_k . \tag{5.46}$$

To complete the recursion, the first term on the right-hand side should be expressed in terms of $\hat{X}_{k|k-1}$ and ϵ_{k-1}. Projecting the state equation (5.42) on the linear subspace spanned by $Y_0, \ldots, Y_{k-1}$ yields

$$\hat{X}_{k+1|k-1} = A_k \hat{X}_{k|k-1} + \hat{U}_{k|k-1} = A_k \hat{X}_{k|k-1} , \tag{5.47}$$

because $\mathrm{E}(U_k Y_j^t) = 0$ for indices $j = 0, \ldots, k-1$. Thus, (5.46) may be written

$$\hat{X}_{k+1|k} = A_k \hat{X}_{k|k-1} + H_k \epsilon_k , \tag{5.48}$$

where H_k, called the *Kalman gain*[1], is a deterministic matrix defined by

$$H_k \stackrel{\text{def}}{=} \mathrm{E}(X_{k+1}\epsilon_k^t) \Gamma_k^{-1} . \tag{5.49}$$

To evaluate the Kalman gain, first note that

$$\epsilon_k = Y_k - B_k \hat{X}_{k|k-1} = B_k (X_k - \hat{X}_{k|k-1}) + S_k V_k . \tag{5.50}$$

---

[1]Readers familiar with the topic will certainly object that we do not comply with the well-established tradition of denoting the Kalman gain by the letter K. We will however meet in Algorithm 5.2.13 below a different version of the Kalman gain for which we reserve the letter K.

Because $\mathrm{E}(V_k(X_k - \hat{X}_{k|k-1})^t) = 0$, (5.50) implies that

$$\Gamma_k = B_k \Sigma_{k|k-1} B_k^t + S_k S_k^t , \tag{5.51}$$

where $\Sigma_{k|k-1}$ is our notation for the covariance of the state prediction error $X_k - \hat{X}_{k|k-1}$. Using the same principle,

$$\begin{aligned} \mathrm{E}(X_{k+1}\epsilon_k^t) &= A_k \, \mathrm{E}(X_k \epsilon_k^t) + R_k \, \mathrm{E}(U_k \epsilon_k^t) \\ &= A_k \Sigma_{k|k-1} B_k^t + R_k \, \mathrm{E}[U_k (X_k - \hat{X}_{k|k-1})^t] B_k^t \\ &= A_k \Sigma_{k|k-1} B_k^t , \end{aligned} \tag{5.52}$$

where we have used the fact that

$$U_k \perp \mathrm{span}(X_0, U_0, \ldots, U_{k-1}, V_0, \ldots, V_{k-1}) \supseteq \mathrm{span}(X_k, Y_0, \ldots, Y_{k-1}) .$$

Combining (5.51) and (5.52) yields the expression of the Kalman gain:

$$H_k = A_k \Sigma_{k|k-1} B_k^t \left\{ B_k \Sigma_{k|k-1} B_k^t + S_k S_k^t \right\}^{-1} . \tag{5.53}$$

As a final step, we now need to evaluate $\Sigma_{k+1|k}$. Because $X_{k+1} = A_k X_k + R_k U_k$ and $\mathrm{E}(X_k U_k^t) = 0$,

$$\mathrm{Cov}(X_{k+1}) = A_k \, \mathrm{Cov}(X_k) A_k^t + R_k R_k^t . \tag{5.54}$$

Similarly, the predicted state estimator follows (5.48) in which $\hat{X}_{k|k-1}$ and ϵ_k also are uncorrelated, as the former is an element of $\mathrm{span}(Y_0, \ldots, Y_{k-1})$. Hence

$$\mathrm{Cov}(\hat{X}_{k+1|k}) = A_k \, \mathrm{Cov}(\hat{X}_{k+1|k}) A_k^t + H_k \Gamma_k H_k^t . \tag{5.55}$$

Using (5.31),

$$\begin{aligned} \Sigma_{k+1|k} &= \mathrm{Cov}(X_{k+1}) - \mathrm{Cov}(\hat{X}_{k+1|k}) \\ &= A_k \Sigma_{k|k-1} A_k^t + R_k R_k^t - H_k \Gamma_k H_k^t , \end{aligned} \tag{5.56}$$

upon subtracting (5.55) from (5.54). Equation (5.56) is known as the *Riccati equation*. Collecting (5.45), (5.48), (5.51), (5.53), and (5.56), we obtain the standard form of the so-called *Kalman filter*, which corresponds to the prediction recursion.

Algorithm 5.2.9 (Kalman Prediction).

Initialization: $\hat{X}_{0|-1} = 0$ and $\Sigma_{0|-1} = \Sigma_\nu$.
Recursion: For $k = 0, \ldots n$,

$$\epsilon_k = Y_k - B_k \hat{X}_{k|k-1} , \qquad \text{innovation} \tag{5.57}$$

$$\Gamma_k = B_k \Sigma_{k|k-1} B_k^t + S_k S_k^t , \qquad \text{innovation cov.} \tag{5.58}$$

$$H_k = A_k \Sigma_{k|k-1} B_k^t \Gamma_k^{-1} , \qquad \text{Kalman Gain} \tag{5.59}$$

$$\hat{X}_{k+1|k} = A_k \hat{X}_{k|k-1} + H_k \epsilon_k , \qquad \text{predict. state estim.} \tag{5.60}$$

$$\Sigma_{k+1|k} = (A_k - H_k B_k) \Sigma_{k|k-1} A_k^t + R_k R_k^t . \qquad \text{predict. error cov.} \tag{5.61}$$

It is easily checked using (5.59) that (5.61) and (5.56) are indeed equivalent, the former being more suited for practical implementation, as it requires fewer matrix multiplications. Equation (5.61) however dissimulates the fact that $\Sigma_{k+1|k}$ indeed is a symmetric matrix. One can also check by simple substitution that Algorithm 5.2.9 is also equivalent to the application of the recursion derived in Proposition 5.2.3 for Gaussian models.

Remark 5.2.10. Evaluating the likelihood function for general linear state-space models is a complicated task. For Gaussian models however, ϵ_k and Γ_k entirely determine the first two moments, and hence the full conditional probability density function of Y_k given the previous observations $Y_0, \ldots, Y_{k-1}$, in the form

$$(2\pi)^{-d_y/2}|\Gamma_k|^{-1/2} \exp\left\{-\frac{1}{2}\epsilon_k^t \Gamma_k^{-1} \epsilon_k\right\} \tag{5.62}$$

where d_y is the dimension of the observations. As a consequence, the log-likelihood of observations up to index n may be computed as

$$\ell_n = -\frac{(n+1)d_y}{2}\log(2\pi) - \frac{1}{2}\sum_{k=0}^{n}\left\{\log|\Gamma_k| + \epsilon_k^t \Gamma_k^{-1}\epsilon_k\right\} , \tag{5.63}$$

which may be evaluated recursively (in n) using Algorithm 5.2.9. Equation (5.63), which is very important in practice for parameter estimation in state-space models, is easily recognized as a particular form of the general relation (3.29). ■

Example 5.2.11 (Random Walk Plus Noise Model). To illustrate Algorithm 5.2.9 on a simple example, consider the scalar random walk plus noise model defined by

$$\begin{aligned} X_{k+1} &= X_k + \sigma_u U_k , \\ Y_k &= X_k + \sigma_v V_k , \end{aligned}$$

where all variables are scalar. Applying the Kalman prediction equations yields, for $k \geq 1$,

$$\begin{aligned} \hat{X}_{k+1|k} &= \hat{X}_{k|k-1} + \frac{\Sigma_{k|k-1}}{\Sigma_{k|k-1} + \sigma_v^2}\left(Y_k - \hat{X}_{k|k-1}\right) \qquad (5.64) \\ &= (1 - a_k)\hat{X}_{k|k-1} + a_k Y_k , \\ \Sigma_{k+1|k} &= \Sigma_{k|k-1} + \sigma_u^2 - \frac{\Sigma_{k|k-1}^2}{\Sigma_{k|k-1} + \sigma_v^2} \\ &= \frac{\Sigma_{k|k-1}\sigma_v^2}{\Sigma_{k|k-1} + \sigma_v^2} + \sigma_u^2 \stackrel{\text{def}}{=} f(\Sigma_{k|k-1}) , \qquad (5.65) \end{aligned}$$

with the notation $a_k = \Sigma_{k|k-1}/(\Sigma_{k|k-1} + \sigma_v^2)$. This recursion is initialized by setting $\hat{X}_{0|-1} = 0$ and $\Sigma_{0|-1} = \Sigma_\nu$. For such a state-space model with time-independent parameters, it is interesting to consider the *steady-state solutions* for the prediction error covariance, that is, to solve for Σ in the equation

$$\Sigma = f(\Sigma) = \frac{\Sigma \sigma_v^2}{\Sigma + \sigma_v^2} + \sigma_u^2 \, .$$

Solving this equation for $\Sigma \geq 0$ yields

$$\Sigma_\infty = \frac{1}{2}\left(\sigma_u^2 + \sqrt{\sigma_u^4 + 4\sigma_u^2\sigma_v^2}\right) .$$

Straightforward calculations show that, for any $M < \infty$, $\sup_{0 \leq \Sigma \leq M} |\dot{f}(\Sigma)| < 1$. In addition, for $k \geq 1$, $(\Sigma_{k+1|k} - \Sigma_\infty)(\Sigma_{k|k-1} - \Sigma_\infty) \geq 0$. These remarks imply that $\Sigma_{k+1|k}$ always falls between $\Sigma_{k|k-1}$ and Σ_∞, and in particular that $\Sigma_{k+1|k} \leq \max(\Sigma_{1|0}, \Sigma_\infty)$. Because f is strictly contracting on any compact subset of $\mathbb{R}^+$, regardless of the value of Σ_ν, the coefficients $a_k = \Sigma_{k|k-1}/(\Sigma_{k|k-1} + \sigma_v^2)$ converge to

$$a_\infty = \frac{\Sigma_\infty}{\Sigma_\infty + \sigma_v^2} \, ,$$

and the mean squared error of the observation predictor $(Y_{k+1} - \hat{Y}_{k+1|k})$ converges to $\Sigma_\infty + \sigma_v^2$. ∎

Remark 5.2.12 (Algebraic Riccati Equation). The equation obtained by assuming that the model parameters A_k, B_k, $S_k S_k^t$, and $R_k R_k^t$ are time invariant, that is, do not depend on the index k, and then dropping indices in (5.56), is the so-called *algebraic Riccati equation* (ARE). Using (5.51) and (5.53), one finds that the ARE may be written

$$\Sigma = A\Sigma A^t + A\Sigma B^t(B\Sigma B^t + SS^t)^{-1} B\Sigma A^t + RR^t .$$

Conditions for the existence of a symmetric positive semi-definite solution to this equation, and conditions under which the recursive form (5.56) converges to such a solution can be found, for instance, in (Caines, 1988). ∎

5.2.3.2 Kalman Filtering

Algorithm 5.2.9 is primarily intended to compute the state predictor $\hat{X}_{k|k-1}$ and the covariance $\Sigma_{k|k-1}$ of the associated prediction error. It is of course possible to obtain a similar recursion for the filtered state estimator $\hat{X}_{k|k}$ and associated covariance matrix $\Sigma_{k|k}$.

Let us start once again with (5.40), applied with $Z = X_k$, to obtain

$$\hat{X}_{k|k} = \hat{X}_{k|k-1} + \mathrm{E}(X_k \epsilon_k^t)\Gamma_k^{-1}\epsilon_k = \hat{X}_{k|k-1} + K_k \epsilon_k \qquad (5.66)$$

where, this time, $K_k \stackrel{\text{def}}{=} \text{Cov}(X_k, \epsilon_k)\Gamma_k^{-1}$ is the filter version of the Kalman gain. The first term on the right-hand side of (5.66) may be rewritten as

$$\hat{X}_{k|k-1} = A_{k-1}\hat{X}_{k-1|k-1} + R_{k-1}\hat{U}_{k-1|k-1} = A_{k-1}\hat{X}_{k-1|k-1}, \tag{5.67}$$

where we have used

$$U_{k-1} \perp \text{span}(X_0, U_0, \ldots, U_{k-2}) \supseteq \text{span}(Y_0, \ldots, Y_{k-1}) .$$

Likewise, the second term on the right-hand side of (5.66) reduces to

$$K_k = \Sigma_{k|k-1} B_k^t \Gamma_k^{-1} , \tag{5.68}$$

because $\epsilon_k = B_k(X_k - \hat{X}_{k|k-1}) + S_k V_k$ with $\text{E}(X_k V_k^t) = 0$.

The only missing piece is the relationship between the error covariance matrices $\Sigma_{k|k}$ and $\Sigma_{k|k-1}$. The state equation $X_k = A_{k-1}X_{k-1} + R_{k-1}U_{k-1}$ and the state prediction equation $\hat{X}_{k|k-1} = A_{k-1}\hat{X}_{k-1|k-1}$ imply that

$$\begin{aligned} \text{Cov}(X_k) &= A_{k-1}\,\text{Cov}(X_{k-1})A_{k-1}^t + R_{k-1}R_{k-1}^t , \\ \text{Cov}(\hat{X}_{k|k-1}) &= A_{k-1}\,\text{Cov}(\hat{X}_{k-1|k-1})A_{k-1}^t , \end{aligned}$$

which, combined with (5.31), yield

$$\Sigma_{k|k-1} = A_{k-1}\Sigma_{k-1|k-1}A_{k-1}^t + R_{k-1}R_{k-1}^t . \tag{5.69}$$

By the same argument, the state recursion $X_k = A_{k-1}X_{k-1} + R_{k-1}U_{k-1}$ and the filter update $\hat{X}_{k|k} = A_{k-1}\hat{X}_{k-1|k-1} + K_k\epsilon_k$ imply that

$$\Sigma_{k|k} = A_{k-1}\Sigma_{k-1|k-1}A_{k-1}^t + R_{k-1}R_{k-1}^t - K_k\Gamma_k K_k^t . \tag{5.70}$$

These relations are summarized in the form of an algorithm.

Algorithm 5.2.13 (Kalman Filtering). For $k = 0, \ldots n$, do the following.

- If $k = 0$, set $\hat{X}_{k|k-1} = 0$ and $\Sigma_{k|k-1} = \Sigma_\nu$; otherwise, set

$$\begin{aligned} \hat{X}_{k|k-1} &= A_{k-1}\hat{X}_{k-1|k-1} , \\ \Sigma_{k|k-1} &= A_{k-1}\Sigma_{k-1|k-1}A_{k-1}^t + R_{k-1}R_{k-1}^t . \end{aligned}$$

- Compute

$\epsilon_k = Y_k - B_k\hat{X}_{k|k-1} ,$	innovation	(5.71)
$\Gamma_k = B_k\Sigma_{k|k-1}B_k^t + S_kS_k^t ,$	innovation cov.	(5.72)
$K_k = \Sigma_{k|k-1}B_k^t\Gamma_k^{-1} ,$	Kalman (filter.) gain	(5.73)
$\hat{X}_{k|k} = \hat{X}_{k|k-1} + K_k\epsilon_k ,$	filter. state estim.	(5.74)
$\Sigma_{k|k} = \Sigma_{k|k-1} - K_kB_k\Sigma_{k|k-1} .$	filter. error cov.	(5.75)

There are several different ways in which Algorithm 5.2.13 may be equivalently rewritten. In particular, it is possible to completely omit the prediction variables $\hat{X}_{k|k-1}$ and $\Sigma_{k|k-1}$ (Kailath *et al.*, 2000).

Remark 5.2.14. As already mentioned in Remark 5.2.5, the changes needed to adapt the filtering and prediction recursions to the case where the state and measurement noises are not assumed to be zero-mean are straightforward. The basic idea is to convert the state-space model by defining properly centered states and measurement variables. Define $X_k^* = X_k - \mathrm{E}[X_k]$, $U_k^* = U_k - \mathrm{E}[U_k]$, $Y_k^* = Y_k - \mathrm{E}[Y_k]$, and $V_k^* = V_k - \mathrm{E}[V_k]$; the expectations of the state and measurement variables can be computed recursively using

$$\begin{aligned} \mathrm{E}[X_{k+1}] &= A_k \,\mathrm{E}[X_k] + R_k \,\mathrm{E}[U_k] \;, \\ \mathrm{E}[Y_k] &= B_k \,\mathrm{E}[X_k] + S_k \,\mathrm{E}[V_k] \;. \end{aligned}$$

It is obvious that

$$\begin{aligned} X_{k+1}^* &= X_{k+1} - \mathrm{E}[X_{k+1}] = A_k(X_k - \mathrm{E}[X_k]) + R_k(U_k - \mathrm{E}[U_k]) \\ &= A_k X_k^* + R_k U_k^* \end{aligned}$$

and, similarly,

$$Y_k^* = Y_k - \mathrm{E}[Y_k] = B_k X_k^* + S_k V_k^* \;.$$

Thus $\{X_k^*, Y_k^*\}_{k\geq 0}$ follows the model defined by (5.42)–(5.43) with $X_0^* = 0$, $\mathrm{E}[U_k^*] = 0$ and $\mathrm{E}[V_k^*] = 0$. The Kalman recursions may be applied directly to compute for instance $\hat{X}^*_{k|k-1}$, the best linear estimate of X_k^* in terms of $Y_0^*, \ldots, Y_{k-1}^*$. The best linear estimate of X_k in terms of $Y_0, \ldots, Y_{k-1}$ is then given by

$$\hat{X}_{k|k-1} = \hat{X}^*_{k|k-1} + \mathrm{E}[X_k] \;.$$

All other quantities of interest can be treated similarly. ∎

5.2.4 Disturbance Smoothing

After revisiting Proposition 5.2.3, we are now ready to derive an alternative solution to the smoothing problem that will share the general features of Algorithm 5.2.4 (RTS smoothing) but operate only on the disturbance vectors U_k rather than on the states X_k. This second form of smoothing, which is more efficient in situations discussed at the beginning of Section 5.2.2, has been popularized under the name of *disturbance smoothing* by De Jong (1988), Kohn and Ansley (1989), and Koopman (1993). It is however a rediscovery of a technique known, in the engineering literature, as Bryson-Frazier (or BF) smoothing, named after Bryson and Frazier (1963)—see also (Kailath *et al.*, 2000, Section 10.2.2). The original arguments invoked by Bryson and Frazier (1963) were however very different from the ones discussed here and the use of the innovation approach to obtain smoothing estimates was initiated by Kailath and Frost (1968).

Recall that for $k = 0, \ldots, n-1$ we denote by $\hat{U}_{k|n}$ the smoothed disturbance estimator, i.e., the best linear prediction of the disturbance U_k in terms of the observations $Y_0, \ldots, Y_n$. The additional notation

$$\Xi_{k|n} \stackrel{\text{def}}{=} \operatorname{Cov}(U_k - \hat{U}_{k|n})$$

will also be used. We first state the complete algorithm before proving that it is actually correct.

Algorithm 5.2.15 (Disturbance Smoother).

Forward filtering: Run the Kalman filter (Algorithm 5.2.9) and store for $k = 0, \ldots, n$ the innovation ϵ_k, the inverse innovation covariance Γ_k^{-1}, the state prediction error covariance $\Sigma_{k|k-1}$, and

$$\Lambda_k \stackrel{\text{def}}{=} A_k - H_k B_k ,$$

where H_k is the Kalman (prediction) gain.

Backward smoothing: For $k = n-1, \ldots, 0$, compute

$$p_k = \begin{cases} B_n^t \Gamma_n^{-1} \epsilon_n & \text{for } \quad k = n-1, \\ B_{k+1}^t \Gamma_{k+1}^{-1} \epsilon_{k+1} + \Lambda_{k+1}^t p_{k+1} & \text{otherwise,} \end{cases} \tag{5.76}$$

$$C_k = \begin{cases} B_n^t \Gamma_n^{-1} B_n & \text{for } \quad k = n-1, \\ B_{k+1}^t \Gamma_{k+1}^{-1} B_{k+1} + \Lambda_{k+1}^t C_{k+1} \Lambda_{k+1} & \text{otherwise,} \end{cases} \tag{5.77}$$

$$\hat{U}_{k|n} = R_k^t p_k , \tag{5.78}$$

$$\Xi_{k|n} = I - R_k^t C_k R_k . \tag{5.79}$$

Initial Smoothed State Estimator: Compute

$$\hat{X}_{0|n} = \Sigma_\nu \left(B_0^t \Gamma_0^{-1} \epsilon_0 + \Lambda_0^t p_0 \right) , \tag{5.80}$$

$$\Sigma_{0|n} = \Sigma_\nu - \Sigma_\nu \left[B_0^t \Gamma_0^{-1} B_0 + \Lambda_0^t C_0 \Lambda_0 \right] \Sigma_\nu . \tag{5.81}$$

Smoothed State Estimator: For $k = 0, \ldots n-1$,

$$\hat{X}_{k+1|n} = A_k \hat{X}_{k|n} + R_k \hat{U}_{k|n} , \tag{5.82}$$

$$\begin{aligned} \Sigma_{k+1|n} &= A_k \Sigma_{k|n} A_k^t + R_k \Xi_{k|n} R_k^t \\ &\quad - A_k \Sigma_{k|k-1} \Lambda_k^t C_k R_k R_k^t - R_k R_k^t C_k \Lambda_k \Sigma_{k|k-1} A_k^t . \end{aligned} \tag{5.83}$$

Algorithm 5.2.15 is quite complex, starting with an application of the Kalman prediction recursion, followed by a backward recursion to obtain the smoothed disturbances and then a final forward recursion needed to evaluate the smoothed states. The proof below is split into two parts that concentrate on each of the two latter aspects of the algorithm.

Proof (Backward Smoothing). We begin with the derivation of the equations needed for computing the smoothed disturbance estimator $\hat{U}_{k|n}$ for $k = n - 1$ down to 0. As previously, it is advantageous to use the innovation sequence $\{\epsilon_0, \ldots, \epsilon_n\}$ instead of the correlated observations $\{Y_0, \ldots, Y_n\}$. Using (5.40), we have

$$\hat{U}_{k|n} = \sum_{i=0}^{n} \mathrm{E}(U_k \epsilon_i^t) \Gamma_i^{-1} \epsilon_i = \sum_{i=k+1}^{n} \mathrm{E}(U_k \epsilon_i^t) \Gamma_i^{-1} \epsilon_i \;, \tag{5.84}$$

where the fact that

$$U_k \perp \operatorname{span}\{Y_0, \ldots Y_k\} = \operatorname{span}\{\epsilon_0, \ldots, \epsilon_k\} \;,$$

has been used to obtain the second expression. We now prove by induction that for any $i = k+1, \ldots, n$,

$$\mathrm{E}[U_k (X_i - \hat{X}_{i|i-1})^t] = \begin{cases} R_k^t \;, & i = k+1 \;, \\ R_k^t \Lambda_{k+1}^t \; \Lambda_{k+2}^t \; \ldots \; \Lambda_{i-1}^t \;, & i \geq k+2 \;, \end{cases} \tag{5.85}$$

$$\mathrm{E}(U_k \epsilon_i^t) = \begin{cases} R_k^t B_{k+1}^t \;, & i = k+1 \;, \\ R_k^t \Lambda_{k+1}^t \; \Lambda_{k+2}^t \; \ldots \; \Lambda_{i-1}^t B_i^t \;, & i \geq k+2 \;. \end{cases} \tag{5.86}$$

First note that

$$\begin{aligned} \mathrm{E}(U_k \epsilon_{k+1}^t) &= \mathrm{E}[U_k (X_{k+1} - \hat{X}_{k+1|k})^t] B_{k+1}^t \\ &= \mathrm{E}(U_k X_{k+1}^t) B_{k+1}^t = R_k^t B_{k+1}^t \;, \end{aligned}$$

using (5.45) and the orthogonality relations $U_k \perp V_{k+1}$, $U_k \perp \operatorname{span}(Y_0, \ldots, Y_k)$ and $U_k \perp X_k$. Now assume that (5.85)–(5.86) hold for some $i \geq k+1$. Combining the state equation (5.42) and the prediction update equation (5.48), we obtain

$$X_{i+1} - \hat{X}_{i+1|i} = \Lambda_i (X_i - \hat{X}_{i|i-1}) + R_i U_i - H_i S_i V_i \;. \tag{5.87}$$

Because $\mathrm{E}(U_k U_i^t) = 0$ and $\mathrm{E}(U_k V_i^t) = 0$, the induction assumption implies that

$$\mathrm{E}[U_k (X_{i+1} - \hat{X}_{i+1|i})^t] = \mathrm{E}[U_k (X_i - \hat{X}_{i|i-1})^t] \Lambda_i^t = R_k^t \Lambda_{k+1}^t \; \Lambda_{k+2}^t \; \cdots \; \Lambda_i^t \;. \tag{5.88}$$

Proceeding as in the case $i = k$ above,

$$\mathrm{E}(U_k \epsilon_{i+1}^t) = \mathrm{E}[U_k (X_{i+1} - \hat{X}_{i+1|i})^t] B_{i+1}^t = R_k^t \Lambda_{k+1}^t \; \Lambda_{k+2}^t \; \cdots \; \Lambda_i^t B_{i+1}^t \;, \tag{5.89}$$

which, by induction, shows that (5.85)–(5.86) hold for all indices $i \geq k+1$. Plugging (5.86) into (5.84) yields

$$\hat{U}_{k|n} = R_k^t \left(B_{k+1}^t \Gamma_{k+1}^{-1} \epsilon_{k+1} + \sum_{i=k+2}^{n} \Lambda_{k+1}^t \ldots \Lambda_{i-1}^t B_i^t \Gamma_i^{-1} \epsilon_i \right) , \tag{5.90}$$

where the term between parentheses is easily recognized as p_k defined recursively by (5.76), thus proving (5.78).

To compute the smoothed disturbance error covariance $\Xi_{k|n}$, we apply once again (5.41) to obtain

$$\begin{aligned}\Xi_{k|n} &= \operatorname{Cov}(U_k) - \operatorname{Cov}\left(\hat{U}_{k|n}\right) \\ &= I - \sum_{i=k+1}^{n} \mathrm{E}(U_k \epsilon_i^t) \Gamma_i^{-1} \mathrm{E}(\epsilon_i U_k^t) \\ &= I - R_k^t \left(B_{k+1}^t \Gamma_{k+1}^{-1} B_{k+1} \right. \\ &\qquad \left. + \sum_{i=k+2}^{n} \Lambda_{k+1}^t \ \ldots \ \Lambda_{i-1}^t B_i^t \Gamma_i^{-1} B_i \Lambda_{i-1} \ \ldots \ \Lambda_{k+1} \right) R_k \ ,\end{aligned} \tag{5.91}$$

where I is the identity matrix with dimension that of the disturbance vector and (5.89) has been used to obtain the last expression. The term in parentheses in (5.91) is recognized as C_k defined by (5.77), and (5.79) follows. □

Proof (Smoothed State Estimation). The key ingredient here is the following set of relations:

$$\mathrm{E}[X_k(X_i - \hat{X}_{i|i-1})^t] = \begin{cases} \Sigma_{k|k-1} \ , & i = k \ , \\ \Sigma_{k|k-1} \Lambda_k^t \ \Lambda_{k+1}^t \ \ldots \ \Lambda_{i-1}^t \ , & i \geq k+1 \ , \end{cases} \tag{5.92}$$

$$\mathrm{E}(X_k \epsilon_i^t) = \begin{cases} \Sigma_{k|k-1} B_k^t \ , & i = k \ , \\ \Sigma_{k|k-1} \Lambda_k^t \ \Lambda_{k+1}^t \ \ldots \ \Lambda_{i-1}^t B_i^t \ , & i \geq k+1 \ , \end{cases} \tag{5.93}$$

which may be proved by induction exactly like (5.85)–(5.86).

Using (5.40) as usual, the minimum mean squared error linear predictor of the initial state X_0 in terms of the observations $Y_0, \ldots, Y_n$ may be expressed as

$$\hat{X}_{0|n} = \sum_{i=0}^{n} \mathrm{E}(X_0 \epsilon_i^t) \Gamma_i^{-1} \epsilon_i \ . \tag{5.94}$$

Hence by direct application of (5.93),

$$\hat{X}_{0|n} = \Sigma_\nu \left(B_0^t \Gamma_0^{-1} \epsilon_0 + \sum_{i=1}^{n} \Lambda_0^t \ \ldots \ \Lambda_{i-1}^t B_i^t \Gamma_i^{-1} \epsilon_i \right) \ , \tag{5.95}$$

proving (5.80). Proceeding as for (5.91), the expression for the smoothed initial state error covariance in (5.81) follows from (5.41).

The update equation (5.82) is a direct consequence of the linearity of the projection operator applied to the state equation (5.42). Finally, to prove (5.83), first combine the state equation (5.42) with (5.82) to obtain

$$\operatorname{Cov}(X_{k+1} - \hat{X}_{k+1|n}) = \operatorname{Cov}[A_k(X_k - \hat{X}_{k|n}) + R_k(U_k - \hat{U}_{k|n})] =$$
$$A_k \Sigma_{k|n} A_k^t + R_k \Xi_{k|n} R_k^t - A_k \operatorname{E}(X_k \hat{U}_{k|n}^t) R_k^t - R_k \operatorname{E}(\hat{U}_{k|n} X_k^t) A_k^t \,, \quad (5.96)$$

where the remark that $\operatorname{E}[\hat{X}_{k|n}(U_k - \hat{U}_{k|n})^t] = 0$, because $\hat{X}_{k|n}$ belongs to $\operatorname{span}(Y_0, \ldots, Y_n)$, has been used to obtain the second expression. In order to compute $\operatorname{E}(X_k \hat{U}_{k|n}^t)$ we use (5.90), writing

$$\operatorname{E}(X_k \hat{U}_{k|n}^t) = \operatorname{E}(X_k \epsilon_{k+1}^t) \Gamma_{k+1}^{-1} B_{k+1} R_k +$$
$$\sum_{i=k+2}^{n} \operatorname{E}(X_k \epsilon_i^t) \Gamma_i^{-1} B_i \Lambda_{i-1} \ldots \Lambda_{k+1} R_k \,. \quad (5.97)$$

Finally, invoke (5.93) to obtain

$$\operatorname{E}(X_k \hat{U}_{k|n}^t) = \Sigma_{k|k-1} \Lambda_k^t B_{k+1}^t \Gamma_{k+1}^{-1} B_{k+1} R_k +$$
$$\sum_{i=k+2}^{n} \Sigma_{k|k-1} \Lambda_k^t \Lambda_{k+1}^t \ldots \Lambda_{i-1}^t B_i^t \Gamma_i^{-1} B_i \Lambda_{i-1} \ldots \Lambda_{k+1} R_k \,,$$

which may be rewritten as

$$\operatorname{E}(X_k \hat{U}_{k|n}^t) = \Sigma_{k|k-1} \Lambda_k^t C_k R_k \,. \quad (5.98)$$

Equation (5.83) then follows from (5.96). □

Remark 5.2.16. There are a number of situations where computing the best linear prediction of the state variables is the only purpose of the analysis, and computation of the error covariance $\operatorname{Cov}(X_k - \hat{X}_{k|n})$ is not required. Algorithm 5.2.15 may then be substantially simplified because (5.77), (5.79), (5.81), and (5.83) can be entirely skipped. Storage of the prediction error covariance matrices $\Sigma_{k|k-1}$ during the initial Kalman filtering pass is also not needed anymore. ■

Remark 5.2.17. An important quantity in the context of parameter estimation (to be discussed in Section 10.4 of Chapter 10) is the one-step posterior cross-covariance

$$C_{k,k+1|n} \stackrel{\text{def}}{=} \operatorname{E}\left[\left(X_k - \hat{X}_{k|n}\right) \left(X_{k+1} - \hat{X}_{k+1|n}\right)^t \middle| Y_{0:n} \right] . \quad (5.99)$$

This is a quantity that can readily be evaluated during the final forward recursion of Algorithm 5.2.15. Indeed, from (5.42)–(5.82),

$$X_{k+1} - \hat{X}_{k+1|n} = A_k \left(X_k - \hat{X}_{k|n}\right) + R_k \left(U_k - \hat{U}_{k|n}\right) .$$

Hence

$$C_{k,k+1|n} = \Sigma_{k|n} A_k^t - \operatorname{E}\left(X_k \hat{U}_{k|n}^t\right) R_k^t \,,$$

where the fact that $E(X_k U_k^t) = 0$ has been used. Using (5.98) then yields

$$C_{k,k+1|n} = \Sigma_{k|n} A_k^t - \Sigma_{k|k-1} \Lambda_k^t C_k R_k R_k^t \,. \quad (5.100)$$

■

5.2.5 The Backward Recursion and the Two-Filter Formula

Notice that up to now, we have not considered the backward functions $\beta_{k|n}$ in the case of Gaussian linear state-space models. In particular, and although the details of both approaches differ, the smoothing recursions discussed in Sections 5.2.1 and 5.2.4 are clearly related to the general principle of backward Markovian smoothing discussed in Section 3.3.2 and do not rely on the forward-backward decomposition discussed in Section 3.2.

A first terminological remark is that although major sources on Gaussian linear models never mention the forward-backward decomposition, it is indeed known under the name of *two-filter formula* (Fraser and Potter, 1969; Kitagawa, 1996; Kailath *et al.*, 2000, Section 10.4). A problem however is that, as noted in Chapter 3, the backward function $\beta_{k|n}$ is not directly interpretable as a probability distribution (recall for instance that the initialization of the backward recursion is $\beta_{n|n}(x) = 1$ for all $x \in \mathsf{X}$). A first approach consists in introducing some additional assumptions on the model that ensure that $\beta_{k|n}(x)$, suitably normalized, can indeed be interpreted as a probability density function. The backward recursion can then be interpreted as the Kalman prediction algorithm, applied backwards in time, starting from the end of the data record (Kailath *et al.*, 2000, Section 10.4).

A different option, originally due to Mayne (1966) and Fraser and Potter (1969), consists in deriving the backward recursion using a reparameterization of the backward functions $\beta_{k|n}$, which is robust to the fact that $\beta_{k|n}(x)$ may not be integrable over X. This solution has the advantage of being generic in that it does not require any additional assumptions on the model, other than $S_k S_k^t$ being invertible. The drawback is that we cannot simply invoke a variant of Algorithm 5.2.3 but need to derive a specific form of the backward recursion using a different parameterization. This implementation of the backward recursion (which could also be used, with some minor modifications, for usual forward prediction) is referred to as the *information form* of the Kalman filtering and prediction recursions (Anderson and Moore, 1979, Section 6.3; Kailath *et al.*, 2000, Section 9.5.2). In the time series literature, this method is also sometimes used as a tool to compute the smoothed estimates when using so-called *diffuse priors* (usually for X_0), which correspond to the notion of improper flat distributions to be discussed below.

5.2.5.1 The Information Parameterization

The main ingredient of what follows consists in revisiting the calculation of the posterior distribution of the unobserved component X in the basic Gaussian linear model

$$Y = BX + V \; .$$

Indeed, in order to prove Proposition 5.2.2, we could have followed a very different route: assuming that both Σ_V and $\mathrm{Cov}(Y) = B^t \Sigma_X B + \Sigma_V$ are full rank matrices, the posterior probability density function of X given Y, which we denote by $p(x|y)$, is known by Bayes' rule to be proportional to the product of the prior $p(x)$ on X and the conditional probability density function $p(y|x)$ of Y given X, that is,

$$p(x|y) \propto \exp\left\{-\frac{1}{2}\left[(y-Bx)^t \Sigma_V^{-1}(y-Bx) + (x-\mu_X)^t \Sigma_X^{-1}(x-\mu_X)\right]\right\} , \tag{5.101}$$

where the symbol $\propto$ indicates proportionality up to a constant that does not depend on the variable x. Note that this normalizing constant could easily be determined in the current case because we know that $p(x|y)$ corresponds to a multivariate Gaussian probability density function. Hence, to fully determine $p(x|y)$, we just need to rewrite (5.101) as a quadratic form in x:

$$p(x|y) \propto \exp\left\{\{-\frac{1}{2}\left[x^t(B^t \Sigma_V^{-1} B + \Sigma_X^{-1})x - x^t(B^t \Sigma_V^{-1} y + \Sigma_X^{-1}\mu_X)\right.\right.$$
$$\left.\left. - (B^t \Sigma_V^{-1} y + \Sigma_X^{-1}\mu_X)^t x\right]\right\} , \tag{5.102}$$

that is,

$$p(x|y) \propto \exp\left\{-\frac{1}{2}\left[(x-\mu_{X|Y})^t \Sigma_{X|Y}^{-1}(x-\mu_{X|Y})\right]\right\} , \tag{5.103}$$

where

$$\mu_{X|Y} = \Sigma_{X|Y}^{-1}\left(B^t \Sigma_V^{-1} y + \Sigma_X^{-1}\mu_X\right) , \tag{5.104}$$
$$\Sigma_{X|Y} = \left(B^t \Sigma_V^{-1} B + \Sigma_X^{-1}\right)^{-1} . \tag{5.105}$$

Note that in going from (5.102) to (5.104), we have used once again the fact that $p(x|y)$ only needs be determined up to a normalization factor, whence terms that do not depend on x can safely be ignored.

As a first consequence, (5.105) and (5.104) are alternate forms of equations (5.17) and (5.16), respectively, which we first met in Proposition 5.2.2. The fact that (5.17) and (5.105) coincide is a well-known result from matrix theory known as the *matrix inversion lemma* that we could have invoked directly to obtain (5.104) and (5.105) from Proposition 5.2.2. This simple rewriting of the conditional mean and covariance in the Gaussian linear model is however not the only lesson that can be learned from (5.104) and (5.105). In particular, a very natural parameterization of the Gaussian distribution in this context consists in considering the inverse of the covariance matrix $\Pi = \Sigma^{-1}$ and the vector $\kappa = \Pi\mu$ rather than the covariance Σ and the mean vector μ. Both of these parameterizations are of course fully equivalent when the covariance matrix Σ is invertible. In some contexts, the inverse covariance matrix Π is referred to as the *precision matrix*, but in the filtering context the use of this parameterization is generally associated with the word *information* (in reference to the fact that in a Gaussian experiment, the inverse of the covariance matrix is precisely the Fisher information matrix associated with the estimation of the mean). We shall adopt this terminology and refer to the use of κ and Π as parameters of the Gaussian distribution as the *information parameterization.* Note that because a Gaussian probability density function $p(x)$ with mean μ and covariance Σ may be written

$$\begin{aligned} p(x) &\propto \exp\left\{-\frac{1}{2}\left[x^t\Sigma^{-1}x - 2x^t\Sigma^{-1}\mu\right]\right\} \\ &= \exp\left\{-\frac{1}{2}\left[\operatorname{trace}\left(xx^t\Sigma^{-1}\right) - 2x^t\Sigma^{-1}\mu\right]\right\} , \end{aligned}$$

$\Pi = \Sigma^{-1}$ and $\kappa = \Pi\mu$ also form the *natural parameterization* of the multivariate normal, considered as a member of the exponential family of distributions (Lehmann and Casella, 1998).

5.2.5.2 The Gaussian Linear Model (Again!)

We summarize our previous findings—Eqs. (5.104) and (5.105)—in the form of the following alternative version of Proposition 5.2.2,

Proposition 5.2.18 (Conditioning in Information Parameterization). *Let*

$$Y = BX + V ,$$

where X and V are two independent Gaussian random vectors such that, in information parameterization, $\kappa_X = \operatorname{Cov}(X)^{-1}\operatorname{E}(X)$, $\Pi_X = \operatorname{Cov}(X)^{-1}$, $\Pi_V = \operatorname{Cov}(V)^{-1}$ and $\kappa_V = \operatorname{E}(V) = 0$, B being a deterministic matrix. Then

$$\kappa_{X|Y} = \kappa_X + B^t\Pi_V Y , \tag{5.106}$$

$$\Pi_{X|Y} = \Pi_X + B^t\Pi_V B , \tag{5.107}$$

where $\kappa_{X|Y} = \operatorname{Cov}(X|Y)^{-1}\operatorname{E}(X|Y)$ and $\Pi_{X|Y} = \operatorname{Cov}(X|Y)^{-1}$.

If the matrices Π_X, Π_V, or $\Pi_{X|Y}$ are not full rank matrices, (5.106) *and* (5.107) *can still be interpreted in a consistent way using the concept of improper (flat) distributions.*

Equations (5.106) and (5.107) deserve no special comment as they just correspond to a restatement of (5.104) and (5.105), respectively. The last sentence of Proposition 5.2.18 is a new element, however. To understand the point, consider (5.101) again and imagine what would happen if $p(x)$, for instance, was assumed to be constant. Then (5.102) would reduce to

$$p(x|y) \propto \exp\left\{-\frac{1}{2}\left[x^t(B^t\Sigma_V^{-1}B)x - x^t(B^t\Sigma_V^{-1}y) - (B^t\Sigma_V^{-1}y)^t x\right]\right\} , \tag{5.108}$$

which corresponds to a perfectly valid Gaussian distribution, when viewed as a function of x, at least when $B^t \Sigma_V^{-1} B$ has full rank. The only restriction is that there is of course no valid probability density function $p(x)$ that is constant on X. This practice is however well established in Bayesian estimation (to be discussed in Chapter 13.1.1) where such a choice of $p(x)$ is referred to as using an *improper* flat prior. The interpretation of (5.108) is then that under an (improper) flat prior on Y, the posterior mean of X given Y is

$$\left(B^t \Sigma_V^{-1} B\right)^{-1} B^t \Sigma_V^{-1} Y , \tag{5.109}$$

which is easily recognized as the (deterministic) optimally weighted least-squares estimate of x in the linear regression model $Y = Bx + V$. The important message here is that (5.109) can be obtained direct from (5.106) by assuming that Π_X is the null matrix and κ_X the null vector. Hence Proposition 5.2.18 also covers the case where X has an improper flat distribution, which is handled simply by setting the precision matrix Π_X and the vector κ_X equal to 0. A more complicated situation is illustrated by the following example.

Example 5.2.19. Assume that the linear model is such that X is bivariate Gaussian and the observation Y is scalar with

$$B = \begin{pmatrix} 1 & 0 \end{pmatrix} \qquad \text{and} \qquad \operatorname{Cov}(V) = \sigma^2 .$$

Proposition 5.2.18 asserts that the posterior parameters are then given by

$$\kappa_{X|Y} = \kappa_X + \begin{pmatrix} \sigma^{-2} Y \\ 0 \end{pmatrix} , \tag{5.110}$$

$$\Pi_{X|Y} = \Pi_X + \begin{pmatrix} \sigma^{-2} & 0 \\ 0 & 0 \end{pmatrix} . \tag{5.111}$$

In particular, if the prior on X is improper flat, then (5.110) and (5.111) simply mean that the posterior distribution of the first component of X given Y is Gaussian with mean Y and variance σ^2, whereas the posterior on the second component is also improper flat. ■

In the above example, what is remarkable is not the result itself, which is obvious, but the fact that it can be obtained by application of a single set of formulas that are valid irrespectively of the fact that some distributions are improper. In more general situations, directions that are in the null space of $\Pi_{X|Y}$ form a subspace where the resulting posterior is improper flat, whereas the posterior distribution of X projected on the image $\Pi_{X|Y}$ is a valid Gaussian distribution.

The information parameterization is ambivalent because it can be used both as a Gaussian prior density function as in Proposition 5.2.18 but also as an observed likelihood. There is nothing magic here but simply the observation that as we (i) allow for improper distributions and (ii) omit the normalization factors, Gaussian priors and likelihood are equivalent. The following lemma is a complement to Proposition 5.2.18, which will be needed below.

Lemma 5.2.20. *Up to terms that do not depend on x,*

$$\int \exp\left\{-\frac{1}{2}\left[(y-Bx)^t\Sigma^{-1}(y-Bx)\right]\right\}\exp\left\{-\frac{1}{2}\left[(y^t\Pi y-2y^t\kappa)\right]\right\}dy$$
$$\propto \exp\left\{-\frac{1}{2}\left[x^tB^t(I+\Pi\Sigma)^{-1}\Pi Bx-2x^tB^t(I+\Pi\Sigma)^{-1}\kappa\right]\right\}, \quad (5.112)$$

where I denotes the identity matrix of suitable dimension.

Proof. The left-hand side of (5.112), which we denote by $p(x)$, may be rewritten as

$$p(x) = \exp\left\{-\frac{1}{2}xB^t\Sigma^{-1}Bx\right\}\times$$
$$\int \exp -\frac{1}{2}\left[y^t(\Pi+\Sigma^{-1})y-2y^t(\kappa+\Sigma^{-1}Bx)\right]dy\,. \quad (5.113)$$

Completing the square, the bracketed term in the integrand of (5.113) may be written

$$\left\{y-(\Pi+\Sigma^{-1})^{-1}(\kappa+\Sigma^{-1}Bx)\right\}^t(\Pi+\Sigma^{-1})$$
$$\times\left\{y-(\Pi+\Sigma^{-1})^{-1}(\kappa+\Sigma^{-1}Bx)\right\}$$
$$-(\kappa+\Sigma^{-1}Bx)^t(\Pi+\Sigma^{-1})^{-1}(\kappa+\Sigma^{-1}Bx)\,. \quad (5.114)$$

The exponent of $-1/2$ times the first two lines of (5.114) integrates to a constant (or, rather, a number not depending on x), as it is recognized as a Gaussian probability density function. Thus

$$p(x)\propto \exp -\frac{1}{2}\left\{\left[-2x^tB^t\Sigma^{-1}(\Pi+\Sigma^{-1})^{-1}\kappa\right.\right.$$
$$\left.+x^tB^t\left(\Sigma^{-1}-\Sigma^{-1}(\Pi+\Sigma^{-1})^{-1}\Sigma^{-1}\right)Bx\right\}, \quad (5.115)$$

where terms that do not depend on x have been dropped. Equation (5.112) follows from the equalities $\Sigma^{-1}(\Pi+\Sigma^{-1})^{-1}=(I+\Pi\Sigma)^{-1}$ and

$$\Sigma^{-1}-\Sigma^{-1}(\Pi+\Sigma^{-1})^{-1}\Sigma^{-1}$$
$$=\Sigma^{-1}(\Pi+\Sigma^{-1})^{-1}\left[(\Pi+\Sigma^{-1})-\Sigma^{-1}\right]=(I+\Pi\Sigma)^{-1}\Pi\,.$$

Note that the last identity is the matrix inversion lemma that we already met, as $(I+\Pi\Sigma)^{-1}\Pi=(\Pi^{-1}+\Sigma)^{-1}$. Using this last form however is not a good idea in general, however, as it obviously does not apply in cases where Π is non-invertible. □

5.2.5.3 The Backward Recursion

The question now is, what is the link between our original problem, which consists in implementing the backward recursion in Gaussian linear state-space models, and the information parameterization discussed in the previous section? The connection is the fact that the backward functions defined by (3.16) do not correspond to probability measures. More precisely, $\beta_{k|n}(X_k)$ defined by (3.16) is the conditional density of the "future" observations $Y_{k+1}, \ldots, Y_n$ given X_k. For Gaussian linear models, we know from Proposition 5.2.18 that this density is Gaussian and hence that $\beta_{k|n}(x)$ has the form of a Gaussian likelihood,

$$p(y|x) \propto \exp -\frac{1}{2} \left[(y - Mx)^t \Sigma^{-1} (y - Mx) \right] ,$$

for some M and Σ given by (5.16) and (5.17). Proceeding as previously, this equation can be put in the same form as (5.108) (replacing B and Σ_V by M and Σ, respectively). Hence, a possible interpretation of $\beta_{k|n}(x)$ is that it corresponds to the posterior distribution of X_k given $Y_{k+1}, \ldots, Y_n$ *in the pseudo-model where X_k is assumed to have an improper flat prior distribution.* According to the previous discussion, $\beta_{k|n}(x)$ itself may not correspond to a valid Gaussian distribution unless one can guarantee that $M^t \Sigma^{-1} M$ is a full rank matrix. In particular, recall from Section 3.2.1 that the backward recursion is initialized by setting $\beta_{n|n}(x) = 1$, and hence $\beta_{n|n}$ never is a valid Gaussian distribution.

The route from now on is clear: in order to implement the backward recursion, one needs to define a set of information parameters corresponding to $\beta_{k|n}$ and derive (backward) recursions for these parameters based on Proposition 5.2.18. We will denote by $\kappa_{k|n}$ and $\Pi_{k|n}$ the information parameters (precision matrix times mean and precision matrix) corresponding to $\beta_{k|n}$ for $k = n$ down to 0 where, by definition, $\kappa_{n|n} = 0$ and $\Pi_{n|n} = 0$. It is important to keep in mind that $\kappa_{k|n}$ and $\Pi_{k|n}$ define the backward function $\beta_{k|n}$ *only up to an unknown constant.* The best we can hope to determine is

$$\frac{\beta_{k|n}(x)}{\int \beta_{k|n}(x)\, dx} ,$$

by computing the Gaussian normalization factor in situations where $\Pi_{k|n}$ is a full rank matrix. But this normalization is not more legitimate or practical than other ones, and it is preferable to consider that $\beta_{k|n}$ will be determined up to a constant only. In most situations, this will be a minor concern, as formulas that take into account this possible lack of normalization, such as (3.21), are available.

Proposition 5.2.21 (Backward Information Recursion). *Consider the Gaussian linear state-space model (5.11)–(5.12) and assume that $S_k S_k^t$ has full rank for all $k \geq 0$. The information parameters $\kappa_{k|n}$ and $\Pi_{k|n}$, which determine $\beta_{k|n}$ (up to a constant), may be computed by the following recursion.*

Initialization: Set $\kappa_{n|n} = 0$ *and* $\Pi_{n|n} = 0$.
Backward Recursion: For $k = n - 1$ *down to 0,*

$$\tilde{\kappa}_{k+1|n} = B_{k+1}^t \left(S_{k+1} S_{k+1}^t\right)^{-1} Y_{k+1} + \kappa_{k+1|n} \,, \tag{5.116}$$

$$\tilde{\Pi}_{k+1|n} = B_{k+1}^t \left(S_{k+1} S_{k+1}^t\right)^{-1} B_{k+1} + \Pi_{k+1|n} \,, \tag{5.117}$$

$$\kappa_{k|n} = A_k^t \left(I + \tilde{\Pi}_{k+1|n} R_k R_k^t\right)^{-1} \tilde{\kappa}_{k+1|n} \,, \tag{5.118}$$

$$\Pi_{k|n} = A_k^t \left(I + \tilde{\Pi}_{k+1|n} R_k R_k^t\right)^{-1} \tilde{\Pi}_{k+1|n} A_k \,. \tag{5.119}$$

Proof. The initialization of Proposition 5.2.21 has already been discussed and we just need to check that (5.116)–(5.119) correspond to an implementation of the general backward recursion (Proposition 3.2.1).

We split this update in two parts and first consider computing

$$\tilde{\beta}_{k+1|n}(x) \propto g_{k+1}(x) \beta_{k+1|n}(x) \tag{5.120}$$

from $\beta_{k+1|n}$. Equation (5.120) may be interpreted as the posterior distribution of X in the pseudo-model in which X has a (possibly improper) prior distribution $\beta_{k+1|n}$ (with information parameters $\kappa_{k+1|n}$ and $\Pi_{k+1|n}$) and

$$Y = B_{k+1} X + S_{k+1} V$$

is observed, where V is independent of X. Equations(5.116)–(5.117) thus correspond to the information parameterization of $\tilde{\beta}_{k+1|n}$ by application of Proposition 5.2.18.

From (3.19) we then have

$$\beta_{k|n}(x) = \int Q_k(x, dx') \tilde{\beta}_{k+1|n}(x') \,, \tag{5.121}$$

where we use the notation Q_k rather than Q to emphasize that we are dealing with possibly non-homogeneous models. Given that Q_k is a Gaussian transition density function corresponding to (5.12), (5.121) may be computed explicitly by application of Lemma 5.2.20 which gives (5.118) and (5.119). □

While carrying out the backward recursion according to Proposition 5.2.21, it is also possible to simultaneously compute the marginal smoothing distribution by use of (3.21).

Algorithm 5.2.22 (Forward-Backward Smoothing).

Forward Recursion: Perform Kalman filtering according to Algorithm 5.2.13 and store the values of $\hat{X}_{k|k}$ and $\Sigma_{k|k}$.

Backward Recursion: Compute the backward recursion, obtaining for each k the mean and covariance matrix of the smoothed estimate as

$$\hat{X}_{k|n} = \hat{X}_{k|k} + \Sigma_{k|k} \left(I + \Pi_{k|n} \Sigma_{k|k}\right)^{-1} \left(\kappa_{k|n} - \Pi_{k|n} \hat{X}_{k|k}\right) \,, \tag{5.122}$$

$$\Sigma_{k|n} = \Sigma_{k|k} - \Sigma_{k|k} \left(I + \Pi_{k|n} \Sigma_{k|k}\right)^{-1} \Pi_{k|n} \Sigma_{k|k} \,. \tag{5.123}$$

Proof. These two equations can be obtained exactly as in the proof of Lemma 5.2.20, replacing $(y - Bx)^t \Sigma^{-1}(y - Bx)$ by $(x - \mu)^t \Sigma^{-1}(x - \mu)$ and applying the result with $\mu = \hat{X}_{k|k}$, $\Sigma = \Sigma_{k|k}$, $\kappa = \kappa_{k|n}$ and $\Pi = \Pi_{k|n}$. If $\Pi_{k|n}$ is invertible, (5.122) and (5.123) are easily recognized as the application of Proposition 5.2.2 with $B = I$, $\mathrm{Cov}(V) = \Pi_{k|n}^{-1}$, and an equivalent observed value of $Y = \Pi_{k|n}^{-1}\kappa_{k|n}$. □

Remark 5.2.23. In the original work by Mayne (1966), the backward information recursion is carried out on the parameters of $\tilde{\beta}_{k|n}$, as defined by (5.120), rather than on $\beta_{k|n}$. It is easily checked using (5.116)–(5.119) that, except for this difference of focus, Proposition 5.2.21 is equivalent to the Mayne (1966) formulas—see also Kailath *et al.* (2000, Section 10.4) on this point. Of course, in the work of Mayne (1966), $\tilde{\beta}_{k|n}$ has to be combined with the predictive distribution $\phi_{k|k-1}$ rather than with the filtering distribution ϕ_k, as $\tilde{\beta}_{k|n}$ already incorporates the knowledge of the observation Y_k. Proposition 5.2.21 and Algorithm 5.2.22 are here stated in a form that is compatible with our general definition of the forward-backward decomposition in Section 3.2. ■

5.2.6 Application to Marginal Filtering and Smoothing in CGLSSMs

The algorithms previously derived for linear state-space models also have important implications for conditionally Gaussian linear state-space models (CGLSSMs). According to Definition 2.2.6, a CGLSSM is such that conditionally on $\{C_k\}_{k\geq 0}$,

$$
\begin{aligned}
W_{k+1} &= A(C_{k+1})W_k + R(C_{k+1})U_k \ , \quad W_0 \sim \mathrm{N}(\mu_\nu, \Sigma_\nu) \ , \\
Y_k &= B(C_k)W_k + S(C_k)V_k \ ,
\end{aligned}
$$

where the indicator process $\{C_k\}_{k\geq 0}$ is a Markov chain on a finite set X, with some transition matrix Q_C.

We follow the general principle outlined in Section 4.2.3 and consider the computation of the posterior distribution of the indicator variables $C_{0:k}$ given the observations $Y_{0:k}$, marginalizing with respect to the continuous component of the state $W_{0:k}$. The key remark—see (4.11)—is that one may evaluate the conditional distribution of W_k given the observations $Y_{0:k-1}$ and the indicator variables $C_{0:k}$. For CGLSSMs, this distribution is Gaussian with mean $\hat{W}_{k|k-1}(C_{0:k})$ and covariance $\Sigma_{k|k-1}(C_{0:k})$—the dependence on the measurement, here $Y_{0:k-1}$, is implicit and we emphasize only the dependence with respect to the indicator variables in the following. Both of these quantities may be evaluated using the Kalman filter recursion (Algorithm 5.2.13), which we briefly recall here.

Given $\hat{W}_{k-1|k-1}(C_{0:k-1})$ and $\Sigma_{k-1|k-1}(C_{0:k-1})$, the filtered partial state estimator and the filtered partial state error covariance at time $k-1$, evaluate the predicted partial state and the associated predicted partial state error covariance as

$$
\begin{aligned}
\hat{W}_{k|k-1}(C_{0:k}) &= A(C_k)\hat{W}_{k-1|k-1}(C_{0:k-1}) , \\
\Sigma_{k|k-1}(C_{0:k}) &= A(C_k)\Sigma_{k-1|k-1}(C_{0:k-1})A^t(C_k) + R(C_k)R^t(C_k) .
\end{aligned}
\tag{5.124}
$$

From these quantities, determine in a second step the innovation and the covariance of the innovation given the indicator variables,

$$
\begin{aligned}
\epsilon_k(C_{0:k}) &= Y_k - B(C_k)\hat{W}_{k|k-1}(C_{0:k}) , \\
\Gamma_k(C_{0:k}) &= B(C_k)\Sigma_{k|k-1}(C_{0:k})B^t(C_k) + S(C_k)S^t(C_k) .
\end{aligned}
\tag{5.125}
$$

In a third and last step, evaluate the filtered partial state estimation and filtered partial state error covariance from the innovation and the innovation covariance,

$$
\begin{aligned}
K_k(C_{0:k}) &= \Sigma_{k|k-1}(C_{0:k})B(C_k)\Gamma_k^{-1}(C_{0:k}) , \\
\hat{W}_{k|k}(C_{0:k}) &= \hat{W}_{k|k-1}(C_{0:k}) + K_k(C_{0:k})\epsilon_k(C_{0:k}) , \\
\Sigma_{k|k}(C_{0:k}) &= \{I - K_k(C_{0:k})B(C_k)\}\ \Sigma_{k|k-1}(C_{0:k}) .
\end{aligned}
\tag{5.126}
$$

As a by-product of the above recursion, one may also determine the conditional probability of C_k given the history of the indicator process $C_{0:k-1}$ and the observations $Y_{0:k}$ up to index k. Indeed, by Bayes' rule,

$$
\frac{\mathrm{P}_\nu(C_k = c \,|\, C_{0:k-1}, Y_{0:k})}{\mathrm{P}_\nu(C_k = c' \,|\, C_{0:k-1}, Y_{0:k})} = \frac{\mathrm{L}_\nu(Y_{0:k} \,|\, C_{0:k-1}, C_k = c)Q_C(C_{k-1}, c)}{\mathrm{L}_\nu(Y_{0:k} \,|\, C_{0:k-1}, C_k = c')Q_C(C_{k-1}, c')} , \tag{5.127}
$$

where L_ν denotes the conditional likelihood of the observations given the indicator variables. Both the numerator and the denominator can be evaluated, following Remark (5.2.10), by applying the Kalman recursions (5.125)–(5.126) for the two values $C_k = c$ and $C_k = c'$. Using (5.62) and (5.127) then yields

$$
\begin{aligned}
\mathrm{P}_\nu(C_k = c \,|\, C_{0:k-1}, Y_{0:k}) &\propto |\Gamma_k(C_{0:k-1}, c)|^{-1/2} \times \\
\exp\left\{-\frac{1}{2}\epsilon_k^t(C_{0:k-1}, c)\Gamma_k^{-1}(C_{0:k-1}, c)\epsilon_k(C_{0:k-1}, c)\right\} &Q_C(C_{k-1}, c) ,
\end{aligned}
\tag{5.128}
$$

where the normalization factor may be evaluated by summation of (5.128) over all $c \in \mathsf{C}$. At the expense of computing r times (5.125)–(5.126), where r is the cardinality of C, it is thus possible to evaluate the conditional distribution of C_k given the history of the indicator process $C_{0:k-1}$, where the continuous variables $W_{0:k}$ have been fully marginalized out. To be applicable however, (5.128) implies that the history of the indicator process before index k be exactly known. This is hardly conceivable except in simulation-based smoothing approximations where one imputes values of the unknown sequence of indicators $\{C_k\}_{k\geq 0}$. The application of (5.125)–(5.126) and (5.128) for this purpose will be fully described in Chapter 8.

A similar remark holds regarding the computation of the conditional distribution of C_k given both the history $C_{0:k-1}$ and future $C_{k+1:n}$ of the indicator sequence and the corresponding observations $Y_{0:n}$. The principle that we follow here is an instance of the generalized forward-backward decomposition (4.13) which, in the case of CGLSSMs, amounts to adapting Algorithm 5.2.22 as follows.

1. Use the backward information recursion of Proposition 5.2.21 to compute $\kappa_{k|n}(C_{k+1:n})$ and $\Pi_{k|n}(C_{k+1:n})$[2].
2. Use the filtering recursion of Algorithm 5.2.13—restated above as (5.124)–(5.126)—to compute $\hat{W}_{k-1|k-1}(C_{0:k-1})$ and $\Sigma_{k-1|k-1}(C_{0:k-1})$.
3. For all values of $c \in \mathsf{C}$, evaluate $\epsilon_k(C_{0:k-1}, c)$, $\Gamma_k(C_{0:k-1}, c)$, as well as $\hat{W}_{k|k}(C_{0:k-1}, c)$, $\Sigma_{k|k}(C_{0:k-1}, c)$ using one step of Algorithm 5.2.13. Then apply (5.122) and (5.123) to obtain $\hat{W}_{k|n}(C_{0:k-1}, c, C_{k+1:n})$ and $\Sigma_{k|n}(C_{0:k-1}, c, C_{k+1:n})$.

The most difficult aspect then consists in computing the likelihood of the observations $Y_{0:n}$ given the indicator sequence, where all indicators variables but c_k are fixed and c_k takes all possible values in C. The lemma below provides a simple formula for this task.

Lemma 5.2.24. *Assume that $\epsilon_k(c_k)$, $\Gamma_k(c_k)$, $\hat{W}_{k|k}(c_k)$, $\Sigma_{k|k}(c_k)$, $\hat{W}_{k|n}(c_k)$, and $\Sigma_{k|n}(c_k)$ are available, where we omit dependence with respect to the indicator variables c_l for $l \neq k$, which is implicit in the following.*

The likelihood of the observations $Y_{0:n}$ given the indicator sequence $C_{0:n} = c_{0:n}$ is then proportional to the quantity

$$\frac{1}{|\Gamma_k(c_k)|^{1/2}} \exp\left[-\frac{1}{2}\epsilon_k^t(c_k)\Gamma_k^{-1}(c_k)\epsilon_k(c_k)\right] \\ \times \frac{1}{|\Sigma_{k|k}(c_k)|^{1/2}} \exp\left[-\frac{1}{2}\hat{W}_{k|k}^t(c_k)\Sigma_{k|k}^{-1}(c_k)\hat{W}_{k|k}(c_k))\right] \\ \times \left\{\frac{1}{|\Sigma_{k|n}(c_k)|^{1/2}} \exp\left[-\frac{1}{2}\hat{W}_{k|n}^t(c_k)\Sigma_{k|n}^{-1}(c_k)\hat{W}_{k|n}(c_k))\right]\right\}^{-1}, \quad (5.129)$$

where the proportionality constant does not depend on the value of c_k.

Before actually proving this identity, we give a hint of the fundamental argument behind (5.129). If X and Y are jointly Gaussian variables (with non-singular covariance matrices), Bayes' rule implies that

---

[2]We do not repeat Proposition 5.2.21 with the notations appropriate for CGLSSMs as we did for (5.124)–(5.126).

$$p(x|y) = \frac{p(y|x)p(x)}{\int p(y|x')p(x')\,dx'} .$$

In particular, the denominator on the right-hand side equals $p(y|x)p(x)/p(x|y)$ for any value of x. For instance, in the linear model of Proposition 5.2.2, applying this identity for $x = 0$ yields

$$\begin{aligned}\int p(Y|x)p(x)\,dx \propto & \frac{1}{|\Sigma_V|^{1/2}} \exp\left[-\frac{1}{2} Y^t \Sigma_V^{-1} Y\right] \\ & \times \frac{1}{|\Sigma_X|^{1/2}} \exp\left[-\frac{1}{2}\mu_X^t \Sigma_X^{-1} \mu_X\right] \\ & \times \left\{\frac{1}{|\Sigma_{X|Y}|^{1/2}} \exp\left[-\frac{1}{2}\mu_{X|Y}^t \Sigma_{X|Y} \mu_{X|Y}\right]\right\}^{-1} , \end{aligned} \tag{5.130}$$

where $\mu_{X|Y} \stackrel{\text{def}}{=} \mathrm{E}(X|Y)$ and $\Sigma_{X|Y} \stackrel{\text{def}}{=} \mathrm{Cov}(X|Y)$ and constants have been omitted. It is tedious but straightforward to check from (5.16) and (5.17) using the matrix inversion lemma that (5.130) indeed coincides with what we know to be the correct result:

$$\int p(Y|x)p(x)\,dx = p(Y) \propto$$
$$\frac{1}{|\Sigma_V + B\Sigma_X B^t|^{1/2}} \exp\left[-\frac{1}{2}(Y - B\mu_X)^t(\Sigma_V + B\Sigma_X B^t)^{-1}(Y - B\mu_X)\right] .$$

Equation (5.130) is certainly not the most efficient way of computing $p(Y)$ but it is one that does not necessitate any other knowledge than that of the prior $p(x)$, the conditional $p(y|x)$, and the posterior $p(x|y)$. Lemma 5.2.24 will now be proved by applying the same principle to the conditional smoothing distribution in a CGLSSM.

Proof (Conditional Smoothing Lemma). The forward-backward decomposition provides a simple general expression for the likelihood of the observations $Y_{0:n}$ in the form

$$\mathrm{L}_n = \int \alpha_k(dw)\beta_{k|n}(w) \tag{5.131}$$

for any $k = 0, \ldots, n$. Recall that our focus is on the likelihood of the observations conditional on a given sequence of indicator variables $C_{0:n} = c_{0:n}$, and more precisely on the evaluation of the likelihood for all values of c_k in C, the other indicator variables c_l, $l \neq k$, being held fixed. In the following, every expression should be understood as being conditional on $C_{0:n} = c_{0:n}$, where only the dependence with respect to c_k is of interest (terms that do not depend on the value of c_k will cancel out by normalization). This being said, (5.131) may be rewritten as

$$\mathrm{L}_n^{(c_k)} = \iint \alpha_{k-1}(dw_{k-1})Q^{(c_k)}(w_{k-1}, dw_k)g_k^{(c_k)}(w_k)\beta_{k|n}(w_k) \tag{5.132}$$

using the forward recursion (3.17), where the superscript (c_k) is used to highlight quantities that depend on this variable. Because the first term of the integrand does not depend on c_k, it may be replaced by its normalized version ϕ_{k-1} to obtain

$$\mathrm{L}_n^{(c_k)} \propto \iint \phi_{k-1}(dw_{k-1}) Q^{(c_k)}(w_{k-1}, dw_k) g_k^{(c_k)}(w_k) \beta_{k|n}(w_k) \,, \tag{5.133}$$

where the proportionality constant does not depend on c_k. Now, using the prediction and filtering relations (see Proposition 3.2.5 and Remark 3.2.6), the right-hand side of (5.133) may be rewritten as the product

$$\int \phi_{k|k-1}^{(c_k)}(dw) g_k^{(c_k)}(w) \times \int \phi_k^{(c_k)}(dw) \beta_{k|n}(w) \,. \tag{5.134}$$

Finally note that in the case of conditionally Gaussian linear state-space models: (i) the first integral in (5.134) may be computed from the innovation ϵ_k as the first line of (5.129)—a remark that was already used in obtaining (5.128); (ii) $\phi_k^{(c_k)}$ is a Gaussian probability density function with parameters $\hat{W}_{k|k}(c_k)$ and $\Sigma_{k|k}(c_k)$; (iii) $\beta_{k|n}$ is a Gaussian likelihood defined, up to a constant, by the information parameters $\kappa_{k|n}$ and $\Pi_{k|n}$;

$$\text{(iv)} \quad \phi_{k|n}^{(c_k)}(dw) = \frac{\phi_k^{(c_k)}(dw) \beta_{k|n}(w)}{\int \phi_k^{(c_k)}(dw') \beta_{k|n}(w')}$$

is the Gaussian distribution with parameters $\hat{X}_{k|n}$ and $\Sigma_{k|n}$ given by (5.122) and (5.123), respectively. The last two factors of (5.129) are now easily recognized as an instance of (5.130) applied to the second integral term in (5.134), where the factor $\beta_{k|n}(0)$ has been ignored because it does not depend on the value of c_k. Note that as a consequence, the fact that $\kappa_{k|n}$ and $\Pi_{k|n}$ define $\beta_{k|n}$ up to an unknown constant only is not detrimental. □

Once again, the context in which Lemma 5.2.24 will be useful is not entirely obvious at this point and will be fully discussed in Section 6.3.2 when reviewing Monte Carlo methods. From the proof of this result, it should be clear however that (5.129) is deeply connected to the smoothing approach discussed in Section 5.2.5 above.

6

Monte Carlo Methods

This chapter takes a different path to the study of hidden Markov models in that it abandons the pursuit of closed-form formulas and exact algorithms to cover instead simulation-based techniques. This change of perspective allows for a much broader coverage of HMMs, which is not restricted to the specific cases discussed in Chapter 5. In this chapter, we consider *sampling* the unknown sequence of states $X_0, \ldots, X_n$ *conditionally on the observed sequence* $Y_0, \ldots Y_n$. In subsequent chapters, we will also use simulation to do inference about the parameters of HMMs, either using simulation-based stochastic algorithms that optimize the likelihood (Chapter 11) or in the context of Bayesian joint inference on the states and parameters (Chapter 13). But even the sole simulation of the missing states may prove itself a considerable challenge in complex settings like continuous state-space HMMs. Therefore, and although these different tasks are presented in separate chapters, simulating hidden states in a model whose parameters are assumed to be known is certainly not disconnected from parameter estimation to be discussed in Chapters 11 and 13.

6.1 Basic Monte Carlo Methods

Although we will not go into a complete description of simulation methods in this book, the reader must be aware that recent developments of these methods have offered new opportunities for inference in complex models like hidden Markov models and their generalizations. For a more in-depth covering of these simulation methods and their implications see, for instance, the books by Chen and Shao (2000), Evans and Swartz (2000), Liu (2001), and Robert and Casella (2004).

6.1.1 Monte Carlo Integration

Integration, in general, is most useful for computing probabilities and expectations. Of course, when given an expectation to compute, the first thing is to try to compute the integral analytically. When analytic evaluation is impossible, numerical integration is an option. However, especially when the dimension of the space is large, numerical integration can become numerically involved: the number of function evaluations required to achieve some degree of approximation increases exponentially in the dimension of the problem (this is often called the *curse of dimensionality*).

Thus it is useful to consider other methods for evaluating integrals. Fortunately, there are methods that do not suffer so directly from the curse of dimensionality, and Monte Carlo methods belong to this group. In particular, recall that, by the strong law of large numbers, if $\xi^1, \xi^2, \ldots$ is a sequence of i.i.d. X-valued random variables with common probability distribution π, then the estimator

$$\hat{\pi}_N^{\mathrm{MC}}(f) = N^{-1} \sum_{i=1}^{N} f(\xi^i)$$

converges almost surely to $\pi(f)$ for all π-integrable functions f. Obviously this *Monte Carlo estimate* of the expectation is not exact, but generating a sufficiently large number of random variables can render this approximation error arbitrarily small, in a suitable probabilistic sense. It is even possible to assess the size of this error. If

$$\pi(|f|^2) = \int |f(x)|^2 \, \pi(dx) < \infty ,$$

the central limit theorem shows that $\sqrt{N}\left[\hat{\pi}_N^{\mathrm{MC}}(f) - \pi(f)\right]$ has an asymptotic normal distribution, which can be used to construct asymptotic confidence regions for $\pi(f)$. For instance, if f is real-valued, a confidence interval with asymptotic probability of coverage α is given by

$$\left[\hat{\pi}_N^{\mathrm{MC}}(f) - c_\alpha N^{-1/2} \sigma_N(\pi, f),\ \hat{\pi}_N^{\mathrm{MC}}(f) + c_\alpha N^{-1/2} \sigma_N(\pi, f)\right] , \tag{6.1}$$

where

$$\sigma_N^2(\pi, f) \stackrel{\mathrm{def}}{=} N^{-1} \sum_{i=1}^{N} \left[f(\xi^i) - \hat{\pi}_N^{\mathrm{MC}}(f)\right]^2$$

and c_α is the $\alpha/2$ quantile of the standard Gaussian distribution. If generating a sequence of i.i.d. samples from π is practicable, one can make the confidence interval as small as desired by increasing the sample size N. When compared to univariate numerical integration and quasi-Monte Carlo methods (Niederreiter, 1992), the convergence rate is not fast. In practical terms, (6.1) implies that an extra digit of accuracy on the approximation requires 100 times as many replications, where the rate $1/\sqrt{N}$ cannot be improved. On the other hand, it is possible to derive methods to reduce the asymptotic variance of the Monte Carlo estimate by allowing a certain amount of dependence among the random variables $\xi^1, \xi^2, \ldots$ Such methods include antithetic variables, control variates, stratified sampling, etc. These techniques are not discussed here (see for instance Robert and Casella, 2004, Chapter 4). A remarkable fact however is that the rate of convergence of $1/\sqrt{N}$ in (6.1) remains the same whatever the dimension of the space X is, which leaves some hope of effectively using the Monte Carlo approach in large-dimensional settings.

6.1.2 Monte Carlo Simulation for HMM State Inference

6.1.2.1 General Markovian Simulation Principle

We now turn to the specific task of simulating the unobserved sequence of states in a hidden Markov model, given some observations. The main result has already been discussed in Section 3.3: given some observations, the unobserved sequence of states constitutes a non-homogeneous Markov chain whose transition kernels may be evaluated, either from the backward functions for the forward chain (with indices increasing as usual) or from the forward measures—or equivalently filtering distributions—for the backward chain (with indices in reverse order). Schematically, both available options are rather straightforward to implement.

Backward Recursion/Forward Sampling: First compute (and store) the backward functions $\beta_{k|n}$ by backward recursion, for $k = n, n-1$ down to 0 (Proposition 3.2.1). Then, simulate X_{k+1} given X_k from the forward transition kernels $\mathrm{F}_{k|n}$ specified in Definition 3.3.1.

Forward Recursion/Backward Sampling: First compute and store the forward measures $\alpha_{\nu,k}$ by forward recursion, according to Proposition 3.2.1. As an alternative, one may evaluate the normalized versions of the forward measures, which coincide with the filtering distributions $\phi_{\nu,k}$, following Proposition 3.2.5. Then X_k is simulated conditionally on X_{k+1} (starting from X_n) according to the backward transition kernel $\mathrm{B}_{\nu,k}$ defined by (3.38).

Despite its beautiful simplicity, the method above will obviously be of no help in cases where an exact implementation of the forward-backward recursion is not available.

6.1.2.2 Models with Finite State Space

In the case where the state space X is finite, the implementation of the forward-backward recursions is feasible and has been fully described in Section 5.1. The second method described above is a by-product of Algorithm 5.1.3.

Algorithm 6.1.1 (Markovian Backward Sampling). Given the stored values of $\phi_0, \ldots, \phi_n$ computed by forward recursion according to Algorithm 5.1.1, do the following.

Final State: Simulate X_n from ϕ_n.

Backward Simulation: For $k = n-1$ down to 0, compute the backward transition kernel according to (5.7) and simulate X_k from $\mathrm{B}_k(X_{k+1}, \cdot)$.

The numerical complexity of this sampling algorithm is thus equivalent to that of Algorithm 5.1.3, whose computational cost depends most importantly on the cardinal r of X and on the difficulty of evaluating the function $g(x, Y_k)$ for all $x \in \mathsf{X}$ and $k = 0, \ldots, n$ (see Section 5.1). The backward simulation pass in Algorithm 6.1.1 is simpler than its smoothing counterpart in Algorithm 5.1.3, as one only needs to evaluate $\mathrm{B}_k(X_{k+1}, \cdot)$ for the simulated value of X_{k+1} rather than $\mathrm{B}_k(i,j)$ for all $(i,j) \in \{1, \ldots, r\}^2$.

6.1.2.3 Gaussian Linear State-Space Models

As discussed in Section 5.2, Rauch-Tung-Striebel smoothing (Algorithm 5.2.4) is the exact counterpart of Algorithm 5.1.3 in the case of Gaussian linear state-space models. Not surprisingly, to obtain the smoothing means and covariance matrices in Algorithm 5.2.4, we explicitly constructed the backward Gaussian transition density, whose mean and covariance are given by (5.23) and (5.24), respectively. We simply reformulate this observation in the form of an algorithm as follows.

Algorithm 6.1.2 (Gaussian Backward Markovian State Sampling). Assume that the filtering moments $\hat{X}_{k|k}$ and $\Sigma_{k|k}$ have been computed using Proposition 5.2.3. Then do the following.

Final State: Simulate

$$X_n \sim \mathrm{N}(\hat{X}_{n|n}, \Sigma_{n|n}).$$

Backward Simulation: For $k = n-1$ down to 0, simulate X_k from a Gaussian distribution with mean and covariance matrix given by (5.23) and (5.24), respectively.

The limitations discussed in the beginning of Section 5.2.2 concerning RTS smoothing (Algorithm 5.2.4) also apply here. In some models, Algorithm 6.1.2 is far from being computationally efficient (Frühwirth-Schnatter, 1994; Carter and Kohn, 1994). With these limitations in mind, De Jong and Shephard (1995) described a sampling algorithm inspired by disturbance (or Bryson-Frazier) smoothing (Algorithm 5.2.15) rather than by RTS smoothing. The method of De Jong and Shephard (1995) is very close to Algorithm 5.2.15 and proceeds by sampling the disturbance vectors U_k backwards (for $k = n-1, \ldots, 0$) and then the initial state X_0, from which the complete sequence $X_{0:n}$ may be obtained by repeated applications of the dynamic equation (5.11). Because the sequence of disturbance vectors $\{U_k\}_{k=n-1,\ldots,0}$ does not however have a backward Markovian structure, the method of De Jong and Shephard (1995) is not a simple by-product of disturbance smoothing (as was the case for Algorithms 5.2.4 and 6.1.2). Durbin and Koopman (2002) described an approach that is conceptually simpler and usually about as efficient as the disturbance sampling method of De Jong and Shephard (1995).

The basic remark is that if X and Y are jointly Gaussian variables, the conditional distribution of X given Y is Gaussian with mean vector $\mathrm{E}[X \mid Y]$ and covariance matrix $\mathrm{Cov}(X \mid Y)$, where $\mathrm{Cov}(X \mid Y)$ equals $\mathrm{Cov}(X - \mathrm{E}[X \mid Y])$ and, in addition, does not depend on Y (Proposition 5.2.2). In particular, if (X^*, Y^*) is another independent pair of Gaussian distributed random vectors with the same (joint) distribution, $X - \mathrm{E}[X \mid Y]$ and $X^* - \mathrm{E}[X^* \mid Y^*]$ are independent and both are $\mathrm{N}(0, \mathrm{Cov}(X \mid Y))$ distributed. In summary, to simulate ξ from the distribution of X given Y, one may

1. Simulate an independent pair of Gaussian variables (X^*, Y^*) with the same distribution as (X, Y) and compute $X^* - \mathrm{E}[X^* \mid Y^*]$;
2. Given Y, compute $\mathrm{E}[X \mid Y]$, and set

$$\xi = \mathrm{E}[X \mid Y] + X^* - \mathrm{E}[X^* \mid Y^*] .$$

This simulation approach only requires the ability to compute conditional expectations and to simulate from the prior joint distribution of X and Y. When applied to the particular case of Gaussian linear state-space models, this general principle yields the following algorithm.

Algorithm 6.1.3 (Sampling with Dual Smoothing). Given a Gaussian linear state-space model following (5.11)–(5.12) and observations $Y_0, \ldots, Y_n$, do the following.

1. Simulate a fictitious independent sequence $\{X_k^*, Y_k^*\}_{k=0,\ldots,n}$ of both states and observations using the model equations.
2. Compute $\{\hat{X}_{k|n}\}_{k=0,\ldots,n}$ and $\{\hat{X}_{k|n}^*\}_{k=0,\ldots,n}$ using Algorithm 5.2.15 for the two sequences $\{Y_k\}_{k=0,\ldots,n}$ and $\{Y_k^*\}_{k=0,\ldots,n}$.

Then $\{\hat{X}_{k|n} + X_k^* - \hat{X}_{k|n}^*\}_{k=0,\ldots,n}$ is distributed according to the posterior distribution of the states given $Y_0, \ldots, Y_n$.

Durbin and Koopman (2002) list a number of computational simplifications that are needed to make the above algorithm competitive with the disturbance sampling approach. As already noted in Remark 5.2.16, the backward recursion of Algorithm 5.2.15 may be greatly simplified when only the best linear estimates (and not their covariances) are to be computed. During the forward Kalman prediction recursion, it is also possible to save on computations by noting that all covariance matrices (state prediction error, innovation) will be common for the two sequences $\{Y_k\}$ and $\{Y_k^*\}$, as these matrices do not depend on the observations but only on the model. The same remark should be used when the purpose is not only to simulate *one* sequence but N sequences of states conditional on the same observations, which will be the standard situation in a Monte Carlo approach. Further improvement can be gained by carrying out simultaneously the simulation and Kalman prediction tasks, as both of them are implemented recursively (Durbin and Koopman, 2002).

6.2 A Markov Chain Monte Carlo Primer

As we have seen above, the general task of simulating the unobserved $X_{0:n}$ given observations $Y_{0:n}$ is non-trivial except when X in finite or the model is a Gaussian linear state-space model. In fact, in such models, analytic integration with respect to (low-dimensional marginals of) the conditional distribution of $X_{0:n}$ given observations is most often feasible, whence there is generally no true need for simulation of the unobserved Markov chain. The important and more difficult challenge is rather to explore methods to carry out this task in greater generality, and this is the object of the current section. We start by describing the accept-reject algorithm, which is a general approach to simulation of i.i.d. samples from a prescribed distribution, and then turn to so-called Markov chain Monte Carlo methods, which are generally more successful in large-dimensional settings.

6.2.1 The Accept-Reject Algorithm

For specific distributions such as the Gaussian, Poisson, or Gamma distributions, there are efficient tailor-made simulation procedures; however, we shall not discuss here the most basic (but nonetheless essential) aspects of random variate generation for which we refer, for instance, to the books by Devroye (1986), Ripley (1987), or Gentle (1998). We are rather concerned with methods that can provide i.i.d. samples from any pre-specified distribution π, not just for specific choices of this distribution. It turns out that there are only a limited number of options for this task, which include the accept-reject algorithm discussed here and the sampling importance resampling approach to be discussed in Section 7.1 (although the latter only provides an *approximate* i.i.d. sample).

The accept-reject algorithm, first described by von Neumann, is important both for its direct applications and also because its principle is at the core of many of the more advanced methods to be discussed in the following (for general references on the accept-reject method, see Devroye, 1986, Chapter 2, Ripley, 1987, p. 60–62, or Robert and Casella, 2004, Chapter 2). It is easier to introduce the key concepts using probability densities, and we assume that π has a density with respect to a measure λ; because this assumption will be adopted all through this section, we shall indeed use the notation π for this density as well. The key requirement of the method is the availability of another probability density function (with respect to λ) r whose functional form is known and from which i.i.d. sampling is readily feasible. We also

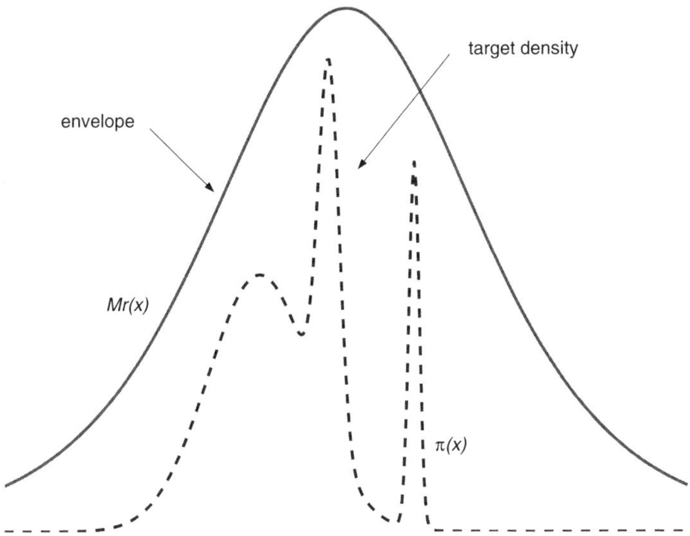

Fig. 6.1. Illustration of the accept-reject method. Random points are drawn uniformly under the bold curve and rejected if the ordinate exceeds $\pi(x)$ (dashed curve).

assume that for some constant $M > 1$, $Mr(x) \geq \pi(x)$, for all $x \in \mathsf{X}$, as illustrated by Figure 6.1.

Proposition 6.2.2 below asserts that abscissas of i.i.d. random points in $\mathsf{X} \times \mathbb{R}^+$ that are generated uniformly under the graph of $\pi(x)$ are distributed according to π. Of course, it is not easier to sample uniformly under the graph of $\pi(x)$ in $\mathsf{X} \times \mathbb{R}^+$ than it is to sample directly from π, but one may instead sample uniformly under the graph of the envelope $Mr(x)$ and accept only those samples that fall under the graph of π. To do this, first generate a candidate, say ξ according to the density r and compute $\pi(\xi)$ as well as the height of the envelope $Mr(\xi)$. A uniform $\mathrm{U}([0,1])$ random variable U is then generated independently from ξ, and the pair is accepted if $UMr(\xi) \leq \pi(\xi)$. In case of rejection, the whole procedure is started again until one eventually obtains a pair ξ, U which is accepted. The algorithm is summarized below,

Algorithm 6.2.1 (Accept-Reject Algorithm).

Repeat: Generate two independent random variables: $\xi \sim r$ and $U \sim \mathrm{U}([0,1])$.
Until: $U \leq \pi(\xi)/(Mr(\xi))$.

The correctness of the accept-reject method can be deduced from the following two simple results.

Proposition 6.2.2. *Let ξ be a random variable with density π with respect to a measure λ on X and U be an independent real random variable uniformly*

distributed on the interval $[0, M]$. *Then the pair* $(\xi, U\pi(\xi))$ *of random variables is uniformly distributed on*

$$\mathscr{S}_{\pi,M} = \left\{(x, u) \in \mathsf{X} \times \mathbb{R}^+ : 0 < u < M\pi(x)\right\} ,$$

with respect to $\lambda \otimes \lambda^{\text{Leb}}$, *where* λ^{Leb} *denotes Lebesgue measure.*

Conversely, if a random vector (ξ, U) *of* $\mathsf{X} \times \mathbb{R}^+$ *is uniformly distributed on* $\mathscr{S}_{\pi,M}$, *then* ξ *admits* π *as marginal probability density function.*

Proof. Obviously, if Proposition 6.2.2 is to be true for some value of M_0, then both claims also hold for all values of $M > 0$ simply by scaling the ordinate by M/M_0. In the following, we thus consider the case where M equals one. For the first statement, take a measurable subset $B \subseteq \mathscr{S}_{\pi,1}$ and let B_x denote the section of B in x, that is, $B_x = \{u : (x, u) \in B\}$. Then

$$\mathrm{P}\left\{(\xi, U\pi(\xi)) \in B\right\} = \\ \int_{x\in\mathsf{X}} \int_{u\in B_x} \frac{1}{\pi(x)} \lambda^{\text{Leb}}(du)\pi(x)\, \lambda(dx) = \iint_B \lambda^{\text{Leb}}(du)\, \lambda(dx) .$$

For the second statement, consider a measurable subset $A \subseteq \mathsf{X}$ and set $\bar{A} = \{(x, u) \in A \times \mathbb{R}^+ : 0 \leq u \leq \pi(x)\}$. Then

$$\mathrm{P}(\xi \in A) = \mathrm{P}\left((\xi, U) \in \bar{A}\right) = \frac{\iint_{\bar{A}} \lambda^{\text{Leb}}(du)\, \lambda(dx)}{\iint_{\mathscr{S}_{\pi,1}} \lambda^{\text{Leb}}(du)\, \lambda(dx)} = \int_A \pi(x)\, \lambda(dx) .$$

□

Lemma 6.2.3. *Let* $V_1, V_2, \ldots$ *be a sequence of i.i.d. random variables taking values in a measurable space* $(\mathsf{V}, \mathcal{V})$ *and* $B \in \mathcal{V}$ *a set such that* $\mathrm{P}(V_1 \in B) = p > 0$.

The integer-valued random variable $\sigma = \inf\{k \geq 1, V_k \in B\}$ *(with the convention that* $\inf \emptyset = \infty$*) is geometrically distributed with parameter* p, *i.e., for all* $i \geq 0$,

$$\mathrm{P}(\sigma = i) = (1 - p)^{i-1} p . \tag{6.2}$$

The random variable $V = V_\sigma \mathbb{1}_{\sigma<\infty}$ *is distributed according to*

$$\mathrm{P}(V \in A) = \frac{\mathrm{P}(V \in A \cap B)}{p} . \tag{6.3}$$

Proof. First note that

$$\mathrm{P}(\sigma = i) = \mathrm{P}(V_1 \notin B, \ldots, V_{i-1} \notin B, V_i \in B) = (1 - p)^{i-1} p ,$$

showing (6.2), which implies in particular that the waiting time σ is finite with probability one. For $A \in \mathcal{V}$,

$$
\begin{aligned}
\mathrm{P}(V \in A) &= \sum_{i=1}^{\infty} \mathrm{P}(V_1 \notin B, \ldots, V_{i-1} \notin B, V_i \in A \cap B) \\
&= \sum_{i=1}^{\infty} (1-p)^{i-1}\, \mathrm{P}(V_1 \in A \cap B) \\
&= \mathrm{P}(V_1 \in A \cap B) \frac{1}{1-(1-p)} .
\end{aligned}
$$

□

Hence by Proposition 6.2.2, the intermediate pairs (ξ_i, U_i) generated in Algorithm 6.2.1 are such that $(\xi_i, MU_i r(\xi_i))$ are uniformly distributed under the graph of $Mr(x)$. By Lemma 6.2.3, the accepted pair (ξ, U) is then uniformly distributed under the graph of $\pi(x)$ and, using Proposition 6.2.2, ξ is marginally distributed according to π. The probability p of acceptance is equal to

$$
\mathrm{P}\left\{ U_1 \le \frac{\pi(\xi_1)}{Mr(\xi_1)} \right\} = \mathrm{P}\left\{ (\xi_1, MU_1 r(\xi_1)) \in \mathscr{S}_{\pi,M} \right\} = \frac{\int_{\mathsf{X}} \pi(x)\, \lambda(dx)}{\int_{\mathsf{X}} Mr(x)\, \lambda(dx)} = \frac{1}{M} .
$$

Remark 6.2.4. The same algorithm can be applied also in cases where the densities π or r are known only up to a constant. In that case, denote by $C_\pi = \int \pi(x)\, \lambda(dx)$ and $C_r = \int r(x)\, \lambda(dx)$ the normalizing constants. The condition $\pi(x) \le Mr(x)$ can be equivalently written as $\bar{\pi}(x) \le M(C_r/C_\pi)\bar{r}(x)$, where $\bar{\pi}(x) = \pi(x)/C_\pi$ and $\bar{r}(x) = r(x)/C_r$ denote the actual probability density functions. Because the two stopping conditions $\bar{\pi}(x) \le M(C_r/C_\pi)\bar{r}(x)$ and $\pi(x) \le Mr(x)$ are equivalent, using the accept-reject algorithm with π, r, and M amounts to using it with $\bar{\pi}$, $\bar{r}$ and MC_r/C_π. Therefore, the knowledge of the normalizing constants C_π and C_r is not required. Note however that when either C_π or C_r differs from one, it is not possible anymore to interpret $1/M$ as the acceptance probability, and the actual acceptance probability $C_\pi/(C_r M)$ is basically unknown. In that case, the complexity of the accept-reject algorithm (typically how many intermediate draws are required on average before accepting a single one) cannot be determined in advance and may only be estimated empirically. ■

Of course, the assumption $\pi(x) \le Mr(x)$ puts some stringent constraints on the choice of the density r from which samples are drawn. The density r should have both heavier tails and sharper infinite peaks than π. The efficiency of the algorithm is the ratio of the areas under the two graphs of $\pi(x)$ and $Mr(x)$, which equals $1/M$. Therefore, it is essential to keep M as close to one as possible. The optimal choice of M for a given r is $M_r = \sup_{x \in \mathsf{X}} \pi(x)/r(x)$, as it maximizes the acceptance probability and therefore minimizes the average required computational effort. Determining a proposal density r such that M_r is small and evaluating M_r (or a tight upper bound for it) are the two key ingredients for practical application of the accept-reject method. In many situations, and especially in multi-dimensional settings, both of these tasks are often equally difficult (see Robert and Casella, 2004, for examples).

6.2.2 Markov Chain Monte Carlo

The remarks above highlight that although accept-reject is often a viable approach in low-dimensional problems, it has serious drawbacks in large-dimensional ones. Most fortunately, there exists a class of alternatives that allow us to handle arbitrary distributions, on large-dimensional sets, without a detailed study of them. This class of simulation methods is called Markov chain Monte Carlo (or MCMC) methods, as they rely on Markov-dependent simulations. It should be stressed at this point that the "Markov" in "Markov chain Monte Carlo" has nothing to do with the "Markov" in "hidden Markov models". These MCMC methods are generic/universal and, while they naturally apply in HMM settings, they are by no means restricted to those.

The original MCMC algorithm was introduced by Metropolis *et al.* (1953) for the purpose of optimization on a discrete state space, in connection with statistical physics: the paper was actually published in the *Journal of Chemical Physics*. The Metropolis algorithm was later generalized by Hastings (1970) and Peskun (1973, 1981) to statistical simulation. Despite several other papers that highlighted its usefulness in specific settings (see, for example, Geman and Geman, 1984; Tanner and Wong, 1987; Besag, 1989), the starting point for an intensive use of MCMC methods by the statistical community can be traced to the presentation of the *Gibbs sampler* by Gelfand and Smith (1990). The MCMC approach is now well-known in many scientific domains, which include physics and statistics but also biology, engineering, etc.

Returning for a while to the general case where π is a distribution, the tenet of MCMC methods is the remark that simulating an i.i.d. sequence $\xi^1, \ldots, \xi^n$ with common probability distribution π is not the only way to approximate π in the sense of being able to approximate the expectation of *any* π-integrable function f. In particular, one may consider Markov-dependent sequences $\{\xi^i\}_{i \geq 1}$ rather than i.i.d. sequences. The ergodic theorem for Markov chains asserts that, under suitable conditions (discussed in Section 14.2.6 of Chapter 14),

$$\hat{\pi}_N^{\text{MCMC}}(f) = \frac{1}{N} \sum_{i=1}^{N} f(\xi^i) \tag{6.4}$$

is a reasonable estimate of the expectation of f under the stationary distribution of the chain $\{\xi^i\}_{i \geq 1}$, for all integrable functions f. In addition, the rate of convergence is identical to that of standard (independent) Monte Carlo, that is, $1/\sqrt{N}$. To make this idea practicable however requires simulation schemes that guarantee

(i) that simulating the chain $\{\xi^i\}_{i \geq 1}$ given an arbitrary initial value ξ^1 is an easily implementable process;

(ii) that the stationary distribution of $\{\xi^i\}_{i\geq 1}$ indeed coincides with the desired distribution π;
(iii) that the chain $\{\xi^i\}_{i\geq 1}$ satisfies conditions needed to guarantee the convergence towards π, irrespectively of the initial value ξ^1.

We will introduce below two major classes of such algorithms, and we refer the reader to Robert and Casella (2004) and Roberts and Tweedie (2005) for an appropriate detailed coverage of these MCMC methods.

In this context, the specific distribution of interest is generally referred to as the *target distribution*. To keep the presentation simple, we will also assume that all distributions and conditional distributions arising are dominated by a common measure λ. The target distribution in particular is assumed to have a probability density function, as above denoted by π, with respect to λ.

6.2.3 Metropolis-Hastings

The (very limited) assumption underlying the Metropolis-Hastings algorithm, besides the availability of π, is that one can simulate from a transition density function r (with respect to the same measure λ), called the *proposal distribution*, whose functional form is also known.

Algorithm 6.2.5 (The Metropolis-Hastings Algorithm). Simulate a sequence of values $\{\xi^i\}_{i\geq 1}$, which forms a Markov chain on X, with the following mechanism: given ξ^i,

1. Generate $\xi \sim r(\xi^i, \cdot)$;
2. Set

$$\xi^{i+1} = \begin{cases} \xi & \text{with probability } \alpha(\xi^i, \xi) \stackrel{\text{def}}{=} \dfrac{\pi(\xi)\, r(\xi, \xi^i)}{\pi(\xi^i)\, r(\xi^i, \xi)} \wedge 1 \\ \xi^i & \text{otherwise} \end{cases} \tag{6.5}$$

The initial value ξ^1 may be chosen arbitrarily.

In practice, (6.5) is carried out by drawing an independent U([0, 1]) variable U and accepting ξ only if $U \leq A(\xi^i, \xi)$, where

$$A(\xi^i, \xi) = \frac{\pi(\xi)\, r(\xi, \xi^i)}{\pi(\xi^i)\, r(\xi^i, \xi)} ,$$

is generally referred to as the Metropolis-Hastings *acceptance ratio*.

The reason for this specific choice of acceptance probability in (6.5), whose name follows from Metropolis *et al.* (1953) and Hastings (1970), is that the associated Markov chain $\{\xi_t\}$ satisfies the *detailed balance equation* (2.12) discussed in Chapter 2.

Proposition 6.2.6 (Reversibility of the Metropolis-Hastings Kernel). *The chain $\{\xi^i\}_{i\geq 1}$ generated by Algorithm 6.2.5 is reversible and π is its stationary probability density function.*

Proof. The transition kernel K associated with Algorithm 6.2.5 is such that for a function $f \in \mathcal{F}_b(\mathsf{X})$,

$$K(x, f) = \int f(x') \left[\alpha(x, x')r(x, x')\,\lambda(dx') + p_R(x)\,\delta_x(dx')\right] ,$$

where $p_R(x)$ is the probability of remaining in the state x, given by

$$p_R(x) = 1 - \int \alpha(x, x')r(x, x')\,\lambda(dx') .$$

Hence

$$\iint f_1(x)f_2(x')\pi(x)\,\lambda(dx)\,K(x, dx') = \iint f_1(x)f_2(x')\pi(x)\alpha(x, x')r(x, x')\,\lambda(dx)\,\lambda(dx') + \int f_1(x)f_2(x)\pi(x)p_R(x)\,\lambda(dx) \tag{6.6}$$

for all functions $f_1, f_2 \in \mathcal{F}_b(\mathsf{X})$. According to (6.5),

$$\pi(x)\alpha(x, x')r(x, x') = \pi(x')r(x', x) \wedge \pi(x)r(x, x') ,$$

which is symmetric in x and x', and thus K satisfies the detailed balance condition (2.12), as we may swap the functions f_1 and f_2 in both terms on the right-hand side of (6.6). This implies in particular that π is a stationary density for the kernel K. □

The previous result is rather weak as there is no guarantee that the chain $\{\xi^i\}_{i\geq 1}$ indeed converges in distribution to π, whatever the choice of the initialization ξ^1. We postpone the study of such questions until Chapter 14, where we show that such results can be obtained under weak additional conditions (see for instance Theorem 14.2.37). We refer to the books by Robert and Casella (2004) and Roberts and Tweedie (2005) for further discussion of convergence issues and focus, in the following, on the practical aspects of MCMC.

Remark 6.2.7. An important feature of the Metropolis-Hastings algorithm is that it can be applied also when π or r is known only through the ratio $\pi(x')/\pi(x)$ or $r(x', x)/r(x, x')$. This allows the algorithm to be used without knowing the normalizing constants: evaluating π and/or r only up to a constant scale factor, or even the ratio π/r, is sufficient to apply Algorithm 6.2.5. This fact is instrumental when the algorithm is to be used to simulate from posterior distributions in Bayesian models (see Chapter 13 for examples), as these distributions are most often defined though Bayes theorem as the product of the likelihood and the prior density, where the normalization is not computable (or else one would not consider using MCMC...).

In hidden Markov models, this feature is very useful for simulating from the posterior distribution of an unobservable sequence of states $X_{0:n}$ given the corresponding observations $Y_{0:n}$. Indeed, the functional form of the conditional distribution of $X_{0:n}$ given $Y_{0:n}$ is given in (3.13), which is fully explicit except for the normalization factor $\mathrm{L}_{\nu,n}$. For MCMC approaches, there is no point in trying to evaluate this normalization factor $\mathrm{L}_{\nu,n}$, and it suffices to know that the desired joint target distribution is proportional to

$$\phi_{0:n|n}(x_{0:n}) \propto \nu(x_0) g_0(x_0) \prod_{k=1}^{n} q(x_{k-1}, x_k) g_k(x_k) , \tag{6.7}$$

where we assume that the model is fully dominated in the sense of Definition (2.2.3) and hence that ν and q denote, respectively, a probability density function and a transition density function (with respect to λ). The target distribution $\phi_{0:n|n}$ defined by (6.7) is thus perfectly suitable for MCMC simulation. ■

We now consider two important classes of Metropolis-Hastings algorithms.

6.2.3.1 Independent Metropolis-Hastings

A first option for the choice of the proposal transition density function $r(x, \cdot)$ is to select a fixed—that is, independent of x—distribution over X, like the uniform distribution if X is compact, or more likely some other distribution that is related to π. This method, as first proposed by Hastings (1970), appears to be an alternative to importance sampling and the accept-reject algorithms[1]. To stress this special case, we denote the independent proposal density by $r_{\mathrm{ind}}(x)$. The Metropolis-Hastings acceptance probability then reduces to

$$\alpha(x, x') = \frac{\pi(x')/r_{\mathrm{ind}}(x')}{\pi(x)/r_{\mathrm{ind}}(x)} \wedge 1 .$$

In particular, in the case of a uniform proposal r_{ind}, the acceptance probability is nothing but the ratio $\pi(x')/\pi(x)$ (a feature shared with the random walk Metropolis-Hastings algorithm below). Intuitively, the transition from $X_n = x$ to $X_{n+1} = x'$ is accomplished by generating an independent sample from a proposal distribution r_{ind} and then thinning it down based on a comparison of the corresponding importance ratios $\pi(x)/r_{\mathrm{ind}}(x)$ and $\pi(x')/r_{\mathrm{ind}}(x')$.

One can notice the connection with the importance sampling method (see Section 7.1.1) in that the Metropolis-Hastings acceptance probability is also based on the importance weight $\pi(\xi')/r_{\mathrm{ind}}(\xi')$. A major difference is

[1] The importance sampling algorithm is conceptually simpler than MCMC methods. For coherence reasons however, the former will be discussed later in the book, when considering sequential Monte Carlo methods. Readers not familiar with the concept of importance sampling may want to go through Section 7.1.1 at this point.

that importance sampling preserves all the simulations while the independent Metropolis-Hastings algorithm only accepts moving to new values ξ' with sufficiently large importance ratio. It can thus be seen as an approximation to sampling importance resampling of Section 7.1.2 in that it also replicates the points with the highest importance weights.

As reported in Mengersen and Tweedie (1996), the performance of an independent Metropolis-Hastings algorithm will vary widely, depending on, in particular, whether or not the importance ratio $\pi(\xi)/r_{\text{ind}}(\xi)$ is bounded (which is also the condition required for applying the accept-reject algorithm). In Mengersen and Tweedie (1996, Theorem 2.1), it is proved that the algorithm is uniformly ergodic (see definition 4.3.15) if there exists $\beta > 0$ such that

$$\pi\left\{x \in \mathsf{X} : \frac{r_{\text{ind}}(x)}{\pi(x)} \geq \beta\right\} = 1 , \tag{6.8}$$

and then, for any $x \in \mathsf{X}$,

$$\|K^n(x,\cdot) - \pi\|_{\text{TV}} \leq (1-\beta)^n .$$

Conversely, if for every $\beta > 0$ the set on which (6.8) fails has positive π-measure, then the algorithm is not even geometrically ergodic. The practical implication is that the chain may tend to "get stuck" in regions with low values of π. This happens when the proposal has lighter tails than the target distribution. To ensure robust performance, it is thus advisable to let r_{ind} be a relatively heavy-tailed distribution (such as the t-distribution for example).

Example 6.2.8 (Squared and Noisy Autoregression). Consider the following model where the hidden Markov chain is from a regular AR(1) model,

$$X_{k+1} = \phi X_k + U_k$$

with $U_k \sim \mathrm{N}(0, \tau^2)$, and where the observable is

$$Y_k = X_k^2 + V_k$$

with $V_k \sim \mathrm{N}(0, \sigma^2)$. The conditional distribution of X_k given X_{k-1}, X_{k+1} and $Y_{0:n}$ is, by Remark 6.2.7, equal to the conditional distribution of X_k given X_{k-1}, X_{k+1} and Y_k, with density proportional to

$$\exp\left[-\frac{1}{2\tau^2}\left\{(x_k - \phi x_{k-1})^2 + (x_{k+1} - \phi x_k)^2 + \frac{\tau^2}{\sigma^2}(y_k - x_k^2)^2\right\}\right] . \tag{6.9}$$

Obviously, the difficulty with this distribution is the $(y_k - x_k^2)^2$ term in the exponential. A naive resolution of this difficulty is to ignore the term in the proposal distribution, which is then a $\mathrm{N}(\mu_k, \rho_k^2)$ distribution with

$$\mu_k = \phi\frac{x_{k-1} + x_{k+1}}{1+\phi^2} \quad \text{and} \quad \rho_k^2 = \frac{\tau^2}{1+\phi^2} .$$

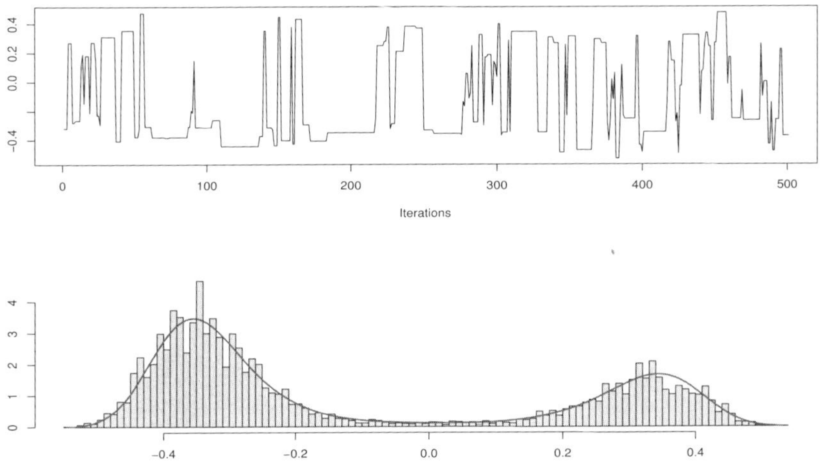

Fig. 6.2. Illustration of Example 6.2.8. Top: plot of the last 500 realizations of the chain $\{\xi_i\}_{i\geq 1}$ produced by the independent Metropolis-Hastings algorithm associated with the $\mathrm{N}(\mu_k, \rho_k^2)$ proposal over 10,000 iterations. Bottom: histogram of a chain of length 10,000 compared with the target distribution (normalized by numerical integration).

The ratio $\pi(x)/r_{\text{ind}}(x)$ is then equal to $\exp -(y_k - x_k^2)^2/2\sigma^2$, which is bounded.

Figure 6.2 *(bottom)* shows how the Markov chain produced by Algorithm 6.2.5 does converge to the proper posterior distribution, even though the target is bimodal (because of the ambiguity on the sign of x_t resulting from the square in the observation equation). Figure 6.2 *(top)* also illustrates the fact that, to jump from one mode to another, the chain has to remain in a given state for several iterations before jumping to the alternative modal region. ■

When the ratio $\pi(x)/r_{\text{ind}}(x)$ is not bounded, the consequences may be very detrimental on the convergence of the algorithm, as shown by the following elementary counterexample.

Example 6.2.9 (Cauchy Meets Normal). Consider a Cauchy $\mathrm{C}(0,1)$ target distribution with a Gaussian $\mathrm{N}(0,1)$ proposal. The ratio $\pi(x)/r_{\text{ind}}(x)$ is then $\exp\{x^2/2\}/(1+x^2)$, which is unbounded and can produce very high values. Quite obviously, the simulation of a sequence of normal proposals to achieve simulation from a Cauchy $\mathrm{C}(0,1)$ distribution is bound to fail, as the normal distribution, whatever its scale, cannot reach the tails of the Cauchy distribution: this failure is illustrated in Figure 6.3. To stress the importance of this requirement (that the ratio $\pi(x)/r_{\text{ind}}(x)$ be bounded), it is important to remember that we can diagnose the failure in Figure 6.3 only because we are cheating and know what the target distribution is, including its normalization. In real practical uses of the method, it would be very difficult in such a case to detect that the sampling algorithm is not doing what it is expected to. ■

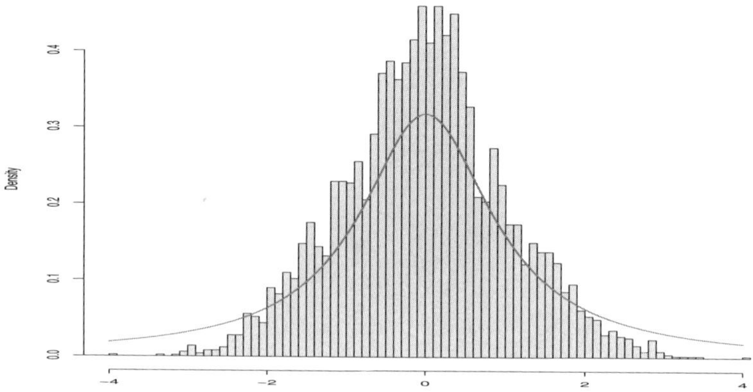

Fig. 6.3. Illustration of Example 6.2.9. Histogram of a independent Metropolis-Hastings chain of length 5,000, based on a N(0, 1) proposal, compared with the target C(0, 1) distribution.

6.2.3.2 Random Walk Metropolis-Hastings

Given that the derivation of an acceptable independent proposal becomes less realistic as the dimension of the problem increases, another option for the choice of $r(x, \cdot)$ is to propose local moves around x with the hope that, by successive jumps, the Markov chain will actually explore the whole range of the target distribution. The most natural (and historically first) proposal in a continuous state space X is the random walk proposal,

$$r(x, x') = h(x' - x) ,$$

where h is a symmetric density. The Metropolis-Hastings acceptance probability is then

$$\alpha(x, x') = \frac{\pi(x')}{\pi(x)} \wedge 1 ,$$

due to the symmetry assumption on h. Once again, the chain $\{\xi^i\}_{i \geq 1}$ thus visits each state x in proportion to $\pi(x)$.

Example 6.2.10 (Squared and Noisy Autoregression, Continued). The conditional distribution of X_k given X_{k-1}, X_{k+1} and Y_k (6.9) is generally bimodal as in Figure 6.2. For some occurrences of X_{k-1}, X_{k+1} and Y_k, the zone located in between the modes has a very low probability under the conditional distribution. If we use a Gaussian random walk, i.e., $h = \mathrm{N}(0, \rho^2)$, with a scale ρ that is too small, the random walk will never jump to the other mode. This is illustrated in Figure 6.4 for $\rho = 0.1$. On the opposite, if the scale ρ is sufficiently large, the corresponding Markov chain will explore both

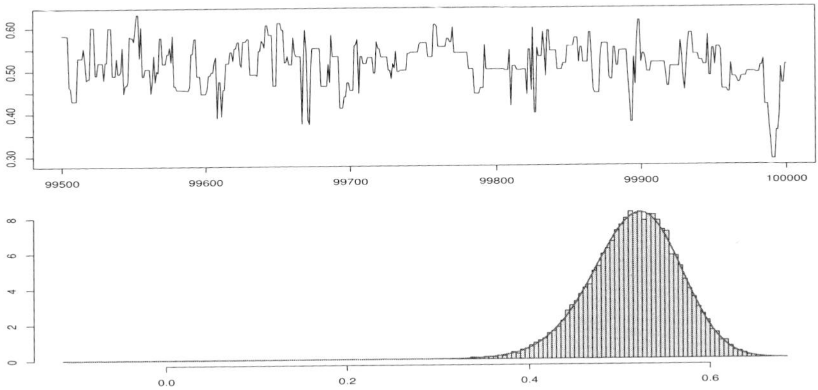

Fig. 6.4. Illustration of Example 6.2.10. Same legend as Figure 6.2 but for a different outcome of (X_{t-1}, X_{t+1}, Y_t) and with the Markov chain based on a random walk with scale $\rho = 0.1$.

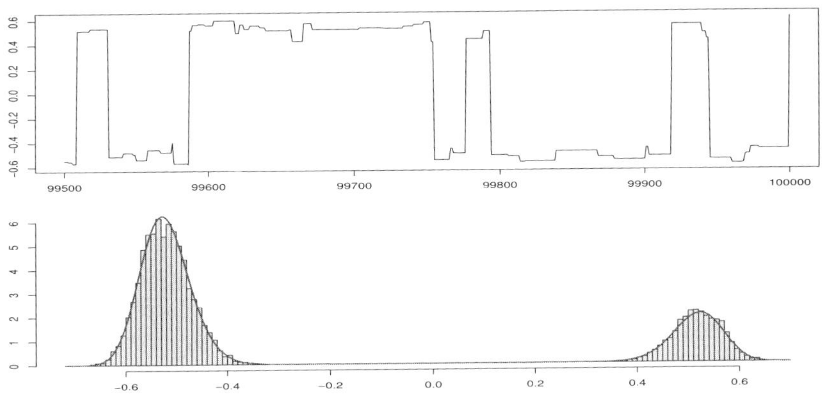

Fig. 6.5. Illustration of Example 6.2.10. Same legend and data set (X_{t-1}, X_{t+1}, Y_t) as Figure 6.4 but with the Markov chain based on a random walk with scale $\rho = 0.5$.

modes and give a satisfactory approximation of the target distribution, as shown by Figure 6.5 for $\rho = 0.5$.

Comparing Figures 6.4 and 6.5 also confirms that a higher acceptance rate does not necessarily imply, by far, a better performance of the Metropolis-Hastings algorithm (in Figure 6.4, the acceptance rate is about 50% and it drops to 13% in the case of Figure 6.5). Especially with random walk proposals, it is normal to observe a fair amount of rejections when the algorithm is properly tuned. ■

Even though the choice of a symmetric density h seems to offer less opportunities for misbehaving, there are two levels at which the algorithm may err: one is related to tail behavior, namely that the tail of h must be heavy enough if geometric convergence is to occur (Mengersen and Tweedie, 1996); and the other is the scale of the random walk. From a theoretical point of view, note that the random walk Metropolis-Hastings kernel is never uniformly ergodic in unbounded state spaces X (Robert and Casella, 2004, Section 7.5). Depending on which scale is chosen, the Markov chain may be very slow to converge either because it moves too cautiously (if the scale is too small) or too wildly (if the scale is too large). Based on time-scaling arguments (i.e., continuous-time limits for properly rescaled random walk Metropolis-Hastings chains), Roberts and Rosenthal (2001) recommend setting the acceptance rate in the range 0.2–0.35, which can be used as a guideline to select the scale of the random walk. In cases similar to the one considered in Example 6.2.10, with well-separated modes, it is customary to observe that the "best" scaling of the proposal (in terms of the empirical correlation of the MCMC chain for instance) corresponds to an acceptance rate that is even lower than these numbers. Unexpected multimodality really is a very significant difficulty in this respect: if the target distribution has several separated modes that are not expected, a random walk with too small a scale can miss those modes without detecting a problem with convergence, as the exploration of the known modes may well be very satisfactory, as exemplified in Figure 6.4.

Example 6.2.11 (Cauchy Meets Normal, Continued). To keep up with the spirit of this toy example, we also try to use in this case a Gaussian random walk as a proposal. The corresponding acceptance probability is then

$$\alpha(x, x') = \frac{1 + x^2}{1 + (x')^2} \wedge 1 .$$

Figure 6.6 illustrates the performance of the algorithm in this setting. The graphic fit of the Cauchy density by the histogram is good but, if we follow Roberts and Tweedie (2005) and look at the chain in more detail, it appears that after 10,000 iterations the range of the chain is $(-14.44, 15.57)$, which shows that the chain fails to explore in a satisfactory fashion the tails of the Cauchy distribution. In fact, the 99% quantile of the Cauchy $\mathrm{C}(0, 1)$ distribution is 31, implying that on average 200 points out of the 10,000 first values of the Markov chain should be above 31 in absolute value! Roberts and Tweedie (2005) show in essence that, when the density of the random walk has tails that are not heavy enough, the corresponding Markov chain is not geometrically ergodic.

■

The two previous categories are the most common choices for the proposals density r, but they are by no means the only or best choices. For instance, in a large-dimension compact state space with a concentrated *target* distribution π, the uniform proposal is very inefficient in that it leads to a very low average acceptance probability; this translates, in practice, to the chain $\{\xi^i\}_{i\geq 1}$ being essentially constant. Similarly, using the random walk proposal with a small scale parameter while the target is multimodal with a very low density in between the modes may result in the chain never leaving its initial mode.

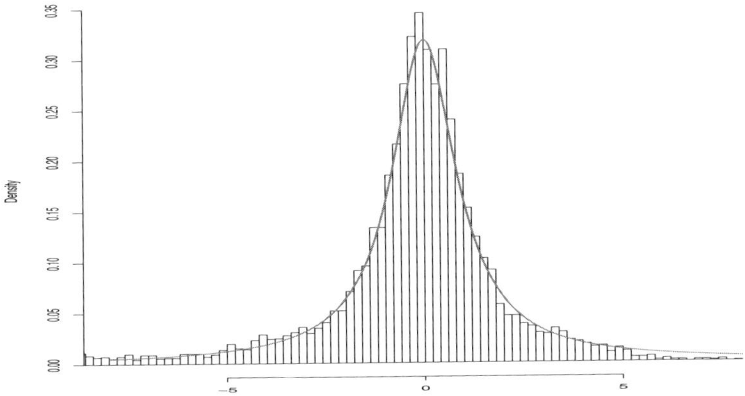

Fig. 6.6. Illustration of Example 6.2.11. Histogram of the 10,000 first steps of a random walk Metropolis-Hastings Markov chain using a Gaussian proposal with scale 1 and Cauchy target distribution.

6.2.4 Hybrid Algorithms

Although the Metropolis-Hastings rule of Algorithm 6.2.5 is our first effective approach for constructing MCMC samplers, we already have a number of available options, as we may freely choose the proposal distribution r. A natural question to ask in this context is to know whether it is possible to build new samplers from existing ones. It turns out that there are two generic and easily implemented ways of combining several MCMC samplers into a new one, which we shall refer to as a *hybrid sampler*. The following lemma is easy to prove from the corresponding definitions of Chapters 2 and 14.

Lemma 6.2.12 (Hybrid Kernels). *Assume that $K_1, \ldots, K_m$ are Markov transition kernels that all admit π as stationary distribution. Then*

(a) $K_{\text{syst}} = K_1 K_2 \cdots K_m$ and
(b) $K_{\text{rand}} = \sum_{i=1}^{m} \alpha_i K_i$, with $\alpha_i > 0$ for $i = 1, \ldots, m$ and $\sum_{i=1}^{m} \alpha_i = 1$,

also admit π as stationary distribution. If in addition $K_1, \ldots, K_m$ are π reversible, K_{rand} also is π reversible but K_{syst} need not be.

Both of these constructions are easily implemented in practice: in (a), each iteration of the hybrid sampler consists in systematically cycling through the m available MCMC kernels; in (b), at each iteration we first toss an m-ary coin with probability of turning i equal to α_i and then apply the MCMC kernel K_i. The additional warning that K_{syst} may not be π reversible (even if all the individual kernels K_i are) is not a problem *per se*. Reversibility is not a necessary condition for MCMC, it is only prevalent because it is easier to devise rules that enforce the (strongest) detailed balance condition. Note also that it is always possible to induce reversibility by appropriate modifications of the cycling strategy. For instance, the symmetric combination $K_{\text{syst}}K_{\text{rev}}$ with $K_{\text{rev}} = K_m K_{m-1} \cdots K_1$ is easily checked to be π reversible. In practice, it means that the cycle through the various available MCMC kernels K_i has to be done in descending and then ascending order.

Regarding irreducibility, it is clear that the *random scan* kernel K_{rand} is guaranteed to be phi-irreducible if at least one of the kernels K_i is. For the *systematic scan* strategy, the situation is more complex and K_{syst} may fail to be phi-irreducible even in cases where all the individual kernels K_i are phi-irreducible (with common irreducibility measure ϕ). A more useful remark is that if $K_1, \ldots, K_m$ all admit π as stationary distribution but are not phi-irreducible—meaning that they do not yet correspond to fully functional converging MCMC algorithms—there are cases where both K_{syst} and K_{rand} are phi-irreducible. It is thus possible to build viable sampling strategies from individual MCMC transitions that are not in themselves fully functional. The main application of this remark is to break large-dimensional problems into smaller ones by modifying only one part of the state at a time.

6.2.5 Gibbs Sampling

When the distribution of interest is multivariate, it may be the case that for each particular variable, its conditional distribution given all remaining variables has a simple form. This is in particular the case for models specified using conditional independence relations like HMMs and more general latent variable models. In this case, a natural MCMC algorithm is the so-called *Gibbs sampler*, which we now describe. Its name somehow inappropriately stems from its use for the simulation of Gibbs Markov random fields by Geman and Geman (1984).

6.2.5.1 A Generic Conditional Algorithm

Suppose we are given a joint distribution with probability density function π on a space X such that $x \in \mathsf{X}$ may be decomposed into m components $x = (x_1, \ldots, x_m)$, where $x_k \in \mathsf{X}_k$. If k is an index in $\{1, \ldots, m\}$, we shall denote by x_k the kth component of x and by $x_{-k} = \{x_l\}_{l \neq k}$ the collection of remaining components. We further denote by $\pi_k(\cdot|x_{-k})$ the conditional probability density function of X_k given $\{X_l\}_{l \neq k}$ and assume that simulation from this conditional distribution is feasible (for $k = 1, \ldots, m$). Note that x_k is not necessarily scalar but may be itself vector-valued.

Algorithm 6.2.13 (Gibbs Sampler). Starting from an initial arbitrary state ξ^1, update the current state $\xi^i = (\xi_1^i, \ldots, \xi_m^i)$ to a new state ξ^{i+1} as follows.

For $k = 1, 2, \ldots, m$: Simulate ξ_k^{i+1} from $\pi_i(\cdot|\xi_1^{i+1}, \ldots, \xi_{k-1}^{i+1}, \xi_{k+1}^i, \ldots, \xi_m^i)$.

In other words, in the kth round of the cycle needed to simulate ξ^{i+1}, the kth component is updated by simulation from its conditional distribution given all other components (which remain fixed). This new value then supersedes the old one and is used in the subsequent simulation steps. A complete round of m conditional simulations is usually referred to as a *sweep* of the algorithm. Another representation of the Gibbs sampler is to break the complete cycle as a combination of m individual MCMC steps where only one of the m components is modified according to the corresponding conditional distribution. This approach is easily recognized as the combination of type (a)—systematic cycling—in Lemma 6.2.12. Hence we know from Lemma 6.2.12 that the correct behavior of the complete cycle can be inferred from that of the individual updates. The next result is a first step in this direction.

Proposition 6.2.14 (Reversibility of Individual Gibbs Steps). *Each of the m individual steps of the Gibbs sampler (Algorithm 6.2.13) is π reversible and thus admits π as a stationary probability density function.*

Proof. Consider the step that updates the kth component and denote by K_k the corresponding transition kernel. We can always write $\lambda = \lambda_k \otimes \lambda_{-k}$ where λ_k and λ_{-k} are measures on X_k and X_{-k}, respectively, such that λ_k dominates $\pi_k(\cdot|x_{-k})$ for all values of $x_k \in \mathsf{X}_k$. With these notations,

$$K_k(x, dx') = \delta_{\{x_{-k}\}}(dx'_{-k})\pi_k(x'_k|x_{-k})\, \lambda_k(dx'_k)\ .$$

Hence, for any functions $f_1, f_2 \in \mathcal{F}_{\mathrm{b}}\left(\mathsf{X}\right)$,

$$\iint f_1(x) f_2(x')\pi(x)\, \lambda(dx)\, K(x, dx') = \\ \int \left\{ f_1(x)\pi(x)\, \lambda_k(dx_k) \int f_2(x'_k, x_{-k})\pi_k(x'_k|x_{-k})\, \lambda_k(dx'_k) \right\} \lambda_{-k}(dx_{-k})\ ,$$

where (x'_k, x_{-k}) refers to the element u of X such that $u_k = x'_k$ and $u_{-k} = x_{-k}$. Because $\pi(x_k, x_{-k})\pi_k(x'_k|x_{-k}) = \pi_k(x_k|x_{-k})\pi(x'_k, x_{-k})$, we may also write

$$\iint f_1(x) f_2(x')\pi(x)\, \lambda(dx)\, K(x, dx') = \\ \int \Big\{ f_2(x'_k, x_{-k})\pi(x'_k, x_{-k})\, \lambda_k(dx'_k) \\ \times \int f_1(x_k, x_{-k})\pi_k(x_k|x_{-k})\, \lambda_k(dx_k) \Big\} \lambda_{-k}(dx_{-k})\ ,$$

which is the same expression as before where the roles of f_1 and f_2 have been exchanged, thus showing that the detailed balance condition (2.12) holds. □

An insightful interpretation of Proposition 6.2.14 is that each step corresponds to a very special type of Metropolis-Hastings move where the acceptance probability is uniformly equal to 1, due to choice of π_k as the proposal distribution. However, Proposition 6.2.14 does not suffice to establish proper convergence of the Gibbs sampler, as none of the individual steps produces a phi-irreducible chain. Only the combination of the m moves in the complete cycle has a chance of producing a chain with the ability to visit the whole space X from any starting point. Of course, one can also adopt the combination of type (b) in Lemma 6.2.12 to obtain the *random scan Gibbs sampler* as opposed to the *systematic scan Gibbs sampler*, which corresponds to the solution exposed in Algorithm 6.2.13. We refer to (Robert and Casella, 2004) and (Roberts and Tweedie, 2005) for more precise convergence results pertaining to these variants of the Gibbs sampler.

One perspective that is somehow unique to Gibbs sampling is *Rao-Blackwellization*, named after the Rao-Blackwell theorem used in classical statistics (Lehmann and Casella, 1998) and recalled as Proposition A.2.5. It is in essence a variance reduction technique (see Robert and Casella, 2004, Chapter 4) that takes advantage of the conditioning abilities of the Gibbs sampler. If only a part of the vector x is of interest (as is often the case in latent variable models), say x_k, Rao-Blackwellization consists in replacing the empirical average

$$\hat{\pi}_N^{\mathrm{MCMC}}(f) = \frac{1}{N}\sum_{i=1}^{N} f(\xi_k^i) \qquad \text{with} \qquad \hat{\pi}_N^{\mathrm{RB}}(f) = \frac{1}{N}\sum_{i=1}^{N} \mathrm{E}_\pi[f(\xi_k)\,|\,\xi_{-k}^i] ,$$

where $\{\xi^i\}_{i\geq 1}$ denotes the chain produced by Algorithm 6.2.13. This is of course only feasible in cases where the integral of the function of interest f under $\pi_k(\cdot|x_{-k})$ may be easily evaluated for all $x \in \mathsf{X}$. In i.i.d. settings, $\hat{\pi}_N^{\mathrm{MCMC}}(f)$ would be more variable than $\hat{\pi}_N^{\mathrm{RB}}(f)$ by Proposition A.2.5. For Markov chain simulations $\{\xi^i\}_{i\geq 1}$, this is not necessarily the case, and it is only in specific situations (see Robert and Casella, 2004, Sections 9.3 and 10.4.3) that the latter estimate can be shown to be less variable. Another substantial benefit of Rao-Blackwellization is to provide an elegant method for the approximation of probability density functions of the different components of x. Indeed,

$$\frac{1}{N}\sum_{i=1}^{N} \pi_k\left(\cdot\,|\xi_{-k}^i\right)$$

is unbiased and converges to the marginal density of kth component, under the target distribution. If the conditional probability density functions are available in closed form, it is unnecessary (and inefficient) to use nonparametric density estimation methods such as kernel methods for postprocessing the output of Gibbs sampling.

We now discuss a clever use of the Gibbs sampling principle, known as the *slice sampler*, which is of interest in its own right.

6.2.5.2 The Slice Sampler

Proposition 6.2.2 asserts that the bivariate random variable (X, U) whose distribution is uniform on

$$\mathscr{S}_\pi = \left\{(x, u) \in \mathsf{X} \times \mathbb{R}^+ : 0 \le u \le \pi(x)\right\} ,$$

is such that the marginal distribution of X is π. This observation is at the core of the accept-reject algorithm discussed in Section 6.2.1. We will use the letter U to denote uniform distributions on sets, writing, for instance, $(X, U) \sim \mathrm{U}(\mathscr{S}_\pi)$.

From the perspective of MCMC algorithms, we can consider using a *random walk* on $\mathscr{S}_\pi$ to produce a Markov chain with stationary distribution equal to this uniform distribution on $\mathscr{S}_\pi$. There are many ways of implementing a random walk on this set, but a natural solution is to go one direction at a time, that is, to move iteratively along the u-axis and then along the x-axis. Furthermore, we can use uniform moves in both directions; that is, starting from a point (x, u) in $\mathscr{S}_\pi$, the move along the u-axis will correspond to the conditional distribution

$$\mathrm{U}\left(\{u : u \le \pi(\xi)\}\right) , \tag{6.10}$$

resulting in a change from point (x, u) to point (x, u'), still in $\mathscr{S}_\pi$, and then the move along the ξ-axis to the conditional distribution

$$\mathrm{U}\left(\{x : \pi(x) \ge u'\}\right) , \tag{6.11}$$

resulting in a change from point (x, u') to point (x', u'). This set of proposals is the basis chosen for the original *slice sampler* of Damien and Walker (1996), Neal (1997) (published as Neal, 2003), and Damien *et al.* (1999).

Algorithm 6.2.15 (Slice Sampler). Starting from an arbitrary point (ξ^1, U^1) in $\mathscr{S}_\pi$, simulate for $i \ge 1$,

1. $U^{i+1} \sim \mathrm{U}\left([0, \pi(\xi^i)]\right)$;
2. $\xi^{i+1} \sim \mathrm{U}\left(S(U^{i+1})\right)$, with $S(u) = \{x : \pi(x) \ge u\}$.

The important point here is that Algorithm 6.2.15 is validated as a Gibbs sampling method, as steps 1 and 2 in the above are simply the conditional distributions of U and ξ associated with the joint distribution $\mathrm{U}(\mathscr{S}_\pi)$.

Obviously, this does not make the slice sampler a universal generator: in many settings, resolving the simulation from the uniform $\mathrm{U}(S(u))$ is just as hard (and impossible) as to generate directly from π, and extensions are often necessary (Robert and Casella, 2004, Chapter 8). Still, this potential universality shows that Gibbs sampling does not only pertain to a special category of hierarchical models.

Example 6.2.16 (Single Site Conditional Distribution in Stochastic Volatility Model). To illustrate the slice sampler, we consider the stochastic volatility model discussed in Example 1.3.13 whose state-space form is as follows:

$$X_{k+1} = \phi X_k + \sigma U_k ,$$
$$Y_k = \beta \exp(X_k/2) V_k ,$$

where $\{U_k\}_{k\geq 0}$ and $\{V_k\}_{k\geq 0}$ are independent standard Gaussian white noise processes. In this model, $\beta^2 \exp(X_k)$ is referred to as the volatility, and its estimation is one of the purposes of the analysis (see Example 1.3.13 for details). As in Example 6.2.8 above, we consider the conditional distribution of X_k given X_{k-1}, X_{k+1} and Y_k, whose transition density function $\pi_k(x|x_{k-1}, x_k)$ is proportional to

$$\exp\left[-\left\{\frac{(x_{k+1}-\phi x)^2}{2\sigma^2} + \frac{(x-\phi x_{k-1})^2}{2\sigma^2}\right\}\right] \frac{1}{\beta \exp(x/2)} \exp\left[-\frac{y_k^2}{2\beta^2 \exp(x)}\right] , \tag{6.12}$$

ignoring constants. In fact, terms that do not depend on x can be ignored as well, and we may complete the square (in x) to obtain

$$\pi_k(x|x_{k-1}, x_k) \propto \exp\left[-\frac{1+\phi^2}{2\sigma^2}\left\{(x-\mu_k)^2 + \frac{y_k^2 \sigma^2}{(1+\phi^2)\beta^2} \exp(-x)\right\}\right] ,$$

where

$$\mu_k = \frac{\phi(x_{k+1}+x_{k-1}) - \sigma^2/2}{1+\phi^2} . \tag{6.13}$$

Defining

$$\alpha_k = \frac{y_k^2 \sigma^2 \exp(-\mu_k)}{(1+\phi^2)\beta^2} \quad \text{and} \quad \rho = \frac{1+\phi^2}{2\sigma^2} , \tag{6.14}$$

$\pi_k(x|x_{k-1}, x_k)$ is thus proportional to

$$\exp\left[-\rho\left\{(x-\mu_k)^2 + \alpha_k \exp[-(x-\mu_k)]\right\}\right] .$$

The parameter μ_k corresponds to a simple shift that poses no simulation problem. Hence, the general form of the conditional probability density function from which simulation is required is $\exp[-\rho\{x^2 + \alpha \exp(-x)\}]$ for positive values of ρ and α. Shephard and Pitt (1997) (among others) discuss an approach based on accept-reject ideas for carrying out this conditional simulation, but we may also use the slice sampler for this purpose. The second step of Algorithm 6.2.15 then requires simulation from the uniform distribution on the set

$$S(u) = \left\{x : \exp[-\rho\{x^2 + \alpha \exp(-x)\}] \geq u\right\} = \left\{x : x^2 + \alpha \exp(-x) \leq \omega\right\} ,$$

setting $\omega = -(1/\rho)\log u$. Now, while the inversion of $x^2 + \alpha \exp(-x) = \omega$ is not possible analytically, the fact that this function is convex (for $\alpha > 0$) and that the previous value of x belongs to the set $S(u)$ help in solving this equation by numerical trial-and-error or more elaborate zero-finding algorithms.

As pointed out by Neal (2003), there is also no need to solve precisely this equation, as knowledge of an interval that contains the set $S(u)$ is enough to simulate from the uniform distribution on $S(u)$: it then suffices to simulate candidates ξ uniformly from the larger set and accept them only if $\xi \in S(u)$ (which is also the accept-reject method but with a high acceptance rate that is controlled by the accuracy of the zero-finding algorithm). Figure 6.7 (top plot) shows that the fit between the histogram of 10,000 consecutive values produced by the slice sampler and the true distribution is quite satisfactory. In addition, the bottom plot shows that the autocorrelation between successive values of ξ^i is quite modest. This fast mixing of the one-dimensional slice sampler is an appealing feature that has been shown to hold under fairly general assumptions on the target distribution (Roberts and Rosenthal, 1998; Robert and Casella, 2004, Chapter 8). ■

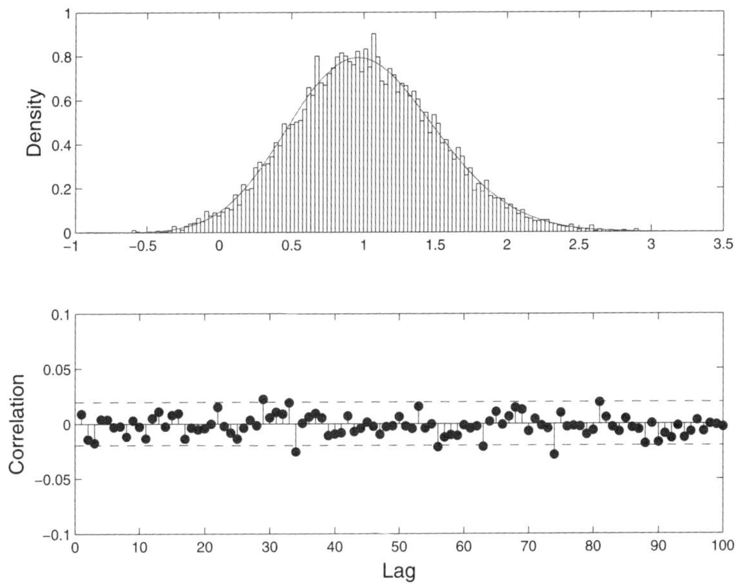

Fig. 6.7. Illustration of Example 6.2.16. Top: histogram of a Markov chain produced by the slice sampler for $\alpha = 5$ and $\rho = 1$ with target distribution in overlay. Bottom: correlogram with 95% confidence interval corresponding to the assumption of white noise.

6.2.6 Stopping an MCMC Algorithm

There is an intrinsic difficulty with using Markov chain Monte Carlo methods for simulation purposes in that, were we to stop the iterations "too early", we would still be influenced by the (arbitrary) starting value of the chain, and were we to stop the iteration "too late", we would be wasting simulation time. In contrast with what happens for independent Monte Carlo where (6.1) may be used to obtain confidence intervals, it is fairly difficult to estimate the accuracy of estimates derived from the MCMC sample because of the unknown correlation structure of the simulated ξ^i. Apart for often useful graphic diagnostics (trace of the samples, correlograms, comparison of histograms obtained with different starting points...), there exist (more or less) empirical rules that provide hints on when an MCMC sampler should be stopped. A branch of MCMC, known as *perfect sampling*, corresponds to a refinement of these rules in which the aim is to guarantee that the Markov chain, when observed at appropriate times, is exactly distributed from the stationary distribution. Not surprisingly, these methods are very difficult to devise and equally costly to implement. Another direction, generally referred to as *computable bounds*, consists in obtaining bounds on the convergence speed of MCMC-generated Markov chains. When available, such results are very powerful, as they do not require any empirical estimation, and the number of required MCMC simulations may be calibrated beforehand. Of course, the drawback here is that for complex samplers, typically hybrid samplers that incorporate several different MCMC sampling steps, such results are simply not available (Robert and Casella, 2004).

6.3 Applications to Hidden Markov Models

This section describes methods that may be used to simulate the unobservable sequence of states $X_{0:n}$ given the corresponding observations $Y_{0:n}$ in HMMs for which the direct (independent) Monte Carlo simulations methods discussed in Section 6.1.2 are not applicable.

We start from the most generic and easily implementable approaches in which each individual hidden state X_k is simulated conditionally on all X_l except itself. We then move to a more specific sampling technique that takes profit of the structure found in conditionally Gaussian linear state-space models (see, in particular, Definition 2.2.6 and Sections 4.2.3 and 5.2.6).

6.3.1 Generic Sampling Strategies

6.3.1.1 Single Site Sampling

We now formalize an argument that was underlying in Examples 6.2.8 and 6.2.16. Starting from the joint conditional distribution of $X_{0:n}$ given $Y_{0:n}$ defined (up to a proportionality constant) by (6.7), the conditional probability density function of a *single* variable in the hidden chain, X_k say, given $Y_{0:n}$ *and* its two neighbors X_{k-1} and X_{k+1} is such that

$$\phi_{k-1:k+1|n}(x_k|x_{k-1}, x_{k+1}) \propto \phi_{k-1:k+1|n}(x_{k-1}, x_k, x_{k+1}) \\ \propto q(x_{k-1}, x_k) q(x_k, x_{k+1}) g_k(x_k) . \qquad (6.15)$$

At the two endpoints $k = 0$ and $k = n$, we have the obvious corrections

$$\phi_{0:1|n}(x_0|x_1) \propto \nu(x_0) q(x_0, x_1)$$

and

$$\phi_{n-1:n|n}(x_n|x_{n-1}) \propto q(x_{n-1}, x_n) g_n(x_n) .$$

Therefore, if we aim at simulating the whole vector $X_{0:n}$ by the most basic Gibbs sampler that simulates one component of the vector at a time, $\phi_{k-1:k+1|n}(x_k|x_{k-1}, x_{k+1})$ is given by (6.15) in a simple closed-form expression. Remember that the expression looks simple only because knowledge of the normalization factor is not required for performing MCMC simulations.

In the case where X is finite, the simulation of $X_{0:n}$ by this Gibbs sampling approach is rather straightforward, as the only operations that are requested (for $k = 0, \ldots, n$) are

- computing $q(x_{k-1}, x) q(x, x_{k+1}) g_k(x)$ for all values of $x \in \mathsf{X}$ and normalizing them to form a probability vector π_k;
- simulating a value of the state according to π_k.

It is interesting to contrast this Gibbs sampling algorithm with the simpler Monte Carlo approach of Algorithm 6.1.1. A complete sweep of the Gibbs sampler is simpler to implement, as each Gibbs simulation step requires that r products be computed (where r is the cardinality of X). Hence, the complete Gibbs sweep requires $\mathrm{O}(r(n+1))$ operations compared to $\mathrm{O}\left(r^2(n+1)\right)$ for Algorithm 6.1.1 due to the necessity of computing all the filtering distributions by Algorithm 5.1.1. On the other hand, the Monte Carlo simulations obtained by Algorithm 6.1.1 are independent, which is not the case for those produced by Gibbs sampling. For a comparable computational effort, we may thus perform r times as many simulations by Gibbs sampling than by independent Monte Carlo. This does not necessarily correspond to a gain though, as the variance of MCMC estimates is most often larger than that of Monte Carlo ones due to the Markov dependence between successive samples. It remains that if the number of possible values of X_k is very large (a case usually found in related models used in applications such as image processing), it may be the case that implementing Monte Carlo simulation is overwhelming while the Gibbs sampler is still feasible.

It is generally true that, apart from this case (finite but very large state space), there are very few examples of hidden Markov models where the Gibbs sampling approach is applicable and the general Monte Carlo approach of Section 6.1.2.1 is not. This has to do with the fact that determining $\phi_{k-1:k+1|n}(\cdot|x_{k-1}, x_{k+1})$ exactly, not only up to a constant, involves exactly the same type of marginalization operation involved in the implementation of the filtering recursion. An important point to stress here is that replacing an exact simulation by a Metropolis-Hastings step in a general MCMC algorithm does not jeopardize its validity as long as the Metropolis-Hastings step is associated with the correct stationary distribution. Hence, the most natural alternative to the Gibbs sampler in cases where sampling from the full conditional distribution is not directly feasible is the *one-at-a-time Metropolis-Hastings* algorithm that combines successive Metropolis-Hastings steps that update only one of the variables. For $k = 0, \ldots, n$, we thus update the kth component x_k^i of the current simulated sequence of states x^i by proposing a new candidate for x_k^{i+1} and accepting it according to (6.5), using (6.15) as the target.

Example 6.3.1 (Single Site Conditional Distribution in Stochastic Volatility Model, Continued). We return to the stochastic volatility model already examined in Example 6.2.16 but with the aim of simulating complete sequences under the posterior distribution rather than just individual states. From the preceding discussion, we may use the algorithm described in Example 6.2.16 for each index ($k = 0, \ldots, n$) in the sequence of states to simulate. Although the algorithm itself applies to all indices, the expression of μ_k, α_k and ρ in (6.13)–(6.14) need to be modified for the two endpoints as follows.

For $k = 0$, the first term in (6.12) should be replaced by

$$\exp -\left\{\frac{(x_1 - \phi x)^2}{2\sigma^2} + \frac{(1-\phi^2)x^2}{2\sigma^2}\right\} , \tag{6.16}$$

as it is sensible to assume that the initial state X_0 is *a priori* distributed as the stationary distribution of the AR(1) process, that is, $\mathrm{N}\left(0, \sigma^2/(1-\phi^2)\right)$. Hence for $k = 0$, (6.13) and (6.14) should be replaced by

$$\begin{cases} \mu_0 = \phi x_1 - \sigma^2/2 , \\ \alpha_0 = Y_0^2 \sigma^2 \exp(-\mu_0)/\beta^2 , \\ \rho_0 = 1/(2\sigma^2) . \end{cases} \tag{6.17}$$

For $k = n$, the first term in (6.12) reduces to

$$\exp -\left\{\frac{(x - x_{n-1})^2}{2\sigma^2}\right\} , \tag{6.18}$$

and thus

$$\begin{cases} \mu_n = \phi x_{n-1} - \sigma^2/2 , \\ \alpha_n = Y_n^2 \sigma^2 \exp(-\mu_n)/\beta^2 , \\ \rho_n = 1/(2\sigma^2) , \end{cases} \tag{6.19}$$

replace (6.13) and (6.14).

An iteration of the complete algorithm thus proceeds by computing, for each index $k = 0, \ldots, n$ in turn, μ_k, α_k and ρ according to (6.13) and (6.14), or (6.17) or (6.19) if $k = 0$ or n. Then one iteration of the slice sampling algorithm discussed in Example 6.2.16 is applied.

For comparison purposes, we also consider a simpler alternative that consists in using a random walk Metropolis-Hastings proposal for the simulation of each individual site. As discussed in Section 6.2.3.2, the acceptance probability of the move at index k is given by

$$\alpha_k(x, x') = \frac{\pi_k(x')}{\pi_k(x)} \wedge 1 ,$$

where π_k is defined in (6.12) with the modifications mentioned in (6.16) and (6.18) for the two particular cases $k = 0$ and $k = n$. Remember that for random walk proposals, we are still free to choose the proposal density itself because, as long as it is of random walk type, it does not affect the acceptance ratio. Because the positive tail of π_k is equivalent to that of a Gaussian distribution with variance $(2\rho)^{-1} = \sigma^2/(1+\phi^2)$ and the negative one decays much faster, it seems reasonable to use a Gaussian random walk proposal with a standard deviation about $2.4 \times \sigma/\sqrt{1+\phi^2}$ based on (Roberts and Rosenthal, 2001)—see also discussion in Section 6.2.3.2 above about setting the scale of random walk proposals.

To compare the relative efficiency of these approaches, we use data simulated from the stochastic volatility model with parameter values corresponding to those fitted by Shephard and Pitt (1997) on log-returns of a historical daily exchange rate series, that is, $\phi = 0.98$, $\sigma = 0.14$, and $\beta = 0.66$. We first consider the case where $n = 20$ for which the simulated state trajectory and the observed data are plotted in Figure 6.8. Because of the highly non-linear nature of the model, comparing the values of daily log-return Y_k and those of the day volatility X_k is not very helpful. To provide a clearer picture, the crosses in Figure 6.8 represent $\hat{\sigma}_k^2 = \log(Y_k^2/\beta^2)$ rather than Y_k itself. Note that $\hat{\sigma}_k^2$ is the maximum likelihood estimate of the daily volatility X_k in the absence of an *a priori* model on the dynamics of the volatility sequence. It is also easily checked from (6.12) and similar expressions that the posterior distribution of the states depend only on the values of Y_k^2/β^2. Figure 6.8 shows that while larger values of $\log(Y_k^2/\beta^2)$ provide a rather good idea of the actual volatility, smaller ones look more like outliers and can be very far from the volatility (beware that the y-scale in Figure 6.8 is reversed). Indeed, a volatility value x rules out observations significantly larger (in magnitude) than, say, three times $\beta \exp(x/2)$, but not observations significantly smaller than $\beta \exp(x/2)$.

Figure 6.9 summarizes the output of 50,000 complete cycles of the single site slice sampling strategy on this data. The initial volatility sequence $x_{0:n}^1$, whose choice is arbitrary, was set to be zero at all sites. Obviously, in this model, the smoothing distributions are very dispersed and do not allow a precise estimation of the actual sequence of states. Note however that there is a possible misinterpretation of Figure 6.9, which would be that the most likely state sequence is the very smooth trajectory connecting the modes of the marginal smoothing distributions displayed here. This is not the case, and typical simulated sequence of states have variations comparable to that of the true sequence. But because of the large dispersion of the marginal

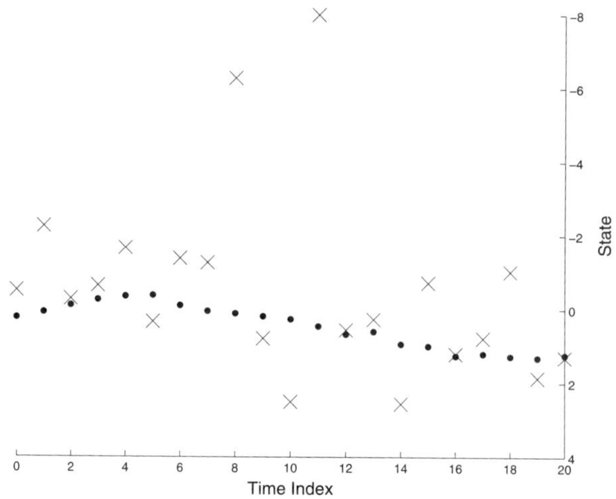

Fig. 6.8. Illustration of Example 6.3.1. Simulated data: values of X_k (black circles) and $\log(Y_k^2/\beta^2)$ (cross). Note that the ordinates (y-axis) run from top to bottom.

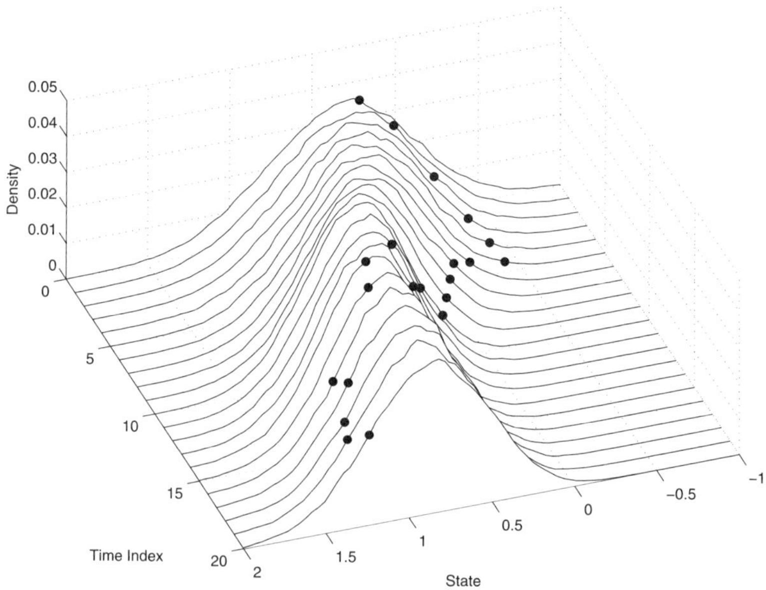

Fig. 6.9. Illustration of Example 6.3.1. Waterfall representation of the marginal smoothing distributions estimated from 50,000 iterations of the single site slice sampler (densities estimated with Epanechnikov kernel, bandwidth 0.05). The bullets show the true simulated state sequence.

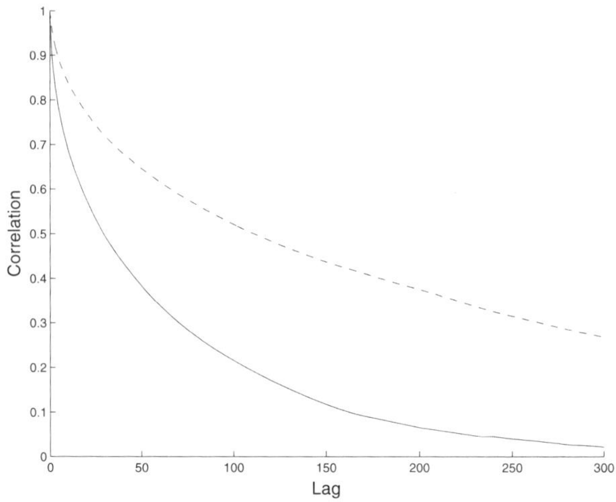

Fig. 6.10. Correlogram of the values simulated at index $k = 10$: solid line, single site slice sampler; dashed line, single site random walk Metropolis-Hastings.

posterior distributions and the absence of clearly marked posterior modes, their marginal averages produce the very smooth curves displayed here.

In this example the efficiency of the simulation algorithm itself is reasonable. To obtain Figure 6.9, for instance, 15,000 iterations would already have been sufficient, in the sense of producing no visible difference, showing that the sampler has converged to the stationary distribution. Figures such as 50,000 or even 15,000 may seem frightening, but they are rather moderate in MCMC applications. Figure 6.10 is the analog of the bottom plot in Figure 6.7, displaying the empirical autocorrelations of the sequence of simulated values for the state with index $k = 10$ (in the center of the sequence). It is interesting to note that while the single site slice sampler (Figure 6.7) produces a sequence of values that are almost uncorrelated, Figure 6.10 exhibits a strong positive correlation due to the interaction between neighboring sites.

Also shown in Figure 6.10 (dashed line) is the autocorrelation for the other algorithm discussed above, based on Gaussian random walk proposals for the simulation of each individual site. This second algorithm has a tuning parameter that corresponds to the standard deviation of the proposals. In the case shown in Figure 6.10, this standard deviation was set to $2.4 \times \sigma/\sqrt{1+\phi^2}$ as previously discussed. With this choice, the acceptance rates are of the order of 50%, ranging from 65% for edge sites ($k = 0$ and $k = n$) to 45% at the center of the simulated sequence. Figure 6.10 shows that this second algorithm produces successive draws that are more correlated (with positive correlation) than the single site slice sampling approach. A frequently used numerical measure of the performance of an MCMC sampler is twice the sum of the autocorrelations, over all the range of indices where the estimation is accurate (counting the value one that corresponds to the index zero only

We conclude this example by noting that slice sampling is obviously not the only available approach to tackle posterior simulation in this model and we do not claim that it is necessary the best one either. Because of its practical importance in econometric applications, MCMC approaches suitable for this model have been considered by several authors including Jacquier *et al.* (1994), Shephard and Pitt (1997) and Kim *et al.* (1998). ■

6.3.1.2 Block Sampling Strategies

In some cases, single site updating can be painfully slow. It is thus of interest to try to speed up the simulation by breaking some of the dependence involved in single site updating. A natural solution is to propose a joint update of a group of X_k, as this induces more variability in the simulated values. This strategy has been shown to be successful in some particular models (Liu *et al.*, 1994). The drawback of this approach however is that when the size of the blocks increases, it is sometimes difficult to imagine efficient proposal strategies in larger dimensional spaces. For the stochastic volatility model discussed above for instance, Shephard and Pitt (1997) discuss the use of approximations based on Gaussian expansions.

There are no general rules here, however, and the eventual improvements in mixing speed have to be gauged at the light of the extra computational efforts required to simulate larger blocks. In the case of multivariate Gaussian distributions for instance, simulating in blocks of size m involves computing the Cholevski factorization of m by m matrices, an operation whose cost is of order m^3. Hence moving to block simulations will be most valuable in cases where single site sampling is pathologically slow.

6.3.2 Gibbs Sampling in CGLSSMs

For the stochastic volatility model, Kim *et al.* (1998) (among others) advocate the use of a specific technique that consists in approximating the behavior of the model by a conditionally Gaussian linear state-space structure. This makes sense as there are simulation techniques specific to CGLSSMs that are usually more efficient than generic simulation methods. This is also the main reason why CGLSSMs are often preferred to less structured (but perhaps more accurate) alternative models in a variety of situations such as "heavy tailed" noise or outliers as in Examples 1.3.11 and 1.3.10, non-Gaussian observation noise (Kim *et al.*, 1998), or signals (Cappé *et al.*, 1999). Not surprisingly, efficient simulation in CGLSSMs is a topic that has been considered by many authors, including Carter and Kohn (1994), De Jong and Shephard (1995), Carter and Kohn (1996), and Doucet and Andrieu (2001).

$k = 10$ is equal to 110, that is, similar to what was observed for the shorter ($n = 20$) sequence.

In this context, the most natural approach to simulation consists in adequately combining the two specific Monte Carlo techniques discussed in Section 6.1.2 (for the finite state space case and Gaussian linear state-space models). Indeed, if we assume knowledge of the indicator sequence $C_{0:n}$, the continuous component of the state, $\{W_k\}_{0\leq k\leq n}$, follows a non-homogeneous Gaussian linear state-space model from which one can sample (block-wise) by Algorithms 6.1.2 or 6.1.3. If we now assume that $W_{0:n}$ is known, Figure 1.6 clearly corresponds to a (non-homogeneous) finite state space hidden Markov model for which we may use Algorithm 6.1.1. To illustrate this conditional two-step block simulation approach, we consider an illustrative example.

Example 6.3.2 (Non-Gaussian Autoregressive Process Observed in Noise). Example 1.3.8 dealt with the case of a Gaussian autoregressive process observed in noise. When the state and/or observation noises are non-Gaussian, a possible solution is to represent the corresponding distributions by mixtures of Gaussians. The model then becomes a CGLSSM according to

$$W_{k+1} = AW_k + \overbrace{\begin{bmatrix} \rho(C_{k+1}) \\ 0 \\ \vdots \\ 0 \end{bmatrix}}^{R(C_{k+1})} U_k \;, \tag{6.20}$$

$$Y_k = \begin{bmatrix} 1 \; 0 \cdots \; 0 \end{bmatrix} W_k + S(C_k) V_k \;, \tag{6.21}$$

where the matrix A is the companion matrix defined in (1.11), which is such that $W_{k+1}(1)$ (the first coordinate of W_{k+1}) is the regression $\sum_{i=1}^{p} \phi_i W_k(i)$, whereas the rest of the vector W_{k+1} is simply a copy of the first $p-1$ coordinates of W_k.

By allowing ρ and S to depend on the indicator sequence C_k, either the state or the observation noise (or both) can be represented as finite scale mixtures of Gaussians. We will assume in the following that $\{C_k\}_{k\geq 0}$ is a Markov chain taking values in the finite set $\{1, \ldots, r\}$; the initial distribution is denoted by ν_C, and the transition matrix by Q_C. In addition, we will assume that W_0 is $\mathrm{N}(0, \Sigma_{W_0})$ distributed where Σ_{W_0} does not depend on the indicator C_0.

The simulation of the continuous component of the state W_k for $k = 0, \ldots, n$, conditionally on $C_{0:n}$, is straightforward: for a specified sequence of indicators, (6.20) and (6.21) are particular instances of a non-homogeneous Gaussian linear state-space model for which Algorithm 6.1.3 applies directly. Recall that due to the particular structure of the matrix A in (6.20), the noisy AR model is typically an example for which the disturbance smoothing (Algorithm 5.2.15) will be more efficient.

For the simulation of the indicator variables given the disturbances $U_{0:n-1}$, two different situations can be distinguished.

Indicators in the Observation Equation: If ρ is constant (does not depend on C_k), only the terms related to the observation equation (6.21) contribute to the posterior joint distribution of the indicators $C_{0:n}$ whose general expression is given in (4.10). Hence the joint posterior distribution of the indicators satisfies

$$\psi_{0:n|n}(c_{0:n}|w_{0:n}, y_{0:n}) \propto \nu_C(c_0) \left\{ \prod_{k=0}^{n-1} Q_C(c_k, c_{k+1}) \right\} \prod_{k=0}^{n} \frac{1}{S(c_k)} \exp\left[-\frac{(y_k - w_k)^2}{2S^2(c_k)} \right] , \quad (6.22)$$

where factors that do not depend on the indicator variables have been omitted. Equation (6.22) clearly has the same structure as the joint distribution of the states in an HMM given by (3.13). Because C_k is finite-valued, we may use Algorithm 5.1.1 for filtering and then Algorithm 6.1.1 for sampling granted that the function g_k be defined as

$$g_k(c) = \frac{1}{S(c)} \exp\left[-\frac{(y_k - w_k)^2}{2S^2(c)} \right] . \quad (6.23)$$

Indicators in the Dynamic Equation: In the opposite case, S is constant but ρ is a function of the indicator variables. The joint distribution of the indicators $C_{0:n}$ given $W_{0:n}$ and $Y_{0:n}$ depends on the quantities defining the dynamic equation (6.20) only, according to

$$\psi_{0:n|n}(c_{0:n}|w_{0:n}, y_{0:n}) \propto \nu_C(c_0) \left\{ \prod_{k=0}^{n-1} Q_C(c_k, c_{k+1}) \right\} \prod_{k=1}^{n} \frac{1}{\rho(c_k)} \exp\left[-\frac{u_{k-1}^2}{2\rho^2(c_k)} \right] , \quad (6.24)$$

where $u_k \stackrel{\text{def}}{=} w_{k+1} - A w_k$. Algorithms 5.1.1 and 6.1.1 once again apply with g_k defined as

$$g_k(c) = \frac{1}{\rho(c)} \exp\left[-\frac{u_{k-1}^2}{2\rho^2(c)} \right] \quad (6.25)$$

for $k = 1, \ldots, n$ and $g_0 = 1$. Note that in this second case, we do not need to condition on the sequence of states $W_{0:n}$, and knowledge of the disturbances $U_{0:n-1}$ is sufficient. In particular, when using Algorithm 6.1.3 (conditionally given $C_{0:n}$), one can omit the last two steps to keep track only of the simulated disturbance sequence

$$\hat{U}_{k|n} + U_k^* - \hat{U}_{k|n}^* \qquad (\text{for } k = 0, \ldots, n-1) ,$$

using the notations introduced in Algorithm 6.1.3.

Of course, in cases where the indicator variables modify the variances of both the state noise and the observation noise, the two cases considered above should be merged, which involves in particular that the functions g_k be defined as the product of the expressions given in (6.23) and (6.25), respectively. ■

In general, the algorithm described above is reasonably successful. However, the rate of convergence of the MCMC sampler typically depends on the values of the parameters and the particular data under consideration: it can be slow in adverse situations, making it difficult to reach general conclusions. There are however a number of cases of practical importance where the algorithm fails. This has to do with the fact that in some situations, there is a very close association between the admissible values of the continuous component $\{W_k\}_{k\geq 0}$ and the indicator variables $\{C_k\}_{k\geq 0}$ leading to a very slow exploration of the space by the MCMC simulations. This happens in particular when using so-called Bernoulli-Gaussian noises (Kormylo and Mendel, 1982; Lavielle, 1993; Doucet and Andrieu, 2001). In the model of Example 6.3.2 for instance, if we just want to model outlying values—a model of interest in audio restoration applications (Ó Ruanaidh and Fitzgerald, 1996; Godsill and Rayner, 1998)—we could set $S = 0$ in the absence of outliers (say if $C_k = 1$) and $S = \sigma$, where σ^2 is large compared to the variance of $\{W_k\}_{k\geq 0}$, in the opposite case ($C_k = 2$). In this case, however, it is easily seen from (6.23) that $C_k = 1$ is only possible if $W_k = Y_k$ and, conversely, $W_k = Y_k$ has zero probability (remember that it is a continuous variable) unless $C_k = 1$. Hence the above algorithm would be fully stuck in that case. Not surprisingly, if $S^2(1)$ is not exactly equal to 0 but still very small (compared to the variance of $\{W_k\}_{k\geq 0}$), the Gibbs sampling approach, which simulates $W_{0:n}$ and then $C_{0:n}$ conditionally on each other, both block-wise, is not very efficient. We illustrate this situation with a very simple instance of Example 6.3.2.

Example 6.3.3 (Gaussian AR Process with Outliers). We consider again (6.20) and (6.21) in the AR(1) case, that is, when all variables in the models are scalar. For the state equation, the parameters are set as

$$A \stackrel{\text{def}}{=} \phi = 0.98 \qquad \text{and} \qquad R = \sqrt{1-\phi^2}\ ,$$

so that the stationary distribution of $\{W_k\}_{k\geq 0}$ is Gaussian with unit variance. We first assume that $S = 3$ in the presence of outliers and 0.2 otherwise, corresponding to a moderately noisy signal in the absence of outliers. By convention, $C_k = 2$ will correspond to the presence of an outlier at index k and we set $C_k = 1$ otherwise.

The light curve in the top plot of Figure 6.13 displays the corresponding simulated observations, where outliers have been generated at (arbitrarily selected) indices 25, 50, and 75. For modeling purposes, we assume that outliers occur independently of each other and with probability 0.95. The alternating block sampling algorithm discussed above is applied by initially setting $C_k^1 = 1$ for $k = 0, \ldots, n$, thus assuming that there are no outliers. Then $W_{0:n}^i$

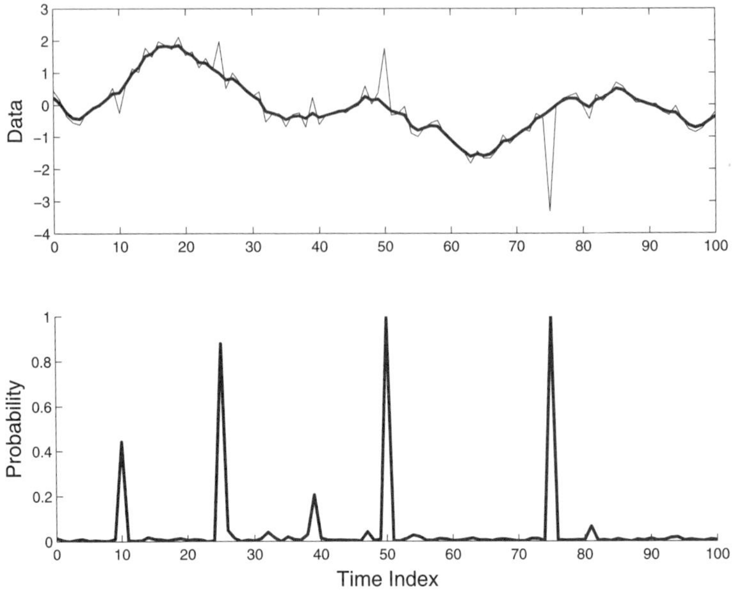

Fig. 6.13. Top plot: observed signal (light curve) and estimated state sequence (bold curve) as estimated after 500 iterations of alternating block sampling from $C_{0:n}$ and $W_{0:n}$. Bottom plot: estimated probability of presence of an outlier; $S(1) = 0.2$ in this case.

is simulated (as a block) conditionally on $C_{0:n}^i$, and $C_{0:n}^{i+1}$ conditionally on $W_{0:n}^i$, for $i = 1$ to 500. The bottom plot in Figure 6.13 displays estimates of the probability of the presence of an outlier at index k obtained by counting the number of times where $C_k^i = 1$. Not surprisingly, the three outliers are clearly localized, although there are two or three other points that could also be considered as outliers given the model, with some degree of plausibility. The bold curve in the top plot of Figure 6.13 shows the average of the simulated state sequences $W_{0:n}^i$. This is in fact a very good approximation of the actual state sequence that is not shown here because it is would be nearly indiscernible from the estimated state sequence in this case.

We now keep the same sequence of states and observation noises but consider the case where $S(1) = 0.02$, that is, ten times smaller than before. In some sense, the task is easier now because there is almost no observation noise except for the outliers, so that localizing them should be all the more easy. Figure 6.14, which is the analog of the top plot in Figure 6.13, shows that it is indeed not the case as the outlier located at index 25 is visibly not detected, resulting in a grossly incorrect estimation of the underlying state at index 25. The source of the problem is transparent: because initially $C_k^1 = 1$ for all indices, simulated values of W_k are very close to the observation Y_k because

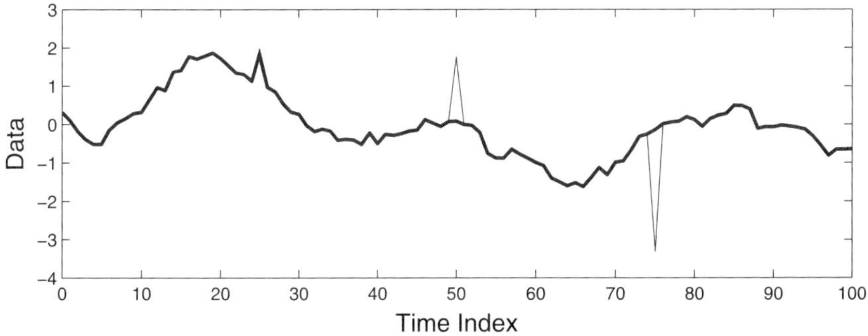

Fig. 6.14. Observed signal (light curve) and estimated state sequence (bold curve) as estimated after 500 iterations of alternating block sampling from $C_{0:n}$ and $W_{0:n}$ when $S(1) = 0.02$.

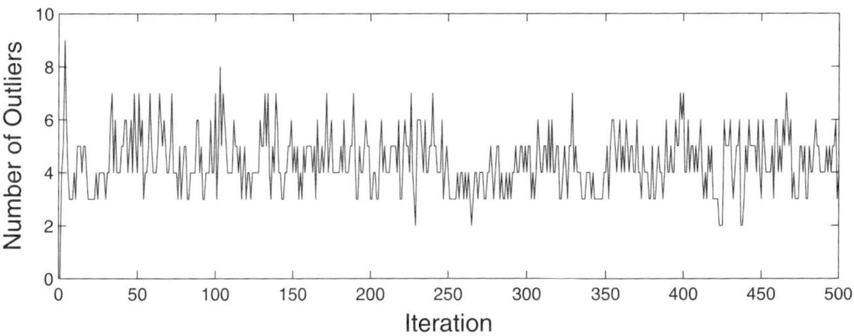

Fig. 6.15. Number of outliers as a function of the iteration index when $S(1) = 0.2$.

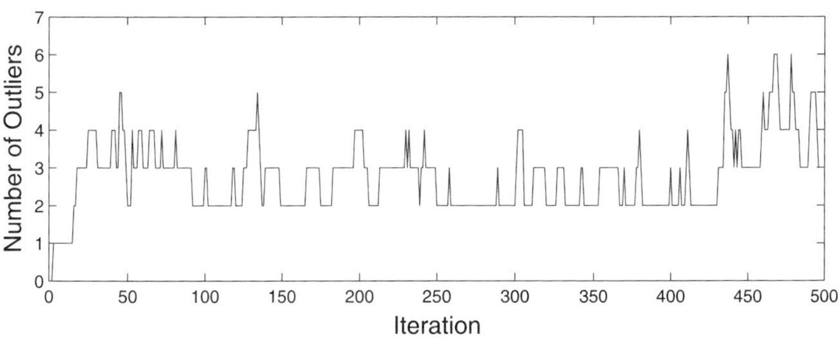

Fig. 6.16. Number of outliers as a function of the iteration index when $S(1) = 0.02$.

$S(1)$ is very small, in turn making it very difficult to reach configurations with $C_k = 2$.

This lack of convergence when $S(1) = 0.02$ is also patent when comparing Figures 6.15 and 6.16: both figures show the simulated number of outliers, that is, the number of indices k for which $C_k^i = 2$, as a function of the iteration index i. In Figure 6.15, this number directly jumps from 0 initially to reach the most likely values (between 3 and 6) and move very quickly in subsequent iterations. In contrast, in Figure 6.16 the estimated number of outliers varies only very slowly with very long steady period. A closer examination of the output reveals that in the second case, it is only after 444 iterations that C_{26} is finally simulated as 2, which explains why the estimated sequence of states is still grossly wrong after 500 simulations. ■

The moral of Example 6.3.3 is by no means that the case where $S(1) = 0.02$ is desperate. Running the simulation for much longer than 500 iterations—and once again, 500 is not considered as a big number in the MCMC world—does produce the expected results. On the other hand, the observation that the same sampling algorithm performs significantly worse in a task that is arguably easier is not something that can easily be swept under the carpet. At the risk of frightening newcomers to the field, it is important to underline that this is not an entirely lonely observation, as it is often difficult to sample efficiently from very concentrated distributions. In Example 6.3.3, the subsets of $(\mathsf{C} \times \mathsf{W})^{n+1}$ that have non-negligible probability under the posterior distribution are very narrow (in some suitable sense) and thus hard to explore with generic MCMC approaches.

To overcome the limits of the method used so far, we can however take profit of the remark that in CGLSSMs, the conditional distribution of the continuous component of the state, $W_{0:n}$, given *both* the observations $Y_{0:n}$ *and* the sequence of indicators $C_{0:n}$, is multivariate Gaussian and can be fully characterized using the algorithms discussed in Section 5.2. Hence the idea to devise MCMC algorithms that target the conditional distribution of $C_{0:n}$ given $Y_{0:n}$, where the continuous part $W_{0:n}$ is marginalized out, rather than the joint distribution of $C_{0:n}$ and $W_{0:n}$. This is the principle of the approaches proposed by Carter and Kohn (1996) and Doucet and Andrieu (2001). The specific contribution of Doucet and Andrieu (2001) was to remark that using the information form of the backward smoothing recursion (discussed in Section 5.2.5) is preferable because it is more generally applicable.

The main tool here is Lemma 5.2.24, which makes it possible to evaluate the likelihood of the observations, marginalizing with respect to the continuous part of the state sequence, where all indicators except one are fixed. Combined with the information provided by the prior distribution of the sequence of indicators, this is all we need for sampling an indicator given all its neighbors, which is the Gibbs sampling strategy discussed in full generality in Section 6.2.5. There is however one important detail concerning the application Lemma 5.2.24 that needs to be clarified. To apply Lemma 5.2.24

at index k, it is required that the results of both the filtering recursion for index $k-1$, $\{\hat{W}_{k-1|k-1}(C_{0:k-1}), \Sigma_{k-1|k-1}(C_{0:k-1})\}$, as well as those of the backward information recursion at index k, $\{\kappa_{k|n}(C_{k+1:n}), \Pi_{k|n}(C_{k+1:n})\}$, be available. None of these two recursions is particularly simple as each step of each recursion involves in particular the inversion of a square matrix whose dimension is that of the continuous component of the state. The important point noted by Carter and Kohn (1996) and Doucet and Andrieu (2001) is that because the forward quantities at index k depend on indicators C_l for $l \le k$ only and, conversely, the backward quantities depend on indicators C_l such that $l > k$ only, it is advantageous to use a systematic scan Gibbs sampler that simulates C_k given its neighbors for $k = 0, \ldots, n$ (or in reverse order) so as to avoid multiple evaluation of identical quantities. This however makes the overall algorithm somewhat harder to describe because it is necessary to carry out the Gibbs simulations and the forward (or backward) recursion simultaneously. The overall computational complexity of a complete sweep of the Gibbs sampler is then only of the order of what it takes to implement Algorithm 5.2.13 or Proposition 5.2.21 for all indices k between 0 and n, times the number r of possible values of the indicator, as these need to be enumerated exhaustively at each index. We now describe the version of the systematic scan Gibbs sampler that uses the result previously obtained in Section 5.2.6.

Algorithm 6.3.4 (Gibbs Sampler for Indicators in Conditional Gaussian Linear State-Space Model). Consider a conditionally Gaussian linear state-space model (Definition 2.2.6) with indicator-dependent matrices A, R, B, and S for which the covariance of the initial state Σ_ν may depend on C_0 and denote by ν_C and Q_C, respectively, the initial distribution and transition matrix of $\{C_k\}_{k \ge 0}$.

Assuming that a current simulated sequence of indicators $C^i_{0:n}$ is available, draw $C^{i+1}_{0:n}$ as follows.

Backward Recursion: Apply Proposition 5.2.21 for $k = n$ down to 0 with $A_k = A(C^i_{k+1})$, $R_k = R(C^i_{k+1})$, $B_k = B(C^i_k)$, and $S_k = S(C^i_k)$. Store the computed quantities $\kappa_{k|n}$ and $\Pi_{k|n}$ for $k = n$ down to 0.

Initial State: For $c = 1, \ldots, r$, compute

$$
\begin{aligned}
&\epsilon_0 = Y_0 , \\
&\Gamma_0(c) = B(c)\Sigma_\nu(c)B^t(c) + S(c)S^t(c) , \\
&\hat{W}_{0|0}(c) = \Sigma_\nu(c)B^t(c)\Gamma_0^{-1}(c)\epsilon_0 , \\
&\Sigma_{0|0}(c) = \Sigma_\nu(c) - \Sigma_\nu(c)B^t(c)\Gamma_0^{-1}(c)B^t(c)\Sigma_\nu(c) , \\
&\ell_0(c) = -\left[\log|\Gamma_0(c)| + \epsilon_0^t\Gamma_0^{-1}(c)\epsilon_0\right]/2 \\
&\hat{W}_{0|n}(c) = \hat{W}_{0|0}(c) + \Sigma_{0|0}(c)\left[I + \Pi_{0|n}\Sigma_{0|0}(c)\right]^{-1}\left[\kappa_{0|n} - \Pi_{0|n}\hat{W}_{0|0}(c)\right] , \\
&\Sigma_{0|n}(c) = \Sigma_{0|0}(c) - \Sigma_{0|0}(c)\left[I + \Pi_{0|n}\Sigma_{0|0}(c)\right]^{-1}\Sigma_{0|0}(c) ,
\end{aligned}
$$

$$m_0(c) = -\left[\log|\Sigma_{0|0}(c)| + \hat{W}^t_{0|0}(c)\Sigma^{-1}_{0|0}(c)\hat{W}_{0|0}(c)\right]/2$$
$$+\left[\log|\Sigma_{0|n}(c)| + \hat{W}^t_{0|n}(c)\Sigma^{-1}_{0|n}(c)\hat{W}_{0|n}(c)\right]/2\,,$$
$$p_0(c) = \exp\left[\ell_0(c) + m_0(c)\right]\nu_C(c)Q_C(c, C^i_1)\,.$$

Normalize the vector p_0 by computing $\bar{p}_0(c) = p_0(c)/\sum_{c'=1}^r p_0(c')$ for $c = 1,\ldots,r$ and sample C^{i+1}_0 from the probability distribution $\bar{p}_0$ on $\{1,\ldots,r\}$. Then store the Kalman filter variables corresponding to $c = C^{i+1}_0$ ($\hat{W}_{0|0}(C^{i+1}_0)$ and $\Sigma_{0|0}(C^{i+1}_0)$, that is) for the next iteration.

For $k = 1,\ldots,n$: for $c = 1,\ldots,r$, compute

$$\hat{W}_{k|k-1}(c) = A(c)\hat{W}_{k-1|k-1}(C^{i+1}_{k-1})\,,$$
$$\Sigma_{k|k-1}(c) = A(c)\Sigma_{k-1|k-1}(C^{i+1}_{k-1})A^t(c) + R(c)R^t(c)\,,$$
$$\epsilon_k(c) = Y_k - B(c)\hat{W}_{k|k-1}(c)\,,$$
$$\Gamma_k(c) = B(c)\Sigma_{k|k-1}(c)B^t(c) + S(c)S^t(c)\,,$$
$$\hat{W}_{k|k}(c) = \hat{W}_{k|k-1}(c) + \Sigma_{k|k-1}(c)B^t(c)\Gamma_k^{-1}(c)\epsilon_k(c)\,,$$
$$\Sigma_{k|k}(c) = \Sigma_{k|k-1}(c) - \Sigma_{k|k-1}(c)B^t(c)\Gamma_k^{-1}(c)B^t(c)\Sigma_{k|k-1}(c)\,,$$
$$\ell_k(c) = -\left[\log|\Gamma_k(c)| + \epsilon^t_k\Gamma_k^{-1}(c)\epsilon_k\right]/2\,,$$
$$\hat{W}_{k|n}(c) = \hat{W}_{k|k}(c) + \Sigma_{k|k}(c)\left[I + \Pi_{k|n}\Sigma_{k|k}(c)\right]^{-1}\left[\kappa_{k|n} - \Pi_{k|n}\hat{W}_{k|k}(c)\right]\,,$$
$$\Sigma_{k|n}(c) = \Sigma_{k|k}(c) - \Sigma_{k|k}(c)\left[I + \Pi_{k|n}\Sigma_{k|k}(c)\right]^{-1}\Sigma_{k|k}(c)\,,$$
$$m_k(c) = -\left[\log|\Sigma_{k|k}(c)| + \hat{W}^t_{k|k}(c)\Sigma^{-1}_{k|k}(c)\hat{W}_{k|k}(c)\right]/2$$
$$+\left[\log|\Sigma_{k|n}(c)| + \hat{W}^t_{k|n}(c)\Sigma^{-1}_{k|n}(c)\hat{W}_{k|n}(c)\right]/2\,,$$
$$p_k(c) = \begin{cases} \exp\left[\ell_k(c) + m_k(c)\right]Q_C(C^{i+1}_{k-1}, c)Q_C(c, C^i_{k+1}) & \text{for } k < n \\ \exp\left[\ell_n(c) + m_n(c)\right]Q_C(C^{i+1}_{n-1}, c) & \text{for } k = n \end{cases}.$$

Set $\bar{p}_k(c) = p_k(c)/\sum_{c'=1}^r p_k(c')$ (for $c = 1,\ldots,r$) and sample C^{i+1}_k from $\bar{p}_k$. If $k < n$, the corresponding Kalman filter variables $\hat{W}_{k|k}(C^{i+1}_k)$ and $\Sigma_{k|k}(C^{i+1}_k)$ are stored for the next iteration.

Despite the fact that it is perhaps the most complex algorithm that is to be met in this book, Algorithm 6.3.4 deserves no special comment as it simply combines the results obtained in Chapter 5 (Algorithms 5.2.13 and 5.2.22, Lemma 5.2.24) with the principle of the Gibbs sampler exposed in Section 6.2.5 and the clever remark that using a systematic scanning order of the simulation sites (here in ascending order) greatly reduces the computation load. Algorithm 6.3.4 is similar to the method described by Doucet and Andrieu (2001), but here the expression used for evaluating $m_k(c)$ has been made more transparent by use of the smoothing moments $\hat{W}_{k|n}(c)$ and $\Sigma_{k|n}(c)$.

Remark 6.3.5. Note that in Algorithm 6.3.4, the quantities $\ell_k(c)$ and, most importantly, $m_k(c)$ are evaluated on a log-scale. Only when computation of the probabilities $\bar{p}_k(c)$ is necessary are those converted back to the linear scale using the exponential function. Although rarely explicitly mentioned, this remark is of some importance in many practical applications of MCMC methods (and particularly those to be discussed in Section 13.2) that involve ratios of likelihood terms, each of which may well exceed the machine precision. In the case of Algorithm 6.3.4, remember that these terms need to be evaluated for all possible values of c and hence their range of variations is all the more important that some of these indicator configurations may be particularly unlikely. ■

To illustrate the behavior of Algorithm 6.3.4, we consider again the noisy AR(1) models with outliers in the case where $S(1) = 0.02$, which led to the poor mixing illustrated in Figure 6.16.

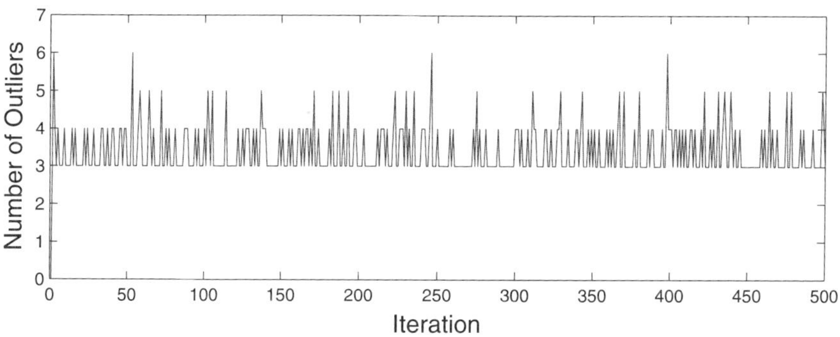

Fig. 6.17. Number of outliers as a function of the iteration index when $S(1) = 0.02$ for the systematic scan Gibbs sampler.

Example 6.3.6 (Gaussian AR Process with Outliers, Continued). Applying Algorithm 6.3.4 to the model of Example 6.3.3 provides the result shown in Figure 6.17, this figure being the exact analog Figure 6.16. Figure 6.17 shows that with Algorithm 6.3.4, only configurations with at least three outliers are ever visited. This is logical, as with such a low value of the observation noise ($S(1) = 0.02$), the values observed at indices 25, 50, and 75 can only correspond to outliers. A closer examination of the simulation shows that all simulated sequences $C^i_{0:n}$ except the initial one—we are still initializing the sampler with the configuration such that $C^1_k = 1$ for all $k = 0, \dots, n$—are such that $C^i_{25} = C^i_{50} = C^i_{75} = 2$. From the simulation, we are thus as certain as we can be that there are indeed outliers at these locations and most probably there are no others (the configuration with exactly these three outliers is selected about 67% of the time and no individual site—other than those with indices 25, 50, and 75—is selected more than 15 times out of the 500 iterations). Figure 6.17 also shows that the other configurations that are explored are visited rather rapidly rather than with long idle periods as in Figure 6.16, which also suggests good mixing properties of the Gibbs sampling algorithm. ■

To conclude this section with a more realistic example of the use of CGLSSMs and MCMC techniques, we consider the change point model and the well-log data already discussed in Example 1.3.10.

Example 6.3.7 (Gibbs Sampler for the Well-Log Data). To analyze the well-log data shown in Figure 1.7, we consider the conditionally Gaussian state-space model

$$W_{k+1} = A(C_{k+1,1})W_k + R(C_{k+1,1})\ , U_k \quad U_k \sim \mathrm{N}(0,1)\ ,$$
$$Y_k = \mu_Y(C_{k,2}) + B(C_{k,2})X_k + S(C_{k,2})\ , V_k \quad V_k \sim \mathrm{N}(0,1)\ ,$$

where $C_{k,1} \in \{1,2\}$ and $C_{k,2} \in \{1,2\}$ are indicator variables indicating, respectively, the presence of a jump in the level of the underlying signal and that of an outlier in the measurement noise, as discussed in Examples 1.3.10 and 1.3.11. For comparison purposes, we use exactly the same model specification as the one advocated by Fearnhead and Clifford (2003).

- The data shown in Figure 1.7 is first centered (approximately) by subtracting $\mu = 115,000$ from each observation; in the following, Y_k refers to the data with this average level μ subtracted.
- When $C_{k,1} = 1$ the underlying signal level is constant and we set $A(1) = 1$ and $R(1) = 0$. When $C_{k,1} = 2$, the occurrence of a jump is modeled by $A(2) = 0$, $R(2) =$ 10,000, which is an informative prior on the size of the jump. Though as explained in the introduction this is presumably an oversimplified assumption, we assume a constant probability for the presence of a jump, or, equivalently, that $\{C_{k,1}\}_{k\geq 0}$ is an i.i.d. sequence of Bernoulli random variables with constant probability of success p. The jump positions then form a discrete renewal sequence whose increment distribution is geometric with expectation $1/p$. Because there are about 16 jumps in a sequence of 4,000 samples, the average of the increment distribution is about 250, suggesting $p = 1/250$.
- When $C_{k,2} = 1$, the observation is modeled as the true state corrupted by additive noise, so that $B(1) = 1$, where $S(1)$ is set to 2,500 based on the empirical standard deviation of the median filter residual shown in the right plot of Figure 1.7. When $C_{k,2} = 2$, the occurrence of an outlier is modeled by a Gaussian random variable whose parameters are independent of the true state, so that $B(2) = 0$, and the outlier is assumed to have mean $\mu_Y(2) =$ -30,000 and standard deviation $S(2) =$ 12,500. The outliers appear to be clustered in time, with a typical cluster size of four samples. Visual inspection shows that there are about 16 clusters of noise, which suggests to model the sequence $\{C_{k,2}\}_{k\geq 0}$ as a Markov

chain with transition probabilities $\mathrm{P}(C_{k,2} = 2 \,|\, C_{k,2} = 1) = 1/250$ and $\mathrm{P}(C_{k,2} = 1 \,|\, C_{k,2} = 2) = 1/4$. The initial $C_{0,2}$ is assumed to be distributed according to the stationary distribution of this chain, which is $\mathrm{P}(C_{0,2} = 1) = 125/127$.

- The initial distribution of W_0 is assumed to have zero mean with a very large variance, which corresponds to an approximation of the so-called diffuse (or improper flat, following the terminology of Section 5.2.5) prior. Note that because $B(C_0)$ may be null (when $C_0 = 2$), using a truly diffuse prior (with "infinite" covariance matrix) cannot be done in this case by simply computing $\hat{W}_{0|0}$ as in (5.109), which is customary. In the case under consideration, however, the prior on W_0 is non-essential because the initial state is very clearly identified from the data anyway.

Note that in the model above, the presence of outliers induces non-zero means in the observation equation. As discussed in Remark 5.2.14, however, this does not necessitate significant modifications, and we just need to apply Algorithm 6.3.4 using as "observation" $Y_k - \mu_Y(c_k)$ rather than Y_k, where $\mu_Y(1) = 0$ and $\mu_Y(2) =$ -30,000.

Because $R(1) = 0$ implies that the continuous component of the state W_k stays exactly constant between two jump points, this model belongs to the category discussed earlier for which the alternating block sampling algorithm cannot be applied at all. We thus consider the result of the Gibbs sampler that operates on the indicator variables only. Figure 6.18 displays the results obtained by application of Algorithm 6.3.4 after 5,000 iterations, one iteration referring to a complete cycle of the Gibbs sampler through all the $n+1$ sites. Initially, $C^1_{k,1}$ and $C^1_{k,2}$ are both set to 1 for all sites $k = 0, \ldots, n$, which corresponds to the (very improbable) configuration in which there are neither jumps nor outliers. Clearly, after 5,000 iterations both the jump and outlier positions are located very clearly. There is however a marked difference, which is that whereas the outliers (middle plot) are located with posterior probabilities very close to 1, the jumps are only located with probabilities between 0.3 and 0.6. There are two reasons for this behavior, the second being more fundamental. First, the model for the distribution of outliers is more precise and incorporate in particular the fact that outliers systematically induce a downward bias. The second reason is a slightly deficient modeling of the occurrence of jumps. For the outliers, the selected Markov transition kernel implies that outlier periods are infrequent (occurring 2/127 of the time on average) but have durations that are exponential with average duration 4. This is a crucial feature, as a closer examination of the data reveals that some of these periods of outliers last for 10 or even 20 consecutive samples. In contrast, our model for jumps implies that jumps are infrequent (occurring in one sample out of 250 on average) and isolated. For instance, a sequence of four consecutive jumps is, *a priori*, judged as being 6.2×10^7 times less probable than the occurrence of just one jump in one of these four positions. The real data however, cf. Figure 6.19, shows that the actual jumps are not abrupt and involve at least

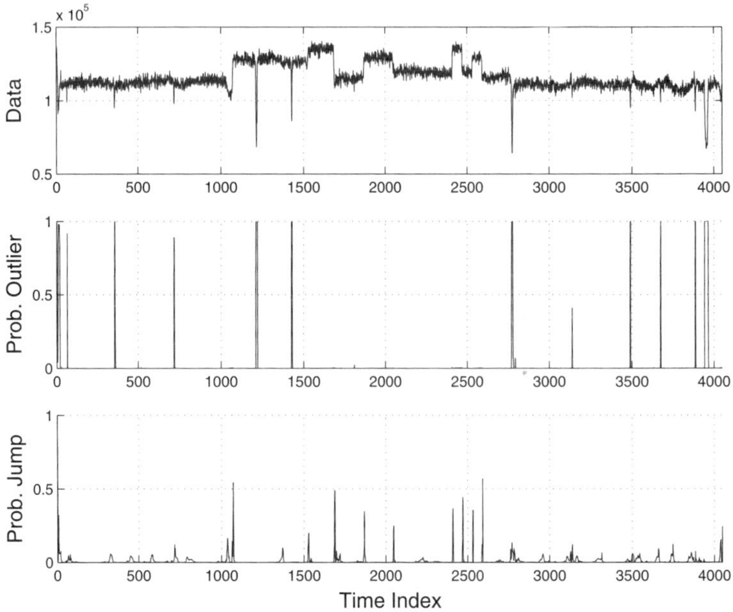

Fig. 6.18. From top to bottom: original data, posterior probability of the presence of outliers, and jumps estimated from 5,000 iterations of Algorithm 6.3.4.

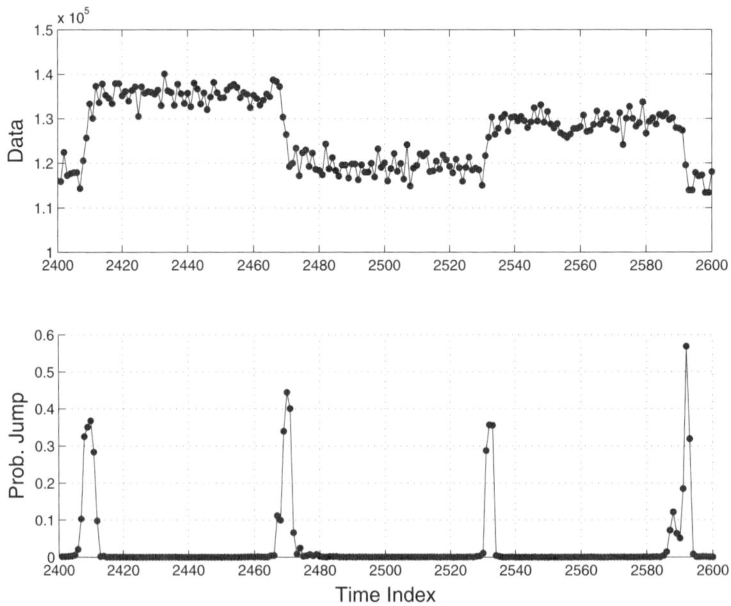

Fig. 6.19. From top to bottom: original data and posterior probability of the presence of jumps (zoom on a detail of Figure 6.18).

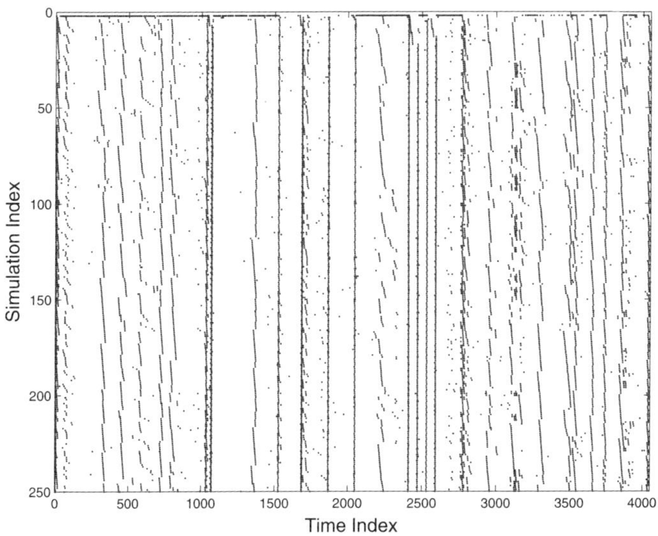

Fig. 6.20. Jump detection indicators (indices such that $C_k^i = 2$) for the first 250 iterations.

two and sometimes as many as five consecutive points. Because the modeling assumptions do not allow all of these points to be marked as jumps, the result tends to identify one of these only as the preferred jump location, whence the larger uncertainty (lower posterior probability) concerning which one is selected. Interestingly, the picture will be very different when we consider the filtering distributions (that is, the distribution of C_k given data up to index k only) in Example 8.2.10 of Chapter 8.

Figure 6.20 gives an idea of the way the simulation visits the configurations of indicators (for the jumps), showing that the algorithm almost instantaneously forgets its erroneous initial state. Consequently, the configurations change at a rather fast pace, suggesting good mixing behavior of the sampler. Note that those time indices for which jumps are detected in the bottom plot of Figure 6.18 correspond to abscissas for which the indicators of jump stay "on" very systematically through the simulation. ■

To conclude this section on MCMC sampling in conditionally Gaussian linear state-space models, we note that there is an important and interesting literature that discusses the "best" use of simulations for the purpose of estimating the unobservable state sequence $\{W_k, C_k\}_{k \geq 0}$. To estimate a function f of the unobserved sequence of states $W_{0:n}$, the most natural options are the straightforward MCMC estimate

$$\frac{1}{N} \sum_{i=1}^{N} f(W_{0:n}^i) ,$$

directly available with alternating block sampling (as in Example 6.3.3), or its Rao-Blackwellized version

$$\frac{1}{N}\sum_{i=1}^{N} \mathrm{E}[f(W_{0:n}) \mid C^i_{0:n}] ,$$

which can easily be computed when using Algorithm 6.3.4, at least for linear and quadratic functions f, as the smoothing moments $\hat{W}_{k|n}(C^{i+1}_{0:k}, C^i_{k+1:n})$ and $\Sigma_{k|n}(C^{i+1}_{0:k}, C^i_{k+1:n})$ are evaluated at each iteration i and for all sites k. But both of these alternatives are estimates of $\mathrm{E}[f(W_{0:n}) \mid Y_{0:n}]$, which, in some applications, is perhaps not what is regarded as the "best" estimate of the states. In the change point application discussed in Example 6.3.7 in particular, $\mathrm{E}\left[f(W_{0:n}) \mid Y_{0:n}\right]$ does not correspond to a piecewise constant trajectory, especially if some jump locations are only detected with some ambiguity. If one really believes that the model is correct, it may thus make more sense to estimate first the best sequence of indicators $\hat{c}_{0:n}$, that is, the one that maximizes $\mathrm{P}(C_{0:n} = c_{0:n} \mid Y_{0:n})$, and then use $\mathrm{E}[f(W_{0:n}) \mid Y_{0:n}, C_{0:n} = \hat{c}_{0:n}]$ as the estimate of the continuous part of the state sequence. In the change point model, this third way of proceeding is guaranteed to return a piecewise constant sequence. This is not an easy task, however, because finding the indicator sequence $\hat{c}_{0:n}$ that maximizes the posterior probability is a difficult combinatorial optimization problem, especially given the fact that we cannot evaluate $\mathrm{P}(C_{0:n} = c_{0:n} \mid Y_{0:n})$ directly. We refer to Lavielle and Lebarbier (2001), Doucet and Andrieu (2001), and references therein for further reading on this issue.

7

Sequential Monte Carlo Methods

The use of Monte Carlo methods for non-linear filtering can be traced back to the pioneering contributions of Handschin and Mayne (1969) and Handschin (1970). These early attempts were based on sequential versions of the *importance sampling* paradigm, a technique that amounts to simulating samples under an instrumental distribution and then approximating the target distributions by weighting these samples using appropriately defined *importance weights*. In the non-linear filtering context, importance sampling algorithms can be implemented sequentially in the sense that, by defining carefully a sequence of instrumental distributions, it is not needed to regenerate the population of samples from scratch upon the arrival of each new observation. This algorithm is called *sequential importance sampling*, often abbreviated SIS. Although the SIS algorithm has been known since the early 1970s, its use in non-linear filtering problems was rather limited at that time. Most likely, the available computational power was then too limited to allow convincing applications of these methods. Another less obvious reason is that the SIS algorithm suffers from a major drawback that was not clearly identified and properly cured until the seminal paper by Gordon *et al.* (1993). As the number of iterations increases, the importance weights tend to degenerate, a phenomenon known as *sample impoverishment* or *weight degeneracy*. Basically, in the long run most of the samples have very small normalized importance weights and thus do not significantly contribute to the approximation of the target distribution. The solution proposed by Gordon *et al.* (1993) is to allow rejuvenation of the set of samples by duplicating the samples with high importance weights and, on the contrary, removing samples with low weights.

The *particle filter* of Gordon *et al.* (1993) was the first successful application of sequential Monte Carlo techniques to the field of non-linear filtering. Since then, sequential Monte Carlo (or SMC) methods have been applied in many different fields including computer vision, signal processing, control, econometrics, finance, robotics, and statistics (Doucet *et al.*, 2001a; Ristic *et al.*, 2004). This chapter reviews the basic building blocks that are needed to implement a sequential Monte Carlo algorithm, starting with concepts related to the importance sampling approach. More specific aspects of sequential Monte Carlo techniques will be further discussed in Chapter 8, while convergence issues will be dealt with in Chapter 9.

7.1 Importance Sampling and Resampling

7.1.1 Importance Sampling

Importance sampling is a method that dates back to, at least, Hammersley and Handscomb (1965) and that is commonly used in several fields (for general references on importance sampling, see Glynn and Iglehart, 1989, Geweke, 1989, Evans and Swartz, 1995, or Robert and Casella, 2004.)

Throughout this section, μ will denote a probability measure of interest on a measurable space $(\mathsf{X}, \mathcal{X})$, which we shall refer to as the *target distribution*. As in Chapter 6, the aim is to approximate integrals of the form $\mu(f) = \int_{\mathsf{X}} f(x)\,\mu(dx)$ for real-valued measurable functions f. The Monte Carlo approach exposed in Section 6.1 consists in drawing an i.i.d. sample $\xi^1, \ldots, \xi^N$ from the probability measure μ and then evaluating the sample mean $N^{-1}\sum_{i=1}^N f(\xi^i)$. Of course, this technique is applicable only when it is possible (and reasonably simple) to sample from the target distribution μ.

Importance sampling is based on the idea that in certain situations it is more appropriate to sample from an *instrumental distribution* ν, and then to apply a change-of-measure formula to account for the fact that the instrumental distribution is different from the target distribution. More formally, assume that the target probability measure μ is absolutely continuous with respect to an *instrumental probability measure* ν from which sampling is easily feasible. Denote by $d\mu/d\nu$ the Radon-Nikodym derivative of μ with respect to ν. Then for any μ-integrable function f,

$$\mu(f) = \int f(x)\,\mu(dx) = \int f(x)\,\frac{d\mu}{d\nu}(x)\,\nu(dx)\,. \tag{7.1}$$

In particular, if $\xi^1, \xi^2, \ldots$ is an i.i.d. sample from ν, (7.1) suggests the following estimator of $\mu(f)$:

$$\tilde{\mu}^{\mathrm{IS}}_{\nu,N}(f) = N^{-1}\sum_{i=1}^N f(\xi^i)\frac{d\mu}{d\nu}(\xi^i)\,. \tag{7.2}$$

Because this estimator is the sample mean of independent random variables, there is a range of results to assess the quality of $\tilde{\mu}^{\mathrm{IS}}_{\nu,N}(f)$ as an estimator of $\mu(f)$. First of all, the strong law of large number implies that $\tilde{\mu}^{\mathrm{IS}}_{\nu,N}(f)$ converges to $\mu(f)$ almost surely as N tends to infinity. In addition, the central limit theorem for i.i.d. variables (or deviation inequalities) may serve as a guidance for selecting the proposal distribution ν, beyond the obvious requirement that it should dominate the target distribution μ. We postpone this issue and, more generally, considerations that pertain to the behavior of the approximation for large values of N to Chapter 9.

In many situations, the target probability measure μ or the instrumental probability measure ν is known only up to a normalizing factor. As already discussed in Remark 6.2.7, this is particularly true when applying importance sampling ideas to HMMs and, more generally, in Bayesian statistics. The Radon-Nikodym derivative $d\mu/d\nu$ is then known up to a (constant) scaling factor only. It is however still possible to use the importance sampling paradigm in that case, by adopting the self-normalized form of the importance sampling estimator,

$$\widehat{\mu}^{\mathrm{IS}}_{\nu,N}(f) = \frac{\sum_{i=1}^{N} f(\xi^i)\frac{d\mu}{d\nu}(\xi^i)}{\sum_{i=1}^{N} \frac{d\mu}{d\nu}(\xi^i)} . \tag{7.3}$$

This quantity is obviously free from any scale factor in $d\mu/d\nu$. The self-normalized importance sampling estimator $\widehat{\mu}^{\mathrm{IS}}_{\nu,N}(f)$ is defined as a ratio of the sample means of the functions $f_1 = f \times (d\mu/d\nu)$ and $f_2 = d\mu/d\nu$. The strong law of large numbers thus implies that $N^{-1}\sum_{i=1}^{N} f_1(\xi^i)$ and $N^{-1}\sum_{i=1}^{N} f_2(\xi^i)$ converge almost surely, to $\mu(f_1)$ and $\nu(d\mu/d\nu) = 1$, respectively, showing that $\widehat{\mu}^{\mathrm{IS}}_{\nu,N}(f)$ is a consistent estimator of $\mu(f)$. Again, more precise results on the behavior of this estimator will be given in Chapter 9. In the following, the term *importance sampling* usually refers to the self-normalized form (7.3) of the importance sampling estimate.

7.1.2 Sampling Importance Resampling

Although importance sampling is primarily intended to overcome difficulties with direct sampling from μ when approximating integrals of the form $\mu(f)$, it can also be used for (approximate) sampling from the distribution μ. The latter can be achieved by the *sampling importance resampling* (or SIR) method due to Rubin (1987, 1988). Sampling importance resampling is a two-stage procedure in which importance sampling as discussed below is followed by an additional random sampling step. In the first stage, an i.i.d. sample $(\tilde{\xi}^1, \ldots, \tilde{\xi}^M)$ is drawn from the instrumental distribution ν, and one computes the normalized version of the importance weights,

$$\omega^i = \frac{\frac{d\mu}{d\nu}(\tilde{\xi}^i)}{\sum_{i=1}^{M} \frac{d\mu}{d\nu}(\tilde{\xi}^i)} , \quad i = 1, \ldots, M . \tag{7.4}$$

In the second stage, the resampling stage, a sample of size N denoted by $\xi^1, \ldots, \xi^N$ is drawn from the intermediate set of points $\tilde{\xi}^1, \ldots, \tilde{\xi}^M$, taking into account the weights computed in (7.4). The rationale is that points $\tilde{\xi}^i$ for which ω^i in (7.4) is large are most likely under the target distribution μ and should thus be selected with higher probability during the resampling than

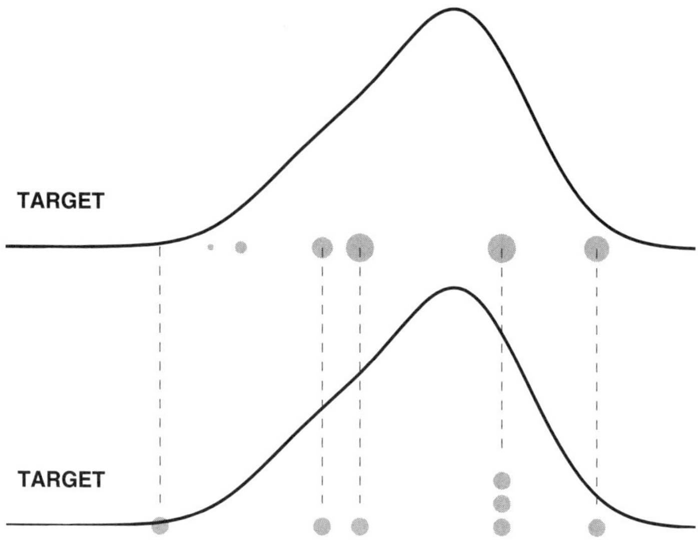

Fig. 7.1. Principle of resampling. Top plot: the sample drawn from ν with associated normalized importance weights depicted by bullets with radii proportional to the normalized weights (the target density corresponding to μ is plotted in solid line). Bottom plot: after resampling, all points have the same importance weight, and some of them have been duplicated ($M = N = 7$).

points with low (normalized) importance weights. This principle is illustrated in Figure 7.1.

There are several ways of implementing this basic idea, the most obvious approach being sampling with replacement with probability of sampling each ξ^i equal to the importance weight ω^i. Hence the number of times N^i each particular point $\tilde{\xi}^i$ in the first-stage sample is selected follows a binomial $\text{Bin}(N, \omega^i)$ distribution. The vector $(N^1, \ldots, N^M)$ is distributed from $\text{Mult}(N, \omega^1, \ldots, \omega^M)$, the multinomial distribution with parameter N and probabilities of success $(\omega^1, \ldots, \omega^M)$. In this resampling step, the points in the first-stage sample that are associated with small normalized importance weights are most likely to be discarded, whereas the best points in the sample are duplicated in proportion to their importance weights. In most applications, it is typical to choose M, the size of the first-stage sample, larger (and sometimes much larger) than N. The SIR algorithm is summarized below.

Algorithm 7.1.1 (SIR: Sampling Importance Resampling).

Sampling: Draw an i.i.d. sample $\tilde{\xi}^1, \ldots, \tilde{\xi}^M$ from the instrumental distribution ν.

Weighting: Compute the (normalized) importance weights

$$\omega^i = \frac{\frac{d\mu}{d\nu}(\tilde{\xi}^i)}{\sum_{j=1}^M \frac{d\mu}{d\nu}(\tilde{\xi}^j)} \quad \text{for } i = 1, \dots, M .$$

Resampling:

- Draw, conditionally independently given $(\tilde{\xi}^1, \dots, \tilde{\xi}^M)$, N discrete random variables $(I^1, \dots, I^N)$ taking values in the set $\{1, \dots, M\}$ with probabilities $(\omega^1, \dots, \omega^M)$, i.e.,

$$\mathrm{P}(I^1 = j) = \omega^j , \quad j = 1, \dots, M . \tag{7.5}$$

- Set, for $i = 1, \dots, N$, $\xi^i = \tilde{\xi}^{I^i}$.

The set $(I^1, \dots, I^N)$ is thus a multinomial trial process. Hence, this method of selection is known as the *multinomial* resampling scheme.

At this point, it may not be obvious that the sample $\xi^1, \dots, \xi^N$ obtained from Algorithm 7.1.1 is indeed (approximately) i.i.d. from μ in any suitable sense. In Chapter 9, it will be shown that the sample mean of the draws obtained using the SIR algorithm,

$$\hat{\mu}^{\mathrm{SIR}}_{\nu,M,N}(f) = \frac{1}{N} \sum_{i=1}^N f(\xi^i) , \tag{7.6}$$

is a consistent estimator of $\mu(f)$ for all functions f satisfying $\mu(|f|) < \infty$. The resampling step might thus be seen as a means to transform the weighted importance sampling estimate $\widehat{\mu}^{\mathrm{IS}}_{\nu,M}(f)$ defined by (7.3) into an unweighted sample average. Recall that N^i is the number of times that the element $\tilde{\xi}^i$ is resampled. Rewriting

$$\hat{\mu}^{\mathrm{SIR}}_{\nu,M,N}(f) = \frac{1}{N} \sum_{i=1}^N f(\xi^i) = \sum_{i=1}^M \frac{N^i}{N} f(\tilde{\xi}^i) ,$$

it is easily seen that the sample mean $\hat{\mu}^{\mathrm{SIR}}_{\nu,M,N}(f)$ of the SIR sample is, conditionally on the first-stage sample $(\tilde{\xi}^1, \dots, \tilde{\xi}^M)$, equal to the importance sampling estimator $\widehat{\mu}^{\mathrm{IS}}_{\nu,M}(f)$ defined in (7.3),

$$\mathrm{E}\left[\hat{\mu}^{\mathrm{SIR}}_{\nu,M,N}(f) \mid \tilde{\xi}^1, \dots, \tilde{\xi}^M\right] = \widehat{\mu}^{\mathrm{IS}}_{\nu,M}(f) .$$

As a consequence, the SIR estimator $\hat{\mu}^{\mathrm{SIR}}_{\nu,M,N}(f)$ is an unbiased estimate of $\mu(f)$, but its mean squared error is always larger than that of the importance sampling estimator (7.3) due to the well-known variance decomposition

$$\begin{aligned}\mathrm{E}&\left[\left(\hat{\mu}^{\mathrm{SIR}}_{\nu,M,N}(f) - \mu(f)\right)^2\right] \\ &= \mathrm{E}\left[\left(\hat{\mu}^{\mathrm{SIR}}_{\nu,M,N}(f) - \widehat{\mu}^{\mathrm{IS}}_{\nu,M}(f)\right)^2\right] + \mathrm{E}\left[\left(\widehat{\mu}^{\mathrm{IS}}_{\nu,M}(f) - \mu(f)\right)^2\right] .\end{aligned}$$

The variance $\mathrm{E}[(\hat{\mu}^{\mathrm{SIR}}_{\nu,M,N}(f) - \widehat{\mu}^{\mathrm{IS}}_{\nu,M}(f))^2]$ may be interpreted as the price to pay for converting the weighted importance sampling estimate into an unweighted approximation.

Showing that the SIR estimate (7.6) is a consistent and asymptotically normal estimator of $\mu(f)$ is not a trivial task, as $\xi^1, \ldots, \xi^N$ are no more independent due to the normalization of the weights followed by resampling. As such, the elementary i.i.d. convergence results that underlie the theory of the importance sampling estimator are of no use, and we refer to Section 9.2 for the corresponding proofs.

Remark 7.1.2. A closer examination of the numerical complexity of Algorithm 7.1.1 reveals that whereas all steps of the algorithm have a complexity that grows in proportion to M and N, this is not quite true for the multinomial sampling step whose numerical complexity is, *a priori*, growing faster than N (about $N \log_2 M$—see Section 7.4.1 below for details). This is very unfortunate, as we know from elementary arguments discussed in Section 6.1 that Monte Carlo methods are most useful when N is large (or more appropriately that the quality of the approximation improves rather slowly as N grows).

A clever use of elementary probabilistic results however makes it possible to devise methods for sampling N times from a multinomial distribution with M possible outcomes using a number of operations that grows only linearly with the maximum of N and M. In order not to interrupt our exposition of sequential Monte Carlo, the corresponding algorithms are discussed in Section 7.4.1 at the end of this chapter. Note that we are here only discussing implementations issues. There are however different motivations, also discussed in Section 7.4.2, for adopting sampling schemes other than multinomial sampling. ■

7.2 Sequential Importance Sampling

7.2.1 Sequential Implementation for HMMs

We now specialize the sampling techniques considered above to hidden Markov models. As in previous chapters, we adopt the hidden Markov model as specified by Definition 2.2.2 where Q denotes the Markov transition kernel of the hidden chain, ν is the distribution of the initial state X_0, and $g(x, y)$ (for $x \in \mathsf{X}, y \in \mathsf{Y}$) denotes the transition density function of the observation given the state, with respect to the measure μ on $(\mathsf{Y}, \mathcal{Y})$. To simplify the mathematical expressions, we will also use the shorthand notation $g_k(\cdot) = g(\cdot, Y_k)$ introduced in Section 3.1.4. We denote the joint smoothing distribution by $\phi_{0:k|k}$, omitting the dependence with respect to the initial distribution ν, which does not play an important role here. According to (4.1), the joint smoothing distribution may be updated recursively in time according to the relations

$$\phi_0(f) = \frac{\int f(x_0)\, g_0(x_0)\, \nu(dx_0)}{\int g_0(x_0)\, \nu(dx_0)} \qquad \text{for all } f \in \mathcal{F}_{\mathrm{b}}(\mathsf{X}) \;,$$

$$\phi_{0:k+1|k+1}(f_{k+1}) = \int \cdots \int f_{k+1}(x_{0:k+1})\, \phi_{0:k|k}(dx_{0:k}) T_k^{\mathrm{u}}(x_k, dx_{k+1}) \qquad \text{for all } f_{k+1} \in \mathcal{F}_{\mathrm{b}}\left(\mathsf{X}^{k+2}\right) \;, \tag{7.7}$$

where T_k^{u} is the transition kernel on $(\mathsf{X}, \mathcal{X})$ defined by

$$T_k^{\mathrm{u}}(x, f) = \left(\frac{\mathrm{L}_{k+1}}{\mathrm{L}_k}\right)^{-1} \int f(x')\, Q(x, dx') g_{k+1}(x') \qquad \text{for all } x \in \mathsf{X}, f \in \mathcal{F}_{\mathrm{b}}(\mathsf{X}) \;. \tag{7.8}$$

The superscript "u" (for "unnormalized") in the notation T_k^{u} is meant to highlight the fact that T_k^{u} is not a probability transition kernel. This distinction is important here because the normalized version $T_k = T_k^{\mathrm{u}} / T_k^{\mathrm{u}}(1)$ of the kernel will play an important role in the following. Note that except in some special cases discussed in Chapter 5, the likelihood ratio $\mathrm{L}_{k+1}/\mathrm{L}_k$ can generally not be computed in closed form, rendering analytic evaluation of T_k^{u} or $\phi_{0:k|k}$ hopeless. The rest of this section reviews importance sampling methods that make it possible to approximate $\phi_{0:k|k}$ recursively in k.

First, because importance sampling can be used when the target distribution is known only up to a scaling factor, the presence of non-computable constants such as $\mathrm{L}_{k+1}/\mathrm{L}_k$ does not preclude the use of the algorithm. Next, it is convenient to choose the instrumental distribution as the probability measure associated with a possibly non-homogeneous Markov chain on X. As seen below, this will make it possible to derive a sequential version of the importance sampling technique. Let $\{R_k\}_{k \geq 0}$ denote a family of Markov transition kernels on $(\mathsf{X}, \mathcal{X})$ and let ρ_0 denote a probability measure on $(\mathsf{X}, \mathcal{X})$. Further denote by $\{\rho_{0:k}\}_{k \geq 0}$ the family of probability measures associated with the inhomogeneous Markov chain with initial distribution ρ_0 and transition kernels $\{R_k\}_{k \geq 0}$,

$$\rho_{0:k}(f_k) \stackrel{\mathrm{def}}{=} \int \cdots \int f_k(x_{0:k})\, \rho_0(dx_0) \prod_{l=0}^{k-1} R_l(x_l, dx_{l+1}) \;.$$

In this context, the kernels R_k will be referred to as the *instrumental kernels.* The term *importance kernel* is also used. The following assumptions will be adopted in the sequel.

Assumption 7.2.1 (Sequential Importance Sampling).

1. *The target distribution* ϕ_0 *is absolutely continuous with respect to the instrumental distribution* ρ_0.
2. *For all* $k \geq 0$ *and all* $x \in \mathsf{X}$, *the measure* $T_k^{\mathrm{u}}(x, \cdot)$ *is absolutely continuous with respect to* $R_k(x, \cdot)$.

Then for any $k \geq 0$ and any function $f_k \in \mathcal{F}_{\mathrm{b}}\left(\mathsf{X}^{k+1}\right)$,

$$\phi_{0:k|k}(f_k) = \int \cdots \int f_k(x_{0:k}) \, \frac{d\phi_0}{d\rho_0}(x_0) \left\{ \prod_{l=0}^{k-1} \frac{dT_l^{\mathrm{u}}(x_l, \cdot)}{dR_l(x_l, \cdot)}(x_{l+1}) \right\} \rho_{0:k}(dx_{0:k}) \,, \tag{7.9}$$

which implies that the target distribution $\phi_{0:k|k}$ is absolutely continuous with respect to the instrumental distribution $\rho_{0:k}$ with Radon-Nikodym derivative given by

$$\frac{d\phi_{0:k|k}}{d\rho_{0:k}}(x_{0:k}) = \frac{d\phi_0}{d\rho_0}(x_0) \prod_{l=0}^{k-1} \frac{dT_l^{\mathrm{u}}(x_l, \cdot)}{dR_i(x_l, \cdot)}(x_{l+1}) \,. \tag{7.10}$$

It is thus legitimate to use $\rho_{0:k}$ as an instrumental distribution to compute importance sampling estimates for integrals with respect to $\phi_{0:k|k}$. Denoting by $\xi_{0:k}^1, \ldots, \xi_{0:k}^N$ N i.i.d. random sequences with common distribution $\rho_{0:k}$, the importance sampling estimate of $\phi_{0:k|k}(f_k)$ for $f_k \in \mathcal{F}_{\mathrm{b}}\left(\mathsf{X}^{k+1}\right)$ is defined as

$$\hat{\phi}_{0:k|k}^{\mathrm{IS}}(f_k) = \frac{\sum_{i=1}^N \omega_k^i f_k(\xi_{0:k}^i)}{\sum_{i=1}^N \omega_k^i} \,, \tag{7.11}$$

where ω_k^i are the unnormalized importance weights defined recursively by

$$\omega_0^i = \frac{d\phi_0}{d\rho_0}(\xi_0^i) \qquad \text{for } i = 1, \ldots, N \,, \tag{7.12}$$

and, for $k \geq 0$,

$$\omega_{k+1}^i = \omega_k^i \frac{dT_k^{\mathrm{u}}(\xi_k^i, \cdot)}{dR_k(\xi_k^i, \cdot)}(\xi_{k+1}^i) \qquad \text{for } i = 1, \ldots, N \,. \tag{7.13}$$

The multiplicative decomposition of the (unnormalized) importance weights in (7.13) implies that these weights may be computed recursively in time as successive observations become available. In the sequential Monte Carlo literature, the update factor dT_k^{u}/dR_k is often called the *incremental weight.* As discussed previously in Section 7.1.1, the estimator in (7.11) is left unmodified if the weights, or equivalently the incremental weights, are evaluated up to a constant only. In particular, one may omit the problematic scaling factor $\mathrm{L}_{k+1}/\mathrm{L}_k$ that we met in the definition of T_k^{u} in (7.8). The practical implementation of sequential importance sampling thus goes as follows.

Algorithm 7.2.2 (SIS: Sequential Importance Sampling).

Initial State: Draw an i.i.d. sample $\xi_0^1, \ldots, \xi_0^N$ from ρ_0 and set

$$\omega_0^i = g_0(\xi_0^i) \frac{d\nu}{d\rho_0}(\xi_0^i) \qquad \text{for } i = 1, \ldots, N .$$

Recursion: For $k = 0, 1, \ldots,$

- Draw $(\xi_{k+1}^1, \ldots, \xi_{k+1}^N)$ conditionally independently given $\{\xi_{0:k}^j, j = 1, \ldots, N\}$ from the distribution $\xi_{k+1}^i \sim R_k(\xi_k^i, \cdot)$. Append ξ_{k+1}^i to $\xi_{0:k}^i$ to form $\xi_{0:k+1}^i = (\xi_{0:k}^i, \xi_{k+1}^i)$.
- Compute the updated importance weights

$$\omega_{k+1}^i = \omega_k^i \times g_{k+1}(\xi_{k+1}^i) \frac{dQ(\xi_k^i, \cdot)}{dR_k(\xi_k^i, \cdot)}(\xi_{k+1}^i), \qquad i = 1, \ldots, N .$$

At any iteration index k importance sampling estimates may be evaluated according to (7.11).

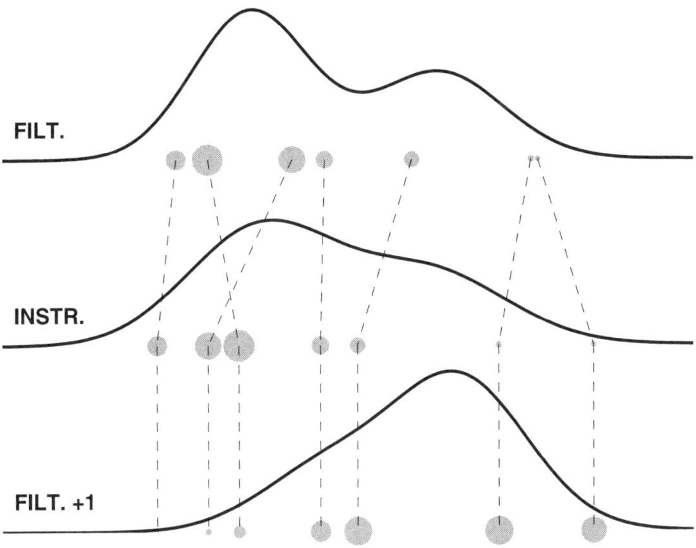

Fig. 7.2. Principle of sequential importance sampling (SIS). Upper plot: the curve represents the filtering distribution, and the particles with weights are represented along the axis by bullets, the radii of which being proportional to the normalized weight of the particle. Middle plot: the instrumental distribution with resampled particle positions. Bottom plot: filtering distribution at the next time index with particle updated weights. The case depicted here corresponds to the choice $R_k = Q$.

An important feature of Algorithm 7.2.2, which corresponds to the method originally proposed in Handschin and Mayne (1969) and Handschin (1970), is that the N trajectories $\xi^1_{0:k}, \ldots, \xi^N_{0:k}$ are independent and identically distributed for all time indices k. Following the terminology in use in the non-linear filtering community, we shall refer to the sample at time index k, $\xi^1_k, \ldots, \xi^N_k$, as the population (or system) of *particles* and to $\xi^i_{0:k}$ for a specific value of the particle index i as the history (or trajectory) of the ith particle. The principle of the method is illustrated in Figure 7.2.

7.2.2 Choice of the Instrumental Kernel

Before discussing in Section 7.3 a serious drawback of Algorithm 7.2.2 that needs to be fixed in order for the method to be applied to any problem of practical interest, we examine strategies that may be helpful in selecting proper instrumental kernels R_k in several models (or families of models) of interest.

7.2.2.1 Prior Kernel

The first obvious and often very simple choice of instrumental kernel R_k is that of setting $R_k = Q$ (irrespectively of k). In that case, the instrumental kernel simply corresponds to the prior distribution of the new state in the absence of the corresponding observation. The incremental weight then simplifies to

$$\frac{dT^{\mathrm{u}}_k(x, \cdot)}{dQ(x, \cdot)}(x') = \frac{\mathrm{L}_k}{\mathrm{L}_{k+1}} g_{k+1}(x') \propto g_{k+1}(x') \qquad \text{for all } (x, x') \in \mathsf{X}^2 \,. \tag{7.14}$$

A distinctive feature of the prior kernel is that the incremental weight in (7.14) *does not depend on* x, that is, on the previous position. The use of the prior kernel $R_k = Q$ is popular because sampling from the prior kernel Q is often straightforward, and computing the incremental weight simply amounts to evaluating the conditional likelihood of the new observation given the current particle position. The prior kernel also satisfies the minimal requirement of importance sampling as stated in Assumption 7.2.1. In addition, because the importance function reduces to g_{k+1}, it is upper-bounded as soon as one can assume that $\sup_{x \in \mathsf{X}, y \in \mathsf{Y}} g(x, y)$ is finite, which (often) is a very mild condition (see also Section 9.1). Despite these appealing properties, the use of the prior kernel can sometimes lead to poor performance, often manifesting itself as a lack of robustness with respect to the values taken by the observed sequence $\{Y_k\}_{k \geq 0}$. The following example illustrates this problem in a very simple situation.

Example 7.2.3 (Noisy AR(1) Model). To illustrate the potential problems associated with the use of the prior kernel, Pitt and Shephard (1999) consider the simple model where the observations arise from a first-order linear autoregression observed in noise,

$$X_{k+1} = \phi X_k + \sigma_U U_k , \qquad U_k \sim \mathrm{N}(0,1) ,$$
$$Y_k = X_k + \sigma_V V_k , \qquad V_k \sim \mathrm{N}(0,1) ,$$

where $\phi = 0.9$, $\sigma_U^2 = 0.01$, $\sigma_V^2 = 1$ and $\{U_k\}_{k \geq 0}$ and $\{V_k\}_{k \geq 0}$ are independent Gaussian white noise processes. The initial distribution ν is the stationary distribution of the Markov chain $\{X_k\}_{k \geq 0}$, that is, normal with zero mean and variance $\sigma_U^2/(1-\phi^2)$.

In the following, we assume that $n = 5$ and simulate the first five observations from the model, whereas the sixth observation is set to the arbitrary value 20. The observed series is

$$(-0.652, -0.345, -0.676, 1.142, 0.721, 20) .$$

The last observation is located 20 standard deviations away from the mean (zero) of the stationary distribution, which definitively corresponds to an aberrant value from the model's point of view. In a practical situation however, we would of course like to be able to handle also data that does not necessarily come from the model under consideration. Note also that in this toy example, one can evaluate the exact smoothing distributions by means of the Kalman filtering recursion discussed in Section 5.2.

Figure 7.3 displays box and whisker plots for the SIS estimate of the posterior mean of the final state X_5 as a function of the number N of particles when using the prior kernel. These plots have been obtained from 125 independent replications of the SIS algorithm. The vertical line corresponds to the true posterior mean of X_5 given $Y_{0:5}$, computed using the Kalman filter. The

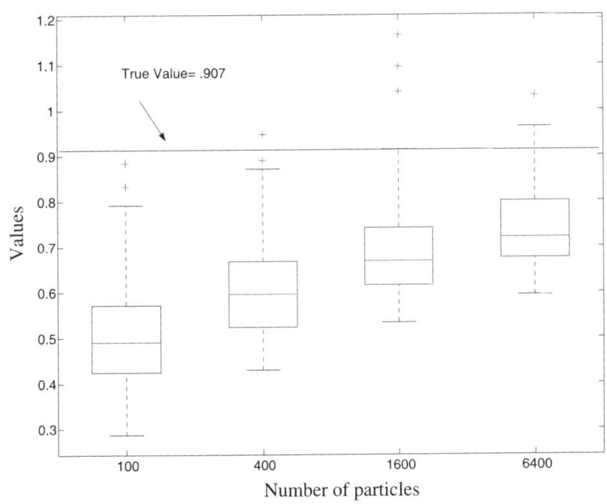

Fig. 7.3. Box and whisker plot of the posterior mean estimate of X_5 obtained from 125 replications of the SIS filter using the prior kernel and increasing numbers of particles. The horizontal line represents the true posterior mean.

figure shows that the SIS algorithm with the prior kernel grossly underestimates the values of the state even when the number of particles is very large. This is a case where there is a conflict between the prior distribution and the posterior distribution: under the instrumental distribution, all particles are proposed in a region where the conditional likelihood function g_5 is extremely low. In that case, the renormalization of the weights used to compute the filtered mean estimate according to (7.11) may even have unexpectedly adverse consequences: a weight close to 1 does not necessarily correspond to a simulated value that is important for the distribution of interest. Rather, it is a weight that is large relative to other, even smaller weights (of particles even less important for the filtering distribution). This is a logical consequence of the fact that the weights must sum to one. ■

7.2.2.2 Optimal Instrumental Kernel

The mismatch between the instrumental distribution and the posterior distribution observed in the previous example is the type of problem that one should try to alleviate by a proper choice of the instrumental kernel. An interesting choice to address this problem is the kernel

$$T_k(x, f) = \frac{\int f(x')\, Q(x, dx') g_{k+1}(x')}{\int Q(x, dx') g_{k+1}(x')} \qquad \text{for } x \in \mathsf{X}, f \in \mathcal{F}_{\mathrm{b}}(\mathsf{X}), \tag{7.15}$$

which is just T_k^{u} defined in (7.8) properly normalized to correspond to a Markov transition kernel (that is, $T_k(x, 1) = 1$ for all $x \in \mathsf{X}$). The kernel T_k may be interpreted as a regular version of the conditional distribution of the hidden state X_{k+1} given X_k and the current observation Y_{k+1}. In the sequel, we will refer to this kernel as the *optimal kernel*, following the terminology found in the sequential importance sampling literature. This terminology dates back probably to Zaritskii *et al.* (1975) and Akashi and Kumamoto (1977) and is largely adopted by authors such as Liu and Chen (1995), Chen and Liu (2000), Doucet *et al.* (2000a), Doucet *et al.* (2001a) and Tanizaki (2003). The word "optimal" is somewhat misleading, and we refer to Chapter 9 for a more precise discussion of optimality of the instrumental distribution in the context of importance sampling (which generally has to be defined for a specific choice of the function f of interest). The main property of T_k as defined in (7.15) is that

$$\frac{dT_k^{\mathrm{u}}(x, \cdot)}{dT_k(x, \cdot)}(x') = \frac{\mathrm{L}_k}{\mathrm{L}_{k+1}} \gamma_k(x) \propto \gamma_k(x) \qquad \text{for } (x, x') \in \mathsf{X}^2, \tag{7.16}$$

where $\gamma_k(x)$ is the denominator of T_k in (7.15):

$$\gamma_k(x) \stackrel{\text{def}}{=} \int Q(x, dx') g_{k+1}(x') \,. \tag{7.17}$$

Equation (7.16) means that the incremental weight in (7.13) now depends on the previous position of the particle only (and not on the new position proposed at index $k+1$). This is the exact opposite of the situation observed previously for the prior kernel. The optimal kernel (7.15) is attractive because it incorporates information both on the state dynamics and on the current observation: the particles move "blindly" with the prior kernel, whereas they tend to cluster into regions where the current local likelihood g_{k+1} is large when using the optimal kernel. There are however two problems with using T_k in practice. First, drawing from this kernel is usually not directly feasible. Second, calculation of the incremental importance weight γ_k in (7.17) may be analytically intractable. Of course, the optimal kernel takes a simple form with easy simulation and explicit evaluation of (7.17) in the particular cases discussed in Chapter 5. It turns out that it can also be evaluated for a slightly larger class of non-linear Gaussian state-space models, as soon as the observation equation is linear (Zaritskii *et al.*, 1975). Indeed, consider the state-space model with non-linear state evolution equation

$$X_{k+1} = A(X_k) + R(X_k)U_k \,, \qquad U_k \sim \mathrm{N}(0, I) \,, \tag{7.18}$$

$$Y_k = BX_k + SV_k \,, \qquad V_k \sim \mathrm{N}(0, I) \,, \tag{7.19}$$

where A and R are matrix-valued functions of appropriate dimensions. By application of Proposition 5.2.2, the conditional distribution of the state vector X_{k+1} given $X_k = x$ and Y_{k+1} is multivariate Gaussian with mean $m_{k+1}(x)$ and covariance matrix $\Sigma_{k+1}(x)$, given by

$$\begin{aligned} K_{k+1}(x) &= R(x)R^t(x)B^t \left[BR(x)R^t(x)B^t + SS^t\right]^{-1} \,, \\ m_{k+1}(x) &= A(x) + K_{k+1}(x)\left[Y_{k+1} - BA(x)\right] \,, \\ \Sigma_{k+1}(x) &= \left[I - K_{k+1}(x)B\right] R(x)R^t(x) \,. \end{aligned}$$

Hence new particles ξ^i_{k+1} need to be simulated from the distribution

$$\mathrm{N}\left(m_{k+1}(\xi^i_k), \Sigma_{k+1}(\xi^i_k)\right) \,, \tag{7.20}$$

and the incremental weight for the optimal kernel is proportional to

$$\gamma_k(x) = \int q(x, x') g_{k+1}(x')\, dx' \propto \\ |\Gamma_{k+1}(x)|^{-1/2} \exp\left\{-\frac{1}{2}\left[Y_{k+1} - BA(x)\right]^t \Gamma^{-1}_{k+1}(x)\left[Y_{k+1} - BA(x)\right]\right\}$$

where

$$\Gamma_{k+1}(x) = BR(x)R^t(x)B^t + SS^t \,.$$

In other situations, sampling from the kernel T_k and/or computing the normalizing constant γ_k is a difficult task. There is no general recipe to solve this problem, but rather a set of possible solutions that should be considered.

Example 7.2.4 (Noisy AR(1) Model, Continued). We consider the noisy AR(1) model of Example 7.2.3 again using the optimal importance kernel, which corresponds to the particular case where all variables are scalar and A and R are constant in (7.18)–(7.19) above. Thus, the optimal instrumental transition density is given by

$$t_k(x,\cdot) = \mathrm{N}\left(\frac{\sigma_U^2\sigma_V^2}{\sigma_U^2+\sigma_V^2}\left\{\frac{\phi x}{\sigma_U^2}+\frac{Y_k}{\sigma_V^2}\right\}, \frac{\sigma_U^2\sigma_V^2}{\sigma_U^2+\sigma_V^2}\right)$$

and the incremental importance weights are proportional to

$$\gamma_k(x) \propto \exp\left[-\frac{1}{2}\frac{(Y_k-\phi x)^2}{\sigma_U^2+\sigma_V^2}\right] .$$

Figure 7.4 is the exact analog of Figure 7.3, also obtained from 125 independent runs of the algorithm, for this new choice of instrumental kernel. The figure shows that whereas the SIS estimate of posterior mean is still negatively biased, the optimal kernel tends to reduce the bias compared to the prior kernel. It also shows that as soon as $N = 400$, there are at least some particles located around the true filtered mean of the state, which means that the method should not get entirely lost as subsequent new observations arrive.

To illustrate the advantages of the optimal kernel with respect to the prior kernel graphically, we consider the model (7.18)–(7.19) again with $\phi = 0.9$, $\sigma_u^2 = 0.4$, $\sigma_v^2 = 0.6$, and $(0, 2.6, 0.6)$ as observed series (of length 3). The initial distribution is a mixture $0.6\,\mathrm{N}(-1, 0.3) + 0.4\,\mathrm{N}(1, 0.4)$ of two Gaussians, for which it is still possible to evaluate the exact filtering distributions as the

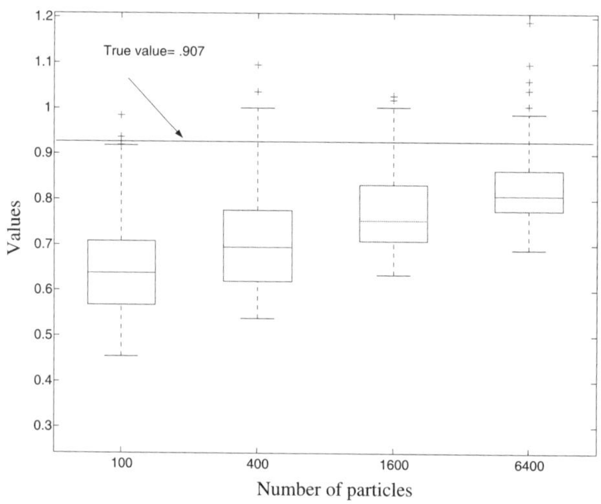

Fig. 7.4. Box and whisker plot of the posterior mean estimate for X_5 obtained from 125 replications of the SIS filter using the optimal kernel and increasing numbers of particles. Same data and axes as Figure 7.3.

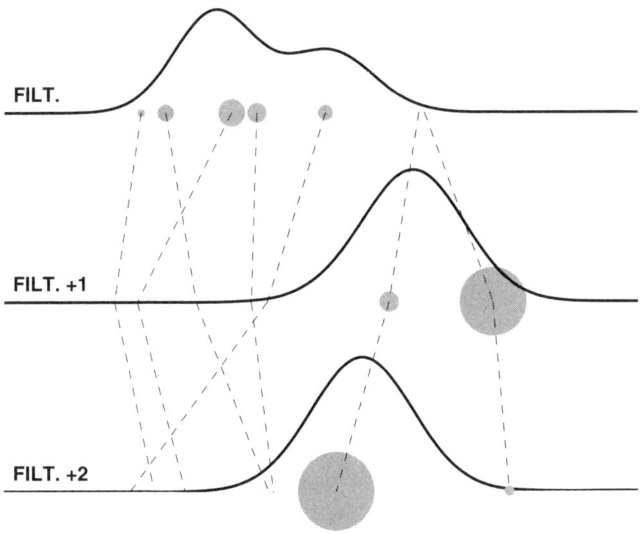

Fig. 7.5. SIS using the prior kernel. The positions of the particles are indicated by circles whose radii are proportional to the normalized importance weights. The solid lines show the filtering distributions for three consecutive time indices.

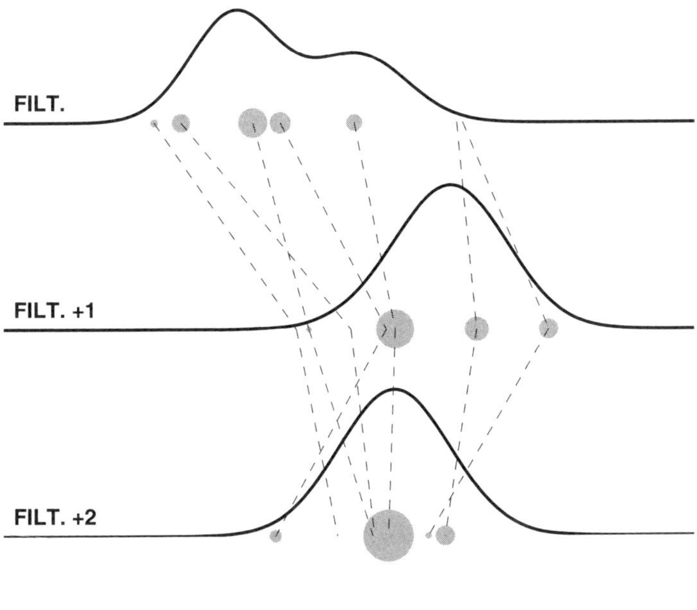

Fig. 7.6. SIS using the optimal kernel (same data and display as in Figure 7.5).

mixture of two Kalman filters using, respectively, $\mathrm{N}(-1, 0.3)$ and $\mathrm{N}(1, 0.4)$ as the initial distribution of X_0. We use only seven particles to allow for an interpretable graphical representation. Figures 7.5 and 7.6 show the positions of the particles propagated using the prior kernel and the optimal kernel, respectively. At time 1, there is a conflict between the prior and the posterior as the observation does not agree with the particle approximation of the predictive distribution. With the prior kernel (Figure 7.5), the mass becomes concentrated on a single particle with several particles lost out in the left tail of the distribution with negligible weights. In contrast, in Figure 7.6 most of the particles stay in high probability regions through the iterations with several distinct particles having non-negligible weights. This is precisely because the optimal kernel "pulls" particles toward regions where the current local likelihood $g_{k+1}(x) = g_{k+1}(x, Y_k)$ is large, whereas the prior kernel does not. ■

7.2.2.3 Accept-Reject Algorithm

Because drawing from the optimal kernel T_k is most often not feasible, a first natural idea consists in trying the accept-reject method (Algorithm 6.2.1), which is a versatile approach to sampling from general distributions. To sample from the optimal importance kernel $T_k(x, \cdot)$ defined by (7.15), one needs an instrumental kernel $R_k(x, \cdot)$ from which it is easy to sample and such that there exists M satisfying $\frac{dQ(x,\cdot)}{dR_k(x,\cdot)}(x')g_k(x') \leq M$ (for all $x \in \mathsf{X}$). Note that because it is generally impossible to evaluate the normalizing constant γ_k of T_k, we must resort here to the unnormalized version of the accept-reject algorithm (see Remark 6.2.4). The algorithm consists in generating pairs (ξ, U) of independent random variables with $\xi \sim R_k(x, \cdot)$ and U uniformly distributed on $[0, 1]$ and accepting ξ if

$$U \leq \frac{1}{M} \frac{dQ(x, \cdot)}{dR_k(x, \cdot)}(\xi) g_k(\xi) .$$

Recall that the distribution of the number of simulations required is geometric with parameter

$$p(x) = \frac{\int Q(x, dx') g_k(x')}{M} .$$

The strength of the accept-reject technique is that, using any instrumental kernel R_k satisfying the domination condition, one can obtain independent samples from the optimal importance kernel T_k. When the conditional likelihood of the observation $g_k(x)$—viewed as a function of x—is bounded, one can for example use the prior kernel Q as the instrumental distribution. In that case

$$\frac{dT_k(x, \cdot)}{dQ(x, \cdot)}(x') = \frac{g_k(x')}{\int g_k(u) \, Q(x, du)} \leq \frac{\sup_{x' \in \mathsf{X}} g_k(x')}{\int g_k(u) \, Q(x, du)} .$$

The algorithm then consists in drawing ξ from the prior kernel $Q(x, \cdot)$, U uniformly on $[0, 1]$ and accepting the draw if $U \leq g_k(\xi)/\sup_{x \in \mathsf{X}} g_k(x)$. The acceptance rate of this algorithm is then given by

$$p(x) = \frac{\int_{\mathsf{X}} Q(x, dx') g_k(x')}{\sup_{x' \in \mathsf{X}} g_k(x')} .$$

Unfortunately, it is not always possible to design an importance kernel $R_k(x, \cdot)$ that is easy to sample from, for which the bound M is indeed finite, *and* such that the acceptance rate $p(x)$ is reasonably large.

7.2.2.4 Local Approximation of the Optimal Importance Kernel

A different option consists in trying to approximate the optimal kernel T_k by a simpler proposal kernel R_k that is handy for simulating. Ideally, R_k should be such that $R_k(x, \cdot)$ both has heavier tails than $T_k(x, \cdot)$ and is close to $T_k(x, \cdot)$ around its modes, with the aim of keeping the ratio $\frac{dT_k(x,\cdot)}{dR_k(x,\cdot)}(x')$ as small as possible. To do so, authors such as Pitt and Shephard (1999) and Doucet *et al.* (2000a) suggest to first locate the high-density regions of the optimal distribution $T_k(x, \cdot)$ and then use an over-dispersed (that is, with sufficiently heavy tails) approximation of $T_k(x, \cdot)$. The first part of this program mostly applies to the case where the distribution $T_k(x, \cdot)$ is known to be unimodal with a mode that can be located in some way. The overall procedure will need to be repeated N times with x corresponding in turn to each of the current particles. Hence the method used to construct the approximation should be reasonably simple if the potential advantages of using a "good" proposal kernel are not to be offset by an unbearable increase in computational cost.

A first remark of interest is that there is a large class of state-space models for which the distribution $T_k(x, \cdot)$ can effectively be shown to be unimodal using convexity arguments. In the remainder of this section, we assume that $\mathsf{X} = \mathbb{R}^d$ and that the hidden Markov model is fully dominated (in the sense of Definition 2.2.3), denoting by q the transition density function associated with the hidden chain. Recall that for a certain form of non-linear state-space models given by (7.18)–(7.19), we were able to derive the optimal kernel and its normalization constant explicitly. Now consider the case where the state evolves according to (7.18), so that

$$q(x, x') \propto \exp\left[-\frac{1}{2}(x' - A(x))^t \left\{R(c)R^t(x)\right\}^{-1} (x' - A(x))\right] ,$$

and $g(x, y)$ is simply constrained to be a log-concave function of its x argument. This of course includes the linear Gaussian observation model considered previously in (7.19) but also many other cases like the non-linear observation considered below in Example 7.2.5. Then the optimal transition density $t_k^{\mathrm{u}}(x, x') = (\mathrm{L}_{k+1}/\mathrm{L}_k)^{-1} q(x, x') g_k(x')$ is also a log-concave function of its x' argument, as its logarithm is the sum of two concave functions (and a constant term). This implies in particular that $x' \mapsto t_k^{\mathrm{u}}(x, x')$ is unimodal and that its mode may be located using computationally efficient techniques such as Newton iterations.

The instrumental transition density function is usually chosen from a parametric family $\{r_\theta\}_{\theta \in \Theta}$ of densities indexed by a finite-dimensional parameter θ. An obvious choice is the multivariate Gaussian distribution with mean m and covariance matrix Γ, in which case $\theta = (\mu, \Gamma)$. A better choice is a multivariate t-distribution with η-degrees of freedom, location m, and scale matrix Γ. Recall that the density of this distribution is proportional to $r_\theta(x) \propto [\eta + (x-m)^t \Gamma^{-1}(x-m)]^{(-\eta+d)/2}$. The choice $\eta = 1$ corresponds to a Cauchy distribution. This is a conservative choice that ensures over-dispersion, but if X is high-dimensional, most draws from a multivariate Cauchy might be too far away from the mode to reasonably approximate the target distribution. In most situations, values such as $\eta = 4$ (three finite moments) are more reasonable, especially if the underlying model does not feature heavy-tailed distributions. Recall also that simulation from the multivariate t-distribution with η degrees of freedom, location m, and scale Σ can easily be achieved by first drawing from a multivariate Gaussian distribution with mean m and covariance Γ and then dividing the outcome by the square root of an independent chi-square draw with η degrees of freedom divided by η.

To choose the parameter θ of the instrumental distribution r_θ, one should try to minimize the supremum of the importance function,

$$\min_{\theta \in \Theta} \sup_{x' \in \mathsf{X}} \frac{q(x, x') g_k(x')}{r_\theta(x')} . \tag{7.21}$$

This is a minimax guarantee by which θ is chosen to minimize an upper bound on the importance weights. Note that if r_θ was to be used for sampling from $t_k(x, \cdot)$ by the accept-reject algorithm, the value of θ for which the minimum is achieved in (7.21) is also the one that would make the acceptance probability maximal (see Section 6.2.1). In practice, solving the optimization problem in (7.21) is often too demanding, and a more generic strategy consists in locating the mode of $x' \mapsto t_k(x, x')$ by an iterative algorithm and evaluating the Hessian of its logarithm at the mode. The parameter θ is then selected in the following way.

Multivariate normal: fit the mean of the normal distribution to the mode of $t_k(x, \cdot)$ and fit the covariance to minus the inverse of the Hessian of $\log t(x, \cdot)$ at the mode.

Multivariate t-distribution: fit the location and scale parameters as the mean and covariance parameters in the normal case; the number of degrees of freedom is usually set arbitrarily (and independently of x) based on the arguments discussed above.

We discuss below an important model for which this strategy is successful.

Example 7.2.5 (Stochastic Volatility Model). We return to the stochastic volatility model introduced as Example 1.3.13 and considered previously in the context of MCMC methods as Example 6.2.16. From the state-space equations that define the model,

$$
\begin{aligned}
X_{k+1} &= \phi X_k + \sigma U_k \ , \\
Y_k &= \beta \exp(X_k/2) V_k \ ,
\end{aligned}
$$

we directly obtain

$$
\begin{aligned}
q(x, x') &= \frac{1}{\sqrt{2\pi\sigma^2}} \exp\left[-\frac{(x' - \phi x)^2}{2\sigma^2}\right] , \\
g_k(x') &= \frac{1}{\sqrt{2\pi\beta^2}} \exp\left[-\frac{Y_k^2}{2\beta^2} \exp(-x') - \frac{1}{2} x'\right] .
\end{aligned}
$$

Simulating from the optimal transition kernel $t_k(x, x')$ is difficult, but the function $x' \mapsto \log(q(x, x') g_k(x'))$ is indeed (strictly) concave. The mode $m_k(x)$ of $x' \mapsto t_k(x, x')$ is the unique solution of the non-linear equation

$$
-\frac{1}{\sigma^2}(x' - \phi x) + \frac{Y_k^2}{2\beta^2} \exp(-x') - \frac{1}{2} = 0 , \tag{7.22}
$$

which can be found using Newton iterations. Once at the mode, the (squared) scale $\sigma_k^2(x)$ is set as minus the inverse of the second-order derivative of $x' \mapsto (\log q(x, x') g_k(x'))$ evaluated at the mode $m_k(x)$. The result is

$$
\sigma_k^2(x) = \left\{ \frac{1}{\sigma^2} + \frac{Y_k^2}{2\beta^2} \exp\left[-m_k(x)\right] \right\}^{-1} . \tag{7.23}
$$

In this example, a t-distribution with $\eta = 5$ degrees of freedom was used, with location $m_k(x)$ and scale $\sigma_k(x)$ obtained as above. The incremental importance weight is then given by

$$
\frac{\exp\left[-\frac{(x' - \phi x)^2}{2\sigma^2} - \frac{Y_k^2}{2\beta^2} \exp(-x') - \frac{x'}{2}\right]}{\sigma_k^{-1}(x) \left\{ \eta + \frac{[x' - m_k(x)]^2}{\sigma_k^2(x)} \right\}^{-(\eta+1)/2}} .
$$

As in the case of Example 6.2.16, the first time index ($k = 0$) is particular, and it is easily checked that $m_0(x)$ is the solution of

$$
-\frac{1 - \phi^2}{\sigma^2} x - \frac{1}{2} + \frac{Y_0^2}{2\beta^2} \exp(-x) = 0 ,
$$

and $\sigma_0(x)$ is given by

$$
\sigma_0^2(x) = \left[\frac{1 - \phi^2}{\sigma^2} + \frac{Y_0^2}{2\beta^2} \exp(-m_0)\right]^{-1} .
$$

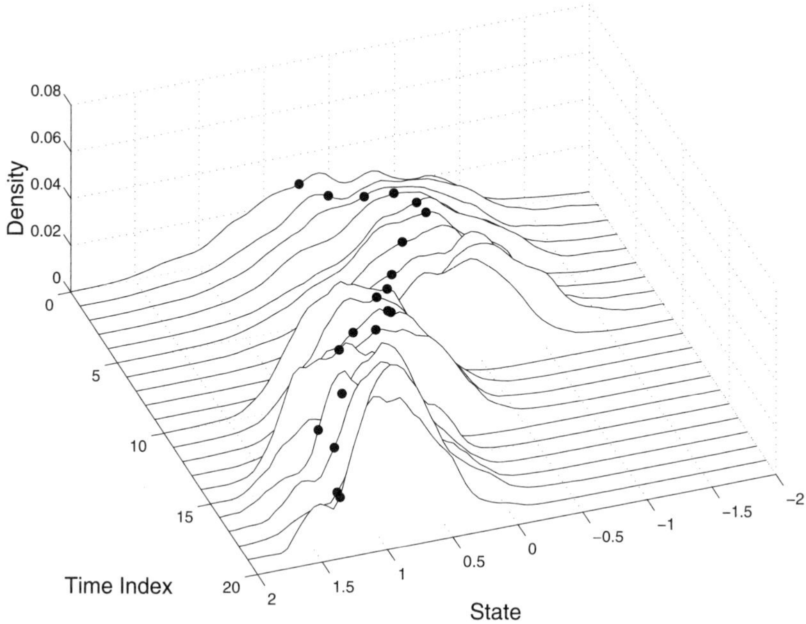

Fig. 7.7. Waterfall representation of filtering distributions as estimated by SIS with $N = 1{,}000$ particles (densities estimated with Epanechnikov kernel, bandwidth 0.2). Data is the same as in Figure 6.8.

Figure 7.7 shows a typical example of the type of fit that can be obtained for the stochastic volatility model with this strategy using 1,000 particles. Note that although the data used is the same as in Figure 6.8, the estimated distributions displayed in both figures are not directly comparable, as the MCMC method in Figure 6.9 approximates the marginal smoothing distribution, whereas the sequential importance sampling approach used for Figure 7.7 provides a (recursive) approximation to the *filtering* distributions. ■

When there is no easy way to implement the local linearization technique, a natural idea explored by Doucet *et al.* (2000a) and Van der Merwe *et al.* (2000) consists in using classical non-linear filtering procedures to approximate t_k. These include in particular the so-called extended Kalman filter (EKF), which dates back to the 1970s (Anderson and Moore, 1979, Chapter 10), as well as the unscented Kalman filter (UKF) introduced by Julier and Uhlmann (1997)—see, for instance, Ristic *et al.* (2004, Chapter 2) for a recent review of these techniques. We illustrate below the use of the extended Kalman filter in the context of sequential importance sampling.

We now consider the most general form of the state-space model with Gaussian noises:

$$X_{k+1} = a(X_k, U_k) , \qquad U_k \sim \mathrm{N}(0, I) , \tag{7.24}$$
$$Y_k = b(X_k, V_k) , \qquad V_k \sim \mathrm{N}(0, I) , \tag{7.25}$$

where a, b are vector-valued measurable functions. It is assumed that $\{U_k\}_{k\geq 0}$ and $\{V_k\}_{k\geq 0}$ are independent white Gaussian noises. As usual, X_0 is assumed to be $\mathrm{N}(0, \Sigma_\nu)$ distributed and independent of $\{U_k\}$ and $\{V_k\}$. The extended Kalman filter proceeds by approximating the non-linear state-space equations (7.24)–(7.25) by a non-linear Gaussian state-space model with linear measurement equation. We are then back to a model of the form (7.18)–(7.19) for which the optimal kernel may be determined exactly using Gaussian formulas. We will adopt the approximation

$$X_k \approx a(X_{k-1}, 0) + R(X_{k-1})U_{k-1} , \tag{7.26}$$
$$Y_k \approx b\left[a(X_{k-1}, 0), 0\right] + B(X_{k-1})\left[X_k - a(X_{k-1}, 0)\right] + S(X_{k-1})V_k , \tag{7.27}$$

where

- $R(x)$ is the $d_x \times d_u$ matrix of partial derivatives of $a(x, u)$ with respect to u and evaluated at $(x, 0)$,
$$[R(x)]_{i,j} \stackrel{\text{def}}{=} \frac{\partial\left[a(x,0)\right]_i}{\partial u_j} \qquad \text{for } i = 1, \ldots, d_x \text{ and } j = 1, \ldots, d_u ;$$
- $B(x)$ and $S(x)$ are the $d_y \times d_x$ and $d_y \times d_v$ matrices of partial derivatives of $b(x, v)$ with respect to x and v respectively and evaluated at $(a(x, 0), 0)$,
$$[B(x)]_{i,j} = \frac{\partial\left\{b\left[a(x,0),0\right]\right\}_i}{\partial x_j} \qquad \text{for } i = 1, \ldots, d_y \text{ and } j = 1, \ldots, d_x ,$$
$$[S(x)]_{i,j} = \frac{\partial\left\{b\left[a(x,0),0\right]\right\}_i}{\partial v_j} \qquad \text{for } i = 1, \ldots, d_y \text{ and } j = 1, \ldots, d_v .$$

It should be stressed that the measurement equation in (7.27) differs from (7.19) in that it depends both on the current state X_k and on the previous one X_{k-1}. The approximate model specified by (7.26)–(7.27) thus departs from the HMM assumptions. On the other hand, *when conditioning* on the value of X_{k-1}, the structure of both models, (7.18)–(7.19) and (7.26)–(7.27), are exactly similar. Hence the posterior distribution of the state X_k given $X_{k-1} = x$ and Y_k is a Gaussian distribution with mean $m_k(x)$ and covariance matrix $\Gamma_k(x)$, which can be evaluated according to

$$K_k(x) = R(x)R^t(x)B^t(x)\left[B(x)R(x)R^t(x)B^t(x) + S(x)S^t(x)\right]^{-1} ,$$
$$m_k(x) = a(x, 0) + K_k(x)\ \left\{Y_k - b\left[a(x, 0), 0\right]\right\} ,$$
$$\Gamma(x) = \left[I - K_k(x)B(x)\right]R(x)R^t(x) .$$

The Gaussian distribution with mean $m_k(x)$ and covariance $\Gamma_k(x)$ may then be used as a proxy for the optimal transition kernel $T_k(x, \cdot)$. To improve the robustness of the method, it is safe to increase the variance, that is, to use $c\Gamma_k(x)$ as the simulation variance, where c is a scalar larger than one. A perhaps more recommendable option consists in using as previously a proposal distribution with tails heavier than the Gaussian, for instance, a multivariate t-distribution with location $m_k(x)$, scale $\Gamma_k(x)$, and four or five degrees of freedom.

Example 7.2.6 (Growth Model). We consider the univariate growth model discussed by Kitagawa (1987) and Polson *et al.* (1992) given, in state-space form, by

$$X_k = a_{k-1}(X_{k-1}) + \sigma_u U_{k-1} \,, \qquad U_k \sim \mathrm{N}(0,1) \,, \tag{7.28}$$

$$Y_k = bX_k^2 + \sigma_v V_k \,, \qquad V_k \sim \mathrm{N}(0,1) \,, \tag{7.29}$$

where $\{U_k\}_{k\geq 0}$ and $\{V_k\}_{k\geq 0}$ are independent white Gaussian noise processes and

$$a_{k-1}(x) = \alpha_0 x + \alpha_1 \frac{x}{1+x^2} + \alpha_2 \cos\left[1.2(k-1)\right] \tag{7.30}$$

with $\alpha_0 = 0.5$, $\alpha_1 = 25$, $\alpha_2 = 8$, $b = 0.05$, and $\sigma_v^2 = 1$ (the value of σ_u^2 will be discussed below). The initial state is known deterministically and set to $X_0 = 0.1$. This model is non-linear both in the state and in the measurement equation. Note that the form of the likelihood adds an interesting twist to the problem: whenever $Y_k \leq 0$, the conditional likelihood function

$$g_k(x) \stackrel{\text{def}}{=} g(x; Y_k) \propto \exp\left[-\frac{b^2}{2\sigma_v^2}\left(x^2 - Y_k/b\right)^2\right]$$

is unimodal and symmetric about 0; when $Y_k > 0$ however, the likelihood g_k is symmetric about 0 with two modes located at $\pm(Y_k/b)^{1/2}$.

The EKF approximation to the optimal transition kernel is a Gaussian distribution with mean $m_k(x)$ and variance $\Gamma_k(x)$ given by

$$K_k(x) = 2\sigma_u^2 b a_{k-1}(x)\left[4\sigma_u^2 b^2 f_{k-1}^2(x) + \sigma_v^2\right]^{-1} \,,$$

$$m_k(x) = a_{k-1}(x) + K_{k-1}(x)\left[Y_k - b a_{k-1}^2(x)\right] \,,$$

$$\Gamma_k(x) = \frac{\sigma_v^2 \sigma_u^2}{4\sigma_u^2 b^2 a_{k-1}^2(x) + \sigma_v^2} \,.$$

In Figure 7.8, the optimal kernel, the EKF approximation to the optimal kernel, and the prior kernel for two different values of the state variance are compared. This figure corresponds to the time index one, and Y_1 is set to 6 (recall that the initial state X_0 is equal to 0.1). In the case where $\sigma_u^2 = 1$ (left plot in Figure 7.8), the prior distribution of the state, $\mathrm{N}(a_0(X_0), \sigma_u^2)$, turns out to be more informative (more peaky, less diffuse) than the conditional likelihood g_1. In other words, the observed Y_1 does not carry a lot of information about the state X_1, compared to the information provided by X_0; this

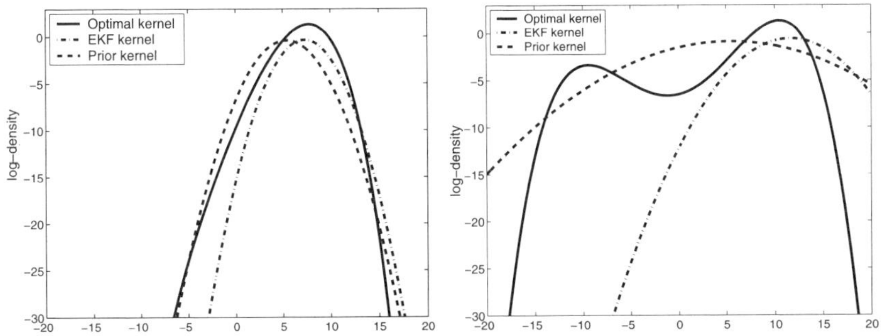

Fig. 7.8. Log-density of the optimal kernel (solid line), EKF approximation of the optimal kernel (dashed-dotted line), and the prior kernel (dashed line) for two different values of the state noise variance σ_u^2: left, $\sigma_u^2 = 1$; right, $\sigma_u^2 = 10$.

is because the measurement variance σ_v^2 is not small compared to σ_u^2. The optimal transition kernel, which does take Y_1 into account, is then very close to the prior kernel, and the differences between the three kernels are minor. In such a situation, one should not expect much improvement with the EKF approximation compared to the prior kernel.

In the case shown in the right plot of Figure 7.8 ($\sigma_u^2 = 10$), the situation is reversed. Now σ_v^2 is relatively small compared to σ_u^2, so that the information about X_1 contained in g_1 is large to that provided by the prior information on X_0. This is the kind of situation where we expect the optimal kernel to improve considerably on the prior kernel. Indeed, because $Y_1 > 0$, the optimal kernel is bimodal, with the second mode far smaller than the first one (recall that the plots are on log-scale); the EKF kernel correctly picks the dominant mode. Figure 7.8 also illustrates the fact that, in contrast to the prior kernel, the EKF kernel does not necessarily dominate the optimal kernel in the tails; hence the need to simulate from an over-dispersed version of the EKF approximation as discussed above. ■

7.3 Sequential Importance Sampling with Resampling

Despite quite successful results for short data records, as was observed in Example 7.2.5, it turns out that the sequential importance sampling approach discussed so far is bound to fail in the long run. We first substantiate this claim with a simple illustrative example before examining solutions to this shortcoming based on the concept of resampling introduced in Section 7.1.2.

7.3.1 Weight Degeneracy

The intuitive interpretation of the importance sampling weight ω_k^i is as a measure of the adequacy of the simulated trajectory $\xi_{0:k}^i$ to the target distribution

$\phi_{0:k|n}$. A small importance weight implies that the trajectory is drawn far from the main body of the posterior distribution $\phi_{0:k|n}$ and will contribute only moderately to the importance sampling estimates of the form (7.11). Indeed, a particle such that the associated weight ω_k^i is orders of magnitude smaller than the sum $\sum_{i=1}^N \omega_k^i$ is practically ineffective. If there are too many ineffective particles, the particle approximation becomes both computationally and statistically inefficient: most of the computing effort is put on updating particles and weights that do not contribute significantly to the estimator; the variance of the resulting estimator will not reflect the large number of terms in the sum but only the small number of particles with non-negligible normalized weights.

Unfortunately, the situation described above is the rule rather than the exception, as the importance weights will (almost always) degenerate as the time index k increases, with most of the normalized importance weights $\omega_k^i / \sum_{j=1}^N \omega_k^j$ close to 0 except for a few ones. We consider below the case of i.i.d. models for which it is possible to show using simple arguments that the large sample variance of the importance sampling estimate can only increase with the time index k.

Example 7.3.1 (Weight Degeneracy in the I.I.D. Case). The simplest case of application of the sequential importance sampling technique is when μ is a probability distribution on $(\mathsf{X}, \mathcal{X})$ and the sequence of target distributions corresponds to the product distributions, that is, the sequence of distributions on $(\mathsf{X}^{k+1}, \mathcal{X}^{\otimes(k+1)})$ defined recursively by $\mu_0 = \mu$ and $\mu_k = \mu_{k-1} \otimes \mu$, for $k \geq 1$. Let ν be another probability distribution on $(\mathsf{X}, \mathcal{X})$ and assume that μ is absolutely continuous with respect to ν and that

$$\int \left[\frac{d\mu}{d\nu}(x) \right]^2 \nu(dx) < \infty \,. \tag{7.31}$$

Finally, let f be a bounded measurable function that is not (μ-a.s.) constant such that its variance under μ, $\mu(f^2) - \mu^2(f)$, is strictly positive.

Consider the sequential importance sampling estimate given by

$$\widehat{\mu}_{k,N}^{\mathrm{IS}}(f) = \sum_{i=1}^N f(\xi_k^i) \frac{\prod_{l=0}^k \frac{d\mu}{d\nu}(\xi_k^i)}{\sum_{j=1}^N \prod_{l=0}^k \frac{d\mu}{d\nu}(\xi_k^j)} \,, \tag{7.32}$$

where the random variables $\{\xi_l^j\}$, $l = 1, \ldots, k$, $j = 1, \ldots, N$ are i.i.d. with common distribution ν. As discussed in Section 7.2, the unnormalized importance weights may be computed recursively and hence (7.32) really corresponds to an estimator of the form (7.11) in the particular case of a function f_k that depends on the last component only. This is of course a rather convoluted and very inefficient way of constructing an estimate of $\mu(f)$ but still constitutes a valid instance of the sequential importance sampling approach (in a very particular case).

Now let k be fixed and write

$$N^{1/2}\left\{\widehat{\mu}^{\mathrm{IS}}_{k,N}(f)-\mu(f)\right\}=\frac{N^{-1/2}\sum_{i=1}^{N}\prod_{l=0}^{k}\left\{f(\xi_k^i)-\mu(f)\right\}\frac{d\mu}{d\nu}(\xi_l^i)}{N^{-1}\sum_{i=1}^{N}\prod_{l=0}^{k}\frac{d\mu}{d\nu}(\xi_l^i)}\,. \tag{7.33}$$

Because

$$\mathrm{E}\left[\prod_{l=0}^{k}\frac{d\mu}{d\nu}(\xi_l^i)\right]=1\,,$$

the weak law of large numbers implies that the denominator of the right-hand side of (7.33) converges to 1 in probability as N increases. Likewise, under (7.31), the central limit theorem shows that the numerator of the right-hand side of (7.33) converges in distribution to the normal $\mathrm{N}(0,\sigma_k^2(f))$ distribution, where

$$\begin{aligned}\sigma_k^2(f)&=\mathrm{E}\left(\left\{\prod_{l=0}^{k}\left[f(\xi_k^1)-\mu(f)\right]^2\frac{d\mu}{d\nu}(\xi_l^1)\right\}^2\right) \qquad (7.34)\\&=\left[\int\left(\frac{d\mu}{d\nu}(x)\right)^2\nu(dx)\right]^k\int\left[\frac{d\mu}{d\nu}(x)\right]^2\left[f(x)-\mu(f)\right]^2\nu(dx)\,.\end{aligned}$$

Slutsky's lemma then implies that (7.33) also converges in distribution to the same $\mathrm{N}(0,\sigma_k^2(f))$ limit as N grows. Now Jensen's inequality implies that

$$1=\left[\int\frac{d\mu}{d\nu}(x)\nu(dx)\right]^2\le\int\left[\frac{d\mu}{d\nu}(x)\right]^2\nu(dx)\,,$$

with equality if and only if $\mu=\nu$. Therefore, if $\mu\neq\nu$, the asymptotic variance $\sigma_k^2(f)$ grows exponentially with the iteration index k for all functions f such that

$$\int\left[\frac{d\mu}{d\nu}(x)\right]^2\left[f(x)-\mu(f)\right]^2\nu(dx)=\int\frac{d\mu}{d\nu}(x)\left[f(x)-\mu(f)\right]^2\mu(dx)\neq 0\,.$$

Because μ is absolutely continuous with respect to ν, $\mu\{x\in\mathsf{X}:d\mu/d\nu(x)=0\}=0$ and the last integral is null if and only if f has zero variance under μ.

Thus in the i.i.d. case, the asymptotic variance of the importance sampling estimate (7.32) increases exponentially with the time index k as soon as the proposal and target differ (except for constant functions). ∎

It is more difficult to characterize the degeneracy of the weights for general target and instrumental distributions. There have been some limited attempts to study more formally this phenomenon in some specific scenarios. In particular, Del Moral and Jacod (2001) have shown the degeneracy of the sequential importance sampling estimator of the posterior mean in Gaussian linear models when the instrumental kernel is the prior kernel. Such results are in general difficult to derive (even in the Gaussian linear models where most of the derivations can be carried out explicitly) and do not provide much additional insight. Needless to say, in practice, weight degeneracy is a prevalent and serious problem making the vanilla sequential importance sampling method discussed so far almost useless. The degeneracy can occur after a very limited number of iterations, as illustrated by the following example.

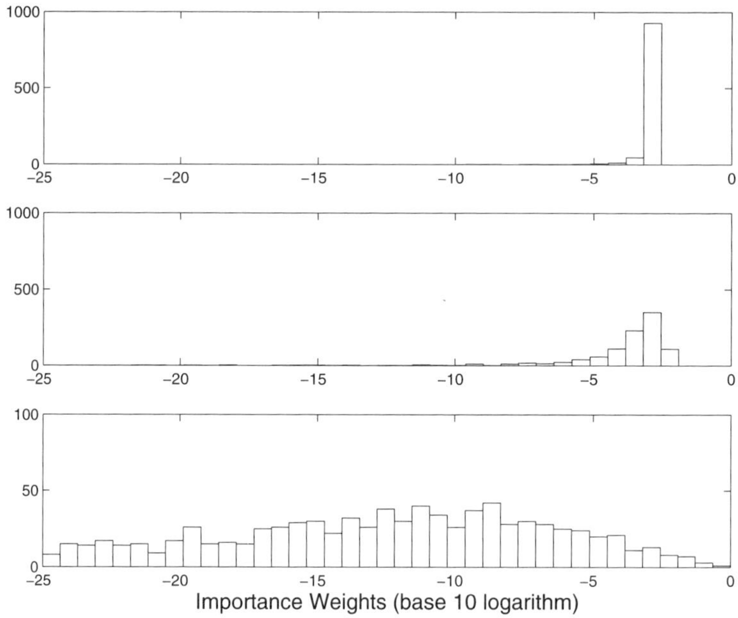

Fig. 7.9. Histograms of the base 10 logarithm of the normalized importance weights after (from top to bottom) 1, 10, and 100 iterations for the stochastic volatility model of Example 7.2.5. Note that the vertical scale of the bottom panel has been multiplied by 10.

Example 7.3.2 (Stochastic Volatility Model, Continued). Figure 7.9 displays the histogram of the base 10 logarithm of the normalized importance weights after 1, 10, and 100 time indices for the stochastic volatility model considered in Example 7.2.5 (using the same instrumental kernel). The number of particles is set to 1,000. Figure 7.9 shows that, despite the choice of a reasonably good approximation to the optimal importance kernel, the normalized importance weights quickly degenerate as the number of iterations of the SIS algorithm increases. Clearly, the results displayed in Figure 7.7 still are reasonable for $k = 20$ but would be disastrous for larger time horizons such as $k = 100$. ■

Because the weight degeneracy phenomenon is so detrimental, it is of great practical significance to set up tests that can detect this phenomenon. A simple criterion is the coefficient of variation of the normalized weights used by Kong *et al.* (1994), which is defined by

$$\mathrm{CV}_N = \left[\frac{1}{N} \sum_{i=1}^{N} \left(N \frac{\omega^i}{\sum_{j=1}^{N} \omega^j} - 1 \right)^2 \right]^{1/2} . \tag{7.35}$$

The coefficient of variation is minimal when the normalized weights are all equal to $1/N$, and then $\mathrm{CV}_N = 0$. The maximal value of CV_N is $\sqrt{N-1}$, which corresponds to one of the normalized weights being one and all others being null. Therefore, the coefficient of variation is often interpreted as a measure of the number of ineffective particles (those that do not significantly contribute to the estimate). A related criterion with a simpler interpretation is the so-called *effective sample size* N_{eff} (Liu, 1996), defined as

$$N_{\mathrm{eff}} = \left[\sum_{i=1}^{N} \left(\frac{\omega^i}{\sum_{j=1}^{N} \omega^j} \right)^2 \right]^{-1} , \tag{7.36}$$

which varies between 1 (all weights null but one) and N (equal weights). It is straightforward to verify the relation

$$N_{\mathrm{eff}} = \frac{N}{1 + \mathrm{CV}_N^2} .$$

Some additional insights and heuristics about the coefficient of variation are given by Liu and Chen (1995).

Yet another possible measure of the weight imbalance is the Shannon entropy of the importance weights,

$$\mathrm{Ent} = - \sum_{i=1}^{N} \frac{\omega^i}{\sum_{j=1}^{N} \omega^j} \log_2 \left(\frac{\omega^i}{\sum_{j=1}^{N} \omega^j} \right) . \tag{7.37}$$

When all the normalized importance weights are null except for one of them, the entropy is null. On the contrary, if all the weights are equal to $1/N$, then the entropy is maximal and equal to $\log_2 N$.

Example 7.3.3 (Stochastic Volatility Model, Continued). Figure 7.10 displays the coefficient of variation (left) and Shannon entropy (right) as a function of the time index k under the same conditions as for Figure 7.9, that is, for the stochastic volatility model of 7.2.5. The figure shows that the distribution of the weights steadily degenerates: the coefficient of variation increases and the entropy of the importance weights decreases. After 100 iterations, there are less than 50 particles (out 1,000) significantly contributing to

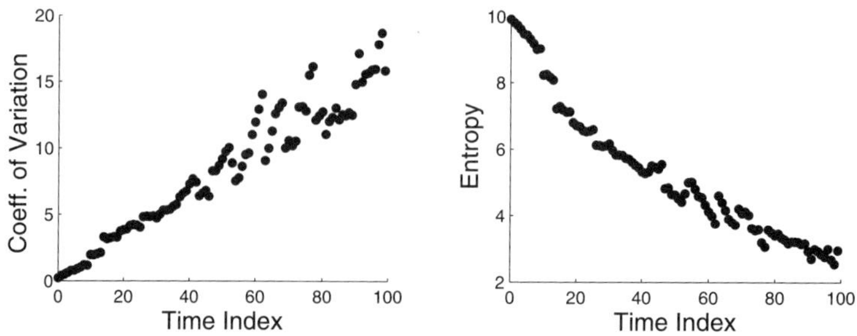

Fig. 7.10. Coefficient of variation (left) and entropy of the normalized importance weights as a function of the number of iterations for the stochastic volatility model of Example 7.2.5. Same model and data as in Figure 7.9.

the importance sampling estimator. Most particles have importance weights that are zero to machine precision, which is of course a tremendous waste in computational resource.

■

7.3.2 Resampling

The solution proposed by Gordon *et al.* (1993) to reduce the degeneracy of the importance weights is based on the concept of *resampling* already discussed in the context of importance sampling in Section 7.1.2. The basic method consists in resampling in the current population of particles using the normalized weights as probabilities of selection. Thus, trajectories with small importance weights are eliminated, whereas those with large importance weights are duplicated. After resampling, all importance weights are reset to one. Up to the first instant when resampling occurs, the method can really be interpreted as an instance of the sampling importance resampling (SIR) technique discussed in Section 7.1.2. In the context of sequential Monte Carlo, however, the main motivation for resampling is to avoid future weight degeneracy by reseting (periodically) the weights to equal values. The resampling step has a drawback however: as emphasized in Section 7.1.2, resampling introduces additional variance in Monte Carlo approximations. In some situations, the additional variance may be far from negligible: when the importance weights already are nearly equal for instance, resampling can only reduce the number of distinct particles, thus degrading the accuracy of the Monte Carlo approximation. The one-step effect of resampling is thus negative but, in the long term, resampling is required to guarantee a stable behavior of the algorithm. This interpretation suggests that it may be advantageous to restrict the use of resampling to cases where the importance weights are becoming very uneven. The criteria defined in (7.35), (7.36), or (7.37) are of course helpful for that purpose. The resulting algorithm, which is generally known under the name of *sequential importance sampling with resampling* (SISR), is summarized below.

Algorithm 7.3.4 (SISR: Sequential Importance Sampling with Resampling). Initialize the particles as in Algorithm 7.2.2, optionally applying the resampling step below. For subsequent time indices $k \geq 0$, do the following.

Sampling:

- Draw $(\tilde{\xi}^1_{k+1}, \ldots, \tilde{\xi}^N_{k+1})$ conditionally independently given $\{\xi^j_{0:k}, j = 1, \ldots, N\}$ from the instrumental kernel: $\tilde{\xi}^i_{k+1} \sim R_k(\xi^i_k, \cdot)$, $i = 1, \ldots, N$.
- Compute the updated importance weights

$$\omega^i_{k+1} = \omega^i_k g_{k+1}(\tilde{\xi}^i_{k+1}) \frac{dQ(\xi^i_k, \cdot)}{dR_k(\xi^i_k, \cdot)}(\tilde{\xi}^i_{k+1}) \,, \qquad i = 1, \ldots, N \,.$$

Resampling (Optional):

- Draw, conditionally independently given $\{(\xi^i_{0:k}, \tilde{\xi}^j_{k+1}),\ i, j = 1, \ldots, N\}$, the multinomial trial $(I^1_{k+1}, \ldots I^N_{k+1})$ with probabilities of success

$$\frac{\omega^1_{k+1}}{\sum_j^N \omega^j_{k+1}}, \ldots, \frac{\omega^N_{k+1}}{\sum_j^N \omega^j_{k+1}} \,.$$

- Reset the importance weights ω^i_{k+1} to a constant value for $i = 1, \ldots, N$.

If resampling is not applied, set for $i = 1, \ldots, N$, $I^i_{k+1} = i$.

Trajectory update: for $i = 1, \ldots, N$,

$$\xi^i_{0:k+1} = \left(\xi^{I^i_{k+1}}_{0:k}, \tilde{\xi}^{I^i_{k+1}}_{k+1} \right) \,. \tag{7.38}$$

As discussed previously the resampling step in the algorithm above may be used systematically (for all indices k), but it is often preferable to perform resampling from time to time only. Usually, resampling is either used systematically but at a lower rate (for one index out of m, where m is fixed) or at random instants based on the values of the coefficient of variation or the entropy criteria defined in (7.35) and (7.37), respectively. Note that in addition to arguments based on the variance of the Monte Carlo approximation, there is usually also a computational incentive for limiting the use of resampling; indeed, except in models where the evaluation of the incremental weights is costly (think of large-dimensional multivariate observations for instance), the computational cost of the resampling step is not negligible. Both Sections 7.4.1 and 7.4.2 discuss several implementations and variants of the resampling step that may render the latter argument less pregnant.

The term *particle filter* is often used to refer to Algorithm 7.3.4 although the terminology SISR is preferable, as particle filtering is sometimes also used more generically for any sequential Monte Carlo method. Gordon *et al.* (1993) actually proposed a specific instance of Algorithm 7.3.4 in which resampling is done systematically at each step and the instrumental kernel is chosen as the prior kernel $R_k = Q$. This particular algorithm, commonly known as the *bootstrap filter*, is most often very easy to implement because it only involves simulating from the transition kernel Q of the hidden chain and evaluation of the conditional likelihood function g.

There is of course a whole range of variants and refinements of Algorithm 7.3.4, many of which will be covered in some detail in the next chapter. A simple remark though is that, as in the case of the simplest SIR method discussed in Section 7.1.2, it is possible to resample N times from a larger population of M intermediate samples. In practice, it means that Algorithm 7.3.4 should be modified as follows at indices k for which resampling is to be applied.

SIS: For $i = 1, \ldots, N$, draw α candidates $\tilde{\xi}_{k+1}^{i,1}, \ldots, \tilde{\xi}_{k+1}^{i,\alpha}$ from each proposal distribution $R_k(\xi_k^i, \cdot)$.

Resampling: Draw $(N_{k+1}^{1,1}, \ldots, N_{k+1}^{1,\alpha}, \ldots, N_{k+1}^{N,1}, \ldots, N_{k+1}^{N,\alpha})$ from the multinomial distribution with parameter N and probabilities

$$\frac{\omega_{k+1}^{i,j}}{\sum_{l=1}^{N} \sum_{m=1}^{\alpha} \omega_{k+1}^{l,m}} \qquad \text{for } i = 1, \ldots, N,\ j = 1, \ldots, \alpha \ .$$

Hence, while this form of resampling keeps the number of particles fixed and equal to N after resampling, the intermediate population (before resampling) has size $M = \alpha \times N$. Although obviously heavier to implement, the use of α larger than one may be advantageous in some models. In particular, we will show in Chapter 9 that using α larger than one effectively reduces the variance associated with the resampling operation in a proportion that may be significant.

Remark 7.3.5 (Marginal Interpretation of SIS and SISR). Both Algorithms 7.2.2 and 7.3.4 have been introduced as methods to simulate whole *trajectories* $\{\xi_{0:k}^i\}_{1 \le i \le N}$ that approximate the joint smoothing distribution $\phi_{0:k|k}$. This was done quite easily in the case of sequential importance sampling (Algorithm 7.2.2), as the trajectories are simply extended independently of one another as new samples arrive. When using resampling however, the process is more involved because it becomes necessary to duplicate or discard some trajectories according to (7.38).

This presentation of the SIS and SISR methods has been adopted because it is the most natural way to introduce sequential Monte Carlo methods. It does not mean that, when implementing the SISR algorithm, storing the whole trajectories is required. Neither do we claim that for large k, the approximation of the complete joint distribution $\phi_{0:k|k}$ provided by the particle trajectories $\{\xi_{0:k}^i\}_{1 \le i \le N}$ is accurate (this point will be discussed in detail in Section 8.3). Most often, Algorithm 7.3.4 is implemented storing only the current generation of particles $\{\xi_k^i\}_{1 \le i \le N}$, and (7.38) simplifies to

$$\xi_{k+1}^i = \tilde{\xi}_{k+1}^{I_{k+1}^i} \qquad i = 1, \ldots, N \ .$$

In that case, the system of particles $\{\xi_k^i\}_{1\le i\le N}$ with associated weights $\{\omega_k^i\}_{1\le i\le N}$, provides an approximation to the filtering distribution ϕ_k, which is the *marginal* of the joint smoothing distribution $\phi_{0:k|k}$.

The notation ξ_k^i could be ambiguous when resampling is applied, as the first $k+1$ elements of the ith trajectory $\xi_{0:k+1}^i$ at time $k+1$ do not necessarily coincide with the ith trajectory $\xi_{0:k}^i$ at time k. By convention, ξ_k^i *always refers to the last point in the ith trajectory, as simulated at index k*. Likewise, $\xi_{l:k}^i$ is the portion of the same trajectory that starts at index l and ends at the last index (that is, k). When needed, we will use the notation $\xi_{0:k}^i(l)$ for the element of index l in the ith particle trajectory at time k to avoid ambiguity. ■

To conclude this section on the SISR algorithm, we briefly revisit two of the examples already considered previously to contrast the results obtained with the SIS and SISR approaches.

Example 7.3.6 (Stochastic Volatility Model, Continued). To illustrate the effectiveness of the resampling strategy, we consider once again the stochastic volatility model introduced in Example 7.2.5, for which the weight degeneracy phenomenon (in the basic SIS approach) was patent in Figures 7.9 and 7.10.

Figures 7.11 and 7.12 are the counterparts of Figs. 7.10 and 7.9, respectively, when resampling is applied whenever the coefficient of variation (7.35) of the normalized weights exceeds one. Note that Figure 7.11 displays the coefficient of variation and Shannon entropy computed, for each index k, *before resampling*, at indices for which resampling do occur. Contrary to what happened in plain importance sampling, the histograms of the normalized importance weights shown in Figure 7.12 are remarkably similar, showing that the weight degeneracy phenomenon is now under control. Another important remark in this example is that both criteria (the coefficient of variation and

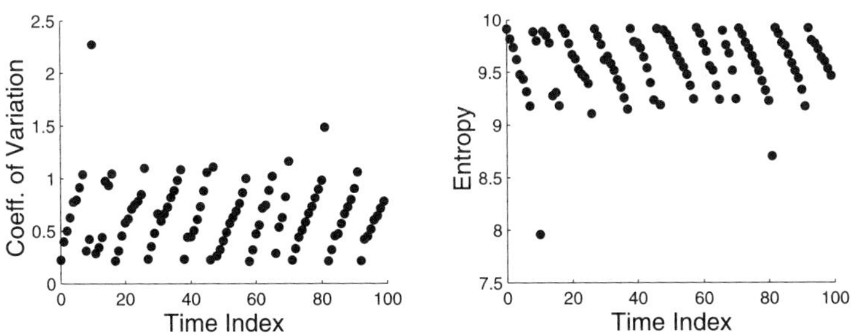

Fig. 7.11. Coefficient of variation (left) and entropy of the normalized importance weights as a function of the number of iterations in the stochastic volatility model of Example 7.2.5. Same model and data as in Figure 7.10. Resampling occurs when the coefficient of variation gets larger than 1.

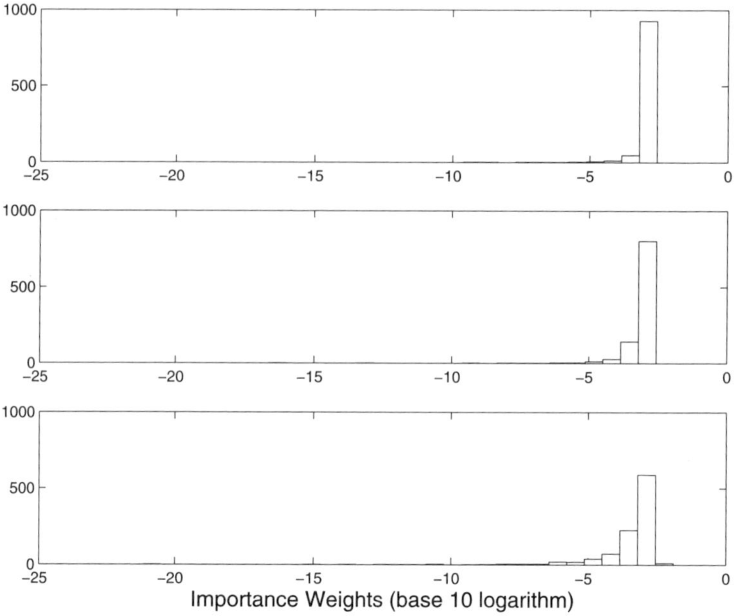

Fig. 7.12. Histograms of the base 10 logarithm of the normalized importance weights after (from top to bottom) 1, 10, and 100 iterations in the stochastic volatility model of Example 7.2.5. Same model and data as in Figure 7.9. Resampling occurs when the coefficient of variation gets larger than 1.

entropy) are strongly correlated. Triggering resampling whenever the entropy gets below, say 9.2, would thus be nearly equivalent with resampling occurring, on average, once every tenth time indices. The Shannon entropy of the normalized importance weights evolves between 10 and 9, suggesting that there are at least 500 particles that are significantly contributing to the importance sampling estimate (out of 1,000). ■

Example 7.3.7 (Growth Model, Continued). Consider again the nonlinear state-space model of Example 7.2.6, with the variance σ_u^2 of the state noise set to 10; this makes the observations very informative relative to the prior distribution on the hidden states. Figures 7.13 and 7.14 display the filtering distributions estimated for the first 31 time indices when using the SIS method with the prior kernel Q as instrumental kernel (Figure 7.13), and the corresponding SISR algorithm with systematic resampling—that is, the bootstrap filter—in Figure 7.14. Both algorithms use 500 particles.

For each time index, the top plots of Figures 7.13 and 7.14 show the highest posterior density (HPD) regions corresponding to the estimated filtering distribution, where the lighter grey zone contains 95% of the probability mass and the darker area corresponds to 50% of the probability mass. These HPD

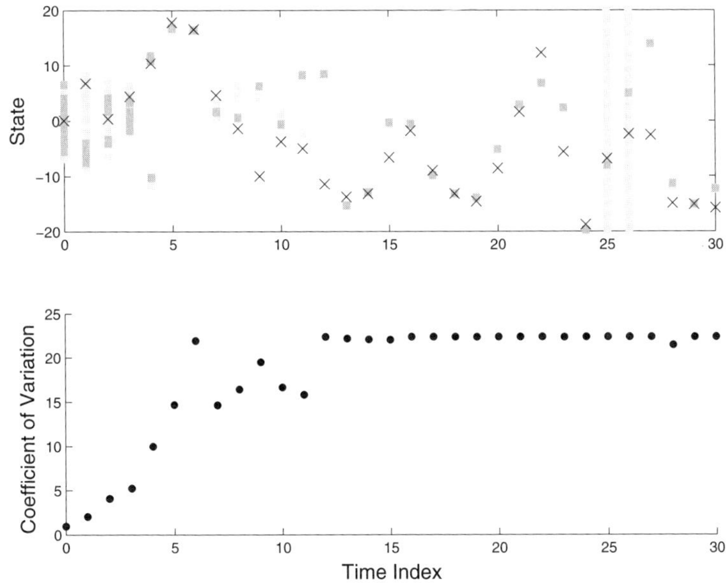

Fig. 7.13. SIS estimates of the filtering distributions in the growth model with instrumental kernel being the prior one and 500 particles. Top: true state sequence ($\times$) and 95%/50% HPD regions (light/dark grey) of estimated filtered distribution. Bottom: coefficient of variation of the normalized importance weights.

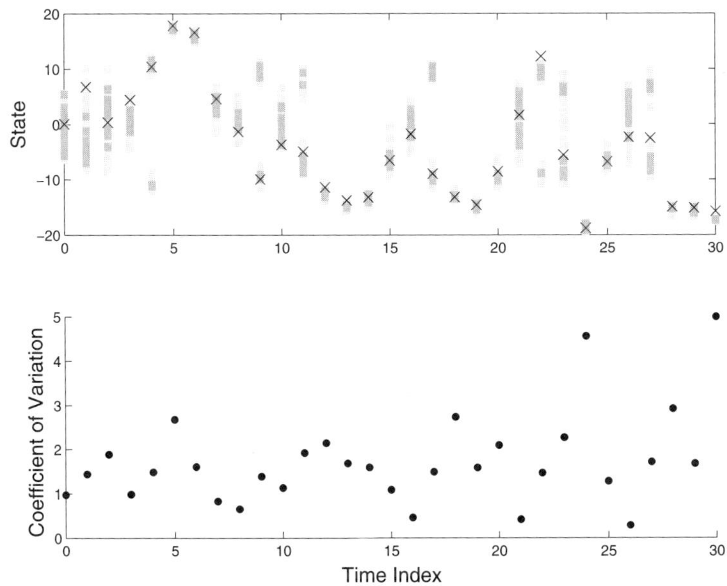

p]

Fig. 7.14. Same legend for Figure 7.13, but with results for the corresponding bootstrap filter.

regions are based on a kernel density estimate (using the Epanechnikov kernel with bandwidth 0.2) computed from the weighted particles (that is, before resampling in the case of the bootstrap filter). Up to $k = 8$, the two methods yield very similar results. With the SIS algorithm however, the bottom panel of Figure 7.13 shows that the weights degenerate quickly. Remember that the maximal value of the coefficient of variation (7.35) is $\sqrt{N-1}$, that is about 22.3 in the case of Figure 7.13. Hence for $k = 6$ and for all indices after $k = 12$, the bottom panel of Figure 7.13 indeed means that almost all normalized weights but one are null: the filtered estimate is concentrated at one point, which sometimes severely departs from the actual state trajectory shown by the crosses. In contrast, the bootstrap filter (Figure 7.14) appears to be very stable and provides reasonable state estimates even at indices for which the filtering distribution is strongly bimodal (see Example 7.2.6 for an explanation of this latter feature). ■

7.4 Complements

As discussed above, resampling is a key ingredient of the success of sequential Monte Carlo techniques. We discuss below two separate aspects related to this issue. First, we show that there are several schemes based on clever probabilistic results that may be exploited to reduce the computational load associated with multinomial resampling. Next, we examine some variants of resampling that achieves lower conditional variance than multinomial resampling. In this latter case, the aim is of course to be able to decrease the number of particles without losing too much on the quality of the approximation.

Throughout this section, we will assume that it is required to draw N samples $\xi^1, \ldots, \xi^N$ out of a, usually larger, set $\{\tilde{\xi}^1, \ldots, \tilde{\xi}^M\}$ according to the *normalized* importance weights $\{\omega^1, \ldots, \omega^M\}$. We denote by $\mathcal{G}$ a σ-field such that both $\omega^1, \ldots, \omega^M$ and $\tilde{\xi}^1, \ldots, \tilde{\xi}^M$ are $\mathcal{G}$-measurable.

7.4.1 Implementation of Multinomial Resampling

Drawing from the multinomial distribution is equivalent to drawing N random indices $I^1, \ldots, I^N$ conditionally independently given $\mathcal{G}$ from the set $\{1, \ldots, M\}$ and such that $\mathrm{P}(I^j = i \,|\, \mathcal{G}) = \omega^i$. This is of course the simplest example of use of the *inversion method*, and each index may be obtained by first simulating a random variable U with uniform distribution on $[0, 1]$ and then determining the index I such that $U \in (\sum_{j=1}^{I-1} \omega^j, \sum_{j=1}^{I} \omega^j]$ (see Figure 7.15). Determining the appropriate index I thus requires on the average $\log_2 M$ comparisons (using a simple binary tree search). Therefore, the naive technique to implement multinomial resampling requires the simulation of N independent uniform random variables and, on the average, of the order $N \log_2 M$ comparisons.

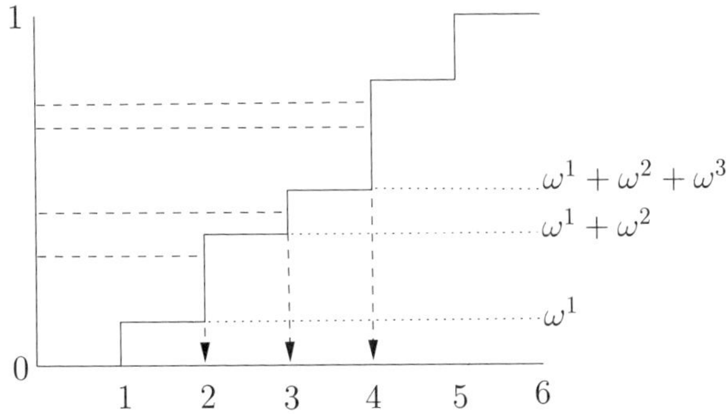

Fig. 7.15. Multinomial sampling from uniform distribution by the inversion method.

A nice solution to avoid the repeated sorting operations consists in pre-sorting the uniform variables. Because the resampling is to be repeated N times, we need N uniform random variables, which will be denoted by $U_1, \ldots, U_N$ and $U_{(1)} \leq U_{(2)} \leq \cdots \leq U_{(N)}$ denoting the associated order statistics. It is easily checked that applying the inversion method from the *ordered* uniforms $\{U_{(i)}\}$ requires, in the worst case, only M comparisons. The problem is that determining the order statistics from the unordered uniforms $\{U_i\}$ by sorting algorithms such as Heapsort or Quicksort is an operation that requires, at best, of the order $N \log_2 N$ comparisons (Press *et al.*, 1992, Chapter 8). Hence, except in cases where $N \ll M$, we have not gained anything yet by pre-sorting the uniform variables prior to using the inversion method. It turns out however that two distinct algorithms are available to sample directly the ordered uniforms $\{U_{(i)}\}$ with a number of operations that scales linearly with N.

Both of these methods are fully covered in by Devroye (1986, Chapter 5), and we only cite here the appropriate results, referring to Devroye (1986, pp. 207–215) for proofs and further references on the methods.

Proposition 7.4.1 (Uniform Spacings). *Let $U_{(1)} \leq \ldots \leq U_{(N)}$ be the order statistics associated with an i.i.d. sample from the* $\mathrm{U}([0,1])$ *distribution. Then the increments*

$$S_i = U_{(i)} - U_{(i-1)} \; , \quad i = 1, \ldots, N \; , \tag{7.39}$$

(where by convention $S_1 = U_{(1)}$) are called the uniform spacings and distributed as

$$\frac{E_1}{\sum_{i=1}^{N+1} E_i}, \ldots, \frac{E_N}{\sum_{i=1}^{N+1} E_i} \; ,$$

where $E_1, \ldots, E_{N+1}$ is a sequence of i.i.d. exponential random variables.

Proposition 7.4.2 (Malmquist, 1950). *Let $U_{(1)} \leq \ldots \leq U_{(N)}$ be the order statistics of $U_1, U_2, \ldots, U_N$—a sequence of i.i.d. uniform $[0,1]$ random vari-*

ables. Then $U_N^{1/N}$, $U_N^{1/N} U_{N-1}^{1/(N-1)}$, ..., $U_N^{1/N} U_{N-1}^{1/(N-1)} \cdots U_1^{1/1}$ *is distributed as* $U_{(N)}, \ldots, U_{(1)}$.

The two sampling algorithms associated with these probabilistic results may be summarized as follows.

Algorithm 7.4.3 (After Proposition 7.4.1).

For $i = 1, \ldots, N+1$: Simulate $U_i \sim \mathrm{U}([0,1])$ and set $E_i = -\log U_i$.
Set $G = \sum_{i=1}^{N+1} E_i$ and $U_{(1)} = E_1/G$.
For $i = 2, \ldots, n$: $U_{(i)} = U_{(i-1)} + E_i/G$.

Algorithm 7.4.4 (After Proposition 7.4.2).

Generate $V_N \sim \mathrm{U}([0,1])$ and set $U_{(N)} = V_N^{1/N}$.
For $i = N-1$ down to 1: Generate $V_i \sim \mathrm{U}([0,1])$ and set $U_{(i)} = V_i^{1/i} U_{(i+1)}$.

Note that Devroye (1986) also discusses a third, slightly more complicated algorithm—the bucket sort method of Devroye and Klincsek (1981)—which also has an expected computation time of order N. Using any of these methods, the computational cost of multinomial resampling scales only linearly in N and M, which makes the method practicable even when a large number of particles is used.

7.4.2 Alternatives to Multinomial Resampling

Instead of using the multinomial sampling scheme, it is also possible to use a different resampling (or reallocation) scheme. For $i = 1, \ldots, M$, denote by N^i the number of times the ith element $\tilde{\xi}^i$ is selected. A resampling scheme will be said to be *unbiased with respect to* $\mathcal{G}$ if

$$\sum_{i=1}^{M} N^i = N , \tag{7.40}$$

$$\mathrm{E}\left[N^i \middle| \mathcal{G}\right] = N\omega^i, \qquad i = 1, \ldots, M . \tag{7.41}$$

We focus here on resampling techniques that keep the number of particles constant (see for instance Crisan *et al.*, 1999, for unbiased sampling with a random number of particles). There are many different conditions under which a resampling scheme is unbiased. The simplest unbiased scheme is multinomial resampling, for which $(N^1, \ldots, N^M)$, conditionally on $\mathcal{G}$, has the multinomial distribution $\mathrm{Mult}(N, \omega^1, \ldots, \omega^N)$. Because $I^1, \ldots, I^M$ are conditionally i.i.d. given $\mathcal{G}$, it is easy to evaluate the conditional variance in the multinomial resampling scheme:

$$\operatorname{Var}\left[\left.\frac{1}{N}\sum_{i=1}^{N} f(\xi^{I^i})\right| \mathcal{G}\right] = \frac{1}{N}\sum_{i=1}^{M}\omega^i\left[f(\tilde{\xi}^i) - \sum_{j=1}^{M}\omega^j f(\tilde{\xi}^j)\right]^2$$

$$= \frac{1}{N}\left\{\sum_{i=1}^{M}\omega^i f^2(\tilde{\xi}^i) - \left[\sum_{i=1}^{M}\omega^i f(\tilde{\xi}^i)\right]^2\right\} . \quad (7.42)$$

A sensible objective is to try to construct resampling schemes for which the conditional variance $\operatorname{Var}(\sum_{i=1}^{N}\frac{N^i}{N}f(\tilde{\xi}^i)\,|\,\mathcal{G})$ is as small as possible and, in particular, smaller than (7.42), preferably for any choice of the function f.

7.4.2.1 Residual Resampling

Residual resampling, or *remainder resampling*, is mentioned by Whitley (1994) (see also Liu and Chen, 1998) as a simple means to decrease the variance incurred by the sampling step. In this scheme, for $i = 1, \ldots, M$ we set

$$N^i = \lfloor N\omega^i \rfloor + \bar{N}^i , \quad (7.43)$$

where $\bar{N}^1, \ldots, \bar{N}^M$ are distributed, conditionally on $\mathcal{G}$, according to the multinomial distribution $\operatorname{Mult}(N - R, \bar{\omega}^1, \ldots, \bar{\omega}^N)$ with $R = \sum_{i=1}^{M}\lfloor N\omega^i\rfloor$ and

$$\bar{\omega}^i = \frac{N\omega^i - \lfloor N\omega^i\rfloor}{N - R}, \quad i = 1, \ldots, M . \quad (7.44)$$

This scheme is obviously unbiased with respect to $\mathcal{G}$. Equivalently, for any measurable function f, the residual sampling estimator is

$$\frac{1}{N}\sum_{i=1}^{N} f(\xi^i) = \sum_{i=1}^{M}\frac{\lfloor N\omega^i\rfloor}{N} f(\tilde{\xi}^i) + \frac{1}{N}\sum_{i=1}^{N-R} f(\tilde{\xi}^{J^i}) , \quad (7.45)$$

where $J^1, \ldots, J^{N-R}$ are conditionally independent given $\mathcal{G}$ with distribution $\mathrm{P}(J^i = k\,|\,\mathcal{G}) = \bar{\omega}^k$ for $i = 1, \ldots, N-R$ and $k = 1, \ldots, M$. Because the residual resampling estimator is the sum of one term that, given $\mathcal{G}$, is deterministic and one term that involves conditionally i.i.d. labels, the variance of residual resampling is given by

$$\frac{1}{N^2}\operatorname{Var}\left[\left.\sum_{i=1}^{N-R} f(\tilde{\xi}^{J^i})\right|\mathcal{G}\right] = \frac{N-R}{N^2}\operatorname{Var}\left[\left. f(\tilde{\xi}^{J^1})\right|\mathcal{G}\right] \quad (7.46)$$

$$= \frac{(N-R)}{N^2}\sum_{i=1}^{M}\bar{\omega}^i\left\{f(\tilde{\xi}^i) - \sum_{j=1}^{M}\bar{\omega}^j f(\tilde{\xi}^j)\right\}^2$$

$$= \frac{1}{N}\sum_{i=1}^{M}\omega^i f^2(\tilde{\xi}^i) - \sum_{i=1}^{M}\frac{\lfloor N\omega^i\rfloor}{N^2} f^2(\tilde{\xi}^i) - \frac{N-R}{N^2}\left\{\sum_{i=1}^{M}\bar{\omega}^i f(\tilde{\xi}^i)\right\}^2 .$$

Residual sampling dominates multinomial sampling also in the sense of having smaller conditional variance. Indeed, first write

$$\sum_{i=1}^{M} \omega^i f(\tilde{\xi}^i) = \sum_{i=1}^{M} \frac{\lfloor N\omega^i \rfloor}{N} f(\tilde{\xi}^i) + \frac{N-R}{N} \sum_{i=1}^{M} \bar{\omega}^i f(\tilde{\xi}^i) \; .$$

Then note that the sum of the M numbers $\lfloor N\omega^i \rfloor / N$ plus $(N-R)/N$ equals one, whence this sequence of $M+1$ numbers can be viewed as a probability distribution. Thus Jensen's inequality applied to the square of the right-hand side of the above display yields

$$\left\{ \sum_{i=1}^{M} \omega^i f(\tilde{\xi}^i) \right\}^2 \le \sum_{i=1}^{M} \frac{\lfloor N\omega^i \rfloor}{N} f^2(\tilde{\xi}^i) + \frac{N-R}{N} \left\{ \sum_{i=1}^{M} \bar{\omega}^i f(\tilde{\xi}^i) \right\}^2 \; .$$

Combining with (7.46) and (7.42), this shows that the conditional variance of residual sampling is always smaller than that of multinomial sampling.

7.4.2.2 Stratified Resampling

The inversion method for sampling a multinomial sequence of trials maps uniform $(0, 1)$ random variables $U^1, \ldots, U^N$ into indices $I^1, \ldots, I^N$ through a deterministic function. For any function f,

$$\sum_{i=1}^{N} f(\tilde{\xi}^{I^i}) = \sum_{i=1}^{N} \Phi_f(U^i) \; ,$$

where the function Φ_f (which depends on both f and $\{\tilde{\xi}^i\}$) is defined, for any $u \in (0, 1]$, by

$$\Phi_f(u) \stackrel{\text{def}}{=} f(\tilde{\xi}^{I(u)}), \qquad I(u) = \sum_{i=1}^{M} i \mathbb{1}_{(\sum_{j=1}^{i-1} \omega^j, \sum_{j=1}^{i} \omega^j]}(u) \; . \tag{7.47}$$

Note that, by construction, $\int_0^1 \Phi_f(u)\, du = \sum_{i=1}^{M} \omega^i f(\tilde{\xi}^i)$. To reduce the conditional variance of $\sum_{i=1}^{N} f(\tilde{\xi}^{I^i})$, we may change the way in which the sample $U^1, \ldots, U^N$ is drawn. A possible solution, commonly used in survey sampling, is based on *stratification* (see Kitagawa, 1996, and Fearnhead, 1998, Section 5.3, for discussion of the method in the context of particle filtering). The interval $(0, 1]$ is partitioned into different *strata*, assumed for simplicity to be intervals $(0, 1] = (0, 1/N] \cup (1/N, 2/N] \cup \cdots \cup (\{N-1\}/N, 1]$. More general partitions could have been considered as well; in particular, the number of partitions does not have to equal N, and the interval lengths could be made dependent on the ω^i. One then draws a sample $\tilde{U}^1, \ldots, \tilde{U}^N$ conditionally independently given $\mathcal{G}$ from the distribution $\tilde{U}^i \sim \mathrm{U}\left(\left(\{i-1\}/N, i/N\right]\right)$

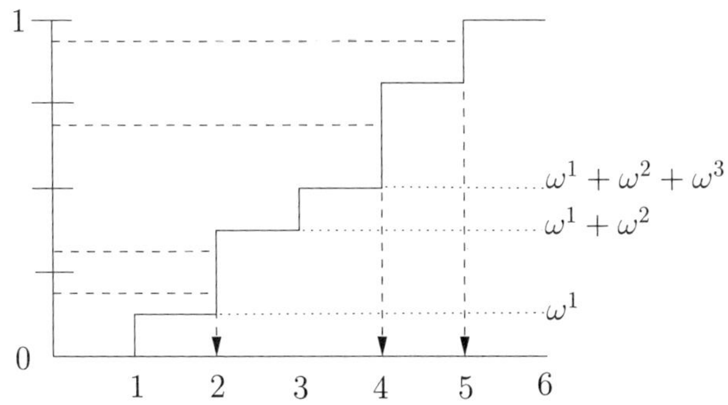

Fig. 7.16. Stratified sampling: the interval $(0,1]$ is divided into N intervals $((i-1)/N, i/N]$. One sample is drawn uniformly from each interval, independently of samples drawn in the other intervals.

(for $i = 1, \ldots, N$) and let $\tilde{I}^i = I(\tilde{U}^i)$ with I as in (7.47) (see Figure 7.16). By construction, the difference between $\tilde{N}^i = \sum_{j=1}^N \mathbb{1}_{\{\tilde{I}^j = i\}}$ and the target (non-integer) value $N\omega^i$ is less than one in absolute value. It also follows that

$$\mathrm{E}\left[\sum_{i=1}^N f(\tilde{\xi}^{\tilde{I}^i}) \,\middle|\, \mathcal{G}\right] = \mathrm{E}\left[\sum_{i=1}^N \Phi_f(\tilde{U}^i) \,\middle|\, \mathcal{G}\right]$$
$$= N \sum_{i=1}^N \int_{(i-1)/N}^{i/N} \Phi_f(u)\, du = N \int_0^1 \Phi_f(u)\, du = N \sum_{i=1}^M \omega^i f(\tilde{\xi}^i)\,,$$

showing that the stratified sampling scheme is unbiased. Because $\tilde{U}^1, \ldots, \tilde{U}^N$ are conditionally independent given $\mathcal{G}$,

$$\mathrm{Var}\left[\frac{1}{N}\sum_{i=1}^N f(\tilde{\xi}^{\tilde{I}^i}) \,\middle|\, \mathcal{G}\right] = \mathrm{Var}\left[\frac{1}{N}\sum_{i=1}^N \Phi_f(\tilde{U}^i) \,\middle|\, \mathcal{G}\right]$$
$$= \frac{1}{N^2}\sum_{i=1}^N \mathrm{Var}\left[\Phi_f(\tilde{U}^i) \,\middle|\, \mathcal{G}\right]$$
$$= \frac{1}{N}\sum_{i=1}^M \omega^i f^2(\tilde{\xi}^i) - \frac{1}{N}\sum_{i=1}^N \left[N \int_{(i-1)/N}^{i/N} \Phi_f(u) du\right]^2 ;$$

here we used that $\int_0^1 \Phi_f^2(u)\, du = \int_0^1 \Phi_{f^2}(u)\, du = \sum_{i=1}^M \omega^i f^2(\tilde{\xi}^i)$. By Jensen's inequality,

$$\frac{1}{N}\sum_{i=1}^N \left[N \int_{(i-1)/N}^{i/N} \Phi_f(u) du\right]^2 \geq \left[\sum_{i=1}^N \int_{(i-1)/N}^{i/N} \Phi_f(u) du\right]^2$$
$$= \left[\sum_{i=1}^M \omega^i f(\tilde{\xi}^i)\right]^2 ,$$

showing that the conditional variance of stratified sampling is always smaller than that of multinomial sampling.

Remark 7.4.5. Note that stratified sampling may be coupled with the residual sampling method discussed previously: the proof above shows that using stratified sampling on the R residual indices that are effectively drawn randomly can only decrease the conditional variance. ■

7.4.2.3 Systematic Resampling

Stratified sampling aims at reducing the *discrepancy*

$$D_N^\star(U^1, \ldots, U^N) \stackrel{\text{def}}{=} \sup_{a \in (0,1]} \left| \frac{1}{N} \sum_{i=1}^N \mathbb{1}_{(0,a]}(U^i) - a \right|$$

of the sample U from the uniform distribution function on $(0, 1]$. This is simply the Kolmogorov-Smirnov distance between the empirical distribution function of the sample and the distribution function of the uniform distribution. The Koksma-Hlawka inequality (Niederreiter, 1992) shows that for any function f having bounded variation on $[0, 1]$,

$$\left| \frac{1}{N} \sum_{i=1}^N f(u^i) - \int_0^1 f(u)\, du \right| \leq C(f) D_N^\star(u^1, \ldots, u^N) ,$$

where $C(f)$ is the variation of f. This inequality suggests that it is desirable to design random sequences $U^1, \ldots, U^N$ whose expected discrepancy is as low as possible. This provides another explanation of the improvement brought by stratified resampling (compared to multinomial resampling).

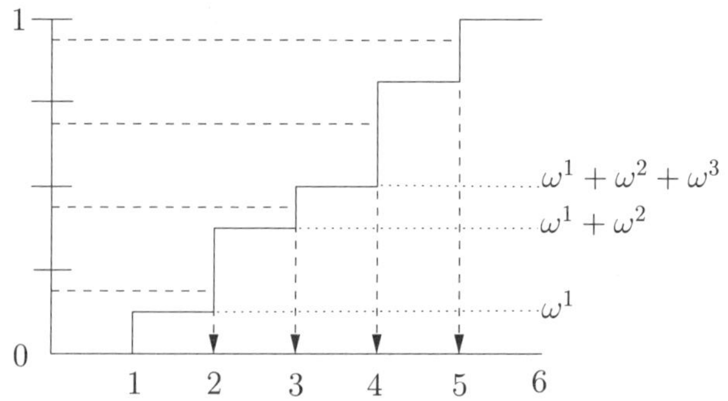

Fig. 7.17. Systematic sampling: the unit interval is divided into N intervals $((i-1)/N, i/N]$ and one sample is drawn from each of them. Contrary to stratified sampling, each sample has the same relative position within its stratum.

Pursuing in this direction, it makes sense to look for sequences with even smaller average discrepancy. One such sequence is $U^i = U + (i-1)/N$, where U is drawn from a uniform $\mathrm{U}((0, 1/N])$ distribution. In survey sampling, this method is known as *systematic sampling*. It was introduced in the particle filter literature by Carpenter *et al.* (1999) but is mentioned by Whitley (1994) under the name of *universal* sampling. The interval $(0, 1]$ is still divided into N sub-intervals $(\{i-1\}/N, i/N]$ and one sample is taken from each of them, as in stratified sampling. However, the samples are no longer independent, as they have the same relative position within each stratum (see Figure 7.17). This sampling scheme is obviously still unbiased. Because the samples are not taken independently across strata, it is however not possible to obtain simple formulas for the conditional variance (Künsch, 2003). It is often conjectured that the conditional variance of systematic resampling is always lower than that of multinomial resampling. This is not correct, as demonstrated by the following example.

Example 7.4.6. Consider the case where the initial population of particles $\{\tilde{\xi}^i\}_{1 \le i \le N}$ is composed of the interleaved repetition of only two distinct values x_0 and x_1, with identical multiplicities (assuming N to be even). In other words,

$$\{\tilde{\xi}^i\}_{1 \le i \le N} = \{x_0, x_1, x_0, x_1, \ldots, x_0, x_1\} .$$

We denote by $2\omega/N$ the common value of the normalized weight ω^i associated to the $N/2$ particles $\tilde{\xi}^i$ that satisfy $\tilde{\xi}^i = x_1$, so that the remaining ones (which are such that $\tilde{\xi}^i = x_0$) share a common weight of $2(1-\omega)/N$. Without loss of generality, we assume that $1/2 \le \omega < 1$ and that the function of interest f is such that $f(x_0) = 0$ and $f(x_1) = F$.

Under multinomial resampling, (7.42) shows that the conditional variance of the estimate $N^{-1} \sum_{i=1}^N f(\xi^i)$ is given by

$$\mathrm{Var}\left[\frac{1}{N} \sum_{i=1}^N f(\xi^i_{\mathrm{mult}}) \middle| \mathcal{G} \right] = \frac{1}{N} (1-\omega)\omega F^2 . \tag{7.48}$$

Because the value $2\omega/N$ is assumed to be larger than $1/N$, it is easily checked that systematic resampling deterministically sets $N/2$ of the ξ^i to be equal to x_1. Depending on the draw of the initial shift, *all* the $N/2$ remaining particles are either set to x_1, with probability $2\omega - 1$, or to x_0, with probability $2(1-\omega)$. Hence the variance is that of a *single* Bernoulli draw scaled by $N/2$, that is,

$$\mathrm{Var}\left[\frac{1}{N} \sum_{i=1}^N f(\xi^i_{\mathrm{syst}}) \middle| \mathcal{G} \right] = (\omega - 1/2)(1-\omega) F^2 .$$

Note that in this case, the conditional variance of systematic resampling is not only larger than (7.48) for most values of ω (except when ω is very close to $1/2$), but it does not even decrease to zero as N grows! Clearly, this observation is very dependent on the order in which the initial population of particles

ω	0.51	0.55	0.6	0.65	0.70	0.75
Multinomial	0.050	0.049	0.049	0.048	0.046	0.043
Residual, stratified	0.010	0.021	0.028	0.032	0.035	0.035
Systematic	0.070	0.150	0.200	0.229	0.245	0.250
Systematic with prior random shuffling	0.023	0.030	0.029	0.029	0.028	0.025

Table 7.1. Standard deviations of various resampling methods for $N = 100$ and $F = 1$. The bottom line has been obtained by simulations, averaging 100,000 Monte Carlo replications.

is presented. Interestingly, this feature is common to the systematic and stratified sampling schemes, whereas the multinomial and residual approaches are unaffected by the order in which the particles are labelled. In this particular example, it is straightforward to verify that residual and stratified resampling are equivalent—which is not the case in general—and amount to deterministically setting $N/2$ particles to the value x_1, whereas the $N/2$ remaining ones are drawn by $N/2$ *conditionally independent* Bernoulli trials with probability of picking x_1 equal to $2\omega - 1$. Hence the conditional variance, for both the residual and stratified schemes, is equal to $N^{-1}(2\omega - 1)(1 - \omega)F^2$. It is hence always smaller than (7.48), as expected from the general study of these two methods.

Once again, the failure of systematic resampling in this example is entirely due to the specific order in which the particles are labelled: it is easy to verify, at least empirically, that the problem vanishes upon randomly permuting the initial particles before applying systematic resampling. Table 7.1 also shows that a common feature of both the residual, stratified, and systematic resampling procedures is to become very efficient in some particular configurations of the weights such as when $\omega = 0.51$ for which the probabilities of selecting the two types of particles are almost equal and the selection becomes quasi-deterministic. Note also that prior random shuffling does somewhat compromise this ability in the case of systematic resampling. ■

In practical applications of sequential Monte Carlo methods, residual, stratified, and systematic resampling are generally found to provide comparable results. Despite the lack of complete theoretical analysis of its behavior, systematic resampling is often preferred because it is the simplest method to implement. Note that there are specific situations, to be discussed in Section 8.2, where more subtle forms of resampling (which do not necessarily bring back all the weights to equal values) are advisable.

8

Advanced Topics in Sequential Monte Carlo

This chapter deals with three disconnected topics that correspond to variants and extensions of the sequential Monte Carlo framework introduced in the previous chapter. Remember that we have already examined in Section 7.2 a first and very important degree of freedom in the application of sequential Monte Carlo methods, namely the choice of the instrumental kernel R_k used to simulate the trajectories of the particles. We now consider solutions that depart, more or less significantly, from the sequential importance sampling with resampling (SISR) method of Algorithm 7.3.4.

The first section covers a far-reaching revision of the principles behind the SISR algorithm in which sequential Monte Carlo is interpreted as a repeated sampling task. This reinterpretation suggests several other sequential Monte Carlo schemes that differ, sometimes significantly, from the SISR approach. Section 8.2 reviews methods that exploit the specific hierarchical structure found in some hidden Markov models, and in particular in conditionally Gaussian linear state-space models (CGLSSMs). The algorithms to be considered there combine the sequential simulation approach presented in the previous chapter with the Kalman filtering recursion discussed in Chapter 5. Finally, Section 8.3 discusses the use of sequential Monte Carlo methods for approximating smoothed quantities of the form introduced in Section 4.1.

8.1 Alternatives to SISR

We first present a reinterpretation of the objectives of the sequential importance sampling with resampling (SISR) algorithm in Section 7.3. This new interpretation suggests a whole range of different approaches that combines more closely the sampling (trajectory update) and resampling (weight reset) operators involved in the SISR algorithm.

In the basic SISR approach (Algorithm 7.3.4), we expect that after a resampling step, say at index k, the particle trajectories $\xi^1_{0:k}, \ldots, \xi^N_{0:k}$ approximately form an i.i.d. sample of size N from the distribution $\phi_{0:k|k}$. We will discuss more precisely in Chapter 9 the degree to which this assertion is correct but assume for the moment that the general intuition is justifiable. Even in the absence of resampling at index k, in which case the weights $\omega_k^1, \ldots, \omega_k^N$ are not identical, the expectation of any function $f_k \in \mathcal{F}_{\mathrm{b}}\left(\mathsf{X}^{k+1}\right)$ under $\phi_{0:k|k}$ may be approximated, following (7.11), by

$$\sum_{i=1}^{N} \frac{\omega_k^i}{\sum_{j=1}^{N} \omega_k^j} f_k(\xi_{0:k}^i) \, .$$

This behavior may indeed be adopted as a general principle for sequential Monte Carlo techniques, considering that a valid algorithm is such that it is recursive and guarantees that the *weighted empirical distribution,*

$$\hat{\phi}_{0:k|k} = \sum_{i=1}^{N} \frac{\omega_k^i}{\sum_{j=1}^{N} \omega_k^j} \, \delta_{\xi_{0:k}^i} \, , \tag{8.1}$$

is a consistent approximation to $\phi_{0:k|k}$, in some suitable sense, as the number N of particles increases (the symbol δ denotes Dirac measures).

The particular feature of the sequence of target distributions encountered in the HMM filtering application is the relatively simple recursive form recalled by (7.7):

$$\phi_{0:k+1|k+1}(f_{k+1}) = \int \cdots \int f_{k+1}(x_{0:k+1}) \phi_{0:k|k}(dx_{0:k}) \, T_k^{\mathrm{u}}(x_k, dx_{k+1}) \, ,$$

for all functions $f_{k+1} \in \mathcal{F}_{\mathrm{b}}\left(\mathsf{X}^{k+2}\right)$, where T_k^{u} is the (unnormalized) kernel defined in (7.8). This relation may be rewritten replacing T_k^{u} by its normalized version T_k defined in (7.15), the so-called optimal importance kernel, to obtain

$$\begin{aligned} \phi_{0:k+1|k+1}(f_{k+1}) = \int \cdots \int f_{k+1}(x_{0:k+1}) \, \phi_{0:k|k}(dx_{0:k}) & \\ \times \frac{\mathrm{L}_k}{\mathrm{L}_{k+1}} \gamma_k(x_k) \, T_k(x_k, dx_{k+1}) \, , & \end{aligned} \tag{8.2}$$

where γ_k is the normalizing function defined in (7.17). Because the likelihoods L_k and L_{k+1} are precisely the type of quantities that are non-evaluable in contexts where sequential Monte Carlo is useful, it is preferable to rewrite (8.2) in the equivalent auto-normalized form

$$\phi_{0:k+1|k+1}(f_{k+1}) = \frac{\int \cdots \int f_{k+1}(x_{0:k+1}) \, \phi_{0:k|k}(dx_{0:k}) \gamma_k(x_k) \, T_k(x_k, dx_{k+1})}{\int \cdots \int \phi_{0:k|k}(dx_{0:k}) \gamma_k(x_k)} \, . \tag{8.3}$$

A natural idea in the context of sequential Monte Carlo is to plug the approximate empirical distribution defined in (8.1) into the recursive update formula (8.3), which yields

$$\tilde{\phi}_{0:k+1|k+1}(f_{k+1}) \stackrel{\text{def}}{=} \sum_{i=1}^{N} \frac{\omega_k^i \, \gamma_k(\xi_k^i)}{\sum_{j=1}^{N} \omega_k^j \, \gamma_k(\xi_k^j)} \int f_{k+1}(\xi_{0:k}^i, x) \, T_k(\xi_k^i, dx) \,. \quad (8.4)$$

This equation defines a probability distribution $\tilde{\phi}_{0:k+1|k+1}$ on X^{k+2}, which is a finite mixture distribution and which also has the particularity that its restriction to the first $k+1$ component is a weighted empirical distribution with support $\xi_{0:k}^1, \ldots, \xi_{0:k}^N$ and weights proportional to $\omega_k^i \gamma_k(\xi_k^i)$. Following this argument, the updated empirical approximation $\hat{\phi}_{0:k|k}$ should approximate the distribution defined in (8.4) as closely as possible, but with the constraint that it is supported by N points only. The simplest idea of course consists in trying to obtain a (conditionally) i.i.d. sample from this mixture distribution. This interpretation opens a range of new possibilities, as we are basically faced with a sampling problem for which several methods, including those discussed in Chapter 6, are available.

8.1.1 I.I.D. Sampling

As discussed above, the first obvious idea is to simulate, if possible, the new particle trajectories as N i.i.d. draws from the distribution defined by (8.4). Note that the term "i.i.d." is used somewhat loosely here, as the statement obviously refers to the conditional distribution of the new particle trajectories $\xi_{0:k+1}^1, \ldots, \xi_{0:k+1}^N$ given the current state of the system as defined by the particle trajectories $\xi_{0:k}^1, \ldots, \xi_{0:k}^N$ and the weights $\omega_k^1, \ldots, \omega_k^N$. The algorithm obtained when following this principle is distinct from Algorithm 7.3.4, although it is very closely related to SISR when the optimal importance kernel T_k is used as the instrumental kernel.

Algorithm 8.1.1 (I.I.D. Sampling or Selection/Mutation Algorithm).

Weight computation: For $i = 1, \ldots, N$, compute the (unnormalized) importance weights

$$\alpha_k^i = \omega_k^i \, \gamma_k(\xi_k^i) \,. \quad (8.5)$$

Selection: Draw $I_{k+1}^1, \ldots, I_{k+1}^N$ conditionally i.i.d. given $\{\xi_{0:k}^i\}_{1 \le i \le N}$, with probabilities $\mathrm{P}(I_{k+1}^1 = j)$ proportional to α_k^j, $j = 1, \ldots, N$.

Sampling: Draw $\tilde{\xi}_{k+1}^1, \ldots, \tilde{\xi}_{k+1}^N$ conditionally independently given $\{\xi_{0:k}^i\}_{1 \le i \le N}$ and $\{I_{k+1}^i\}_{1 \le i \le N}$, with distribution $\tilde{\xi}_{k+1}^i \sim T_k(\xi_k^{I_{k+1}^i}, \cdot)$. Set $\xi_{0:k+1}^i = (\xi_{0:k}^{I_{k+1}^i}, \tilde{\xi}_{k+1}^i)$ and $\omega_{k+1}^i = 1$ for $i = 1, \ldots, N$.

Comparing the above algorithm with Algorithm 8.1.1 for the particular choice $R_k = T_k$ reveals that they differ only by the order in which the sampling and selection operations are performed. Algorithm 8.1.1 prescribes that each trajectory be first extended by setting $\xi_{0:k+1}^i = (\xi_{0:k}^i, \tilde{\xi}_{k+1}^i)$ with $\tilde{\xi}_{k+1}^i$ drawn from $T_k(\xi_k^i, \cdot)$. Then resampling is performed in the population of extended trajectories, based on weights given by (8.5) when $R_k = T_k$. In contrast, Algorithm 8.1.1 first selects the trajectories based on the weights α_k^i and then simulates an independent extension for each selected trajectory. This is of course possible only because the optimal importance kernel T_k is used as instrumental kernel, rendering the incremental weights independent of the position of the particle at index $k+1$ and thus allowing for early selection. Intuitively, Algorithm 8.1.1 is preferable because it does not simply duplicate trajectories with high weights but rather selects the most promising trajectories at index k using independent extensions (at index $k+1$) for each selected trajectory. Following the terminology in use in genetic algorithms[1], Algorithm 8.1.1 is a selection/mutation algorithm, whereas the SISR approach is based on mutation/selection. Recall that the latter is more general, as it does not require that the optimal kernel T_k be used, although we shall see later, in Section 8.1.2, that the i.i.d. sampling approach can be modified to allow for general instrumental kernels.

Remark 8.1.2. In Chapter 7 as well as in the exposition above, we considered that the quantity of interest is the joint smoothing measure $\phi_{0:k|k}$. It is important however to understand that this focus on the joint smoothing measure $\phi_{0:k|k}$ is unessential as all the algorithms presented so far only rely on the recursive structure observed in (8.4). Of course, in the case of the joint smoothing measure $\phi_{0:k|k}$, the kernel T_k and the function γ_k that appear in (8.4) have a specific form given by (7.15) and (7.17):

$$\int f(x')\, \gamma_k(x) T_k(x, dx') = \int f(x')\, g_{k+1}(x') Q(x, dx') \tag{8.6}$$

for functions $f \in \mathcal{F}_{\mathrm{b}}(\mathsf{X})$, where $\gamma_k(x)$ equals the above expression evaluated for $f = 1$. However, any of the sequential Monte Carlo algorithms discussed so far can be used for generic choices of the kernel T_k and the function γ_k provided the expression for the incremental weights is suitably modified. The core of SMC techniques is thus the structure observed in (8.4), whose connection with the methods exposed here is worked out in detail in the recent book by Del Moral (2004).

As an example, recall from Chapter 3 that the distribution $\phi_{0:k|k-1}$ differs from $\phi_{0:k-1|k-1}$ only by an application of the prior (or state transition) kernel Q and hence satisfies a recursion similar to (8.4) with the kernel T_k and the function γ_k replaced by Q and g_k, respectively:

$$\phi_{0:k+1|k}(f_{k+1}) = \frac{\int \cdots \int f_{k+1}(x_{0:k+1})\, \phi_{0:k|k-1}(dx_{0:k}) g_k(x_k)\, Q(x_k, dx_{k+1})}{\int \cdots \int \phi_{0:k|k-1}(dx_{0:k}) g_k(x_k)}\,, \tag{8.7}$$

[1]Genetic algorithms (see, e.g., Whitley, 1994) have much in common with sequential Monte Carlo methods. Their purpose is different however, with an emphasis on optimization rather than, as for SMC, simulation. Both fields do share a lot of common terminology.

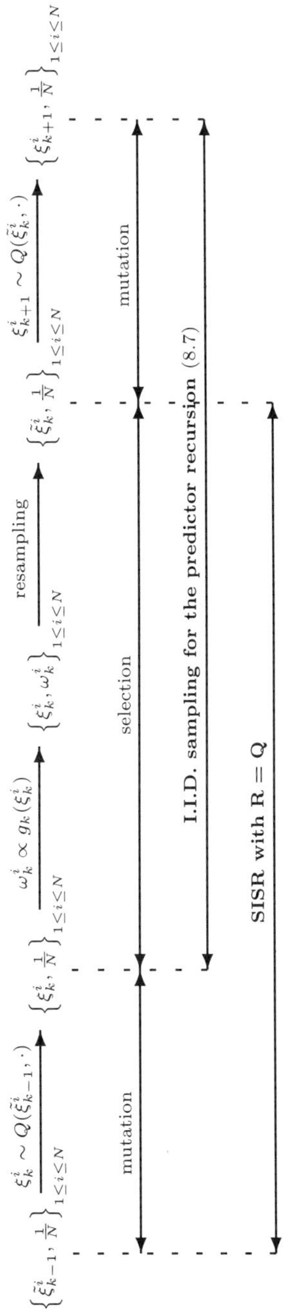

Fig. 8.1. The bootstrap filter decomposed into elementary mutation/selection steps.

where the denominator could be written more compactly as $\phi_{k|k-1}(g_k)$. The recursive update formula obtained for the (joint) predictive distribution is much simpler than (8.4), as (8.7) features the prior kernel Q—from which we generally assume that sampling is feasible—and the conditional likelihood function g_k—whose analytical expression is known. In particular, it is straightforward to apply Algorithm 8.1.1 in this case by selecting with weights $g_k(\xi_k^1), \ldots, g_k(\xi_k^N)$ and mutating the selected particles using the kernel Q. This is obviously equivalent to the bootstrap filter (Algorithm 7.3.4 with Q as the instrumental kernel) viewed at a different stage: just after the selection step for Algorithm 7.3.4 and just after the mutation step for Algorithm 8.1.1 *applied to the predictive distribution* (see Figure 8.1 for an illustration). The previous interpretation however suggests that the bootstrap filter operates very differently on the filtering and predictive approximations, either according to Algorithm 7.3.4 or to Algorithm 8.1.1. We shall see in the next chapter (Section 9.4) that this observation has important implications when it comes to evaluating the asymptotic (for large N) performance of the method. ∎

Coming back to the joint smoothing distribution $\phi_{0:k|k}$, Algorithm 8.1.1 is generally not applicable directly as it involves sampling from T_k and evaluation of the normalization function γ_k (see also the discussion in Section 7.2.2 on this point). In the remainder of this section, we will examine a number of more practicable options that keep up with the general objective of sampling from the distribution defined in (8.4). The first section below presents a method that is generally known under the name *auxiliary particle filter* after Pitt and Shephard (1999) (see also Liu and Chen, 1998). The way it is presented here however differs notably from the exposition of Pitt and Shephard (1999), whose original argument will be discussed in Section 8.1.3.

8.1.2 Two-Stage Sampling

We now consider using the sampling importance resampling method introduced in Section 7.1.2 to sample approximately from $\tilde{\phi}_{0:k+1|k+1}$. Recall that SIR sampling proceeds in two steps: in a first step, a new population is drawn according to an instrumental distribution, say $\rho_{0:k+1}$; then, in a second step, the points are selected with probabilities proportional to the importance ratio between the target (here $\tilde{\phi}_{0:k+1|k+1}$) and the instrumental distribution $\rho_{0:k+1}$.

Our aim is to find an instrumental distribution $\rho_{0:k+1}$ that is as close as possible to $\tilde{\phi}_{0:k+1|k+1}$ as defined in (8.4), yet easy to sample from. A sensible option is provided by mixture distributions such that for all functions $f_{k+1} \in \mathcal{F}_\mathrm{b}\left(\mathsf{X}^{k+2}\right)$,

$$\rho_{0:k+1}(f_{k+1}) = \sum_{i=1}^{N} \frac{\omega_k^i \, \tau_k^i}{\sum_{j=1}^{N} \omega_k^j \, \tau_k^j} \int f_{k+1}(\xi_{0:k}^i, x) \, R_k(\xi_k^i, dx) \; . \tag{8.8}$$

Here, $\tau_{k+1}^1, \ldots, \tau_{k+1}^N$ are positive numbers, called *adjustment multiplier weights* by Pitt and Shephard (1999), and R_k is a transition kernel on X. Both the adjustment multiplier weights and the instrumental kernel may depend on the new observation Y_{k+1} although, as always, we do not explicitly mention it in our notation. To ensure that the importance ratio is well-defined, we require that the adjustment multiplier weights be strictly positive and that $T_k(x,\cdot)$, or equivalently $T_k^{\mathrm{u}}(x,\cdot)$, be absolutely continuous with respect to $R_k(x,\cdot)$, for all $x \in \mathsf{X}$.

These assumptions imply that the target distribution $\tilde{\phi}_{0:k+1|k+1}$ defined in (8.4) is dominated by the instrumental distribution $\rho_{0:k+1}$ with importance function given by the Radon-Nikodym derivative

$$\frac{d\tilde{\phi}_{0:k+1|k+1}}{d\rho_{0:k+1}}(x_{0:k+1}) = C_k \sum_{i=1}^{N} \mathbb{1}_{\{\xi_{0:k}^i\}}(x_{0:k}) \frac{\gamma_k(\xi_k^i)}{\tau_k^i} \frac{dT_k(\xi_k^i,\cdot)}{dR_k(\xi_k^i,\cdot)}(x_{k+1}) , \tag{8.9}$$

where

$$C_k = \frac{\sum_{i=1}^N \omega_k^i \, \tau_k^i}{\sum_{i=1}^N \omega_k^i \, \gamma_k(\xi_k^i)} .$$

Because the factor C_k is a normalizing constant that does not depend on $x_{0:k+1}$, it is left here only for reference; its evaluation is never required when using the SIR approach. In order to obtain (8.9), we used the fundamental observation that a set $A_{k+1} \in \mathcal{X}^{\otimes(k+2)}$ can have non-null probability under *both* (8.4) and (8.8) only if there exists an index i and a measurable set $A \subseteq \mathsf{X}$ such that $\{\xi_0^i\} \times \cdots \times \{\xi_k^i\} \times A \subseteq A_{k+1}$, that is, A_{k+1} must contain (at least) one of the current particle trajectories. Recall that

$$\gamma_k(\xi_k^i)\, T_k(\xi_k^i, dx) = g_{k+1}(x)\, Q(\xi_k^i, dx) ,$$

and hence (8.9) may be rewritten as

$$\frac{d\tilde{\phi}_{0:k+1|k+1}}{d\rho_{0:k+1}}(x_{0:k+1}) = C_k \sum_{i=1}^{N} \mathbb{1}_{\{\xi_{0:k}^i\}}(x_{0:k}) \frac{g_{k+1}(x_{k+1})}{\tau_k^i} \frac{dQ(\xi_k^i,\cdot)}{dR_k(\xi_k^i,\cdot)}(x_{k+1}) , \tag{8.10}$$

Thanks to the relatively simple expression of the importance function in (8.10), the complete SIR algorithm is straightforward provided that we can simulate from the instrumental kernel R_k.

Algorithm 8.1.3 (Two-Stage Sampling).

First-Stage Sampling:

- Draw $I_k^1, \ldots, I_k^M$ conditionally i.i.d. given $\{\xi_{0:k}^i\}_{1\le i\le N}$, with probabilities $\mathrm{P}(I_k^i = j)$ proportional to the (unnormalized) *first-stage weights* $\omega_k^j \tau_k^j$, $j = 1, \ldots, M$.
- Draw $\tilde{\xi}_{k+1}^1, \ldots, \tilde{\xi}_{k+1}^M$ conditionally independently given $\{\xi_{0:k}^l\}_{1\le l\le N}$ and $\{I_k^i\}_{1\le i\le M}$, with distribution $\tilde{\xi}_{k+1}^i \sim R_k(\xi_k^{I_k^i}, \cdot)$. Set $\tilde{\xi}_{0:k+1}^i = (\xi_{0:k}^{I_k^i}, \tilde{\xi}_{k+1}^i)$ for $i = 1, \ldots, M$.

Weight computation: For $i = 1, \ldots, M$, compute the (unnormalized) *second-stage weights*

$$\alpha_k^i = \frac{g_{k+1}(\tilde{\xi}_{k+1}^i)}{\tau_k^{I_k^i}} \frac{dQ(\xi_k^{I_k^i}, \cdot)}{dR_k(\xi_k^{I_k^i}, \cdot)}(\tilde{\xi}_{k+1}^i) . \tag{8.11}$$

Second-Stage Resampling:

- Draw $J_{k+1}^1, \ldots, J_{k+1}^N$ conditionally i.i.d. given $\{\tilde{\xi}_{0:k+1}^i\}_{1 \le i \le M}$, with probabilities $\mathrm{P}(J_{k+1}^1 = j)$ proportional to the second-stage weights α_k^j, $j = 1, \ldots, M$.
- For $i = 1, \ldots, N$, set $\xi_{0:k+1}^i = \tilde{\xi}_{0:k+1}^{J_{k+1}^i}$ and $\omega_{k+1}^i = 1$.

The adjustment multiplier weights $\{\tau_k^i\}_{1 \le i \le n}$ should be chosen to sample preferentially (in the first stage) the particle trajectories that are most likely under $\tilde{\phi}_{0:k+1|k+1}$. Usually the multiplier weight τ_k^i depends on the new observation Y_{k+1} and on the position of the particle at index k, ξ_k^i, but more general conditions can be considered as well. If one can guess, based on the new observation, which particle trajectories are most likely to survive or die, the resampling stage may be anticipated by increasing (or decreasing) the importance weights. As such, the use of adjustment multiplier weights is a mechanism to prevent sample impoverishment.

The expression for the second-stage weights in (8.11) provides additional insights on how to choose the adjustment multiplier weights. The efficiency of the SIR procedure is best when the importance weights are well-balanced, that is, when the total mass is spread over a large number of particles. The multiplier adjustment weights τ_k^i should thus be chosen to render the second-stage weights as evenly distributed as possible. In the particular case where sampling is done from the prior (or state transition) kernel, that is if $R_k = Q$, the expression of the second-stage weight simplifies to

$$\alpha_k^i = g_{k+1}(\tilde{\xi}_{k+1}^i)/\tau_k^{I_{k+1}^i} .$$

Although it is not possible to equate this expression to one, as τ_k^i cannot depend on $\tilde{\xi}_{k+1}^i$, it is easy to imagine strategies that reach this objective on average. Pitt and Shephard (1999) suggest that the adjustment multiplier weights be set as the likelihood of the mean of the predictive distribution corresponding to each particle,

$$\tau_k^i = g_{k+1}\left(\int x \, Q(\xi_k^i, dx)\right) . \tag{8.12}$$

In particular, in examples where Q corresponds to a random walk move, the adjustment multiplier weight τ_k^i is thus equal to $g_{k+1}(\xi_k^i)$, the conditional likelihood of the new observation given the current position, which is quite natural. In general situations, the success of this approach depends on our ability to choose the adjustment multiplier weights in a way that the first sampling stage is effective.

N	Ref.	Bootstrap filter			Auxiliary particle filter		
		$M = 100$	1,000	10,000	$M = 100$	1,000	10,000
100	0.91	0.49 (0.12)	0.57 (0.10)	0.61 (0.11)	0.56 (0.11)	0.62 (0.11)	0.62 (0.10)
1,000	0.91	-	0.64 (0.10)	0.71 (0.09)	0.59 (0.11)	0.71 (0.10)	0.74 (0.09)
10,000	0.91	-	-	0.75 (0.09)	0.60 (0.12)	0.73 (0.09)	0.80 (0.08)

Table 8.1. Approximations of the posterior mean $\hat{X}_{5|5}$ in the noisy AR(1) model, obtained using the bootstrap filter and auxiliary particle filter. The model and observations $Y_{0:5}$ are given in Example 7.2.3. Results are reported for different values of M (size of the first stage sample) and N (number of particle retained in the second stage). The figures are means and standard errors from 500 independent replications for each pair of M and N. The column "ref" displays the true posterior mean computed by Kalman filtering.

Example 8.1.4 (Noisy AR(1) Model, Continued). To illustrate the behavior of the method, we consider again the simple noisy AR(1) model of Example 7.2.3, which has the advantage that exact filtering quantities may be computed by the Kalman recursions. In Example 7.2.3, we approximated the posterior mean of X_5 given the observed $Y_{0:5}$ using sequential importance sampling with the prior kernel Q as instrumental kernel and found that this approximation grossly underestimates the true posterior mean, which evaluates (by Kalman filtering) to 0.91. The situation improves somewhat when using the optimal kernel T_k (Example 7.2.4). Because there are only six observations, the differences between the results of SIS and SISR are small, as the weights do not have the time to degenerate (given that, in addition, the outlier occurs at the last time index).

In Table 8.1, we compare the results of the SISR algorithm with Q as the instrumental kernel (also known as the bootstrap filter) and the two-stage algorithm. Following (8.12), the adjustment multiplier weights were set to

$$\tau_k^i = \mathrm{N}(Y_{k+1};\ \phi\xi_k^i, \sigma_V^2)\ ;$$

see Example 7.2.3 for details on the notation. This second algorithm is usually referred to as the (or an) auxiliary particle filter. The table shows that for all values of M (the size of the first stage sampling population) and N (the number of particles retained in the second stage), the auxiliary article filter outperforms the bootstrap filter. The auxiliary filter effectively reduces the bias to a level that is, in this case, comparable (albeit slightly larger) to that obtained when using the optimal kernel T_k as instrumental kernel (see Figure 7.4).

For the bootstrap filter (Algorithm 7.3.4), only values of M larger than N have been considered. Indeed, because the algorithm operates by first extending the trajectories and then resampling, it does not apply directly when $M < N$. Note however that the examination of the figures obtained for the auxiliary filter (Algorithm 8.1.3), for which both M and N may be chosen freely, suggests that it is more efficient to use M larger than N than the converse. The payoff for using M larger than N, compared to the base situation where $M = N$, is also much more significant in the case of the bootstrap filter—whose baseline performance is worse—than for the auxiliary particle filter. ■

8.1.3 Interpretation with Auxiliary Variables

We now discuss another interpretation of Algorithm 8.1.3, which is more in the spirit of (Pitt and Shephard, 1999). This alternative perspective on Algorithm 8.1.3 is based on the observation that although we generally consider our target distributions to be the joint smoothing distribution $\phi_{0:k|k}$, the obtained algorithms are directly applicable for approximating the filtering distribution ϕ_k simply by dropping the history of the particles (Remark 7.3.5).

In particular, if we now consider that only the current system of particles $\{\xi_k^i\}_{1\le i\le N}$, with associated weights $\{\omega_k^i\}_{1\le i\le N}$ is available, (8.3) should be replaced by the marginal relation

$$\tilde{\phi}_{k+1}(f) \stackrel{\text{def}}{=} \sum_{i=1}^{N} \frac{\omega_k^i\, \gamma_k(\xi_k^i)}{\sum_{j=1}^{N} \omega_k^j\, \gamma_k(\xi_k^j)} \int f(x)\, T_k(\xi_k^i, dx)\;, \qquad f \in \mathcal{F}_{\mathrm{b}}(\mathsf{X})\;, \tag{8.13}$$

which thus defines our target distribution for updating the system of particles.

For the same reason as above, it makes sense to select a proposal distribution (this time on X) closely related to (8.13). Indeed, we consider the N component mixture

$$\rho_{k+1}(f) = \sum_{i=1}^{N} \frac{\omega_k^i \tau_k^i}{\sum_{j=1}^{N} \omega_k^j \tau_k^j} \int f(x)\, R_k(\xi_k^i, dx)\;. \tag{8.14}$$

Proceeding as in (8.9)–(8.10), the Radon-Nikodym derivative is now given by

$$\frac{d\tilde{\phi}_{\nu,k+1}}{d\rho_{k+1}}(x) = C_k \frac{d\left\{\sum_{i=1}^{N} \omega_k^i\, T_k(\xi_k^i, \cdot)\right\}}{d\left\{\sum_{i=1}^{N} \omega_k^i \tau_k^j\, R_k(\xi_k^i, \cdot)\right\}}(x)\;. \tag{8.15}$$

Compared to (8.10), this marginal importance ratio would be costly to evaluate as such, as both its numerator and its denominator involve summing over N terms. This difficulty can be overcome by *data augmentation*, introducing an *auxiliary variable* that corresponds to the mixture component that is selected when drawing the new particle position. Consider the following distribution ϕ_{k+1}^{aux} on the product space $\{1, \ldots, N\} \times \mathsf{X}$:

$$\phi_{k+1}^{\text{aux}}(\{i\} \times A) = \frac{\omega_k^i \int_A g_{k+1}(x)\, Q(\xi_k^i, dx)}{\sum_{j=1}^{N} \omega_k^j \gamma_k(\xi_k^j)}\;, \qquad A \in \mathcal{X},\; i = 1, \ldots, N\;. \tag{8.16}$$

Because

$$\phi^{\text{aux}}_{k+1}(\{1,\ldots,N\} \times A) = \sum_{i=1}^{N} \phi^{\text{aux}}_{k+1}(\{i\} \times A) = \tilde{\phi}_{k+1}(A) , \qquad A \in \mathcal{X} ,$$

$\tilde{\phi}_{k+1}$ is the marginal distribution of ϕ^{aux}_{k+1} and we may sample from $\tilde{\phi}_{k+1}$ by sampling from ϕ^{aux}_{k+1} and discarding the auxiliary index. To sample from ϕ^{aux}_{k+1} using the SIR method, we can then use the following instrumental distribution on the product space $\{1,\ldots,N\} \times \mathsf{X}$:

$$\rho^{\text{aux}}_{k+1}(\{i\} \times A) = \frac{\omega^i_k \, \tau^i_k}{\sum_{j=1}^{N} \omega^j_k \, \tau^j_k} R_k(\xi^i_k, A) , \qquad A \in \mathcal{X} . \tag{8.17}$$

This distribution may be interpreted as the joint distribution of the selection index I^i_k and the proposed new particle position $\tilde{\xi}^i_{k+1}$ in Algorithm 8.1.3. This time, the importance function is very simple and similar to (8.10),

$$\frac{d\phi^{\text{aux}}_{k+1}}{d\rho^{\text{aux}}_{k+1}}(i,x) = C_k \frac{g_{k+1}(x)}{\tau^i_k} \frac{dQ(\xi^i_k,\cdot)}{dR_k(\xi^i_k,\cdot)}(x) , \qquad i = 1,\ldots,N,\ x \in \mathsf{X} , \tag{8.18}$$

Hence Algorithm 8.1.3 may also be understood in terms of auxiliary sampling.

8.1.4 Auxiliary Accept-Reject Sampling

Rather than using the SIR method, simulating from (8.17) and using the importance ratio defined in (8.18), we may consider other methods for simulating directly from (8.16). An option, already discussed in the context of sequential importance sampling in Section 7.2.2, consists in using the accept-reject method (defined in Section 6.2.1).

The accept-reject method may be used to generate a truly i.i.d. sample from the target distribution. The price to pay compared to the SIR algorithm is a typically higher computational cost, especially when the acceptance probability is low. In addition, the number of simulations needed is itself random and the computation time cannot be predicted beforehand, especially when there are unknown normalizing constants (Remark 6.2.4). The method has nonetheless been studied for sequential simulation by several authors including Tanizaki (1996), Tanizaki and Mariano (1998), and Hürzeler and Künsch (1998) (see also Pitt and Shephard, 1999, and Liu and Chen, 1998).

In auxiliary accept-reject the idea is to find an instrumental distribution ρ^{aux}_{k+1} that dominates the target ϕ^{aux}_{k+1} and is such that the Radon-Nikodym derivative $d\phi^{\text{aux}}_{k+1}/d\rho^{\text{aux}}_{k+1}$ is bounded. Indeed, proposals of the form given in (8.8) still constitute an appropriate choice granted that we strengthen somewhat the assumptions that were needed for applying the SIR method.

Assumption 8.1.5. *For any $k \geq 0$ and $x \in \mathsf{X}$,*

$$\sup_{x' \in \mathsf{X}} g_{k+1}(x') \frac{dQ(x,\cdot)}{dR_k(x,\cdot)}(x') < \infty . \tag{8.19}$$

Because the index i runs over a finite set $\{1, \ldots, N\}$, we may define

$$M_k = \max_{1 \le i \le N} \frac{A_k^i}{\tau_k^i}, \qquad \text{where} \quad A_k^i \ge \sup_{x \in \mathsf{X}} g_{k+1}(x) \frac{dQ(\xi_k^i, \cdot)}{dR_k(\xi_k^i, \cdot)}(x) . \tag{8.20}$$

With these definitions, the Radon-Nikodym derivative $d\phi_{k+1}^{\text{aux}} / d\rho_{k+1}^{\text{aux}}$ given by (8.18) is bounded by

$$\frac{d\phi_{k+1}^{\text{aux}}}{d\rho_{k+1}^{\text{aux}}}(i, x) \le M_k \, \frac{\sum_{i=1}^N \omega_k^i \, \tau_k^i}{\sum_{i=1}^N \omega_k^i \, \gamma_k(\xi_k^i)} , \tag{8.21}$$

and hence the use of the accept-reject algorithm is valid. The complete algorithm proceeds as follows.

Algorithm 8.1.6 (Auxiliary Accept-Reject). For $i = 1, \ldots, N$,

Repeat:

- Draw an index $I_k^i \in \{1, \ldots, N\}$ with probabilities proportional to the first-stage weights $\omega_k^1 \tau_k^1, \ldots, \omega_k^N \tau_k^N$.
- Conditionally on I_k^i, draw a proposal $\tilde{\xi}_{k+1}^i$ from the instrumental transition kernel $R_k(\xi_k^{I_k^i}, \cdot)$ and U^i from a uniform distribution on $[0, 1]$.

Until:

$$U^i \le \frac{1}{M_k} \frac{g_{k+1}(\tilde{\xi}_{k+1}^i)}{\tau_k^{I_k^i}} \frac{dQ(\xi_k^{I_k^i}, \cdot)}{dR_k(\xi_k^{I_k^i}, \cdot)}(\tilde{\xi}_{k+1}^i) .$$

Update: Set $\xi_{k+1}^i = \tilde{\xi}_{k+1}^i$.

When done, reset all weights $\{\omega_{k+1}^i\}_{1 \le i \le N}$ to a (common) constant value.

Because the joint distribution of the accepted pairs is ϕ_{k+1}^{aux}, as defined by (8.16), the marginal distribution of the accepted draws (forgetting about the index) is (8.13) as required. One should typically try to increase the acceptance rate by proper choices of the adjustment multiplier weights τ_k^i and, whenever possible, by also choosing the instrumental kernel R_k in an appropriate fashion. The user should also determine the upper bounds A_k^i in (8.20) as tightly as possible. The following lemma, due to Künsch (2003), gives some indications on how the multiplier weights should be chosen to maximize the acceptance ratio.

Lemma 8.1.7. *For a given choice of instrumental kernels R_k and upper bounds A_k^i, the average acceptance probability is maximal when the incremental adjustment weights τ_k^i are proportional to A_k^i for $i = 1, \ldots, N$.*

Proof. Recall from Remark 6.2.4 that because of the presence of unknown normalization constants, the acceptance probability of the accept-reject method is not $1/M_k$ but rather the inverse of the upper bound on the importance function, that is, the right-hand side of (8.21). Because

$$\frac{\sum_{i=1}^N \omega_k^i \, \tau_k^i}{\sum_{i=1}^N \omega_k^i \, \gamma_k(\xi_k^i)} M_k \geq \frac{\sum_{i=1}^N \omega_k^i \, \tau_k^i \, \frac{A_k^i}{\tau_k^i}}{\sum_{i=1}^N \omega_k^i \, \gamma_k(\xi_k^i)} = \frac{\sum_{i=1}^N \omega_k^i \, A_k^i}{\sum_{i=1}^N \omega_k^i \, \gamma_k(\xi_k^i)} ,$$

the acceptance probability is bounded by

$$\frac{\sum_{i=1}^N \omega_k^i \, \gamma_k(\xi_k^i)}{\sum_{i=1}^N \omega_k^i \, A_k^i} . \tag{8.22}$$

The bound is attained when $A_k^i / \tau_k^i = M_k$ for all i. □

Tanizaki and Mariano (1998) and Hürzeler and Künsch (1998) both consider the particular choice $R_k = Q$. Lemma 8.1.7 shows that the optimal adjustment multiplier weights are then constant, $\tau_k^i = 1$ for all i. This is somewhat surprising in light of the discussion in Section 8.1.2, as one could conjecture heuristically that it is more appropriate to favor particles that agree with the next observations. Lemma 8.1.7 however shows that the only means to improve the acceptance rate is, whenever possible, to properly optimize the instrumental kernel.

8.1.5 Markov Chain Monte Carlo Auxiliary Sampling

Rather than using the accept-reject algorithm to sample exactly from (8.16), Berzuini *et al.* (1997) suggest that a few iterations of a Markov chain Monte Carlo sampler with target distribution (8.16) be used. The algorithm proposed by Berzuini *et al.* (1997) is based on the independent Metropolis-Hastings sampler discussed in Section 6.2.3.1. Once again, we use a distribution ρ_{k+1}^{aux} of the form defined in (8.8) as the proposal, but this time the chain moves from (i, x) to (i', x') with a probability given by $A[(i, x), (i', x')] \wedge 1$ where

$$A\left[(i, x), (i', x')\right] = \left[\frac{g_{k+1}(x')}{\tau_k^{i'}} \frac{dQ(\xi_k^i, \cdot)}{dR_k(\xi_k^i, \cdot)}(x') \right] \left[\frac{g_{k+1}(x)}{\tau_k^i} \frac{dQ(\xi_k^i, \cdot)}{dR_k(\xi_k^i, \cdot)}(x) \right]^{-1} . \tag{8.23}$$

In case of rejection, the chain stays in (i, x). This update step is then repeated independently N times.

Algorithm 8.1.8 (Auxiliary MCMC). For $i = 1, \ldots, N$,

Initialization: Draw an index $I_k^{i,1} \in \{1, \ldots, N\}$ with probabilities proportional to the first-stage weights $\omega_k^1 \tau_k^1, \ldots, \omega_k^N \tau_k^N$, and $\tilde{\xi}_{k+1}^{i,1}$ from the instrumental transition kernel $R_k(\xi_k^{I_k^{i,1}}, \cdot)$. Set $\xi_{k+1}^i = \tilde{\xi}_{k+1}^{i,1}$ and $I_k^i = I_k^{i,1}$.

For $j = 2$ to $J_{\max}$: Draw an index $I_k^{i,j} \in \{1, \ldots, N\}$ with probabilities proportional to the first-stage weights $\omega_k^1 \tau_k^1, \ldots, \omega_k^N \tau_k^N$, draw $\tilde{\xi}_{k+1}^{i,j}$ from the instrumental transition kernel $R_k(\xi_k^{I_k^{i,j}}, \cdot)$ and a $\mathrm{U}([0,1])$ variable U^j. If

$$U^j \leq A[(I_k^i, \xi_{k+1}^i), (I_k^{i,j}, \tilde{\xi}_{k+1}^{i,j})] ,$$

set $\xi_{k+1}^i = \tilde{\xi}_{k+1}^{i,j}$ and $I_k^i = I_k^{i,j}$.

When done, all weights $\{\omega_{k+1}^i\}_{1\le i\le N}$ are reset to a constant value.

In the above algorithm, ρ_{k+1}^{aux} is used both as proposal distribution for the independent Metropolis-Hastings sampler and for generating the initial values $I_k^{i,1}$ and $\tilde{\xi}_{k+1}^{i,1}$. Compared to the accept-reject approach of the previous section, Algorithm 8.1.8 is appealing, as it is associated with a deterministic computation time that scales like the product of N and $J_{\max}$. On the other hand, the method can only be useful if $J_{\max}$ is "small" which in turn is legitimate only if the independent Metropolis-Hastings chain is fast mixing. As discussed in Section 6.2.3.1, the mixing of each individual chain is governed by the behavior of the quantity

$$M_k^i = \sup_{x\in\mathsf{X}} \frac{g_{k+1}(x)}{\tau_k^i} \frac{dQ(\xi_k^i,\cdot)}{dR_k(\xi_k^i,\cdot)}(x) ,$$

and the chain is uniformly (geometrically) ergodic, at rate $(1 - 1/M_k^i)$, only if M_k^i is finite. Not surprisingly, this approach thus shares many common features and properties with the accept-reject algorithm discussed in the previous section. It is of course possible to combine both methods (Tanizaki, 2003) or to resort to other type of MCMC samplers. We refer to Berzuini and Gilks (2001) for a full discussion of this approach together with some examples where it is particularly useful.

8.2 Sequential Monte Carlo in Hierarchical HMMs

In Section 4.2, we examined a general class of HMMs, referred to as hierarchical HMMs, for which the state can be partitioned into two components, one of which can be analytically integrated out—or *marginalized*—conditionally on the other component. When marginalization is feasible, one may derive computationally efficient sampling procedures that focus their full attention on a state space whose dimension is smaller—and in most applications, much smaller—than the original one. As a result, when marginalization is feasible, it usually significantly improves the performance of particle filtering, allowing in particular a drastic reduction of the number of particles needed to achieve a given level of accuracy of the estimates (Akashi and Kumamoto, 1977; Liu and Chen, 1998; MacEachern *et al.*, 1999; Doucet *et al.*, 2000a,b). One should however keep in mind that marginalization requires the use of rather sophisticated algorithms, and that the computations necessary to update each marginal particle can be much more demanding than for an unstructured particle that lives in the complete state space.

Marginalizing out some of the variables is an example of a classical technique in computational statistics referred to as the *Rao-Blackwellization*, because it is related to the Rao-Blackwell risk reduction principle in statistics. Rao-Blackwellization is an important ingredient of simulation-based methods that we already met in the context of MCMC methods in Chapter 6.

In the hierarchical hidden Markov model introduced in Section 1.3.4, the state variable X_k can be decomposed in two parts (C_k, W_k), where C_k is called the indicator variable or the regime and W_k is the partial state, which can be marginalized out conditionally on the regime. We will focus on the special case where the indicator variables are discrete and finite. Although it is possible to use the marginalization principle in a more general setting (see, e.g., Doucet *et al.*, 2001b, or Andrieu *et al.*, 2003), the case of discrete indicator variables remains the most important in practical applications.

8.2.1 Sequential Importance Sampling and Global Sampling

Assume that the indicator variables take their values in the finite set $\mathsf{C} = \{1, \ldots, r\}$. We consider here, as previously, that the goal is to simulate from the sequence of joint probability measures $\{\psi_{0:k|k}\}_{k \geq 0}$ of $C_{0:k}$ given $Y_{0:k}$. For the moment, the details of the structure of $\psi_{0:k|k}$ do not matter and we simply assume that there exists an (unnormalized) transition kernel $T_k^{\mathrm{u}} : \mathsf{C}^{k+1} \times \mathsf{C} \to \mathbb{R}^+$ such that

$$\psi_{0:k+1|k+1}(c_{0:k+1}) = \psi_{0:k|k}(c_{0:k}) T_k^{\mathrm{u}}(c_k, c_{k+1}) \ . \tag{8.24}$$

Note that as usual for probabilities on discrete spaces, we use the notation $\psi_{0:k|k}(c_{0:k})$ rather than $\psi_{0:k|k}(\{c_{0:k}\})$. This definition should be compared to (7.8). Indeed, T_k^{u} is an unnormalized kernel similar to that which appears in (7.8), although it does not depend—as a function of $c_{0:k}$—on c_k only. This modification is due to the fact that the structure of the joint smoothing distribution in hierarchical HMMs, *when marginalizing with respect to the intermediate component* $\{W_k\}$, is more complex than in the models that we have met so far in this chapter (see Section 4.2.3). Once again, these considerations are not important for the moment, and the reader should consider (8.24) as the definition of a (generic) sequence of probability distributions over increasing spaces.

8.2.1.1 Sequential Importance Sampling

In the sequential importance sampling framework, the target distribution at time k is approximated by independent path particles denoted, as previously, by $\xi_{0:k}^1, \ldots, \xi_{0:k}^N$, associated with non-negative (normalized) weights $(\omega_k^1, \ldots, \omega_k^N)$ such that

$$\hat{\psi}_{0:k|k}(c_{0:k}) = \sum_{i=1}^{N} \omega_k^i \mathbb{1}_{\xi_{0:k}^i}(c_{0:k}) \ . \tag{8.25}$$

These particles and weights are updated sequentially by drawing from an instrumental distribution over sequences in $\mathsf{C}^{\mathbb{N}}$ defined by an initial probability distribution $\rho_{0:0}$ on C and a family of transition kernels $R_k : \mathsf{C}^{k+1} \times \mathsf{C} \to \mathbb{R}^+$, for $k \geq 0$, such that

$$\rho_{0:k+1}(c_{0:k+1}) = \rho_{0:k}(c_{0:k}) R_k(c_{0:k}, c_{k+1}) , \tag{8.26}$$

where $\rho_{0:k}$ denotes the joint distribution of $\xi^1_{0:k}$. It is assumed that for each k, the instrumental kernel R_k dominates the transition T^{u}_k in the sense that for any $c_{0:k}$ and any $c = 1, \ldots, r$, the condition $T^{\mathrm{u}}_k(c_{0:k}, c) > 0$ implies $R_k(c_{0:k}, c) > 0$. In words, all transitions that are permitted (have positive probability) under the model are permitted also under the instrumental kernel. In the sequential importance sampling procedure, one draws exactly one successor for each path particle $\xi^i_{0:k}$, $i = 1, \ldots, N$. More precisely, an N-uplet $I^1_{k+1}, \ldots, I^N_{k+1}$ is drawn conditionally independently given the past and with probabilities proportional to the weights

$$R_k(\xi^i_{0:k}, 1), \ldots, R_k(\xi^i_{0:k}, r) .$$

The particle system is then updated according to $\xi^i_{0:k+1} = (\xi^i_{0:k}, I^i_{k+1})$. If $\xi^1_0, \ldots, \xi^N_0$ are drawn independently from a probability distribution $\rho_{0:0}$, the particle system $\xi^1_{0:k}, \ldots, \xi^N_{0:k}$ consists of N independent draws from the instrumental distribution $\rho_{0:k}$. As in (7.13), the associated (unnormalized) importance weights can be written as a product of incremental weights

$$\omega^i_{k+1} = \omega^i_k \frac{T^{\mathrm{u}}_k(\xi^i_{0:k}, I^i_{k+1})}{R_k(\xi^i_{0:k}, I^i_{k+1})} . \tag{8.27}$$

The instrumental transition kernel that minimizes the variance of the importance weights conditionally on the history of the particle system will be denoted by T_k and is given by the analog of (7.15):

$$T_k(c_{0:k}, c) = \frac{T^{\mathrm{u}}_k(c_{0:k}, c)}{T^{\mathrm{u}}_k(c_{0:k}, \mathsf{C})}, \quad c_{0:k} \in \mathsf{C}^{k+1},\ c \in \mathsf{C} . \tag{8.28}$$

This kernel is referred to as the *optimal instrumental kernel*. The importance weights (8.27) associated with this kernel are updated according to

$$\omega^i_{k+1} = \omega^i_k T^{\mathrm{u}}_k(\xi^i_{0:k}, \mathsf{C}) . \tag{8.29}$$

As before, these incremental importance weights do not depend on the descendant of the particle. The SIS algorithm using the optimal importance kernel is equivalent to the random sampling algorithm of Akashi and Kumamoto (1977). In this scheme, resampling is stochastic with precisely one descendant of each particle at time k being kept. For each particle, a descendant is chosen with probabilities proportional to the descendant's weights $T^{\mathrm{u}}_k(\xi^i_{0:k}, 1), \ldots, T^{\mathrm{u}}_k(\xi^i_{0:k}, r)$. The weight of the chosen particle is set to the product of its parent's weight and the sum $\sum_{c=1}^r T^{\mathrm{u}}_k(\xi^i_{0:k}, c)$.

8.2.1.2 Global Sampling

As in the previous chapter, the particle system produced by sequential importance sampling degenerates, and the way to fight this degeneracy is resampling. Because the state space is finite however, we now can probe the whole state space because each particle has a finite number (r) of possible descendants. The sampling and resampling steps may then be combined into a single random draw. Recall that a natural estimator of the target distribution $\psi_{0:k|k}$ at time k is the empirical distribution of the particles defined in (8.25). Equation (8.24) suggests to estimate the probability distribution Π_{k+1} by

$$\tilde{\psi}_{0:k+1|k+1}(c_{0:k+1}) = \frac{\sum_{i=1}^N \omega_k^i \delta_{\xi_{0:k}^i}(c_{0:k}) T_k^{\mathrm{u}}(\xi_{0:k}^i, c_{k+1})}{\sum_{i=1}^N \omega_k^i T_k^{\mathrm{u}}(\xi_{0:k}^i, \mathsf{C})} . \tag{8.30}$$

This equation corresponds to (8.4) in the current discrete setting. The support of this distribution is included in the set of all the possible descendants of the current system of particles. Each particle has at most r possible descendants and thus the support of this distribution has at most $N \times r$ points. A straightforward solution (see for instance Fearnhead and Clifford, 2003) to sample from this distribution is as follows.

Algorithm 8.2.1 (Global Sampling).

Weighting: For $i = 1, \ldots, N$ and $j = 1, \ldots, r$, compute the (normalized) weights

$$\omega_{k+1}^{i,j} = \frac{\omega_k^i T_k^{\mathrm{u}}(\xi_{0:k}^i, j)}{\sum_{l=1}^N \sum_{c=1}^r \omega_k^l T_k^{\mathrm{u}}(\xi_{0:k}^i, c)} . \tag{8.31}$$

Sampling : Conditionally independently from the particle system history, draw N identically distributed pairs $(I_k^i, J_{k+1}^i) \in \{1, \ldots, N\} \times \{1, \ldots, r\}$, for $i = 1, \ldots N$, such that $\mathrm{P}[(I_k^1, J_{k+1}^1) = (i,j) \,|\, \mathcal{G}_k] = \omega_{k+1}^{i,j}$, where $\mathcal{G}_k$ is the σ-field generated by the history of the particle system up to time k.

Update: Set $\xi_{0:k+1}^i = (\xi_{0:k}^{I_k^i}, J_{k+1}^i)$ and $\omega_{k+1}^i = 1/N$ for $i = 1, \ldots, N$.

Remark 8.2.2. There are several closely related algorithms that have appeared in the literature, in particular the detection estimation algorithm of Tugnait (1984). In this algorithm, the resampling stage is deterministic, with the N particles having largest weights being kept. The application of such ideas has been especially investigated in digital communication applications and is discussed, for instance, by Punskaya *et al.* (2002) and Bertozzi *et al.* (2003). ■

8.2.2 Optimal Sampling

As stressed in Section 7.4.2, there are other options to draw the reallocation variables such as residual, stratified, or systematic resampling. Although these can certainly be useful in this context, the discrete nature of the state space has an unexpected consequence that is not addressed properly by the resampling techniques discussed so far. For problems in which the state space is continuous, having multiple copies of particles is not detrimental. After resampling, each copy of a given duplicated particle will evolve independently from the others. Therefore, a particle with a large importance weight that is replicated many times in the resampling stage may, in the future, have a large number of distinct descendants. When the state space is finite however, each particle can probe *all* its possible descendants $(\xi^i_{0:k}, j)$ such that $T^{\mathrm{u}}_k(\xi^i_{0:k}, j) > 0$. Hence, if the resampling procedure replicates a particle at time k, the replications of this particle will probe exactly the same configurations in the future. Having multiple copies of the same path particle in finite state space models is thus particularly wasteful.

A possible solution to this problem has been suggested by Fearnhead and Clifford (2003) under the name *optimal sampling*. Instead of drawing reallocation variables $\{(I^i_k, J^i_{k+1})\}_{1\le i\le N}$, we sample non-negative importance weights $\{W^{i,j}_{k+1}\}$ satisfying the constraints

$$\sum_{i=1}^{N}\sum_{j=1}^{r} \mathbb{1}_{\{W^{i,j}_{k+1}>0\}} \le N \tag{8.32}$$

$$\mathrm{E}[W^{i,j}_{k+1} \,|\, \mathcal{G}_k] = \omega^{i,j}_{k+1}, \quad i=1,\dots,N,\ j=1,\dots,r, \tag{8.33}$$

where the weights $\omega^{i,j}_{k+1}$ are defined in (8.31). The first constraint is that there are at most N particles with non-zero weights. The second constraint is that the importance weights be unbiased—in the terminology of Liu and Chen (1998) or Liu *et al.* (2001), the new sample is the said to be *properly weighted.* A word of caution is needed here: despite the fact that the unbiasedness condition is very sensible in the context of resampling, it does not, in itself, guarantee a proper behavior of the algorithm (more on this will be said in Chapter 9). Conversely, *exact* unbiasedness is not absolutely necessary, and it is perfectly possible to consider algorithms that exhibit a low, and controllable, bias. The problem reduces to that of approximating a probability distribution having $M = N \times r$ points of support by a *random* probability distribution having at most N points of support. Resampling is equivalent to assigning a new, random weight to each of the $M = N \times r$ particles. If the weight is zero the particle is removed, whereas if the weight is non-zero the particle is kept; the non-zero random variables $W^{i,j}_{k+1}$ represent the new weights of the descendants of the particle system.

In a more general perspective, the problem can be formulated as follows. Let ω be a discrete probability distribution with M points of support

$$\omega = (\omega_1, \dots, \omega_M), \quad \omega_i \ge 0, \quad \sum_{i=1}^{M} \omega_i = 1\,. \tag{8.34}$$

We want to find a *random* probability distribution $W = (W_1, \ldots, W_M)$ on $\{1, \ldots, M\}$ with at most $N \leq M$ points of support,

$$W_i \geq 0\,, \quad \sum_{i=1}^{M} W_i = 1\,, \quad \sum_{i=1}^{M} \mathbb{1}_{\{W_i > 0\}} \leq N\,, \tag{8.35}$$

satisfying

$$\mathrm{E}[W_i] = \omega_i, \quad i = 1, \ldots, M\,. \tag{8.36}$$

There are of course a number of different ways to achieve (8.35) and (8.36). In particular, all the resampling methods discussed in Section 7.4.2 (as well as multinomial resampling) draw integer counts N_i, which are such that $W_i = N_i/N$ satisfy the above requirements, with equality for the last condition in (8.35). The "optimal" solution is the one that guarantees that the random distribution W is close, in some suitable sense, to the target distribution ω. We follow the suggestion of Fearnhead and Clifford (2003) and use the average L^2 distance. The problem then becomes equivalent to finding a random probability distribution $W = (W_1, \ldots, W_M)$ that minimizes

$$\sum_{i=1}^{M} \mathrm{E}(W_i - \omega_i)^2 \tag{8.37}$$

subject to (8.35) and (8.36). To compute the solution we rely on two lemmas.

Lemma 8.2.3. *Let $\omega \geq 0$ and $p \in (0, 1]$. If W is a non-negative random variable satisfying*

$$\mathrm{E}[W] = \omega \quad \text{and} \quad \mathrm{P}(W > 0) = p\,, \tag{8.38}$$

then

$$\mathrm{E}(W - \omega)^2 \geq \frac{1-p}{p}\omega^2\,. \tag{8.39}$$

The lower bound is attained by any random variable W such that W equals ω/p on the subset of the sample space where $W > 0$.

Proof. By decomposing the sample space into $\{W > 0\}$ and $\{W = 0\}$, we obtain

$$\omega = \mathrm{E}[W] = \mathrm{E}[W \mid W > 0]\,\mathrm{P}(W > 0) = \mathrm{E}[W \mid W > 0]p\,, \tag{8.40}$$

and by a similar decomposition,

$$\mathrm{E}(W - \omega)^2 = \mathrm{E}[(W - \omega)^2 \mid W > 0]p + \omega^2(1 - p)\,. \tag{8.41}$$

A bias-variance decomposition of $\mathrm{E}[(W - \omega)^2 \mid W > 0]$ then gives

$$
\begin{aligned}
&\mathrm{E}[(W-\omega)^2 \mid W>0] \\
&\quad = \mathrm{E}[(W-\mathrm{E}[W|W>0])^2 \mid W>0] + (\mathrm{E}[W \mid W>0]-\omega)^2 \\
&\quad = \mathrm{E}[(W-\mathrm{E}[W|W>0])^2 \mid W>0] + \omega^2 \frac{(1-p)^2}{p^2} ,
\end{aligned}
$$

where we used (8.40) to obtain the second equality. The right-hand side of this display is bounded from below by $\omega^2(1-p)^2/p^2$, and inserting this into the right-hand side of (8.41) we obtain (8.39). Using the last display once again, we also see that the bound is attained if and only if W equals $\mathrm{E}[W|W>0] = \omega/p$ on the set where $W>0$. □

Lemma 8.2.4. *Let $N < M$ be integers and let $\beta_1, \ldots, \beta_M$ be non-negative numbers. Consider the problem*

$$
\begin{aligned}
&\textit{minimize} && \sum_{j=1}^{M} \frac{\beta_j}{p_j} \\
&\textit{subject to} && \sum_{j=1}^{M} p_j \le N , \\
& && 0 \le p_j \le 1, \quad j=1,\ldots,M .
\end{aligned}
$$

This problem has a unique solution given by

$$
p_j = \mu\sqrt{\beta_j} \wedge 1, \quad j=1,\ldots,M , \tag{8.42}
$$

where the constant μ is the unique solution of the equation

$$
\sum_{i=1}^{M} \mu\sqrt{\beta_j} \wedge 1 = N . \tag{8.43}
$$

Proof. Denote by λ and λ_i the Lagrange multipliers associated respectively with the inequality constraints $\sum_{i=1}^{M} p_i \le N$ and $p_i \le 1$, $i=1,\ldots,M$. The Karush-Kuhn-Tucker conditions (see Boyd and Vandenberghe, 2004) for the primal $p_1,\ldots,p_M$ and dual $\lambda, \lambda_1, \ldots, \lambda_M$ optimal points are given by

$$
\sum_{i=1}^{M} p_i \le N, \qquad p_i \le 1, \qquad i=1,\ldots,M , \tag{8.44}
$$

$$
\lambda \ge 0, \qquad \lambda_i \ge 0, \qquad i=1,\ldots,M , \tag{8.45}
$$

$$
\lambda\left(\sum_{i=1}^{M} p_i - N\right) = 0, \qquad \lambda_i(p_i-1)=0, \qquad i=1,\ldots,M , \tag{8.46}
$$

$$
-\frac{\beta_i}{p_i^2} + \lambda + \lambda_i = 0, \qquad i=1,\ldots,M . \tag{8.47}
$$

The complementary slackness condition (8.46) implies that for all indices i such that $p_i < 1$, the corresponding multiplier λ_i is zero. Hence, using (8.47),

$$p_i = \sqrt{\frac{\beta_i}{\lambda}} \wedge 1, \quad i = 1, \ldots, M . \tag{8.48}$$

From this we see that if $\lambda = 0$, then $p_i = 1$ for all i and (8.44) cannot be satisfied. Thus $\lambda > 0$, and the complementary slackness condition (8.46) therefore implies that $\sum_{i=1}^M p_i = N$. Plugging (8.48) into this equation determines the multiplier λ by solving for λ in the equation $\sum_1^M \sqrt{\beta_i/\lambda} \wedge 1 = N$. □

By combining these two lemmas, we readily obtain a characterization of the random distribution achieving the minimal average divergence (8.37) subject to the support constraint $\sum_{i=1}^M \mathrm{P}(W_i > 0) \leq N$ and the unbiasedness constraint (8.36).

Proposition 8.2.5. *Let $W = (W_1, \ldots, W_M)$ be a random vector with non-negative entries. This vector is a solution to the problem*

$$\begin{aligned} &\textit{minimize} && \sum_{i=1}^M \mathrm{E}(W_i - \omega_i)^2 \\ &\textit{subject to} && \sum_{i=1}^M \mathrm{P}(W_i > 0) \leq N , \\ & && \mathrm{E}[W_i] = \omega_i, \quad i = 1, \ldots, M , \end{aligned}$$

if and only if for any $i = 1, \ldots, M$,

$$W_i = \begin{cases} \omega_i/p_i & \textit{with probability } p_i \stackrel{\text{def}}{=} \mu\omega_i \wedge 1 , \\ 0 & \textit{otherwise} , \end{cases} \tag{8.49}$$

where μ is the unique solution of the equation

$$\sum_{i=1}^M \mu\omega_i \wedge 1 = N . \tag{8.50}$$

Proof. Put $p_i = \mathrm{P}(W_i > 0)$. By Lemma 8.2.3,

$$\sum_{i=1}^M \mathrm{E}(W_i - \omega_i)^2 \geq \sum_{i=1}^M \frac{\omega_i^2}{p_i} - \sum_{i=1}^M \omega_i^2 . \tag{8.51}$$

The proof follows from Lemma 8.2.4. □

Remark 8.2.6. Note that if $\mu\omega_i \geq 1$, then $p_i = 1$ and $\omega_i/p_i = \omega_i$. Thus (8.49) implies that weights exceeding a given threshold (depending on the weights themselves) are left unchanged. For a particle i whose weight falls below this threshold, the algorithm proceeds as follows. With probability $1 - p_i > 0$, the weight is set to zero; otherwise it is set (and thus increased) to $\omega_i/p_i = 1/\mu$ in order to satisfy the unbiasedness condition. The algorithm is related to the procedure proposed in Liu *et al.* (2001) under the name *partial rejection control.* ■

The above proposition describes the marginal distribution of the W_i that solves (8.37). The following result proposes a simple way to draw random weights $(W_1, \ldots, W_M)$ that satisfy (8.49) with $\sum_{i=1}^M \mathbb{1}_{\{W_i>0\}} = N$.

Proposition 8.2.7. *Let* μ *be the solution of* (8.50),

$$\mathsf{S} \stackrel{\text{def}}{=} \{i \in \{1, \ldots, M\} : \mu\omega_i \geq 1\} \tag{8.52}$$

and $p_i = \mu\omega_i \wedge 1$. *Let* U *be a uniform random variable on* $(0, 1)$ *and set*

$$N_i = \left\lfloor \textstyle\sum_{j \notin \mathsf{S},\, j \leq i} p_j + U \right\rfloor - \left\lfloor \textstyle\sum_{j \notin \mathsf{S},\, j < i} p_j + U \right\rfloor, \quad i = 1, \ldots, M,$$

with $\lfloor \cdot \rfloor$ *being the integer part. Define the random vector* $W = (W_1, \ldots, W_M)$ *by*

$$W_i = \begin{cases} \omega_i & \text{if } i \in \mathsf{S}\ , \\ 1/\mu & \text{if } i \notin \mathsf{S} \text{ and } N_i > 0\ , \\ 0 & \text{if } i \notin \mathsf{S} \text{ and } N_i = 0\ . \end{cases} \tag{8.53}$$

Then W *satisfies* (8.49) *and*

$$\sum_{i=1}^M \mathbb{1}_{\{W_i>0\}} = N\ , \tag{8.54}$$

$$\sum_{i=1}^M W_i = 1\ . \tag{8.55}$$

Proof. We first show that $\mathrm{P}(W_i > 0) = p_i$. For $i \in \mathsf{S}$ this is immediate, with $p_i = 1$. Thus pick $i \notin \mathsf{S}$. Then

$$N_i \leq \sup_{x \geq 0}(\lfloor x + p_i \rfloor - \lfloor x \rfloor) \leq 1\ .$$

Therefore $N_i = \mathbb{1}_{\{W_i>0\}}$, which implies $\mathrm{P}(W_i > 0) = \mathrm{P}(N_i > 0) = \mathrm{E}[N_i]$. It is straightforward to check that the expectation of N_i is the difference of the two sums involved in its definition, whence $\mathrm{E}[N_i] = p_i$. Thus $\mathrm{P}(W_i > 0) = p_i$, showing that (8.49) is satisfied.

Next observe that $\sum_1^M \mathbb{1}_{\{W_i>0\}} = |\mathsf{S}| + \sum_{i \notin \mathsf{S}} N_i$. The sum of N_i over all $i \notin \mathsf{S}$ is a telescoping one, whence

$$\sum_{i=1}^{M} \mathbb{1}_{\{W_i>0\}} = |\mathsf{S}| + \left\lfloor \sum_{i\notin\mathsf{S}} p_i + U \right\rfloor - \lfloor U \rfloor$$
$$= |\mathsf{S}| + \lfloor N - |\mathsf{S}| + U \rfloor - \lfloor U \rfloor = |\mathsf{S}| - (N - |\mathsf{S}|) = N ,$$

where we used $\sum_{i\notin\mathsf{S}} p_i = \sum_1^M p_i - \sum_{i\in\mathsf{S}} p_i = N - |\mathsf{S}|$ for the second equality. Thus we have (8.54).

Finally,

$$\sum_{i=1}^{M} W_i = \sum_{i\in\mathsf{S}} \omega_i + \sum_{i\in\mathsf{S}} N_i/\mu .$$

From the above, we know that the second sum on the right-hand side equals $(N - |\mathsf{S}|)/c$. Because, by definition, $\omega_i/p_i = 1/\mu$ for $i \notin \mathsf{S}$, the first sum is

$$\sum_{i\in\mathsf{S}} \omega_i = 1 - \sum_{i\notin\mathsf{S}} \omega_i = 1 - \mu^{-1} \sum_{i\notin\mathsf{S}} p_i = 1 - \frac{N - |\mathsf{S}|}{\mu} .$$

We conclude that $\sum_1^M W_i = 1$, that is, (8.55) holds. □

Back to our original problem, Proposition 8.2.7 suggests the following sampling algorithm.

Algorithm 8.2.8 (Optimal Sampling).

Weighting : For $i = 1, \dots, N$ and $j = 1, \dots, r$, compute the weights

$$\omega_{k+1}^{i,j} = \frac{\omega_k^i T_k^{\mathrm{u}}(\xi_{0:k}^i, j)}{\sum_{l=1}^N \sum_{c=1}^r \omega_k^l T_k^{\mathrm{u}}(\xi_{0:k}^i, c)} . \tag{8.56}$$

Sampling:

- Determine the solution μ_{k+1} of the equation
$$\sum_{i=1}^{N} \sum_{j=1}^{r} \mu_{k+1} \omega_{k+1}^{i,j} \wedge 1 = N .$$
- Draw $U \sim \mathrm{U}([0,1])$ and set $S = 0$.
- For $i = 1, \dots, N$ and $j = 1, \dots, r$,
  - If $\mu_{k+1}\omega_{k+1}^{i,j} \geq 1$, then set $W_{k+1}^{i,j} = \omega_{k+1}^{i,j}$.
  - If $\mu_{k+1}\omega_{k+1}^{i,j} < 1$, then set
  $$W_{k+1}^{i,j} = \begin{cases} \mu_{k+1}^{-1} & \text{if } \lfloor \mu_{k+1}(S + \omega_{k+1}^{i,j}) + U \rfloor - \lfloor \mu_{k+1} S + U \rfloor > 0 , \\ 0 & \text{otherwise} , \end{cases}$$

  and set $S = S + \omega_{k+1}^{i,j}$.

Update: For $i = 1, \dots, N$ and $j = 1, \dots, r$, if $W_{k+1}^{i,j} > 0$ set

$$\xi_{0:k+1}^{\mathfrak{I}(i,j)} = (\xi_{0:k}^i, j) ,$$
$$\omega_{k+1}^{\mathfrak{I}(i,j)} = W_{k+1}^{i,j} , \qquad \text{where} \quad \mathfrak{I}(i,j) = \sum_{l=1}^{i} \sum_{c=1}^{j-1} \mathbb{1}_{\{W_{k+1}^{l,c}>0\}} .$$

8.2.3 Application to CGLSSMs

In this section, we consider conditionally Gaussian linear state-space models (CGLSSMs), introduced in Section 1.3.4 and formally defined in Section 2.2.3. Recall that a CGLSSM is such that

$$\begin{aligned} W_{k+1} &= A(C_{k+1})W_k + R(C_{k+1})U_k \;, \\ Y_k &= B(C_k)W_k + S(C_k)V_k \;, \end{aligned} \tag{8.57}$$

where

- $\{C_k\}_{k\geq 0}$ is a Markov chain on the finite set $\mathsf{C} = \{1, \ldots, r\}$, with transition kernel Q_C and initial distribution ν_C;
- the state noise $\{U_k\}_{k\geq 0}$ and measurement noise $\{V_k\}_{k\geq 0}$ are independent multivariate Gaussian white noises with zero mean and identity covariance matrices;
- the initial partial state W_0 is assumed to be independently $\mathrm{N}(\mu_\nu, \Sigma_\nu)$ distributed;
- A, B, R, and S are known matrix-valued functions of appropriate dimensions.

Efficient recursive procedures, presented in Section 5.2.6, are available to compute the filtered or predicted estimate of the partial state and the associated error covariance matrix conditionally on the indicator variables and observations. By embedding these algorithms in the sequential importance sampling resampling framework, it is possible to derive computationally efficient sampling procedures that operate in the space of indicator variables (Doucet *et al.*, 2000a; Chen and Liu, 2000). Recall in particular that the kernel T_k^{u} in (8.24) has an expression given by (4.11), which we repeat below.

$$\begin{aligned} T_k^{\mathrm{u}}(c_{0:k}, c_{k+1}) = \left(\frac{\mathrm{L}_{k+1}}{\mathrm{L}_k}\right)^{-1} Q_C(c_k, c_{k+1}) \times \\ \int_{\mathsf{W}} g_{k+1}(c_{k+1}, w_{k+1}) \varphi_{k+1|k}(c_{0:k+1}, w_{k+1})\, dw_{k+1} \;, \end{aligned} \tag{8.58}$$

for $c_{0:k+1} \in \mathsf{C}^{k+2}$, where

- L_k is the likelihood of the observations up to time k;
- $g_{k+1}(c_{k+1}, w_{k+1}) = g\left[(c_{k+1}, w_{k+1}), Y_{k+1}\right]$ is the value of the transition density function of the observation Y_{k+1} given the state and indicator variables, that is,

$$g_{k+1}(c_{k+1}, w_{k+1}) = \mathrm{N}(Y_{k+1}; B(c_{k+1})w_{k+1}, S(c_{k+1})S^t(c_{k+1})) \;, \tag{8.59}$$

  with $\mathrm{N}(\cdot; \mu, \Sigma)$ being the density of the Gaussian multivariate distribution with mean μ and covariance matrix Σ;

- $\varphi_{k+1|k}(c_{0:k+1}, w_{k+1})$ is the density of the predictive distribution of the partial state W_{k+1} given the observations up to time k and the indicator variables up to time $k+1$:

$$\varphi_{k+1|k}(c_{0:k+1}, w_{k+1}) = \mathrm{N}\left(w_{k+1}; \hat{W}_{k+1|k}(c_{0:k+1}), \Sigma_{k+1|k}(c_{0:k+1})\right) , \tag{8.60}$$

where $\hat{W}_{k+1|k}(c_{0:k+1})$ and $\Sigma_{k+1|k}(c_{0:k+1})$ denote respectively the conditional mean and error covariance matrix of the prediction of the partial state W_{k+1} in terms of the observations $Y_{0:k}$ and indicator variables $C_{0:k+1} = c_{0:k+1}$—these quantities can be computed recursively using the Kalman one-step prediction/correction formula (see Section 5.2.3).

As discussed in Section 4.2.3, the distribution of the partial state W_n conditional on the observations up to time n is a mixture of r^{n+1} components—here, Gaussian components—with weights given by $\psi_{0:n|n}$. In the particle approximation, each particle $\xi^i_{0:n}$ relates to a single term in this mixture. Particle approximation of the filtering distribution $\varphi_{n|n}$ of the partial state W_n thus consists in recursively choosing N components out of a growing mixture of r^{n+1} components and adjusting accordingly the weights of the components which are kept; hence the name *mixture Kalman filter* proposed by Chen and Liu (2000) to describe this approach.

Algorithm 8.2.9 (Mixture Kalman Filter).

Initialization: For $i = 1, \ldots, r$, compute

$$\begin{aligned}
\xi^i_0 &= i , \\
\omega^i_0 &= \mathrm{N}(Y_0; B(i)\mu_\nu, B(i)\Sigma_\nu B^t(i) + S(i)S^t(i))\, \nu_C(c_0) , \\
K_0(\xi^i_0) &= B^t(i)\Sigma_\nu \left[B(i)\Sigma_\nu B^t(i) + S(i)S^t(i)\right]^{-1} , \\
\hat{W}_{0|0}(\xi^i_0) &= \mu_\nu + K_0(\xi^i_0)\left[Y_0 - B(i)\mu_\nu\right] , \\
\Sigma_{0|0}(\xi^i_0) &= \Sigma_\nu - K_0(\xi^i_0)B(i)\Sigma_\nu .
\end{aligned}$$

Recursion:

Computation of weights: For $i = 1, \ldots, N$ and $j = 1, \ldots, r$, compute

$$\begin{aligned}
\hat{W}_{k+1|k}(\xi^i_{0:k}, j) &= A(j)\hat{W}_{k|k}(\xi^i_{0:k}) , \\
\Sigma_{k+1|k}(\xi^i_{0:k}, j) &= A(j)\Sigma_{k|k}(\xi^i_{0:k})A^t(j) + R(j)R^t(j) , \\
\hat{Y}_{k+1|k}(\xi^i_{0:k}, j) &= B(j)\hat{W}_{k+1|k}(\xi^i_{0:k}, j) , \\
\Gamma_{k+1}(\xi^i_{0:k}, j) &= B(j)\Sigma_{k+1|k}(\xi^i_{0:k}, j)B^t(j) + S(j)S^t(j) , \\
\tilde{\omega}^{i,j}_{k+1} &= \omega^i_k\, \mathrm{N}(Y_{k+1}\, ; \hat{Y}_{k+1|k}(\xi^i_{0:k}, j), \Gamma_{k+1}(\xi^i_{0:k}, j))\, Q_C(\xi^i_k, j) .
\end{aligned}$$

(First Option) Importance Sampling Step: For $i = 1, \ldots, N$, draw J^i_{k+1} in $\{1, \ldots, r\}$ with probabilities proportional to $\tilde{\omega}^{i,1}_k, \ldots, \tilde{\omega}^{i,r}_k$, conditionally independently of the particle history, and set

$$\xi^i_{0:k+1} = (\xi^i_{0:k}, J^i_{k+1}) ,$$

$$\omega^i_{k+1} = \sum_{j=1}^{r} \tilde{\omega}^{i,j}_{k+1} \Big/ \sum_{i=1}^{N}\sum_{j=1}^{r} \tilde{\omega}^{i,j}_{k+1} ,$$

$$K_{k+1}(\xi^i_{0:k+1}) = \Sigma_{k+1|k}(\xi^i_{0:k}, J^i_{k+1}) B^t(J^i_{k+1}) \Gamma^{-1}_{k+1}(\xi^i_{0:k+1}, J^i_{k+1}) ,$$

$$\hat{W}_{k+1|k+1}(\xi^i_{0:k+1}) = \hat{W}_{k+1|k}(\xi^i_{0:k}, J^i_{k+1}) + K_{k+1}(\xi^i_{0:k+1}) \left[Y_{k+1} - \hat{Y}_{k+1|k}(\xi^i_{0:k}, J^i_{k+1}) \right] ,$$

$$\Sigma_{k+1|k+1}(\xi^i_{0:k+1}) = \left[I - K_{k+1}(\xi^i_{0:k+1}) B(J_{k+1}) \right] \Sigma_{k+1|k}(\xi^i_{0:k}, J^i_{k+1}) .$$

(Second Option) Optimal Sampling Step:

- Draw importance weights $W^{i,j}_k$ for $i = 1, \ldots, N$ and $j = 1, \ldots, r$ using Algorithm 8.2.8.
- Set $\mathfrak{I} = 0$. For $i = 1, \ldots, N$ and $j = 1, \ldots, r$, if $W^{i,j}_{k+1} > 0$ then

$$\xi^{\mathfrak{I}}_{0:k+1} = (\xi^i_{0:k}, j) ,$$

$$\omega^{\mathfrak{I}}_{k+1} = W^{i,j}_{k+1} ,$$

$$K_{k+1}(\xi^{\mathfrak{I}}_{0:k+1}) = \Sigma_{k+1|k}(\xi^i_{0:k}, j) B^t(j) \Gamma^{-1}_{k+1}(\xi^i_{0:k}, j) ,$$

$$\hat{W}_{k+1|k+1}(\xi^{\mathfrak{I}}_{0:k+1}) = \hat{W}_{k+1|k}(\xi^i_{0:k}, j) + K_{k+1}(\xi^{\mathfrak{I}}_{0:k+1}) \left[Y_{k+1} - \hat{Y}_{k+1|k}(\xi^i_{0:k}, j) \right] ,$$

$$\Sigma_{k+1|k+1}(\xi^{\mathfrak{I}}_{0:k+1}) = \left[I - K_{k+1}(\xi^{\mathfrak{I}}_{0:k+1}) B(j) \right] \Sigma_{k+1|k}(\xi^i_{0:k}, j) ,$$

$$\mathfrak{I} = \mathfrak{I} + 1 .$$

Note that in the algorithm above, $\{W^{i,j}_k\}$ are the weights drawn according to Algorithm 8.2.8. These have nothing to do with the state variable W_k and should not be mistaken with the corresponding predictor denoted by $\hat{W}_{k+1|k}(\xi^i_{0:k}, j)$. The first option corresponds to the basic importance sampling strategy—without resampling—and is thus analogous to the SIS approach of Algorithm (7.2.2). As usual, after several steps without resampling, the particle system quickly degenerates into a situation where the discrepancy between the weights $\{\omega^i_k\}_{1 \le i \le N}$ is more and more pronounced as k grows. The second option corresponds to a resampling step based on Algorithm 8.2.8, which avoids particle duplication in the situation where C_k is finite-valued.

Example 8.2.10. To illustrate the previous algorithm, we consider once more the well-log data of Example 1.3.10 using the same modeling assumptions as in Example 6.3.7. In contrast to Example 6.3.7 however, we now consider sequential approximation of the filtering (or fixed-lag smoothing) distributions of the jump and outlier indicators rather than the block (non-sequential) approximation of the joint smoothing distributions of these variables.

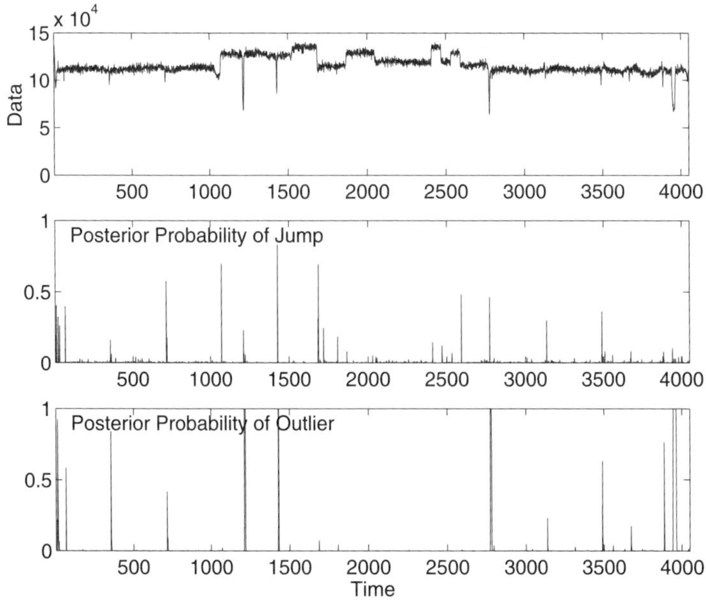

Fig. 8.2. On-line analysis of the well-log data, using 100 particles with detection delay $\Delta = 0$. Top: data; middle: posterior probability of a jump; bottom: posterior probability of an outlier.

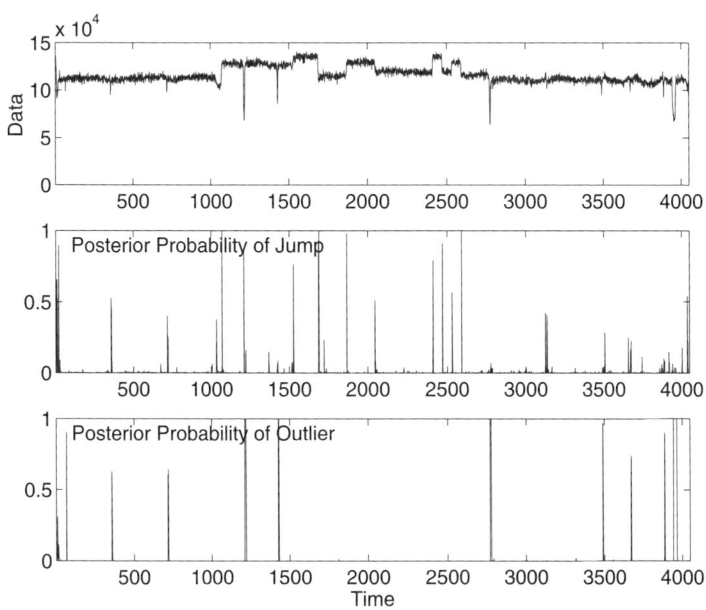

Fig. 8.3. On-line analysis of the well-log data, using 100 particles with detection delay $\Delta = 5$ (same display as above).

The main aim of analyzing well-log data is the on-line detection of abrupt changes in the level of the response. The detection delay, defined as the number of samples that are processed before a decision is taken, should be kept as small as possible. Here the detection delay has been set to $\Delta = 0$ and $\Delta = 5$: after processing each observation Y_k, the probability of a jump having occurred at time $k - \Delta$ was estimated by averaging the values of $\{\xi^i_{0:k}(k-\Delta)\}_{1 \le i \le N}$ (see Example 6.3.7 for the detail of the parameterization used in this example).

The results of a single on-line analysis of the well-log data using the optimal sampling strategy (at each step) are shown in Figures 8.2 ($\Delta = 0$) and 8.3 ($\Delta = 5$). In both cases, $N = 100$ particles are used. For $\Delta = 0$, the particle filter has performed reasonably well: most of the obvious jumps in the level of the data have a posterior probability close to 1, although some of them are obviously missing (around time index 2000 for instance). In addition, differentiating jumps from outliers is particularly difficult in this case and the filter has misclassified outliers into change points (at time index 700 for instance). On Figure 8.3 ($\Delta = 5$), most of the misclassification errors have disappeared and the overall result is quite good (although some points are still detected both as change points and outliers as in index 1200). Because the typical length of an outlier is about four, five samples are usually enough to tell whether a change in the level has occurred. ■

8.3 Particle Approximation of Smoothing Functionals

As emphasized in Section 4.1, it is often of interest to approximate the expectation of some statistic $t_n(x_{0:n})$ under the joint smoothing distribution $\phi_{0:n|n}$,

$$\int \cdots \int t_n(x_{0:n})\, \phi_{0:n|n}(dx_{0:n}) \; .$$

This difficult problem admits a computationally simpler solution in cases where the statistic has the specific form—which we called a *smoothing functional* in Section 4.1—given by (see Definition 4.1.2):

$$t_{n+1}(x_{0:n+1}) = m_n(x_n, x_{n+1}) t_n(x_{0:n}) + s_n(x_n, x_{n+1}) \; , \quad n \ge 0 \; , \qquad (8.61)$$

for all $x_{0:n+1} \in \mathsf{X}^{n+2}$. Here $\{m_n\}_{n \ge 0}$ and $\{s_n\}_{n \ge 0}$ are two sequences of real measurable functions on $\mathsf{X} \times \mathsf{X}$. Examples include the sample mean $t_n(x_{0:n}) = (n+1)^{-1} \sum_{k=0}^n x_k$, the first-order sample autocovariance coefficient $t_n(x_{0:n}) = n^{-1} \sum_{k=1}^n x_{k-1} x_k$, etc. Other important examples of smoothing functionals arise in parameter estimation when using the EM algorithm or when computing the gradient of the log-likelihood function (see Chapters 10 and 11 for details).

Define the finite signed measure τ_n on $(\mathsf{X}, \mathcal{X})$ by

$$\tau_n(f) \stackrel{\text{def}}{=} \int \cdots \int f(x_n)\, t_n(x_{0:n})\, \phi_{0:n|n}(dx_{0:n}) \; , \quad f \in \mathcal{F}_{\mathrm{b}}(\mathsf{X}) \; . \qquad (8.62)$$

Note that by construction, $\tau_n(\mathsf{X}) = \phi_{0:n|n}(t_n)$, that is, the quantity of interest. By Proposition 4.1.3, the measures $\{\tau_n\}_{n\geq 0}$ may be updated recursively according to

$$\tau_0(f) = \{\nu(g_0)\}^{-1} \int f(x_0)\, t_0(x_0) g_0(x_0)\, \nu(dx_0)$$

and

$$\tau_{n+1}(f) = c_{n+1}^{-1} \iint f(x_{n+1}) \Big[\tau_n(dx_n) Q(x_n, dx_{n+1}) g_{n+1}(x_{n+1}) m_n(x_n, x_{n+1}) \\ + \phi_n(dx_n)\, Q(x_n, dx_{n+1}) g_{n+1}(x_{n+1}) s_n(x_n, x_{n+1}) \Big] , \tag{8.63}$$

where the normalizing constant c_{n+1} is given by (3.22) as $c_{n+1} = \phi_n Q g_{n+1}$. It is easily seen that τ_n is absolutely continuous with respect to the filtering measure ϕ_n. Hence (8.63) may be rewritten as

$$\tau_{n+1}(f) = \iint f(x_{n+1}) \times \\ \left\{ \frac{d\tau_n}{d\phi_n}(x_n) m_n(x_n, x_{n+1}) + s_n(x_n, x_{n+1}) \right\} \phi_{n:n+1|n+1}(dx_{n:n+1}) \,. \tag{8.64}$$

In SISR algorithms, the joint smoothing distribution $\phi_{0:n+1|n+1}$ at time $n+1$ is approximated by a set $\{\xi^i_{0:n+1}\}_{1\leq i\leq N}$ of particles with associated importance weights $\{\omega^i_{n+1}\}_{1\leq i\leq N}$. Due to the sequential update of the particle trajectories, there exist indices $I^1_{n+1}, \ldots, I^N_{n+1}$ (see Algorithm 7.3.4) such that

$$\xi^i_{0:n+1} = (\xi^{I^i_{n+1}}_{0:n}, \xi^i_{n+1}) \,,$$

meaning that the first $n+1$ coordinates of the path are simply copied from the previous generation of particles. Because τ_n is absolutely continuous with respect to ϕ_n for any n, it seems reasonable to approximate τ_n using the same system of particles as that used to approximate ϕ_n. That is, for any n we approximate τ_n by

$$\widehat{\tau}_n = \sum_{i=1}^N \frac{\omega^i_n}{\sum_{j=1}^N \omega^j_n} \gamma^i_n \delta_{\xi^i_n} \,, \tag{8.65}$$

where γ^i_n, $i = 1, \ldots, N$, are signed weights. Such approximations have been considered in different settings by Cappé (2001a), Cérou *et al.* (2001), Doucet and Tadić (2003), and Fichou *et al.* (2004). This approximation of τ_n yields the following estimator of $\phi_{0:n|n}(t_n) = \tau_n(\mathsf{X})$:

$$\hat{\phi}_{0:n|n}(t_n) = \sum_{i=1}^N \frac{\omega^i_n}{\sum_{j=1}^N \omega^j_n} \gamma^i_n \,. \tag{8.66}$$

The two measures $\widehat{\tau}_n$ and $\hat{\phi}_n$ have the same support, which implies that $\widehat{\tau}_n$ is absolutely continuous with respect to $\hat{\phi}_n$; in addition, for any $x \in \{\xi_n^1, \ldots, \xi_n^N\}$,

$$\frac{d\widehat{\tau}_n}{d\hat{\phi}_n}(x) = \frac{\sum_{j \in \mathsf{I}_n(x)} \omega_n^j \gamma_n^j}{\sum_{j \in \mathsf{I}_n(x)} \omega_n^j} , \tag{8.67}$$

where $\mathsf{I}_n(x) \stackrel{\text{def}}{=} \{j = 1, \ldots, N : \xi_n^j = x\}$. In cases where there are no ties (all particle locations are distinct), we simply have

$$\frac{d\widehat{\tau}_n}{d\hat{\phi}_n}(\xi_n^i) = \gamma_n^i . \tag{8.68}$$

To derive a recursive approximation of $\widehat{\tau}_n$, it is only needed to derive update equations for the signed weights γ_n^i. Plugging the particle approximation $\hat{\phi}_{n:n+1|n+1} \propto \sum_{i=1}^N \omega_{n+1}^i \delta_{\xi_{n:n+1}^i}$ of the retrospective smoothing distribution $\phi_{n:n+1|n+1}$ into the update equation (8.64) yields the following approximation of the measure τ_{n+1}:

$$\sum_{i=1}^N \frac{\omega_{n+1}^i}{\sum_{j=1}^N \omega_{n+1}^j} \left\{ \frac{d\tau_n}{d\phi_n}(\xi_n^{I_{n+1}^i}) m_n(\xi_n^{I_{n+1}^i}, \xi_{n+1}^i) + s_n(\xi_n^{I_{n+1}^i}, \xi_{n+1}^i) \right\} \delta_{\xi_{n+1}^i} . \tag{8.69}$$

Using the approximation (8.68) of $\frac{d\tau_n}{d\phi_n}(\xi_n^j)$, the latter relation suggests the following recursion for the weights $\{\gamma_n^i\}_{1 \le i \le N}$:

$$\gamma_0^i = t_0(\xi_0^i) , \tag{8.70}$$

$$\gamma_{n+1}^i = \gamma_n^{I_{n+1}^i} m_n(\xi_n^{I_{n+1}^i}, \xi_{n+1}^i) + s_n(\xi_n^{I_{n+1}^i}, \xi_{n+1}^i) . \tag{8.71}$$

This relation, originally derived by Cappé (2001a)[2], is computationally attractive because the approximation uses the same set particles and weights as those used to approximate the filtering distribution; only the incremental signed weights need to be computed recursively. Also, it mimics the exact recursion for τ_n and therefore seems like a good way to approximate this sequence of measures.

To get a better understanding of the behavior of the algorithm, we will derive the recursion (8.71) from a different (admittedly more elementary) perspective. The sequential importance sampling approximation of the joint smoothing distribution $\phi_{0:n|n}$ amounts to approximate, for any statistic $t_n(x_{0:n})$, $\phi_{0:n|n}(t_n)$ by

$$\hat{\phi}_{0:n|n}(t_n) = \sum_{i=1}^N \frac{\omega_n^i}{\sum_{j=1}^N \omega_n^j} t_n(\xi_{0:n}^i) . \tag{8.72}$$

---

[2]The recursion obtained by Cérou *et al.* (2001) is based on a very different argument but turns out to be equivalent in the case where the functional of interest corresponds to the gradient of the log-likelihood function (see Section 10.2.4 for details).

If the statistic t_n is a smoothing functional as defined in (8.61), this quantity can be evaluated sequentially so that storing the whole particle paths is avoided. Denote by $\{t_n^i\}_{1\leq i\leq N}$ the current value of the smoothing functional t_n along the particle path $\xi_{0:n}^i$: $t_n^i = t_n(\xi_{0:n}^i)$. This quantity may be updated according to the recursion $t_0^i = t_0(\xi_0^i)$ and

$$t_{n+1}^i = t_n^{I_{n+1}^i} m_n(\xi_n^{I_{n+1}^i}, \xi_{n+1}^i) + s_n(\xi_n^{I_{n+1}^i}, \xi_{n+1}^i), \qquad i = 1, \ldots, N . \tag{8.73}$$

Perhaps surprisingly, because the two approximations have been derived from two different perspectives, (8.73) and (8.71) are identical. This means that both equations are recursive ways to compute the approximation (8.72) of expectations with respect to the joint smoothing distribution. The second reasoning, which led to recursion (8.73), however, raises some concern about the practical use of this approximation. Because the path particles $\{(\xi_{0:n}^i, \omega_n^i)\}_{1\leq i\leq N}$ are targeted to approximate a probability distribution over the space X^{n+1}, whose dimension grows with n, it is to be expected that the curse of dimensionality can only be fought by increasing the number N of path particles as n increases (Del Moral, 2004). A worst case analysis suggests that the number N of path particles should grow exponentially with n, which is of course unrealistic. This assertion should however be taken with some care because we are in general interested only in low-dimensional statistical summaries of the particle paths. Hence, the situation usually is more contrasted, as illustrated below on an example.

Example 8.3.1. We consider here the stochastic volatility model of Example 7.2.5:

$$\begin{aligned} X_{k+1} &= \phi X_k + \sigma U_k , & U_k &\sim \mathrm{N}(0,1) , \\ Y_k &= \beta \exp(X_k/2) V_k , & V_k &\sim \mathrm{N}(0,1) . \end{aligned}$$

Here the observations $\{Y_k\}_{k\geq 0}$ are the log-returns, $\{X_k\}_{k\geq 0}$ is the log-volatility, and $\{U_k\}_{k\geq 0}$ and $\{V_k\}_{k\geq 0}$ are independent sequences of standard white Gaussian noise. We use the SISR algorithm with systematic resampling and instrumental kernel being a t-distribution with 5 degrees of freedom and mode and scale adjusted to the mode and curvature of the optimal instrumental kernel (see Example 7.2.5). We consider the daily log-returns, that is, the difference of the log of the series, on the British pound/US dollar exchange rate from October, 1 1981, to June, 28 1985 (the data is scaled by 100 and mean-corrected—see Kim *et al.*, 1998, and Shephard and Pitt, 1997 for details). The number of samples is $n = 945$, and we used the stochastic volatility model with parameters $\phi = 0.975$, $\beta = 0.63$, and $\sigma = 0.16$; these are the maximum likelihood estimates reported by Sandmann and Koopman (1998) on this data set.

The path particles after 70 iterations are plotted in Figure 8.4. The figure clearly shows that the selection mechanism implies that for any given time index $k \leq n$, the number of ancestors, at that time, of the particle trajectories

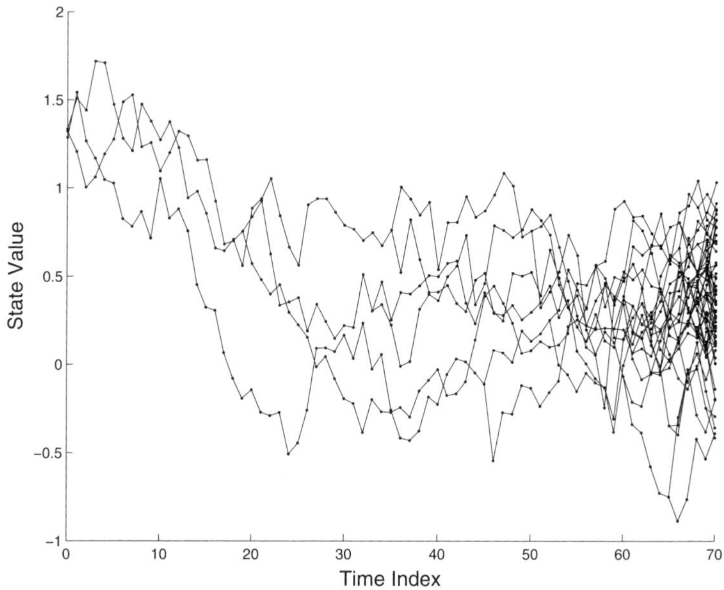

Fig. 8.4. Particle trajectories at time $n = 70$ for the stochastic volatility model using the algorithm of Example 7.2.5 with $N = 100$ particles and systematic resampling.

ending in index n becomes small as the difference between n and k grows. It is therefore to be expected that estimation of the expectation under the joint smoothing distribution of statistics involving the first time lags will typically display large fluctuations and that these fluctuations will get larger when n increases.

This behavior is indeed illustrated in Figure 8.5, which shows particle estimates of $\int x^2 \, \phi_{0|n}(dx)$ for different values of n and N. The variance of the particle estimate steadily increases with n for all values of N. In addition, a fairly large number N of particles is needed to obtain reliable estimates for larger values of n although the value to be estimated does not change much when n gets larger than, say, $n = 20$.

It is interesting to contrast the results of particle methods with those that can be obtained with the (non-sequential) Markov chain Monte Carlo (MCMC) methods of Chapter 6. For MCMC methods, because the target distribution is static and equal to the joint distribution $\phi_{0:n|n}$, we simply ran 100 instances of the sampler of Example 6.3.1 for each value of n and recorded the averaged value of the first component (squared) in each sample. Here a "sweep" refers to the successive updates of each of the $n + 1$ sites of the simulated sequence $X^i_{0:n}$ (see Example 6.3.1 for details). The computation cost of the MCMC and particles approaches, with comparable values of n and N, are thus roughly the same. Remember however that in the particle approach, estimated values of $\int x^2 \, \phi_{0|n}(dx)$ for different values of n may be

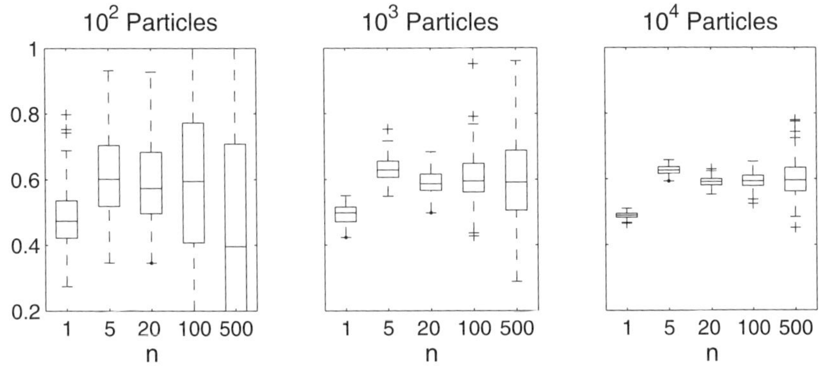

Fig. 8.5. Box and whisker plots of particle estimates of $\int x^2 \phi_{0|n}(dx)$ for $n = 1$, 5, 20, 30, and 500, and particle population sizes $N = 10^2$, 10^3, and 10^4. The plots are based on 100 independent replications.

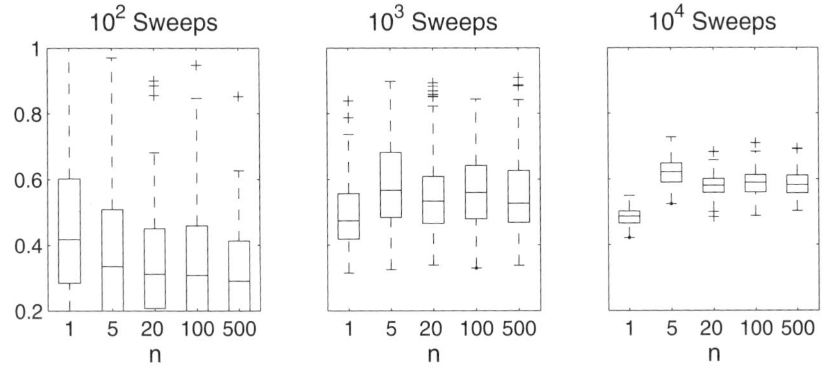

Fig. 8.6. Same figure as above for MCMC estimates of $\int x^2 \phi_{0|n}(dx)$, where N refers to the number of MCMC sweeps though the data, using the MCMC sampler of Example 6.3.1.

obtained in a single run of the algorithm due to the sequential nature of the computations. Observe first on the leftmost display of Figure 8.6 that the MCMC estimates obtained with just $N = 100$ sweeps are severely downward biased: this is due to the fact that the sequence of states $X^1_{0:n}$ is initialized with zero values and $N = 100$ is insufficient to forget this initialization, due to the correlation between successive MCMC simulations (see Figure 6.10). On this data set (and with those parameter values), about 200 iterations are indeed needed to obtain reasonably unbiased estimates. The next important observation about Figure 8.6 is that the variance of the estimate does not vary much with n. This is of course connected to the observation, made in Example 6.3.1, that the correlation between successive MCMC simulations does not change (significantly) as n increases. For smaller values of n, the existence of correlation makes the MCMC approach far less reliable than the particle method. But for larger values of n, the degradation of the results previously observed for the particle method—with a fixed value of N and as n increases—kicks in and the comparison is more balanced (compare the fifth boxes in the rightmost displays of Figures 8.5 and 8.6).

In some sense, the degradation observed on Figure 8.5 as n grows (N being fixed) is all the more disturbing that we expect the result to be nearly independent of n, once it exceeds a given value (which is clearly the case on Figures 8.5 and 8.6). Indeed, the *forgetting property* of the smoothing distributions discussed in Section 4.3 implies that the posterior distribution of the state x_0 depends predominantly on the observations Y_k with time indices close to $k = 0$[3] (see, e.g., Polson *et al.*, 2002, for a related use of the forgetting property). For large values of n, it is thus reasonable to approximate the expectation of $t_{n,0}(x_{0:n}) = x_0$ under $\phi_{0|n}$ by that of the same quantity under $\phi_{0|k}$ for k large enough, but still much smaller than n. Of course it is to be expected that the bias of the approximation decreases when increasing the number of lags k. On the other hand, as mentioned above, the dispersion of the particle estimator of the expectation under the reduced-lag smoothing distribution $\phi_{0|k}(t_{0,n})$ increases with k. We are thus faced with a classical bias-variance trade-off problem; when k is large the bias is small but the dispersion is large, and *vice versa*. Setting k smaller than n is thus an effective way of robustifying the estimator without any modification of the sequential Monte Carlo procedure.

To give an idea of how large k should be for the example under consideration, the difference between the means of the particle estimates (obtained using $N = 10^5$ particles) of $\phi_{0|n}(t_{n,0})$ and $\phi_{0|k}(t_{n,0})$ is less than 10^{-3} for $n = 100$ and $k = 20$. For $k = 1$ and $k = 10$, the corresponding differences are 0.2 and -0.12, respectively. This means that we can safely estimate $\phi_{0|n}(t_{0,n})$ by $\phi_{0|k}(t_{0,n})$ if we take $k \geq 20$. The standard error of the reduced-lag smoothing estimator $\phi_{0|20}(t_{0,n})$ is, at least, three times less than that of $\phi_{0|500}(t_{0,n})$. As a consequence, we can achieve the same level of performance using reduced-lag smoothing with about 10 times less particles (compare, on Figure 8.5, the third box in the second display with the fifth one in the third display).

This naturally raises the question whether the same conclusion can be drawn for other statistics of interest. Suppose that we want to approximate the expectations of $t_{n,1}(x_{0:n}) = \sum_{l=0}^{n-1} x_l^2$ and $t_{n,2}(x_{0:n}) = \sum_{l=1}^{n} x_{l-1}x_l$ under the joint smoothing distribution $\phi_{0:n|n}$[4]. These two statistics may be written as time averages, immediately suggesting the *fixed-lag* approximations $\sum_{l=0}^{n-1} \int x_l^2 \, \phi_{l|(l+k)\wedge n}(dx_l)$ and $\sum_{l=1}^{n} \int x_{l-1}x_l \, \phi_{l-1:l|(l+k)\wedge n}(dx_{l-1:l})$ for some

[3]Note that we invoke here the spirit of Section 4.3 rather than an exact result, as we are currently unable to prove that the forgetting property holds for the stochastic volatility model (see discussion at the end of Section 4.3), although empirical evidence says it does.

[4]These statistics need to be evaluated in order to estimate the intermediate quantity of the Expectation-Maximization algorithm—see Example 11.1.2 for details.

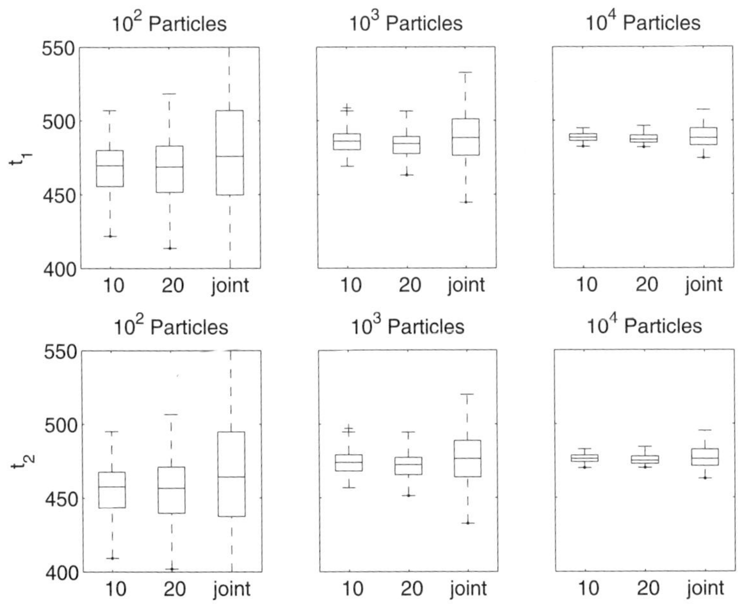

Fig. 8.7. Box and whisker plots of particle estimators of the expectations of the two statistics $t_{n,1}(x_{0:n}) = \sum_{k=0}^{n-1} x_k^2$ (top) and $t_{n,2}(x_{0:n}) = \sum_{k=1}^{n} x_k x_{k-1}$ (bottom) for $n = 945$: from left to right, increasing particle population sizes of $N = 10^2$, 10^3, and 10^4; on each graph, fixed-lag smoothing approximation for smoothing delays $k = 10$ and 20 and full path "joint" particle approximation. The plots are based on 100 independent replications.

lag k—where the term *fixed-lag* refers to the fact that k is fixed and does not vary with n. To approximate both of these sums, one can use a variant of (8.73) in which only the part of the sum that pertains to indices l located less than k lags away from the current time index is updated, while the contribution of indices further back in the past is fixed. A little thought should convince the reader that this can be achieved by storing the cumulative contribution of past sections of the trajectories that do not get resampled anymore $\sum_{i=1}^{N} \sum_{l=0}^{n-k-1} s(\xi_{0:n}^i(l))$ as well as the recent history of the particles $\{\xi_{0:n}^i(l)\}$ for $l = n-k, \ldots, n$ and $i = 1, \ldots, N$; here s is the function of interest, say $s(x) = x^2$ in the case of $t_{n,1}$, and $\xi_{0:n}^i(l)$ denotes the element of index l in the path $\xi_{0:n}^i$. As above, it is expected that increasing the number of lags k will increase the dispersion but decrease the bias. This is confirmed by the results displayed in Figure 8.7. Again, the use of fixed-lag instead of joint smoothing provides more accurate estimators. ■

To conclude this section, we would like to stress again the difference between fixed-dimensional statistics like $t_{n,0}(x_{0:n}) = x_0$ and smoothing functionals, in the sense of Definition (4.1.2), which depend on the complete collection of hidden states up to time n (for instance, $t_{n,1}(x_{0:n}) = \sum_{l=0}^{n-1} x_l^2$). Although the latter case does seem to be more challenging, the averaging effect due to n should not be underestimated: even crude approximations of the individual terms, say $\int x_l^2 \phi_{l|n}(dx_l)$ in the case of $t_{n,1}$, may add up to provide a reliable approximation of the conditional expectation of $t_{n,1}$. In our experience, the strategy discussed above is usually successful with rather moderate values of the lag k and the number N of particles, as will be illustrated in Chapter 11. In the case of fixed-dimensional statistics, more elaborate smoothing algorithms may be more recommendable, particularly in situations where relying on forgetting properties might be questionable (Kitagawa, 1996; Fong *et al.*, 2002; Briers *et al.*, 2004).

9

Analysis of Sequential Monte Carlo Methods

The previous chapters have described many algorithms to approximate prediction, filtering, and smoothing distributions. The development of these algorithms was motivated mainly on heuristic grounds, and the validity of these approximations is of course a question of central interest. In this chapter, we analyze these methods, mainly from an asymptotic perspective. That is, we study the behavior of the estimators in situations where the number of particles gets large. Asymptotic analysis provides approximations that in many circumstances have proved to be relatively robust. Most importantly, asymptotic arguments provide insights in the sampling methodology by verifying that the procedures are sensible, providing a framework for comparing competing procedures, and providing understanding of the impact of different options (choice of importance kernel, etc.) on the overall performance of the samplers.

9.1 Importance Sampling

9.1.1 Unnormalized Importance Sampling

Let $(\mathsf{X}, \mathcal{X})$ be a measurable space. Define on $(\mathsf{X}, \mathcal{X})$ two probability distributions: the *target distribution* μ and the *instrumental distribution* ν.

Assumption 9.1.1. *The target distribution μ is absolutely continuous with respect to the instrumental distribution ν, $\mu \ll \nu$, and $d\mu/d\nu > 0$ ν-a.s.*

Let f be a real-valued measurable function on X such that $\mu(|f|) = \int |f| \, d\mu < \infty$. Denote by $\xi^1, \xi^2, \ldots$ an i.i.d. sample from ν and consider the estimator

$$\tilde{\mu}^{\mathrm{IS}}_{\nu,N}(f) = \frac{1}{N} \sum_{i=1}^{N} f(\xi^i) \frac{d\mu}{d\nu}(\xi^i) \,. \tag{9.1}$$

Because this estimator is the sample average of independent random variables, there is a range of results to assess the accuracy of $\tilde{\mu}^{\mathrm{IS}}_{\nu,N}(f)$ as an estimator of $\mu(f)$. Some of these results are asymptotic in nature, like the law of large numbers (LLN) and the central limit theorem (CLT). It is also possible to derive non-asymptotic bounds like Berry-Esseen bounds, bounds on error moments $\mathrm{E}\,|\tilde{\mu}^{\mathrm{IS}}_{\nu,N}(f) - \mu(f)|^p$ for some $p > 0$ or on the tail probability $\mathrm{P}\left(|\tilde{\mu}^{\mathrm{IS}}_{\nu,N}(f) - \mu(f)| \geq \epsilon\right)$. Instead of covering the full scale of results that can be derived, we establish for the different algorithms presented in the previous chapter a law of large numbers, a central limit theorem, and deviation bounds.

A direct application of the LLN and of the CLT yields the following result.

Theorem 9.1.2. *Let f be a real measurable function such that $\mu(|f|) < \infty$ and $|f|\mu \ll |f|\nu$, and let $\xi^1, \xi^2, \ldots$ be a sequence of i.i.d. random variables from ν. Then the unnormalized importance sampling estimator $\tilde{\mu}^{\mathrm{IS}}_{\nu,N}(f)$ given by (9.1) is strongly consistent, $\lim_{N\to\infty} \tilde{\mu}^{\mathrm{IS}}_{\nu,N}(f) = \mu(f)$ a.s.*

Assume in addition that

$$\int f^2 \left[\frac{d\mu}{d\nu}\right]^2 d\nu < \infty\,. \tag{9.2}$$

Then $\tilde{\mu}^{\mathrm{IS}}_{\nu,N}(f)$ is asymptotically Gaussian,

$$\sqrt{N}(\tilde{\mu}^{\mathrm{IS}}_{\nu,N}(f) - \mu(f)) \xrightarrow{\mathcal{D}} \mathrm{N}\left(0, \mathrm{Var}_\nu\left(f\frac{d\mu}{d\nu}\right)\right) \quad \text{as } N \to \infty\,,$$

where $\mathrm{Var}_\nu\left(f\frac{d\mu}{d\nu}\right)$ is given by

$$\mathrm{Var}_\nu\left(f\frac{d\mu}{d\nu}\right) = \int \left[f\frac{d\mu}{d\nu} - \mu(f)\right]^2 d\nu\,.$$

Obviously, while the importance sampling construction (9.1) is universal, the performance of the importance sampling estimator depends heavily on the relation between the target distribution μ, the instrumental distribution ν, and the function f. It is also worthwhile to note that for a given function f, it is most often possible to find a distribution ν that yields an estimate with a lower variance than when using the Monte Carlo method, that is, taking $\nu = \mu$. In some situations the improvements can be striking: this is in particular the case where the function f is non-zero only for values that are in the tails of the target distribution μ, a situation that occurs for instance when estimating the probability of rare events. The basic idea is to choose the importance distribution ν so that it generates values that are in the region where the integrand $f\frac{d\mu}{d\nu}$ is large, as this region is where the most important contributions are made to the value of the integral.

Notice that

$$\mathrm{Var}_\nu\left(f\frac{d\mu}{d\nu}\right) = [\mu(f)]^2\,\nu\left[\left(\frac{|f|d\mu/d\nu}{\mu(|f|)} - 1\right)^2\right],$$

where the second factor on the right-hand side is the *chi-square* distance between the densities 1 and $|f|\frac{d\mu}{d\nu}/\mu(|f|)$ under ν. This factor is of course in general unknown, but may be estimated consistently by computing the (squared) coefficient of variation CV_N^2, see (7.35), of the importance weights $\omega^i = |f(\xi^i)|\frac{d\mu}{d\nu}(\xi^i)$, $i = 1, \ldots, N$.

Poor selection of the instrumental distribution can induce large variations in the importance weights $d\mu/d\nu$ and thus unreliable approximations of $\mu(f)$. In many settings, an inappropriate choice of the instrumental distribution might lead to an estimator (9.1) whose variance is infinite (and which therefore does not satisfy the assumptions of the CLT). Here is a simple example of this behavior.

Example 9.1.3 (Importance Sampling with Cauchy and Gaussian Variables). In this example, the target $\mu = \mathrm{C}(0,1)$ is a standard Cauchy distribution, and the instrumental distribution $\nu = \mathrm{N}(0,1)$ is a standard Gaussian distribution. The importance weight function, given by

$$\frac{d\mu}{d\nu}(x) = \sqrt{2\pi}\,\frac{\exp(x^2/2)}{\pi\,(1+x^2)}\;,$$

is obviously badly behaved. In particular

$$\frac{1}{\sqrt{2\pi}}\int_{-\infty}^{\infty}\left[\frac{d\mu}{d\nu}(x)\right]^2 \exp(-x^2/2)\,dx = \infty\;.$$

Figure 9.1 illustrates the poor performance of the associated importance sampling estimator for the function $f(x) = \exp(-|x|)$. We have displayed the quantile-quantile plot of the sample quantiles of the unnormalized IS estimator $\tilde{\mu}_{\nu,N}^{\mathrm{IS}}(f)$, obtained from $m = 500$ independent Monte Carlo experiments, versus the quantiles of a standard normal distribution. In the left panel N =100 and in the right panel N =1,000. The quantile-quantile plot shows deviations from the normal distribution in both the lower and the upper tail for both N =100 and N =1,000, indicating that the distribution of $\tilde{\mu}_{\nu,N}^{\mathrm{IS}}(f)$ does not converge in the limit to a Gaussian distribution. ∎

Example 9.1.4. We now switch the roles of the target and instrumental distributions, taking $\mu = \mathrm{N}(0,1)$ and $\nu = \mathrm{C}(0,1)$. The importance weight is bounded by $\sqrt{2\pi/e}$, and this time Theorem 9.1.2 can be applied. Quantile-quantile plots of the sample quantiles of the unnormalized IS estimator $\tilde{\mu}_{\nu,N}^{\mathrm{IS}}(f)$ are shown in Figure 9.2. The fit is good, even when the sample size is small ($N = 100$). It is worthwhile to investigate the impact of the choice of the scale of the Cauchy distribution. Assume now that $\nu = \mathrm{C}(0,\sigma)$ where $\sigma > 0$ is the scale parameter. The importance weight function is bounded by

$$\begin{cases} \frac{\sqrt{2\pi}}{e\sigma}\mathrm{e}^{\sigma^2/2}\,, & \sigma < \sqrt{2}\,, \\ \sigma\sqrt{\pi/2}\,, & \sigma \geq \sqrt{2}\,. \end{cases} \tag{9.3}$$

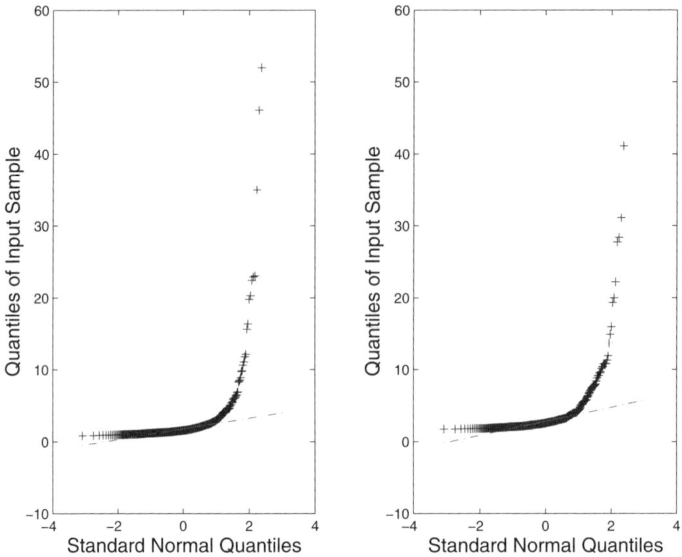

Fig. 9.1. Quantile-quantile plot of the sample quantiles of the unnormalized IS estimator of $\mu(f)$ versus the quantiles of a standard normal distribution. The target and instrumental distributions μ and ν are standard Cauchy and standard Gaussian, respectively, and $f(x) = \exp(-|x|)$. The number of Monte Carlo replications is $m = 500$. Left panel: sample size $N = 100$. Right panel: sample size $N = 1,000$.

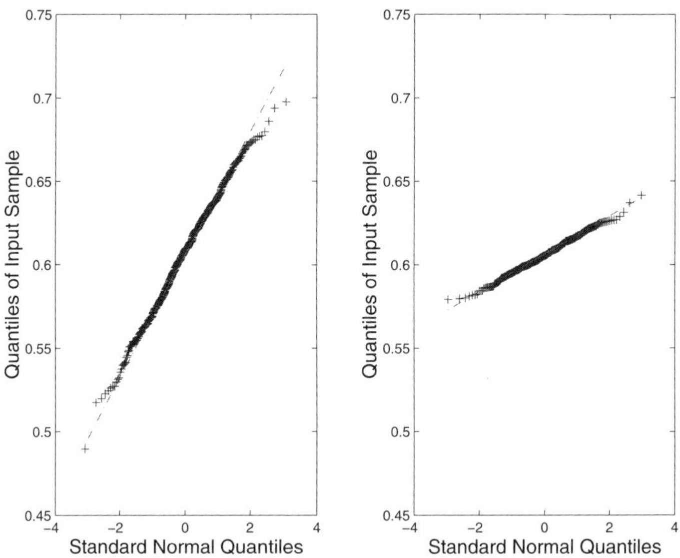

Fig. 9.2. Same figure as above with the roles of μ and ν switched: the target distribution μ is standard Gaussian and the instrumental distribution ν is standard Cauchy.

For $\sigma < \sqrt{2}$, the maximum is attained at $\pm\sqrt{2-\sigma^2}$, while for $\sigma \geq \sqrt{2}$ it is attained at $x = 0$. The upper bound on the importance weight has a minimum at $\sigma = 1$.

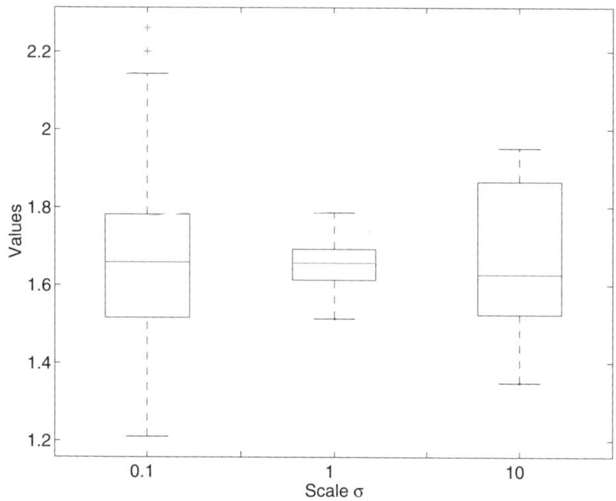

Fig. 9.3. Box-and-whisker plots of the unnormalized IS estimator of $\mu(f)$. The target and instrumental distributions μ and ν were standard Gaussian and Cauchy with scale σ, respectively, and $f(x) = \exp(-|x|)$. Left to right: σ: 0.1, 1, and 10. The sample size was $N = 1{,}000$ and the number of Monte Carlo replications for each plot was $m = 500$.

Figure 9.3 displays box-and-whisker plots of the unnormalized IS estimator for three different values of the scale: $\sigma = 0.1$, $\sigma = 1$, and $\sigma = 10$. The choice $\sigma = 1$ leads to estimators that are better behaved than for $\sigma = 0.1$ and $\sigma = 10$. In the first case, the values drawn from the instrumental distribution are typically too small to represent the standard Gaussian distribution around 0. In the second case, the values drawn are typically too large, and many draws fall far in the tail of the Gaussian distribution. ■

9.1.2 Deviation Inequalities

As outlined above, it is interesting to obtain some non-asymptotic control of the fluctuations of the importance sampling estimator. We may either want to compute bounds on moments $\mathrm{E}\,|\tilde{\mu}^{\mathrm{IS}}_{\nu,N}(f) - \mu(f)|^p$, or to control the probability $\mathrm{P}(|\tilde{\mu}^{\mathrm{IS}}_{\nu,N}(f) - \mu(f)| \geq t)$ for some $t > 0$. Because $\tilde{\mu}^{\mathrm{IS}}_{\nu,N}(f)$ is a sum of i.i.d. random variables, there is a variety of probability inequalities that may be applied for this purpose (see Petrov, 1995, Chapter 2). We do not develop this topic in detail, but just mention two inequalities that will be used later in the book.

The first family of inequalities is related to the control on moments of sums of random variables. There are a variety of inequalities of that kind, which are all similar (except for the constants).

Theorem 9.1.5 (Marcinkiewicz-Zygmund Inequality). *If $X_1, \ldots, X_n$ is a sequence of independent random variables and $p \geq 2$, then*

$$\mathrm{E}\left|\sum_{i=1}^{n}(X_i - \mathrm{E}[X_i])\right|^p \leq C(p) n^{p/2-1} \sum_{i=1}^{n} \mathrm{E}\left|X_i - \mathrm{E}(X_i)\right|^p \tag{9.4}$$

for some positive constant $C(p)$ only depending on p.

The second family of inequalities is related to bounding the tail probabilities. There is a large amount of work in this domain too. The archetypal result is the so-called Hoeffding inequality.

Theorem 9.1.6 (Hoeffding Inequality). *Let $X_1, \ldots, X_n$ be independent bounded random variables such that $\mathrm{P}(a_i \leq X_i \leq b_i) = 1$. Then for any $t \geq 0$,*

$$\mathrm{P}\left\{\sum_{i=1}^{n}[X_i - \mathrm{E}(X_i)] \geq t\right\} \leq \mathrm{e}^{-2t^2/\sum_{i=1}^{n}(b_i - a_i)^2}$$

and

$$\mathrm{P}\left\{\sum_{i=1}^{n}[X_i - \mathrm{E}(X_i)] \leq -t\right\} \leq \mathrm{e}^{-2t^2/\sum_{i=1}^{n}(b_i - a_i)^2} .$$

From these inequalities, it is straightforward to derive non-asymptotic bounds on moments and tail probabilities of the importance sampling estimator. Because the importance ratio is formally not defined on sets A that are such that $\nu(A) = 0$, we first need to extend the concept of oscillation—see (4.14)—as follows. For any measurable function f and measure ν, we define the essential oscillation of f with respect to ν by

$$\operatorname{osc}_\nu(f) \stackrel{\text{def}}{=} 2 \inf_{c \in \mathbb{R}} \|f - c\|_{\nu,\infty} , \tag{9.5}$$

where $\|g\|_{\nu,\infty}$ denotes the essential supremum of g (with respect to ν), the smallest number a such that $\{x : g(x) > a\}$ has ν-measure 0. It is easily checked that the above definition implies that for any a and b such that $a \leq f(\xi) \leq b$ ν-a.s., $\operatorname{osc}_\nu(f) \leq (b - a)$.

Theorem 9.1.7. *For $p \geq 2$ and any $N \geq 1$, the estimator $\tilde{\mu}^{\mathrm{IS}}_{\nu,N}(f)$ defined in (9.1) satisfies*

$$\mathrm{E}\left|\tilde{\mu}^{\mathrm{IS}}_{\nu,N}(f) - \mu(f)\right|^p \leq C(p) N^{-p/2} \nu\left(\left|f\frac{d\mu}{d\nu} - \mu(f)\right|^p\right) ,$$

where the constant $C(p) < \infty$ only depends on p. Moreover, for any $N \geq 1$ and any $t \geq 0$,

$$\mathrm{P}\left[\left|\tilde{\mu}^{\mathrm{IS}}_{\nu,N}(f) - \mu(f)\right| \geq t\right] \leq 2\exp\left[-2Nt^2 \big/ \operatorname{osc}^2_\nu(f\, d\mu/d\nu)\right] . \tag{9.6}$$

9.1.3 Self-normalized Importance Sampling Estimator

When the normalizing constant of the target distribution μ is unknown, it is customary to use the self-normalized form of the importance sampling estimator,

$$\widehat{\mu}^{\text{IS}}_{\nu,N}(f) = \frac{\sum_{i=1}^{N} f(\xi^i) \frac{d\mu}{d\nu}(\xi^i)}{\sum_{i=1}^{N} \frac{d\mu}{d\nu}(\xi^i)} . \tag{9.7}$$

This quantity is obviously free from any scale factor in $d\mu/d\nu$. The properties of this estimator are of course closely related to those of the unnormalized importance sampling estimator.

9.1.3.1 Consistency and Asymptotic Normality

Theorem 9.1.8. *Let f be a measurable function such that $\mu(|f|) < \infty$. Assume that $\mu \ll \nu$ and let $\xi^1, \xi^2, \ldots,$ be an i.i.d. sequence with distribution ν. Then*

$$\widehat{\mu}^{\text{IS}}_{\nu,N}(f) \xrightarrow{a.s.} \mu(f) \quad \text{as } N \to \infty .$$

Assume in addition that f satisfies

$$\int [1 + f^2] \left[\frac{d\mu}{d\nu}\right]^2 d\nu < \infty . \tag{9.8}$$

Then the sequence of estimators $\widehat{\mu}^{\text{IS}}_{\nu,N}(f)$ is asymptotically Gaussian,

$$\sqrt{N}\left[\widehat{\mu}^{\text{IS}}_{\nu,N}(f) - \mu(f)\right] \xrightarrow{\mathcal{D}} \text{N}\left(0, \sigma^2(\nu, f)\right) \quad \text{as } N \to \infty ,$$

where

$$\sigma^2(\nu, f) = \int \left[\frac{d\mu}{d\nu}\right]^2 [f - \mu(f)]^2 d\nu . \tag{9.9}$$

Proof. Strong consistency follows from

$$N^{-1} \sum_{i=1}^{N} f(\xi^i) \frac{d\mu}{d\nu}(\xi^i) \xrightarrow{\text{a.s.}} \mu(f) \quad \text{and} \quad N^{-1} \sum_{i=1}^{N} \frac{d\mu}{d\nu}(\xi^i) \xrightarrow{\text{a.s.}} 1 .$$

Write

$$\sqrt{N}\left[\widehat{\mu}^{\text{IS}}_{\nu,N}(f) - \mu(f)\right] = \frac{N^{-1/2} \sum_{i=1}^{N} \frac{d\mu}{d\nu}(\xi^i)\left[f(\xi^i) - \mu(f)\right]}{N^{-1} \sum_{i=1}^{N} \frac{d\mu}{d\nu}(\xi^i)} .$$

By the central limit theorem, the numerator of the right-hand side above converges weakly to $\text{N}(0, \sigma^2(\nu, f))$ as $N \to \infty$, with $\sigma^2(\nu, f)$ given by (9.9), and as noted above the corresponding denominator converges a.s. to 1. The second part of the theorem then follows by Slutsky's theorem (Billingsley, 1995). □

9.1.3.2 Deviation Inequalities

Assessing deviance bounds for (9.7) is not a trivial task, because both the numerator and the denominator of $\widehat{\mu}^{\mathrm{IS}}_{\nu,N}(f)$ are random. The following elementary lemma plays a key role in deriving such bounds.

Lemma 9.1.9. *Let f be a measurable function and assume that $\mu \ll \nu$. Let c be a real constant and define $\bar{f} = f - c$. Then*

$$\left|\widehat{\mu}^{\mathrm{IS}}_{\nu,N}(f) - \mu(f)\right| \le \left|\frac{1}{N}\sum_{i=1}^{N}\left[\frac{d\mu}{d\nu}(\xi^i)\bar{f}(\xi^i) - \mu(\bar{f})\right]\right| \\ + \left\|\bar{f}\right\|_{\nu,\infty}\left|\frac{1}{N}\sum_{i=1}^{N}\left[\frac{d\mu}{d\nu}(\xi^i) - 1\right]\right| \quad \nu\text{-a.s.} \tag{9.10}$$

Proof. First note that $\widehat{\mu}^{\mathrm{IS}}_{\nu,N}(f) - \mu(f) = \widehat{\mu}^{\mathrm{IS}}_{\nu,N}(\bar{f}) - \mu(\bar{f})$. Next consider the decomposition

$$\widehat{\mu}^{\mathrm{IS}}_{\nu,N}(\bar{f}) - \mu(\bar{f}) = \frac{1}{N}\sum_{i=1}^{N}\left[\frac{d\mu}{d\nu}(\xi^i)(\bar{f}(\xi^i) - \mu(\bar{f}))\right] \\ + \frac{\sum_{i=1}^{N}\frac{d\mu}{d\nu}(\xi^i)\bar{f}(\xi^i)}{\sum_{i=1}^{N}\frac{d\mu}{d\nu}(\xi^i)}\left[1 - \frac{1}{N}\sum_{i=1}^{N}\frac{d\mu}{d\nu}(\xi^i)\right] .$$

Finally, use the triangle inequality and maximize over $\bar{f}(\xi^i)$ in the second term. □

From this result we may obtain moment bounds using the Marcinkiewicz-Zygmund inequality or, under more stringent conditions, exponential bounds on tail probabilities.

Theorem 9.1.10. *Assume that $\nu\left[(d\mu/d\nu)^p\right] < \infty$ for some $p \ge 2$. Then there exists a constant $C < \infty$ such that for any $N \ge 1$ and measurable function f,*

$$\mathrm{E}\left|\widehat{\mu}^{\mathrm{IS}}_{\nu,N}(f) - \mu(f)\right|^p \le C N^{-p/2}\operatorname{osc}^p_\nu(f) . \tag{9.11}$$

In addition, for any $t \ge 0$,

$$\mathrm{P}\left[\left|\widehat{\mu}^{\mathrm{IS}}_{\nu,N}(f) - \mu(f)\right| \ge t\right] \le 4\exp\left[-8Nt^2 \Big/ 9\left\|d\mu/d\nu\right\|^2_{\nu,\infty}\operatorname{osc}^2_\nu(f)\right] . \tag{9.12}$$

Proof. The bound (9.11) is a direct consequence of Lemma 9.1.9 and the Marcinkiewicz-Zygmund inequality (Theorem 9.1.5). Note that by minimizing over c in the right-hand side of (9.10), we may replace $\left\|\bar{f}\right\|_{\nu,\infty}$ by $(1/2)\operatorname{osc}_\nu(f)$, which is done here.

For the second part pick $b \in (0,1)$ and write, using Lemma 9.1.9,

$$\mathrm{P}\left[\left|\widehat{\mu}^{\mathrm{IS}}_{\nu,N}(f) - \mu(f)\right| \geq t\right] \leq \mathrm{P}\left[\left|\sum_{i=1}^{N}\left(\frac{d\mu}{d\nu}(\xi^i)\bar{f}(\xi^i) - \mu(\bar{f})\right)\right| \geq Nbt\right] \\ + \mathrm{P}\left[\left|\sum_{i=1}^{N}\left(\frac{d\mu}{d\nu}(\xi^i) - 1\right)\right| \geq N(1-b)t \Big/ \left\|\bar{f}\right\|_{\nu,\infty}\right] .$$

Next apply Hoeffding's inequality (Theorem 9.1.6) to both terms on the right-hand side to obtain

$$\mathrm{P}\left[\left|\widehat{\mu}^{\mathrm{IS}}_{\nu,N}(f) - \mu(f)\right| \geq t\right] \leq 2\exp\left\{-2Nb^2t^2 \Big/ \mathrm{osc}^2_\nu\left[(d\mu/d\nu)\bar{f}\right]\right\} \\ + 2\exp\left[-2N(1-b)^2t^2 \Big/ \left\|d\mu/d\nu\right\|^2_{\nu,\infty}\left\|\bar{f}\right\|^2_{\nu,\infty}\right] , \quad (9.13)$$

where the fact that $\mathrm{osc}_\nu(d\mu/d\nu) \leq \|d\mu/d\nu\|_{\nu,\infty}$ (as $d\mu/d\nu$ is positive) has been used. Now note that when $\bar{f}$ is such that $\left\|\bar{f}\right\|_{\nu,\infty} = (1/2)\,\mathrm{osc}_\nu(f)$, $\mathrm{osc}_\nu\left[(d\mu/d\nu)\bar{f}\right] \leq \|d\mu/d\nu\|_{\nu,\infty}\mathrm{osc}_\nu(f)$. Hence to equate both terms on the right-hand side of (9.13) we set $b = 2/3$ which gives (9.12). □

9.2 Sampling Importance Resampling

9.2.1 The Algorithm

In this section, we study the sampling importance resampling (SIR) technique, introduced by Rubin (1987, 1988). It enables drawing an *asymptotically independent* sample $\xi^1, \ldots, \xi^N$ from a *target distribution* μ. The method requires that we know an *instrumental distribution* ν satisfying $\mu \ll \nu$ and such that the Radon-Nikodym derivative $d\mu/d\nu$ is known up to a normalizing factor. Therefore either μ or ν, or both, may be known up to a normalizing constant only. A tacit assumption is that sampling from the instrumental distribution ν is doable.

The SIR method proceeds in two steps. In the *sampling stage*, we draw an i.i.d. sample $\xi^1, \ldots, \xi^M$ from the instrumental distribution ν. The size M of this intermediate sample is usually taken to be larger, and sometimes much larger, than the size $\tilde{M}$ of the final sample. In the *resampling stage*, we draw a sample $\tilde{\xi}^1, \ldots, \tilde{\xi}^{\tilde{M}}$ of size $\tilde{M}$ from the instrumental sample $\xi^1, \ldots, \xi^M$. There are several ways of implementing this basic idea, the most obvious approach being to sample with replacement with a probability of picking each ξ^i, $i = 1, \ldots, M$, that is proportional to its importance weight $\frac{d\mu}{d\nu}(\xi^i)$. That is, $\tilde{\xi}^i = \xi^{I^i}$ for $i = 1, \ldots, \tilde{M}$, where $I^1, \ldots, I^{\tilde{M}}$ are conditionally independent given the instrumental sample and with distribution

$$\mathrm{P}(I^1 = i \,|\, \xi^1, \ldots, \xi^M) = \frac{\frac{d\mu}{d\nu}(\xi^i)}{\sum_{j=1}^{M}\frac{d\mu}{d\nu}(\xi^j)} .$$

For any measurable real-valued function f, we may associate to this sample an estimator $\hat{\mu}^{\mathrm{SIR}}_{\nu,\tilde{M}}(f)$ of $\mu(f)$, defined as the Monte Carlo estimator of $\mu(f)$ associated to the resampled particles $\tilde{\xi}^1, \ldots, \tilde{\xi}^{\tilde{M}}$,

$$\hat{\mu}^{\mathrm{SIR}}_{\nu,\tilde{M}}(f) = \frac{1}{\tilde{M}} \sum_{i=1}^{\tilde{M}} f(\tilde{\xi}^i) = \frac{1}{\tilde{M}} \sum_{i=1}^{M} N^i f(\xi^i) . \tag{9.14}$$

Here N^i is the total number of times that ξ^i was selected from the instrumental sample. Thus $(N^1, \ldots, N^M)$ have a multinomial distribution with

$$\mathrm{E}[N^i \,|\, \xi^1, \ldots, \xi^M] = \tilde{M} \frac{\frac{d\mu}{d\nu}(\xi^i)}{\sum_{j=1}^M \frac{d\mu}{d\nu}(\xi^i)} , \qquad i = 1, \ldots, M .$$

The conditional expectation of the SIR estimate with respect to the instrumental sample equals the (self-normalized) importance sampling estimator provided by this sample,

$$\mathrm{E}[\hat{\mu}^{\mathrm{SIR}}_{\nu,\tilde{M}}(f) \,|\, \xi^1, \ldots, \xi^M] = \sum_{i=1}^M \frac{\frac{d\mu}{d\nu}(\xi^i)}{\sum_{i=1}^M \frac{d\mu}{d\nu}(\xi^i)} f(\xi^i) .$$

The asymptotic analysis of the SIR estimator involves more sophisticated arguments however, because $\tilde{\xi}^1, \ldots, \tilde{\xi}^{\tilde{M}}$ is *not* an i.i.d. sample from μ. Nevertheless, for any measurable bounded real-valued function f on X and $j = 1, \ldots, \tilde{M}$,

$$\mathrm{E}[f(\tilde{\xi}^j) \,|\, \xi^1, \ldots, \xi^M] = \sum_{i=1}^M \frac{\frac{d\mu}{d\nu}(\xi^i)}{\sum_{j=1}^M \frac{d\mu}{d\nu}(\xi^j)} f(\xi^i) \stackrel{\mathrm{P}}{\longrightarrow} \mu(f) ,$$

where the convergence follows from Theorem 9.1.8. Because the conditional expectation on the left-hand side is bounded by $\|f\|_\infty$, we can take expectations of both sides and appeal to dominated convergence to conclude that $\mathrm{E}[f(\tilde{\xi}^j)] \to \mu(f)$ as $M \to \infty$. This shows that, whereas *marginally* the $\tilde{\xi}^i$ are not distributed according to μ, the distribution of any $\tilde{\xi}^i$ is *asymptotically* correct in the sense that for any i, the marginal distribution of $\tilde{\xi}^i$ converges to the target distribution μ as $M \to \infty$. In the same way, for any $i \neq j$ and $f, g \in \mathcal{F}_{\mathrm{b}}(\mathsf{X})$ we have

$$\begin{aligned}
\mathrm{E}[f(\tilde{\xi}^i) g(\tilde{\xi}^j)] &= \mathrm{E}[\mathrm{E}[f(\tilde{\xi}^i) g(\tilde{\xi}^j) \,|\, \xi^1, \ldots, \xi^M]] \\
&= \mathrm{E}[\mathrm{E}[f(\tilde{\xi}^i) \,|\, \xi^1, \ldots, \xi^M] \; \mathrm{E}[g(\tilde{\xi}^j) \,|\, \xi^1, \ldots, \xi^M]] \\
&= \mathrm{E}[\widehat{\mu}^{\mathrm{IS}}_{\nu,M}(f) \; \widehat{\mu}^{\mathrm{IS}}_{\nu,M}(g)] .
\end{aligned}$$

Repeating the argument above shows that $\mathrm{E}[f(\tilde{\xi}^i) g(\tilde{\xi}^j)] \to \mu(f)\mu(g)$. Thus, whereas the random variables $\tilde{\xi}^i$ and $\tilde{\xi}^j$ for $i \neq j$ are *not* independent for any given sample size M, they are *asymptotically* independent as the sample size M goes to infinity.

The estimation error $\hat{\mu}^{\mathrm{SIR}}_{\nu,M}(f) - \mu(f)$ can be decomposed into two terms,

$$\hat{\mu}^{\mathrm{SIR}}_{\nu,\tilde{M}}(f) - \mu(f) = \hat{\mu}^{\mathrm{SIR}}_{\nu,\tilde{M}}(f) - \widehat{\mu}^{\mathrm{IS}}_{\nu,M}(f) + \widehat{\mu}^{\mathrm{IS}}_{\nu,M}(f) - \mu(f) . \tag{9.15}$$

The first term on the right-hand side is the error associated with the approximation of the importance sampling estimator $\widehat{\mu}^{\mathrm{IS}}_{\nu,M}(f)$ by its sampled version $\hat{\mu}^{\mathrm{SIR}}_{\nu,\tilde{M}}(f)$. The second term is the error associated to the importance sampling estimator. To obtain asymptotic results, we now assume that the instrumental and final sample sizes are non-decreasing sequences of integers, denoted by $\{M_N\}$ and $\{\tilde{M}_N\}$, respectively, both diverging to infinity. As shown in Theorem 9.2.15, when $\mu(|f|) < \infty$, these two error terms go to zero and therefore $\hat{\mu}^{\mathrm{SIR}}_{\nu,\tilde{M}_N}(f)$ is a consistent estimator of $\mu(f)$.

The next question to answer in the elementary asymptotic theory developed in this chapter is to find conditions upon which $a_N\{\hat{\mu}^{\mathrm{SIR}}_{\nu,\tilde{M}_N}(f) - \mu(f)\}$ is asymptotically normal; here $\{a_N\}$, the rate sequence, is a non-decreasing sequence of positive reals. Again we use the decomposition (9.15). First a conditional central limit theorem shows that, for any $f \in L^2(\mathsf{X}, \mu)$,

$$\begin{aligned} &\tilde{M}_N^{1/2}\left[\hat{\mu}^{\mathrm{SIR}}_{\nu,\tilde{M}_N}(f) - \widehat{\mu}^{\mathrm{IS}}_{\nu,M_N}(f)\right] \\ &\quad = \tilde{M}_N^{-1/2}\left\{\sum_{i=1}^{\tilde{M}_N} f(\tilde{\xi}^i) - \mathrm{E}[f(\tilde{\xi}^i) \,|\, \xi^1, \dots, \xi^{M_N}]\right\} \xrightarrow{\mathcal{D}} \mathrm{N}\left(0, \mathrm{Var}_\mu(f)\right) . \end{aligned}$$

Note that $\mathrm{N}(0, \mathrm{Var}_\mu(f))$ is the limiting distribution of the plain Monte Carlo estimator of $\mu(f)$ from an i.i.d. sample from μ. Theorem 9.1.8 shows that if $(1 + f^2)(d\mu/d\nu)^2$ is ν-integrable, then

$$M_N^{1/2}\left\{\widehat{\mu}^{\mathrm{IS}}_{\nu,M_N}(f) - \mu(f)\right\} \xrightarrow{\mathcal{D}} \mathrm{N}\left(0, \mathrm{Var}_\nu\left\{\frac{d\mu}{d\nu}[f - \mu(f)]\right\}\right) .$$

The key result, shown in Theorem 9.2.15, is that $\tilde{M}_N^{1/2}\{\hat{\mu}^{\mathrm{SIR}}_{\nu,\tilde{M}_N}(f) - \widehat{\mu}^{\mathrm{IS}}_{\nu,M_N}(f)\}$ and $M_N^{1/2}\{\widehat{\mu}^{\mathrm{IS}}_{\nu,M_N}(f) - \mu(f)\}$ are asymptotically independent.

In many circumstances, and in particular when studying the resampling step in sequential or iterative applications of the SIR algorithm (such as in the sequential Monte Carlo framework), it is convenient to relax the conditions on the instrumental sample $\xi^1, \dots, \xi^M$. In addition, it is of interest to consider weighted samples $(\xi^1, \omega^1), \dots, (\xi^M, \omega^M)$, where ω^i are non-negative (importance) weights. We now proceed by introducing precise definitions and notations and then present the main results.

9.2.2 Definitions and Notations

Let $\{M_N\}_{N \geq 0}$ be a sequence of positive integers. Throughout this section, we use the word *triangular array* to refer to a system $\{U^{N,i}\}_{1 \leq i \leq M_N}$ of random variables defined on a common probability space $(\Omega, \mathcal{F}, \mathrm{P})$ and organized as follows:

$$\begin{array}{cccccc} U^{1,1} & U^{1,2} & \dots & U^{1,M_1} & & \\ U^{2,1} & U^{2,2} & \dots & \dots & U^{2,M_2} & \\ U^{3,1} & U^{3,2} & \dots & \dots & \dots & U^{3,M_3} \\ \vdots & \vdots & \vdots & \vdots & \vdots & \ddots \end{array}$$

The row index N ranges over $1, 2, 3, \dots$ while the column index i ranges from 1 to M_N, where M_N is a sequence of integers satisfying $\lim_{N \to \infty} M_N = \infty$. It will usually be the case that $M_1 < M_2 < \dots$; hence the term *triangular.* It is not necessary to assume this, however. It is not assumed that the random variables within each row are independent nor that they are identically distributed. We assume nothing about the relation between the random variables on different rows.

Let $\{\mathcal{G}^N\}_{N \geq 0}$ be a sequence of sub-σ-fields of $\mathcal{F}$. We say that a triangular array $\{U^{N,i}\}_{1 \leq i \leq M_N}$ is *measurable* with respect to this sequence if for any N the random variables $U^{N,1}, \dots, U^{N,M_N}$ are $\mathcal{G}^N$-measurable. We say that the triangular array $\{U^{N,i}\}_{1 \leq i \leq M_N}$ is *conditionally independent given* $\{\mathcal{G}^N\}$ if for any N the random variables $U^{N,1}, \dots, U^{N,M_N}$ are conditionally independent given $\mathcal{G}^N$. The term *conditionally i.i.d. given* $\{\mathcal{G}^N\}$ is defined in an entirely similar manner.

In the sequel, we will need a number of technical results regarding triangular arrays. To improve readability of the text, however, these results are gathered at the end of the chapter, in Section 9.5.1.

Definition 9.2.1 (Weighted Sample). *A triangular array of random variables* $\{(\xi^{N,i}, \omega^{N,i})\}_{1 \leq i \leq M_N}$ *is said to be a* weighted sample *if for any* $N \geq 1$, $\omega^{N,i} \geq 0$ *for* $i = 1, \dots, M_N$ *and* $\sum_{i=1}^{M_N} \omega^{N,i} > 0$ *a.s.*

Let us now consider specifically the case when the variables $\xi^{N,i}$ take values in the space X. Assume that the weighted sample $\{(\xi^{N,i}, \omega^{N,i})\}_{1 \leq i \leq M_N}$ approximates the instrumental distribution ν in the sense that for any f in an appropriately defined class of functions, $W_N^{-1} \sum_{i=1}^{M_N} \omega^{N,i} f(\xi^{N,i})$, with $W_N = \sum_{i=1}^{M_N} \omega^{N,i}$ being the normalization factor, converges in an appropriately defined sense to $\nu(f)$ as N tends to infinity. The most elementary way to assess this convergence consists in requiring that $W_N^{-1} \sum_{i=1}^{M_N} \omega^{N,i} f(\xi^{N,i})$ converges to $\nu(f)$ in probability for functions f in some class C of real-valued functions on X.

Definition 9.2.2 (Consistent Weighted Sample). *The weighted sample* $\{(\xi^{N,i}, \omega^{N,i})\}_{1 \leq i \leq M_N}$ *is said to be a* consistent *for the probability measure* ν *and the set* $\mathsf{C} \subseteq L^1(\mathsf{X}, \nu)$ *if for any* $f \in \mathsf{C}$,

$$\sum_{i=1}^{M_N} \frac{\omega^{N,i}}{\sum_{j=1}^{M_N} \omega^{N,j}} f\left(\xi^{N,i}\right) \stackrel{\mathrm{P}}{\longrightarrow} \nu(f) \quad \text{as } N \to \infty .$$

In order to obtain sensible results, we restrict our attention to classes of sets that are sufficiently rich.

Definition 9.2.3 (Proper Set). *A set C of real-valued measurable functions on X is said to be* proper *if the following conditions are satisfied.*

(i) C is a linear space: for any f and g in C and reals α and β, $\alpha f + \beta g \in \mathsf{C}$.
(ii) If $|g| \in \mathsf{C}$ and f is measurable with $|f| \leq |g|$, then $|f| \in \mathsf{C}$.

For any function f, define the positive and negative parts of it by

$$f^+ \stackrel{\text{def}}{=} f \vee 0 \quad \text{and} \quad f^- \stackrel{\text{def}}{=} (-f) \vee 0 ,$$

and note that f^+ and f^- are both dominated by $|f|$. Thus, if $|f| \in \mathsf{C}$, then f^+ and f^- both belong to C and so does $f = f^+ - f^-$. It is easily seen that for any $p \geq 0$ and any measure μ on $(\mathsf{X}, \mathcal{X})$, the set $L_p(\mathsf{X}, \mu)$ is proper.

There are many different ways to obtain a consistent weighted sample. An i.i.d. sample $\{\xi^{N,i}\}_{1 \leq i \leq M_N}$ with common distribution ν is consistent for $(\nu, L^1(\mathsf{X}, \nu))$, and $\{(\xi^{N,i}, \frac{d\mu}{d\nu}(\xi^{N,i})\}_{1 \leq i \leq M_N}$ is consistent for $(\mu, L_1(\mathsf{X}, \mu))$. Of course, when dealing with such elementary situations, the use of triangular arrays can be avoided. Triangular arrays come naturally into play when considering iterated applications of the SIR algorithm, as in sequential importance sampling techniques. In this case, the weighted sample $\{(\xi^{N,i}, \omega^{N,i}\}_{1 \leq i \leq M_N}$ is the result of iterated applications of importance sampling, resampling, and propagation steps. We study several examples of such situations later in this chapter.

The notion of sample consistency is weak but is in practice only moderately helpful, because it does not indicate the rate at which the estimator $W_N^{-1} \sum_{i=1}^N \omega^{N,i} f(\xi^{N,i})$ converges to $\nu(f)$. In particular, this definition does not provide a way to construct an asymptotic confidence interval for $\nu(f)$. A natural way to strengthen it is to consider distributional convergence of the normalized difference $a_N \sum_{i=1}^{M_N} \frac{\omega^{N,i}}{W_N} \{f(\xi^{N,i}) - \nu(f)\}$.

Definition 9.2.4 (Asymptotically Normal Weighted Sample). *Let A be a class of real-valued measurable functions on X, let σ be a real non-negative function on A, and let $\{a_N\}$ be a non-decreasing real sequence diverging to infinity. We say that the weighted sample $\{(\xi^{N,i}, \omega^{N,i})\}_{1 \leq i \leq M_N}$ is* asymptotically normal *for $(\nu, \mathsf{A}, \sigma, \{a_N\})$ if for any function $f \in \mathsf{A}$ it holds that $\nu(|f|) < \infty$, $\sigma^2(f) < \infty$ and*

$$a_N \sum_{i=1}^{M_N} \frac{\omega^{N,i}}{\sum_{j=1}^{M_N} \omega^{N,j}} \left[f(\xi^{N,i}) - \nu(f)\right] \xrightarrow{\mathcal{D}} \mathrm{N}(0, \sigma^2(f)) \quad \text{as } N \to \infty .$$

Of course, if $\{(\xi^{N,i}, \omega^{N,i})\}_{1 \leq i \leq M_N}$ is asymptotically normal for $(\nu, \mathsf{A}, \sigma, \{a_N\})$, then it is also consistent for (ν, A). If $\{\xi^{N,i}\}_{1 \leq i \leq M_N}$ are i.i.d. with common distribution ν then for any function $f \in L^2(\mathsf{X}, \nu)$ and any non-decreasing sequence $\{M_N\}$ such that $\lim_{N\to\infty} M_N = \infty$,

$$\frac{1}{\sqrt{M_N}} \sum_{i=1}^{M_N} \left[f(\xi^{N,i}) - \nu(f)\right] \xrightarrow{\mathcal{D}} \mathrm{N}\left(0, \nu\left\{[f - \nu(f)]^2\right\}\right) .$$

Therefore $\{(\xi^{N,i}, 1)\}_{1\le i\le M_N}$ is an asymptotically normal weighted sample for $\left(\nu, L^2(\mathsf{X}, \nu), \sigma, \{\sqrt{M_N}\}\right)$ with $\sigma^2(f) = \nu\left(\{f - \nu(f)\}^2\right)$. In the context of importance sampling, for each N we draw $\{\xi^{N,i}\}_{1\le i\le M_N}$ independently from the instrumental distribution ν and assign it weights $\{\frac{d\mu}{d\nu}(\xi^{N,i})\}_{1\le i\le M_N}$. Using an argument as in the proof of Theorem 9.1.8, it also follows that $\{(\xi^{N,i}, \frac{d\mu}{d\nu}(\xi^{N,i}))\}_{1\le i\le M_N}$ is an asymptotically normal weighted sample for $(\mu, \mathsf{A}, \sigma, \{\sqrt{M_N}\})$, with

$$\mathsf{A} = \left\{ f \in L^2(\mathsf{X}, \mu) : \nu\left(\left\{\frac{d\mu}{d\nu}[f - \mu(f)]\right\}^2\right) < \infty \right\}$$

and

$$\sigma^2(f) = \nu\left(\left\{\frac{d\mu}{d\nu}[f - \mu(f)]\right\}^2\right), \qquad f \in \mathsf{A} .$$

When the SIR algorithm is applied sequentially, the rate $\{a_N\}$ can be different from $\sqrt{M_N}$ because of the dependence among the random variables $\{\xi^{N,i}\}_{1\le i\le M_N}$ introduced by the resampling procedure.

9.2.3 Weighting and Resampling

Assume that $\{(\xi^{N,i}, 1)\}_{1\le i\le M_N}$ is an i.i.d. sample from the instrumental distribution ν. In the first stage of the SIR procedure, we assign to these samples importance weights $\frac{d\mu}{d\nu}(\xi^{N,i})$, $i = 1, \dots, M_N$, where μ is the target distribution, assumed to be absolutely continuous with respect to ν. We then draw, conditionally independently given $\mathcal{F}^N = \sigma(\{\xi^{N,1}, \dots, \xi^{N,M_N}\})$, random variables $I^{N,1}, \dots, I^{N,\tilde{M}_N}$ with distribution $\mathrm{P}(I^{N,k} = \xi^{N,i} \mid \mathcal{F}^N) = \frac{d\mu}{d\nu}(\xi^{N,i})$ and let $\tilde{\xi}^{N,i} = \xi^{N,I^{N,i}}$ for $i = 1, \dots, \tilde{M}_N$. Proceeding this way, we thus define a weighted sample $\{(\tilde{\xi}^{N,i}, 1)\}_{1\le i\le \tilde{M}_N}$. As outlined in the discussion above, we know that $\{(\xi^{N,i}, 1)\}_{1\le i\le M_N}$ is consistent for $(\nu, L^1(\mathsf{X}, \nu))$. We have already mentioned that $\{(\xi^{N,i}, \frac{d\mu}{d\nu}(\xi^{N,i}))\}_{1\le M_N}$ is consistent for $(\mu, L^1(\mathsf{X}, \mu))$; therefore the weighting operation transforms a weighted sample consistent for $(\nu, L^1(\mathsf{X}, \nu))$ into a weighted sample consistent for $(\mu, L^1(\mathsf{X}, \mu))$. Similarly, in the second step, the resampling operation transforms a weighted sample $\{(\xi^{N,i}, \frac{d\mu}{d\nu}(\xi^{N,i}))\}_{1\le i\le M_N}$ into another one $\{(\tilde{\xi}^{N,i}, 1)\}_{1\le i\le \tilde{M}_N}$. It is a natural question to ask whether the latter one is consistent for μ and, if so, what an appropriately defined class of functions on X might be. Of course, in this discussion it is also sensible to strengthen the requirement of consistency into asymptotic normality and again prove that the weighting and resampling operations transform an asymptotically normal weighted sample for ν into an asymptotically normal sample for μ (for appropriately defined class of functions, normalizing factors, etc.)

The main purpose of this section is to establish such results. Because we apply these results in a sequential context, we start from a weighted sample $\{(\xi^{N,i}, \omega^{N,i}\}_{1 \le i \le M_N}$, with weights $\omega^{N,i}$ that are not necessarily identical. Also, we do not assume that $\{\xi^{N,i}\}_{1 \le i \le M_N}$ are conditionally i.i.d. with distribution ν. In addition, we denote by $\{\mathcal{G}^N\}$ a sequence of sub-σ-fields of $\mathcal{F}$. When studying the single-stage SIR estimator, one may simply set, for any $N \ge 0$, $\mathcal{G}^N$ equal to the trivial σ-field $\{\emptyset, \Omega\}$. Indeed, the use of $\{\mathcal{G}^N\}_{N \ge 0}$ is a provision for situations in which the SIR algorithm is applied sequentially; $\{\mathcal{G}^N\}_{N \ge 0}$ handles the history of the particle system up to the current iteration.

Algorithm 9.2.5 (Weighting and Resampling).

Resampling: Draw random variables $\{I^{N,1}, \ldots, I^{N,\tilde{M}_N}\}$ conditionally independently given

$$\mathcal{F}^N = \mathcal{G}^N \vee \sigma\left\{(\xi^{N,1}, \omega^{N,1}), \ldots, (\xi^{N,M_N}, \omega^{N,M_N})\right\} , \tag{9.16}$$

with probabilities proportional to $\omega^{N,1} \frac{d\mu}{d\nu}(\xi^{N,1}), \ldots, \omega^{N,M_N} \frac{d\mu}{d\nu}(\xi^{N,M_N})$. In other words, for $k = 1, \ldots, \tilde{M}_N$,

$$\mathrm{P}(I^{N,k} = i \,|\, \mathcal{F}^N) = \frac{\omega^{N,i} \frac{d\mu}{d\nu}(\xi^{N,i})}{\sum_{j=1}^{M_N} \omega^{N,j} \frac{d\mu}{d\nu}(\xi^{N,j})} , \qquad i = 1, \ldots, M_N .$$

Assignment: For $i = 1, \ldots, \tilde{M}_N$, set

$$\tilde{\xi}^{N,i} = \xi^{N,I^{N,i}} . \tag{9.17}$$

We now study in which sense the weighted sample $\{(\tilde{\xi}^{N,i}, 1)\}_{1 \le i \le \tilde{M}_N}$ approximates the target distribution μ. Consider the following assumption.

Assumption 9.2.6. *$\{(\xi^{N,i}, \omega^{N,i})\}_{1 \le i \le M_N}$ is consistent for (ν, C), where C is a proper set of functions. In addition, $d\mu/d\nu \in \mathsf{C}$.*

The following theorem is an elementary extension of Theorem 9.1.8. It shows that the if the original weighted sample of Algorithm 9.2.5 is consistent for ν, then the reweighted sample is consistent for μ.

Theorem 9.2.7. *Assume 9.1.1 and 9.2.6. Then*

$$\tilde{\mathsf{C}} \stackrel{\text{def}}{=} \left\{ f \in L^1(\mathsf{X}, \mu) : |f| \frac{d\mu}{d\nu} \in \mathsf{C} \right\} \tag{9.18}$$

is a proper set of functions and $\{(\xi^{N,i}, \omega^{N,i} \frac{d\mu}{d\nu}(\xi^{N,i}))\}_{1 \le i \le M_N}$ is consistent for $(\mu, \tilde{\mathsf{C}})$.

Proof. It is easy to check that $\tilde{\mathsf{C}}$ is proper. Because $\{(\xi^{N,i}, \omega^{N,i})\}_{1 \le i \le M_N}$ is consistent for (ν, C), for any function $h \in \mathsf{C}$ it holds that

$$\sum_{i=1}^{M_N} \frac{\omega^{N,i}}{\sum_{j=1}^{M_N} \omega^{N,j}} h\left(\xi^{N,i}\right) \xrightarrow{\mathrm{P}} \nu(h) .$$

By construction $h \frac{d\mu}{d\nu} \in \mathsf{C}$ for any $h \in \tilde{\mathsf{C}}$. Therefore

$$\sum_{i=1}^{M_N} \frac{\omega^{N,i}}{\sum_{j=1}^{M_N} \omega^{N,j}} \frac{d\mu}{d\nu}(\xi^{N,i}) h(\xi^{N,i}) \xrightarrow{\mathrm{P}} \nu\left(h \frac{d\mu}{d\nu}\right) = \mu(h) . \tag{9.19}$$

The proof is concluded by applying (9.19) with $h \equiv 1$ and $h = f$. □

The next step is to show that the sample $\{\tilde{\xi}^{N,i}\}$, which is the result of the resampling operation, is consistent for μ as well. The key result to proving this is the following theorem, which establishes a conditional weak law of large numbers for conditionally independent random variables under easily verified technical conditions.

Theorem 9.2.8. *Let μ be a probability distribution on $(\mathsf{X}, \mathcal{X})$ and let f be in $L^1(\mathsf{X}, \mu)$. Assume that the triangular array $\{\xi^{N,i}\}_{1 \le i \le M_N}$ is conditionally independent given $\{\mathcal{F}^N\}$ and that for any non-negative C,*

$$\frac{1}{M_N} \sum_{i=1}^{M_N} \mathrm{E}\left[|f|(\xi^{N,i}) \mathbb{1}_{\{|f|(\xi^{N,i}) \ge C\}} \mid \mathcal{F}^N \right] \xrightarrow{\mathrm{P}} \mu\left(|f| \mathbb{1}_{\{|f| \ge C\}} \right) . \tag{9.20}$$

Then

$$\frac{1}{M_N} \sum_{i=1}^{M_N} \left(f(\xi^{N,i}) - \mathrm{E}\left[f(\xi^{N,i}) \mid \mathcal{F}^N \right] \right) \xrightarrow{\mathrm{P}} 0 . \tag{9.21}$$

Proof. We have to check conditions (ii)–(iii) of Proposition 9.5.7. Set $V_{N,i} = M_N^{-1} f(\xi^{N,i})$ for any N and $i = 1, \ldots, M_N$, By construction, the triangular array $\{V_{N,i}\}$ is conditionally independent given $\{\mathcal{F}^N\}$ and $\mathrm{E}[|V_{N,i} \mid \mathcal{F}^N] < \infty$. Equation (9.20) with $C = 0$ shows that

$$\sum_{i=1}^{M_N} \mathrm{E}\left[|V_{N,i}| \mid \mathcal{F}^N \right] \le M_N^{-1} \sum_{i=1}^{M_N} \mathrm{E}\left[|f(\xi^{N,i})| \mid \mathcal{F}^N \right] \xrightarrow{\mathrm{P}} \mu(|f|) < \infty ,$$

whence the sequence $\{\sum_{i=1}^{M_N} \mathrm{E}[|V_{N,i}| \mid \mathcal{F}^N\}_{N \ge 0}$ is bounded in probability [condition (ii)]. Next, for any positive ϵ and C we have for sufficiently large N,

$$\sum_{i=1}^{M_N} \mathrm{E}\left[|V_{N,i}| \mathbb{1}_{\{|V_{N,i}| \geq \epsilon\}} \mid \mathcal{F}^N \right]$$
$$= \frac{1}{M_N} \sum_{i=1}^{M_N} \mathrm{E}\left[|f(\xi^{N,i})| \mathbb{1}_{\{|f|(\xi^{N,i}) \geq \epsilon M_N\}} \mid \mathcal{F}^N \right]$$
$$\leq M_N^{-1} \sum_{i=1}^{M_N} \mathrm{E}\left[|f(\xi^{N,i})| \mathbb{1}_{\{|f|(\xi^{N,i}) \geq C\}} \mid \mathcal{F}^N \right] \xrightarrow{\mathrm{P}} \mu(|f| \mathbb{1}_{\{|f| \geq C\}}) \,.$$

By dominated convergence, the right-hand side of this display tends to zero as $C \to \infty$. Thus, the left-hand side of the display converges to zero in probability, which is condition (iii). □

We can now prove that the resampled particles are consistent for μ.

Theorem 9.2.9. *Let* $\{(\tilde{\xi}^{N,i}, 1)\}_{1 \leq i \leq \tilde{M}_N}$ *be as in Algorithm 9.2.5 and let* $\tilde{\mathsf{C}}$ *be as in* (9.18). *Then under Assumptions 9.1.1, and 9.2.6,* $\{(\tilde{\xi}^{N,i}, 1)\}_{1 \leq i \leq \tilde{M}_N}$ *is consistent for* $(\mu, \tilde{\mathsf{C}})$.

Proof. We will apply Theorem 9.2.8 and thus need to verify its assumptions. By construction, $\{\tilde{\xi}^{N,i}\}_{1 \leq i \leq \tilde{M}_N}$ is conditionally independent given $\mathcal{F}^N$. Pick f in $\tilde{\mathsf{C}}$. Because $\tilde{\mathsf{C}}$ is proper, $|f| \mathbb{1}_{\{|f| \geq C\}} \in \tilde{\mathsf{C}}$ for any $C \geq 0$. Therefore

$$\frac{1}{\tilde{M}_N} \sum_{i=1}^{\tilde{M}_N} \mathrm{E}\left[|f|(\tilde{\xi}^{N,i}) \mathbb{1}_{\{|f|(\tilde{\xi}^{N,i}) \geq C\}} \,\Big|\, \mathcal{F}^N \right]$$
$$= \sum_{i=1}^{M_N} \frac{\omega^{N,i} \frac{d\mu}{d\nu}(\xi^{N,i})}{\sum_{j=1}^{M_N} \omega^{N,j} \frac{d\mu}{d\nu}(\xi^{N,j})} |f|(\xi^{N,i}) \mathbb{1}_{\{|f|(\xi^{N,i}) \geq C\}} \xrightarrow{\mathrm{P}} \mu(|f| \mathbb{1}_{\{|f| \geq C\}}) \,,$$

where the convergence follows from Theorem 9.2.7. Thus Theorem 9.2.8 applies, and taking $C = 0$, it allows us to conclude that $\tilde{M}_N^{-1} \sum_1^{\tilde{M}_N} f(\tilde{\xi}^{N,i})$ converges to $\mu(f)$ in probability for any non-negative f. By dividing a general f in $\tilde{\mathsf{C}}$ into its positive and negative parts, we see that the same conclusion holds true for such f. □

Our next objective is to establish asymptotic normality of the resampled particles $\{(\tilde{\xi}^{N,i}, 1)\}$. Consider the following assumption.

Assumption 9.2.10. *The weighted sample* $\{(\xi^{N,i}, \omega^{N,i})\}_{1 \leq i \leq M_N}$ *is asymptotically normal for* $(\nu, \mathsf{A}, \sigma, \{a_N\})$, *where* A *is a proper set of functions,* σ *is a non-negative function on* A, *and* $\{a_N\}$ *is a non-decreasing sequence of positive constants diverging to infinity. In addition,* $\frac{d\mu}{d\nu} \in \mathsf{A}$.

We proceed in two steps. In a first step, we strengthen the conclusions of Theorem 9.1.8 to show that the reweighted sample $\{(\xi^{N,i}, \omega^{N,i} \frac{d\mu}{d\nu}(\xi^{N,i}))\}_{1 \leq i \leq M_N}$ is asymptotically normal. Then we show that the sampling operation preserves asymptotic normality.

Theorem 9.2.11. *Assume 9.1.1, 9.2.6, and 9.2.10 and define*

$$\bar{\mathsf{A}} \stackrel{\text{def}}{=} \left\{ f \in L^2(\mathsf{X}, \mu) \;:\; |f| \frac{d\mu}{d\nu} \in \mathsf{A} \right\} .$$

Then $\bar{\mathsf{A}}$ is a proper set and the weighted sample $\{(\xi^{N,i}, \omega^{N,i} \frac{d\mu}{d\nu}(\xi^{N,i}))\}_{1 \le i \le M_N}$ is asymptotically normal for $(\mu, \bar{\mathsf{A}}, \bar{\sigma}, \{a_N\})$ with

$$\bar{\sigma}^2(f) = \sigma^2 \left\{ \frac{d\mu}{d\nu} [f - \mu(f)] \right\} .$$

Proof. Once again it is easy to see that A is proper. Pick f in $\bar{\mathsf{A}}$. Under the stated assumptions, $\frac{d\mu}{d\nu} \in \mathsf{A}$ and $f \frac{d\mu}{d\nu} \in \mathsf{A}$. Therefore $\mu(|f|) = \nu(|f| \frac{d\mu}{d\nu}) < \infty$, showing that $f \in L^1(\mathsf{X}, \mu)$. In addition, again as A is a proper, $h = \frac{d\mu}{d\nu}\{f - \mu(f)\} \in \mathsf{A}$. By construction, $\nu(h) = 0$. Write

$$a_N \left\{ \sum_{i=1}^{M_N} \frac{\omega^{N,i} \frac{d\mu}{d\nu}(\xi^{N,i})}{\sum_{j=1}^{M_N} \omega^{N,j} \frac{d\mu}{d\nu}(\xi^{N,j})} \left[f(\xi^{N,i}) - \mu(f) \right] \right\} = \frac{a_N \sum_{i=1}^{M_N} \omega^{N,i} h(\xi^{N,i})}{\sum_{i=1}^{N} \omega^{N,i} \frac{d\mu}{d\nu}(\xi^{N,i})} .$$

Because the weighted sample $\{(\xi^{N,i}, \omega^{N,i}\}_{1 \le i \le M_N}$ is asymptotically normal for $(\nu, \mathsf{A}, \sigma, \{a_N\})$, $h \in \mathsf{A}$, and $\nu(h) = 0$, we conclude that

$$a_N \sum_{i=1}^{M_N} \frac{\omega^{N,i}}{\sum_{j=1}^{M_N} \omega^{N,j}} h(\xi^{N,i}) \xrightarrow{\mathcal{D}} \mathrm{N}\left(0, \sigma^2(h)\right)$$

and note that $\sigma^2(h) = \bar{\sigma}^2(f)$. Moreover, because the same weighted sample is consistent for ν,

$$\sum_{i=1}^{M_N} \frac{\omega^{N,i}}{\sum_{j=1}^{N} \omega^{N,j}} \frac{d\mu}{d\nu}(\xi^{N,i}) \xrightarrow{\mathrm{P}} \nu\left(\frac{d\mu}{d\nu}\right) = 1 .$$

The proof now follows by Slutsky's theorem (Billingsley, 1995). □

In order to proceed to asymptotic normality after resampling, we need some preparatory results. The following proposition establishes a conditional CLT for triangular arrays of conditionally independent random variables. It is an almost direct application of Theorem 9.5.13, which is stated and proved in Section 9.5.1.

Proposition 9.2.12. *Assume 9.1.1 and 9.2.6. Then for any $u \in \mathbb{R}$ and any function f such that $f^2 \frac{d\mu}{d\nu} \in \mathsf{C}$,*

$$\mathrm{E}\left[\exp\left(\mathrm{i}u \, \tilde{M}_N^{-1/2} \sum_{i=1}^{\tilde{M}_N} \{ f(\tilde{\xi}^{N,i}) - \mathrm{E}[f(\tilde{\xi}^{N,i}) \,|\, \mathcal{F}^N] \} \right) \,\middle|\, \mathcal{F}^N \right]$$
$$\xrightarrow{\mathrm{P}} \exp\left(-(u^2/2) \operatorname{Var}_\mu(f) \right) , \quad (9.22)$$

where $\{\mathcal{F}^N\}$ and $\{\tilde{\xi}^{N,i}\}_{1 \le i \le \tilde{M}_N}$ are defined in (9.16) *and* (9.17), *respectively.*

Corollary 9.2.13. *Assume 9.1.1 and 9.2.6. Then*

$$\tilde{M}_N^{-1/2} \sum_{i=1}^{\tilde{M}_N} \{f(\tilde{\xi}^{N,i}) - \mathrm{E}[f(\tilde{\xi}^{N,i}) \,|\, \mathcal{F}^N]\} \xrightarrow{\mathcal{D}} \mathrm{N}(0, \mathrm{Var}_\mu(f)) \,. \tag{9.23}$$

Proof (of Proposition 9.2.12). We will appeal to Theorem 9.5.13 and hence need to check that its conditions (ii) and (iii) are satisfied. First,

$$\mathrm{Var}[f(\tilde{\xi}^{N,1}) \,|\, \mathcal{F}^N] =$$
$$\sum_{i=1}^{M_N} \frac{\omega^{N,i} \frac{d\mu}{d\nu}(\xi^{N,i})}{\sum_{j=1}^{M_N} \omega^{N,j} \frac{d\mu}{d\nu}(\xi^{N,j})} f^2(\xi^{N,i}) - \left\{ \sum_{i=1}^{M_N} \frac{\omega^{N,i} \frac{d\mu}{d\nu}(\xi^{N,i})}{\sum_{j=1}^{M_N} \omega^{N,j} \frac{d\mu}{d\nu}(\xi^{N,j})} f(\xi^{N,i}) \right\}^2 \,.$$

The assumptions say that $\{(\xi^{N,i}, \omega^{N,i})\}_{1 \le i \le M_N}$ is consistent for (ν, C). Because $\frac{d\mu}{d\nu} \in \mathsf{C}$ and $f^2 \frac{d\mu}{d\nu} \in \mathsf{C}$, the inequality $|f| \frac{d\mu}{d\nu} \le \mathbb{1}_{\{|f| \le 1\}} \frac{d\mu}{d\nu} + f^2 \frac{d\mu}{d\nu}$ shows that $|f| \frac{d\mu}{d\nu} \in \mathsf{C}$. Theorem 9.2.7 then implies that

$$\mathrm{Var}[f(\tilde{\xi}^{N,1}) \,|\, \mathcal{F}^N] \xrightarrow{\mathrm{P}} \mu(f^2) - \{\mu(f)\}^2 = \mathrm{Var}_\mu(f) \,.$$

Condition (ii) follows. Moreover, for any positive constant C,

$$\tilde{M}_N^{-1} \sum_{i=1}^{\tilde{M}_N} \mathrm{E}[f^2(\tilde{\xi}^{N,i}) \mathbb{1}_{\{|f|(\tilde{\xi}^{N,i}) \ge C\}} \,|\, \mathcal{F}^N]$$
$$= \sum_{i=1}^{M_N} \frac{\omega^{N,i} \frac{d\mu}{d\nu}(\xi^{N,i})}{\sum_{j=1}^{M_N} \omega^{N,j} \frac{d\mu}{d\nu}(\xi^{N,j})} f^2(\xi^{N,i}) \mathbb{1}_{\{|f|(\xi^{N,i}) \ge C\}} \,.$$

Because $f^2 \frac{d\mu}{d\nu}$ belongs to the proper set C, we have $f^2 \mathbb{1}_{\{|f| \ge C\}} \frac{d\mu}{d\nu} \in \mathsf{C}$. This implies that the right-hand side of the above display converges in probability to $\mu(f^2 \mathbb{1}_{|f| \ge C})$. Hence condition (iii) also holds. □

Applying successively Theorem 9.2.11 and Proposition 9.2.12 yields the following result, showing that the resampling preserves asymptotic normality.

Theorem 9.2.14. *Assume 9.1.1, 9.2.6, and 9.2.10, and that $a_N^2 / \tilde{M}_N$ has a limit, α say, possibly infinite. Define*

$$\tilde{\mathsf{A}} \stackrel{\text{def}}{=} \left\{ f \in L^2(\mathsf{X}, \mu) : |f| \frac{d\mu}{d\nu} \in \mathsf{A},\, f^2 \frac{d\mu}{d\nu} \in \mathsf{C} \right\} , \tag{9.24}$$

where A and C are as in Assumptions 9.2.10 and 9.2.6, respectively. Then $\tilde{\mathsf{A}}$ is a proper set and the following holds true for the resampled system $\{(\tilde{\xi}^{N,i}, 1)\}_{1 \le i \le \tilde{M}_N}$ defined as in Algorithm 9.2.5.

(i) If $\alpha < 1$, then $\{(\tilde{\xi}^{N,i}, 1)\}$ is asymptotically normal for $(\mu, \tilde{\mathsf{A}}, \tilde{\sigma}, \{a_N\})$ with

$$\tilde{\sigma}^2(f) = \alpha \operatorname{Var}_\mu(f) + \sigma^2\left(\frac{d\mu}{d\nu}\{f - \mu(f)\}\right), \qquad f \in \tilde{\mathsf{A}}. \tag{9.25}$$

(ii) If $\alpha \geq 1$, then $\{(\tilde{\xi}^{N,i}, 1)\}$ is asymptotically normal for $(\mu, \tilde{\mathsf{A}}, \tilde{\sigma}, \{\tilde{M}_N^{1/2}\})$ with

$$\tilde{\sigma}^2(f) = \operatorname{Var}_\mu(f) + \alpha^{-1}\sigma^2\left(\frac{d\mu}{d\nu}\{f - \mu(f)\}\right), \qquad f \in \tilde{\mathsf{A}}. \tag{9.26}$$

Thus, we see that if $\tilde{M}_N$ increases much slower than a_N, so that $\alpha = \infty$, then the rate of convergence is $\tilde{M}_N^{1/2}$ and the limiting variance is the basic Monte Carlo variance $\operatorname{Var}_\mu(f)$. This means that a_N is so large compared to $\tilde{M}_N$ that the weighted sample $\{(\xi^{N,i}, \omega^{N,i}\frac{d\mu}{d\nu}(\xi^{N,i}))\}$ approximates μ with negligible error, and the resampled particles can effectively be thought of as an i.i.d. sample from μ. On the other hand, when $\tilde{M}_N$ increases much faster than a_N, so that $\alpha = 0$, then the rate of convergence is a_N and the limiting variance is that associated with the weighted sample $\{(\xi^{N,i}, \omega^{N,i}\frac{d\mu}{d\nu}(\xi^{N,i}))\}$ alone (see Theorem 9.2.11). This means that the size of the resample is so large that the error associated with this part of the overall procedure can be disregarded.

Proof (Theorem 9.2.14). Pick $f \in \tilde{\mathsf{A}}$ and write $\tilde{M}_N^{-1}\sum_{i=1}^{\tilde{M}_N} f(\tilde{\xi}^{N,i}) - \mu(f) = A_N + B_N$ with

$$A_N = \sum_{i=1}^{M_N} \frac{\omega^{N,i}\frac{d\mu}{d\nu}(\xi^{N,i})}{\sum_{j=1}^{M_N}\omega^{N,j}\frac{d\mu}{d\nu}(\xi^{N,j})}\{f(\xi^{N,i}) - \mu(f)\},$$

$$B_N = \tilde{M}_N^{-1}\sum_{i=1}^{\tilde{M}_N}\{f(\tilde{\xi}^{N,i}) - \mathrm{E}[f(\tilde{\xi}^{N,i}) \mid \mathcal{F}^N]\}.$$

Under the stated assumptions, Proposition 9.2.11 shows that

$$a_N A_N \xrightarrow{\mathcal{D}} \mathrm{N}\left(0, \sigma^2\left\{\frac{d\mu}{d\nu}[f - \mu(f)]\right\}\right).$$

Combining this with Proposition 9.2.12, we find that for any real numbers u and v,

$$\begin{aligned}
&\mathrm{E}\left[\exp(\mathrm{i}(u\tilde{M}_N^{1/2}B_N + va_N A_N)\right] \\
&= \mathrm{E}\left[\mathrm{E}\left[\exp(\mathrm{i}u\tilde{M}_N^{1/2}B_N)\,\middle|\,\mathcal{F}^N\right]\exp(\mathrm{i}va_N A_N)\right] \\
&\to \exp\left[-(u^2/2)\operatorname{Var}_\mu(f)\right]\exp\left(-(v^2/2)\sigma^2\left\{\frac{d\mu}{d\nu}[f - \mu(f)]\right\}\right).
\end{aligned}$$

Thus the bivariate characteristic function converges to the characteristic function of a bivariate normal, implying that

$$\begin{pmatrix} a_N A_N \\ \tilde{M}_N^{1/2} B_N \end{pmatrix} \xrightarrow{\mathcal{D}} \mathrm{N}\left(0, \begin{bmatrix} \sigma^2\left(\frac{d\mu}{d\nu}\{f - \mu[f]\}\right) & 0 \\ 0 & \mathrm{Var}_\mu\{f\} \end{bmatrix}\right) .$$

Put $b_N = a_N$ if $\alpha < 1$ and $b_N = \tilde{M}_N^{1/2}$ if $\alpha \geq 1$. The proof follows from

$$b_N(A_N + B_N) = (b_N a_N^{-1}) a_N A_N + (b_N \tilde{M}_N^{-1/2}) \tilde{M}_N^{1/2} B_N .$$

□

9.2.4 Application to the Single-Stage SIR Algorithm

We now apply the above results to the single-stage SIR algorithm, sampling from an instrumental distribution ν and then weighting and resampling to obtain an approximately i.i.d. sample from μ. The procedure is illustrated in Figure 9.4. Thus $\{\xi^{N,i}\}_{1 \leq i \leq M_N}$ is an i.i.d. sample from ν and the weights are set to 1; $\omega^{N,i} \equiv 1$. The LLN shows that Assumption 9.2.6 is satisfied with $\mathsf{C} = L^1(\mathsf{X}, \nu)$. Theorem 9.2.9 shows that for any $f \in \tilde{\mathsf{C}} = L^1(\mathsf{X}, \mu)$ (see the definition in (9.18)),

$$\frac{1}{\tilde{M}_N} \sum_{i=1}^{\tilde{M}_N} f(\tilde{\xi}^{N,i}) \xrightarrow{\mathrm{P}} \mu(f) .$$

Moreover, the weighted sample $\{(\xi^{N,i}, 1)\}_{1 \leq i \leq M_N}$ satisfies Assumption 9.2.10 with $\mathsf{A} = L^2(\mathsf{X}, \nu)$, $\sigma^2(f) = \nu\left(\{f - \nu(f)\}^2\right)$, and $a_N = M_N^{1/2}$, provided $d\mu/d\nu \in L^2(\mathsf{X}, \nu)$. Thus Theorem 9.2.14 shows that $\{(\tilde{\xi}^{N,i}, 1)\}_{1 \leq i \leq \tilde{M}_N}$ is asymptotically normal for μ. We summarize this in the following result.

Theorem 9.2.15. *Assume 9.1.1 and let $\{\xi^{N,i}\}_{1 \leq i \leq M_N}$ be i.i.d. random variables with distribution ν. Then $\{(\tilde{\xi}^{N,i}, 1)\}_{1 \leq i \leq \tilde{M}_N}$ given by Algorithm 9.2.5 is consistent for $\left(\mu, L^1(\mathsf{X}, \mu)\right)$.*

Assume in addition that $\lim_{N \to \infty} M_N / \tilde{M}_N = \alpha$ for some $\alpha \in [0, \infty]$ and that $\frac{d\mu}{d\nu} \in L^2(\mathsf{X}, \nu)$. Define $\tilde{\mathsf{A}} = \{f \in L^2(\mathsf{X}, \mu) : f \frac{d\mu}{d\nu} \in L^2(\mathsf{X}, \nu)\}$. Then the following holds true.

(i) If $\alpha < 1$, then $\{(\tilde{\xi}^{N,i}, 1))\}_{1 \leq i \leq \tilde{M}_N}$ is asymptotically normal for $(\nu, \tilde{\mathsf{A}}, \tilde{\sigma}, \{M_N^{1/2}\})$ with

$$\tilde{\sigma}^2(f) \stackrel{\text{def}}{=} \alpha \,\mathrm{Var}_\mu(f) + \mathrm{Var}_\nu \left\{ \frac{d\mu}{d\nu} [f - \mu(f)] \right\}, \qquad f \in \tilde{\mathsf{A}} .$$

(ii) If $\alpha \geq 1$, then $\{(\tilde{\xi}^{N,i}, 1))\}_{1 \leq i \leq \tilde{M}_N}$ is asymptotically normal for $(\nu, \tilde{\mathsf{A}}, \tilde{\sigma}, \{\tilde{M}_N^{1/2}\})$ with

$$\tilde{\sigma}^2(f) \stackrel{\text{def}}{=} \mathrm{Var}_\mu(f) + \alpha^{-1} \,\mathrm{Var}_\nu \left\{ \frac{d\mu}{d\nu} [f - \mu(f)] \right\}, f \in \tilde{\mathsf{A}} .$$

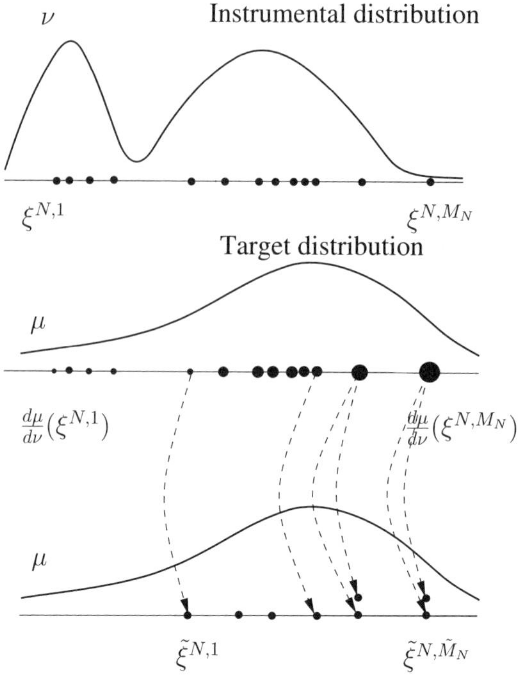

Fig. 9.4. The single-stage SIR algorithm.

Without loss of generality, we may assume here that $\tilde{M}_N = N$. To obtain a rate $\sqrt{N}$ asymptotically normal sample for the target distribution μ, the cardinality M_N of the instrumental sample should grow at least as fast as N, $\lim_{N\to\infty} M_N/N > 0$. If $\lim_{N\to\infty} M_N/N = \infty$, then

$$\sqrt{N}[\hat{\mu}^{\mathrm{SIR}}_{\nu,N}(f) - \mu(f)] \xrightarrow{\mathcal{D}} \mathrm{N}(0, \mathrm{Var}_\mu(f)) ,$$

that is, the SIR estimator and the plain Monte Carlo estimator $\hat{\mu}^{\mathrm{MC}}_N(f)$ of $\mu(f)$ (the estimator of $\mu(f)$ obtained by computing the sample average $N^{-1}\sum_{i=1}^N f(\xi^i)$ with $\{\xi^i\}$ being an i.i.d. sample from the target distribution μ) have the same limiting Gaussian distribution. In practice, this means that large values for the instrumental sample should be used when one is asking for a sample that behaves as an i.i.d. sample from μ.

We conclude this section with some elementary deviations inequalities. These inequalities are non-asymptotic and allow evaluating the performance of the SIR estimator for finite sample sizes.

Theorem 9.2.16. *Assume 9.1.1 and let $\{\xi^{N,i}\}_{1\le i\le M_N}$ be i.i.d. random variables with distribution ν. Then for any $t > 0$, $f \in \mathcal{F}_{\mathrm{b}}(\mathsf{X})$, $a \in (0,1)$, and $N \geq 0$,*

$$\mathrm{P}\left[\left|M_N^{-1}\sum_{i=1}^{M_N} f(\tilde{\xi}^{N,i}) - \mu(f)\right| \geq t\right]$$
$$\leq 2\exp\left[-2\tilde{M}_N a^2 t^2 \,/\mathrm{osc}^2(f)\right]$$
$$+ 4\exp\left[-8M_N(1-a)^2 t^2 \Big/ 9\,\mathrm{osc}^2(f)\,\|d\mu/d\nu\|^2_{\nu,\infty}\right] .$$

Proof. Decompose $\tilde{M}_N^{-1}\sum_{i=1}^{\tilde{M}_N}\{f(\tilde{\xi}^{N,i}) - \mu(f)\}$ as a sum $A^N + B^N$ of the two terms

$$A^N(f) = \tilde{M}_N^{-1}\sum_{i=1}^{\tilde{M}_N}\{f(\tilde{\xi}^{N,i}) - \mathrm{E}[f(\tilde{\xi}^{N,i}) \,|\, \xi^{N,1},\ldots,\xi^{N,M_N}]\} ,$$
$$B^N(f) = \tilde{M}_N^{-1}\sum_{i=1}^{\tilde{M}_N}\{\mathrm{E}[f(\tilde{\xi}^{N,i}) \,|\, \xi^{N,1},\ldots,\xi^{N,M_N}] - \mu(f)\}$$
$$= \frac{\sum_{i=1}^{M_N}\frac{d\mu}{d\nu}(\xi^{N,i})f(\xi^{N,i})}{\sum_{i=1}^{M_N}\frac{d\mu}{d\nu}(\xi^{N,i})} - \frac{\nu\left(\frac{d\mu}{d\nu}f\right)}{\nu\left(\frac{d\mu}{d\nu}\right)} .$$

Hoeffding's inequality implies that

$$\mathrm{P}\left(|A^N(f)| \geq at \,|\, \xi^{N,1},\ldots,\xi^{N,M_N}\right) \leq 2\exp\left[-2\tilde{M}_N a^2 t^2 \,/\mathrm{osc}^2(f)\right] .$$

The result also holds unconditionally by taking the expectation of the left-hand side. For $\mathrm{P}(|B^N(f)| \geq (1-a)t)$, use the bound (9.12) of Theorem 9.1.10. □

Example 9.2.17 (Importance Sampling with Cauchy and Gaussian Variables, Continued). In this continuation of Example 9.1.3, the target distribution μ is standard Gaussian and the instrumental distribution ν is standard Cauchy. In this case $d\mu/d\nu$ is bounded by some finite M, so that

$$\nu\left[f^2\left(\frac{d\mu}{d\nu}\right)^2\right] \leq M\,\nu\left(f^2\frac{d\mu}{d\nu}\right) \leq M\,\mu(f^2) .$$

Hence Theorem 9.2.15 applies to functions f that are square integrable with respect to the standard Gaussian distribution. This condition is also required to establish asymptotic normality of the importance sampling estimator. We set N =1,000 and investigate the impact of the size M of the instrumental sample on the accuracy of the SIR estimator for $f(x) = \exp(-x)$. Figure 9.5 displays the box-and-whisker plot obtained from 500 independent Monte Carlo replications of the IS and SIR estimators of $\mu(f)$, for instrumental sample sizes M =100, 1,000, 10,000, and 100,000. As expected, the fluctuations of the SIR estimate decrease as the ratio M/N increases. Not surprisingly, when

M =100 ($\alpha = 0.1$) the fluctuation of $\widehat{\mu}^{\mathrm{IS}}_{\nu,M}(f) - \mu(f)$ dominates the resampling fluctuation $\hat{\mu}^{\mathrm{SIR}}_{\nu,M,N}(f) - \widehat{\mu}^{\mathrm{IS}}_{\nu,M}(f)$. On the contrary, when $M = 10,000$ ($\alpha = 10$), the resampling fluctuation is much larger than the error associated with the importance sampling estimate. Likewise, for this M the variance of the SIR estimator is not significantly different from the variance of the plain Monte Carlo estimator using an i.i.d. sample of size N =1,000 from the target distribution μ. To judge the ability of the SIR sample to mimic the distribution of an independent sample from μ, we applied a goodness-of-fit test.

Figure 9.5 displays observed p-values and observed rejection probabilities for the Kolmogorov-Smirnov (KS) goodness-of-fit test of the null hypothesis that the distribution is standard Gaussian (with significance level 5%). For M =100 and 1,000, the p-values are small and the rejection probabilities are large, meaning that the KS test detects a deviation from the null hypothesis of Gaussianity. For $M = 10,000$ and $100,000$ the p-values are much higher and the probabilities of rejection are much smaller. ■

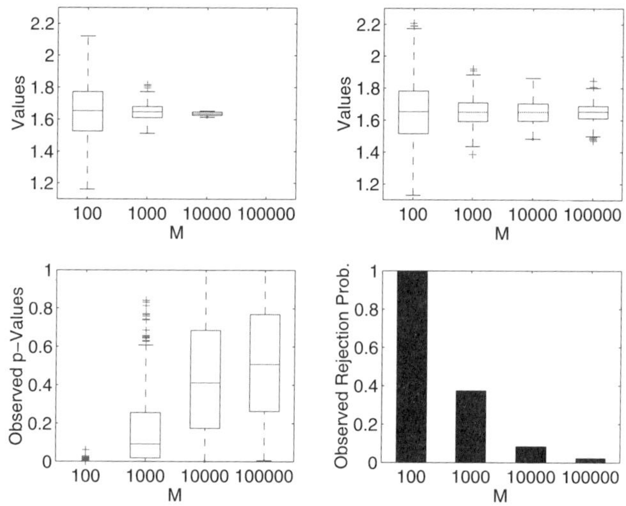

Fig. 9.5. Simulation results for estimation of the integral $\mu(f)$ with $f(x) = \exp(-x)$ and sample size N =1,000, using importance sampling (IS) and sampling importance resampling (SIR) estimators. The instrumental distribution ν was standard Cauchy and target distribution μ was standard Gaussian. The number of Monte Carlo replications was 500 and the instrumental sample sizes were M = 100, 1,000, 10,000, and 100,000. Top left: Box-and-whisker plot of the IS estimates. Top right: Box-and-whisker plot of the SIR estimates. Bottom left: Observed p-values of the Kolmogorov-Smirnov goodness-of-fit test of the null hypothesis that the distribution after resampling is standard Gaussian. Bottom right: Observed rejection probabilities of the null hypothesis at significance level 5%.

9.3 Single-Step Analysis of SMC Methods

We now carry the analysis one step forward to encompass elementary steps of (some of) the sequential Monte Carlo methods discussed in the previous chapters. To do that, we need to consider transformations of the weighted sample that are more sophisticated than weighting and sampling. As outlined in the previous chapter, many different actions might be considered, and it is out of the scope of this chapter to investigate all possible variants. We focus in the following on the SISR approach (Algorithm 7.3.4) and on the variant that we called i.i.d. sampling (Algorithm 8.1.1). As discussed in Section 8.1.1, each iteration of both of these algorithms is composed of two simple procedures—selection and mutation—which we consider separately below.

9.3.1 Mutation Step

To study SISR algorithms, we need first to show that when moving the particles using a Markov transition kernel and then assigning them appropriately defined importance weights, we transform a weighted sample consistent (or asymptotically normal) for one distribution into a weighted sample consistent (or asymptotically normal) for another appropriately defined distribution. As before, we let ν be a probability measure on $(\mathsf{X}, \mathcal{X})$, L be a finite transition kernel on $(\mathsf{X}, \mathcal{X})$, and R be a probability kernel on $(\mathsf{X}, \mathcal{X})$. Define the probability measure μ on $(\mathsf{X}, \mathcal{X})$ by

$$\mu(A) = \frac{\int_{\mathsf{X}} \nu(dx)\, L(x, A)}{\int_{\mathsf{X}} \nu(dx)\, L(x, \mathsf{X})} . \tag{9.27}$$

We then wish to construct a sample consistent for μ, given a weighted sample $\{(\xi^{N,i}, 1)\}_{1 \le i \le M_N}$ from ν. To do so, we move the particles using R as an instrumental kernel and then assign them suitable importance weights. Before writing down the algorithm, we introduce some assumptions.

Assumption 9.3.1. $\nu L(\mathsf{X}) = \int_{\mathsf{X}} \nu(dx)\, L(x, \mathsf{X})$ *is positive and finite.*

Assumption 9.3.2. $\{(\xi^{N,i}, 1)\}_{1 \le i \le M_N}$ *is consistent for* (ν, C), *where* C *is a proper set. In addition, the function* $x \mapsto L(x, \mathsf{X})$ *belongs to* C.

Assumption 9.3.3. *For any* $x \in \mathsf{X}$, $L(x, \cdot)$ *is absolutely continuous with respect to* $R(x, \cdot)$ *and there exists a (strictly) positive version of* $dL(x, \cdot)/dR(x, \cdot)$.

Now let $\{\alpha_N\}$ be a sequence of integers and put $\tilde{M}_N = \alpha_N M_N$. Consider the following algorithm.

Algorithm 9.3.4 (Mutation). Draw $\tilde{\xi}^{N,1}, \ldots, \tilde{\xi}^{N,\tilde{M}_N}$ conditionally independently given $\mathcal{F}^N = \mathcal{G}^N \vee \sigma(\xi^{N,1}, \ldots, \xi^{N,M_N})$ with distribution

$$\mathrm{P}(\tilde{\xi}^{N,j} \in A \,|\, \mathcal{F}^N) = R(\xi^{N,i}, A)$$

for $i = 1, \ldots, M_N$, $j = \alpha_N(i-1)+1, \ldots, \alpha_N i$, and $A \in \mathcal{X}$, and assign $\tilde{\xi}^{N,j}$ the weight

$$\tilde{\omega}^{N,j} = \frac{dL(\xi^{N,i}, \cdot)}{dR(\xi^{N,i}, \cdot)}(\tilde{\xi}^{N,j}) .$$

Thus each particle gives birth to α_N offspring. In many cases, we set $\alpha_N = 1$; then each particle is propagated forward only once. Increasing the number α_N of offspring increases the particle diversity before the resampling step and is thus a practical means for contending particle degeneracy. This of course increases the computational complexity of the algorithm.

Theorem 9.3.5. *Assume 9.3.1, 9.3.2 and 9.3.3, and define*

$$\tilde{\mathsf{C}} \stackrel{\text{def}}{=} \{f \in L^1(\mathsf{X}, \mu) : x \mapsto L(x, |f|) \in \mathsf{C}\} , \tag{9.28}$$

where μ is given by (9.27). *Then $\tilde{\mathsf{C}}$ is a proper set and $\{(\tilde{\xi}^{N,i}, \tilde{\omega}^{N,i})\}_{1 \le i \le \tilde{M}_N}$ defined by Algorithm 9.3.4 is consistent for $(\mu, \tilde{\mathsf{C}})$.*

Proof. Checking that $\tilde{\mathsf{C}}$ is proper is straightforward, so we turn to the consistency. We prove this by showing that for any $f \in \tilde{\mathsf{C}}$,

$$\frac{1}{\tilde{M}_N} \sum_{j=1}^{\tilde{M}_N} \tilde{\omega}^{N,j} f(\tilde{\xi}^{N,j}) \xrightarrow{\text{P}} \nu L(f) . \tag{9.29}$$

Under the assumptions made, the function $x \mapsto L(x, \mathsf{X})$ belongs to C, implying that the constant function 1 belongs to $\tilde{\mathsf{C}}$; hence $\tilde{M}_N^{-1} \sum_{j=1}^{\tilde{M}_N} \tilde{\omega}^{N,j}$ converges to $\nu L(\mathsf{X})$ in probability. Then for any $f \in \tilde{\mathsf{C}}$, the ratio of the two sample means considered tends to $\nu L(f)/\nu L(\mathsf{X}) = \mu(f)$ in probability. This is consistency.

To prove (9.29), pick f in $\tilde{\mathsf{C}}$ and note that $\mathrm{E}[\tilde{\omega}^{N,j} f(\tilde{\xi}^{N,j}) \,|\, \mathcal{F}^N] = L(\xi^{N,i}, f)$ for j and i as in Algorithm 9.3.4. Hence

$$\tilde{M}_N^{-1} \sum_{j=1}^{\tilde{M}_N} \mathrm{E}[\tilde{\omega}^{N,j} f(\tilde{\xi}^{N,j}) \,|\, \mathcal{F}^N] = M_N^{-1} \sum_{i=1}^{M_N} L(\xi^{N,i}, f) \xrightarrow{\text{P}} \nu L(f) ,$$

so that it is sufficient to show that

$$\tilde{M}_N^{-1} \sum_{j=1}^{\tilde{M}_N} \tilde{\omega}^{N,j} f(\tilde{\xi}^{N,j}) - \tilde{M}_N^{-1} \sum_{j=1}^{\tilde{M}_N} \mathrm{E}[\tilde{\omega}^{N,j} f(\tilde{\xi}^{N,j}) \,|\, \mathcal{F}^N] \xrightarrow{\text{P}} 0 . \tag{9.30}$$

For that purpose, we put $V_{N,j} = \tilde{M}_N^{-1} \tilde{\omega}^{N,j} f(\tilde{\xi}^{N,j})$ and appeal to Proposition 9.5.7; we need to check its conditions (i)–(iii). The triangular array $\{V_{N,j}\}_{1 \le j \le \tilde{M}_N}$ is conditionally independent given $\{\mathcal{F}^N\}$; this is condition (i). Next, just as above,

$$\sum_{j=1}^{\tilde{M}_N} \mathrm{E}[|V_{N,j}| \mid \mathcal{F}^N] = M_N^{-1} \sum_{i=1}^{M_N} L(\xi^{N,i}, |f|) \xrightarrow{\mathrm{P}} \nu L(|f|) ,$$

showing condition (ii). We finally need to show that for any positive C,

$$A_N = \sum_{j=1}^{\tilde{M}_N} \mathrm{E}[|V_{N,j}| \mathbb{1}_{\{|V_{N,j}| \geq C\}} \mid \mathcal{F}^N] \xrightarrow{\mathrm{P}} 0 .$$

Put $h(x, x') = \frac{dL(x,\cdot)}{dR(x,\cdot)}(x')|f|(x')$. For any positive C, we then have

$$\int R(x, dx')\, h(x, x') \mathbb{1}_{\{h(x,x') \geq C\}} \leq \int R(x, dx')\, h(x, x') = L(x, |f|) .$$

Because the function $x \mapsto L(x, |f|) \in \mathsf{C}$ and the set C is proper, this shows that the left-hand side of the above display is in C. Hence for large enough N,

$$\begin{aligned} A_N \leq M_N^{-1} \sum_{i=1}^{M_N} \int R(\xi^{N,i}, dx')\, h(\xi^{N,i}, x') \mathbb{1}_{\{h(\xi^{N,i},x') \geq C\}} \\ \xrightarrow{\mathrm{P}} \iint \nu(dx)\, R(x, dx')\, h(x, x') \mathbb{1}_{\{h(x,x') \geq C\}} . \end{aligned}$$

The right-hand side of this inequality is bounded by $\nu L(|f|) < \infty$ (cf. above), so that, by dominated convergence, the right-hand side can be made arbitrarily small by letting $C \to \infty$. This shows that A_N tends to zero in probability, which is condition (iii). Thus Proposition 9.5.7 applies, (9.30) holds, and the proof is complete. □

To establish asymptotic normality of the estimators, we must strengthen Assumption 9.3.2 as follows.

Assumption 9.3.6. *The weighted sample* $\{(\xi^{N,i}, 1)\}_{1 \leq i \leq M_N}$ *is asymptotically normal for* $(\nu, \mathsf{A}, \sigma, \{M_N^{1/2}\})$, *where* A *is a proper set and* σ *is a non-negative function on* A.

Theorem 9.3.7. *Assume 9.3.1, 9.3.2, 9.3.3, and 9.3.6, and that* $\{\alpha_N\}$ *has a limit* α, *possibly infinite. Define*

$$\begin{aligned} \tilde{\mathsf{A}} \stackrel{\text{def}}{=} \Big\{ f \in L^2(\mathsf{X}, \mu) : x \mapsto L(x, f) \in \mathsf{A} \text{ and} \\ x \mapsto \int_{\mathsf{X}} R(x, dx') \left[\frac{dL(x,\cdot)}{dR(x,\cdot)}(x') f(x') \right]^2 \in \mathsf{C} \Big\} . \quad (9.31) \end{aligned}$$

Then $\tilde{\mathsf{A}}$ *is a proper set and* $\{(\tilde{\xi}^{N,i}, \tilde{\omega}^{N,i})\}_{1 \leq i \leq \tilde{M}_N}$ *given by Algorithm 9.3.4 is asymptotically normal for* $(\mu, \tilde{\mathsf{A}}, \tilde{\sigma}, \{M_N^{1/2}\})$ *with*

$$\tilde{\sigma}^2(f) \stackrel{\text{def}}{=} \frac{\sigma^2\{L[f-\mu(f)]\} + \alpha^{-1}\eta^2[f-\mu(f)]}{[\nu L(\mathsf{X})]^2}, \qquad f \in \tilde{\mathsf{A}}\,, \tag{9.32}$$

and η^2 *defined by*

$$\eta^2(f) \stackrel{\text{def}}{=} \iint \nu(dx)\, R(x, dx') \left[\frac{dL(x,\cdot)}{dR(x,\cdot)}(x') f(x')\right]^2 - \int \nu(dx)\, [L(x,f)]^2\,. \tag{9.33}$$

Proof. First we note that by definition, α is necessarily at least 1. Checking that $\tilde{\mathsf{A}}$ is proper is straightforward, so we turn to the asymptotic normality. Pick $f \in \tilde{\mathsf{A}}$ and assume, without loss of generality, that $\mu(f) = 0$. Write

$$\sum_{i=1}^{\tilde{M}_N} \frac{\tilde{\omega}^{N,i}}{\sum_{j=1}^{\tilde{M}_N} \tilde{\omega}^{N,j}} f(\tilde{\xi}^{N,i}) = \frac{\tilde{M}_N}{\sum_{j=1}^{\tilde{M}_N} \tilde{\omega}^{N,j}} (A_N + B_N)\,,$$

with

$$A_N = \tilde{M}_N^{-1} \sum_{i=1}^{\tilde{M}_N} \mathrm{E}[\tilde{\omega}^{N,i} f(\tilde{\xi}^{N,i}) \,|\, \mathcal{F}^N] = M_N^{-1} \sum_{i=1}^{M_N} L(\xi^{N,i}, f)\,,$$

$$B_N = \tilde{M}_N^{-1} \sum_{i=1}^{\tilde{M}_N} \{\tilde{\omega}^{N,i} f(\tilde{\xi}^{N,i}) - \mathrm{E}[\tilde{\omega}^{N,i} f(\tilde{\xi}^{N,i}) \,|\, \mathcal{F}^N]\}\,.$$

Because $\tilde{M}_N / \sum_{i=1}^{\tilde{M}_N} \tilde{\omega}^{N,i}$ converges to $1/\nu L(\mathsf{X})$ in probability (cf. the proof of Theorem 9.3.5), the conclusion of the theorem follows from Slutsky's theorem if we prove that $M_N^{1/2}(A_N + B_N)$ converges weakly to $\mathrm{N}(0, \sigma^2(Lf) + \alpha^{-1}\eta^2(f))$.

In order to do that, we first note that as the function $x \mapsto L(x,f)$ belongs to A and $\{(\xi^{N,i}, 1)\}_{1 \le i \le M_N}$ is asymptotically normal for $(\nu, \mathsf{A}, \sigma, \{M_N^{1/2}\})$,

$$M_N^{1/2} A_N \xrightarrow{\mathcal{D}} \mathrm{N}(0, \sigma^2(Lf))\,.$$

Next we prove that for any real u,

$$\mathrm{E}\left[\exp(\mathrm{i}u\tilde{M}_N^{1/2} B_N) \,\middle|\, \mathcal{F}^N\right] \xrightarrow{\mathrm{P}} \exp\left[-(u^2/2)\eta^2(f)\right]\,.$$

For that purpose, we use Proposition 9.5.12, and we thus need to check its conditions (i)–(iii). Set $V_{N,i} = \tilde{M}_N^{-1/2} \tilde{\omega}^{N,i} f(\tilde{\xi}^{N,i})$. The triangular array $\{V_{N,i}\}_{1 \le i \le \tilde{M}_N}$ is conditionally independent given $\{\mathcal{F}^N\}$ [condition (i)]. Moreover, the function $x \mapsto \int R(x, dx')\, h^2(x,x')$ with $h(x,x') = \frac{dL(x,\cdot)}{dR(x,\cdot)}(x') f(x')$ belongs to C. Therefore

$$\sum_{i=1}^{\tilde{M}_N} \mathrm{E}[V_{N,i}^2 \,|\, \mathcal{F}^N] \xrightarrow{\mathrm{P}} \iint \nu(dx)\, R(x, dx')\, h^2(x,x')\,,$$

$$\sum_{i=1}^{\tilde{M}_N} (\mathrm{E}[V_{N,i} \,|\, \mathcal{F}^N])^2 \xrightarrow{\mathrm{P}} \int \nu(dx) \left[\int R(x, dx')\, h(x, x') \right]^2 .$$

These displays imply that condition (ii) holds.

It remains to verify (iii), the Lindeberg condition. For any positive C, the inequality

$$\int_{\mathsf{X}} R(x, dx')\, h^2(x, x') \mathbb{1}_{\{|h(x,x')| \geq C\}} \leq \int_{\mathsf{X}} R(x, dx')\, h^2(x, x')$$

shows that the function $x \mapsto \int_{\mathsf{X}} R(x, dx') h^2(x, x') \mathbb{1}_{\{|h(x,x')| \geq C\}}$ belongs to C. This yields

$$M_N^{-1} \sum_{i=1}^{M_N} \int R(\xi^{N,i}, dx')\, h^2(\xi^{N,i}, x') \mathbb{1}_{\{h(\xi^{N,i}, x') \geq C\}} \xrightarrow{\mathrm{P}} \iint \nu(dx) R(x, dx')\, h^2(x, x') \mathbb{1}_{\{h(x,x') \geq C\}} .$$

Because $\iint \nu(dx) R(x, dx') h^2(x') < \infty$, the right-hand side of this display can be made arbitrarily small by letting $C \to \infty$. Therefore

$$\sum_{i=1}^{\tilde{M}_N} \mathrm{E}[V_{N,i}^2 \mathbb{1}_{\{|V_{N,i}| \geq \epsilon\}} \,|\, \mathcal{F}^N] \xrightarrow{\mathrm{P}} 0 ,$$

and this is condition (iii).

Thus Proposition 9.5.12 applies, and just as in the proof of Theorem 9.2.14 it follows that

$$\begin{pmatrix} M_N^{1/2} A_N \\ \tilde{M}_N^{1/2} B_N \end{pmatrix} \xrightarrow{\mathcal{D}} \mathrm{N}\left(0, \begin{bmatrix} \sigma^2(Lf) & 0 \\ 0 & \eta^2(f) \end{bmatrix}\right) .$$

The proof is now concluded upon writing $M_N^{1/2}(A_N + B_N) = M_N^{1/2} A_N + \alpha_N^{-1/2} \tilde{M}_N^{1/2} B_N$. □

9.3.2 Description of Algorithms

It is now time to combine the mutation step and the resampling step. This can be done in two different orders, mutation first or selection first, leading to two different algorithms that we call mutation/selection and selection/mutation, respectively. In the mutation/selection algorithm, we first apply the mutation algorithm, 9.3.4, to obtain a weighted sample $\{(\tilde{\xi}^{N,i}, \tilde{\omega}^{N,i})\}_{1 \leq i \leq \tilde{M}_N}$, and then resample according to the importance weights. The selection/mutation algorithm on the other hand is based on a particular decomposition of μ, namely

$$\mu(A) = \frac{\int \nu(dx)\, L(x, A)}{\nu L(\mathsf{X})} = \int \bar{\mu}(dx) \frac{L(x, A)}{L(x, \mathsf{X})} , \tag{9.34}$$

where

$$\bar{\mu}(A) \stackrel{\text{def}}{=} \frac{\int_A \nu(dx)\, L(x, \mathsf{X})}{\nu L(\mathsf{X})} , \qquad A \in \mathcal{X} . \tag{9.35}$$

From a sample $\{(\xi^{N,i}, \omega^{N,i})\}_{1 \le i \le M_N}$, we compute importance weights as $L(\xi^{N,i}, \mathsf{X})$, resample, and finally mutate the resampled system using the Markov kernel $(x, A) \mapsto L(x, A)/L(x, \mathsf{X})$. We now describe the algorithms formally.

Let $\{\alpha_N\}$ be a sequence of integers and set $\tilde{M}_N = \alpha_N M_N$.

Algorithm 9.3.8 (Mutation/Selection).

Mutation: Draw $\tilde{\xi}^{N,1}, \ldots, \tilde{\xi}^{N,\tilde{M}_N}$ conditionally independently given $\mathcal{F}^N = \mathcal{G}^N \vee \sigma(\xi^{N,1}, \ldots, \xi^{N,M_N})$, with distribution $\mathrm{P}(\tilde{\xi}^{N,j} \in \cdot \mid \mathcal{F}^N) = R(\xi^{N,i}, \cdot)$ for $i = 1, \ldots, M_N$ and $j = \alpha_N(i-1)+1, \ldots, \alpha_N i$. Assign $\tilde{\xi}^{N,j}$ the weight $\tilde{\omega}^{N,j} = \frac{dL(\xi^{N,i}, \cdot)}{dR(\xi^{N,i}, \cdot)}(\tilde{\xi}^{N,j})$.

Sampling: Draw M_N random variables $I^{N,1}, \ldots, I^{N,M_N}$ conditionally independently given $\tilde{\mathcal{F}}^N = \mathcal{F}^N \vee \sigma(\tilde{\xi}^{N,1}, \ldots, \tilde{\xi}^{N,\tilde{M}_N})$, with the probability of outcome j, $1 \le j \le \tilde{M}_N$, being proportional to $\tilde{\omega}^{N,j}$. Set $\check{\xi}^{N,i} = \tilde{\xi}^{N,I^{N,i}}$ for $i = 1, \ldots, M_N$.

To avoid notational explosion, it is assumed here that the sample size after the resampling stage is identical to the size of the initial sample. Extensions to general sample sizes are straightforward. The algorithm is illustrated in Figure 9.6.

For the selection/mutation algorithm, we have to strengthen the assumption on the transition kernel L.

Assumption 9.3.9. *For any $x \in \mathsf{X}$, $L(x, \mathsf{X}) > 0$.*

Algorithm 9.3.10 (Selection/Mutation).

Selection: Draw random variables $I^{N,1}, \ldots, I^{N,M_N}$ conditionally independently given $\mathcal{F}^N = \mathcal{G}^N \vee \sigma(\xi^{N,1}, \ldots, \xi^{N,M_N})$, with the probability of outcome j, $1 \le j \le M_N$, being proportional to $L(\xi^{N,j}, \mathsf{X})$. Set $\bar{\xi}^{N,i} = \xi^{N,I^{N,i}}$ for $i = 1, \ldots, M_N$.

Mutation: Draw $\check{\xi}^{N,1}, \ldots, \check{\xi}^{N,M_N}$ conditionally independently given $\bar{\mathcal{F}}^N = \mathcal{F}^N \vee \sigma(I^{N,1}, \ldots, I^{N,M_N})$, with distribution $\mathrm{P}(\check{\xi}^{N,i} \in \cdot \mid \bar{\mathcal{F}}^N) = \frac{L(\bar{\xi}^{N,i}, \cdot)}{L(\bar{\xi}^{N,i}, \mathsf{X})}$.

The algorithm is illustrated in Figure 9.7. As described above, the selection/mutation algorithm requires evaluation of, for any $x \in \mathsf{X}$, the normalizing constant $L(x, \mathsf{X})$, and then sampling from the Markov transition kernel $L(x, \cdot)/L(x, \mathsf{X})$. As emphasized in Chapter 7, these steps are not always easy to carry out. In this sense, this algorithm is in general less widely applicable

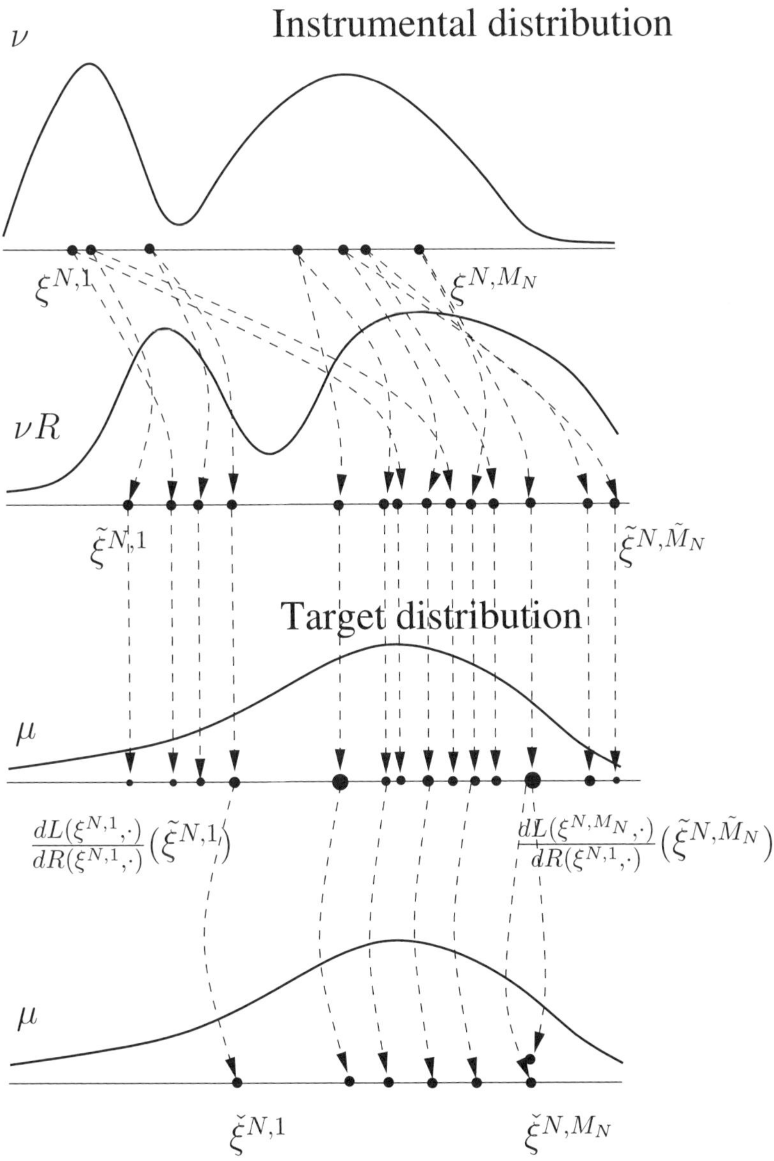

Fig. 9.6. The mutation/selection algorithm. The figure depicts the transformation of the particle system by application of a mutation step followed by a resampling step. In the first stage, an intermediate sample is generated using an instrumental kernel R. Each individual particle of the original system has exactly α_N offspring. In a second step, importance weights taking into account the initial and final positions of the particles are computed. A resampling step, in accordance with these importance weights, is then applied.

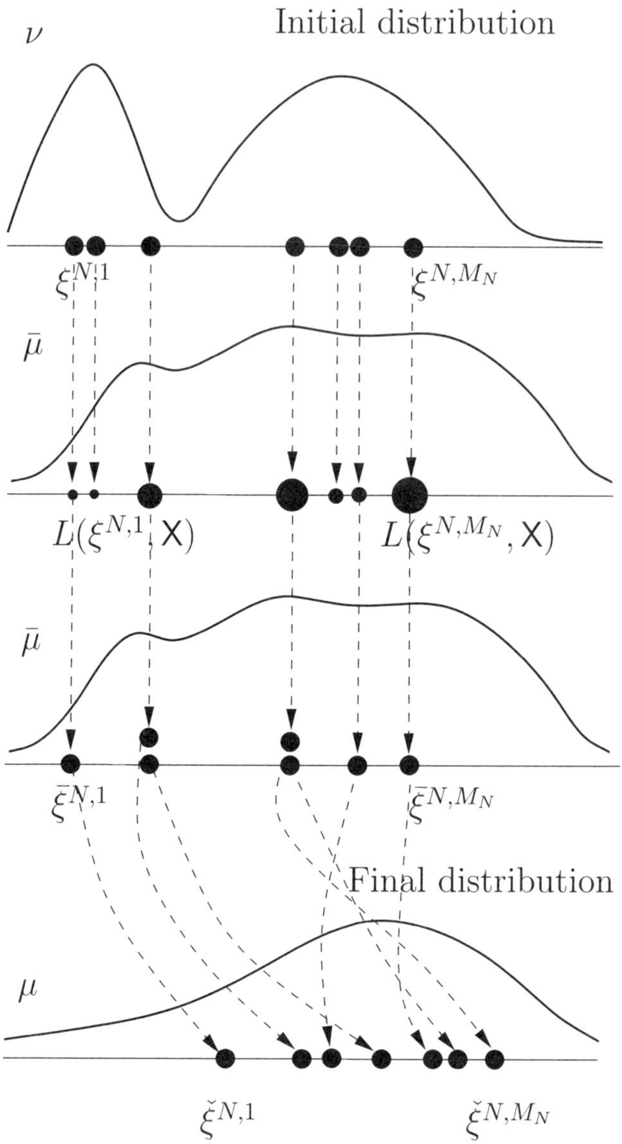

Fig. 9.7. The selection/mutation algorithm. The figure depicts the transformation of the particle system by application of a selection step followed by a mutation step. In the first stage, the importance weights $\{L(\xi^{N,i}, \mathsf{X})\}_{1 \leq i \leq M_N}$ are computed and the system of particles is resampled according to these importance weights. In the second stage, each resampled particle $\{\bar{\xi}^{N,i}\}_{1 \leq i \leq M_N}$ is mutated using the kernel $L(\bar{\xi}^{N,i}, \cdot)/L(\bar{\xi}^{N,i}, \mathsf{X})$.

than mutation/selection. However, it is worthwhile to note that the random variables $\check{\xi}^{N,1}, \ldots, \check{\xi}^{N,M_N}$ are conditionally independent given $\mathcal{F}^N$ and distributed according to the mixture of probability kernels

$$\sum_{i=1}^{M_N} \frac{L(\xi^{N,i}, \mathsf{X})}{\sum_{j=1}^{M_N} L(\xi^{N,j}, \mathsf{X})} \frac{L(\xi^{N,i}, A)}{L(\xi^{N,i}, \mathsf{X})} .$$

As pointed out in Section 8.1.4, it is possible to draw from this distribution without having to follow the selection/mutation steps.

9.3.3 Analysis of the Mutation/Selection Algorithm

Using the tools derived above we establish the consistency and asymptotic normality of the mutation/selection algorithm, 9.3.8. A direct application of Theorems 9.3.5 and 9.2.9 yields the following result.

Theorem 9.3.11. *Assume 9.3.1, 9.3.2, and 9.3.3, and define*

$$\tilde{\mathsf{C}} \stackrel{\text{def}}{=} \{f \in L^1(\mathsf{X}, \mu) : x \mapsto L(x, |f|) \in \mathsf{C}\} . \tag{9.36}$$

where μ is given by (9.27). Then $\tilde{\mathsf{C}}$ is a proper set and

(i) $\{(\tilde{\xi}^{N,i}, \tilde{\omega}^{N,i})\}_{1 \leq i \leq \tilde{M}_N}$ given by Algorithm 9.3.8 is consistent for $(\mu, \tilde{\mathsf{C}})$;
(ii) $\{(\check{\xi}^{N,i}, 1)\}_{1 \leq i \leq M_N}$ given by Algorithm 9.3.8 is consistent for $(\mu, \tilde{\mathsf{C}})$.

Moreover, Theorems 9.3.7 and 9.2.14 imply the following.

Theorem 9.3.12. *Assume 9.3.1, 9.3.2, 9.3.3, and 9.3.6, and that $\{\alpha_N\}$ has a limit, possibly infinite. Define*

$$\tilde{\mathsf{A}} \stackrel{\text{def}}{=} \left\{ f \in L^2(\mathsf{X}, \mu) : x \mapsto L(x, |f|) \in \mathsf{A} \text{ and } \right. \\ \left. x \mapsto \int R(x, dx') \left[\frac{dL(x, \cdot)}{dR(x, \cdot)}(x') f(x') \right]^2 \in \mathsf{C} \right\} .$$

Then $\tilde{\mathsf{A}}$ is a proper set and

(i) $\{(\tilde{\xi}^{N,i}, \tilde{\omega}^{N,i})\}_{1 \leq i \leq \tilde{M}_N}$ given by Algorithm 9.3.8 is asymptotically normal for $(\mu, \tilde{\mathsf{A}}, \tilde{\sigma}, \{M_N^{1/2}\})$ with

$$\tilde{\sigma}^2(f) \stackrel{\text{def}}{=} \frac{\sigma^2\{L[f - \mu(f)]\} + \alpha^{-1}\eta^2[f - \mu(f)]}{[\nu L(\mathsf{X})]^2}, \qquad f \in \tilde{\mathsf{A}} ,$$

and η^2 being defined in (9.33);
(ii) $\{(\check{\xi}^{N,i}, 1)\}_{1 \leq i \leq M_N}$ given by Algorithm 9.3.8 is asymptotically normal for $(\mu, \tilde{\mathsf{A}}, \check{\sigma}, \{M_N^{1/2}\})$ with $\check{\sigma}^2(f) = \mathrm{Var}_\mu(f) + \tilde{\sigma}^2(f)$ for $f \in \tilde{\mathsf{A}}$.

9.3.4 Analysis of the Selection/Mutation Algorithm

We now analyze the selection/mutation algorithm, 9.3.10.

Theorem 9.3.13. *Assume 9.3.2 and 9.3.9. Then*

(i) $\{(\xi^{N,i}, L(\xi^{N,i}, \mathsf{X}))\}_{1\le i\le M_N}$ *given by Algorithm 9.3.10 is consistent for* $(\bar{\mu}, \bar{C})$, *where* $\bar{\mu}$ *is defined in* (9.35) *and*

$$\bar{\mathsf{C}} \stackrel{\text{def}}{=} \{f \in L^1(\mathsf{X}, \bar{\mu}) : x \mapsto |f(x)|L(x, \mathsf{X}) \in \mathsf{C}\} ;$$

(ii) $\{(\check{\xi}^{N,i}, 1)\}_{1\le i\le M_N}$ *given by Algorithm 9.3.10 is consistent for* $(\mu, \check{\mathsf{C}})$, *where*

$$\check{\mathsf{C}} = \{f \in L^1(\mathsf{X}, \mu) : x \mapsto L(x, |f|) \in \mathsf{C}\} .$$

Proof. By construction, $\bar{\mu}$ is absolutely continuous with respect to ν and

$$\frac{d\bar{\mu}}{d\nu}(x) = \frac{L(x, \mathsf{X})}{\nu L(\mathsf{X})}, \quad x \in \mathsf{X} . \tag{9.37}$$

The first assertion follows from Theorem 9.2.7. Theorem 9.2.9 shows that the weighted sample $\{(\bar{\xi}^{N,i}, 1)\}_{1\le i\le M_N}$ is consistent for $(\bar{\mu}, \bar{\mathsf{C}})$. Assertion (ii) then follows from the representation (9.34) of μ and Theorem 9.3.5. □

We may similarly formulate conditions under which the selection/mutation scheme transforms an asymptotically normal sample from the distribution ν into an asymptotically normal sample from μ.

Assumption 9.3.14. $\{(\xi^{N,i}, 1)\}_{1\le i\le M_N}$ *is asymptotically normal for* $(\nu, \mathsf{A}, \sigma, \{M_N^{1/2}\})$, *where* A *is a proper set and* σ *is a non-negative function on* A. *In addition the function* $x \mapsto L(x, \mathsf{X})$ *belongs to* A.

Theorem 9.2.11, Theorem 9.2.14, and Theorem 9.3.7 lead to the following result.

Theorem 9.3.15. *Assume 9.3.2, 9.3.9, and 9.3.14. Then*

(i) $\{(\xi^{N,i}, L(\xi^{N,i}, \mathsf{X}))\}_{1\le i\le M_N}$ *given by Algorithm 9.3.10 is asymptotically normal for* $(\bar{\mu}, \bar{\mathsf{A}}, \bar{\sigma}, \{M_N^{1/2}\})$, *where* $\bar{\mu}$ *is defined in* (9.35),

$$\bar{\mathsf{A}} = \{f \in L^2(\mathsf{X}, \bar{\mu}) : x \mapsto |f(x)|L(x, \mathsf{X}) \in \mathsf{A}\}$$

and

$$\bar{\sigma}^2(f) = \frac{\sigma^2\left\{L(\cdot, \mathsf{X})[f - \bar{\mu}(f)]\right\}}{[\nu L(\mathsf{X})]^2}, \qquad f \in \bar{\mathsf{A}} ;$$

(ii) $\left\{(\check{\xi}^{N,i}, 1)\right\}_{1\le i\le M_N}$ *given by Algorithm 9.3.10 is asymptotically normal for* $(\mu, \check{\mathsf{A}}, \check{\sigma}, \{M_N^{1/2}\})$, *where*

$$\check{\mathsf{A}} = \{f \in L^2(\mathsf{X}, \mu) : x \mapsto L(x, |f|) \in \mathsf{A} \text{ and } x \mapsto L(x, f^2) \in \mathsf{C}\}$$

and

$$\check{\sigma}^2(f) = \mathrm{Var}_\mu(f) + \frac{\sigma^2\left\{L[f - \mu(f)]\right\}}{[\nu L(\mathsf{X})]^2}, \qquad f \in \check{\mathsf{A}} .$$

9.4 Sequential Monte Carlo Methods

We are now ready to evaluate the performance of repeated applications of the basic procedures studied in the previous section. We begin with the mutation/selection or SISR variant.

9.4.1 SISR

Sequential importance sampling with resampling amounts to successively applying the mutation/selection procedure in order to construct a sample approximating the marginal filtering distribution. In this case, the initial and final probability distributions are the marginal filtering distributions $\phi_{\nu,k}$ at two successive time instants. As discussed in Chapter 7, these two distributions are related by (7.8), which we recall here:

$$\phi_{\nu,0}(A) = \frac{\int_A \nu(dx')\, g_0(x')}{\int_{\mathsf{X}} \nu(dx')\, g_0(x')} , \qquad A \in \mathcal{X} , \tag{9.38}$$

$$\phi_{\nu,k+1}(A) = \frac{\int_{\mathsf{X}} \phi_{\nu,k}(dx)\, T_k^{\mathrm{u}}(x, A)}{\int_{\mathsf{X}} \phi_{\nu,k}(dx)\, T_k^{\mathrm{u}}(x, \mathsf{X})} , \qquad A \in \mathcal{X},\ k \geq 0 , \tag{9.39}$$

$$T_k^{\mathrm{u}}(x, A) = \int_A Q(x, dx')\, g_{k+1}(x') , \qquad A \in \mathcal{X} , \tag{9.40}$$

where, as usual, Q stands for the transition kernel of the hidden chain and g_k for the likelihood of the current observation, $g_k(x) = g(x, Y_k)$[1].

The instrumental distributions are defined by a sequence $\{R_k\}_{k\geq 0}$ of instrumental transition kernels on $(\mathsf{X}, \mathcal{X})$ and a probability distribution ρ_0 on $(\mathsf{X}, \mathcal{X})$. In addition, let $\{\alpha_N\}$ denote a sequence of positive integers that control the size of the intermediate populations of particles (see below). We require the following assumptions.

Assumption 9.4.1.

(i) $\nu(g_0) > 0$.
(ii) $\int_{\mathsf{X}} Q(x, dx') g_k(x') > 0$ *for all* $x \in \mathsf{X}$ *and* $k \geq 0$.
(iii) $\sup_{x \in \mathsf{X}} g_k(x) < \infty$ *for all* $k \geq 0$.

Assumption 9.4.2. *The instrumental distribution ρ_0 for the initial state dominates the filtering distribution $\phi_{\nu,0}$, $\phi_{\nu,0} \ll \rho_0$.*

Assumption 9.4.3. *For any $k \geq 0$ and all $x \in \mathsf{X}$, the instrumental kernel $R_k(x, \cdot)$ dominates $T_k^{\mathrm{u}}(x, \cdot)$, $T_k^{\mathrm{u}}(x, \cdot) \ll R_k(x, \cdot)$. In addition, for any x there exists a version of the Radon-Nikodym derivative $\frac{dT_k^{\mathrm{u}}(x,\cdot)}{dR_k(x,\cdot)}$ that is (strictly) positive and such that* $\sup_{(x,x') \in \mathsf{X}\times\mathsf{X}} \frac{dT_k^{\mathrm{u}}(x,\cdot)}{dR_k(x,\cdot)}(x') < \infty$.

---

[1]Note that in Chapter 7 we defined T_k^{u} with a different scale factor—see (7.8). As mentioned several times, however, this scale factor plays no role in approaches based on (self-normalized) importance sampling and SIR. For notational simplicity, we thus ignore this scale factor here.

These conditions are not minimal but are most often satisfied in practice. The first assumption, 9.4.1, implies that for any positive integer k,

$$0 < \int \cdots \int \nu(dx_0)\, g_0(x_0) \prod_{i=1}^{k} Q(x_{i-1}, dx_i)\, g_i(x_i) \leq \prod_{i=0}^{k} \sup_{x \in \mathsf{X}} g_i(x) < \infty ,$$

so that in particular

$$0 < \phi_{\nu,k} T_k^{\mathrm{u}}(\mathsf{X}) < \infty . \tag{9.41}$$

The SISR approach under study has already been described in Algorithm 7.3.4, which we rephrase below in a more mathematical fashion to underline the conditioning arguments to be used in the following.

Algorithm 9.4.4 (SISR).

Mutation: Draw $\{\tilde{\xi}_{k+1}^{N,i}\}_{1 \leq i \leq \tilde{M}_N}$ conditionally independently given $\mathcal{F}_k^N$, with distribution

$$\mathrm{P}(\tilde{\xi}_{k+1}^{N,j} \in A \,|\, \mathcal{F}_k^N) = R_k(\xi_k^{N,i}, A)$$

for $i = 1, \ldots, M_N$, $j = \alpha_N(i-1)+1, \ldots, \alpha_N i$ and $A \in \mathcal{X}$, and compute the importance weights

$$\tilde{\omega}_{k+1}^{N,j} = g_{k+1}(\tilde{\xi}_{k+1}^{N,j}) \frac{dQ(\xi_k^{N,i}, \cdot)}{dR_k(\xi_k^{N,i}, \cdot)}(\tilde{\xi}_{k+1}^{N,j})$$

for j and i as above.

Selection: Draw $I_{k+1}^{N,1}, \ldots, I_{k+1}^{N,M_N}$ conditionally independently given $\tilde{\mathcal{F}}_k^N = \mathcal{F}_k^N \vee \sigma(\tilde{\xi}_{k+1}^{N,1}, \ldots, \tilde{\xi}_{k+1}^{N,\tilde{M}_N})$, with distribution

$$\mathrm{P}(I_{k+1}^{N,i} = j \,|\, \tilde{\mathcal{F}}_k^N) = \frac{\tilde{\omega}_{k+1}^{N,j}}{\sum_{j=1}^{\tilde{M}_N} \tilde{\omega}_{k+1}^{N,j}} ,$$

and set $\xi_{k+1}^{N,i} = \tilde{\xi}_{k+1}^{N, I_{k+1}^{N,i}}$ and $\mathcal{F}_{k+1}^N = \tilde{\mathcal{F}}_k^N \vee \sigma(\xi_{k+1}^{N,1}, \ldots, \xi_{k+1}^{N,M_N})$.

Two choices, among many others, of the instrumental kernel are the following.

Prior kernel: $R_k = Q$. For any $(x, x') \in \mathsf{X} \times \mathsf{X}$, $[dT_k^{\mathrm{u}}(x, \cdot)/dQ(x, \cdot)](x') = g_{k+1}(x')$, showing that the importance weights $\tilde{\omega}_{k+1}^{N,j} = g_{k+1}(\tilde{\xi}_{k+1}^{N,j})$ only depend on the mutated particle positions. Provided Assumption 9.4.1 holds true, so does Assumption 9.4.3 as soon as $g_{k+1}(x) > 0$ for all $x \in \mathsf{X}$. Note that for the prior kernel, $\{(\tilde{\xi}_{k+1}^{N,i}, 1)\}_{1 \leq i \leq \tilde{M}_N}$ is a sample approximating the marginal predictive distribution $\phi_{k+1|k} = \phi_k Q$.

Optimal kernel: $R_k = T_k$, defined by

$$T_k(x, A) = \frac{T_k^{\mathrm{u}}(x, A)}{T_k^{\mathrm{u}}(x, \mathsf{X})} .$$

For all $(x, x') \in \mathsf{X} \times \mathsf{X}$, $[dT_k^{\mathrm{u}}(x, \cdot)/dT_k(x, \cdot)](x') = T_k^{\mathrm{u}}(x, \mathsf{X})$, which implies that the importance weights $\tilde{\omega}_{k+1}^{N,j} = T_k^{\mathrm{u}}(\xi^{N,i}, \mathsf{X})$, with j and i as above, only depend on the current particle positions. Provided Assumption 9.4.1 holds true, so does Assumption 9.4.3 because, for all $(x, x') \in \mathsf{X} \times \mathsf{X}$, $[dT_k^{\mathrm{u}}(x, \cdot)/dT_k(x, \cdot)] > 0$ and

$$\sup_{(x,x') \in \mathsf{X} \times \mathsf{X}} \frac{dT_k^{\mathrm{u}}(x, \cdot)}{dT_k(x, \cdot)}(x') = \sup_{x \in \mathsf{X}} \int_{\mathsf{X}} Q(x, dx')\, g_{k+1}(x') \leq \sup_{x \in \mathsf{X}} g_{k+1}(x) < \infty \, .$$

For all other instrumental kernels, the importance weights depend on the initial and final positions of the particles.

Theorem 9.4.5. *Assume 9.4.1, 9.4.2, and 9.4.3. Then the following holds true.*

(i) If $\{(\xi_0^{N,i}, 1)\}_{1 \leq i \leq M_N}$ is consistent for $(\phi_{\nu,0}, L^1(\mathsf{X}, \phi_{\nu,0}))$ then for any $k > 0$, $\{(\xi_k^{N,i}, 1)\}_{1 \leq i \leq M_N}$ is consistent for $(\phi_{\nu,k}, L^1(\mathsf{X}, \phi_{\nu,k}))$.
(ii) If in addition $\{(\xi_0^{N,i}, 1)\}_{1 \leq i \leq M_N}$ is asymptotically normal for $(\phi_{\nu,0}, L^2(\mathsf{X}, \phi_{\nu,0}), \sigma_0, \{M_N^{1/2}\})$ then for any $k > 0$, $\{(\xi_k^{N,i}, 1)\}_{1 \leq i \leq M_N}$ is asymptotically normal for $(\phi_{\nu,k}, L^2(\mathsf{X}, \phi_{\nu,k}), \sigma_k, \{M_N^{1/2}\})$, where the sequence $\{\sigma_k\}$ of functions is defined recursively, for $f \in L^2(\mathsf{X}, \phi_{\nu,k})$, by

$$\sigma_{k+1}^2(f) = \mathrm{Var}_{\phi_{\nu,k+1}}(f)$$
$$+ \frac{\sigma_k^2(T_k^{\mathrm{u}}\{f - \phi_{\nu,k+1}(f)\}) + \alpha^{-1}\eta_k^2(\{f - \phi_{\nu,k+1}(f)\}^2)}{(\phi_{\nu,k} T_k^{\mathrm{u}}(\mathsf{X}))^2}$$

with

$$\eta_k^2(f) = \iint \phi_{\nu,k}(dx) R_k(x, dx') \left\{ \frac{dT_k^{\mathrm{u}}(x, \cdot)}{dR_k(x, \cdot)}(x') f(x') \right\}^2 - \iint \phi_{\nu,k}(dx) \{T_k^{\mathrm{u}}(x, f)\}^2 \, .$$

Proof. The proof is by induction over k. Starting with (i), we hence assume that for some $k \geq 0$, $\{(\xi^{N,i}, 1)\}_{1 \leq i \leq M_N}$ is consistent for $(\phi_{\nu,k}, L^1(\mathsf{X}, \phi_{\nu,k}))$. To prove that consistency then holds for $k+1$ as well, we shall employ Theorem 9.3.11 and hence need to verify its underlying assumptions with $\nu = \phi_{\nu,k}$ and $L = T_k^{\mathrm{u}}$. To start with, Assumption 9.3.1 is (9.41) and Assumption 9.3.3 is implied by Assumption 9.4.3. Assumption 9.3.2 follows from the induction hypothesis plus the bound $T_k^{\mathrm{u}}(x, \mathsf{X}) \leq \|g_{k+1}\|_\infty < \infty$ for all x. Finally, to check that consistency applies over $L^1(\mathsf{X}, \phi_{\nu,k+1})$, we need to verify that for any $f \in L^1(\mathsf{X}, \phi_{\nu,k+1})$ the function $x \mapsto T_k^{\mathrm{u}}(x, |f|)$ belongs to $L^1(\mathsf{X}, \phi_{\nu,k})$. This is indeed true, as

$$(\phi_{\nu,k}T_k^{\mathrm{u}})(|f|) = (\phi_{\nu,k}T_k^{\mathrm{u}})(\mathsf{X}) \times \phi_{\nu,k+1}(|f|).$$

Assertion (i) now follows from Theorem 9.3.11 and induction.

We proceed to part (ii), modify the induction hypothesis accordingly, and use Theorem 9.3.12 to propagate it from k to $k+1$. The additional assumption we then need to verify is Assumption 9.3.6, which is the induction hypothesis. Finally, we need to check that asymptotic normality applies over $L^2(\mathsf{X}, \phi_{\nu,k+1})$. Pick $f \in L^2(\mathsf{X}, \phi_{\nu,k+1})$. Then by Jensen's inequality,

$$\begin{aligned}
\phi_{\nu,k}\left[(T_k^{\mathrm{u}}|f|)^2\right] &= \phi_{\nu,k}\left([Q(g_{k+1}|f|)]^2\right) \\
&\leq \phi_{\nu,k}Q(g_{k+1}^2 f^2) \\
&= (\phi_{\nu,k}T_k^{\mathrm{u}})(\mathsf{X})\,\phi_{\nu,k+1}(g_{k+1}f^2) \\
&\leq (\phi_{\nu,k}T_k^{\mathrm{u}})(\mathsf{X})\,\|g_{k+1}\|_\infty\,\phi_{\nu,k+1}(f^2) < \infty\,,
\end{aligned}$$

saying that $T_k^{\mathrm{u}}(|f|)$ is in $L^2(\mathsf{X}, \phi_{\nu,k})$. Similarly

$$\begin{aligned}
&\int_{\mathsf{X}} \phi_{\nu,k}(dx) \int_{\mathsf{X}} R_k(x, dx') \left[\frac{dT_k^{\mathrm{u}}(x,\cdot)}{dR_k(x,\cdot)}(x') f(x')\right]^2 \\
&\qquad \leq \sup_{(x,x')\in\mathsf{X}\times\mathsf{X}} \frac{dT_k^{\mathrm{u}}(x,\cdot)}{dR_k(x,\cdot)}(x')\,(\phi_{\nu,k}T_k^{\mathrm{u}})(\mathsf{X})\,\phi_{\nu,k+1}(f^2) < \infty\,,
\end{aligned}$$

so that the function that $\phi_{\nu,k}$ is acting on in the left-hand side belongs to $L^1(\mathsf{X}, \phi_{\nu,k})$. Assertion (ii) now follows from Theorem 9.3.12 and induction.

□

9.4.2 I.I.D. Sampling

We now consider successive applications of the selection/mutation procedure. The resulting algorithm, referred to as i.i.d. sampling in Section 8.1.1, is recalled below. Because the mathematical analysis of this algorithm is somewhat simpler, we consider below two additional types of results: uniform (in time) convergence results under appropriate forgetting conditions (as discussed in Section 4.3) and exponential tail inequalities. Recall that although the emphasis is here put on filtering estimates, the selection/mutation algorithm may also be applied to approximate the predictive distributions, in which case it is known as the *bootstrap filter* (Figure 8.1). Hence all results below also apply to the analysis of the predictive estimates produced by the bootstrap filter, with only minor adjustments.

Algorithm 9.4.6 (I.I.D. Sampling).

Selection: Assign to the particle $\xi_k^{N,i}$ the importance weight

$$\omega_{k+1}^{N,i} = T_k^{\mathrm{u}}(\xi_k^{N,i}, \mathsf{X}) = \int_{\mathsf{X}} Q(\xi_k^{N,i}, dx')\,g_{k+1}(x')\,.$$

Draw $I_{k+1}^{N,1}, \ldots, I_{k+1}^{N,M_N}$ conditionally independently given $\mathcal{F}_k^N$, with distribution

$$\mathrm{P}(I_{k+1}^{N,i} = j \mid \mathcal{F}_k^N) = \frac{\omega_k^{N,j}}{\sum_{j=1}^{M_N} \omega_k^{N,j}} , \qquad i, j = 1, \ldots, M_N ,$$

and set $\bar{\xi}_k^{N,i} = \xi_k^{N, I_{k+1}^{N,i}}$.

Mutation: Draw $\xi_{k+1}^{N,1}, \ldots, \xi_{k+1}^{N,M_N}$ conditionally independently given $\bar{\mathcal{F}}_k^N = \mathcal{F}_k^N \vee \sigma(I_{k+1}^{N,1}, \ldots, I_{k+1}^{N,M_N})$, with distribution

$$\mathrm{P}(\xi_{k+1}^{N,i} \in A \mid \bar{\mathcal{F}}_k^N) = \frac{T_k^{\mathrm{u}}(\bar{\xi}_k^{N,i}, A)}{T_k^{\mathrm{u}}(\bar{\xi}_k^{N,i}, \mathsf{X})} = \frac{\int_A Q(\bar{\xi}_k^{N,i}, dx')\, g_{k+1}(x')}{\int_{\mathsf{X}} Q(\bar{\xi}_k^{N,i}, dx')\, g_{k+1}(x')} .$$

9.4.2.1 Consistency and Asymptotic Normality

Theorem 9.4.7. *Assume 9.4.1, 9.4.2, and 9.4.3. Then the following holds true.*

(i) If $\{(\xi_0^{N,i}, 1)\}_{1 \le i \le M_N}$ is consistent for $(\phi_{\nu,0}, L^1(\mathsf{X}, \phi_{\nu,0}))$ then for any $k > 0$, $\{(\xi_k^{N,i}, 1)\}_{1 \le i \le M_N}$ is consistent for $(\phi_{\nu,k}, L^1(\mathsf{X}, \phi_{\nu,k}))$.

(ii) If $\{(\xi_0^{N,i}, 1)\}_{1 \le i \le M_N}$ is asymptotically normal for $(\phi_{\nu,0}, L^2(\mathsf{X}, \phi_{\nu,0}), \sigma_0, \{M_N^{1/2}\})$, then for any $k > 0$, $\{(\xi_k^{N,i}, 1)\}_{1 \le i \le M_N}$ is asymptotically normal for $(\phi_{\nu,k}, L^2(\mathsf{X}, \phi_{\nu,k}), \sigma_k, \{M_N^{1/2}\})$, where the sequence $\{\sigma_k\}$ of functions is defined recursively by

$$\sigma_{k+1}^2(f) = \mathrm{Var}_{\phi_{\nu,k+1}}(f) + \frac{\sigma_k^2 \{T_k^{\mathrm{u}}[f - \phi_{\nu,k+1}(f)]\}}{[\phi_{\nu,k} T_k^{\mathrm{u}}(\mathsf{X})]^2} , \quad f \in L^2(\mathsf{X}, \phi_{\nu,k+1}) . \tag{9.42}$$

Proof. Again the proof is by induction. Hence assume that for some $k \ge 0$, $\{(\xi_k^{N,i}, 1)\}_{1 \le i \le M_N}$ is consistent for $(\phi_{\nu,k}, L^1(\mathsf{X}, \phi_{\nu,k}))$. To carry the induction hypothesis from k to $k+1$, we shall employ Theorem 9.3.13 and thus need to check its underlying assumptions. Assumption 9.3.2 was verified in the proof of Theorem 9.4.5, and (9.3.9) is Assumption 9.4.1(ii). What remains to check is that consistency holds over the whole of $L^1(\mathsf{X}, \phi_{\nu,k+1})$, and for that we must verify that for every f in this space, the function $T_k^{\mathrm{u}}(|f|)$ belongs to $L^1(\mathsf{X}, \phi_{\nu,k})$. This was also done in the proof of Theorem 9.4.5. Hence assertion (i) follows from Theorem 9.3.13 and induction.

We proceed to part (ii), modify the induction hypothesis accordingly, and use Theorem 9.3.15 to propagate it from k to $k+1$. The additional assumption we then need to verify is Assumption 9.3.14, which follows from the induction hypothesis and the bound $T_k^{\mathrm{u}}(x, \mathsf{X}) \le \|g_{k+1}\|_\infty$. Finally, we establish that asymptotic normality applies over $L^2(\mathsf{X}, \phi_{\nu,k+1})$, which amounts to verifying that for any $f \in L^2(\mathsf{X}, \phi_{\nu,k+1})$, the function $T_k^{\mathrm{u}}(|f|)$ belongs to $L^2(\mathsf{X}, \phi_{\nu,k})$ and the function $T_k^{\mathrm{u}}(f^2)$ belongs to $L^1(\mathsf{X}, \phi_{\nu,k+1})$. The first of these requirements is part of the proof of Theorem 9.4.5, and the proof of the second requirement is entirely analogous. Assertion (ii) now follows from Theorem 9.3.15 and induction. □

It is worthwhile to note that the asymptotic variance of the i.i.d. sampling algorithm is always lower than that of SISR, whatever choice of instrumental kernel for the latter. This indicates that whenever possible, i.i.d. sampling should be preferred. By iterating (9.42), one can obtain an analytic expression for the asymptotic variance.

Proposition 9.4.8. *Assume 9.4.1 and 9.4.3 and that* $\{(\xi_0^{N,i}, 1)\}_{1 \leq i \leq M_N}$ *is asymptotically normal for* $(\phi_{\nu,0}, L^2(\mathsf{X}, \phi_{\nu,0}), \sigma_0, \{M_N^{1/2}\})$. *Then for any* $k \geq 0$ *and* $f \in L^2(\mathsf{X}, \phi_{\nu,k})$,

$$\sigma_k^2(f) = \sum_{l=1}^{k} \frac{\mathrm{Var}_{\phi_{\nu,l}} \{T_l^{\mathrm{u}} \cdots T_{k-1}^{\mathrm{u}}[f - \phi_{\nu,k}(f)]\}}{\left[\phi_{\nu,l} T_l^{\mathrm{u}} \cdots T_{k-1}^{\mathrm{u}}(\mathsf{X})\right]^2} + \frac{\sigma_0^2 \{T_0^{\mathrm{u}} \cdots T_{k-1}^{\mathrm{u}}[f - \phi_{\nu,k}(f)]\}}{\left[\phi_{\nu,0} T_0^{\mathrm{u}} \cdots T_{k-1}^{\mathrm{u}}(\mathsf{X})\right]^2} ,$$

where, by convention $T_i^{\mathrm{u}} \cdots T_j^{\mathrm{u}}(x, A)$ *is the identity transition kernel* $\delta_x(A)$ *for* $i > j$.

Proof. The proof is by induction on k. The result holds true for $k = 0$. Assume now that the result holds true for some $k \geq 0$. We evaluate the right-hand side of (9.42) with the claimed formula for σ_k^2. Doing this, we first note that $T_k^{\mathrm{u}}[f - \phi_{\nu,k+1}(f)] - \phi_{\nu,k} T_k^{\mathrm{u}}[f - \phi_{\nu,k+1}(f)] = T_k^{\mathrm{u}}[f - \phi_{\nu,k+1}(f)]$, because $\phi_{\nu,k} T_k^{\mathrm{u}}$ equals $\phi_{\nu,k+1}$ up to a multiplicative constant. Thus the right-hand side of (9.42) evaluates to

$$\mathrm{Var}_{\phi_{\nu,k+1}}(f) + \sum_{l=1}^{k} \frac{\mathrm{Var}_{\phi_{\nu,l}} \{T_l^{\mathrm{u}} \cdots T_k^{\mathrm{u}}[f - \phi_{\nu,k+1}(f)]\}}{\left[\phi_{\nu,l} T_l^{\mathrm{u}} \cdots T_{k-1}^{\mathrm{u}}(\mathsf{X})\right]^2 [\phi_{\nu,k} T_k^{\mathrm{u}}(\mathsf{X})]^2} + \frac{\sigma_0^2 \{T_0^{\mathrm{u}} \cdots T_k^{\mathrm{u}}[f - \phi_{\nu,k}(f)]\}}{\left[\phi_{\nu,0} T_0^{\mathrm{u}} \cdots T_{k-1}^{\mathrm{u}}(\mathsf{X})\right]^2 [\phi_{\nu,k} T_k^{\mathrm{u}}(\mathsf{X})]^2} .$$

Comparing this with the claimed expression for $\sigma_{k+1}^2(f)$, we see that what remains to verify is that the denominators of the above ratios equal the square of $\phi_{\nu,l} T_l^{\mathrm{u}} \cdots T_k^{\mathrm{u}}(\mathsf{X})$.

To do that, we observe that the definition of the filtering distribution—see for instance (3.13)—shows that for any $l \leq k - 1$,

$$
\begin{aligned}
\phi_{\nu,k}(h) &= \mathrm{L}_{\nu,k}^{-1} \int \cdots \int \nu(dx_0)\, g_0(x_0) \prod_{i=1}^{k} Q(x_{i-1}, dx_i)\, g_i(x_i) h(x_k) \\
&= \mathrm{L}_{\nu,l} \mathrm{L}_{\nu,k}^{-1} \int \cdots \int \phi_{\nu,l}(dx_l) \prod_{i=l+1}^{k} Q(x_{i-1}, dx_i)\, g_i(x_i) h(x_k) \\
&= \frac{\phi_{\nu,l} T_l^{\mathrm{u}} \cdots T_{k-1}^{\mathrm{u}} f}{\phi_{\nu,l} T_l^{\mathrm{u}} \cdots T_{k-1}^{\mathrm{u}}(\mathsf{X})} .
\end{aligned}
$$

Setting $h = T_k^{\mathrm{u}}(\mathsf{X})$ yields $[\phi_{\nu,k} T_k^{\mathrm{u}}(\mathsf{X})]\, \phi_{\nu,l} T_l^{\mathrm{u}} \cdots T_{k-1}^{\mathrm{u}}(\mathsf{X}) = \phi_{\nu,l} T_l^{\mathrm{u}} \cdots T_k^{\mathrm{u}}(\mathsf{X})$. The proof now follows by induction. □

The expression for the asymptotic variance is rather involved, and it is difficult in general to make simple statements on this quantity. There is however a situation in which some interesting conclusions can be drawn. Consider the following assumption (cf. Lemma 4.3.25).

Assumption 9.4.9. *There exist positive constants σ^- and σ^+ and a probability distribution λ such that $0 < \sigma^- \lambda(A) \le Q(x, A) \le \sigma^+ \lambda(A) < \infty$ for all $x \in \mathsf{X}$ and $A \in \mathcal{X}$.*

Also recall the notation $\rho \stackrel{\mathrm{def}}{=} 1 - \sigma^- / \sigma^+$.

Under this condition, it has been shown that the posterior chain is uniformly geometrically mixing, that is, it forgets its initial condition uniformly and at a geometric (or exponential) rate. Exponential forgetting allows us to prove that the asymptotic variance of the selection/mutation algorithm remains bounded.

Proposition 9.4.10. *Assume 9.4.1, 9.4.3, and 9.4.9. Then for any $f \in \mathcal{F}_{\mathrm{b}}(\mathsf{X})$, it holds that $\sup_{k \ge 0} \sigma_k^2(f) < \infty$, where σ_k^2 is defined in* (9.42).

Proof. Consider the numerators of the ratios of the expression for σ_k in Proposition 9.4.8. Proposition 3.3.2 shows that for any integers $l < k$,

$$
T_l^{\mathrm{u}} \cdots T_{k-1}^{\mathrm{u}}(x, A) = \beta_{l|k}(x) \mathrm{F}_{l|k} \cdots \mathrm{F}_{k-1|k}(x, A) , \quad x \in \mathsf{X},\ A \in \mathcal{X} ,
$$

where the $\mathrm{F}_{l|k}$ are forward smoothing kernels (see Definition 3.3.1) and $\beta_{l|k}$ is the backward function (see Definition 3.1.6). Therefore

$$
\begin{aligned}
T_l^{\mathrm{u}} \cdots T_{k-1}^{\mathrm{u}} f(x) - T_l^{\mathrm{u}} \cdots T_{k-1}^{\mathrm{u}}(x, \mathsf{X}) \phi_{\nu,k}(f) \\
= \beta_{l|k}(x) \left[\mathrm{F}_{l|k} \cdots \mathrm{F}_{k-1|k} f(x) - \phi_{\nu,k}(f) \right] . \quad (9.43)
\end{aligned}
$$

Next we consider the denominators of the expression for σ_k. We have $\phi_{\nu,l} T_l^{\mathrm{u}} \cdots T_{k-1}^{\mathrm{u}}(\mathsf{X}) = \phi_{\nu,l}(\beta_{l|k}) = \prod_{j=l+1}^{k} c_{\nu,j}$, where the first equality follows from the above and the second one from Proposition 3.2.5, and where the constants $c_{\nu,j}$ are defined recursively in (3.22). Moreover, by (3.26) $\mathrm{L}_{\nu,k} = \prod_{j=0}^{k} c_{\nu,j}$, and hence

$$\phi_{\nu,l} T_l^{\mathrm{u}} \cdots T_{k-1}^{\mathrm{u}}(\mathsf{X}) = \frac{\mathrm{L}_{\nu,k}}{\mathrm{L}_{\nu,l}} . \tag{9.44}$$

Combining (9.43) and (9.44) yields for any integers $l \leq k$,

$$\frac{\operatorname{Var}_{\phi_{\nu,l}} \left\{ T_l^{\mathrm{u}} \cdots T_{k-1}^{\mathrm{u}} [f - \phi_{\nu,k}(f)] \right\}}{\left[\phi_{\nu,l} T_l^{\mathrm{u}} \cdots T_{k-1}^{\mathrm{u}}(\mathsf{X}) \right]^2} \\ = \operatorname{Var}_{\phi_{\nu,l}} \left(\beta_{l|k} \frac{\mathrm{L}_{\nu,l}}{\mathrm{L}_{\nu,k}} \{ \mathrm{F}_{l|k} \cdots \mathrm{F}_{k-1|k} f - \phi_{\nu,k}(f) \} \right) . \tag{9.45}$$

In order to bound this variance, we first notice that Lemma 4.3.22(ii) shows that

$$\beta_{l|k}(x) \frac{\mathrm{L}_{\nu,l}}{\mathrm{L}_{\nu,k}} = \frac{\int Q(x, dx')\, g_{l+1}(x') \beta_{l+1|k}(x')}{\iint \phi_{\nu,l}(dx)\, Q(x, dx')\, g_{l+1}(x') \beta_{l+1|k}(x')} \leq \frac{\sigma^+}{\sigma^-} = \frac{1}{1-\rho} . \tag{9.46}$$

Next, Proposition 3.3.4 shows that $\phi_{\nu,k}(f) = \phi_{\nu,l|k} \mathrm{F}_{l|k} \cdots \mathrm{F}_{k-1|k} f$, where $\phi_{\nu,l|k}$ is a smoothing distribution. In addition, by Lemma 4.3.22 again, for any probability measures ξ and ξ' on $(\mathsf{X}, \mathcal{X})$,

$$\left\| \xi \mathrm{F}_{l|k} \cdots \mathrm{F}_{k-1|k} - \xi' \mathrm{F}_{l|k} \cdots \mathrm{F}_{k-1|k} \right\|_{\mathrm{TV}} \leq \rho^{k-l} \left\| \xi - \xi' \right\|_{\mathrm{TV}} .$$

Applying this bound with $\xi = \delta_x$ and $\xi' = \phi_{\nu,l|k}$ shows that

$$\left| \mathrm{F}_{l|k} \cdots \mathrm{F}_{k-1|k}(x, f) - \phi_{\nu,l|k}(f) \right| \leq 2\rho^{k-l} \left\| f \right\|_\infty .$$

Finally, combining with (9.45) and (9.46) shows that

$$\frac{\operatorname{Var}_{\phi_{\nu,l}} \left\{ T_l^{\mathrm{u}} \cdots T_{k-1}^{\mathrm{u}} [f - \phi_{\nu,k}(f)] \right\}}{\left[\phi_{\nu,l} T_l^{\mathrm{u}} \cdots T_{k-1}^{\mathrm{u}}(\mathsf{X}) \right]^2} \leq 4(1-\rho)^{-2} \rho^{2(k-l)} \left\| f \right\|_\infty^2 .$$

This bound together with Proposition 9.4.8 completes the proof. □

9.4.2.2 Exponential Inequalities

The induction argument previously used for the central limit theorem may also be used to derive exponential inequalities for the tail probabilities.

Theorem 9.4.11. *Assume 9.4.1 and that there exist some constants $a(0)$ and $b(0)$ such that for any $t \geq 0$ and $f \in \mathcal{F}_{\mathrm{b}}(\mathsf{X})$,*

$$\mathrm{P} \left[\left| M_N^{-1} \sum_{i=1}^{M_N} f(\xi_0^{N,i}) - \phi_{\nu,0}(f) \right| \geq t \right] \leq a(0) \exp\left[- \frac{2 M_N t^2}{b(0)^2 \operatorname{osc}^2(f)} \right] . \tag{9.47}$$

Then for any $k > 0$, $t > 0$ and $f \in \mathcal{F}_{\mathrm{b}}(\mathsf{X})$,

$$\mathrm{P}\left[\left|M_N^{-1}\sum_{i=1}^{M_N} f(\xi_k^{N,i}) - \phi_{\nu,k}(f)\right| \geq t\right] \leq a(k)\exp\left[-\frac{2M_N t^2}{b(k)^2 \operatorname{osc}^2(f)}\right], \quad (9.48)$$

where the constants $a(k)$ and $b(k)$ are defined recursively through

$$a(k+1) = 2\,(1+a(k))\ ,$$
$$b(k+1) = \frac{(3/2)\,\|g_{k+1}\|_\infty\, b(k) + \phi_{\nu,k} T_k^{\mathrm{u}}(\mathsf{X})}{\phi_{\nu,k} T_k^{\mathrm{u}}(\mathsf{X})}\ .$$

Proof. The proof is by induction; assume that the claim is true for some $k \geq 0$. Decompose $M_N^{-1}\sum_{k=1}^{M_N} f(\xi_{k+1}^{N,i}) - \phi_{\nu,k+1}(f)$ in two terms $A_{k+1}^N(f) + B_{k+1}^N(f)$, where

$$A_{k+1}^N(f) = M_N^{-1}\sum_{i=1}^{M_N}(f(\xi_{k+1}^{N,i}) - \mathrm{E}[f(\xi_{k+1}^{N,i})\,|\,\bar{\mathcal{F}}_k^N])$$
$$B_{k+1}^N(f) = M_N^{-1}\sum_{i=1}^{M_N} \mathrm{E}[f(\xi_{k+1}^{N,i})\,|\,\bar{\mathcal{F}}_k^N] - \phi_{\nu,k+1}(f)$$
$$= \frac{\sum_{k=1}^{M_N} T_k^{\mathrm{u}} f(\xi_k^{N,i})}{\sum_{k=1}^{M_N} T_k^{\mathrm{u}}(\xi_k^{N,i},\mathsf{X})} - \frac{\phi_{\nu,k} T_k^{\mathrm{u}} f}{\phi_{\nu,k} T_k^{\mathrm{u}}(\mathsf{X})}\ .$$

Proceeding like in Theorem 9.2.16, for any $a \in (0,1)$ and $t \geq 0$,

$$\mathrm{P}(|A_{k+1}^N(f)| \geq at) \leq 2\exp\left[-2a^2t^2 M_N/\operatorname{osc}^2(f)\right]\ . \quad (9.49)$$

We now bound $B_{k+1}^N(f)$. First note first for any constant c, $B_{k+1}^N(f) = B_{k+1}^N(f-c)$. We choose c in such a way that $\|f-c\|_\infty = (1/2)\operatorname{osc}(f)$ and set $\bar{f} = f - c$. Writing

$$B_{k+1}^N(f) = \frac{M_N^{-1}\sum_{i=1}^{M_n}\{T_k^{\mathrm{u}}\bar{f}(\xi_k^{N,i}) - \phi_{\nu,k}T_k^{\mathrm{u}}\bar{f}\}}{\phi_{\nu,k}T_k^{\mathrm{u}}(\mathsf{X})}$$
$$- \frac{\sum_{i=1}^{M_N} T_k^{\mathrm{u}}\bar{f}(\xi_k^{N,i})}{\sum_{i=1}^{M_N} T_k^{\mathrm{u}}(\xi_k^{N,i},\mathsf{X})}\,\frac{M_N^{-1}\sum_{i=1}^{M_N}\{T_k^{\mathrm{u}}(\xi_k^{N,i},\mathsf{X}) - \phi_{\nu,k}T_k^{\mathrm{u}}(\mathsf{X})\}}{\phi_{\nu,k}T_k^{\mathrm{u}}(\mathsf{X})} \quad (9.50)$$

and using the induction assumption, it holds that for any $b \in (0,1)$,

$$\mathrm{P}\left[|B_{k+1}^N(f)| \geq (1-a)t\right] \leq a(k)\exp\left\{-\frac{2M_N(1-a)^2b^2t^2[\phi_{\nu,k}T_k^{\mathrm{u}}(\mathsf{X})]^2}{b^2(k)\operatorname{osc}^2\left(T_k^{\mathrm{u}}\bar{f}\right)}\right\}$$
$$+ a(k)\exp\left\{-\frac{2M_N(1-a)^2(1-b)^2t^2[\phi_{\nu,k}T_k^{\mathrm{u}}(\mathsf{X})]^2}{b^2(k)\,\|\bar{f}\|_\infty^2\operatorname{osc}^2(T_k^{\mathrm{u}}\mathbb{1})}\right\}\ .$$

By Lemma 4.3.4, for any $(x,x') \in \mathsf{X}\times\mathsf{X}$,

$$
\begin{aligned}
|T^{\mathrm{u}}_k \bar{f}(x) - T^{\mathrm{u}}_k \bar{f}(x')| &= |Q(x, g_{k+1}\bar{f}) - Q(x', g_{k+1}\bar{f})| \\
&\leq (1/2) \left\| Q(x,\cdot) - Q(x',\cdot) \right\|_{\mathrm{TV}} \operatorname{osc}\left(g_{k+1}\bar{f}\right) \\
&\leq \left\| g_{k+1} \right\|_\infty \operatorname{osc}(f) ,
\end{aligned}
$$

and similarly,

$$
|T^{\mathrm{u}}_k(x, \mathsf{X}) - T^{\mathrm{u}}_k(x', \mathsf{X})| = |Q(x, g_{k+1}) - Q(x', g_{k+1})| \leq \left\| g_{k+1} \right\|_\infty .
$$

Thus, $\operatorname{osc}\left(T^{\mathrm{u}}_k \bar{f}\right)$ and $\operatorname{osc}\left(T^{\mathrm{u}}_k \mathbb{1}\right)$ are bounded by $\left\| g_{k+1} \right\|_\infty \operatorname{osc}(f)$ and $\left\| g_{k+1} \right\|_\infty$, respectively. The result follows by choosing $b = 2/3$ as in the proof of Theorem 9.1.10 and then setting a to equate the bounds on $A^N_{k+1}(f)$ and $B^N_{k+1}(f)$. □

The bound is still of Hoeffding type, but at each iteration the constants $a(k)$ and $b(k)$ increase. Hence, the obtained bound is almost useless in practice for large k, except when the number of iterations is small or the number of particles is large (compared to the iteration index). It would of course be more appropriate to derive an exponential bound with constants that do not depend on the iteration index. Such results hold true when Q satisfies the strong mixing condition.

Theorem 9.4.12. *Assume 9.4.1, 9.4.9, and* (9.47). *Then there exist constants a and b such that for any $n \geq 0$, $t \geq 0$ and $f \in \mathcal{F}_{\mathrm{b}}(\mathsf{X})$,*

$$
\mathrm{P}\left[\left| M_N^{-1} \sum_{i=1}^{M_N} f(\xi^{N,i}_n) - \phi_{\nu,n}(f) \right| \geq t \right] \leq a \exp\left[-\frac{2 M_N t^2}{b^2 \operatorname{osc}^2(f)} \right] .
$$

Proof. Define $\hat{\phi}^N_k = M_N^{-1} \sum_{i=1}^{M_N} \delta_{\xi^{N,i}_k}$. The difference $\hat{\phi}^N_n(f) - \phi_{\nu,n}(f)$ may be expressed as the telescoping sum

$$
\hat{\phi}^N_n(f) - \phi_{\nu,n}(f) = \frac{\hat{\phi}^N_0 T^{\mathrm{u}}_0 \cdots T^{\mathrm{u}}_{n-1} f}{\hat{\phi}^N_0 T^{\mathrm{u}}_0 \cdots T^{\mathrm{u}}_{n-1}(\mathsf{X})} - \phi_{\nu,n}(f) + \\
\sum_{k=1}^{n} \left\{ \frac{\hat{\phi}^N_k T^{\mathrm{u}}_k \cdots T^{\mathrm{u}}_{n-1} f}{\hat{\phi}^N_k T^{\mathrm{u}}_k \cdots T^{\mathrm{u}}_{n-1}(\mathsf{X})} - \frac{\hat{\phi}^N_{k-1} T^{\mathrm{u}}_{k-1} \cdots T^{\mathrm{u}}_{n-1} f}{\hat{\phi}^N_{k-1} T^{\mathrm{u}}_{k-1} \cdots T^{\mathrm{u}}_{n-1}(\mathsf{X})} \right\} , \tag{9.51}
$$

with the convention that $T^{\mathrm{u}}_k \cdots T^{\mathrm{u}}_{n-1}$ is the identity mapping when $k = n$. We shall show that the tail probabilities of each of the terms on the right-hand side of (9.51) are exponentially small. Put

$$
A^N_n(f) = \frac{\hat{\phi}^N_0 T^{\mathrm{u}}_0 \cdots T^{\mathrm{u}}_{n-1} f}{\hat{\phi}^N_0 T^{\mathrm{u}}_0 \cdots T^{\mathrm{u}}_{n-1}(\mathsf{X})} - \phi_{\nu,n}(f) \tag{9.52}
$$

$$
= \frac{\sum_{i=1}^{M_N} \beta_{0|n}(\xi^{N,i}_0) \{ \mathrm{F}_{0|n} \cdots \mathrm{F}_{n-1|n} f(\xi^{N,i}_0) - \phi_{\nu,n}(f) \}}{\sum_{i=1}^{M_N} \beta_{0|n}(\xi^{N,i}_0)} , \tag{9.53}
$$

where $\phi_{\nu,n}(f)$ could also be rewritten as $\phi_{\nu,0}T_0^{\mathrm{u}}\cdots T_{n-1}^{\mathrm{u}}(f)$ (see Section 3.3.1). Thus by Lemma 4.3.4 and Proposition 4.3.23(i),

$$\left\|\mathrm{F}_{0|n}\cdots\mathrm{F}_{n-1|n}(\cdot,f)-\phi_{\nu,n}(f)\right\|_{\infty}\leq\rho^{n}\operatorname{osc}(f) \tag{9.54}$$

and

$$\operatorname{osc}\left(\mathrm{F}_{0|n}\cdots\mathrm{F}_{n-1|n}(\cdot,f)\right)\leq\rho^{n}\operatorname{osc}(f)\ . \tag{9.55}$$

In addition

$$\frac{\phi_{\nu,0}(\beta_{0|n})}{\operatorname{osc}\left(\beta_{0|n}(\cdot)\right)}\geq\frac{\phi_{\nu,0}(\beta_{0|n})}{2\left\|\beta_{0|n}(\cdot)\right\|_{\infty}}\geq\frac{\sigma^{-}}{\sigma^{+}}=1-\rho\ , \tag{9.56}$$

where Lemma 4.3.22(ii) was used for the second inequality. Writing

$$\begin{aligned}A_n^N(f)=M_N^{-1}&\frac{\sum_{i=1}^{M_N}\beta_{0|n}(\xi_0^{N,i})\{\mathrm{F}_{0|n}\cdots\mathrm{F}_{n-1|n}f(\xi_0^{N,i})-\phi_{\nu,n}(f)\}}{\phi_{\nu,0}(\beta_{0|n})}\\&+\frac{\sum_{i=1}^{M_N}\beta_{0|n}(\xi_0^{N,i})\{\mathrm{F}_{0|n}\cdots\mathrm{F}_{n-1|n}f(\xi_0^{N,i})-\phi_{\nu,n}(f)\}}{\sum_{i=1}^{M_N}\beta_{0|n}(\xi_0^{N,i})}\\&\qquad\times\left[1-M_N^{-1}\frac{\sum_{i=1}^{M_N}\beta_{0|n}(\xi_0^{N,i})}{\phi_{\nu,0}(\beta_{0|n})}\right]\end{aligned}$$

we have, using (9.54) and the triangle inequality,

$$\begin{aligned}A_n^N(f)\leq M_N^{-1}&\left|\frac{\sum_{i=1}^{M_N}\beta_{0|n}(\xi_0^{N,i})\{\mathrm{F}_{0|n}\cdots\mathrm{F}_{n-1|n}f(\xi_0^{N,i})-\phi_{\nu,n}(f)\}}{\phi_{\nu,0}(\beta_{0|n})}\right|\\&+\rho^n\operatorname{osc}(f)\,M_N^{-1}\left|\frac{\sum_{i=1}^{M_N}\beta_{0|n}(\xi_0^{N,i})-\phi_{\nu,0}(\beta_{0|n})}{\phi_{\nu,0}(\beta_{0|n})}\right|\ .\end{aligned}$$

Using (9.56) as well as (9.47) twice (for the functions $\mathrm{F}_{0|n}\cdots\mathrm{F}_{n-1|n}f$ and $\beta_{0|n}$) shows that for any $t\geq 0$,

$$\mathrm{P}\left[|A_n^N(f)|\geq t\right]\leq 2a(0)\exp\left[-\frac{M_N t^2(1-\rho)^2}{2b^2(0)\operatorname{osc}^2(f)\,\rho^{2n}}\right]\ . \tag{9.57}$$

For $1\leq k\leq n$, put

$$\Delta_{k,n}^N(f)=\frac{\hat{\phi}_k^N T_k^{\mathrm{u}}\cdots T_{n-1}^{\mathrm{u}}f}{\hat{\phi}_k^N T_k^{\mathrm{u}}\cdots T_{n-1}^{\mathrm{u}}(\mathsf{X})}-\frac{\hat{\phi}_{k-1}^N T_{k-1}^{\mathrm{u}}\cdots T_{n-1}^{\mathrm{u}}f}{\hat{\phi}_{k-1}^N T_{k-1}^{\mathrm{u}}\cdots T_{n-1}^{\mathrm{u}}(\mathsf{X})}\ . \tag{9.58}$$

Proposition 3.3.2 shows that $T_k^{\mathrm{u}}\cdots T_{n-1}^{\mathrm{u}}(x,A)=\beta_{k|n}(x)\mathrm{F}_{k|n}\cdots\mathrm{F}_{n-1|n}(x,A)$. Pick $x_0\in\mathsf{X}$. Then

$$\frac{\hat{\phi}_k^N T_k^{\mathrm{u}}\cdots T_{n-1}^{\mathrm{u}}f}{\hat{\phi}_k^N T_k^{\mathrm{u}}\cdots T_{n-1}^{\mathrm{u}}(\mathsf{X})}-\mathrm{F}_{k|n}\cdots\mathrm{F}_{n-1|n}(x_0)=\frac{\beta_{k|n}(\xi_k^{N,i})\psi_{k|n}(\xi_k^{N,i})}{\sum_{i=1}^{M_N}\beta_{k|n}(\xi_k^{N,i})}\ , \tag{9.59}$$

where $\psi_{k|n}(x) = \mathrm{F}_{k|n} \cdots \mathrm{F}_{k-1|n} f(x) - \mathrm{F}_{k|n} \cdots \mathrm{F}_{k-1|n} f(x_0)$. Set

$$\tilde{\phi}_k^N = \frac{\hat{\phi}_{k-1}^N T_{k-1}^{\mathrm{u}}}{\hat{\phi}_{k-1}^N T_{k-1}^{\mathrm{u}}(\mathsf{X})} \quad \text{and} \quad \tilde{\mu}_{k|n}^N(A) = \frac{\int_A \tilde{\phi}_k^N(dx)\,\beta_{k|n}(x)}{\int_{\mathsf{X}} \tilde{\phi}_k^N(dx)\,\beta_{k|n}(x)} .$$

Then $\tilde{\mu}_{k|n}^N \ll \tilde{\phi}_k^N$, with Radon-Nikodym derivative

$$\frac{d\tilde{\mu}_{k|n}^N}{d\tilde{\phi}_k^N}(x) = \frac{\beta_{k|n}(x)}{\tilde{\phi}_k^N(\beta_{k|n})} .$$

Using these notations,

$$\frac{\hat{\phi}_{k-1}^N T_{k-1}^{\mathrm{u}} \cdots T_{n-1}^{\mathrm{u}} f}{\hat{\phi}_{k-1}^N T_{k-1}^{\mathrm{u}} \cdots T_{n-1}^{\mathrm{u}}(\mathsf{X})} - \mathrm{F}_{k|n} \cdots \mathrm{F}_{n-1|n} f(x_0) =$$
$$\frac{\tilde{\phi}_k^N\left[\beta_{k|n}\{\mathrm{F}_{k|n} \cdots \mathrm{F}_{n-1|n} f - \mathrm{F}_{k|n} \cdots \mathrm{F}_{n-1|n} f(x_0)\}\right]}{\tilde{\phi}_k^N(\beta_{k|n})} = \tilde{\mu}_{k|n}^N(\psi_{k|n}) . \quad (9.60)$$

Combining (9.59) and (9.60), we may express $\Delta_{k,n}^N(f)$ as

$$\Delta_{k,n}^N(f) = \frac{\sum_{i=1}^{\tilde{M}_N} \frac{d\tilde{\mu}_{k|n}^N}{d\tilde{\phi}_k^N}(\xi_k^{N,i})\psi_{k|n}(\xi_k^{N,i})}{\sum_{i=1}^{M_N} \frac{d\tilde{\mu}_{k|n}^N}{d\tilde{\phi}_k^N}(\xi_k^{N,i})} - \tilde{\mu}_{k|n}^N(\psi_{k|n}) .$$

Because $\{\xi_k^{N,i}\}_{1 \le i \le M_N}$ are conditionally i.i.d. given $\mathcal{F}_{k-1}^N$ with common distribution $\tilde{\phi}_k^N$, the first term in the above expression may be seen as an importance sampling estimator of $\tilde{\mu}_{k|n}^N(\psi_{k|n})$. By Lemma 4.3.22(ii), the Radon-Nikodym derivative $d\tilde{\mu}_{k|n}^N / d\tilde{\phi}_k^N(x)$ is bounded uniformly in k, N and x as

$$\frac{d\tilde{\mu}_{k|n}^N}{d\tilde{\phi}_k^N}(x) \le \frac{\sigma^+}{\sigma^-} = \frac{1}{1-\rho} .$$

Proceeding as above, the Hoeffding inequality implies that for any $t \ge 0$,

$$\mathrm{P}\left[\left|\Delta_{k,n}^N(f)\right| \ge t\right] \le 2\exp\left[-\frac{M_N t^2(1-\rho)^2}{2\,\mathrm{osc}^2(f)\,\rho^{2(n-k)}}\right] .$$

Hence the probability that the sum on the right-hand side of (9.51) is (in absolute value) at least t is bounded by

$$2\sum_{k=0}^{n-1} \exp\left[-\frac{M_N t^2(1-\rho)^2 b_k^2}{2\,\mathrm{osc}^2(f)\,\rho^{2k}}\right] \quad (9.61)$$

for any sequence $\{b_k\}_{0 \le k \le n-1}$ of positive numbers summing to one. To obtain a bound that does not depend on n, take $b_k = \theta^k(1-\theta)/(1-\theta^n)$ with $\rho < \theta < 1$. This choice proves that (9.61) is bounded by

$$a \exp\left[-\frac{M_N t^2 (1-\rho)^2 (1-\theta^2)}{2\,\mathrm{osc}^2(f)}\right],$$

where a is a constant that depends only on θ and ρ. □

9.5 Complements

9.5.1 Weak Limits Theorems for Triangular Array

This section summarizes various basic results on the asymptotics of triangular arrays that are used in the proofs of this chapter.

9.5.1.1 Law of Large Numbers

Throughout this section, $\{M_N\}_{N\geq 0}$ denotes a sequence of integers. All random variables are assumed to be defined on a common probability space $(\Omega, \mathcal{F}, \mathrm{P})$.

Proposition 9.5.1. *Let $\{U_{N,i}\}_{1\leq i\leq M_N}$ be a triangular array of random variables and let $\{\mathcal{F}^N\}_{N\geq 0}$ be a sequence of sub-σ-fields of $\mathcal{F}$. Assume that the following conditions hold true.*

(i) The triangular array is conditionally independent given $\{\mathcal{F}^N\}$ and for any N and $i = 1,\ldots,M_N$, $\mathrm{E}[|U_{N,i}|\,|\,\mathcal{F}^N] < \infty$ and $\mathrm{E}[U_{N,i}\,|\,\mathcal{F}^N] = 0$.
(ii) For some positive ϵ,

$$\sum_{i=1}^{M_N} \mathrm{E}[U_{N,i}^2 \mathbb{1}_{\{|U_{N,i}|<\epsilon\}}\,|\,\mathcal{F}^N] \stackrel{\mathrm{P}}{\longrightarrow} 0\,, \tag{9.62}$$

$$\sum_{i=1}^{M_N} \mathrm{E}[|U_{N,i}| \mathbb{1}_{\{|U_{N,i}|\geq\epsilon\}}\,|\,\mathcal{F}^N] \stackrel{\mathrm{P}}{\longrightarrow} 0\,. \tag{9.63}$$

Then

$$\sum_{i=1}^{M_N} U_{N,i} \stackrel{\mathrm{P}}{\longrightarrow} 0\,.$$

Proof. Consider the truncated random variable $\bar{U}_{N,i} = U_{N,i}\mathbb{1}_{\{|U_{N,i}|<\epsilon\}}$. Using (9.63) and $\mathrm{E}[U_{N,i}\,|\,\mathcal{F}^N] = 0$, we find that

$$\sum_{i=1}^{M_N} \mathrm{E}[\bar{U}_{N,i}\,|\,\mathcal{F}^N] \stackrel{\mathrm{P}}{\longrightarrow} 0\,. \tag{9.64}$$

By Chebyshev's inequality, it follows that for any $\delta > 0$,

$$A_N(\delta) = \mathrm{P}\left(\left|\sum_{i=1}^{M_N} \bar{U}_{N,i} - \sum_{i=1}^{M_N} \mathrm{E}[\bar{U}_{N,i} \,|\, \mathcal{F}^N]\right| \geq \delta \,\middle|\, \mathcal{F}^N\right)$$
$$\leq \delta^{-2} \operatorname{Var}\left(\sum_{i=1}^{M_N} \bar{U}_{N,i} \,\middle|\, \mathcal{F}^N\right) ,$$

and hence (9.62) shows that $A_N(\delta) \to 0$ in probability. Because $A_N(\delta)$ is obviously bounded, we also have $\mathrm{E}[A_N(\delta)] \to 0$, that is,

$$\sum_{i=1}^{M_N} \bar{U}_{N,i} - \sum_{i=1}^{M_N} \mathrm{E}[\bar{U}_{N,i} \,|\, \mathcal{F}^N] \xrightarrow{\mathrm{P}} 0 . \tag{9.65}$$

Moreover, for any $\delta > 0$,

$$\mathrm{P}\left(\left|\sum_{i=1}^{M_N} U_{N,i} - \sum_{i=1}^{M_N} \bar{U}_{N,i}\right| \geq \delta \,\middle|\, \mathcal{F}^N\right) \leq \mathrm{P}\left(\sum_{i=1}^{M_N} |U_{N,i}| \mathbb{1}_{\{|U_{N,i}| \geq \epsilon\}} \geq \delta \,\middle|\, \mathcal{F}^N\right)$$
$$\leq \delta^{-1} \sum_{i=1}^{M_N} \mathrm{E}[|U_{N,i}| \mathbb{1}_{\{|U_{N,i}| \geq \epsilon\}} \,|\, \mathcal{F}^N] \xrightarrow{\mathrm{P}} 0 .$$

Thus, $\sum_{i=1}^{M_N} U_{N,i} - \sum_{i=1}^{M_N} \bar{U}_{N,i} \to 0$ in probability. Combining with (9.64) and (9.65), the proof is complete. □

Definition 9.5.2 (Bounded in Probability). *A sequence $\{Z_N\}_{N \geq 0}$ of random variables is said to be* bounded in probability *if*

$$\lim_{C \to \infty} \sup_{N \geq 0} \mathrm{P}(|Z_N| \geq C) = 0.$$

Often the term *tight*, or *asymptotically tight*, is used instead of "bounded in probability". We recall without proof the following elementary properties.

Lemma 9.5.3.

1. *Let $\{U_N\}_{N \geq 0}$ and U be random variables. If $U_N \xrightarrow{\mathcal{D}} U$, then $\{U_N\}$ is bounded in probability.*
2. *Let $\{U_N\}_{N \geq 0}$ and $\{V_N\}_{N \geq 0}$ be two sequences of random variables. If $\{V_N\}$ is bounded in probability and $|U_N| \leq |V_N|$ for any N, then $\{U_N\}$ is bounded in probability.*
3. *Let $\{U_N\}_{N \geq 0}$ and $\{V_N\}_{N \geq 0}$ be two sequences of random variables. If $\{U_N\}_{N \geq 0}$ is bounded in probability and $V_N \to 0$ in probability, then $U_N V_N \to 0$ in probability.*
4. *Let $\{U_N\}_{N \geq 0}$ be a sequence of random variables and let $\{M_N\}_{N \geq 0}$ be a non-decreasing deterministic sequence diverging to infinity. If $\{U_N\}$ is bounded in probability, then $\mathbb{1}_{\{U_N \geq M_N\}} \to 0$ in probability.*

The following elementary lemma is repeatedly used in the sequel.

Lemma 9.5.4. *Let $\{U_N\}_{N\geq 0}$ and $\{V_N\}_{N\geq 0}$ be two sequences of random variables such that $\{V_N\}$ is bounded in probability. Assume that for any positive η there exists a sequence $\{W_N(\eta)\}_{N\geq 0}$ of random variables such that $W_N(\eta) \xrightarrow{\mathrm{P}} 0$ as $N \to \infty$ and*

$$|U_N| \leq \eta V_N + W_N(\eta) .$$

Then $U_N \xrightarrow{\mathrm{P}} 0$.

Proof. For any $\delta > 0$,

$$\mathrm{P}(|U_N| \geq \delta) \leq \mathrm{P}[V_N \geq \delta/(2\eta)] + \mathrm{P}[W_N(\eta) \geq \delta/2] .$$

This implies that for any $\eta > 0$,

$$\limsup_{N\to\infty} \mathrm{P}(|U_N| \geq \delta) \leq \sup_{N\geq 0} \mathrm{P}[V_N \geq \delta/(2\eta)] .$$

Because the right-hand side can be made arbitrarily small by letting $\eta \to 0$, the result follows. □

Proposition 9.5.5. *Let $\{U_{N,i}\}_{1\leq i\leq M_N}$ be a triangular array of random variables and let $\{\mathcal{F}^N\}_{N\geq 0}$ be a sequence of sub-σ-fields of $\mathcal{F}$. Assume that the following conditions hold true.*

(i) The triangular array is conditionally independent given $\{\mathcal{F}^N\}$ and for any N and $i = 1,\ldots,M_N$, $\mathrm{E}[|U_{N,i}|\,|\,\mathcal{F}^N] < \infty$ and $\mathrm{E}[U_{N,i}\,|\,\mathcal{F}^N] = 0$.
(ii) The sequence of random variables

$$\left\{\sum_{i=1}^{M_N} \mathrm{E}[|U_{N,i}|\,|\,\mathcal{F}^N]\right\}_{N\geq 0} \tag{9.66}$$

is bounded in probability.
(iii) For any positive η,

$$\sum_{i=1}^{M_N} \mathrm{E}[|U_{N,i}|\mathbb{1}_{\{|U_{N,i}|\geq\eta\}}\,|\,\mathcal{F}^N] \xrightarrow{\mathrm{P}} 0 . \tag{9.67}$$

Then

$$\sum_{i=1}^{M_N} U_{N,i} \xrightarrow{\mathrm{P}} 0 .$$

Proof. We employ Proposition 9.5.1 and then need to check its condition (ii). The current condition (iii) is (9.63), so it suffices to prove that (9.62) holds for some (arbitrary) $\epsilon > 0$. To do that, note that for any $\eta \in (0, \epsilon)$,

$$\sum_{i=1}^{M_N} \mathrm{E}[U_{N,i}^2 \mathbb{1}_{\{|U_{N,i}|<\epsilon\}} \,|\, \mathcal{F}^N]$$
$$\leq \sum_{i=1}^{M_N} \mathrm{E}[U_{N,i}^2 \mathbb{1}_{\{|U_{N,i}|<\eta\}} \,|\, \mathcal{F}^N] + \sum_{i=1}^{M_N} \mathrm{E}[U_{N,i}^2 \mathbb{1}_{\{\eta\leq|U_{N,i}|<\epsilon\}} \,|\, \mathcal{F}^N]$$
$$\leq \eta \sum_{i=1}^{M_N} \mathrm{E}[|U_{N,i}| \,|\, \mathcal{F}^N] + \epsilon \sum_{i=1}^{M_N} \mathrm{E}[|U_{N,i}| \mathbb{1}_{\{|U_{N,i}|\geq\eta\}} \,|\, \mathcal{F}^N] \,.$$

Now (9.62) follows from Lemma 9.5.4. □

In the special case where the random variables $\{U_{N,i}\}_{1\leq i\leq M_N}$, for any N, are conditionally i.i.d. given $\{\mathcal{F}^N\}$, Proposition 9.5.5 admits a simpler formulation.

Corollary 9.5.6. *Let $\{V_{N,i}\}_{1\leq i\leq M_N}$ be a triangular array of random variables and let $\{\mathcal{F}^N\}_{N\geq 0}$ be a sequence of sub-σ-fields of $\mathcal{F}$. Assume that the following conditions hold true.*

(i) The triangular array is conditionally i.i.d. given $\mathcal{F}^N$ and for any N, $\mathrm{E}[|V_{N,1}| \,|\, \mathcal{F}^N] < \infty$ and $\mathrm{E}[V_{N,1} \,|\, \mathcal{F}^N] = 0$.
(ii) The sequence $\{\mathrm{E}[|V_{N,1}| \,|\, \mathcal{F}^N]\}_{N\geq 0}$ is bounded in probability.
(iii) For any positive η, $\mathrm{E}[|V_{N,1}| \mathbb{1}_{\{|V_{N,1}|\geq \eta M_N\}} \,|\, \mathcal{F}^N] \to 0$ in probability.

Then

$$M_N^{-1} \sum_{i=1}^{M_N} V_{N,i} \xrightarrow{\mathrm{P}} 0 \,.$$

Proposition 9.5.7. *Let $\{V_{N,i}\}_{1\leq i\leq M_N}$ be a triangular array of random variables and let $\{\mathcal{F}^N\}$ be a sequence of sub-σ-fields of $\mathcal{F}$. Assume that the following conditions hold true.*

(i) The triangular array is conditionally independent given $\{\mathcal{F}^N\}$ and for any N and $i = 1, \ldots, M_N$, $\mathrm{E}[|V_{N,i}| \,|\, \mathcal{F}^N] < \infty$.
(ii) The sequence $\{\sum_{i=1}^{M_N} \mathrm{E}[|V_{N,i}| \,|\, \mathcal{F}^N]\}_{N\geq 0}$ is bounded in probability,
(iii) For any positive ϵ,

$$\sum_{i=1}^{M_N} \mathrm{E}[|V_{N,i}| \mathbb{1}_{\{|V_{N,i}|\geq\epsilon\}} \,|\, \mathcal{F}^N] \xrightarrow{\mathrm{P}} 0 \,. \tag{9.68}$$

Then

$$\sum_{i=1}^{M_N} \{V_{N,i} - \mathrm{E}[V_{N,i} \,|\, \mathcal{F}^N]\} \xrightarrow{\mathrm{P}} 0 \,.$$

Proof. We check that the triangular array $U_{N,i} = V_{N,i} - \mathrm{E}[V_{N,i} \,|\, \mathcal{F}^N]$ satisfies conditions (i)–(iii) of Proposition 9.5.5. This triangular array is conditionally independent given $\mathcal{F}^N$, and for any N and any $i = 1, \ldots, M_N$, $\mathrm{E}[|U_{N,i}| \,|\, \mathcal{F}^N] \leq 2\,\mathrm{E}[|V_{N,i}| \,|\, \mathcal{F}^N] < \infty$ and $\mathrm{E}[U_{N,i} \,|\, \mathcal{F}^N] = 0$, showing condition (i). In addition

$$\sum_{i=1}^{M_N} \mathrm{E}[|U_{N,i}| \,|\, \mathcal{F}^N] \leq 2 \sum_{i=1}^{M_N} \mathrm{E}[|V_{N,i}| \,|\, \mathcal{F}^N] ,$$

showing that the sequence $\{\sum_{i=1}^{M_N} \mathrm{E}[|U_{N,i}| \,|\, \mathcal{F}^N]\}_{N \geq 0}$ is bounded in probability. Hence condition (ii) holds.

We now turn to the final condition of Proposition 9.5.5, (9.67). With the bounds $|U_{N,i}| \leq |V_{N,i}| + \mathrm{E}[|V_{N,i}| \,|\, \mathcal{F}^N]$ and $\mathbb{1}_{\{|U_{N,i}| \geq \epsilon\}} \leq \mathbb{1}_{\{|V_{N,i}| \geq \epsilon/2\}} + \mathbb{1}_{\{\mathrm{E}[|V_{N,i}| \,|\, \mathcal{F}^N] \geq \epsilon/2\}}$ and in view of the assumed condition (iii), it suffices to prove that for any positive ϵ,

$$A_N = \sum_{i=1}^{M_N} \mathrm{E}[|V_{N,i}| \,|\, \mathcal{F}^N]\, \mathrm{P}(|V_{N,i}| \geq \epsilon \,|\, \mathcal{F}^N) \xrightarrow{\mathrm{P}} 0 , \tag{9.69}$$

$$B_N = \sum_{i=1}^{M_N} \mathrm{E}[|V_{N,i}| \,|\, \mathcal{F}^N] \mathbb{1}_{\{\mathrm{E}[|V_{N,i}| \,|\, \mathcal{F}^N] \geq \epsilon\}} \xrightarrow{\mathrm{P}} 0 . \tag{9.70}$$

Bound A_N as

$$A_N \leq \mathrm{P}\left(\max_{1 \leq i \leq M_N} |V_{N,i}| \geq \epsilon \,\middle|\, \mathcal{F}^N \right) \sum_{i=1}^{M_N} \mathrm{E}[|V_{N,i}| \,|\, \mathcal{F}^N] .$$

Considering the assumed condition (ii), it is sufficient to prove that the conditional probability of the display tends to zero in probability. To do that, notice that

$$\max_{1 \leq i \leq M_N} |V_{N,i}| \leq \epsilon/2 + \sum_{i=1}^{M_N} |V_{N,i}| \mathbb{1}_{\{|V_{N,i}| \geq \epsilon/2\}} ,$$

whence, using condition (iii),

$$\begin{aligned} \mathrm{P}\left(\max_{1 \leq i \leq M_N} |V_{N,i}| \geq \epsilon \,\middle|\, \mathcal{F}^N \right) &\leq \mathrm{P}\left(\sum_{i=1}^{M_N} |V_{N,i}| \mathbb{1}_{\{|V_{N,i}| \geq \epsilon/2\}} \geq \epsilon/2 \,\middle|\, \mathcal{F}^N \right) \\ &\leq (2/\epsilon) \sum_{i=1}^{M_N} \mathrm{E}[|V_{N,i}| \mathbb{1}_{\{|V_{N,i}| \geq \epsilon/2\}} \,|\, \mathcal{F}^N] \xrightarrow{\mathrm{P}} 0 . \end{aligned}$$

Thus (9.69) holds. Now bound B_N as

$$B_N \leq \mathbb{1}_{\{\max_{1 \leq i \leq M_N} \mathrm{E}[|V_{N,i}| \,|\, \mathcal{F}^N] \geq \epsilon\}} \sum_{i=1}^{M_N} \mathrm{E}[|V_{N,i}| \,|\, \mathcal{F}^N] .$$

To show that $B_N \to 0$ in probability, it is again sufficient to prove that so does the first factor. In a similar fashion as above we have

$$\mathbb{1}\left\{\max_{1\le i\le M_N} \mathrm{E}[|V_{N,i}|\,|\,\mathcal{F}^N] \ge \epsilon\right\}$$
$$\le \mathbb{1}\left\{\sum_{i=1}^{M_N} \mathrm{E}[|V_{N,i}|\mathbb{1}_{\{|V_{N,i}|\ge\epsilon/2\}}\,|\,\mathcal{F}^N] \ge \epsilon/2\right\}$$
$$\le (2/\epsilon)\sum_{i=1}^{M_N} \mathrm{E}[|V_{N,i}|\mathbb{1}_{\{|V_{N,i}|\ge\epsilon/2\}}\,|\,\mathcal{F}^N] \stackrel{\mathrm{P}}{\longrightarrow} 0\,.$$

Thus (9.70) holds. By combining (9.68), (9.69), and (9.70) we find that (9.67) holds, concluding the proof. □

9.5.1.2 Central Limit Theorems

Lemma 9.5.8. *Let $z_1,\dots,z_m$ and $z'_1,\dots,z'_m$ be complex numbers of modulus at most* 1. *Then*

$$|z_1\cdots z_m - z'_1\cdots z'_m| \le \sum_{i=1}^m |z_i - z'_i|\,.$$

Proof. This follows by induction from

$$z_1\cdots z_m - z'_1\cdots z'_m = (z_1 - z'_1)z_2\cdots z_m + z'_1(z_2\cdots z_m - z'_1\cdots z'_m)\,.$$

□

In the investigation of the central limit theorem for triangular arrays, the so-called Lindeberg condition plays a fundamental role.

Proposition 9.5.9. *Let $\{U_{N,i}\}_{1\le i\le M_N}$ be a triangular array of random variables and let $\{\mathcal{F}^N\}_{N\ge 0}$ be a sequence of sub-σ-fields of $\mathcal{F}$. Assume that the following conditions hold true.*

(i) The triangular array is conditionally independent given $\{\mathcal{F}^N\}$ and for any N and $i=1,\dots,M_N$, $\mathrm{E}[U^2_{N,i}\,|\,\mathcal{F}^N]<\infty$, and $\mathrm{E}[U_{N,i}\,|\,\mathcal{F}^N]=0$.
(ii) There exists a positive constant σ^2 such that with $\sigma^2_{N,i}=\mathrm{E}[U^2_{N,i}\,|\,\mathcal{F}^N]$,

$$\sum_{i=1}^{M_N} \sigma^2_{N,i} \stackrel{\mathrm{P}}{\longrightarrow} \sigma^2\,. \tag{9.71}$$

(iii) For all $\epsilon>0$,

$$\sum_{i=1}^{M_N} \mathrm{E}[U^2_{N,i}\mathbb{1}_{\{|U_{N,i}|\ge\epsilon\}}\,|\,\mathcal{F}^N] \stackrel{\mathrm{P}}{\longrightarrow} 0. \tag{9.72}$$

Then for any real u,

$$\mathrm{E}\left[\exp\left(\mathrm{i}u\sum_{i=1}^{M_N} U_{N,i}\right) \middle| \mathcal{F}^N\right] \xrightarrow{\mathrm{P}} \exp\left(-\sigma^2 u^2/2\right) . \tag{9.73}$$

Remark 9.5.10. The condition (9.72) is often referred to as the Lindeberg condition. If this condition is satisfied, then the triangular array also satisfies the *uniform smallness condition*, $\max_{1\le i\le M_N} \mathrm{E}[U_{N,i}^2 \,|\, \mathcal{F}^N] \to 0$ in probability. Indeed, for any $\epsilon > 0$,

$$\begin{aligned}\sigma_{N,i}^2 &= \mathrm{E}[U_{N,i}^2 \mathbb{1}_{\{|U_{N,i}|<\epsilon\}} \,|\, \mathcal{F}^N] + \mathrm{E}[U_{N,i}^2 \mathbb{1}_{\{|U_{N,i}|\ge\epsilon\}} \,|\, \mathcal{F}^N] \\ &\le \epsilon^2 + \mathrm{E}[U_{N,i}^2 \mathbb{1}_{\{|U_{N,i}|\ge\epsilon\}} \,|\, \mathcal{F}^N] ,\end{aligned}$$

which implies that

$$\max_{1\le i\le M_N} \mathrm{E}[U_{N,i}^2 \,|\, \mathcal{F}^N] \le \epsilon^2 + \sum_{i=1}^{M_N} \mathrm{E}[U_{N,i}^2 \mathbb{1}_{|U_{N,i}|\ge\epsilon} \,|\, \mathcal{F}^N] .$$

Because ϵ is arbitrary, the uniform smallness condition is satisfied. The Lindeberg condition guarantees that large values (of the same order as the square root of the variance of the sum) have a negligible influence in the central limit theorem. Such extremely large values have a small influence both on the variance and on the distribution of the sum we investigate. ∎

Proof (of Proposition 9.5.9). The proof is adapted from Billingsley (1995, Theorem 27.1). Because $\sum_{i=1}^N \sigma_{N,i}^2 \xrightarrow{\mathrm{P}} \sigma^2$,

$$\exp\left(-(u^2/2)\sum_{i=1}^{M_N} \sigma_{N,i}^2\right) \xrightarrow{\mathrm{P}} \exp\left(-\sigma^2 u^2/2\right) .$$

Thus it suffices to prove that

$$\mathrm{E}\left[\exp\left(\mathrm{i}u\sum_{i=1}^{M_N} U_{N,i}\right) \middle| \mathcal{F}^N\right] - \exp\left(-\frac{u^2}{2}\sum_{i=1}^{M_N} \sigma_{N,i}^2\right) \xrightarrow{\mathrm{P}} 0 . \tag{9.74}$$

To start with, using the conditional independence of the triangular array and Lemma 9.5.8, it follows that the left-hand side of this display is bounded by

$$\sum_{i=1}^{M_N} \left|\mathrm{E}[\exp\left(\mathrm{i}uU_{N,i}\right) \,|\, \mathcal{F}^N] - \exp(-u^2\sigma_{N,i}^2/2)\right| .$$

From here we proceed in two steps, showing that both

$$A_N = \sum_{i=1}^{M_N} \left|\mathrm{E}[\exp\left(\mathrm{i}uU_{N,i}\right) \,|\, \mathcal{F}^N] - (1 - u^2\sigma_{N,i}^2/2)\right| \xrightarrow{\mathrm{P}} 0$$

and

$$B_N = \sum_{i=1}^{M_N} \left|\exp(-u^2\sigma_{N,i}^2/2) - (1 - u^2\sigma_{N,i}^2/2)\right| \xrightarrow{\mathrm{P}} 0 .$$

These two result suffice to finish the proof.

Now, by Taylor's inequality,

$$\left|\mathrm{e}^{\mathrm{i}tx} - \left(1 + \mathrm{i}tx - \frac{1}{2}t^2x^2\right)\right| \le \min\left\{|tx|^2, |tx|^3\right\} ,$$

so that the characteristic function of $U_{N,i}$ satisfies

$$\left|\mathrm{E}[\exp(\mathrm{i}uU_{N,i}) \,|\, \mathcal{F}^N] - (1 - u^2\sigma_{N,i}^2/2)\right| \le \mathrm{E}[\min(|uU_{N,i}|^2, |uU_{N,i}|^3) \,|\, \mathcal{F}^N] .$$

Note that this expectation is finite. For positive ϵ, the right-hand side of the inequality is at most

$$\mathrm{E}[|uU_{N,i}|^3 \,\mathbb{1}_{\{|U_{N,i}|<\epsilon\}} \,|\, \mathcal{F}^N] + \mathrm{E}[|uU_{N,i}|^2 \,\mathbb{1}_{\{|U_{N,i}|\ge\epsilon\}} \,|\, \mathcal{F}^N] \\ \le \epsilon|u|^3\sigma_{N,i}^2 + u^2\,\mathrm{E}[|U_{N,i}|^2 \,\mathbb{1}_{\{|U_{N,i}|\ge\epsilon\}} \,|\, \mathcal{F}^N] .$$

Summing up the right-hand side over $1 \le i \le M_N$, using the assumed conditions (ii) and (iii) and recalling that ϵ was arbitrary, we find that $A_N \to 0$ in probability. We now turn to B_N. For positive x, $|\mathrm{e}^{-x} - 1 + x| \le x^2/2$. Thus

$$B_N \le \frac{u^4}{8} \sum_{i=1}^{M_N} \sigma_{N,i}^4 \le \frac{u^4}{8} \max_{1\le i\le M_N} \sigma_{N,i}^2 \sum_{i=1}^{M_N} \sigma_{N,i}^2 .$$

Here the sum on the right-hand side converges in probability and, as remarked above, the maximum tends to zero in probability (the uniform smallness condition). Thus $B_N \to 0$ in probability and the proof is complete. □

In the special case where the random variables $\{U_{N,i}\}_{1\le i\le M_N}$, for any N, are conditionally i.i.d. given $\{\mathcal{F}^N\}$, Proposition 9.5.9 admits a simpler formulation.

Corollary 9.5.11. *Let $\{V_{N,i}\}_{1\le i\le M_N}$ be a triangular array of random variables and let $\{\mathcal{F}^N\}_{N\ge 0}$ be a sequence of sub-σ-fields of $\mathcal{F}$. Assume that the following conditions hold true.*

(i) The triangular array is conditionally i.i.d. given $\{\mathcal{F}^N\}$ and for any N, $\mathrm{E}[V_{N,1}^2 \,|\, \mathcal{F}^N] < \infty$ and $\mathrm{E}[V_{N,1} \,|\, \mathcal{F}^N] = 0$.

(ii) There exists a positive constant σ^2 such that $\mathrm{E}[V_{N,1}^2 \,|\, \mathcal{F}^N] \xrightarrow{\mathrm{P}} \sigma^2$.

(iii) For any positive ϵ, $\mathrm{E}[V_{N,1}^2 \mathbb{1}_{\{|V_{N,1}|\ge\epsilon M_N\}} \,|\, \mathcal{F}^N] \xrightarrow{\mathrm{P}} 0$.

Then for any real u,

$$\mathrm{E}\left[\exp\left(\mathrm{i}uM_N^{-1/2}\sum_{i=1}^{M_N} V_{N,i}\right) \,\middle|\, \mathcal{F}^N\right] \xrightarrow{\mathrm{P}} \exp(-\sigma^2u^2/2) . \tag{9.75}$$

Proposition 9.5.12. *Let $\{V_{N,i}\}_{1\leq i\leq M_N}$ be a triangular array of random variables and let $\{\mathcal{F}^N\}_{N\geq 0}$ be a sequence of sub-σ-fields of $\mathcal{F}$. Assume that the following conditions hold true.*

(i) The triangular array is conditionally independent given $\{\mathcal{F}^N\}$ and for any N and $i = 1,\ldots,M_N$, $\mathrm{E}[V_{N,i}^2 \,|\, \mathcal{F}^N] < \infty$.
(ii) There exists a constant $\sigma^2 > 0$ such that

$$\sum_{i=1}^{M_N}\{\mathrm{E}[V_{N,i}^2 \,|\, \mathcal{F}^N] - (\mathrm{E}[V_{N,i} \,|\, \mathcal{F}^N])^2\} \stackrel{\mathrm{P}}{\longrightarrow} \sigma^2 .$$

(iii) For all $\epsilon > 0$,

$$\sum_{i=1}^{M_N} \mathrm{E}[V_{N,i}^2 \mathbb{1}_{\{|V_{N,i}|\geq\epsilon\}} \,|\, \mathcal{F}^N] \stackrel{\mathrm{P}}{\longrightarrow} 0 .$$

Then for any real u,

$$\mathrm{E}\left[\exp\left(\mathrm{i}u\sum_{i=1}^{M_N}\{V_{N,i} - \mathrm{E}[V_{N,i} \,|\, \mathcal{F}^N]\}\right)\,\middle|\, \mathcal{F}^N\right] \stackrel{\mathrm{P}}{\longrightarrow} \exp(-(u^2/2)\sigma^2) .$$

Proof. We check that the triangular array $U_{N,i} = V_{N,i} - \mathrm{E}[V_{N,i} \,|\, \mathcal{F}^N]$ satisfies conditions (i)–(iii) of Proposition 9.5.9. This triangular array is conditionally independent given $\{\mathcal{F}^N\}$ and by construction $\mathrm{E}[U_{N,i} \,|\, \mathcal{F}^N] = 0$ and $\mathrm{E}[U_{N,i}^2 \,|\, \mathcal{F}^N] = \mathrm{E}[V_{N,i}^2 \,|\, \mathcal{F}^N] - \{\mathrm{E}[V_{N,i} \,|\, \mathcal{F}^N]\}^2$. Therefore, conditions (i) and (ii) are fulfilled. It remains to check that for any $\epsilon > 0$, (9.72) holds true. By Jensen's inequality,

$$U_{N,i}^2 \leq 2(V_{N,i}^2 + \mathrm{E}[V_{N,i}^2 \,|\, \mathcal{F}^N]) ,$$
$$\mathbb{1}_{\{|U_{N,i}|\geq\epsilon\}} \leq \mathbb{1}_{\{V_{N,i}^2\geq\epsilon^2/4\}} + \mathbb{1}_{\{\mathrm{E}[V_{N,i}^2 \,|\, \mathcal{F}^N]\geq\epsilon^2/4\}} ,$$

so that the left-hand side of (9.72) is bounded by

$$2\sum_{i=1}^{M_N} \mathrm{E}[V_{N,i}^2 \mathbb{1}_{\{V_{N,i}^2\geq\epsilon^2/4\}} \,|\, \mathcal{F}^N] + 2\sum_{i=1}^{M_N} \mathrm{E}[V_{N,i}^2 \,|\, \mathcal{F}^N]\, \mathrm{P}(V_{N,i}^2 \geq \epsilon^2/4 \,|\, \mathcal{F}^N)$$
$$+ 4\sum_{i=1}^{M_N} \mathrm{E}[V_{N,i}^2 \,|\, \mathcal{F}^N] \mathbb{1}_{\{\mathrm{E}[V_{N,i}^2 \,|\, \mathcal{F}^N]\geq\epsilon^2/4\}} .$$

The proof is concluded using the same arguments as in the proof of Proposition 9.5.7. □

Theorem 9.5.13. *Let $\{\xi^{N,i}\}_{1\leq i\leq M_N}$ be a triangular array of X-valued random variables, let $\{\mathcal{F}^N\}_{N\geq 0}$ be a sequence of sub-σ-fields of $\mathcal{F}$, and let f be a real-valued function on X. Assume that the following conditions hold true.*

(i) The triangular array is conditionally independent given $\{\mathcal{F}^N\}$ and for any N and $i = 1,\ldots,M_N$, $\mathrm{E}[f^2(\xi^{N,i}) \,|\, \mathcal{F}^N] < \infty$,

(ii) There exists a constant $\sigma^2 > 0$ *such that*

$$M_N^{-1} \sum_{i=1}^{M_N} \{\mathrm{E}[f^2(\xi^{N,i}) \,|\, \mathcal{F}^N] - (\mathrm{E}[f(\xi^{N,i}) \,|\, \mathcal{F}^N])^2\} \xrightarrow{\mathrm{P}} \sigma^2 .$$

(iii) There exists a probability measure μ *on* $(\mathsf{X}, \mathcal{X})$ *such that* $f \in L^2(\mathsf{X}, \mu)$ *and for any positive* C*,*

$$M_N^{-1} \sum_{i=1}^{M_N} \mathrm{E}[f^2(\xi^{N,i}) \mathbb{1}_{\{|f(\xi^{N,i})| \geq C\}} \,|\, \mathcal{F}^N] \xrightarrow{\mathrm{P}} \mu(f^2 \mathbb{1}_{\{|f| \geq C\}}) .$$

Then for any real u*,*

$$\mathrm{E}\left[\exp\left(iu M_N^{-1/2} \sum_{i=1}^{M_N} \{f(\xi^{N,i}) - \mathrm{E}[f(\xi^{N,i}) \,|\, \mathcal{F}^N]\} \right) \middle| \, \mathcal{F}^N \right] \xrightarrow{\mathrm{P}} \exp(-\sigma^2 u^2/2) . \quad (9.76)$$

Proof. Set $V_{N,i} = M_N^{-1/2} f(\xi^{N,i})$. We prove the theorem by checking conditions (i)–(iii) of Proposition 9.5.12. Of these conditions, the first two are immediate, so it remains to verify the Lindeberg condition (iii). Pick $\epsilon > 0$. Then for any positive C

$$\sum_{i=1}^{M_N} \mathrm{E}[V_{N,i}^2 \mathbb{1}_{\{|V_{N,i}| \geq \epsilon\}} \,|\, \mathcal{F}^N] \\ \leq M_N^{-1} \sum_{i=1}^{M_N} \mathrm{E}[f^2(\xi^{N,i}) \mathbb{1}_{\{|f(\xi^{N,i})| \geq C\}} \,|\, \mathcal{F}^N] \xrightarrow{\mathrm{P}} \mu(f^2 \mathbb{1}_{\{|f| \geq C\}}) ,$$

where the inequality holds for sufficiently large N. Because $f \in L^2(\mathsf{X}, \mu)$ the right-hand side of this display tends to zero as $C \to \infty$, so that the Lindeberg condition is satisfied. □

9.5.2 Bibliographic Notes

Convergence of interacting particle systems has been considered by many authors in the last decade, triggered by the seminal papers of Del Moral (1996, 1998). Most of the results presented in this chapter have already appeared in the literature, perhaps in a slightly different form. We have focused here on the most elementary convergence properties, the law of large numbers, and the central limit theorem. More sophisticated convergence results are available, covering for instance large deviations (Del Moral and Guionnet, 1998), empirical processes (Del Moral and Ledoux, 2000), propagation of chaos (Del

Moral and Miclo, 2001), and rate of convergence in the central limit theorem. The ultimate reference for convergence analysis of interacting particle systems is Del Moral (2004), which summarizes most of these efforts. An elementary but concise survey of available results is given in Crisan and Doucet (2002). The approach developed here has been inspired by Künsch (2003).

Part II

Parameter Inference

10

Maximum Likelihood Inference, Part I: Optimization Through Exact Smoothing

In previous chapters, we have focused on structural results and methods for HMMs, considering in particular that the models under consideration were always perfectly known. In most situations, however, the model cannot be fully specified beforehand, and some of its parameters need to be calibrated based on observed data. Except for very simplistic instances of HMMs, the structure of the model is sufficiently complex to prevent the use of direct estimators such as those provided by moment or least squares methods. We thus focus in the following on computation of the *maximum likelihood estimator.*

Given the specific structure of the likelihood function in HMMs, it turns out that the key ingredient of any optimization method applicable in this context is the ability to compute smoothed functionals of the unobserved sequence of states. Hence the methods discussed in the second part of the book for evaluating smoothed quantities are instrumental in devising parameter estimation strategies.

This chapter only covers the class of HMMs discussed in Chapter 5, for which the smoothing recursions described in Chapters 3 and 4 may effectively be implemented on computers. For such models, the likelihood function is computable, and hence our main task will be to optimize a possibly complex but entirely known function. The topic of this chapter thus relates to the more general field of numerical optimization. For models that do not allow for exact numerical computation of smoothing distributions, this chapter provides a framework from which numerical approximations can be built. Those will be discussed in Chapter 11.

10.1 Likelihood Optimization in Incomplete Data Models

To describe the methods as concisely as possible, we adopt a very general viewpoint in which we only assume that the likelihood function of interest may be written as the marginal of a higher dimensional function. In the terminology introduced by Dempster *et al.* (1977), this higher dimensional function is described as the *complete data* likelihood; in this framework, the term *incomplete data* refers to the actual observed data while the *complete data* is a (not fully observable) higher dimensional random variable. In Section 10.2, we will exploit the specific structure of the HMM, and in particular the fact that it corresponds to a *missing data model* in which the observations simply are a subset of the complete data. We ignore these specifics for the moment however and consider the general likelihood optimization problem in incomplete data models.

10.1.1 Problem Statement and Notations

Given a σ-finite measure λ on $(\mathsf{X}, \mathcal{X})$, we consider a family $\{f(\cdot; \theta)\}_{\theta \in \Theta}$ of non-negative λ-integrable functions on X. This family is indexed by a parameter $\theta \in \Theta$, where Θ is a subset of $\mathbb{R}^{d_\theta}$ (for some integer d_θ). The task under consideration is the maximization of the integral

$$\mathrm{L}(\theta) \stackrel{\text{def}}{=} \int f(x\,; \theta)\, \lambda(dx) \tag{10.1}$$

with respect to the parameter θ. The function $f(\cdot\,; \theta)$ may be thought of as an *unnormalized probability density* with respect to λ. Thus $\mathrm{L}(\theta)$ is the normalizing constant for $f(\cdot\,; \theta)$. In typical examples, $f(\cdot\,; \theta)$ is a relatively simple function of θ. In contrast, the quantity $\mathrm{L}(\theta)$ usually involves high-dimensional integration and is therefore sufficiently complex to prevent the use of simple maximization approaches; even the direct evaluation of the function might turn out to be non-feasible.

In Section 10.2, we shall consider more specifically the case where f is the joint probability density function of two random variables X and Y, the latter being observed while the former is not. Then X is referred to as the *missing data*, f is the *complete data likelihood*, and L is the density of Y alone, that is, the *likelihood* available for estimating θ. Note however that thus far, the dependence on Y is not made explicit in the notation; this is reminiscent of the implicit conditioning convention discussed in Section 3.1.4 in that the observations do not appear explicitly. Having sketched these statistical ideas, we stress that we feel it is actually easier to understand the basic mechanisms at work without relying on the probabilistic interpretation of the above quantities. In particular, it is not required that L be a likelihood, as any function satisfying (10.1) is a valid candidate for the methods discussed here (cf. Remark 10.2.1).

In the following, we will assume that $\mathrm{L}(\theta)$ is positive, and thus maximizing $\mathrm{L}(\theta)$ is equivalent to maximizing

$$\ell(\theta) \stackrel{\text{def}}{=} \log \mathrm{L}(\theta)\ . \tag{10.2}$$

In a statistical setting, ℓ is the *log-likelihood*. We also associate to each function $f(\cdot\,; \theta)$ the probability density function $p(\cdot\,; \theta)$ (with respect to the dominating measure λ) defined by

$$p(x\,;\theta) \stackrel{\text{def}}{=} f(x\,;\theta)/\mathrm{L}(\theta)\,. \tag{10.3}$$

In the statistical setting sketched above, $p(x;\theta)$ is the conditional density of X given Y.

10.1.2 The Expectation-Maximization Algorithm

The most popular method for solving the general optimization problem outlined above is the EM (for *expectation-maximization*) algorithm introduced, in its full generality, by Dempster *et al.* (1977) in their landmark paper. Given the literature available on the topic, our aim is not to provide a comprehensive review of all the results related to the EM algorithm but rather to highlight some of its key features and properties in the context of hidden Markov models.

10.1.2.1 The Intermediate Quantity of EM

The central concept in the framework introduced by Dempster *et al.* (1977) is an auxiliary function (or, more precisely, a family of auxiliary functions) known as the intermediate quantity of EM.

Definition 10.1.1 (Intermediate Quantity of EM). *The* intermediate quantity of EM *is the family* $\{\mathcal{Q}(\cdot\,;\theta')\}_{\theta'\in\Theta}$ *of real-valued functions on* Θ*, indexed by* θ' *and defined by*

$$\mathcal{Q}(\theta\,;\theta') \stackrel{\text{def}}{=} \int \log f(x\,;\theta) p(x\,;\theta')\,\lambda(dx)\,. \tag{10.4}$$

Remark 10.1.2. To ensure that $\mathcal{Q}(\theta\,;\theta')$ is indeed well-defined for all values of the pair (θ,θ'), one needs regularity conditions on the family of functions $\{f(\cdot\,;\theta)\}_{\theta\in\Theta}$, which will be stated below (Assumption 10.1.3). To avoid trivial cases however, we use the convention $0\log 0 = 0$ in (10.4) and in similar relations below. In more formal terms, for every measurable set N such that *both* $f(x\,;\theta)$ and $p(x\,;\theta')$ vanish λ-a.e. on N, set

$$\int_N \log f(x\,;\theta) p(x\,;\theta')\,\lambda(dx) \stackrel{\text{def}}{=} 0\,.$$

With this convention, $\mathcal{Q}(\theta\,;\theta')$ stays well-defined in cases where there exists a non-empty set N such that *both* $f(x\,;\theta)$ and $f(x\,;\theta')$ vanish λ-a.e. on N. ■

The intermediate quantity $\mathcal{Q}(\theta\,;\theta')$ of EM may be interpreted as the expectation of the function $\log f(X\,;\theta)$ when X is distributed according to the probability density function $p(\cdot\,;\theta')$ indexed by a, possibly different, value θ' of the parameter. Using (10.2) and (10.3), one may rewrite the intermediate quantity of EM in (10.4) as

$$\mathcal{Q}(\theta\,;\theta') = \ell(\theta) - \mathcal{H}(\theta\,;\theta')\,, \tag{10.5}$$

where

$$\mathcal{H}(\theta\,;\theta') \stackrel{\text{def}}{=} -\int \log p(x\,;\theta) p(x\,;\theta')\,\lambda(dx)\,. \tag{10.6}$$

Equation (10.5) states that the intermediate quantity $\mathcal{Q}(\theta\,;\theta')$ of EM differs from (the log of) the objective function $\ell(\theta)$ by a quantity that has a familiar form. Indeed, $\mathcal{H}(\theta'\,;\theta')$ is recognized as the *entropy* of the probability density function $p(\cdot\,;\theta')$ (see for instance Cover and Thomas, 1991). More importantly, the increment of $\mathcal{H}(\theta\,;\theta')$,

$$\mathcal{H}(\theta\,;\theta') - \mathcal{H}(\theta'\,;\theta') = -\int \log \frac{p(x\,;\theta)}{p(x\,;\theta')} p(x\,;\theta')\,\lambda(dx)\,, \tag{10.7}$$

is recognized as the *Kullback-Leibler divergence* (or *relative entropy*) between the probability density functions p indexed by θ and θ', respectively.

The last piece of notation needed is the following: the gradient and Hessian of a function, say L, at θ' will be denoted by $\nabla_\theta \mathrm{L}(\theta')$ and $\nabla_\theta^2 \mathrm{L}(\theta')$, respectively. To avoid ambiguities, the gradient of $\mathcal{H}(\cdot\,;\theta')$ with respect to its first argument, evaluated at θ'', will be denoted by $\nabla_\theta \mathcal{H}(\theta\,;\theta')|_{\theta=\theta''}$ (where the same convention will also be used, if needed, for the Hessian).

We conclude this introductory section by stating a minimal set of assumptions that guarantee that all quantities introduced so far are indeed well-defined.

Assumption 10.1.3.

(i) The parameter set Θ is an open subset of $\mathbb{R}^{d_\theta}$ (for some integer d_θ).
(ii) For any $\theta \in \Theta$, $\mathrm{L}(\theta)$ is positive and finite.
(iii) For any $(\theta, \theta') \in \Theta \times \Theta$, $\int |\nabla_\theta \log p(x\,;\theta)| p(x\,;\theta')\,\lambda(dx)$ is finite.

Assumption 10.1.3(iii) implies in particular that the probability distributions in the family $\{p(\cdot\,;\theta)\,d\lambda\}_{\theta\in\Theta}$ are all absolutely continuous with respect to one another. Any individual distribution $p(\cdot\,;\theta)\,d\lambda$ can only vanish on sets that are assigned null probability by all other probability distributions in the family. Thus both $\mathcal{H}(\theta\,;\theta')$ and $\mathcal{Q}(\theta\,;\theta')$ are well-defined for all pairs of parameters.

10.1.2.2 The Fundamental Inequality of EM

We are now ready to state the fundamental result that justifies the standard construction of the EM algorithm.

Proposition 10.1.4. *Under Assumption 10.1.3, for any $(\theta, \theta') \in \Theta \times \Theta$,*

$$\ell(\theta) - \ell(\theta') \geq \mathcal{Q}(\theta\,;\theta') - \mathcal{Q}(\theta'\,;\theta')\,, \tag{10.8}$$

where the inequality is strict unless $p(\cdot\,;\theta)$ and $p(\cdot\,;\theta')$ are equal λ-a.e.

Assume in addition that

(a) $\theta \mapsto \mathrm{L}(\theta)$ *is continuously differentiable on* Θ*;*
(b) for any $\theta' \in \Theta$, $\theta \mapsto \mathcal{H}(\theta\,;\theta')$ *is continuously differentiable on* Θ.

Then for any $\theta' \in \Theta$, $\theta \mapsto \mathcal{Q}(\theta\,;\theta')$ *is continuously differentiable on* Θ *and*

$$\nabla_\theta \ell(\theta') = \nabla_\theta \mathcal{Q}(\theta\,;\theta')|_{\theta=\theta'} \;. \tag{10.9}$$

Proof. The difference between the left-hand side and the right-hand side of (10.8) is the quantity defined in (10.7), which we already recognized as a Kullback-Leibler distance. Under Assumption 10.1.3(iii), this latter term is well-defined and known to be strictly positive (by direct application of Jensen's inequality) unless $p(\cdot\,;\theta)$ and $p(\cdot\,;\theta')$ are equal λ-a.e. (Cover and Thomas, 1991; Lehmann and Casella, 1998).

For (10.9), first note that $\mathcal{Q}(\theta\,;\theta')$ is a differentiable function of θ, as it is the difference of two functions that are differentiable under the additional assumptions (a) and (b). Next, the previous discussion implies that $\mathcal{H}(\theta\,;\theta')$ is minimal for $\theta = \theta'$, although this may not be the only point where the minimum is achieved. Thus its gradient vanishes at θ', which proves (10.9).

□

10.1.2.3 The EM Algorithm

The essence of the EM algorithm, which is suggested by (10.5), is that $\mathcal{Q}(\theta\,;\theta')$ may be used as a surrogate for $\ell(\theta)$. Both functions are not necessarily comparable but, in view of (10.8), any value of θ such that $\mathcal{Q}(\theta\,;\theta')$ is increased over its baseline $\mathcal{Q}(\theta'\,;\theta')$ corresponds to an increase of ℓ (relative to $\ell(\theta')$) that is at least as large.

The EM algorithm as proposed by Dempster *et al.* (1977) consists in iteratively building a sequence $\{\theta^i\}_{i\geq 1}$ of parameter estimates given an initial guess θ^0. Each iteration is classically broken into two steps as follows.

E-Step: Determine $\mathcal{Q}(\theta\,;\theta^i)$;
M-Step: Choose θ^{i+1} to be the (or any, if there are several) value of $\theta \in \Theta$ that maximizes $\mathcal{Q}(\theta\,;\theta^i)$.

It is certainly not obvious at this point that the M-step may be in practice easier to perform than the direct maximization of the function of interest ℓ itself. We shall return to this point in Section 10.1.2.4 below.

Proposition 10.1.4 provides the two decisive arguments behind the EM algorithm. First, an immediate consequence of (10.8) is that, by the very definition of the sequence $\{\theta^i\}$, the sequence $\{\ell(\theta^i)\}_{i\geq 0}$ of log-likelihood values is non-decreasing. Hence EM is a monotone optimization algorithm. Second, if the iterations ever stop at a point $\theta_\star$, then $\mathcal{Q}(\theta\,;\theta_\star)$ has to be maximal at $\theta_\star$ (otherwise it would still be possible to improve over $\theta_\star$), and hence $\theta_\star$ is such that $\nabla_\theta \mathrm{L}(\theta_\star) = 0$, that is, this is a *stationary point of the likelihood.*

Although this picture is largely correct, there is a slight flaw in the second half of the above intuitive reasoning in that the if part *(if the iterations ever stop at a point)* may indeed never happen. Stronger conditions are required to ensure that the sequence of parameter estimates produced by EM from any starting point indeed converges to a limit $\theta_\star \in \Theta$. However, it is actually true that when convergence to a point takes place, the limit has to be a stationary point of the likelihood. In order not to interrupt our presentation of the EM framework, convergence results pertaining to the EM algorithm are deferred to Section 10.5 at the end of this chapter; see in particular Theorems 10.5.3 and 10.5.4.

10.1.2.4 EM in Exponential Families

The EM algorithm defined in the previous section will only be helpful in situations where the following general conditions hold.

E-Step: It is possible to compute, at reasonable computational cost, the intermediate quantity $\mathcal{Q}(\theta\,;\theta')$ given a value of θ'.
M-Step: $\mathcal{Q}(\theta\,;\theta')$, considered as a function of its first argument θ, is sufficiently simple to allow closed-form maximization.

A rather general context in which both of these requirements are satisfied, or at least are equivalent to easily interpretable necessary conditions, is when the functions $\{f(\cdot\,;\theta)\}$ belong to an *exponential family.*

Definition 10.1.5 (Exponential Family). *The family $\{f(\cdot\,;\theta)\}_{\theta\in\Theta}$ defines an* exponential family *of positive functions on* X *if*

$$f(x\,;\theta) = \exp\{\psi(\theta)^t S(x) - c(\theta)\}h(x)\ , \tag{10.10}$$

where S and ψ are vector-valued functions (of the same dimension) on X and Θ respectively, c is a real-valued function on Θ and h is a non-negative real-valued function on X.

Here $S(x)$ is known as the vector of *natural sufficient statistics*, and $\eta = \psi(\theta)$ is the *natural parameterization.* If $\{f(\cdot\,;\theta)\}_{\theta\in\Theta}$ is an exponential family and if $\int |S(x)|f(x\,;\theta)\,\lambda(dx)$ is finite for any $\theta \in \Theta$, the intermediate quantity of EM reduces to

$$\mathcal{Q}(\theta\,;\theta') = \psi(\theta)^t \left[\int S(x)p(x\,;\theta')\,\lambda(dx)\right] - c(\theta) + \int p(x\,;\theta')\log h(x)\,\lambda(dx)\ . \tag{10.11}$$

Note that the right-most term does not depend on θ and thus plays no role in the maximization. It may as well be ignored, and in practice it is not required to compute it. Except for this term, the right-hand side of (10.11) has an explicit form as soon as it is possible to evaluate the expectation of the vector of sufficient statistics S under $p(\cdot\,;\theta')$. The other important feature of (10.11), ignoring the rightmost term, is that $\mathcal{Q}(\theta\,;\theta')$, viewed as a function of θ, is similar to the logarithm of (10.10) for the particular value $S_{\theta'} = \int S(x)p(x\,;\theta')\,\lambda(dx)$ of the sufficient statistic.

In summary, if $\{f(\cdot\,;\theta)\}_{\theta\in\Theta}$ is an exponential family, the two above general conditions needed for the EM algorithm to be practicable reduce to the following.

E-Step: The expectation of the vector of sufficient statistics $S(X)$ under $p(\cdot\,;\theta')$ must be computable.

M-Step: Maximization of $\psi(\theta)^t s - c(\theta)$ with respect to $\theta \in \Theta$ must be feasible in closed form for any s in the convex hull of $S(\mathsf{X})$ (that is, for any valid value of the expected vector of sufficient statistics).

For the sake of completeness, it should be mentioned that there are variants of the EM algorithm that are handy in cases where the maximization required in the M-step is not directly feasible (see Section 10.5.3 and further references in Section 10.5.4). In the context of HMMs, the main limitation of the EM algorithm rather appears in cases where the E-step is not feasible. This latter situation is the rule rather than the exception in models for which the state space X is not finite. For such cases, approaches that build on the EM concepts introduced in the current chapter will be fully discussed in Chapter 11.

10.1.3 Gradient-based Methods

A frequently ignored observation is that in any model where the EM strategy may be applied, it is also possible to evaluate derivatives of the objective function $\ell(\theta)$ with respect to the parameter θ. This is obvious from (10.9), and we will expand on this matter below. As a consequence, instead of resorting to a specific algorithm such as EM, one may borrow tools from the (comprehensive and well-documented) toolbox of gradient-based optimization methods.

10.1.3.1 Computing Derivatives in Incomplete Data Models

A first remark is that in cases where the EM algorithm is applicable, the objective function $\ell(\theta)$ is actually computable: because the EM requires the computation of expectations under the conditional density $p(\cdot\,;\theta)$, it is restricted to cases where the normalizing constant $\mathrm{L}(\theta)$—and hence $\ell(\theta) = \log \mathrm{L}(\theta)$—is available. The two equalities below show that it is indeed also the case for the first- and second-order derivatives of $\ell(\theta)$.

Proposition 10.1.6 (Fisher's and Louis' Identities). *Assume 10.1.3 and that the following conditions hold.*

(a) $\theta \mapsto \mathrm{L}(\theta)$ is twice continuously differentiable on Θ.

(b) For any $\theta' \in \Theta$, $\theta \mapsto \mathcal{H}(\theta\,;\theta')$ is twice continuously differentiable on Θ. In addition, $\int |\nabla_\theta^k \log p(x\,;\theta)| p(x\,;\theta')\,\lambda(dx)$ is finite for $k = 1, 2$ and any $(\theta, \theta') \in \Theta \times \Theta$, and

$$\nabla_\theta^k \int \log p(x\,;\theta) p(x\,;\theta')\,\lambda(dx) = \int \nabla_\theta^k \log p(x\,;\theta) p(x\,;\theta')\,\lambda(dx)\;.$$

Then the following identities hold:

$$\nabla_\theta \ell(\theta') = \int \nabla_\theta \log f(x\,;\theta)|_{\theta=\theta'}\, p(x\,;\theta')\,\lambda(dx)\;, \tag{10.12}$$

$$- \nabla^2_\theta \ell(\theta') = - \int \nabla^2_\theta \log f(x\,;\theta)\big|_{\theta=\theta'}\, p(x\,;\theta')\,\lambda(dx) \\ + \int \nabla^2_\theta \log p(x\,;\theta)\big|_{\theta=\theta'}\, p(x\,;\theta')\,\lambda(dx)\;. \tag{10.13}$$

The second equality may be rewritten in the equivalent form

$$\nabla^2_\theta \ell(\theta') + \{\nabla_\theta \ell(\theta')\}\,\{\nabla_\theta \ell(\theta')\}^t = \int \Big[\nabla^2_\theta \log f(x\,;\theta)\big|_{\theta=\theta'} \\ + \{\nabla_\theta \log f(x\,;\theta)|_{\theta=\theta'}\}\,\{\nabla_\theta \log f(x\,;\theta)|_{\theta=\theta'}\}^t \Big] p(x\,;\theta')\,\lambda(dx)\;. \tag{10.14}$$

Equation (10.12) is sometimes referred to as *Fisher's identity* (see the comment by B. Efron in the discussion of Dempster *et al.*, 1977, p. 29). In cases where the function L may be interpreted as the likelihood associated with some statistical model, the left-hand side of (10.12) is the *score function* (gradient of the log-likelihood). Equation (10.12) shows that the score function may be evaluated by computing the expectation, under $p(\cdot\,;\theta')$, of the function $\nabla_\theta \log f(X\,;\theta)|_{\theta=\theta'}$. This latter quantity, in turn, is referred to as the *complete score function* in a statistical context, as $\log f(x\,;\theta)$ is the joint log-likelihood of the complete data (X, Y); again we remark that at this stage, Y is not explicit in the notation.

Equation (10.13) is usually called the *missing information principle* after Louis (1982) who first named it this way, although it was mentioned previously in a slightly different form by Orchard and Woodbury (1972) and implicitly used in Dempster *et al.* (1977). In cases where L is a likelihood, the left-hand side of (10.13) is the associated *observed information matrix*, and the second term on the right-hand side is easily recognized as the (negative of the) Fisher information matrix associated with the probability density function $p(\cdot\,;\theta')$.

Finally (10.14), which is here written in a form that highlights its symmetry, was also proved by Louis (1982) and is thus known as *Louis' identity*. Together with (10.12), it shows that the first- and second-order derivatives of ℓ may be evaluated by computing expectations under $p(\cdot\,;\theta')$ of quantities derived from $f(\cdot\,;\theta)$. We now prove these three identities.

Proof (of Proposition 10.1.6). Equations (10.12) and (10.13) are just (10.5) where the right-hand side is differentiated once, using (10.9), and then twice under the integral sign.

To prove (10.14), we start from (10.13) and note that the second term on its right-hand side is the negative of an information matrix for the parameter

θ associated with the probability density function $p(\cdot\,;\theta)$ and evaluated at θ'. We rewrite this second term using the well-known information matrix identity

$$\int \nabla_\theta^2 \log p(x\,;\theta)\big|_{\theta=\theta'}\, p(x\,;\theta')\,\lambda(dx) \\ = -\int \left\{\nabla_\theta \log p(x\,;\theta)|_{\theta=\theta'}\right\}\left\{\nabla_\theta \log p(x\,;\theta)|_{\theta=\theta'}\right\}^t p(x\,;\theta')\,\lambda(dx)\;.$$

This is again a consequence of assumption (b) and the fact that $p(\cdot\,;\theta)$ is a probability density function for all values of θ, implying that

$$\int \nabla_\theta \log p(x\,;\theta)|_{\theta=\theta'}\, p(x\,;\theta')\,\lambda(dx) = 0\;.$$

Now use the identity $\log p(x\,;\theta) = \log f(x\,;\theta) - \ell(\theta)$ and (10.12) to conclude that

$$\int \left\{\nabla_\theta \log p(x\,;\theta)|_{\theta=\theta'}\right\}\left\{\nabla_\theta \log p(x\,;\theta)|_{\theta=\theta'}\right\}^t p(x\,;\theta')\,\lambda(dx) \\ = \int \left\{\nabla_\theta \log f(x\,;\theta)|_{\theta=\theta'}\right\}\left\{\nabla_\theta \log f(x\,;\theta)|_{\theta=\theta'}\right\}^t p(x\,;\theta')\,\lambda(dx) \\ - \left\{\nabla_\theta \ell(\theta')\right\}\left\{\nabla_\theta \ell(\theta')\right\}^t\;,$$

which completes the proof. □

Remark 10.1.7. As was the case for the intermediate quantity of EM, Fisher's and Louis' identities only involve expectations under $p(\cdot\,;\theta')$ of quantities derived from $f(\cdot\,;\theta)$. In particular, when the functions $f(\cdot\,;\theta)$ belong to an exponential family (see Definition 10.1.5), Fisher's identity, for instance, may be rewritten as

$$\nabla_\theta \ell(\theta') = \left\{\nabla_\theta \psi(\theta')\right\}^t \left(\int S(x) p(x\,;\theta')\,\lambda(dx)\right) - \nabla_\theta c(\theta')\;,$$

with the convention that $\nabla_\theta \psi(\theta')$ is the $d_\theta \times d_\theta$ matrix containing the partial derivatives $[\nabla_\theta \psi(\theta')]_{ij} = \partial \psi_i(\theta')/\partial \theta_j$. As a consequence, the only practical requirement for using Fisher's and Louis' identities is the ability to compute expectations of the sufficient statistic $S(x)$ under $p(\cdot\,;\theta)$ for any $\theta \in \Theta$. ■

10.1.3.2 The Steepest Ascent Algorithm

We briefly discuss the main features of gradient-based iterative optimization algorithms, starting with the simplest, but certainly not most efficient, approach. We restrict ourselves to the case where the optimization problem is *unconstrained* in the sense that $\Theta = \mathbb{R}^{d_\theta}$, so that any parameter value produced by the algorithms below is valid. For an in-depth coverage of the subject, we recommend the monographs by Luenberger (1984) and Fletcher (1987).

The simplest method is the *steepest ascent* algorithm in which the current value of the estimate θ^i is updated by adding a multiple of the gradient $\nabla_\theta \ell(\theta^i)$, referred to as the *search direction*:

$$\theta^{i+1} = \theta^i + \gamma_i \nabla_\theta \ell(\theta^i) \; . \tag{10.15}$$

Here the multiplier γ_i is a non-negative scalar that needs to be adjusted at each iteration to ensure, *a minima*, that the sequence $\{\ell(\theta^i)\}$ is non-decreasing—as was the case for EM. The most sensible approach consists in choosing γ_i as to maximize the objective function in the search direction:

$$\gamma_i = \arg\max_{\gamma \geq 0} \ell[\theta^i + \gamma \nabla_\theta \ell(\theta^i)] \; . \tag{10.16}$$

It can be shown (Luenberger, 1984, Chapter 7) that under mild assumptions, the steepest ascent method with multipliers (10.16) is globally convergent, with a set of limit points corresponding to the stationary points of ℓ (see Section 10.5 for precise definitions of these terms and a proof that this property holds for the EM algorithm).

It remains that the use of the steepest ascent algorithm is not recommended, particularly in large-dimensional parameter spaces. The reason for this is that its speed of convergence *linear* in the sense that if the sequence $\{\theta^i\}_{i\geq 0}$ converges to a point $\theta_\star$ such that the Hessian $\nabla_\theta^2 \ell(\theta_\star)$ is negative definite (see Section 10.5.2), then

$$\lim_{i\to\infty} \frac{|\theta^{i+1}(k) - \theta_\star(k)|}{|\theta^i(k) - \theta_\star(k)|} = \rho_k < 1 \; ; \tag{10.17}$$

here $\theta(k)$ denotes the kth coordinate of the parameter vector. For large-dimensional problems it frequently occurs that, at least for some components k, the factor ρ_k is close to one, resulting in very slow convergence of the algorithm. It should be stressed however that the same is true for the EM algorithm, which also exhibits speed of convergence that is linear, and often very poor (Dempster *et al.*, 1977; Jamshidian and Jennrich, 1997; Meng, 1994; Lange, 1995; Meng and Van Dyk, 1997). For gradient-based methods however, there exists a whole range of approaches, based on the second-order properties of the objective function, to guarantee faster convergence.

10.1.3.3 Newton and Second-order Methods

The prototype of second-order methods is the Newton, or Newton-Raphson, algorithm:

$$\theta^{i+1} = \theta^i - H^{-1}(\theta^i) \nabla_\theta \ell(\theta^i) \; , \tag{10.18}$$

where $H(\theta^i) = \nabla_\theta^2 \ell(\theta^i)$ is the Hessian of the objective function. The Newton iteration is based on the second-order approximation

$$\ell(\theta) \approx \ell(\theta') + \nabla \ell(\theta') \, (\theta - \theta') + \frac{1}{2} (\theta - \theta')^t \, H(\theta') \, (\theta - \theta') \; .$$

If the sequence $\{\theta^i\}_{i\geq 0}$ produced by the algorithm converges to a point $\theta_\star$ at which the Hessian is negative definite, the convergence is, at least, quadratic in the sense that for sufficiently large i there exists a positive constant β such that $\|\theta^{i+1} - \theta_\star\| \leq \beta\|\theta^i - \theta_\star\|^2$. Therefore the procedure can be very efficient.

The practical use of the Newton algorithm is however hindered by two serious difficulties. The first is analogous to the problem already encountered for the steepest ascent method: there is no guarantee that the algorithm meets the minimal requirement to provide a final parameter estimate that is at least as good as the starting point θ^0. To overcome this difficulty, one may proceed as for the steepest ascent method and introduce a multiplier γ_i controlling the step-length in the search direction, so that the method takes the form

$$\theta^{i+1} = \theta^i - \gamma_i H^{-1}(\theta^i)\nabla_\theta \ell(\theta^i) \ . \tag{10.19}$$

Again, γ_i may be set to maximize $\ell(\theta^{i+1})$. In practice, it is most often impossible to obtain the exact maximum point called for by the ideal line-search, and one uses approximate directional maximization procedures. Generally speaking, a *line-search algorithm* is an algorithm to find a reasonable multiplier γ_i in a step of the form (10.19). A frequently used algorithm consists in determining the (approximate) maximum based on a polynomial interpolation of $\ell(\theta)$ along the line-segment between the current point θ^i and the proposed update given by (10.18).

A more serious problem is that except in the particular case where the function $\ell(\theta)$ is strictly concave, the direct implementation of (10.18) is prone to numerical instabilities: there may well be whole regions of the parameter space where the Hessian $H(\theta)$ is either non-invertible (or at least very badly conditioned) or not negative semi-definite (in which case $-H^{-1}(\theta^i)\nabla_\theta \ell(\theta^i)$ is not necessarily an ascent direction). To combat this difficulty, Quasi-Newton methods[1] use the modified recursion

$$\theta^{i+1} = \theta^i + \gamma_i W^i \nabla \ell(\theta^i) \ ; \tag{10.20}$$

here W^i is a weight matrix that may be tuned at each iteration, just like the multiplier γ_i. The rationale is that if W^i becomes close to $-H^{-1}(\theta^i)$ when convergence occurs, the modified algorithm will share the favorable convergence properties of the Newton algorithm. On the other hand, by using a weight matrix W^i different from $-H^{-1}(\theta^i)$, numerical issues associated with the matrix inversion may be avoided. We again refer to Luenberger (1984) and Fletcher (1987) for a more precise discussion of the available approaches and simply mention here the fact that usually the methods only take profit of gradient information to construct W^i, for instance using finite difference calculations, without requiring the direct evaluation of the Hessian $H(\theta)$.

In some contexts, it may be possible to build explicit strategies that are not as good as the Newton algorithm—failing in particular to reach quadratic

[1] *Conjugate gradient* methods are another alternative approach that we do not discuss here.

convergence rates—but yet significantly faster at converging than the basic steepest ascent approach. For incomplete data models, Lange (1995) suggested to use in (10.20) a weight matrix $I_c^{-1}(\theta^i)$ given by

$$I_c(\theta') = -\int \nabla_\theta^2 \log f(x\,;\theta)\big|_{\theta=\theta'} p(x\,;\theta')\,\lambda(dx)\,. \tag{10.21}$$

This is the first term on the right-hand side of (10.13). In many models of interest, this matrix is positive definite for all $\theta' \in \Theta$, and thus its inversion is not subject to numerical instabilities. Based on (10.13), it is also to be expected that in some circumstances, $I_c(\theta')$ is a reasonable approximation to the Hessian $\nabla_\theta^2 \ell(\theta')$ and hence that the weighted gradient algorithm converges faster than the steepest ascent or EM algorithms (see Lange, 1995, for further results and examples). In a statistical context, where $f(x\,;\theta)$ is the joint density of two random variables X and Y, $I_c(\theta')$ is the conditional expectation given Y of the observed information matrix of associated with this pair.

10.1.4 Pros and Cons of Gradient-based Methods

A quick search through the literature shows that for HMMs in particular and incomplete data models in general, the EM algorithm is much more popular than are gradient-based alternatives. A first obvious reason for this is that the EM approach is more generally known than its gradient-based counterparts. We list below a number of additional significant differences between both approaches, giving first the arguments in favor of the EM algorithm.

- *The EM algorithm is usually very simple to implement from scratch.* This is not the case for gradient-based methods, which require several specialized routines, for Hessian approximation, line-search, etc. This argument is however made less pregnant by the wide availability of generic numerical optimization code, so that implementing a gradient-based method usually only requires the computation of the objective function ℓ and its gradient. In most situations, this is not more complicated than is implementing EM.
- *The EM algorithm often deals with parameter constraints implicitly.* It is generally the case that the M-step equations are so simple that they can be solved even for parameters that are subject to constraints (see the case of normal HMMs in Section 10.3 for an example). For gradient-based methods this is not the case, and parameter constraints have to be dealt with explicitly, either through reparameterization (see Example 10.3.2) or using constrained optimization routines.
- *The EM algorithm is parameterization independent.* Because the M-step is defined by a maximization operation, it is independent of the way the parameters are represented, as is the maximum likelihood estimator for instance. Thus any (invertible) transformation of the parameter vector θ leaves the EM recursion unchanged. This is obviously not the case for gradient-based methods for which reparameterization will change the gradient and Hessian, and hence the convergence behavior of the algorithm.

In contrast, gradient-based methods may be preferred for the following reasons.

- *Gradient-based methods do not require the M-step.* Thus they may be applied to models for which the M-step does not lead to simple closed-form solutions.
- *Gradient-based methods converge faster.* As discussed above, gradient-based methods can reach quadratic convergence whereas EM usually converges only linearly, following (10.17)—see Example 10.3.2 for an illustration and further discussion of this aspect.

10.2 Application to HMMs

We now return to our primary focus and discuss the application of the previous methods to the specific case of hidden Markov models.

10.2.1 Hidden Markov Models as Missing Data Models

HMMs correspond to a sub-category of incomplete data models known as missing data models. In missing data models, the observed data Y is a subset of some not fully observable *complete data* (X, Y). We here assume that the joint distribution of X and Y, for a given parameter value θ, admits a joint probability density function $f(x, y\,; \theta)$ with respect to the product measure $\lambda \otimes \mu$. As mentioned in Section 10.1.1, the function f is sometimes referred to as the *complete data likelihood*. It is important to understand that f is a probability density function only when considered as a function of both x and y. For a fixed value of y and considered as a function of x only, f is a positive integrable function. Indeed, the actual *likelihood* of the observation, which is defined as the probability density function of Y with respect to μ, is obtained by marginalization as

$$\mathrm{L}(y\,; \theta) = \int f(x, y\,; \theta)\, \lambda(dx)\;. \tag{10.22}$$

For a given value of y this is of course a particular case of (10.1), which served as the basis for developing the EM framework in Section 10.1.2. In missing data models, the family of probability density functions $\{p(\cdot\,; \theta)\}_{\theta \in \Theta}$ defined in (10.3) may thus be interpreted as

$$p(x|y\,; \theta) = \frac{f(x, y\,; \theta)}{\int f(x, y\,; \theta)\, \lambda(dx)}\;, \tag{10.23}$$

the conditional probability density function of X given Y.

In the last paragraph, slightly modified versions of the notations introduced in (10.1) and (10.3) were used to reflect the fact that the quantities of interest now depend on the observed variable Y. This is obviously mostly a change regarding terminology, with no impact on the contents of Section 10.1.2, except that we may now think of integrating with respect to $p(\cdot\,;\theta)\,d\lambda$ as taking the conditional expectation with respect to the *missing data* X, given the observed data Y, in the model indexed by the parameter value θ.

Remark 10.2.1. Applying the EM algorithm defined in Section 10.1.2 in the case of (10.22) yields a sequence of parameter values $\{\theta^i\}_{i\geq 0}$ whose likelihoods $\mathrm{L}(y\,;\theta^i)$ cannot decrease with the iteration index i. Obviously, this connects to *maximum likelihood estimation*. Another frequent use of the EM algorithm is for *maximum a posteriori (MAP)* estimation, in which the objective function to be maximized is a *Bayesian posterior* (Dempster *et al.*, 1977). Indeed, we may replace (10.22) by

$$\mathrm{L}(y\,;\theta) = \pi(\theta) \int f(x, y\,;\theta)\,\lambda(dx)\;, \tag{10.24}$$

where π is a positive function on Θ. In the Bayesian framework (see Section 13.1 for a brief presentation of the Bayesian approach), π is usually selected to be a probability density function (with respect to some measure on Θ) and (10.24) is then interpreted as being proportional, up to a factor that depends on y only, to the *posterior* probability density function of the unknown parameter θ, conditional on the observation Y. In that case, π is referred to as a *prior density* on the parameter θ. But π in (10.24) may also be thought of as a *regularization functional* (sometimes also called a penalty) that may not have a probabilistic interpretation (Green, 1990).

Whether L is defined according to (10.22) or to (10.24) does not modify the definition of $p(\cdot\,;\theta)$ in (10.23), as the factor $\pi(\theta)$ cancels out in the renormalization. Thus the E-step in the EM algorithm is left unchanged and only the M-step depends on the precise choice of π. ∎

10.2.2 EM in HMMs

We now consider more specifically hidden Markov models using the notations introduced in Section 2.2, assuming that observations Y_0 to Y_n (or, in short, $Y_{0:n}$) are available. Because we only consider HMMs that are fully dominated in the sense of Definition 2.2.3, we will use the notations ν and $\phi_{k|n}$ to refer to the probability density functions of these distributions (of X_0 and of X_k given $Y_{0:n}$) with respect to the dominating measure λ. The joint probability density function of the hidden states $X_{0:n}$ and associated observations $Y_{0:n}$, with respect to the product measure $\lambda^{\otimes(n+1)} \otimes \mu^{\otimes(n+1)}$, is given by

$$\begin{aligned} f_n(x_{0:n}, y_{0:n}\,;\theta) = \nu(x_0\,;\theta) g(x_0, y_0\,;\theta) q(x_0, x_1\,;\theta) g(x_1, y_1\,;\theta) \\ \cdots q(x_{n-1}, x_n\,;\theta) g(x_n, y_n\,;\theta)\;, \end{aligned} \tag{10.25}$$

where we used the same convention as above to indicate dependence with respect to the parameter θ.

Because we mainly consider estimation of the HMM parameter vector θ from a single sequence of observations, it does not make much sense to consider ν as an independent parameter. There is no hope to estimate ν consistently, as there is only one random variable X_0 (that is not even observed!) drawn from this density. In the following, we shall thus consider that ν is either fixed (and known) or fully determined by the parameter θ that appears in q and g. A typical example of the latter consists in assuming that ν is the stationary distribution associated with the transition function $q(\cdot,\cdot\,;\theta)$ (if it exists). This option is generally practicable only in very simple models (see Example 10.3.3 below for an example) because of the lack of analytical expressions relating the stationary distribution of $q(\cdot,\cdot\,;\theta)$ to θ for general parameterized hidden chains. Irrespective of whether ν is fixed or determined by θ, it is convenient to omit dependence with respect to ν in our notations, writing, for instance, E_θ for expectations under the model parameterized by (θ, ν).

Note that for left-to-right HMMs (discussed Section 1.4), the case is rather different as the model is trained from several independent sequences and the initial distribution is often a key parameter. Handling the case of multiple training sequences is straightforward as the quantities corresponding to different sequences simply need to be added together due to the independence assumption (see Section 10.3.2 below for the details in the normal HMM case).

The likelihood of the observations $\mathrm{L}_n(y_{0:n}\,;\theta)$ is obtained by integrating (10.25) with respect to the x (state) variables under the measure $\lambda^{\otimes(n+1)}$. Note that here we use yet another slight modification of the notations adopted in Section 10.1 to acknowledge that both the observations and the hidden states are indeed sequences with indices ranging from 0 to n (hence the subscript n). Upon taking the logarithm in (10.25),

$$\log f_n(x_{0:n}, y_{0:n}\,;\theta) = \log \nu(x_0\,;\theta) + \sum_{k=0}^{n-1} \log q(x_k, x_{k+1}\,;\theta) + \sum_{k=0}^{n} \log g(x_k, y_k\,;\theta)\;,$$

and hence the intermediate quantity of EM has the additive structure

$$\mathcal{Q}(\theta\,;\theta') = \mathrm{E}_{\theta'}[\log \nu(X_0\,;\theta) \mid Y_{0:n}] + \sum_{k=0}^{n-1} \mathrm{E}_{\theta'}[\log q(X_k, X_{k+1}\,;\theta) \mid Y_{0:n}] + \sum_{k=0}^{n} \mathrm{E}_{\theta'}[\log g(X_k, Y_k\,;\theta) \mid Y_{0:n}]\;.$$

In the following, we will adopt the "implicit conditioning" convention that we have used extensively from Section 3.1.4 and onwards, writing $g_k(x\,;\theta)$ instead of $g(x, Y_k\,;\theta)$. With this notation, the intermediate quantity of EM may be rewritten as

$$
\begin{aligned}
\mathcal{Q}(\theta\,;\theta') = \mathrm{E}_{\theta'}[\log \nu(X_0\,;\theta) \mid Y_{0:n}] + \sum_{k=0}^{n} \mathrm{E}_{\theta'}[\log g_k(X_k\,;\theta) \mid Y_{0:n}] \\
+ \sum_{k=0}^{n-1} \mathrm{E}_{\theta'}[\log q(X_k, X_{k+1}\,;\theta) \mid Y_{0:n}] \,. \quad (10.26)
\end{aligned}
$$

Equation (10.26) shows that in great generality, evaluating the intermediate quantity of EM only requires the computation of expectations under the marginal $\phi_{k|n}(\cdot\,;\theta')$ and bivariate $\phi_{k:k+1|n}(\cdot\,;\theta')$ smoothing distributions, given the parameter vector θ'. The required expectations may thus be computed using either any of the variants of the forward-backward approach presented in Chapter 3 or the recursive smoothing approach discussed in Section 4.1. To make the connection with the recursive smoothing approach of Section 4.1, we simply rewrite (10.26) as $\mathrm{E}_{\theta'}[t_n(X_{0:n}\,;\theta) \mid Y_{0:n}]$, where

$$
t_0(x_0\,;\theta) = \log \nu(x_0\,;\theta) + \log g_0(x_0\,;\theta) \quad (10.27)
$$

and

$$
t_{k+1}(x_{0:k+1}\,;\theta) = t_k(x_{0:k}\,;\theta) + \{\log q(x_k, x_{k+1}\,;\theta) + \log g_{k+1}(x_{k+1}\,;\theta)\} \,. \quad (10.28)
$$

Proposition 4.1.3 may then be applied directly to obtained the smoothed expectation of the sum functional t_n.

Although the exact form taken by the M-step will obviously depend on the way g and q depend on θ, the EM update equations follow a very systematic scheme that does not change much with the exact model under consideration. For instance, all discrete state space models for which the transition matrix q is parameterized by its $r \times r$ elements and such that g and q do not share common parameters (or parameter constraints) give rise to the same update equations for q, given in (10.43) below. Several examples of the EM update equations will be reviewed in Sections 10.3 and 10.4.

10.2.3 Computing Derivatives

Recall that the Fisher identity—(10.12)—provides an expression for the gradient of the log-likelihood $\ell_n(\theta)$ with respect to the parameter vector θ, closely related to the intermediate quantity of EM. In the HMM context, (10.12) reduces to

$$
\begin{aligned}
\nabla_\theta \ell_n(\theta) = \mathrm{E}_\theta[\nabla_\theta \log \nu(X_0\,;\theta) \mid Y_{0:n}] + \sum_{k=0}^{n} \mathrm{E}_\theta[\nabla_\theta \log g_k(X_k\,;\theta) \mid Y_{0:n}] \\
+ \sum_{k=0}^{n-1} \mathrm{E}_\theta[\nabla_\theta \log q(X_k, X_{k+1}\,;\theta) \mid Y_{0:n}] \,. \quad (10.29)
\end{aligned}
$$

Hence the gradient of the log-likelihood may also be evaluated using either the forward-backward approach or the recursive technique discussed in Chapter 4. For the latter, we only need to redefine the functional of interest, replacing (10.27) and (10.28) by their gradients with respect to θ.

Louis' identity (10.14) gives rise to more complicated expressions, and we only consider here the case where g does depend on θ, whereas the state transition density q and the initial distribution ν are assumed to be fixed and known (the opposite situation is covered in detail in a particular case in Section 10.3.4). In this case, (10.14) may be rewritten as

$$\begin{aligned}
&\nabla_\theta^2 \ell_n(\theta) + \{\nabla_\theta \ell_n(\theta)\}\{\nabla_\theta \ell_n(\theta)\}^t \\
&\quad = \sum_{k=0}^{n} \mathrm{E}_\theta[\nabla_\theta^2 \log g_k(X_k\,;\theta) \mid Y_{0:n}] \\
&\qquad + \sum_{k=0}^{n}\sum_{j=0}^{n} \mathrm{E}_\theta\left[\{\nabla_\theta \log g_k(X_k\,;\theta)\}\{\nabla_\theta \log g_j(X_j\,;\theta)\}^t \,\middle|\, Y_{0:n}\right] .
\end{aligned} \tag{10.30}$$

The first term on the right-hand side of (10.30) is obviously an expression that can be computed proceeding as for (10.29), replacing first- by second-order derivatives. The second term is however more tricky because it (seemingly) requires the evaluation of the joint distribution of X_k and X_j given the observations $Y_{0:n}$ for all pairs of indices k and j, which is not obtainable by the smoothing approaches based on some form of the forward-backward decomposition. The rightmost term of (10.30) is however easily recognized as a squared sum functional similar to (4.4), which can thus be evaluated recursively (in n) proceeding as in Example 4.1.4. Recall that the trick consists in observing that if

$$\tau_{n,1}(x_{0:n}\,;\theta) \stackrel{\text{def}}{=} \sum_{k=0}^{n} \nabla_\theta \log g_k(x_k\,;\theta) ,$$

$$\tau_{n,2}(x_{0:n}\,;\theta) \stackrel{\text{def}}{=} \left\{\sum_{k=0}^{n} \nabla_\theta \log g_k(x_k\,;\theta)\right\}\left\{\sum_{k=0}^{n} \nabla_\theta \log g_k(x_k\,;\theta)\right\}^t ,$$

then

$$\begin{aligned}
\tau_{n,2}(x_{0:n}\,;\theta) = \tau_{n-1,2}(x_{0:n-1}\,;\theta) &+ \{\nabla_\theta \log g_n(x_n\,;\theta)\}\{\nabla_\theta \log g_n(x_n\,;\theta)\}^t \\
&+ \tau_{n-1,1}(x_{0:n-1}\,;\theta)\{\nabla_\theta \log g_n(x_n\,;\theta)\}^t \\
&+ \nabla_\theta \log g_n(x_n\,;\theta)\{\tau_{n-1,1}(x_{0:n-1}\,;\theta)\}^t .
\end{aligned}$$

This last expression is of the general form given in Definition 4.1.2, and hence Proposition 4.1.3 may be applied to update recursively in n

$$\mathrm{E}_\theta[\tau_{n,1}(X_{0:n}\,;\theta) \mid Y_{0:n}] \qquad \text{and} \qquad \mathrm{E}_\theta[\tau_{n,2}(X_{0:n}\,;\theta) \mid Y_{0:n}] .$$

To make this approach more concrete, we will describe below, in Section 10.3.4, its application to a very simple finite state space HMM.

10.2.4 Connection with the Sensitivity Equation Approach

The method outlined above for evaluating the gradient of the likelihood is coherent with the general approach of Section 4.1. There is however a (seemingly) distinct approach for evaluating the same quantity, which does not require the use of Fisher's identity, and has been used for a very long time in the particular case of Gaussian linear state-space models. The method, known under the name of *sensitivity equations* (see for instance Gupta and Mehra, 1974), postulates that since the log-likelihood can be computed recursively based on the Kalman prediction recursion, its derivatives can also be computed by a recursion—the so-called *sensitivity equations*—which is obtained by differentiating the Kalman relations with respect to the model parameters. For such models, the remark that the gradient of the log-likelihood may also be obtained using Fisher's identity was made by Segal and Weinstein (1989); see also Weinstein *et al.* (1994).

The sensitivity equations approach is in no way limited to Gaussian linear state-space models but may be applied to HMMs in general. This remark, put forward by Campillo and Le Gland (1989) and Le Gland and Mevel (1997), has been subsequently used for finite state-space HMMs (Cappé *et al.*, 1998; Collings and Rydén, 1998) as well as for general HMMs (Cérou *et al.*, 2001; Doucet and Tadić, 2003). In the latter case, it is necessary to resort to some form of sequential Monte Carlo approach discussed in Chapter 7 because exact filtering is not available. It is interesting that the sequential Monte Carlo approximation method used by both Cérou *et al.* (2001) and Doucet and Tadić (2003) has also been derived by Cappé (2001a) using Fisher's identity and the smoothing framework discussed in Section 4.1. Indeed, we show below that the sensitivity equation approach is *exactly* equivalent to the use of Fisher's identity.

Recall that the log-likelihood may be written according to (3.29) as a sum of terms that only involve the prediction density,

$$\ell_n(\theta) = \sum_{k=0}^{n} \log \underbrace{\int \phi_{k|k-1}(x_k\,;\theta) g_k(x_k\,;\theta)\, \lambda(dx_k)}_{c_k(\theta)} , \tag{10.31}$$

where the integral is also the normalizing constant that appears in the prediction and filtering recursion (Remark 3.2.6), which we denoted by $c_k(\theta)$. The filtering recursion as given by (3.27) implies that

$$\phi_{k+1}(x_{k+1}\,;\theta) = c_{k+1}^{-1}(\theta) \int \phi_k(x_k\,;\theta) q(x_k, x_{k+1}\,;\theta) g_{k+1}(x_{k+1}\,;\theta)\, \lambda(dx_k) . \tag{10.32}$$

To differentiate (10.32) with respect to θ, we assume that $c_{k+1}(\theta)$ does not vanish and we use the obvious identity

$$\nabla_\theta \frac{u(\theta)}{v(\theta)} = v^{-1}(\theta) \nabla_\theta u(\theta) - \frac{u(\theta)}{v(\theta)} \nabla_\theta \log v(\theta)$$

to obtain

$$\nabla_\theta \phi_{k+1}(x_{k+1}\,;\theta) = \rho_{k+1}(x_{k+1}\,;\theta) - \phi_{k+1}(x_{k+1}\,;\theta)\,\nabla_\theta \log c_{k+1}(\theta)\;, \tag{10.33}$$

where

$$\rho_{k+1}(x_{k+1}\,;\theta) \stackrel{\text{def}}{=} c_{k+1}^{-1}(\theta)\nabla_\theta \int \phi_k(x_k\,;\theta)q(x_k, x_{k+1}\,;\theta)g_{k+1}(x_{k+1}\,;\theta)\,\lambda(dx_k)\;. \tag{10.34}$$

We further assume that as in Proposition 10.1.6, we may interchange integration with respect to λ and differentiation with respect to θ. Because $\phi_{k+1}(\cdot\,;\theta)$ is a probability density function, $\int \phi_{k+1}(x_{k+1}\,;\theta)\,\lambda(dx_{k+1}) = 1$ and $\nabla_\theta \int \phi_{k+1}(x_{k+1}\,;\theta)\,\lambda(dx_{k+1}) = \int \nabla_\theta \phi_{k+1}(x_{k+1}\,;\theta)\,\lambda(dx_{k+1}) = 0$. Therefore, integration of both sides of (10.33) with respect to $\lambda(dx_{k+1})$ yields

$$0 = \int \rho_{k+1}(x_{k+1}\,;\theta)\,\lambda(dx_{k+1}) - \nabla_\theta \log c_{k+1}(\theta)\;.$$

Hence, we may evaluate the gradient of the incremental log-likelihood in terms of ρ_{k+1} according to

$$\nabla_\theta \log c_{k+1}(\theta) \stackrel{\text{def}}{=} \nabla_\theta(\ell_{k+1}(\theta) - \ell_k(\theta)) = \int \rho_{k+1}(x_{k+1}\,;\theta)\,\lambda(dx_{k+1})\;. \tag{10.35}$$

Now we evaluate the derivative in (10.34) assuming also that q and g_k are non-zero to obtain

$$\begin{aligned}\rho_{k+1}(x_{k+1}\,;\theta) = c_{k+1}^{-1}(\theta) \int \Big\{ &[\nabla_\theta \log q(x_k, x_{k+1}\,;\theta) + \nabla_\theta \log g_{k+1}(x_{k+1}\,;\theta)] \\ &\times \phi_k(x_k\,;\theta) + \nabla_\theta \phi_k(x_k\,;\theta)\Big\} q(x_k, x_{k+1}\,;\theta)g_{k+1}(x_{k+1}\,;\theta)\,\lambda(dx_k)\;.\end{aligned}$$

Plugging (10.33) into the above equation yields an update formula for ρ_{k+1},

$$\begin{aligned}\rho_{k+1}(x_{k+1}\,;\theta) = c_{k+1}^{-1}(\theta) \int \Big\{ &[\nabla_\theta \log q(x_k, x_{k+1}\,;\theta) + \nabla_\theta \log g_{k+1}(x_{k+1}\,;\theta)] \\ &\times \phi_k(x_k\,;\theta) + \rho_k(x_k\,;\theta)\Big\} q(x_k, x_{k+1}\,;\theta)g_{k+1}(x_{k+1}\,;\theta)\,\lambda(dx_k) \\ &- \phi_{k+1}(x_{k+1}\,;\theta)\nabla_\theta \log c_k(\theta)\;, \end{aligned} \tag{10.36}$$

where (10.32) has been used for the last term on the right-hand side. We collect these results in the form of the algorithm below.

Algorithm 10.2.2 (Sensitivity Equations). In addition to the usual filtering recursions, do:

Initialization: Compute

$$\rho(x_0) = [\nabla_\theta \log \nu(x_0\,;\theta) + \nabla_\theta \log q_0(x_0\,;\theta)]\,\phi_0(x_0\,;\theta)$$

and $\nabla_\theta \ell_0(\theta) = \int \rho(x_0)\,\lambda(dx_0)$.

Recursion: For $k = 0, 1 \ldots$, use (10.36) to compute ρ_{k+1} and (10.35) to evaluate $\nabla_\theta \ell_{k+1}(\theta) - \nabla_\theta \ell_k(\theta)$.

Algorithm 10.2.2 updates the intermediate function $\rho_k(\cdot\,;\theta)$, defined in (10.34), whose integral is the quantity of interest $\nabla_\theta \log c_k(\theta)$. Obviously, one can equivalently use as intermediate quantity the derivative of the filtering probability density function $\nabla_\theta \phi_k(\cdot\,;\theta)$, which is directly related to $\rho_k(\cdot\,;\theta)$ by (10.33). The quantity $\nabla_\theta \phi_k(\cdot\,;\theta)$, which is referred to as the *tangent filter* by Le Gland and Mevel (1997), is also known as the *filter sensitivity* and may be of interest in its own right. Using $\nabla_\theta \phi_k(\cdot\,;\theta)$ instead of $\rho_k(\cdot\,;\theta)$ does not however modify the nature of algorithm, except for slightly more involved mathematical expressions.

It is interesting to contrast Algorithm 10.2.2 with the smoothing approach based on Fisher's identity (10.29). Recall from Section 4.1 that in order to evaluate (10.29), we recursively define a sequence of functions by

$$t_0(x_0) = \nabla_\theta \log \nu(x_0\,;\theta) + \nabla_\theta \log g_0(x_0\,;\theta)\ ,$$

and

$$t_{k+1}(x_{0:k+1}) = t_k(x_{0:k}) + \nabla_\theta \log q(x_k, x_{k+1}\,;\theta) + \nabla_\theta \log g_{k+1}(x_k\,;\theta)$$

for $k \geq 0$.

Proposition 4.1.3 asserts that $\mathrm{E}_\theta\left[t_k(X_{0:k})\,|\,Y_{0:k}\right] = \int \tau_k(x_k\,;\theta)\,\lambda(dx_k)$, where τ_k may be updated according to the recursion

$$\begin{aligned}\tau_{k+1}(x_{k+1}\,;\theta) = c_{k+1}^{-1}(\theta) \int \Big\{ &[\nabla_\theta \log q(x_k, x_{k+1}\,;\theta) + \nabla_\theta \log g_{k+1}(x_{k+1}\,;\theta)] \\ &\times \phi_k(x_k\,;\theta) + \tau_k(x_k\,;\theta)\Big\} q(x_k, x_{k+1}\,;\theta) g_{k+1}(x_{k+1}\,;\theta)\,\lambda(dx_k) \quad (10.37)\end{aligned}$$

for $k \geq 0$, where $\tau_0(x_0\,;\theta) = c_0(\theta)^{-1}\nu(x_0)t_0(x_0)g_0(x_0)$.

Comparing (10.37) and (10.36), it is easily established by recurrence on k that $\rho_0(\cdot\,;\theta) = \tau_0(\cdot\,;\theta)$ and

$$\rho_k(\cdot\,;\theta) = \tau_k(\cdot\,;\theta) - \left(\sum_{l=0}^{k-1} \nabla_\theta \log c_l(\theta)\right) \phi_k(\cdot\,;\theta) \qquad (10.38)$$

for $k \geq 1$. Hence, whereas $\int \tau_k(x_k\,;\theta)\,\lambda(dx_k)$ gives access to $\nabla_\theta \ell_k(\theta)$, the gradient of the log-likelihood up to index k, $\int \rho_k(x_k\,;\theta)\,\lambda(dx_k)$ equals the gradient of the increment $\ell_k(\theta) - \ell_{k-1}(\theta)$, where the second term is decomposed into the telescoping sum $\ell_{k-1}(\theta) = \sum_{l=0}^{k-1} \nabla_\theta \log c_l(\theta)$ of increments.

The sensitivity equations and the use of Fisher's identity combined with the recursive smoothing algorithm of Proposition 4.1.3 are thus completely equivalent. The fundamental reason for this rather surprising observation is that whereas the log-likelihood may be written, according to (10.31), as a sum of integrals under the successive prediction distributions, the same is no more true when differentiating with respect to θ. To compute the gradient of (10.31), one needs to evaluate $\rho_k(\cdot\,;\theta)$—or, equivalently, $\nabla_\theta \phi_k(\cdot\,;\theta)$—which depends on all the previous values of $c_l(\theta)$ through the sum $\sum_{l=0}^{k-1} \nabla_\theta \log c_l(\theta)$.

To conclude this section, let us stress again that there are only two different options for computing the gradient of the log-likelihood.

Forward-backward algorithm: based on Fisher's identity (10.29) and forward-backward smoothing.

Recursive algorithm: which can be equivalently derived either through the sensitivity equations or as an application of Proposition 4.1.3 starting from Fisher's identity. Both arguments give rise to the same algorithm.

These two options only differ in the way the computations are organized, as both evaluate *exactly* the sum of terms appearing in (10.29). In considering several examples below, we shall observe that the former solution is generally more efficient from the computational point of view.

10.3 The Example of Normal Hidden Markov Models

In order to make the general principles outlined in the previous section more concrete, we now work out the details on selected examples of HMMs. We begin with the case where the state space is finite and the observation transition function g corresponds to a (univariate) Gaussian distribution. Only the most standard case where the parameter vector is split into two sub-components that parameterize, respectively, g and q, is considered.

10.3.1 EM Parameter Update Formulas

In the widely used normal HMM discussed in Section 1.3.2, X is a finite set, identified with $\{1, \ldots, r\}$, $\mathsf{Y} = \mathbb{R}$, and g is a Gaussian probability density function (with respect to Lebesgue measure) given by

$$g(x, y\,; \theta) = \frac{1}{\sqrt{2\pi\upsilon_x}} \exp\left\{-\frac{(y-\mu_x)^2}{2\upsilon_x}\right\} .$$

By definition, $g_k(x\,;\theta)$ is equal to $g(x, Y_k\,;\theta)$. We first assume that the initial distribution ν is known and fixed, before examining the opposite case briefly in Section 10.3.2 below. The parameter vector θ thus encompasses the transition probabilities q_{ij} for $i, j = 1, \ldots, r$ as well as the means μ_i and variances υ_i for $i = 1, \ldots, r$. Note that in this section, because we will often need to differentiate with respect to υ_i, it is simpler to use the variances $\upsilon_i = \sigma_i^2$ rather than the standard deviations σ_i as parameters. The means and variances are unconstrained, except for the positivity of the latter, but the transition probabilities are subject to the equality constraints $\sum_{j=1}^{r} q_{ij} = 1$

for $i = 1, \ldots, r$ (in addition to the obvious constraint that q_{ij} should be non-negative). When considering the parameter vector denoted by θ', we will denote by μ'_i, υ'_i, and q'_{ij} its various elements.

For the model under consideration, (10.26) may be rewritten as

$$\mathcal{Q}(\theta\,;\theta') = C^{st} - \frac{1}{2}\sum_{k=0}^{n} \mathrm{E}_{\theta'}\left[\sum_{i=1}^{r} \mathbb{1}\{X_k = i\}\left(\log \upsilon_i + \frac{(Y_k - \mu_i)^2}{\upsilon_i}\right) \middle|\, Y_{0:n}\right] \\ + \sum_{k=1}^{n} \mathrm{E}_{\theta'}\left[\sum_{i=1}^{r}\sum_{j=1}^{r} \mathbb{1}\{(X_{k-1}, X_k) = (i,j)\} \log q_{ij} \,\middle|\, Y_{0:n}\right] ,$$

where the leading term does not depend on θ. Using the notations introduced in Section 3.1 for the smoothing distributions, we may write

$$\mathcal{Q}(\theta\,;\theta') = C^{st} - \frac{1}{2}\sum_{k=0}^{n}\sum_{i=1}^{r} \phi_{k|n}(i\,;\theta')\left[\log \upsilon_i + \frac{(Y_k - \mu_i)^2}{\upsilon_i}\right] \\ + \sum_{k=1}^{n}\sum_{i=1}^{r}\sum_{j=1}^{r} \phi_{k-1:k|n}(i,j\,;\theta') \log q_{ij} \,. \quad (10.39)$$

In the above expression, we use the same convention as in Chapter 5 and denote the smoothing probability $\mathrm{P}_{\theta'}(X_k = i \,|\, Y_{0:n})$ by $\phi_{k|n}(i\,;\theta')$ rather than by $\phi_{k|n}(\{i\}\,;\theta')$. The variable θ' is there to recall the dependence of the smoothing probability on the unknown parameters.

Now, given the initial distribution ν and parameter θ', the smoothing distributions appearing in (10.39) can be evaluated by any of the variants of forward-backward smoothing discussed in Chapter 3. As already explained above, the E-step of EM thus reduces to solving the smoothing problem. The M-step is specific and depends on the model parameterization: the task consists in finding a global optimum of $\mathcal{Q}(\theta\,;\theta')$ that satisfies the constraints mentioned above. For this, simply introduce the Lagrange multipliers $\lambda_1, \ldots, \lambda_r$ that correspond to the equality constraints $\sum_{j=1}^{r} q_{ij} = 1$ for $i = 1, \ldots, r$ (Luenberger, 1984, Chapter 10). The first-order partial derivatives of the Lagrangian

$$\mathfrak{L}(\theta, \lambda\,;\theta') = \mathcal{Q}(\theta\,;\theta') + \sum_{i=1}^{r} \lambda_i \left(1 - \sum_{j=1}^{r} q_{ij}\right)$$

are given by

$$\frac{\partial}{\partial \mu_i}\mathfrak{L}(\theta, \lambda\,;\theta') = \frac{1}{\upsilon_i}\sum_{k=0}^{n} \phi_{k|n}(i\,;\theta')(Y_k - \mu_i) \,,$$

$$\frac{\partial}{\partial \upsilon_i}\mathfrak{L}(\theta, \lambda\,;\theta') = -\frac{1}{2}\sum_{k=0}^{n} \phi_{k|n}(i\,;\theta')\left[\frac{1}{\upsilon_i} - \frac{(Y_k - \mu_i)^2}{\upsilon_i^2}\right] ,$$

$$\frac{\partial}{\partial q_{ij}} \mathfrak{L}(\theta, \lambda ; \theta') = \sum_{k=1}^{n} \frac{\phi_{k-1:k|n}(i, j ; \theta')}{q_{ij}} - \lambda_i ,$$

$$\frac{\partial}{\partial \lambda_i} \mathfrak{L}(\theta, \lambda ; \theta') = 1 - \sum_{j=1}^{r} q_{ij} . \tag{10.40}$$

Equating all expressions in (10.40) to zero yields the parameter vector

$$\theta^* = \left[(\mu_i^*)_{i=1,\ldots,r}, (\upsilon_i^*)_{i=1,\ldots,r}, (q_{ij}^*)_{i,j=1,\ldots,r} \right]$$

which achieves the maximum of $\mathcal{Q}(\theta ; \theta')$ under the applicable parameter constraints:

$$\mu_i^* = \frac{\sum_{k=0}^{n} \phi_{k|n}(i ; \theta') Y_k}{\sum_{k=0}^{n} \phi_{k|n}(i ; \theta')} , \tag{10.41}$$

$$\upsilon_i^* = \frac{\sum_{k=0}^{n} \phi_{k|n}(i ; \theta')(Y_k - \mu_i^*)^2}{\sum_{k=0}^{n} \phi_{k|n}(i ; \theta')} , \tag{10.42}$$

$$q_{ij}^* = \frac{\sum_{k=1}^{n} \phi_{k-1:k|n}(i, j ; \theta')}{\sum_{k=1}^{n} \sum_{l=1}^{r} \phi_{k-1:k|n}(i, l ; \theta')} \tag{10.43}$$

for $i, j = 1, \ldots, r$, where the last equation may be rewritten more concisely as

$$q_{ij}^* = \frac{\sum_{k=1}^{n} \phi_{k-1:k|n}(i, j ; \theta')}{\sum_{k=1}^{n} \phi_{k-1|n}(i ; \theta')} . \tag{10.44}$$

Equations (10.41)–(10.43) are emblematic of the intuitive form taken by the parameter update formulas derived though the EM strategy. These equations are simply the maximum likelihood equations for the *complete model* in which both $\{X_k\}_{0 \le k \le n}$ and $\{Y_k\}_{0 \le k \le n}$ would be observed, except that the functions $\mathbb{1}\{X_k = i\}$ and $\mathbb{1}\{X_{k-1} = i, X_k = j\}$ are replaced by their conditional expectations, $\phi_{k|n}(i ; \theta')$ and $\phi_{k-1:k|n}(i, j ; \theta')$, given the actual observations $Y_{0:n}$ and the available parameter estimate θ'. As discussed in Section 10.1.2.4, this behavior is fundamentally due to the fact that the probability density functions associated with the complete model form an exponential family. As a consequence, the same remark holds more generally for all discrete HMMs for which the conditional probability density functions $g(i, \cdot ; \theta)$ belong to an exponential family. A final word of warning about the way in which (10.42) is written: in order to obtain a concise and intuitively interpretable expression, (10.42) features the value of μ_i^* as given by (10.41). It is of course possible to rewrite (10.42) in a way that only contains the current parameter value θ' and the observations $Y_{0:n}$ by combining (10.41) and (10.42) to obtain

$$\upsilon_i^* = \frac{\sum_{k=0}^{n} \phi_{k|n}(i ; \theta') Y_k^2}{\sum_{k=0}^{n} \phi_{k|n}(i ; \theta')} - \left[\frac{\sum_{k=0}^{n} \phi_{k|n}(i ; \theta') Y_k}{\sum_{k=0}^{n} \phi_{k|n}(i ; \theta')} \right]^2 . \tag{10.45}$$

For normal HMMs, the M-step thus reduces to computing averages and ratios of simple expressions that involve the marginal and bivariate smoothing probabilities evaluated during the E-step. The number of operations associated with the implementation of these expressions scales with respect to r and n like $r^2 \times n$, which is similar to the complexity of forward-backward smoothing (see Chapter 5). In practice however, the M-step is usually faster than the E-step because operations such as sums, products, or squares are carried out faster than the exponential (recall that forward-backward smoothing requires the computation of $g_{\theta'}(i, y_k)$ for all $i = 1, \ldots, r$ and $k = 0, \ldots, n$). Although the difference may not be very significant for scalar models, it becomes more and more important for high-dimensional multivariate generalizations of the normal HMM, such as those used in speech recognition.

10.3.2 Estimation of the Initial Distribution

As mentioned above, in this chapter we generally assume that the initial distribution ν, that is, the distribution of X_0, is fixed and known. There are cases when one wants to treat this as an unknown parameter however, and we briefly discuss below this issue in connection with the EM algorithm for the normal HMM. We shall assume that $\nu = (\nu_i)_{1 \leq i \leq r}$ is an unknown probability vector (that is, with non-negative entries summing to unity), which we accommodate within the parameter vector θ. The complete log-likelihood will then be as above, where the initial term

$$\log \nu_{X_0} = \sum_{i=1}^{r} \mathbb{1}\{X_0 = i\} \log \nu_i$$

goes into $\mathcal{Q}(\theta\,;\theta')$ as well, giving the additive contribution

$$\sum_{i=1}^{r} \phi_{0|n}(i\,;\theta') \log \nu_i$$

to (10.39). This sum is indeed part of (10.39) already, but hidden within C^{st} when ν is not a parameter to be estimated. Using Lagrange multipliers as above, it is straightforward to show that the M-step update of ν is $\nu_i^* = \phi_{0|n}(i\,;\theta')$.

It was also mentioned above that sometimes it is desirable to link ν to q_θ as being the stationary distribution of q_θ. Then there is an additive contribution to $\mathcal{Q}(\theta\,;\theta')$ as above, with the difference that ν can now not be chosen freely but is a function of q_θ. As there is no simple formula for the stationary distribution of q_θ, the M-step is no longer explicit. However, once the sums (over k) in (10.39) have been computed for all i and j, we are left with an optimization problem over the q_{ij} for which we have an excellent initial guess, namely the standard update (ignoring ν) (10.43). A few steps of a standard numerical optimization routine (optimizing over the q_{ij}) is then often enough to find the maximum of $\mathcal{Q}(\cdot\,;\theta')$ under the stationarity assumption. Variants of the basic EM strategy, to be discussed in Section 10.5.3, may also be useful in this situation.

10.3.3 Recursive Implementation of E-Step

An important observation about (10.41)–(10.43) is that all expressions are ratios in which both the numerator and the denominator may be interpreted as smoothed expectations of simple additive functionals. As a consequence, the recursive smoothing techniques discussed in Chapter 4 may be used to evaluate separately the numerator and denominator of each expression. The important point here is that to implement the E-step of EM, forward-backward smoothing is not strictly required and it may be replaced by a purely recursive evaluation of the quantities involved in the M-step update.

As an example, consider the case of the first update equation (10.41) that pertains to the means μ_i. For *each pre-specified state* i, say $i = i_0$, one can devise a recursive filter to compute the quantities needed to update μ_{i_0} as follows. First define the two functionals

$$t_{n,1}(X_{0:n}) = \sum_{k=0}^{n} \mathbb{1}\{X_k = i_0\} Y_k \; ,$$
$$t_{n,2}(X_{0:n}) = \sum_{k=0}^{n} \mathbb{1}\{X_k = i_0\} \; . \tag{10.46}$$

Comparing with the general form considered in Chapter 4, the two functionals above are clearly of additive type. Hence the multiplicative functions $\{m_k\}_{0 \le k \le n}$ that appear in Definition 4.1.2 are constant and equal to one in this case. Proceeding as in Chapter 4, we associate with the functionals defined in (10.46) the sequence of signed measures

$$\tau_{n,1}(i\,;\theta') = \mathrm{E}_{\theta'}[\mathbb{1}\{X_n = i\} t_{n,1}(X_{0:n}) \mid Y_{0:n}] \; ,$$
$$\tau_{n,2}(i\,;\theta') = \mathrm{E}_{\theta'}[\mathbb{1}\{X_n = i\} t_{n,2}(X_{0:n}) \mid Y_{0:n}] \; , \tag{10.47}$$

for $i = 1, \ldots, r$. Note that we adopt here the same convention as for the smoothing distributions, writing $\tau_{n,1}(i\,;\theta')$ rather than $\tau_{n,1}(\{i\}\,;\theta')$. In this context, the expression "signed measure" is also somewhat pompous because the state space X is finite and $\tau_{n,1}$ and $\tau_{n,2}$ can safely be identified with vectors in $\mathbb{R}^r$. The numerator and denominator of (10.41) for the state $i = i_0$ are given by, respectively,

$$\sum_{i=1}^{r} \tau_{n,1}(i\,;\theta') \quad \text{and} \quad \sum_{i=1}^{r} \tau_{n,2}(i\,;\theta') \; ,$$

which can also be checked directly from (10.47), as $\sum_{i=1}^{r} \mathbb{1}\{X_n = i\}$ is identically equal to one. Recall from Chapter 4 that $\tau_{n,1}$ and $\tau_{n,2}$ are indeed quantities that may be recursively updated following the general principle of Proposition 4.1.3. Algorithm 10.3.1 below is a restatement of Proposition 4.1.3 in the context of the finite normal hidden Markov model.

Algorithm 10.3.1 (Recursive Smoothing for a Mean).

Initialization: Compute the first filtering distribution according to

$$\phi_0(i\,;\theta') = \frac{\nu(i)g_0(i\,;\theta')}{c_0(\theta')} \,,$$

for $i = 1, \ldots, r$, where $c_0(\theta') = \sum_{j=1}^{r} \nu(j)g_0(j\,;\theta')$. Then

$$\tau_{0,1}(i_0\,;\theta') = \phi_0(i_0\,;\theta')Y_0 \quad \text{and} \quad \tau_{0,2}(i_0\,;\theta') = \phi_0(i_0\,;\theta') \,,$$

and both $\tau_{0,1}(i\,;\theta')$ and $\tau_{0,2}(i\,;\theta')$ are set to zero for $i \neq i_0$.

Recursion: For $k = 0, \ldots, n-1$, update the filtering distribution

$$\phi_{k+1}(j\,;\theta') = \frac{\sum_{i=1}^{r} \phi_k(i\,;\theta')\, q'_{ij}\, g_{k+1}(j\,;\theta')}{c_{k+1}(\theta')}$$

for $j = 1, \ldots, r$, where

$$c_{k+1}(\theta') = \sum_{j=1}^{r}\sum_{i=1}^{r} \phi_k(i\,;\theta')\, q'_{ij}\, g_{k+1}(j\,;\theta') \,.$$

Next,

$$\tau_{k+1,1}(j\,;\theta') = \frac{\sum_{i=1}^{r} \tau_{k,1}(i\,;\theta')\, q'_{ij}\, g_{k+1}(j\,;\theta')}{c_{k+1}(\theta')} + Y_{k+1}\phi_{k+1}(i_0\,;\theta')\delta_{i_0}(j) \tag{10.48}$$

for $j = 1, \ldots, r$, where $\delta_{i_0}(j)$ is equal to one when $j = i_0$ and zero otherwise. Likewise,

$$\tau_{k+1,2}(j\,;\theta') = \frac{\sum_{i=1}^{r} \tau_{k,2}(i\,;\theta')\, q'_{ij}\, g_{k+1}(j\,;\theta')}{c_{k+1}(\theta')} + \phi_{k+1}(i_0\,;\theta')\delta_{i_0}(j) \tag{10.49}$$

for $j = 1, \ldots, r$.

Parameter Update: When the final observation index n is reached, the updated mean $\mu^*_{i_0}$ is obtained as

$$\mu^*_{i_0} = \frac{\sum_{i=1}^{r} \tau_{n,1}(i\,;\theta')}{\sum_{i=1}^{r} \tau_{n,2}(i\,;\theta')} \,.$$

It is clear that one can proceed similarly for parameters other than the means. For the same given state $i = i_0$, the alternative form of the variance update equation given in (10.45) shows that, in addition to $t_{n,1}$ and $t_{n,2}$ defined in (10.46), the functional

$$t_{n,3}(X_{0:n}) = \sum_{k=0}^{n} \mathbb{1}\{X_k = i_0\}Y_k^2$$

is needed to compute the updated variance $\upsilon_{i_0}^*$. The recursive smoother associated with this quantity is updated as prescribed by Algorithm 10.3.1 for $t_{n,1}$ by simply replacing Y_k by Y_k^2.

In the case of the transition probabilities, considering *a fixed pair* of states (i_0, j_0), (10.44) implies that in addition to evaluating $\tau_{n-1,2}$, one needs to derive a smoother for the functional

$$t_{n,4}(X_{0:n}) = \sum_{k=1}^{n} \mathbb{1}\{X_{k-1} = i_0, X_k = j_0\} , \tag{10.50}$$

where $t_{0,4}(X_0)$ is defined to be null. Following Proposition 4.1.3, the associated smoothed quantity

$$\tau_{n,4}(i\,;\theta') = \mathrm{E}_{\theta'}[\mathbb{1}\{X_n = i\} t_{n,4}(X_{0:n}) \,|\, Y_{0:n}]$$

may be updated recursively according to

$$\tau_{k+1,4}(j\,;\theta') = \frac{\sum_{i=1}^{r} \tau_{k,4}(i\,;\theta')\, q'_{ij}\, g_{k+1}(j\,;\theta')}{c_{k+1}(\theta')} + \frac{\phi_k(i_0\,;\theta')\, q'_{i_0 j_0}\, g_{k+1}(j_0\,;\theta') \delta_{j_0}(j)}{c_{k+1}(\theta')} , \tag{10.51}$$

where $\delta_{j_0}(j)$ equal to one when $j = j_0$ and zero otherwise, and c_{k+1} and ϕ_k should be computed recursively as in Algorithm 10.3.1. Because $\tau_{0,4}$ is null, the recursion is initialized by setting $\tau_{0,4}(i\,;\theta') = 0$ for all states $i = 1, \ldots, r$.

The case of the transition probabilities clearly illustrates the main weakness of the recursive approach, namely that a specific recursive smoother must be implemented for each statistic of interest. Indeed, for each time index k, (10.48), (10.49), or (10.51) require of the order of r^2 operations, which is comparable with the computational cost of the (normalized) forward or filtering recursion (Algorithm 5.1.1). The difference is that after application of the complete forward-backward recursions, one may compute *all the statistics* involved in the EM re-estimation equations (10.41)–(10.43). In contrast, the recursive smoothing recursion only provides the smoothed version of *one particular statistic*: in the case of (10.51) for instance, this is (10.50) with a fixed choice of the pair i_0, j_0. Hence implementing the EM algorithm with recursive smoothing requires the order of $r^2 \times (n+1) \times \dim(\theta)$ operations, where $\dim(\theta)$ refers to the number of parameters. In the case of the complete (scalar) normal HMM, $\dim(\theta)$ equals $2r$ for the means and variances, plus $r \times (r-1)$ for the transition probabilities. Hence recursive smoothing is clearly not competitive with approaches based on the forward-backward decomposition.

To make it short, the recursive smoothing approach is not a very attractive option in finite state space HMMs and normal HMMs in particular. More precisely, both the intermediate quantity of EM in (10.26) and the gradient of the log-likelihood in (10.29) are additive. In the terminology used in Section 4.1.2, they both correspond to additive functionals of the form $t_{n+1}(x_{0:n+1}) = t_n(x_{0:n}) + s_n(x_n, x_{n+1})$. In such cases, smoothing approaches based on the forward-backward decompositions such as Algorithms 5.1.2 or 5.1.3 that evaluate the bivariate smoothing distributions $\phi_{k:k+1|n}$ for $k = 0, \ldots, n-1$ are more efficient because they do not require that the functions $\{s_k\}_{k=0,\ldots,n-1}$ be specified. There are however some situations in which the recursive smoothing approach developed in Section 4.1 and illustrated above in the case of normal HMMs may be useful.

- First, because it is recursive, it does not require that the intermediate computation results be stored, which is in sharp contrast with the other smoothing approaches where either the forward or backward variables need to be stored. This is of course of interest when processing very large data sets.
- When the functional whose conditional expectation is to be evaluated is not of the additive type, approaches based on the evaluation of bivariate smoothing distributions are not applicable anymore. In contrast, recursive smoothing stays feasible as long as the functional follows the general pattern of Definition 4.1.2. The most significant functional of practical interest that is not additive is the second-order derivative of the log-likelihood function. The use of recursive smoothing for this purpose will be illustrated on a simple example in Section 10.3.4.

Finally, another different motivation for computing either the intermediate quantity of EM or the gradient of the log-likelihood recursively has to do with recursive estimation. As noted by several authors, including Le Gland and Mevel (1997), Collings and Rydén (1998), and Krishnamurthy and Yin (2002), being able to compute recursively the intermediate quantity of EM or the gradient of the log-likelihood is a key step toward efficient recursive (also called on-line or adaptive) parameter estimation approaches. It is important however to understand that recursive computation procedures do not necessarily directly translate into recursive estimation approaches. Algorithm 10.3.1 for instance describes how to compute the EM update of the mean μ_i given some observations $Y_0, \ldots, Y_n$ and a *fixed* current parameter value $\theta = \theta'$. In recursive estimation on the other hand, once a new observation Y_k is collected, the parameter estimate, $\hat{\theta}_k$ say, needs to be updated. Using the equations of Algorithm 10.3.1 with $\hat{\theta}_k$ substituted for θ' is of course a natural idea, but not one that is guaranteed to produce the desired result. This is precisely the objective of works such as Le Gland and Mevel (1997) and Krishnamurthy and Yin (2002), to study if and when such recursive approaches do produce expected results. It is fair to say that, as of today, this remains a largely open issue.

10.3.4 Computation of the Score and Observed Information

For reasons discussed above, computing the gradient of the log-likelihood is not a difficult task in finite state space HMMs and should preferably be done using smoothing algorithms based on the forward-backward decomposition. The only new requirement is to evaluate the derivatives with respect to θ that appear in (10.29). In the case of the normal HMM, we already met the appropriate expressions in (10.40), as Fisher's identity (10.12) implies that the gradient of the intermediate quantity at the current parameter estimate coincides with the gradient of the log-likelihood. Hence

$$\frac{\partial}{\partial \mu_i} \ell_n(\theta) = \frac{1}{\upsilon_i} \sum_{k=0}^{n} \phi_{k|n}(i\,;\theta)(Y_k - \mu_i) ,$$

$$\frac{\partial}{\partial \upsilon_i} \ell_n(\theta) = -\frac{1}{2} \sum_{k=0}^{n} \phi_{k|n}(i\,;\theta) \left[\frac{1}{\upsilon_i} - \frac{(Y_k - \mu_i)^2}{\upsilon_i^2} \right] ,$$

$$\frac{\partial}{\partial q_{ij}} \ell_n(\theta) = \sum_{k=1}^{n} \frac{\phi_{k-1:k|n}(i, j\,;\theta)}{q_{ij}} .$$

Recall also that the log-likelihood itself is directly available from the filtering recursion, following (5.4).

Before considering the computation of the Hessian, we first illustrate the performance of the optimization methods introduced in Section 10.1.3, which only require the evaluation of the log-likelihood and its gradient.

Example 10.3.2 (Binary Deconvolution Model). We consider the simple binary deconvolution model of Cappé *et al.* (1999), which is somewhat related to the channel coding situation described in Example 1.3.2, except that the channel is unknown. This model is of interest in digital communications (see for instance Krishnamurthy and White, 1992; Kaleh and Vallet, 1994; Fonollosa *et al.*, 1997). It is given by

$$Y_k = \sum_{i=0}^{p} h_i B_{k-i} + N_k , \tag{10.52}$$

where $\{Y_k\}_{k\geq 0}$ is the observed sequence, $\{N_k\}_{k\geq 0}$ is a stationary sequence of white Gaussian noise with zero mean and variance υ, and $\{B_k\}_{k\geq 0}$ is a sequence of transmitted symbols. For simplicity, we assume that $\{B_k\}_{k\geq 0}$ is a binary, i.e., $B_k \in \{-1, 1\}$, sequence of i.i.d. fair Bernoulli draws. We consider below that $p = 1$, so that to cast the model into the HMM framework, we only need to define the state as the vector $X_k = (B_k, B_{k-1})^t$, which takes one of the four possible values

$$s_1 \stackrel{\text{def}}{=} \begin{pmatrix} -1 \\ -1 \end{pmatrix} , \quad s_2 \stackrel{\text{def}}{=} \begin{pmatrix} -1 \\ 1 \end{pmatrix} , \quad s_3 \stackrel{\text{def}}{=} \begin{pmatrix} 1 \\ -1 \end{pmatrix} , \quad s_4 \stackrel{\text{def}}{=} \begin{pmatrix} 1 \\ 1 \end{pmatrix} .$$

Hence, upon defining the vector $h \stackrel{\text{def}}{=} (h_0\ h_1)^t$ of filter coefficients, we may view (10.52) as a four-states normal HMM such that $\mu_i = s_i^t h$ and $\upsilon_i = \upsilon$ for $i = 1, \ldots, 4$. The transition matrix Q is entirely fixed by our assumption that the binary symbols are equiprobable, and is given by

$$Q = \begin{pmatrix} 1/2 & 0 & 1/2 & 0 \\ 1/2 & 0 & 1/2 & 0 \\ 0 & 1/2 & 0 & 1/2 \\ 0 & 1/2 & 0 & 1/2 \end{pmatrix} .$$

The model parameters to be estimated are thus the vector h of filter coefficients and the common variance υ. For simplicity, we assume that the distribution of the initial state X_0 is known.

To make the connection with the general (unconstrained) normal hidden Markov model discussed previously, we need only take into account the facts that $\nabla_h \mu_i = s_i$ and $\partial \upsilon_i / \partial \upsilon = 1$, as all variances are equal. Hence, using the chain rule, the gradient of the intermediate quantity of EM may be evaluated from (10.40) as

$$\begin{aligned} \nabla_h \mathcal{Q}(\theta\,;\theta') &= \sum_{i=1}^{4} \frac{\partial}{\partial \mu_i} \mathcal{Q}(\theta\,;\theta') \nabla_h \mu_i \\ &= \frac{1}{\upsilon} \sum_{i=1}^{4} \sum_{k=0}^{n} \phi_{k|n}(i\,;\theta')(Y_k s_i - s_i s_i^t h) \end{aligned} \tag{10.53}$$

and

$$\begin{aligned} \frac{\partial}{\partial \upsilon} \mathcal{Q}(\theta\,;\theta') &= \sum_{i=1}^{4} \frac{\partial}{\partial \upsilon_i} \mathcal{Q}(\theta\,;\theta') \frac{\partial \upsilon_i}{\partial \upsilon} \\ &= -\frac{1}{2} \left[\frac{n}{\upsilon} - \sum_{i=1}^{4} \sum_{k=0}^{n} \phi_{k|n}(i\,;\theta') \frac{(Y_k - s_i^t h)^2}{\upsilon^2} \right] . \end{aligned} \tag{10.54}$$

The M-step update equations (10.41) and (10.42) of the EM algorithm should thus be replaced by

$$h^* = \left[\sum_{i=1}^{4} \sum_{k=0}^{n} \phi_{k|n}(i\,;\theta') s_i s_i^t \right]^{-1} \left[\sum_{i=1}^{4} \sum_{k=0}^{n} \phi_{k|n}(i\,;\theta') Y_k s_i \right] ,$$

$$\begin{aligned} \upsilon^* &= \frac{1}{n} \sum_{i=1}^{4} \sum_{k=0}^{n} \phi_{k|n}(i\,;\theta')(Y_k - s_i^t h^*)^2 \\ &= \frac{1}{n} \left\{ \sum_{k=0}^{n} Y_k^2 - \left[\sum_{i=1}^{4} \sum_{k=0}^{n} \phi_{k|n}(i\,;\theta') Y_k s_i \right]^t h^* \right\} . \end{aligned}$$

For computing the log-likelihood gradient, we may resort to Fisher's identity, setting $\theta = \theta'$ in (10.53) and (10.54) to obtain $\nabla_h \ell_n(\theta')$ and $\frac{\partial}{\partial \upsilon} \ell_n(\theta')$, respectively.

We now compare the results of the EM algorithm and of a quasi-Newton method for this model. In both cases, the forward-backward recursions are used to compute the smoothing probabilities $\phi_{k|n}(i\,;\theta')$ for $k = 0, \ldots, n$ and $i = 1, \ldots, 4$. To avoid parameter constraints, we compute the partial derivative with respect to $\log \upsilon$ rather than with respect to υ, as the parameter $\log \upsilon$ is unconstrained. This modification is not needed for the EM algorithm, which is parameterization independent. The quasi-Newton optimization is performed using the so-called BFGS weight update and cubic line-searches (see Fletcher, 1987, for details concerning the former).

The data set under consideration is the same as in Cappé *et al.* (1999) and consists of 150 synthetic observations generated with the model corresponding to $h_0 = 1.3$, $h_1 = 0.6$ and $\upsilon = (h_0^2 + h_1^2)/4$ (6 dB signal to noise ratio). There are three parameters for this model, and Figures 10.1 and 10.2 show plots of the profile log-likelihood for values of h_0 and h_1 on a regular grid. The profile log-likelihood is $\ell_n(h, \hat{\upsilon}(h))$ with $\hat{\upsilon}(h) = \arg\max_\upsilon \ell_n(h, \upsilon)$, that is, the largest possible log-likelihood for a fixed value of h. The figures show that the profile log-likelihood has a global maximum, the MLE, as well as a local one. The location of the local maximum (or maxima) as well as its presence obviously depends on the particular outcome of the simulated noise $\{N_k\}$. It is

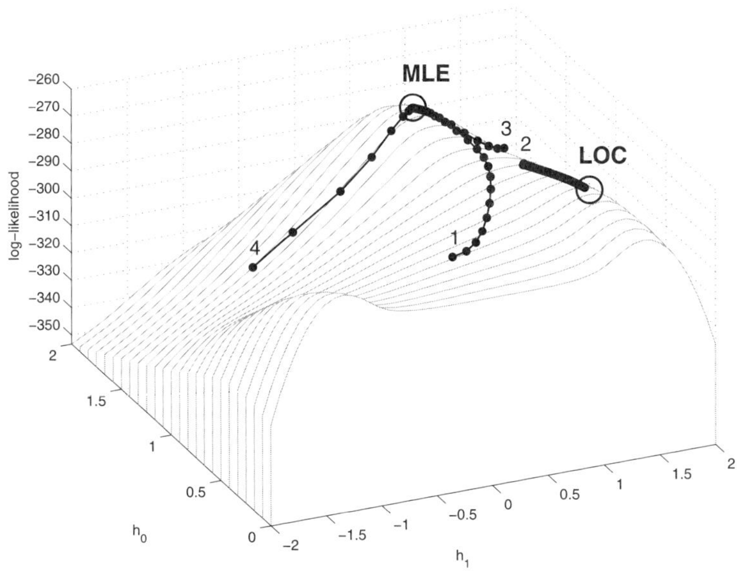

Fig. 10.1. Profile log-likelihood surface over (h_0, h_1) for a particular realization of the binary deconvolution model. The true model parameters are $h_0 = 1.3$ and $h_1 = 0.6$, and 150 observations were taken. The two circled positions labeled MLE and LOC are, respectively, the global maximum of the profile log-likelihood and a local maximum. Also shown are trajectories of 35 iterations of the EM algorithm, initialized at four different points marked 1–4.

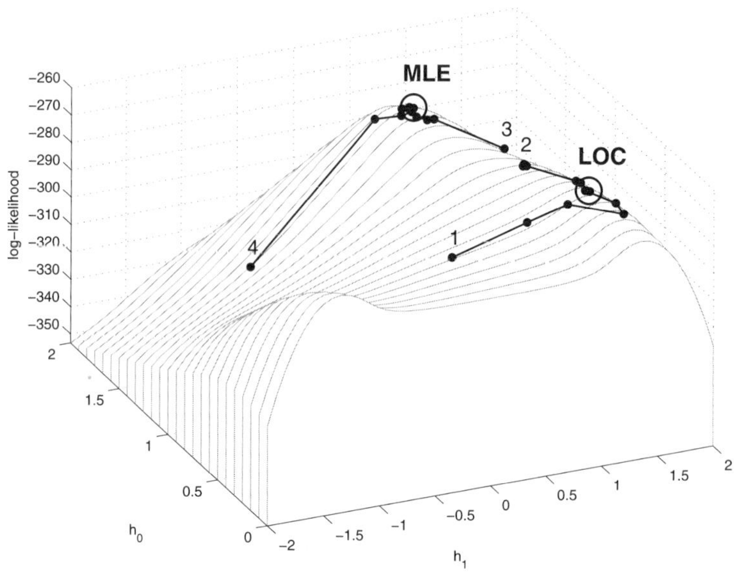

Fig. 10.2. Same profile log-likelihood surface as in Figure 10.1. Also shown are trajectories of 5 iterations of a quasi-Newton algorithm, initialized at the same four different points marked 1–4 as in Figure 10.1.

a fundamental feature of the model however that the parameters $h = (h_0\ h_1)^t$ and $h = (h_1\ h_0)^t$, which govern identical second-order statistical properties of the model, are difficult to discriminate, especially with few observations. Note that as swapping the signs of both h_0 and h_1 leaves the model unchanged, the profile log-likelihood surface is symmetric, and only the half corresponding to positive values of h_0 is shown here.

A first remark is that even in such a simplistic model, there is a local maximum and, depending on the initialization, both algorithms may converge to this point. Because the algorithms operate differently, it may even occur that the EM and quasi-Newton algorithms initialized at the same point eventually converge to different values, as in the case of initialization at point 1. The other important remark is that the EM algorithm (Figure 10.1) shows very different convergence behavior depending on the region of the parameter space where it starts: when initialized at point 4, the algorithm gets real close to the MLE in about seven iterations, whereas when initialized at point 1 or 2, it is still far from having reached convergence after 20 iterations. In contrast, the quasi-Newton method (Figure 10.2) updates the parameter by doing steps that are much larger than those of EM, especially during the first iterations, and provides very accurate parameter estimates with as few as five iterations. It is fair to say that due to the necessity of evaluating the weight matrix (with finite difference computations) and to the cubic line-search procedure, each iteration of the quasi-Newton method requires on average seven evaluations of the log-likelihood and its gradient, which means in particular seven instances of the forward-backward procedure. From a computational point of view, the time needed to run the 5 iterations of the quasi-Newton method in this example is thus roughly equivalent to that required for 35 iterations of the EM algorithm. ∎

As discussed earlier, computing the observed information in HMMs is more involved, as the only computationally feasible option consists in adopting the recursive smoothing framework of Proposition 4.1.3. Rather than embarking into the general normal HMM case, we consider another simpler illustrative example where the parameter of interest is scalar.

Example 10.3.3. Consider a simplified version of the ion channel model (Example 1.3.5) in which the state space X is composed of two states that are (by convention) labeled 0 and 1, and $g(0, y)$ and $g(1, y)$ respectively correspond to the $\mathrm{N}(0, \upsilon)$ and $\mathrm{N}(1, \upsilon)$ distributions. This model may also be interpreted as a state space model in which

$$Y_k = X_k + V_k \ ,$$

where $\{V_k\}$ is an i.i.d. $\mathrm{N}(0, \upsilon)$-distributed sequence, independent of the Markov chain $\{X_k\}$, which takes its values in the set $\{0, 1\}$. The transition matrix Q of $\{X_k\}$ is parameterized in the form

$$Q = \begin{pmatrix} \rho_0 & 1 - \rho_0 \\ 1 - \rho_1 & \rho_1 \end{pmatrix} .$$

It is also most logical in this case to assume that the initial distribution ν of X_0 coincides with the stationary distribution associated with Q, that is, $\nu(0) = \rho_0/(\rho_0 + \rho_1)$ and $\nu(1) = \rho_1/(\rho_0 + \rho_1)$. In this model, the distributions of holding times (number of consecutive steps k for which X_k stays constant) have geometric distributions with expectations $(1 - \rho_0)^{-1}$ and $(1 - \rho_1)^{-1}$ for states 0 and 1, respectively. ∎

We now focus on the computation of the derivatives of the log-likelihood in the model of Example 10.3.3 with respect to the transition parameters ρ_0 and ρ_1. As they play a symmetric role, it is sufficient to consider, say, ρ_0 only. The variance υ is considered as fixed so that the only quantities that depend on the parameter ρ_0 are the initial distribution ν and the transition matrix Q. We will, as usual, use the simplified notation $g_k(x)$ rather than $g(x, Y_k)$ to denote the Gaussian density function $(2\pi\upsilon)^{-1/2} \exp\{-(Y_k - x)^2/(2\upsilon)\}$ for $x \in \{0, 1\}$. Furthermore, in order to simplify the expressions below, we also omit to indicate explicitly the dependence with respect to ρ_0 in the rest of this section. Fisher's identity (10.12) reduces to

$$\frac{\partial}{\partial \rho_0} \ell_n = \mathrm{E}\left[\frac{\partial}{\partial \rho_0} \log \nu(X_0) + \sum_{k=0}^{n-1} \frac{\partial}{\partial \rho_0} \log q_{X_k X_{k+1}} \,\middle|\, Y_{0:n} \right] ,$$

where the notation q_{ij} refers to the element in the $(1+i)$-th row and $(1+j)$-th column of the matrix Q (in particular, q_{00} and q_{11} are alternative notations for ρ_0 and ρ_1). We are thus in the framework of Proposition 4.1.3 with a smoothing functional $t_{n,1}$ defined by

$$
\begin{aligned}
t_{0,1}(x) &= \frac{\partial}{\partial \rho_0} \log \nu(x) , \\
s_{k,1}(x, x') &= \frac{\partial}{\partial \rho_0} \log q_{xx'} \qquad \text{for } k \geq 0 ,
\end{aligned}
$$

where the multiplicative functions $\{m_{k,1}\}_{k \geq 0}$ are equal to 1. Straightforward calculations yield

$$
\begin{aligned}
t_{0,1}(x) &= (\rho_0 + \rho_1)^{-1} \left[\frac{\rho_1}{\rho_0} \delta_0(x) - \delta_1(x) \right] , \\
s_{k,1}(x, x') &= \frac{1}{\rho_0} \delta_{(0,0)}(x, x') - \frac{1}{1 - \rho_0} \delta_{(0,1)}(x, x') .
\end{aligned}
$$

Hence a first recursion, following Proposition 4.1.3.

Algorithm 10.3.4 (Computation of the Score in Example 10.3.3).

Initialization: Compute $c_0 = \sum_{i=0}^{1} \nu(i) g_0(i)$ and, for $i = 0, 1$,

$$
\begin{aligned}
\phi_k(i) &= c_0^{-1} \nu(i) g_0(i) , \\
\tau_{0,1}(i) &= t_{0,1}(i) \phi_0(i) .
\end{aligned}
$$

Recursion: For $k = 0, 1, \ldots$, compute $c_{k+1} = \sum_{i=0}^{1} \sum_{j=0}^{1} \phi_k(i) q_{ij} g_k(j)$ and, for $j = 0, 1$,

$$
\begin{aligned}
\phi_{k+1}(j) &= c_{k+1}^{-1} \sum_{i=0}^{1} \phi_k(i) q_{ij} g_k(j) , \\
\tau_{k+1,1}(j) &= c_{k+1}^{-1} \left\{ \sum_{i=0}^{1} \tau_{k,1}(i) q_{ij} g_{k+1}(j) \right. \\
&\qquad \left. + \phi_k(0) g_{k+1}(0) \delta_0(j) - \phi_k(0) g_{k+1}(1) \delta_1(j) \right\} .
\end{aligned}
$$

At each index k, the log-likelihood is available via $\ell_k = \sum_{l=0}^{k} \log c_l$, and its derivative with respect to ρ_0 may be evaluated as

$$
\frac{\partial}{\partial \rho_0} \ell_k = \sum_{i=0}^{1} \tau_{k,1}(i) .
$$

For the second derivative, Louis' identity (10.14) shows that

$$\frac{\partial^2}{\partial \rho_0^2}\ell_n + \left\{\frac{\partial}{\partial \rho_0}\ell_n\right\}^2 = \mathrm{E}\left[\frac{\partial^2}{\partial \rho_0^2}\log \nu(X_0) + \sum_{k=0}^{n-1}\frac{\partial^2}{\partial \rho_0^2}\log q_{X_k X_{k+1}} \,\middle|\, Y_{0:n}\right]$$
$$+ \mathrm{E}\left[\left(\frac{\partial}{\partial \rho_0}\log \nu(X_0) + \sum_{k=0}^{n-1}\frac{\partial}{\partial \rho_0}\log q_{X_k X_{k+1}}\right)^2 \,\middle|\, Y_{0:n}\right] . \quad (10.55)$$

The first term on the right-hand side of (10.55) is very similar to the case of $\tau_{n,1}$ considered above, except that we now need to differentiate the functions twice, replacing $t_{0,1}$ and $s_{k,1}$ by $\frac{\partial}{\partial \rho_0}t_{0,1}$ and $\frac{\partial}{\partial \rho_0}s_{k,1}$, respectively. The corresponding smoothing functional $t_{n,2}$ is thus now defined by

$$t_{0,2}(x) = -\frac{\rho_1(2\rho_0+\rho_1)}{\rho_0^2(\rho_0+\rho_1)^2}\delta_0(x) + \frac{1}{(\rho_0+\rho_1)^2}\delta_1(x) ,$$
$$s_{k,2}(x,x') = -\frac{1}{\rho_0^2}\delta_{(0,0)}(x,x') - \frac{1}{(1-\rho_0)^2}\delta_{(0,1)}(x,x') .$$

The second term on the right-hand side of (10.55) is more difficult, and we need to proceed as in Example 4.1.4: the quantity of interest may be rewritten as the conditional expectation of

$$t_{n,3}(x_{0:n}) = \left[t_{0,1}(x_0) + \sum_{k=0}^{n-1} s_{k,1}(x_k, x_{k+1})\right]^2 .$$

Expanding the square in this equation yields the update formula

$$t_{k+1,3}(x_{0:k+1}) = t_{k,3}(x_{0:k}) + s_{k,1}^2(x_k,x_{k+1}) + 2t_{k,1}(x_{0:k})s_{k,1}(x_k,x_{k+1}) .$$

Hence $t_{k,1}$ and $t_{k,3}$ jointly are of the form prescribed by Definition 4.1.2 with incremental additive functions $s_{k,3}(x,x') = s_{k,1}^2(x,x')$ and multiplicative updates $m_{k,3}(x,x') = 2s_{k,1}(x,x')$. As a consequence, the following smoothing recursion holds.

Algorithm 10.3.5 (Computation of the Observed Information in Example 10.3.3).

Initialization: For $i = 0, 1$,

$$\tau_{0,2}(i) = t_{0,2}(i)\phi_0(i) .$$
$$\tau_{0,3}(i) = t_{0,1}^2(i)\phi_0(i) .$$

Recursion: For $k = 0, 1, \ldots,$ compute for $j = 0, 1$,

$$\tau_{k+1,2}(j) = c_{k+1}^{-1} \left\{ \sum_{i=0}^{1} \tau_{k,2}(i) q_{ij} g_{k+1}(j) \right.$$
$$\left. - \frac{1}{\rho_0} \phi_k(0) g_{k+1}(0) \delta_0(j) - \frac{1}{(1-\rho_0)} \phi_k(0) g_{k+1}(1) \delta_1(j) \right\} ,$$
$$\tau_{0,3}(j) = c_{k+1}^{-1} \left\{ \sum_{i=0}^{1} \tau_{k,3}(i) q_{ij} g_{k+1}(j) \right.$$
$$+ 2 \left[\tau_{k,1}(0) g_{k+1}(0) \delta_0(j) - \tau_{k,1}(0) g_{k+1}(1) \delta_1(j) \right]$$
$$\left. + \frac{1}{\rho_0} \phi_k(0) g_{k+1}(0) \delta_0(j) + \frac{1}{(1-\rho_0)} \phi_k(0) g_{k+1}(1) \delta_1(j) \right\} .$$

At each index k, the second derivative of the log-likelihood satisfies

$$\frac{\partial^2}{\partial \rho_0^2} \ell_k + \left(\frac{\partial}{\partial \rho_0} \ell_k \right)^2 = \sum_{i=0}^{1} \tau_{k,2}(i) + \sum_{i=0}^{1} \tau_{k,3}(i) ,$$

where the second term on the left-hand side may be evaluated in the same recursion, following Algorithm 10.3.4.

To illustrate the results obtained with Algorithms 10.3.4–10.3.5, we consider the model with parameters $\rho_0 = 0.95$, $\rho_1 = 0.8$, and $\upsilon = 0.1$ (using the notations introduced in Example 10.3.3). Figure 10.3 displays the typical aspect of two sequences of length 200 simulated under slightly different values of ρ_0. One possible use of the output of Algorithms 10.3.4–10.3.5 consists in testing for changes in the parameter values. Indeed, under conditions to be detailed in Chapter 12 (and which hold here), the normalized score $n^{-1/2} \frac{\partial}{\partial \rho_0} \ell_n$ satisfies a central limit theorem with variance given by the limit of the normalized information $-n^{-1}(\partial^2/\partial\rho_0^2)\ell_n$. Hence it is expected that

$$\mathfrak{R}_n = \frac{\frac{\partial}{\partial \rho_0} \ell_n}{\sqrt{-\frac{\partial^2}{\partial \rho_0^2} \ell_n}}$$

be asymptotically N(0, 1)-distributed under the null hypothesis that ρ_0 is indeed equal to the value used for computing the score and information recursively with Algorithms 10.3.4–10.3.5.

Figure 10.4 displays the empirical quantiles of $\mathfrak{R}_n$ against normal quantiles for $n = 200$ and $n =1{,}000$. For the longer sequences ($n =1{,}000$), the result is clearly as expected with a very close fit to the normal quantiles. When $n = 200$, asymptotic normality is not yet reached and there is a significant bias toward high values of $\mathfrak{R}_n$. Looking back at Figure 10.3, even if υ was equal to zero—or in other words, if we were able to identify without ambiguity the 0 and 1 states from the data—there would not be much information about ρ_0 to be gained from runs of length 200: when $\rho_0 = 0.95$ and $\rho_1 = 0.8$, the

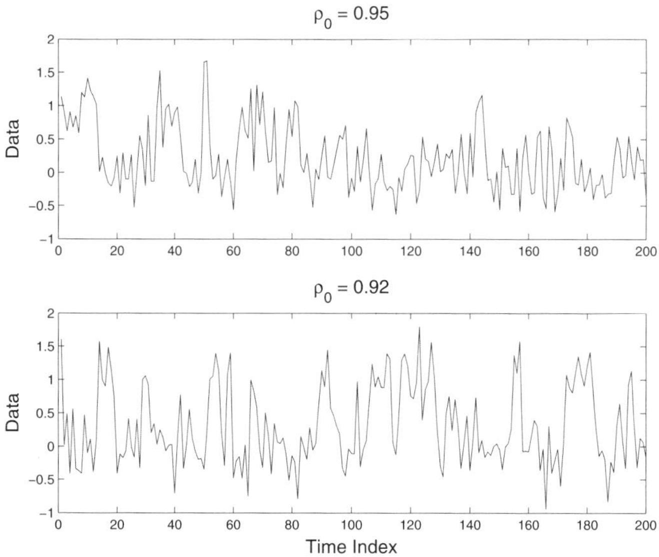

Fig. 10.3. Two simulated trajectories of length $n = 200$ from the simplified ion channel model of Example 10.3.3 with $\rho_0 = 0.95$, $\rho_1 = 0.8$, and $\sigma^2 = 0.1$ (top), and $\rho_0 = 0.92$, $\rho_1 = 0.8$, and $\sigma^2 = 0.1$ (bottom).

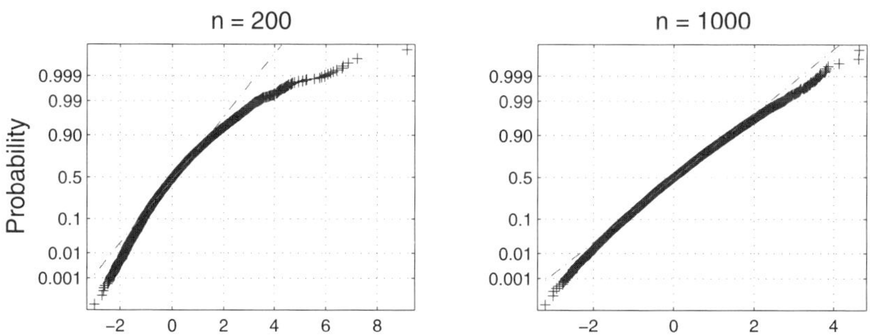

Fig. 10.4. QQ-plot of empirical quantiles of the test statistic $\mathfrak{R}_n$ (abscissas) for the simplified ion channel model of Example 10.3.3 with $\rho_0 = 0.95$, $\rho_1 = 0.8$, and $\sigma^2 = 0.1$ vs. normal quantiles (ordinates). Samples sizes were $n = 200$ (left) and n =1,000 (right), and 10,000 independent replications were used to estimate the empirical quantiles.

average number of distinct runs of 0s that one can observe in 200 consecutive data points is only about $200/(20 + 5) = 8$. To construct a goodness of fit test from $\mathfrak{R}_n$, one can monitor values of $\mathfrak{R}_n^2$, which asymptotically has a chi-square distribution with one degree of freedom. Testing the null hypothesis $\rho_0 = 0.95$ gives p-values of 0.87 and 0.09 for the two sequences in the top and bottom plots, respectively, of Figure 10.3. When testing at the 10% level, both sequences thus lead to the correct decision: no rejection and rejection of the null hypothesis, respectively. Interestingly, testing the other way around, that is, postulating $\rho_0 = 0.92$ as the null hypothesis, gives p-values of 0.20 and 0.55 for the top and bottom sequences of Figure 10.3, respectively. The outcome of the test is now obviously less clear-cut, which reveals an asymmetry in its discrimination ability: it is easier to detect values of ρ_0 that are smaller than expected than the converse. This is because smaller values of ρ_0 means more changes (on average) in the state sequence and hence more usable information about ρ_0 to be obtained from a fixed size record. This asymmetry is connected to the upward bias visible in the left plot of Figure 10.4.

10.4 The Example of Gaussian Linear State-Space Models

We now consider more briefly the case of Gaussian linear state-space models that form the other major class of hidden Markov models for which the methods discussed in Section 10.1 are directly applicable. It is worth mentioning that Gaussian linear state-space models are perhaps the only important subclass of the HMM family for which there exist reasonable simple non-iterative parameter estimation algorithms not based on maximum likelihood arguments but are nevertheless useful in practical applications. These sub-optimal algorithms, proposed by Van Overschee and De Moor (1993), rely on the linear structure of the model and use only eigendecompositions of empirical covariance matrices—a general principle usually referred to under the denomination of *subspace methods* (Van Overschee and De Moor, 1996). Keeping in line with the general topic of this chapter, we nonetheless consider below only algorithms for maximum likelihood estimation in Gaussian linear state-space models.

The Gaussian linear state-space model introduced in Section 1.3.3 is given in so-called state-space form by (1.7)–(1.8), which we recall here:

$$\begin{aligned} X_{k+1} &= AX_k + RU_k \ , \\ Y_k &= BX_k + SV_k \ , \end{aligned}$$

where X_0, $\{U_k\}_{k\geq 0}$ and $\{V_k\}_{k\geq 0}$ are jointly Gaussian. The parameters of the model are the four matrices A, R, B, and S. Note that except for scalar models, it is not possible to estimate R and S because both $\{U_k\}$ and $\{V_k\}$ are unobservable and hence R and S are only identifiable up to an orthonormal matrix. In other words, multiplying R or S by any orthonormal matrix of suitable dimension does not modify the distribution of the observations. Hence the parameters that are identifiable are the covariance matrices $\Upsilon_R = RR^t$ and $\Upsilon_S = SS^t$, which we consider below. Likewise, the matrices A and B are identifiable up to a similarity transformation only. Indeed, setting $X'_k = TX_k$ for some invertible matrix T, that is, making a change of basis for the state process, it is straightforward to check that the joint process $\{(X'_k, Y_k)\}$ satisfies the model assumptions with TAT^{-1}, BT^{-1}, and TR replacing A, B, and R, respectively. Nevertheless, we work with A and B in the algorithm below. If a unique representation is desired, one may use, for instance, the companion form of A given its eigenvalues; this matrix may contain complex entries though. As in the case of finite state space HMMs (Section 10.2.2), it is not sensible to consider the initial covariance matrix Σ_ν as an independent parameter when using a single observed sequence. On the other hand, for such models it is very natural to assume that Σ_ν is associated with the stationary distribution of $\{X_k\}$. Except for the particular case of the scalar AR(1) model however (to be discussed in Example 11.1.2), this option typically renders the EM update equations non-explicit and it is thus standard practice to treat Σ_ν as a fixed matrix unrelated to the parameters (Ghosh, 1989). We shall also assume that both Υ_R and Υ_S are full rank covariance matrices so that all Gaussian distributions admit densities with respect to (multi-dimensional) Lebesgue measure.

10.4.1 The Intermediate Quantity of EM

With the previous notations, the intermediate quantity $\mathcal{Q}(\theta\,;\theta')$ of EM, defined in (10.26), may be expressed as

$$-\frac{1}{2}\,\mathrm{E}_{\theta'}\left[n\log|\Upsilon_R| + \sum_{k=0}^{n-1}(X_{k+1}-AX_k)^t\Upsilon_R^{-1}(X_{k+1}-AX_k) \,\middle|\, Y_{0:n}\right]$$
$$-\frac{1}{2}\,\mathrm{E}_{\theta'}\left[(n+1)\log|\Upsilon_S| + \sum_{k=0}^{n}(Y_k-BX_k)^t\Upsilon_S^{-1}(Y_k-BX_k) \,\middle|\, Y_{0:n}\right], \tag{10.56}$$

up to terms that do not depend on the parameters. In order to elicit the M-step equations or to compute the score, we differentiate (10.56) using elementary perturbation calculus as well as the identity $\nabla_C \log|C| = C^{-t}$ for an invertible matrix C—which is a consequence of the adjoint representation of the inverse (Horn and Johnson, 1985, Section 0.8.2):

$$\nabla_A \mathcal{Q}(\theta\,;\theta') = -\Upsilon_R^{-1}\,\mathrm{E}_{\theta'}\left[\sum_{k=0}^{n-1}(AX_kX_k^t - X_{k+1}X_k^t)\,\middle|\, Y_{0:n}\right], \tag{10.57}$$

$$\nabla_{\Upsilon_R^{-1}} \mathcal{Q}(\theta\,;\theta') = -\frac{1}{2}\left\{-n\Upsilon_R \right. \tag{10.58}$$
$$\left. + \mathrm{E}_{\theta'}\left[\sum_{k=0}^{n-1}(X_{k+1}-AX_k)(X_{k+1}-AX_k)^t \,\middle|\, Y_{0:n}\right]\right\},$$

$$\nabla_B \mathcal{Q}(\theta\,;\theta') = -\Upsilon_S^{-1}\,\mathrm{E}_{\theta'}\left[\sum_{k=0}^{n}(BX_kX_k^t - Y_kX_k^t)\,\middle|\, Y_{0:n}\right], \tag{10.59}$$

$$\nabla_{\Upsilon_S^{-1}} \mathcal{Q}(\theta\,;\theta') = -\frac{1}{2}\Bigg\{ -(n+1)\Upsilon_S + \mathrm{E}_{\theta'}\left[\sum_{k=0}^{n} (Y_k - BX_k)(Y_k - BX_k)^t \,\Bigg|\, Y_{0:n} \right] \Bigg\} . \tag{10.60}$$

Note that in the expressions above, we differentiate with respect to the inverses of Υ_R and Υ_S rather than with respect to the covariance matrices themselves, which is equivalent, because we assume both of the covariance matrices to be positive definite, but yields simpler formulas. Equating all derivatives simultaneously to zero defines the EM update of the parameters. We will denote these updates by A^*, B^*, Υ_R^*, and Υ_S^*, respectively. To write them down, we will use the notations introduced in Chapter 5: $\hat{X}_{k|n}(\theta') = \mathrm{E}_{\theta'}[X_k \,|\, Y_{0:n}]$ and $\Sigma_{k|n}(\theta') = \mathrm{E}_{\theta'}[X_k X_k' \,|\, Y_{0:n}] - \hat{X}_{k|n}(\theta')\hat{X}^t_{k|n}(\theta')$, where we now indicate explicitly that these first two smoothing moments indeed depend on the current estimates of the model parameters (they also depend on the initial covariance matrix Σ_ν, but we ignore this fact here because this quantity is considered as being fixed). We also need to evaluate the conditional covariances

$$\begin{aligned} C_{k,k+1|n}(\theta') &\stackrel{\text{def}}{=} \mathrm{Cov}_{\theta'}[X_k, X_{k+1} \,|\, Y_{0:n}] \\ &= \mathrm{E}_{\theta'}[X_k X_{k+1}^t \,|\, Y_{0:n}] - \hat{X}_{k|n}(\theta')\hat{X}^t_{k+1|n}(\theta') . \end{aligned}$$

For Gaussian models, the latter expression coincides with the definition given in (5.99), and hence one may use expression (5.100) to evaluate $C_{k,k+1|n}(\theta')$ during the final forward recursion of Algorithm 5.2.15.

With these notations, the EM update equations are given by

$$A^* = \left[\sum_{k=0}^{n-1} C_{k,k+1|n}(\theta') + \hat{X}_{k|n}(\theta')\hat{X}^t_{k+1|n}(\theta') \right]^t \left[\sum_{k=0}^{n-1} \Sigma_{k|n}(\theta') + \hat{X}_{k|n}(\theta')\hat{X}^t_{k|n}(\theta') \right]^{-1} , \tag{10.61}$$

$$\Upsilon_R^* = \frac{1}{n} \sum_{k=0}^{n-1} \Big\{ \left[\Sigma_{k+1|n}(\theta') + \hat{X}_{k+1|n}(\theta')\hat{X}^t_{k+1|n}(\theta') \right] - A^* \left[C_{k,k+1|n}(\theta') + \hat{X}_{k|n}(\theta')\hat{X}^t_{k+1|n}(\theta') \right] \Big\} , \tag{10.62}$$

$$B^* = \left[\sum_{k=0}^{n} \hat{X}_{k|n}(\theta') Y_k^t \right]^t \left[\sum_{k=0}^{n} \Sigma_{k|n}(\theta') + \hat{X}_{k|n}(\theta')\hat{X}^t_{k|n}(\theta') \right]^{-1} , \tag{10.63}$$

$$\Upsilon_S^* = \frac{1}{n+1} \sum_{k=0}^{n} \left[Y_k Y_k^t - B^* \hat{X}_{k|n}(\theta') Y_k^t \right] . \tag{10.64}$$

In obtaining the covariance update, we used the same remark that made it possible to rewrite, in the case of normal HMMs, (10.42) as (10.45).

10.4.2 Recursive Implementation

As in the case of finite state space HMMs, it is possible to implement the parameter update equations (10.61)–(10.64) or to compute the gradient (10.57)–(10.60) of the log-likelihood recursively in n. Here we only sketch the general principles and refer to the paper by Elliott and Krishnamurthy (1999) in which the details of the EM re-estimation equations are worked out. Proceeding as in Section 4.1, it is clear that all expressions under consideration may be rewritten term by term as the expectation[2] $\mathrm{E}[t_n(X_{0:n}) \mid Y_{0:n}]$ of well chosen additive functionals t_n. More precisely, the functionals of interest are of the form $t_n(x_{0:n}) = t_0(x_0) + \sum_{k=0}^{n-1} s_k(x_k, x_{k+1})$, where the individual terms in the sum are of one of the types

$$s_{k-1,1}(x_k) = h_k^t x_k , \tag{10.65}$$
$$s_{k-1,2}(x_k) = x_k^t M_k x_k , \tag{10.66}$$
$$s_{k-1,3}(x_{k-1}, x_k) = x_{k-1}^t T_{k-1} x_k , \tag{10.67}$$

and $\{h_k\}_{k\geq 0}$, $\{M_k\}_{k\geq 0}$, and $\{T_k\}_{k\geq 0}$, respectively, denote sequences of vectors and matrices with dimension that of the state vectors (d_x) and which may depend on the model parameters or on the observations.

For illustration purposes, we focus on the example of (10.63): the first factor on the right-hand side of (10.63) is a matrix whose ij elements (ith row, jth column) corresponds to $\mathrm{E}[\sum_{k=0}^{n} h_k^t X_k \mid Y_{0:n}]$ for the particular choice

$$h_k = \begin{pmatrix} 0 \ldots & 0 & Y_k(i) & 0 & \ldots 0 \end{pmatrix}^t . \tag{10.68}$$
$$\qquad\;\; \begin{matrix} 1 \ldots j-1 & j & j+1 \ldots d_x \end{matrix}$$

Likewise, the ij element of the second factor on the right-hand side of (10.63)—before inverting the matrix—corresponds to the expectation of a functional of the second of the three types above with M_k being a matrix of zeros except for a unit entry at position ij.

Let $\tau_{n,1}$ denote the expectation $\mathrm{E}[\sum_{k=0}^{n} h_k^t X_k \mid Y_{0:n}]$ for an additive functional of the first type given in (10.65). To derive a recursion for $\tau_{n,1}$, we use the innovation decomposition (Section 5.2.2) to obtain

[2] Note that in this section, we omit to indicate explicitly the dependence with respect to the model parameters to alleviate the notations.

$$
\begin{aligned}
\tau_{n+1,1} &\stackrel{\text{def}}{=} \mathrm{E}_{\theta'}\left[\sum_{k=0}^{n+1} h_k^t X_k \,\middle|\, Y_{0:n+1}\right] \\
&= h_{n+1}^t \hat{X}_{n+1|n+1} \\
&\quad + \sum_{k=0}^{n} h_k^t \left(\hat{X}_{k|n} + \mathrm{E}[X_k \epsilon_{n+1}^t] \Gamma_{n+1}^{-1} \epsilon_{n+1}\right) \\
&= h_{n+1}^t \hat{X}_{n+1|n+1} + \mathrm{E}\left[\sum_{k=0}^{n} h_k^t X_k \,\middle|\, Y_{0:n}\right] \\
&\quad + \underbrace{\left(\sum_{k=0}^{n} h_k^t \Sigma_{k|k-1} \Lambda_k^t \, \Lambda_{k+1}^t \, \ldots \, \Lambda_n^t\right)}_{r_{n+1}} B^t \Gamma_{n+1}^{-1} \epsilon_{n+1} \;,
\end{aligned}
$$

where (5.93) was used to obtain the last expression, which also features the notation $\Lambda_k = A - H_k B$ with H_k being the Kalman (prediction) gain introduced in the statement of Algorithm 5.2.15. The term that we denoted by r_{n+1} is an intermediate quantity that has some similarities with the variable p_k (or more precisely p_0) that is instrumental in the disturbance smoothing algorithm (Algorithm 5.2.15). The same key remark applies here as r_n can be computed recursively (in n) according to the equations

$$
\begin{aligned}
r_0 &= 0 \;, \\
r_{n+1} &= \left(r_n + h_n \Sigma_{n|n-1}\right) \Lambda_n^t \qquad \text{for } n \geq 0 \;.
\end{aligned}
$$

Hence the following recursive smoothing algorithm, which collects all necessary steps.

Algorithm 10.4.1 (Recursive Smoothing for a Linear Sum Functional).

Initialization: Apply the Kalman filtering recursion for $k = 0$ (Algorithm 5.2.13) and set

$$
\begin{aligned}
r_0 &= 0 \;, \\
\tau_0 &= \mathrm{E}[h_0^t X_0 \,|\, Y_0] = h_0^t \hat{X}_{0|0} \;.
\end{aligned}
$$

Recursion: For $n = 1, 2, \ldots$, run one step of the Kalman filtering and prediction recursions (Algorithms 5.2.9 and 5.2.13) and compute

$$
\begin{aligned}
r_n &= \left(r_{n-1} + h_{n-1} \Sigma_{n-1|n-2}\right) \Lambda_{n-1}^t \;, \\
\tau_n &= \mathrm{E}\left[\sum_{k=0}^{n} h_k^t X_k \,\middle|\, Y_{0:n}\right] = h_n^t \hat{X}_{n|n} + \tau_{n-1} + r_n B^t \Gamma_n^{-1} \epsilon_n \;.
\end{aligned}
$$

Algorithm 10.4.1 illustrates the fact that as in the case of finite state space models, recursive computation is in general less efficient than is forward-backward smoothing from a computational point of view: although Algorithm 10.4.1 capitalizes on a common framework formed by the Kalman filtering and prediction recursions, it does however require the update of a quantity (r_n) that is specific to the choice of the sequence of vectors $\{h_k\}_{k\geq 0}$. To compute the first factor on the right-hand side of (10.63) for instance, one needs to apply the recursion of Algorithm 10.4.1 for the $d_y \times d_x$ possible choices of $\{h_k\}_{k\geq 0}$ given by (10.68). Thus, except for low-dimensional models or particular cases in which the system matrices A, Υ_R, B, and Υ_S are very sparse, recursive computation is usually not the method of choice for Gaussian linear state-space models (see Elliott and Krishnamurthy, 1999, for a discussion of the complexity of the complete set of equations required to carry out the EM parameter update).

10.5 Complements

To conclude this chapter, we briefly return to an issue mentioned in Section 10.1.2 regarding the conditions that ensure that the EM iterations indeed converge to stationary points of the likelihood.

10.5.1 Global Convergence of the EM Algorithm

As a consequence of Proposition 10.1.4, the EM algorithm described in Section 10.1.2 has the property that the log-likelihood function ℓ can never decrease in an iteration. Indeed,

$$\ell(\theta^{i+1}) - \ell(\theta^i) \geq \mathcal{Q}(\theta^{i+1}\,;\theta^i) - \mathcal{Q}(\theta^i\,;\theta^i) \geq 0 \; .$$

This class of algorithms, sometimes referred to as *ascent algorithms* (Luenberger, 1984, Chapter 6), can be treated in a unified manner following a theory developed mostly by Zangwill (1969). Wu (1983) showed that this general theory applies to the EM algorithm as defined above, as well as to some of its variants that he calls generalized EM (or GEM). The main result is a strong stability guarantee known as *global convergence*, which we discuss below.

We first need a mathematical formalism that describes the EM algorithm. This is done by identifying any homogeneous (in the iterations) iterative algorithm with a specific choice of a mapping M that associates θ^{i+1} to θ^i. In the theory of Zangwill (1969), one indeed considers families of algorithms by allowing for *point-to-set* maps M that associate a set $M(\theta') \subseteq \Theta$ to each parameter value $\theta' \in \Theta$. A specific algorithm in the family is such that θ^{i+1} is selected in $M(\theta^i)$. In the example of EM, we may define M as

$$M(\theta') = \left\{\theta \in \Theta \,:\, \mathcal{Q}(\theta\,;\theta') \geq \mathcal{Q}(\tilde{\theta}\,;\theta') \text{ for all } \tilde{\theta} \in \Theta\right\} \,, \tag{10.69}$$

that is, $M(\theta')$ is the set of values θ that maximize $\mathcal{Q}(\theta\,;\theta')$ over Θ. Usually $M(\theta')$ reduces to a singleton, and the mapping M is then simply a point-to-point map (a usual function from Θ to Θ). But the use of point-to-set maps makes it possible to deal also with cases where the intermediate quantity of EM may have several global maxima, without going into the details of what is done in such cases. We next need the following definition before stating the main convergence theorem.

Definition 10.5.1 (Closed Mapping). *A map T from points of Θ to subsets of Θ is said to be* closed *on a set $\mathcal{S} \subseteq \Theta$ if for any converging sequences $\{\theta^i\}_{i\geq 0}$ and $\{\tilde{\theta}^i\}_{i\geq 0}$, the conditions*

(a) $\theta^i \to \theta \in \mathcal{S}$,
(b) $\tilde{\theta}^i \to \tilde{\theta}$ with $\tilde{\theta}^i \in T(\theta^i)$ for all $i \geq 0$,

imply that $\tilde{\theta} \in T(\theta)$.

Note that for point-to-point maps, that is, if $T(\theta)$ is a singleton for all θ, the definition above is equivalent to the requirement that T be continuous on $\mathcal{S}$. Definition 10.5.1 is thus a generalization of continuity for general (point-to-set) maps. We are now ready to state the main result, which is proved in Zangwill (1969, p. 91) or Luenberger (1984, p. 187).

Theorem 10.5.2 (Global Convergence Theorem). *Let Θ be a subset of $\mathbb{R}^{d_\theta}$ and let $\{\theta^i\}_{i\geq 0}$ be a sequence generated by $\theta^{i+1} \in T(\theta^i)$ where T is a point-to-set map on Θ. Let $\mathcal{S} \subseteq \Theta$ be a given "solution" set and suppose that*

(1) the sequence $\{\theta^i\}_{i\geq 0}$ is contained in a compact subset of Θ;
(2) T is closed over $\Theta \setminus \mathcal{S}$ (the complement of $\mathcal{S}$);
(3) there is a continuous "ascent" function s on Θ such that $s(\theta) \geq s(\theta')$ for all $\theta \in T(\theta')$, with strict inequality for points θ' that are not *in $\mathcal{S}$.*

Then the limit of any convergent subsequence of $\{\theta^i\}$ is in the solution set $\mathcal{S}$. In addition, the sequence of values of the ascent function, $\{s(\theta^i)\}_{i\geq 0}$, converges monotonically to $s(\theta_\star)$ for some $\theta_\star \in \mathcal{S}$.

The final statement of Theorem 10.5.2 should not be misinterpreted: that $\{s(\theta^i)\}$ converges to a value that is the image of a point in $\mathcal{S}$ is a simple consequence of the first and third assumptions. It does however not imply that the sequence of parameters $\{\theta^i\}$ is itself convergent in the usual sense, but only that the limit points of $\{\theta^i\}$ have to be in the solution set $\mathcal{S}$. An important property however is that because $\{s(\theta^{i(l)})\}_{l\geq 0}$ converges to $s(\theta_\star)$ for any convergent subsequence $\{\theta^{i(l)}\}$, all limit points of $\{\theta^i\}$ must be in the set $\mathcal{S}_\star = \{\theta \in \Theta : s(\theta) = s(\theta_\star)\}$ (in addition to being in $\mathcal{S}$). This latter statement means that the sequence of iterates $\{\theta^i\}$ will ultimately approach a set of points that are "equivalent" as measured by the ascent function s.

The following general convergence theorem following the proof by Wu (1983) is a direct application of the previous theory to the case of EM.

Theorem 10.5.3. *Suppose that in addition to the hypotheses of Proposition 10.1.4 (Assumptions 10.1.3 as well as parts (a) and (b) of Proposition 10.1.4), the following hold.*

(i) $\mathcal{H}(\theta\,;\theta')$ *is continuous in its second argument,* θ'*, on* Θ.
(ii) For any θ^0*, the level set* $\Theta^0 = \{\theta \in \Theta : \ell(\theta) \geq \ell(\theta^0)\}$ *is compact and contained in the interior of* Θ.

Then all limit points of any instance $\{\theta^i\}_{i\geq 0}$ *of an EM algorithm initialized at* θ^0 *are in* $\mathcal{L}^0 = \{\theta \in \Theta^0 : \nabla_\theta \ell(\theta) = 0\}$*, the set of stationary points of* ℓ *with log-likelihood larger than that of* θ^0*. The sequence* $\{\ell(\theta^i)\}$ *of log-likelihoods converges monotonically to* $\ell_\star = \ell(\theta_\star)$ *for some* $\theta_\star \in \mathcal{L}^0$.

Proof. This is a direct application of Theorem 10.5.2 using $\mathcal{L}^0$ as the solution set and ℓ as the ascent function. The first hypothesis of Theorem 10.5.2 follows from (ii) and the third one from Proposition 10.1.4. The closedness assumption (2) follows from Proposition 10.1.4 and (i): for the EM mapping M defined in (10.69), $\tilde{\theta}^i \in M(\theta^i)$ amounts to the condition

$$\mathcal{Q}(\tilde{\theta}^i\,;\theta^i) \geq \mathcal{Q}(\theta\,;\theta^i) \quad \text{for all } \theta \in \Theta\,,$$

which is also satisfied by the limits of the sequences $\{\tilde{\theta}^i\}$ and $\{\theta^i\}$ (if these converge) by continuity of the intermediate quantity $\mathcal{Q}$, which follows from that of ℓ and $\mathcal{H}$ (note that it is here important that $\mathcal{H}$ be continuous with respect to both arguments). Hence the EM mapping is indeed closed on Θ as a whole and Theorem 10.5.3 follows. □

The assumptions of Proposition 10.1.4 as well as item (i) above are indeed very mild in typical situations. Assumption (ii) however may be restrictive, even for models in which the EM algorithm is routinely used (such as the normal HMMs introduced in Section 1.3.2, for which this assumption does not hold if the variances υ_i are allowed to be arbitrarily small). The practical implication of (ii) being violated is that the EM algorithm may fail to converge to the stationary points of the likelihood for some particularly badly chosen initial points θ^0.

Most importantly, the fact that θ^{i+1} maximizes the intermediate quantity $\mathcal{Q}(\cdot\,;\theta^i)$ of EM does in no way imply that, ultimately, $\ell_\star$ is the global maximum of ℓ over Θ. There is even no guarantee that $\ell_\star$ is a local maximum of the log-likelihood: it may well only be a saddle point (Wu, 1983, Section 2.1). Also, the convergence of the sequence $\ell(\theta^i)$ to $\ell_\star$ does not automatically imply the convergence of $\{\theta^i\}$ to a point $\theta_\star$.

Pointwise convergence of the EM algorithm requires more stringent assumptions that are difficult to verify in practice. As an example, a simple corollary of the global convergence theorem states that if the solution set $\mathcal{S}$ in Theorem 10.5.2 is a single point, $\theta_\star$ say, then the sequence $\{\theta^i\}$ indeed converges to $\theta_\star$ (Luenberger, 1984, p. 188). The sketch of the proof of this corollary is that every subsequence of $\{\theta^i\}$ has a convergent further subsequence because of the compactness assumption (1), but such a subsequence admits s as an ascent function and thus converges to $\theta_\star$ by Theorem 10.5.2 itself. In cases where the solution set is composed of several points, further conditions are needed to ensure that the sequence of iterates indeed converges and does not cycle through different solution points.

In the case of EM, pointwise convergence of the EM sequence may be guaranteed under an additional condition given by Wu (1983) (see also Boyles, 1983, for an equivalent result), stated in the following theorem.

Theorem 10.5.4. *Under the hypotheses of Theorem 10.5.3, if*

(iii) $\|\theta^{i+1} - \theta^i\| \to 0$ *as* $i \to \infty$,

then all limit points of $\{\theta^i\}$ *are in a connected and compact subset of* $\mathcal{L}_\star = \{\theta \in \Theta : \ell(\theta) = \ell_\star\}$, *where* $\ell_\star$ *is the limit of the log-likelihood sequence* $\{\ell(\theta^i)\}$.

In particular, if the connected components of $\mathcal{L}_\star$ *are singletons, then* $\{\theta^i\}$ *converges to some* $\theta_\star$ *in* $\mathcal{L}_\star$.

Proof. The set of limit points of a bounded sequence $\{\theta^i\}$ with $\|\theta^{i+1} - \theta^i\| \to 0$ is connected and compact (Ostrowski, 1966, Theorem 28.1). The proof follows because under Theorem 10.5.2, the limit points of $\{\theta^i\}$ must belong to $\mathcal{L}_\star$. □

10.5.2 Rate of Convergence of EM

Even if one can guarantee that the EM sequence $\{\hat{\theta}^i\}$ converges to some point $\theta_\star$, this limiting point can be either a local maximum, a saddle point, or even a local minimum. The proposition below states conditions under which the stable stationary points of EM coincide with local maxima only (see also Lange, 1995, Proposition 1, for a similar statement). We here consider that the EM mapping M is a point-to-point map, that is, that the maximizer in the M-step is unique.

To understand the meaning of the term "stable", consider the following approximation to the limit behavior of the EM sequence: it is sensible to expect that if the EM mapping M is sufficiently regular in a neighborhood of the limiting fixed point $\theta_\star$, the asymptotic behavior of the EM sequence $\{\theta^i\}$ follows the tangent linear dynamical system

$$(\theta^{i+1} - \theta_\star) = M(\theta^i) - M(\theta_\star) \approx \nabla_\theta M(\theta_\star)(\theta^i - \theta_\star) . \tag{10.70}$$

Here $\nabla_\theta M(\theta_\star)$ is called the *rate matrix* (see for instance Meng and Rubin, 1991). A fixed point $\theta_\star$ is said to be *stable* if the spectral radius of $\nabla_\theta M(\theta_\star)$ is less than 1. In this case, the tangent linear system is asymptotically stable in the sense that the sequence $\{\zeta^i\}$ defined recursively by $\zeta^{i+1} = \nabla_\theta M(\theta_\star)\zeta^i$ tends to zero as n tends to infinity (for any choice of ζ^0). The linear *rate of convergence* of EM is defined as the largest moduli of the eigenvalues of $\nabla_\theta M(\theta_\star)$. This rate is an upper bound on the factors ρ_k that appear in (10.17).

Proposition 10.5.5. *Under the assumptions of Theorem 10.1.6, assume that $\mathcal{Q}(\cdot\,;\theta)$ has a unique maximizer for all $\theta \in \Theta$ and that, in addition,*

$$H(\theta_\star) = -\int \nabla^2_\theta \log f(x\,;\theta)\big|_{\theta=\theta_\star} p(x\,;\theta_\star)\,\lambda(dx) \tag{10.71}$$

and

$$G(\theta_\star) = -\int \nabla^2_\theta \log p(x\,;\theta)\big|_{\theta=\theta_\star} p(x\,;\theta_\star)\,\lambda(dx) \tag{10.72}$$

are positive definite matrices for all stationary points of EM (i.e., such that $M(\theta_\star) = \theta_\star$). Then for all such points, the following hold true.

(i) $\nabla_\theta M(\theta_\star)$ *is diagonalizable and its eigenvalues are positive real numbers.*
(ii) The point $\theta_\star$ is stable for the mapping M if and only if it is a proper maximizer of $\ell(\theta)$ in the sense that all eigenvalues of $\nabla^2_\theta \ell(\theta_\star)$ are negative.

Proof. The EM mapping is defined implicitly through the fact that $M(\theta')$ maximizes $\mathcal{Q}(\cdot\,;\theta')$, which implies that

$$\int \nabla_\theta \log f(x\,;\theta)|_{\theta=M(\theta')}\, p(x\,;\theta')\,\lambda(dx) = 0\,,$$

using assumption (b) of Theorem 10.1.6. Careful differentiation of this relation at a point $\theta' = \theta_\star$, which is such that $M(\theta_\star) = \theta_\star$ and hence $\nabla_\theta\, \ell(\theta)|_{\theta=\theta_\star} = 0$, gives (Dempster *et al.*, 1977; Lange, 1995, see also)

$$\nabla_\theta M(\theta_\star) = [H(\theta_\star)]^{-1}\left[H(\theta_\star) + \nabla^2_\theta \ell(\theta_\star)\right]\,,$$

where $H(\theta_\star)$ is defined in (10.71). The missing information principle—or Louis' formula (see Proposition 10.1.6)—implies that $G(\theta_\star) = H(\theta_\star) + \nabla^2_\theta \ell(\theta_\star)$ is positive definite under our assumptions.

Thus $\nabla_\theta M(\theta_\star)$ is diagonalizable with positive eigenvalues that are the same (counting multiplicities) as those of the matrix $A_\star = I + B_\star$, where $B_\star = [H(\theta_\star)]^{-1/2}\nabla^2_\theta \ell(\theta_\star)[H(\theta_\star)]^{-1/2}$. Thus $\nabla_\theta M(\theta_\star)$ is stable if and only if $B_\star$ has negative eigenvalues only. The Sylvester law of inertia (see for instance Horn and Johnson, 1985) shows that $B_\star$ has the same inertia (number of positive, negative, and zero eigenvalues) as $\nabla^2_\theta \ell(\theta_\star)$. Thus all of $B_\star$'s eigenvalues are negative if and only if the same is true for $\nabla^2_\theta \ell(\theta_\star)$, that is, if $\theta_\star$ is a proper maximizer of ℓ. □

The proof above implies that when $\theta_\star$ is stable, the eigenvalues of $M(\theta_\star)$ lie in the interval $(0, 1)$.

10.5.3 Generalized EM Algorithms

As discussed above, the type of convergence guaranteed by Theorem 10.5.3 is rather weak but, on the other hand, this result is remarkable as it indeed covers not only the original EM algorithm proposed by Dempster *et al.* (1977) but a whole class of variants of the EM approach. One of the most useful extensions of EM is the ECM (for expectation conditional maximization) by Meng and Rubin (1993), which addresses situations where direct maximization of the intermediate quantity of EM is intractable. Assume for instance that the parameter vector θ consists of two sub-components θ_1 and θ_2, which are such that maximization of $\mathcal{Q}((\theta_1, \theta_2); \theta')$ with respect to θ_1 or θ_2 only (the other sub-component being fixed) is easy, whereas joint maximization with respect to $\theta = (\theta_1, \theta_2)$ is problematic. One may then use the following algorithm for updating the parameter estimate at iteration i.

E-step: Compute $\mathcal{Q}((\theta_1, \theta_2); (\theta_1^i, \theta_2^i))$;
CM-step: Determine

$$\theta_1^{i+1} = \arg\max_{\theta_1} \mathcal{Q}((\theta_1, \theta_2^i); (\theta_1^i, \theta_2^i)) ,$$

and then

$$\theta_2^{i+1} = \arg\max_{\theta_2} \mathcal{Q}((\theta_1^{i+1}, \theta_2); (\theta_1^i, \theta_2^i)) .$$

It is easily checked that for this algorithm, (10.8) is still verified and thus ℓ is an ascent function; this implies that Theorem 10.5.3 holds under the same set of assumptions.

The example above is only the simplest case where the ECM approach may be applied, and further extensions are discussed by Meng and Rubin (1993) as well as by Fessler and Hero (1995) and Meng and Van Dyk (1997).

10.5.4 Bibliographic Notes

The EM algorithm was popularized by the celebrated article of Dempster *et al.* (1977). It is generally admitted however that several published works predated this landmark paper by describing applications of the EM principle to some specific cases (Meng and Van Dyk, 1997). Interestingly, the earliest example of a complete EM strategy, which also includes convergence proofs (in addition to describing the forward-backward smoothing algorithm discussed in Chapter 3), is indeed the work by Baum *et al.* (1970) on finite state space HMMs, generalizing the idea put forward by Baum and Eagon (1967). This pioneering contribution has been extended by authors such as Liporace (1982), who showed that the same procedure could be applied to other types of HMMs. The generality of the approach however was not fully recognized until Dempster *et al.* (1977) and Wu (1983) (who made the connection with the theory of global convergence) showed that the convergence of the EM approach (and its generalizations) is guaranteed in great generality.

The fact that the EM algorithm may also be used, with minor modifications, for MAP estimation was first mentioned by Dempster *et al.* (1977). Green (1990) illustrates a number of practical applications where this option plays an important role. Perhaps the most significant of these is speech processing where MAP estimation, as first described by Gauvain and Lee (1994), is commonly used for the model adaptation task (that is, re-retraining from sparse data of some previously trained models).

The ECM algorithm of Meng and Rubin (1993) (discussed Section 10.5.3) was also studied independently by Fessler and Hero (1995) under the name SAGE (space-alternating generalized EM). Fessler and Hero (1995) also introduced the idea that in some settings it is advantageous to use different ways of augmenting the data, that is, different ways of writing the likelihood as in (10.1) depending on the parameter subset that one is trying to re-estimate; see also Meng and Van Dyk (1997) for further developments of this idea.

11

Maximum Likelihood Inference, Part II: Monte Carlo Optimization

This chapter deals with maximum likelihood parameter estimation for models in which the smoothing recursions of Chapter 3 cannot be implemented. The task is then considerably more difficult, as it is not even possible to evaluate the likelihood to be maximized. Most of the methods applicable in such cases are reminiscent of the iterative optimization procedures (EM and gradient methods) discussed in the previous chapter but rely on approximate smoothing computations based on some form of Monte Carlo simulation. In this context, the methods covered in Chapters 6 and 7 for simulating the unobservable sequence of states conditionally on the observations play a prominent role.

It is important to distinguish the topic of this chapter with a distinct—although not entirely disconnected—problem. The methods discussed in the previous chapters were all based on local exploration (also called hill-climbing strategies) of the likelihood function. Such methods are typically unable to guarantee that the point reached at convergence is a global maximum of the function; indeed, it may well be a local maximum only or even a saddle point—see Section 10.5 for details regarding the EM algorithm. Many techniques have been proposed to overcome this significant difficulty, and most of them belong to a class of methods that Geyer (1996) describes as random search optimization. Typical examples are the so-called genetic and simulated annealing algorithms that both involve simulating random moves in the parameter space (see also Section 13.3, which describes a technique related to simulated annealing). In these approaches, the main motivation for using simulations (in parameter space and/or hidden variable space) is the hope to design more robust optimization rules that can avoid local maxima.

The focus of the current chapter is different, however, as we examine below methods that can be considered as simulation-based extensions of approaches introduced in the previous chapter. The primary objective is here to provide tools for maximum likelihood inference also for the class of HMMs in which exact smoothing is not available.

11.1 Methods and Algorithms

11.1.1 Monte Carlo EM

11.1.1.1 The Algorithm

Throughout this section, we use the incomplete data model notations introduced in Section 10.1.2. Recall that the *E-step* of the EM algorithm amounts to evaluating the function $\mathcal{Q}(\theta\,;\theta') = \int \log f(x\,;\theta)p(x\,;\theta')\,\lambda(dx)$ (see Definition 10.1.1). We here consider cases where direct numerical evaluation of this expectation under p is not available. The principle proposed by Wei and Tanner (1991)—see also Tanner (1993)—consists in using the Monte Carlo approach to approximate the intractable E-step with an empirical average based on simulated data:

$$\hat{\mathcal{Q}}_m(\theta\,;\theta') \stackrel{\text{def}}{=} \frac{1}{m}\sum_{j=1}^{m} \log f(\xi^j\,;\theta)\;, \tag{11.1}$$

where $\xi^1,\ldots,\xi^m$ are i.i.d. draws from the density $p(x\,;\theta')$. The subscript m in (11.1) reflects the dependence on the Monte Carlo sample size. The EM algorithm can thus be modified into the Monte Carlo EM (MCEM) algorithm by replacing $\mathcal{Q}(\theta\,;\theta')$ by $\hat{\mathcal{Q}}_m(\theta\,;\theta')$ in the E-step. More formally, the MCEM algorithm consists in iteratively computing a sequence $\{\hat{\theta}^i\}$ of parameter estimates, given an initial guess $\hat{\theta}^0$, by iterating the following two steps.

Algorithm 11.1.1 (MCEM Algorithm). For $i = 1, 2, \ldots,$

Simulation step: Draw $\xi^{i,1},\ldots,\xi^{i,m_i}$ conditionally independently given

$$\mathcal{F}^{i-1} \stackrel{\text{def}}{=} \sigma(\hat{\theta}^0, \xi^{j,l},\; j = 0,\ldots,i-1,\; l = 1,\ldots,m_j) \tag{11.2}$$

from the density $p(x\,;\hat{\theta}^{i-1})$.

M-step: Choose $\hat{\theta}^i$ to be the (or any, if there are several) value of $\theta \in \Theta$ which maximizes $\hat{\mathcal{Q}}_{m_i}(\theta\,;\hat{\theta}^{i-1})$, where $\hat{\mathcal{Q}}_{m_i}(\theta\,;\hat{\theta}^{i-1})$ is as in (11.1) (replacing ξ^j by $\xi^{i,j}$).

The initial point is picked arbitrarily and depends primarily on prior belief about the location of the maximum likelihood estimate. Like the EM algorithm, the MCEM algorithm is particularly well suited to problems in which the parametric model $\{f(x\,;\theta) : \theta \in \Theta\}$ belongs to an exponential family, $f(x\,;\theta) = \exp(\psi^t(\theta)S(x) - c(\theta))h(x)$ (see Definition 10.1.5). In this case, the E-step consists in computing a Monte Carlo approximation

$$\hat{S}^i = \frac{1}{m_i}\sum_{j=1}^{m_i} S(\xi^{i,j}) \tag{11.3}$$

of the expectation $\int S(x)p(x\,;\hat{\theta}^{i-1})\,\lambda(dx)$. The M-step then consists in optimizing the function $\theta \mapsto \psi^t(\theta)\hat{S}^i - c(\theta)$. In many models, this function is convex, and the maximization can be achieved in closed form.

In many situations, the simulation of an i.i.d. sample from the density $p(x\,;\hat{\theta}^{i-1})$ may turn out difficult. One may then use Markov chain Monte Carlo techniques, in which case $\xi^{i,1},\ldots,\xi^{i,m_i}$ is a sequence generated by an ergodic Markov chain whose stationary distribution is $p(x\,;\hat{\theta}^{i-1})$ (see Chapter 6). More precisely,

$$\xi^{i,j} \mid \mathcal{F}^{i,j-1} \sim \Pi_{\hat{\theta}^{i-1}}(\xi^{i,j-1},\cdot), \qquad j = 2,\ldots,m_i,$$

where, for any $\theta \in \Theta$, Π_θ is a Markov transition kernel admitting $p(x\,;\theta)$ as its stationary distribution and $\mathcal{F}^{i,j} = \mathcal{F}^{i-1} \vee \sigma(\xi^{i,1},\ldots,\xi^{i,j-1})$. Using MCMC complicates the control of the MCEM algorithm because of the nested structure of the iterations: an iterative sampling procedure (MCMC) is used in the inner loop of an iterative optimization procedure (MCEM).

Compared to i.i.d. Monte Carlo simulations, MCMC introduces two additional sources of errors. First, for any i and $j = 1,\ldots,m_i$, the distribution of $\xi^{i,j}$ is only approximately equal to the density $p(x\,;\hat{\theta}^{i-1})$, thus inducing a bias in the estimate. To obtain a reasonably accurate sample, it is customary to include a *burn-in period*, whose length should ideally depend on the rate at which the MCMC sampler actually mixes, during which the MCMC samples are not used for computing (11.3). The implementation of such procedures typically requires more or less sophisticated schemes to check for convergence. Second, the successive realizations $\xi^{i,1},\ldots,\xi^{i,m_i}$ of the missing data are not independent. This makes the choice of sample size more involved, because the dependence complicates the estimation of the Monte Carlo error.

11.1.1.2 MCEM for HMMs

The applications of the MCEM algorithm to HMMs is straightforward. We use the same notations and assumptions as in Section 10.2.2. In this context, $\mathrm{L}_n(Y_{0:n}\,;\theta)$ is the likelihood of the observations, $\log f(x_{0:n}\,;\theta)$ is the so-called complete data likelihood (10.25), and $p(x_{0:n}\,;\theta)$ is the conditional density of the state sequence $X_{0:n}$ given the observations $Y_{0:n}$.

In this context, MCEM is (at least conceptually) straightforward to implement: one first simulates m_i trajectories of the hidden states $X_{0:n}$ conditionally on the observations $Y_{0:n}$ and given the current parameter estimate $\hat{\theta}^{i-1}$; (11.1) is then computed using the expression of the intermediate quantity of EM given in (10.26). As discussed above, the M-step is usually straightforward at least in exponential families of distributions. To illustrate the method, we consider the following example, which will serve for illustration purposes throughout this section.

Example 11.1.2 (MCEM in Stochastic Volatility Model). We consider maximum likelihood estimation in the stochastic volatility model of Example 1.3.13,

$$X_{k+1} = \phi X_k + \sigma U_k , \qquad U_k \sim \mathrm{N}(0,1) ,$$
$$Y_k = \beta \exp(X_k/2) V_k , \qquad V_k \sim \mathrm{N}(0,1) ,$$

where the observations $\{Y_k\}_{k\geq 0}$ are the log-returns, $\{X_k\}_{k\geq 0}$ is the log-volatility, and $\{U_k\}_{k\geq 0}$ and $\{V_k\}_{k\geq 0}$ are independent sequences of white Gaussian noise with zero mean and unit variance. We analyze daily log-returns, that is, differences of the log of the series, on the British pound/US dollar exchange rate historical series (from 1 October 1981 to 28 June 1985) already considered in Example 8.3.1. The number of observations is equal to 945.

In our analysis, we will assume that the log-volatility process $\{X_k\}$ is stationary ($|\phi| < 1$) so that the initial distribution ν is given by $X_0 \sim \mathrm{N}(0, \sigma^2/(1-\phi^2))$. For this very simple model, the M-step equations are reasonably simple both for the "exact" likelihood—assuming that the initial state is distributed under the stationary distribution—and for the "conditional" likelihood—assuming that the distribution of X_0 does not depend on the parameters. We use the former approach for illustration purposes, although the results obtained on this data set with both methods are equivalent. The stochastic volatility model can naturally be cast into the framework of exponential families. Define $S(X_{0:n}) = (S_i(X_{0:n}))_{0\leq i\leq 4}$ by

$$S_0(x_{0:n}) = x_0^2 , \qquad S_1(x_{0:n}) = \sum_{k=0}^{n-1} x_k^2 , \qquad S_2(x_{0:n}) = \sum_{k=1}^{n} x_k^2 ,$$
$$S_3(x_{0:n}) = \sum_{k=1}^{n} x_k x_{k-1} , \qquad S_4(x_{0:n}) = \sum_{k=0}^{n} Y_k^2 \exp(-x_k) . \tag{11.4}$$

With these notations, the complete data likelihood may be expressed, up to terms not depending on the parameters, as

$$\log f(X_{0:n}\,; \beta, \phi, \sigma) = F(S(X_{0:n})\,; \beta, \phi, \sigma) ,$$

where the function $s = (s_i)_{0\leq i\leq 4} \mapsto F(s\,; \beta, \phi, \sigma)$ is given by

$$F(s\,; \beta, \phi, \sigma) = -\frac{n+1}{2}\log\beta^2 - \frac{1}{2\beta^2}s_4 - \frac{n+1}{2}\log\sigma^2 + \frac{1}{2}\log(1-\phi^2)$$
$$- \frac{(1-\phi^2)s_0}{2\sigma^2} - \frac{1}{2\sigma^2}\left(s_2 - 2\phi s_3 + \phi^2 s_1\right) .$$

Maximization with respect to β yields the update

$$\beta^* = \sqrt{\frac{s_4}{n+1}} . \tag{11.5}$$

Computing the partial derivative of $F(s\,; \beta, \phi, \sigma)$ with respect to σ^2 yields the relation

$$\begin{aligned}\sigma^2(s\,;\phi) &= \frac{1}{n+1}\left\{(1-\phi^2)s_0 + s_2 - 2\phi s_3 + \phi^2 s_1\right\} \\ &= \frac{1}{n+1}\left\{(s_0+s_2) - 2\phi s_3 + \phi^2(s_1 - s_0)\right\} . \end{aligned} \tag{11.6}$$

Plugging this value into the partial derivative of $F(s\,;\beta,\phi,\sigma)$ with respect to ϕ yields an estimation equation for ϕ:

$$-\frac{\phi}{1-\phi^2} + \frac{\phi s_0}{\sigma^2(s\,;\phi)} + \frac{s_3 - \phi s_1}{\sigma^2(s\,;\phi)} = 0 .$$

The solution of this equation amounts to solving the cubic

$$\begin{aligned}\phi^3[n(s_1 - s_0)] &+ \phi^2[-(n-1)s_3] \\ &+ \phi[-s_2 + ns_0 - (n+1)s_1] + (n+1)s_3 = 0 . \end{aligned} \tag{11.7}$$

Hence the M-step implies the following computations: find ϕ^* as the root of (11.7), selecting the one that is, in absolute value, smaller than one; determine $(\sigma^*)^2$ using (11.6); β^* is given by (11.5).

To implement the MCEM algorithm, we sampled from the joint smoothing distribution of $X_{0:n}$ parameterized by $\hat{\theta}_{i-1}$ using the single-site Gibbs sampler with embedded slice sampler, as described in Example 6.2.16. Initially, the sampler was initialized by setting all $X_k = 0$, and a burn-in period of 200 sweeps (by a *sweep* we mean updating every hidden state X_k once in a linear order from X_0 to X_n) was performed before the computation of the samples averages involved in the statistics S_l (for $l = 0, \ldots, 4$) was initialized. Later E-steps did not reset the state variables like this, but rather started with the final realization $X_{0:n}^{i-1,m_{i-1}}$ of the previous E-step (thus done with different parameters). The statistics $S_l(X_{0:n})$ (for $l = 0, \ldots, 4$) were approximated by averaging over the sampled trajectories letting, for instance, $\hat{S}_3^i = \frac{1}{m_i}\sum_{j=1}^{m_i}\sum_{k=1}^{n} X_k^{i,j} X_{k-1}^{i,j}$. The M-step was carried out as discussed above.

Figure 11.1 shows 400 iterations of the MCEM algorithm with 25,000 MCMC sweeps in each step, started from the parameter values $\beta = 0.8$, $\phi = 0.9$, and $\sigma = 0.3$. Because the number of sweeps at each step is quite large, the MCEM parameter trajectory can be seen as a proxy for the EM trajectory. It should be noted that the convergence of the EM algorithm is in this case quite slow because the eigenvalues of the rate matrix defined in (10.70) are close to one. The final estimates are $\beta = 0.641$, $\phi = 0.975$, and $\sigma = 0.165$, which agrees with figures given by Sandmann and Koopman (1998) up to the second decimal. ■

A key issue, to be discussed in the following, is whether or not such a large number of MCMC simulation is really needed to obtain the results shown on Figure 11.1. In Section 11.1.2, we will see that by a proper choice of the simulation schedule, that is, of the sequence $\{m_i\}_{i\geq 1}$, it is possible to obtain equivalent results with far less computational effort.

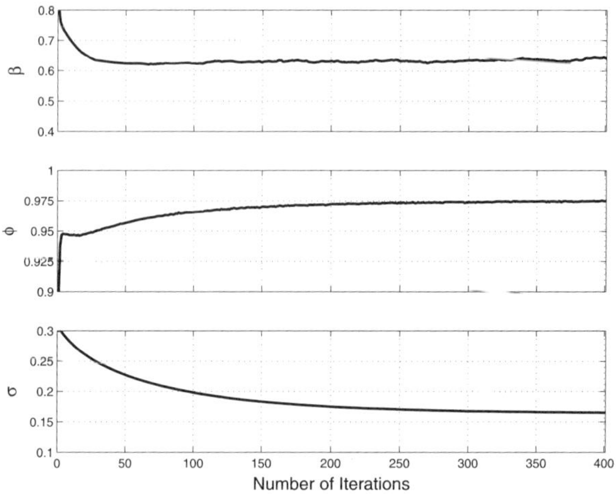

Fig. 11.1. Trajectory of the MCEM algorithm for the stochastic volatility model and GBP/USD exchange rate data. In the E-step, an MCMC algorithm was used to impute the missing data. The plots show 400 EM iterations with 25,000 MCMC sweeps in each iteration.

11.1.1.3 MCEM Based on Sequential Monte Carlo Simulations

The use of Monte Carlo simulations—either Markov chain or i.i.d. ones—is not the only available option for approximating the E-step computations. Another approach, suggested by Gelman (1995) (see also Quintana *et al.*, 1999), consists in approximating the intermediate quantity $\mathcal{Q}(\theta\,;\hat{\theta}^{i-1})$ of EM using importance sampling (see Section 7.1). In this case, we simulate a sample $\tilde{\xi}^{i,1},\ldots,\tilde{\xi}^{i,m_i}$ from an instrumental distribution with density r with respect to the common dominating measure λ and approximate $\mathcal{Q}(\theta\,;\hat{\theta}^{i-1})$ by the weighted sum

$$\hat{\mathcal{Q}}_{m_i}(\theta\,;\hat{\theta}^{i-1}) \stackrel{\text{def}}{=} \sum_{j=1}^{m_i} \omega^{i,j} \log f(\tilde{\xi}^{i,j};\theta) \;, \qquad \omega^{i,j} \stackrel{\text{def}}{=} \frac{\frac{p(\tilde{\xi}^{i,j}\,;\hat{\theta}^{i-1})}{r(\tilde{\xi}^{i,j})}}{\sum_{k=1}^{m_i} \frac{p(\tilde{\xi}^{i,k}\,;\hat{\theta}^{i-1})}{r(\tilde{\xi}^{i,k})}} \;. \tag{11.8}$$

In most implementations of this method reported so far, the instrumental distribution is chosen as the density $p(x\,;\theta^\star)$ for a reference value $\theta^\star$ of the parameter, but other choices can also be valuable. We may keep the same instrumental distribution and therefore the same importance sample during several iterations of the algorithm. Of course, as the iterations go on, the instrumental distribution can become poorly matched to the current target density $p(x;\hat{\theta}^{i-1})$, leading to badly behaved importance sampling estimators. The mismatch between the instrumental and target distributions can be monitored by controlling that the importance weights remain properly balanced.

For HMMs, importance sampling is seldom a sensible choice unless the number of observations is small (see Section 7.3.1). Natural candidates in this context are the sequential Monte Carlo methods based on resampling ideas discussed in Chapters 7 and 8. In Section 8.3, we considered the general problem of estimating quantities of the form $\mathrm{E}(t_n(X_{0:n})|Y_{0:n}\,;\theta)$, when the function t_n complies with Definition 4.1.2, based on sequential Monte Carlo simulations. As discussed in Section 10.2.2, the intermediate quantity of EM is precisely of this form with an additive structure given by (10.26). Recall that the same remark also holds for the gradient of the log-likelihood with respect to the parameter vector θ (Section 10.2.3). For both of these, an approximation of the smoothed expectation can be computed recursively and without storing the complete particle trajectories (see Section 8.3).

For the model of Example 11.1.2, the function t_n is fully determined by the four statistics defined in (11.4). Recursive particle smoothing for the statistics S_0, S_1, and S_3 has already been considered in Example 8.3.1 (see Figures 8.5 and 8.7). The case of the remaining two statistics is entirely similar. Recall from Example 8.3.1 that it is indeed possible to robustify the estimation of such smoothed sum functionals by using fixed-lag approximations. The simple method proposed in Example 8.3.1 consists in replacing the smoothing distributions $\phi_{l|n}$ by the fixed lag-smoothing distribution $\phi_{l|l+k\wedge n}$ for a suitably chosen value of the delay k. The particle approximation to $\sum_{l=0}^{n}\int s(x)_l \phi_{l|l+k\wedge n}(dx_l)$ can be computed recursively using an algorithm that is only marginally more complex than that used for $\sum_{l=0}^{n}\int s(x)_l \phi_{l|n}(dx_l)$. Results obtained following this approach will be discussed in Example 11.1.3 below.

11.1.2 Simulation Schedules

Although the MCEM algorithm provides a solution to intractable E-step, it also raises difficult implementation issues. Intelligent usage of the Monte Carlo simulations is necessary because MCEM can place a huge burden on the user's computational resources.

Heuristically there is no need to use a large number of simulations during the initial stage of the optimization. Even rather crude estimation of $\mathcal{Q}(\theta\,;\hat{\theta}^{i-1})$ might suffice to drive the parameters toward the region of interest. As the EM iterations go on, the number of simulations should be increased however to avoid "zig-zagging" when the algorithm approaches convergence. Thus, in making the trade-off between improving accuracy and reducing the computational cost associated with a large sample size, one should favor increasing the sample size m_i as $\hat{\theta}^i$ approaches its limit. Determining exactly how this increase should be accomplished to produce the "best" possible result is a topic that still attracts much research interest (Booth and Hobert, 1999; Levine and Casella, 2001; Levine and Fan, 2004).

Example 11.1.3 (MCEM with Increasing Simulation Schedule). Results comparable to those of the "brute force" version of the MCEM algorithm

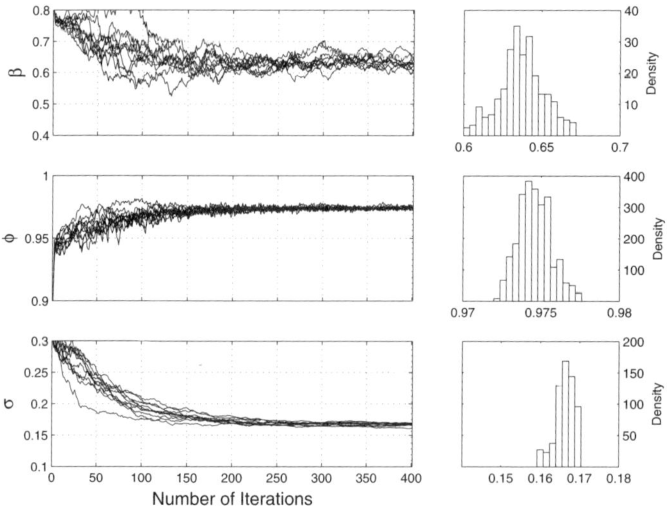

Fig. 11.2. Same model, data, and algorithm as in Figure 11.1, except that the number of MCMC sweeps in the E-step was increased quadratically with the EM iteration number. The plots show results from 400 iterations of the MCEM algorithm with the number of MCMC sweeps ranging from 1 at the first iteration to 374 at iteration 200 and 1,492 at iteration 400; the total number of sweeps was 200,000. Left: 10 independent trajectories of the MCEM algorithm, with identical initial points. Right: histograms, obtained from 50 independent runs, of the final values of the parameters.

considered in Example 11.1.2 can in fact can be achieved with a number of sweeps smaller by an order of magnitude. To allow for comparisons with other methods, we set, in the following, the total number of simulations of the missing data sequence to 200,000. Figure 11.2 shows the results when the number of sweeps of the E-step MCMC sampler increases proportionally to the square of the EM iteration number. This increase is quite slow, because many EM iterations are required to reach convergence (see Figure 11.1). The number of sweeps performed during the final E-step is only about 1500 (compared to the 25,000 for the MCEM algorithm illustrated in Figure 11.1). As a result, the MCEM algorithm is still affected by a significant fraction of simulation noise in its last iteration.

As discussed above, the averaged MCMC simulations may be replaced by time-averages computed from sequential Monte Carlo simulations. To this aim, we consider the SISR algorithm implemented as in Example 8.3.1 with systematic resampling and a t-distribution with 5 degrees of freedom fitted to the mode of the optimal instrumental distribution. The SMC approach requires a minimal number of particles to produce sensible output. Hence we cannot adopt exactly the same simulation schedule as in the case of MCMC above, and the number of particles was set to 250 for the first 100 MCEM

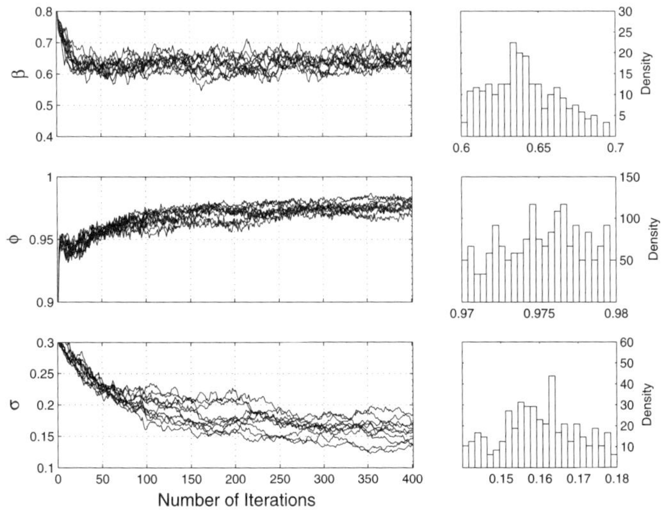

Fig. 11.3. Same model and data as in Figure 11.1. Parameter estimates were computed using an MCEM algorithm employing SISR approximation of the joint smoothing distributions. The plots show results from 400 iterations of the MCEM algorithm. The number of particles was 250 for the first 100 EM iterations, 500 for iterations 101 to 200, and then increased proportionally to the squared iteration number. The contents of the plots are as in Figure 11.2.

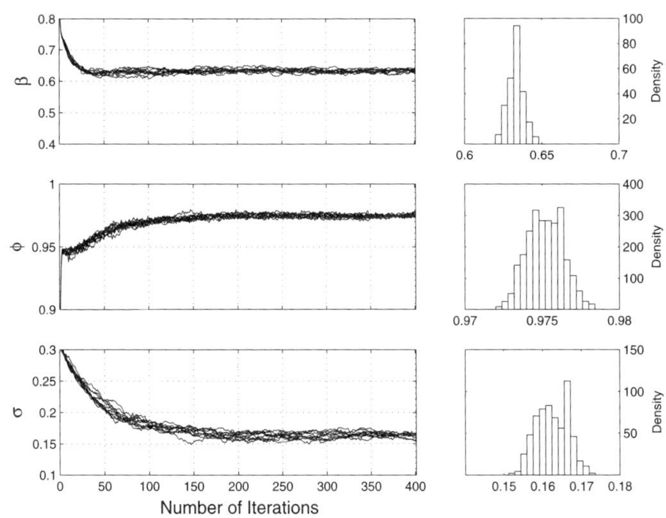

Fig. 11.4. Same model and data as in Figure 11.1. Parameter estimates were computed using an MCEM algorithm employing SISR approximation of fixed-lag smoothing distributions with delay $k = 20$. The plots show results from 400 iterations of the MCEM algorithm. The number of particles was as described in Figure 11.3 and the contents of the plots are as in Figure 11.2.

iterations, 500 for iterations 101 to 200, and then increases proportionally to the square of the MCEM iteration number. The total number of simulations is also equal to 200,000 in this case. The MCEM algorithm was run using both the particle approximation of the joint smoothing distributions and that of the fixed-lag smoothing distributions. Figure 11.3 shows that the implementation based on joint smoothing produces highly variable parameter estimates. This is coherent with the behavior observed in Example 8.3.1. Given that the number of observations is already quite large, it is preferable to use fixed-lag smoothing (here with a lag $k = 20$), as the bias introduced by this approximation is more than compensated by the reduction in the Monte Carlo error variance. As shown in Figure 11.4, the behavior of the resulting algorithm is very close to what is obtained using the MCEM algorithm with MCMC imputation of the missing data. When comparing to Figure 11.2, the level of the Monte Carlo error appears to be reduced in Figure 11.4, and the bias introduced by the fixed-lag smoothing approximation is hardly perceptible. ■

11.1.2.1 Automatic Schedules

From the previous example, it is obvious that it is generally advantageous to vary the precision of the estimate of the intermediate quantity $\mathcal{Q}(\theta\,;\hat{\theta}^{i-1})$ with i, and in particular to increase this precision as i grows and $\hat{\theta}^i$ approaches a limit. In the example above, this was accomplished by increasing the number of sweeps of the MCMC sampler or by increasing the number of particles of the SMC algorithm. So far, the increase was done in a deterministic fashion, and such deterministic schedules may also be given theoretical support (see Section 11.2.3). Deterministic schemes are appealing because of their simplicity, but it is obvious that because there are only few theoretical guidelines on how to choose m_i, finding an appropriate schedule is in general not straightforward.

It has often been advocated that using automatic, or adaptive, procedures to choose m_i would be more appropriate. To do so, it is required to determine, at each iteration, an estimate of the Monte Carlo error $\hat{\mathcal{Q}}_{m_i}(\theta\,;\hat{\theta}^{i-1}) - \mathcal{Q}(\theta\,;\hat{\theta}^{i-1})$. The dependence of this error with respect to m_i should also be known or determined from the output of the algorithm. Such "data-driven" procedures require gauging the Monte Carlo errors, which is, in general, a complicated task. Booth and Hobert (1999) present an automatic method that requires independent Monte Carlo sample in the E-step. Independent simulations allow for computationally inexpensive and straightforward assessment of Monte Carlo error through an application of the central limit theorem.

Such independent sampling routines are often unavailable in practical implementations of the MCEM algorithm however, requiring MCMC or SMC algorithms to obtain relevant Monte Carlo samples. Levine and Casella (2001) present a method for estimating the simulation error of a Monte Carlo E-step using MCMC samples. Their procedure is based on regenerative methods for MCMC simulations and amounts to finding renewal periods across which the

MCMC trajectories are independent (see for instance Hobert *et al.*, 2002). By subsampling the chain between regeneration times, Monte Carlo error may be assessed through the CLT for independent outcomes in a manner analogous to Booth and Hobert (1999). For phi-irreducible Markov chains, such renewal periods can be obtained using the splitting procedure, which requires determining small sets (see Section 14.2 for definitions of the concepts mentioned here). A drawback of this approach is that it may be difficult, if not impossible, to establish the minorization condition necessary for implementing the regenerative simulation procedure. Once such a minorization condition has been established however, implementing the procedure is nearly trivial.

Both of the automatic procedures mentioned above are able to decide when to increase the Monte Carlo sample size, but the choice of sample size at each such instance is arbitrary. Levine and Fan (2004) present a method that overcomes the limitations of the previous algorithm. The Monte Carlo error is gauged directly using a subsampling technique, and the authors use asymptotic results to construct an adaptive rule for updating the Monte Carlo sample size.

Despite their obvious appeal, automatic methods suffer from some drawbacks. First, the estimation of the Monte Carlo error induces a computational overhead that might be non-negligible. Second, because the number of simulations at each iteration is random, the total amount of computation cannot be fixed beforehand; this may be inconvenient. Finally, the convergence of the proposed schemes are based on heuristic arguments and have not been established on firm grounds.

11.1.2.2 Averaging

There is an alternative to automatic selection of the Monte Carlo sample size, developed by Fort and Moulines (2003), which is straightforward to implement and most often useful. This method is inspired by the averaging procedure originally proposed by Polyak (1990) to improve the rate of convergence of stochastic approximation procedures.

To motivate the construction of the averaging procedure, note that provided that the sequence $\{\hat{\theta}^i\}$ converges to a limit $\theta_\star$, each value of $\hat{\theta}^i$ may itself be considered as an estimator of the associated limit $\theta_\star$. Theorem 11.2.14 asserts that the variance of $\hat{\theta}^i - \theta_\star$ is of order $1/m_i$. Thus, in the idealized situation where the random perturbations $\hat{\theta}^i - \theta_\star$ would also be uncorrelated, it is well-known that it is possible to obtain an improved estimator of $\theta_\star$ by combining the individual estimates $\hat{\theta}^i$ in proportion of the inverse of their variance (this is the minimum variance estimate of $\theta_\star$). This optimal linear combination has a variance that decreases as $1/\sum_i m_i$, that is, the total number of simulations rather than the final number of simulations. Although the MCEM perturbations $\hat{\theta}^i - \theta_\star$ are not uncorrelated, even when using i.i.d. Monte Carlo simulation, due to the dependence with respect to θ, Fort and Moulines (2003) suggested using the averaged MCEM estimator

$$\tilde{\theta}_i \stackrel{\text{def}}{=} \sum_{j=i_0}^{i} \frac{m_j}{\sum_{j=i_0}^{i} m_j} \hat{\theta}^j \,, \qquad \text{for } i \geq i_0, \tag{11.9}$$

where i_0 is the iteration index at which computation of the average is started. In general, it is not recommended to start averaging too early, when the algorithm is still in its transient phase.

Example 11.1.4 (Averaging). In Example 11.1.3, the number of sweeps is increased quite slowly and the number of sweeps during the final EM iterations is not large (about 1500). This scheme is advantageous in situations when the EM algorithm is slow, because a large number of iterations can be performed while keeping the total of number of simulations moderate. The problem is rather that the simulation noise at convergence is still significant (see Figure 11.2). This is a typical situation in which averaging can prove to be very helpful. As seen in Figure 11.5, averaging reduces the noise when the parameters are in the neighborhood of their limits. Averaging is also beneficial when the EM statistics are estimated using sequential Monte Carlo (see Figure 11.6). ■

11.1.3 Gradient-based Algorithms

As discussed in Section 10.2.3, computation of the gradient of the log-likelihood is very much related to the E-step of EM as a consequence of Fisher's identity (Proposition 10.1.6). It is thus rather straightforward to derive Monte Carlo versions of the gradient algorithms introduced in Section 10.1.3. At the ith iteration, one may for example approximate the gradient of the log-likelihood $\nabla_\theta \ell(\hat{\theta}^{i-1})$, where $\hat{\theta}^{i-1}$ denotes the current parameter estimate, by

$$\widehat{\nabla_\theta \ell}_{m_i}(\hat{\theta}^{i-1}) = \frac{1}{m_i} \sum_{j=1}^{m_i} \nabla_\theta \log f(\xi^{i,j}\,; \hat{\theta}^{i-1}) \,, \tag{11.10}$$

where $\xi^{i,1}, \ldots, \xi^{i,m_i}$ is an i.i.d. sample from the density $p(x\,; \hat{\theta}^{i-1})$ or a realization of an ergodic Markov chain admitting $p(x\,; \hat{\theta}^{i-1})$ as its stationary density. It is also possible to use importance sampling; if $\tilde{\xi}^{i,1}, \ldots, \tilde{\xi}^{i,m_i}$ is a sample from the instrumental distribution r, then the IS estimate of $\nabla_\theta \ell(\hat{\theta}^{i-1})$ is

$$\widehat{\nabla_\theta \ell}_m(\hat{\theta}^{i-1}) = \sum_{j=1}^{m_i} \omega^{i,j} \nabla_\theta \log f(\tilde{\xi}^{i,j}\,; \hat{\theta}^{i-1}), \quad \omega^{i,j} = \frac{\frac{p(\tilde{\xi}^{i,j};\hat{\theta}^{i-1})}{r(\tilde{\xi}^{i,j})}}{\sum_{k=1}^{m_i} \frac{p(\tilde{\xi}^{i,k};\hat{\theta}^{i-1})}{r(\tilde{\xi}^{i,k})}} \tag{11.11}$$

As in the case of MCEM, it is likely that for HMMs, importance sampling strategies become unreliable when the number of observations increases. To circumvent the problem, one may use sequential Monte Carlo methods such

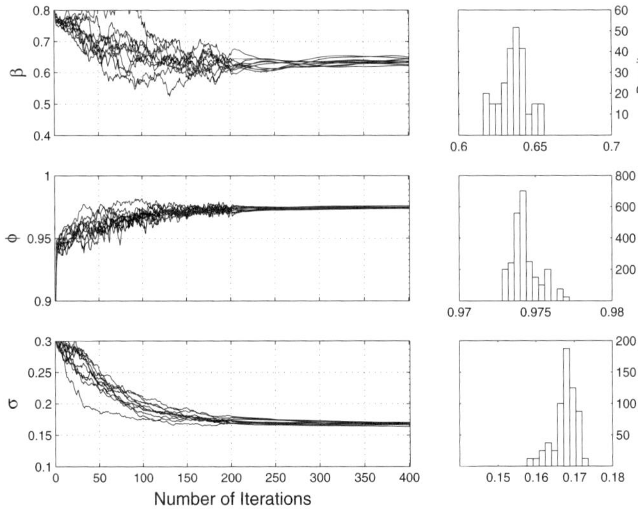

Fig. 11.5. Same model, data, and algorithm as in Figure 11.2, except that averaging according to (11.9) was used to smooth the sufficient statistics of the E-step; averaging was started after $i_0 = 200$ iterations. The plots show results from 400 iterations of the MCEM algorithm. The contents of the plots are as in Figure 11.2.

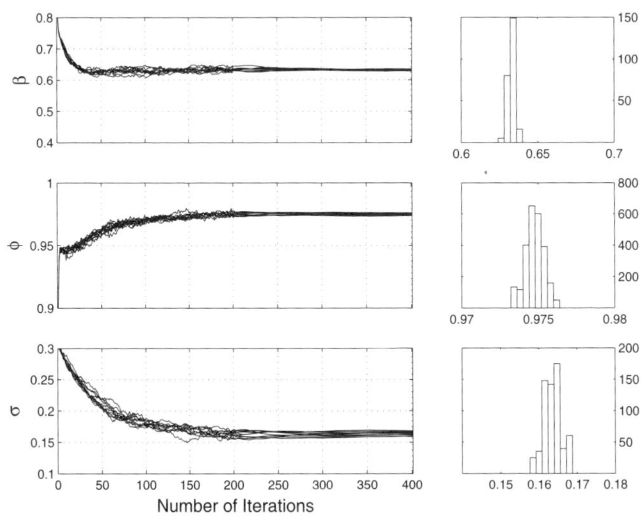

Fig. 11.6. Same model, data, and algorithm as in Figure 11.4, except that averaging according to (11.9) was used to smooth the sufficient statistics of the E-step; averaging was started after $i_0 = 200$ iterations. The plots show results from 400 iterations of the MCEM algorithm. The contents of the plots are as in Figure 11.2.

as SISR where (11.11) is not computed directly but rather constructed recursively (in time) following the approach discussed in Section 8.3 and used in the case of MCEM above. Details are omitted because the gradient of the log-likelihood (10.29) and the intermediate quantity of EM (10.26) are very similar. For models that belong to exponential families, the only quantities that need to be computed in both cases are the smoothed expectation of the sufficient statistics, and hence both computations are exactly equivalent.

Louis's identity (see Proposition 10.1.6) suggests an approximation of the Hessian of $\ell(\theta)$ at $\hat{\theta}^{i-1}$ of the form

$$\hat{J}_{m_i}(\hat{\theta}^{i-1}) = \frac{1}{m_i}\sum_{j=1}^{m_i} \nabla_\theta^2 \log f(\xi^{i,j}\,;\hat{\theta}^{i-1}) + \frac{1}{m_i}\sum_{j=1}^{m_i}\left[\nabla_\theta \log f(\xi^{i,j}\,;\hat{\theta}^{i-1})\right]^{\otimes 2} - \left[\widehat{\nabla_\theta \ell}_{m_i}(\hat{\theta}^{i-1})\right]^{\otimes 2} ,$$

where $\xi^{i,1},\ldots,\xi^{i,m_i}$ are as above, for a vector a we have used the notation $a^{\otimes 2} = aa^t$, and the estimate of the gradient in the final term on the right-hand side may be chosen, for instance, as in (11.10). Using this approximation of the Hessian, it is possible to formulate a Monte Carlo version of the Newton-Raphson procedure. This algorithm was first proposed by Geyer and Thompson (1992) in an exponential family setting and then generalized by Gelfand and Carlin (1993). Gelman (1995) proposed a similar algorithm in which importance sampling is used as the Monte Carlo method.

Now assume that we have, with the help of a Monte Carlo approximation of the gradient and possibly also the Hessian, selected a search direction. The next step is then to determine an appropriate value of the step size γ (see Section 10.1.3). This is not a simple task, because the objective function $\ell(\theta)$ cannot be evaluated analytically, and therefore it is not possible to implement a line search—at least not in an immediate way. A simple option consists in using a step size that is small but fixed (see Dupuis and Simha, 1991), and to let $m_i \to \infty$ as sufficiently fast as $i \to \infty$.

If we want to optimize the step size, we have to approximate the objective function in the search direction. We may for example follow the method proposed by Geyer and Thompson (1992), which consists in approximating (locally) the ratio $\mathrm{L}(\theta)/\mathrm{L}(\hat{\theta}^{i-1})$ by

$$\sum_{j=1}^{m_i} \frac{f(\xi^{i,j}\,;\theta)}{f(\xi^{i,j}\,;\hat{\theta}^{i-1})} ,$$

where $\{\xi^{i,j}\}$ are the samples from $p(x\,;\hat{\theta}^{i-1})$ used to determine the search direction. Under standard assumptions, the sum of this display converges in probability as $m_i \to \infty$ to

$$\int \frac{f(x\,;\theta)}{f(x\,;\hat{\theta}^{i-1})} p(x\,;\hat{\theta}^{i})\,\lambda(dx) = \frac{\mathrm{L}(\theta)}{\mathrm{L}(\hat{\theta}^{i-1})} .$$

This suggests approximating the difference $\ell(\theta) - \ell(\hat{\theta}^{i-1})$ in a neighborhood of $\hat{\theta}^{i-1}$ by

$$\log \left[\frac{1}{m_i} \sum_{j=1}^{m_i} \frac{f(\xi^{i,j}\,;\theta)}{f(\xi^{i,j}\,;\hat{\theta}^{i-1})} \right] . \tag{11.12}$$

This type of approximation nevertheless needs to be considered with some care because the search direction is not necessarily an ascent direction for this approximation of the objective function due to the Monte Carlo errors. To the best of our knowledge, this type of approximation has not been thoroughly investigated in practice.

As for the MCEM algorithm, it is not necessary to estimate the objective function and its gradient with high accuracy during the initial optimization steps. Therefore, the Monte Carlo sample sizes should not be taken large at the beginning of the procedure but should be increased when the algorithm approaches convergence. Procedures to adapt the sample size m_i at each iteration are discussed and analyzed by Sakalauskas (2000, 2002) for gradient algorithms using a (small enough) fixed step size. The suggestion of this author is to increase m_i proportionally to the inverse of the squared norm of the (estimated) gradient at the current parameter estimate. If this proportionality factor is carefully adjusted, it may be shown, under a set of restrictive conditions, that the Monte Carlo steepest ascent algorithm converges almost surely to a stationary point of the objective function.

It is fair to say that in the case of general state space HMMs, gradient-based methods as less popular than their counterparts based on the EM paradigm. An important advantage of EM based methods in this context is that they are parameterization independent (see Section 10.1.4 for further discussion). This property means that the issue of selecting a proper step size γ—which is problematic in simulation-based approaches as discussed above—has no counterpart for EM-based methods, which are scale-free. Remember that it is also precisely the reason why the EM approach sometimes converges much more slowly than gradient-based methods.

11.1.4 Interlude: Stochastic Approximation and the Robbins-Monro Approach

Stochastic approximation is a general term for methods that recursively search for an optimum or zero of a function that can only be observed disturbed by some noise. The original work in the stochastic approximation literature was by Robbins and Monro (1951), who developed and analyzed a recursive procedure for finding the root(s) of the equation $h(\theta) = 0$. If the function h was known, a simple procedure to find a root consists in using the elementary algorithm

$$\theta^i = \theta^{i-1} + \gamma_i h(\theta^{i-1}) , \tag{11.13}$$

where $\{\gamma_i\}$ is a sequence of positive step sizes. In many applications, the evaluation of $h(\theta)$ cannot be performed, either because it is computationally prohibitive or analytical formulas are simply not available, but noise-corrupted observations of the function can be obtained for any value of the parameter $\theta \in \Theta \subseteq \mathbb{R}^{d_\theta}$. One could then, for instance, consider using the procedure (11.13) but with $h(\theta)$ replaced by an accurate estimate of its value obtained by averaging many noisy observations of the function.

It was recognized by Robbins and Monro (1951) that averaging a large number of observations of the function at θ^{i-1} is not always the most efficient solution. Indeed, the value of the function $h(\theta^{i-1})$ is only of interest in so far that it leads us in the right direction, and it is not unreasonable to expect that this happens, at least on the average, even if the estimate is not very accurate. Robbins and Monro (1951) rather proposed the algorithm

$$\hat{\theta}^i = \hat{\theta}^{i-1} + \gamma_i Y^i , \tag{11.14}$$

where γ_i is a deterministic sequence satisfying

$$\gamma_i > 0, \quad \lim_{i\to\infty} \gamma_i = 0, \quad \sum_i \gamma_i = \infty ,$$

and Y^i is a noisy observation of $h(\hat{\theta}^{i-1})$. Although the analysis of the method is certainly simpler when the noise sequence $\{Y^i - h(\hat{\theta}^{i-1})\}_{i\geq 1}$ is i.i.d., in many practical applications the noise $Y^i - h(\hat{\theta}^{i-1})$ depends on $\hat{\theta}^{i-1}$ and sometimes on past values of $\hat{\theta}^j$ and Y^j, for $j \leq i-1$ (see for instance Benveniste *et al.*, 1990; Kushner and Yin, 2003). Using a decreasing step size implies that the parameter sequence $\{\hat{\theta}^i\}$ moves slower as i goes to infinity; the basic idea is that decreasing step sizes provides an averaging of the random errors committed when evaluating the function h.

Ever since the introduction of the now classic Robbins-Monro algorithm, stochastic approximation has been successfully used in many applications and has received wide attention in the literature. The convergence of the stochastic approximation scheme is also a question of importance that has been addressed under a variety of conditions, which cover most of the applications (see for instance Benveniste *et al.*, 1990; Duflo, 1997; Kushner and Yin, 2003).

11.1.5 Stochastic Gradient Algorithms

We now come back to the generic incomplete data model, considering several ways in which the stochastic approximation approach may be put in use. The first obvious option is to apply the Robbins-Monro algorithm to determine the roots of the equations $\nabla_\theta \ell(\theta) = 0$, yielding the following recursions

$$\hat{\theta}^i = \hat{\theta}^{i-1} + \gamma_i \nabla_\theta \log f(\xi^i \,; \hat{\theta}^{i-1}) , \tag{11.15}$$

where ξ^i is a sample from the density $p(x \,; \hat{\theta}^{i-1})$. That is, defining the filtration $\{\mathcal{F}^i\}$ such that $\mathcal{F}^{i-1} = \sigma(\hat{\theta}^0, \xi^0, \ldots, \xi^{i-1})$,

$$\xi^i \,|\, \mathcal{F}^{i-1} \sim p(\cdot\,; \hat{\theta}^{i-1}) \,.$$

Thus $Y^i = \nabla_\theta \log f(\xi^i\,; \hat{\theta}^{i-1})$ can be considered as a noisy measurement of $\nabla_\theta \ell(\hat{\theta}^{i-1})$ because of the Fisher identity, $\mathrm{E}[Y^i \,|\, \mathcal{F}^{i-1}] = \nabla_\theta \ell(\hat{\theta}^{i-1})$. Hence we can write $Y^i = \nabla_\theta \ell(\hat{\theta}^{i-1}) + \zeta^i$, with

$$\zeta^i = \nabla_\theta \log f(\xi^i\,; \hat{\theta}^{i-1}) - \mathrm{E}[\nabla_\theta \log f(\xi^i)\,; \hat{\theta}^{i-1} \,|\, \mathcal{F}^{i-1}] \,;$$

obviously $\{\zeta^i\}$ is an $\{\mathcal{F}^i\}$-adapted martingale difference sequence.

Often it is not possible to sample directly from the density $p(x\,; \hat{\theta}^{i-1})$. One can then replace this draw by iterations from a Markov chain admitting $p(x\,; \hat{\theta}^{i-1})$ as its stationary density. Then $\mathrm{E}[Y^i \,|\, \mathcal{F}^{i-1}]$ does no longer equal $\nabla_\theta \ell(\hat{\theta}^{i-1})$, but rather

$$\xi^i \,|\, \mathcal{F}^{i-1} \sim \Pi_{\hat{\theta}^{i-1}}(\xi^{i-1}, \cdot) \,, \tag{11.16}$$

where for any $\theta \in \Theta$, Π_θ is a transition kernel of an ergodic Markov chain with stationary density $p(x\,;\theta)$. Such algorithms were considered by Younes (1988, 1989) for maximum likelihood estimation in partially observed Gibbs fields. They were later extended by Gu and Kong (1998) to maximum likelihood estimation in general incomplete data problems by (see also Gu and Li, 1998; Delyon *et al.*, 1999, Section 8). In this case, the noise structure is more complicated and analysis and control of the convergence of such algorithms become intricate (see Andrieu *et al.*, 2005, for results in this direction).

Several improvements can be brought to this scheme. First, it is sometimes recommendable to run a certain number, say m, of simulations before updating the value of the parameter. That is,

$$\hat{\theta}^i = \hat{\theta}^{i-1} + \gamma_i \left\{ \frac{1}{m} \sum_{j=1}^m \nabla_\theta \log f(\xi^{i,j}\,; \hat{\theta}^{i-1}) \right\} \,, \tag{11.17}$$

where $\xi^{i,1}, \ldots, \xi^{i,m}$ are draws from $p(x\,; \hat{\theta}^{i-1})$. Choosing $m > 1$ is generally beneficial in that it makes the procedure more stable and saves computational time. The downside is that there are few theoretical guidelines on how to set this number. The above algorithm is very close to the Monte Carlo version of the steepest ascent method. Another possible improvement, much in the spirit of quasi-Newton algorithms, is to modify the search direction by letting

$$\hat{\theta}^i = \hat{\theta}^{i-1} + \gamma_i W^i \left\{ \frac{1}{m} \sum_{j=1}^m \nabla_\theta \log f(\xi^{i,j}\,; \hat{\theta}^{i-1}) \right\} \,, \tag{11.18}$$

where W^i is a properly chosen weight matrix (see for instance Gu and Li, 1998; Gu and Kong, 1998).

One of the main appeals of stochastic approximation is that, at least in principle, the only decision that has to be made is the choice of the step size schedule. Although in theory the method converges for a wide variety of step sizes (see Section 11.3), in practice the choice of step sizes influences the actual number of simulations needed to take the parameter estimate into the neighborhood of the solution (transient regime) and its fluctuations around the solution (misadjustment near convergence). Large step sizes generally speed up convergence to a neighborhood of the solution but fail to mitigate simulation noise. Small step sizes reduce noise but cause slow convergence. Heuristically, it is appropriate to use large step sizes until the algorithm reaches a neighborhood of the solution and then to switch to smaller step sizes (see for instance Gu and Zhu, 2001, for applications to the stochastic gradient algorithm).

A way to alleviate the step size selection problem is to use averaging as in Section 11.1.2. Polyak (1990) (see also Polyak and Juditsky, 1992) showed that if the sequence of step sizes $\{\gamma_i\}$ tends to zero slower than $1/i$, yet fast enough to ensure convergence at a given rate, then the running average

$$\tilde{\theta}^i \stackrel{\text{def}}{=} (i - i_0 + 1)^{-1} \sum_{j=i_0}^{i} \hat{\theta}^j , \qquad i \geq i_0 , \tag{11.19}$$

converges at an *optimal* rate. Here i_0 is an index at which averaging starts, so as to discard the very first steps. This result implies that one should adopt step sizes larger than usual but in conjunction with averaging (to control the increased noise due to use of the larger step sizes). The practical value of averaging has been reported in many different contexts—see (Kushner and Yin, 2003, Chapter 11) for a thorough investigation averaging, as well as (Delyon *et al.*, 1999).

11.1.6 Stochastic Approximation EM

We now consider a variant of the MCEM algorithm that may also be interpreted as a stochastic approximation procedure. Compared to the stochastic gradient approach discussed in the previous section, this algorithm is scale-free in the sense that the step sizes are positive numbers restricted to the interval $[0, 1]$. Compared to the MCEM approach, the E-step involves a weighted average of the approximations of the intermediate quantity of EM obtained in the current as well as in the previous iterations. Hence there is no need to increase the number of replications of the missing data as in MCEM.

Algorithm 11.1.5 (Stochastic Approximation EM). Given an initial parameter estimate $\hat{\theta}^0$ and a decreasing sequence of positive step sizes $\{\gamma_i\}_{i\geq 1}$ such that $\gamma_1 = 1$, do, for $i = 1, 2 \ldots$,

Simulation: Draw $\xi^{i,1}, \ldots, \xi^{i,m}$ from the conditional density $p(x\,; \hat{\theta}^{i-1})$.

Maximization: Compute $\hat{\theta}^i$ as the maximum of the function $\hat{Q}_i(\theta)$ over the feasible set Θ, where

$$\hat{Q}_i(\theta) = \hat{Q}_{i-1}(\theta) + \gamma_i \left\{ \frac{1}{m} \sum_{j=1}^{m} \log f(\xi^{i,j}\,; \theta) - \hat{Q}_{i-1}(\theta) \right\} . \tag{11.20}$$

This algorithm, called the stochastic approximation EM (SAEM) algorithm, was proposed by Cardoso *et al.* (1995) and further analyzed by Delyon *et al.* (1999) and Kuhn and Lavielle (2004). To understand why this algorithm can be cast into the Robbins-Monro framework, consider the simple case where the complete data likelihood is from an exponential family of distributions. In this case, the SAEM algorithm consists in updating, at each iteration, the current estimates $(\hat{S}^i, \hat{\theta}^i)$ of the complete data sufficient statistic and of the parameter. Each iteration of the algorithm is divided into two steps. In a first step, we draw $\xi^{i,1}, \ldots, \xi^{i,m}$ from the conditional density $p(x\,; \hat{\theta}^{i-1})$ and update $\hat{S}^i$ according to

$$\hat{S}^i = \hat{S}^{i-1} + \gamma_i \left[\frac{1}{m} \sum_{j=1}^{m} S(\xi^{i,j}) - \hat{S}^{i-1} \right] . \tag{11.21}$$

In a second step, we compute $\hat{\theta}^i$ as the maximum of the function $\psi^t(\theta)\hat{S}^i - c(\theta)$.

Assume that the function $\psi^t(\theta)s - c(\theta)$ has a single global maximum, denoted $\bar{\theta}(s)$ for all feasible values of $\hat{S}^i$. The difference $m^{-1} \sum_{j=1}^{m} S(\xi^{i,j}) - \hat{S}^{i-1}$ can then be considered as a noisy observation of a function $h(\hat{S}^{i-1})$, where

$$h(s) = \int S(x) p(x\,; \bar{\theta}(s))\, \lambda(dx) - s . \tag{11.22}$$

Thus (11.21) fits into the Robbins-Monro when considering the sufficient statistic s rather than the associated parameter $\bar{\theta}(s)$. This Robbins-Monro procedure searches for the roots of $h(s) = 0$, that is, the values of s satisfying

$$\int S(x) p(x\,; \bar{\theta}(s))\, \lambda(dx) = s .$$

Assume that this equation has a solution $s_\star$ and put $\theta_\star = \bar{\theta}(s_\star)$. Now note that

$$\mathcal{Q}(\theta\,; \theta_\star) = \psi^t(\theta) \int S(x) p(x\,; \theta_\star)\, \lambda(dx) - c(\theta) = \psi^t(\theta) s_\star - c(\theta) ,$$

and by definition the maximum of the right-hand side of this display is obtained at $\theta_\star$. Therefore, an iteration of the EM algorithm started at $\theta_\star$ will stay at $\theta_\star$, and we find that each root $s_\star$ is associated to a fixed point $\theta_\star$ of the EM algorithm.

The SAEM algorithm is simple to implement and has proved to be reasonably successful in different applications. Compared to the stochastic gradient procedure, SAEM inherits from the expectation-maximization algorithm most of the properties that made the success of the EM approach (for instance, the simplicity with which it deals with parameter constraints). One of these properties is invariance with respect to the parameterization. With the SAEM algorithm, the scale of the step sizes $\{\gamma_i\}$ is fixed irrespectively of the parameterization as γ_1 equals 1. As in the case of the stochastic gradient, however, the rate of decrease of the step sizes strongly influences the practical performance of the algorithm. In particular, if the convergence rate of the EM algorithm is already slow, it is unwise to choose fast decreasing step sizes, thereby even further slowing down the method. In contrast, if EM converges fast, then large step sizes introduce an unnecessary amount of extra noise, which should be avoided. Here again, the use of averaging is helpful in reducing the impact of the choice of the rate of decrease of the step sizes.

Example 11.1.6. We implemented the SAEM algorithm for the stochastic volatility model and data described in Example 11.1.2, and the results are displayed in Figure 11.7. In each iteration of the algorithm, a single realization of the missing data was obtained using a sweep of the Gibbs sampler. This draw was used to update the stochastic approximation estimate of the complete data sufficient statistics, which were then used to update the parameter estimate. The only tuning parameter is the sequence of step size γ_n. Here again the theory of stochastic approximation does not tell much about the "optimal" way to choose this sequence. In view of the above discussion, we used slowly decreasing step sizes ($\gamma_n = n^{-0.6}$) to speed up convergence toward the region of interest. As seen in Figure 11.7, the parameters estimates obtained using this implementation of SAEM are rather noisy. In order to reduce the fluctuations, we performed averaging, computing

$$\tilde{\theta}^i = (i - i_0 + 1)^{-1} \sum_{j=i_0}^{i} \hat{\theta}^i , \qquad i \geq i_0 , \tag{11.23}$$

where i_0 was set to 100,000. Averaging is useful only when the parameter approaches convergence and should be turned off during the initial steps of the algorithm. Figure 11.8 shows results for the SAEM algorithm with averaging. Figures 11.7 and 11.8 should be compared with Figures 11.2 and 11.5, respectively, which involve the same sampler and the same overall number of simulations but were obtained using the MCEM strategy. Both procedures (SAEM and MCEM) provides comparable results. ■

11.1.7 Stochastic EM

The stochastic EM (SEM) algorithm is a method that shares many similarities with the stochastic approximation EM algorithm. The SEM algorithm was initially proposed as a means to estimate parameters of mixtures distributions (Celeux and Diebolt, 1985, 1990), but the concept can easily be generalized to cover more general incomplete data models. The basic idea is

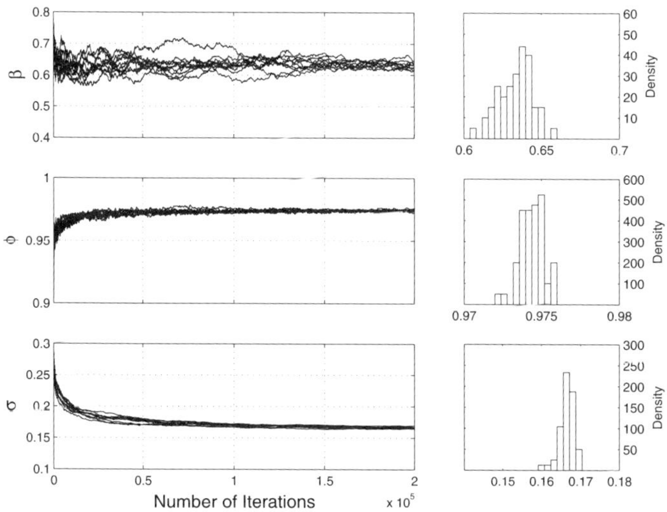

Fig. 11.7. Parameter estimation in the stochastic volatility model with GBP/USD exchange rate data, using the SAEM algorithm with MCMC simulations. The plots show results from 200,000 iterations of the SAEM algorithm with step sizes $\gamma_n = n^{-0.6}$. The contents of the plots are as in Figure 11.2.

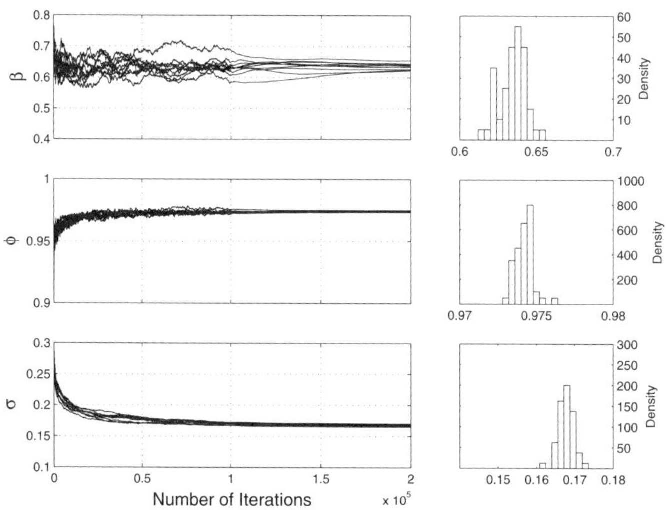

Fig. 11.8. Same model, data, and algorithm as in Figure 11.7, except that averaging was used starting at 100,000 iterations. The plots show results from 200,000 iterations of the SAEM algorithm. The contents of the plots are as in Figure 11.2.

to construct an ergodic homogeneous Markov chain whose stationary distribution is concentrated around the maximum likelihood estimate. SEM is an iterative algorithm in which each iteration proceeds in two steps. In a first step, the stochastic imputation step, the missing data is drawn from the conditional density $p(x\,;\hat{\theta}^{i-1})$, where $\hat{\theta}^{i-1}$ is the current parameter estimate. In a second step, the maximization step, a new parameter estimate $\hat{\theta}^{i}$ is obtained as the maximizer of the complete data likelihood function with the missing data being that imputed in the simulation step. The algorithm thus alternates between simulating (imputing) missing data and computing parameter estimates. In a more general formulation, one may draw several replications of the missing data in the simulation step and use the average of the corresponding complete data log-likelihood functions to obtain a new parameter estimate.

Algorithm 11.1.7 (Stochastic EM Algorithm).

Simulation: Draw $\xi^{i,1},\dots,\xi^{i,m}$ from the conditional density $p(x\,;\hat{\theta}^{i-1})$.

Maximization: Compute $\hat{\theta}^{i}$ as the maximum of the function $\hat{Q}_i(\theta)$ over the feasible set Θ, where

$$\hat{Q}_i(\theta) = \frac{1}{m}\sum_{j=1}^{m} \log f(\xi^{i,j}\,;\theta)\;. \tag{11.24}$$

The main difference between SAEM and SEM is the sequence of decreasing step sizes used in the SAEM approach to smooth the intermediate quantities of EM estimated in successive iterations. In the SEM algorithm, these step sizes are non-decreasing, $\gamma_i = 1$, so there is no averaging of the Monte Carlo error as the iterations progress. The SEM iteration is also obviously identical to the MCEM iteration (see Algorithm 11.1.1) where the difference only lies in the fact that the number of simulated replications of the missing data is not increased with the iteration index.

If $\xi^{i,1},\dots,\xi^{i,m}$ are conditionally independent given $\mathcal{F}^{i-1}$ defined in (11.2), with common density $p(x;\hat{\theta}^{i-1})$, then $\{\theta^i\}$ is a homogeneous Markov chain. Under a set of (rather restrictive) technical conditions, this chain can be shown to be ergodic (Diebolt and Ip, 1996; Nielsen, 2000). Then, as the number of iterations i tends to infinity, the distribution of $\hat{\theta}^{i}$ converges in total variation distance to the distribution of a *random variable* $\hat{\theta}^{\infty}$. The distribution of this random variable is in general difficult to characterize, but, under additional technical assumptions, this stationary distribution may be shown to converge in the sense that as the number of observations increases, it becomes increasingly concentrated around the maximum likelihood estimator (Nielsen, 2000). With SEM, a point estimate can be obtained, for example, by computing sample averages of the simulated parameter trajectories. The theory of the SEM algorithm is difficult even for elementary models, and the available results are far from covering sophisticated setups like continuous state-space HMMs. This is particularly true in situations where imputation of missing data is done using an MCMC algorithm, which clearly adds an addition level of difficulty.

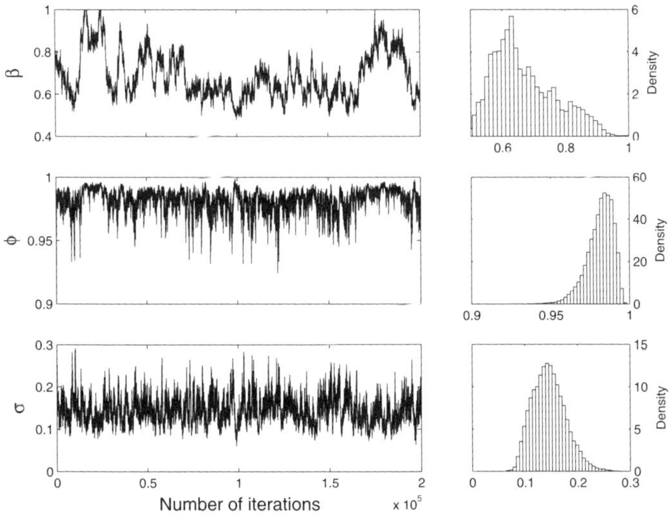

Fig. 11.9. Parameter estimation in the stochastic volatility model with GBP/USD exchange rate data, using an SEM algorithm. The plots show results from 200,000 iterations of the SEM algorithm with a single replication of the missing data imputed in each iteration. Left: 200,000 iterations of a single trajectory of SEM. Right: histograms, computed from the second half of the run, of parameter estimates.

Example 11.1.8. Figure 11.9 displays one trajectory of parameter estimates obtained with the SEM algorithm for the stochastic volatility model and data described in Example 11.1.2, using one sweep of the Gibbs sampler to simulate the unobserved volatility sequence at each iteration.

The histograms of the parameters have a single mode but are highly skewed and show great variability (note that the x-scales are here much larger than in previous figures). The empirical averages for the three parameters are $\beta = 0.687$, $\phi = 0.982$, $\sigma = 0.145$, which do not coincide with the maximum likelihood estimate previously found with other methods (compare with the numbers given at the end of Example 11.1.2). This remains consistent however with the theory developed in Nielsen (2000), as the mismatch is small and, in the current case, probably even less than the order of the random fluctuations due to the use of a finite number of simulations (here 200,000). ■

To conclude this section, we also mention the variant of SEM and MCEM proposed by Doucet *et al.* (2002). This algorithm, which uses concepts borrowed from the Bayesian paradigm, will be presented in Section 13.3.

11.2 Analysis of the MCEM Algorithm

In Section 10.5, the EM algorithm was analyzed by viewing each of its iterations as a mapping M on the parameter space Θ such that the EM sequence of estimates is given by the iterates $\theta^{i+1} = M(\theta^i)$. Under mild conditions, the EM sequence eventually converges to the set of fixed points, $\mathcal{L} = \{\theta \in \Theta : \theta = M(\theta)\}$, of this mapping. EM is an ascent algorithm as each iteration of M increases the observed log-likelihood ℓ, that is, $\ell \circ M(\theta) \geq \ell(\theta)$ for any $\theta \in \Theta$ with equality if and only if $\theta \in \mathcal{L}$. This ascent property is essential in showing that the algorithm converges: it guarantees that the sequence $\{\ell(\theta^i)\}$ is non-decreasing and, hence, convergent if it is bounded.

The MCEM algorithm is an approximation of the EM algorithm. Each iteration of the MCEM algorithm is a perturbed version of an EM iteration, where the "typical size" of the perturbation is controlled by the Monte Carlo error and thus by the number of simulations. The MCEM sequence may thus be written under the form $\hat{\theta}^{i+1} = M(\hat{\theta}^i) + \zeta^{i+1}$, where ζ^{i+1} is the perturbation due to the Monte Carlo approximation. Provided that the number of simulations is increased as the algorithm approaches convergence, the perturbation ζ^i vanishes as $i \to \infty$. Note that the MCEM algorithm is not an ascent algorithm, which prevents us from using the general convergence results of Section 10.5. It is sensible however to expect that the behavior of the MCEM algorithm closely follows that of the EM algorithm, at least for large i, as the random perturbations vanish in the limit.

To prove that this intuition is correct, we first establish in Section 11.2.1 a stability result for deterministically perturbed dynamical systems and then use this result in Section 11.2.2 to deduce a set of conditions implying almost sure convergence of the MCEM algorithm. To avoid entering into too many technicalities, we study convergence under elementary assumptions that do not cover all possible applications of MCEM to maximum likelihood estimation in partially observed models. We feel however that a first exposure to this theory should not be obscured by too many distracting details that will almost inevitably arise when trying to cover more sophisticated cases.

Remark 11.2.1 (Stability in Stochastic Algorithms). One topic of importance that we entirely avoid here is the stability issue. We always assume that it can be independently guaranteed that the sequence of estimates produced by the algorithm deterministically stays in a compact set. Although this will obviously be the case where the parameter space Θ is compact, this assumption may fail to hold in more general settings where the algorithms under study can generate sequences of parameters that either diverge erratically or converge toward the boundary of the parameter space. To circumvent this problem, from both practical and theoretical points of view, it is necessary to modify the elementary recursion of the algorithm, for instance using *re-projections* (Kushner and Yin, 2003; Fort and Moulines, 2003; Andrieu *et al.*, 2005). ■

11.2.1 Convergence of Perturbed Dynamical Systems

Let $T : \Theta \to \Theta$ be a (point-to-point) map on Θ. We study in this section the convergence of the Θ-valued discrete time dynamical system $\theta^{i+1} = T(\theta^i)$

and the perturbed dynamical system $\theta^{i+1} = T(\theta^i) + \zeta^{i+1}$, where $\{\zeta^i\}$ is a deterministic sequence converging to zero. The study of such perturbed dynamical systems was initiated by Kesten (1972), and these results have later been extended by Pierre-Loti-Viaud (1995), Brandière (1998), and Bonnans and Shapiro (1998).

To study the convergence, it is useful to introduce Lyapunov functions associated with the mapping T. A Lyapunov function, as defined below, is equivalent to the concept of ascent function that we met in Section 10.5 when discussing the convergence of EM. The terminology "Lyapunov function" is however more standard, except in numerical optimization texts. Note that Lyapunov functions are traditionally defined as descent functions rather than ascent functions. We reverse this convention to be consistent with the fact that the MLE estimator is defined as the *maximum* of the (log-)likelihood function.

Definition 11.2.2 (Lyapunov Function). *$T : \Theta \to \Theta$ be a map as above and let*

$$\mathcal{L} \stackrel{\text{def}}{=} \{\theta \in \Theta : \theta = T(\theta)\} \tag{11.25}$$

be the set of fixed points of this map. A function $W : \Theta \to \mathbb{R}$ is said to be a Lyapunov function *relative to (T, Θ) if W is continuous and $W \circ T(\theta) \geq W(\theta)$ for all $\theta \in \Theta$, with equality if and only if $\theta \in \mathcal{L}$.*

In other words, the map T is an ascent algorithm for the function W.

Theorem 11.2.3. *Let Θ be an open subset of $\mathbb{R}^{d_\theta}$ and let $T : \Theta \to \Theta$ be a continuous map with set $\mathcal{L}$ of fixed points. Assume that there exists a Lyapunov function W relative to (T, Θ) such that $W(\mathcal{L})$ is a finite set of points. Let $\mathcal{K}$ be a compact set and $\{\theta^i\}$ a $\mathcal{K}$-valued sequence satisfying*

$$\lim_{i\to\infty} |W(\theta^{i+1}) - W \circ T(\theta^i)| = 0 \; . \tag{11.26}$$

Then the set $\mathcal{L} \cap \mathcal{K}$ is non-empty, the sequence $\{W(\theta^i)\}$ converges to a point $w_\star \in W(\mathcal{L} \cap \mathcal{K})$, and the sequence $\{\theta^i\}$ converges to the set $\mathcal{L}_{w_\star} = \{\theta \in \mathcal{L} \cap \mathcal{K} : W(\theta) = w_\star\}$.

The proof of the theorem is based on the following result.

Lemma 11.2.4. *Let $\epsilon > 0$ be a real constant, let $n \geq 1$ be an integer, and let $-\infty < a_1 < b_1 < \ldots < a_n < b_n < \infty$ be real numbers. Let $\{w_j\}$ and $\{e_j\}$ be two sequences such that $\limsup_{j\to\infty} w_j < \infty$, $\lim_{j\to\infty} e_j = 0$ and*

$$w_{j+1} \geq w_j + \epsilon \mathbb{1}_{A^c}(w_j) + e_j \; , \quad \text{where} \quad A \stackrel{\text{def}}{=} \bigcup_{i=1}^{n} [a_i, b_i] \; . \tag{11.27}$$

Then there exists an index $k_\star \in \{1, \ldots, n\}$ such that $a_{k_\star} \leq \liminf w_j \leq \limsup w_j \leq b_{k_\star}$.

Proof. First note that (11.27) implies that the sequence $\{w_j\}$ is infinitely often in the set A (otherwise it would tend to infinity, contradicting the assumptions). Thus it visits infinitely often at least one of the intervals $[a_k, b_k]$ for some k. Choose $\eta < \epsilon \wedge \inf_{1 \le i \le n-1}(a_{i+1} - b_i)/2$ and set j_0 such that $|e_j| \le \eta$ for $j \ge j_0$. Let $p \ge j_0$ such that $w_p \in [a_k, b_k]$. We will show that

$$\text{for any } j \ge p\,, \quad w_j \ge a_k - \eta\,. \tag{11.28}$$

The property is obviously true for $j = p$. Assume now that the property holds true for some $j \ge p$. If $w_j \ge a_k$, then (11.27) shows that $w_{j+1} \ge a_k - \eta$. If $a_k - \eta \le w_j < a_k$, then $w_{j+1} \ge w_j + \epsilon - \eta \ge a_k - \eta$. Therefore $w_{j+1} \ge a_k - \eta$, and (11.28) follows by induction. Because η was arbitrary, we find that $\liminf w_j \ge a_k$. Using a similar induction argument, one may show that $\limsup w_j \le b_k$, which concludes the proof. □

Proof (of Theorem 11.2.3). If $\mathcal{L} \cap \mathcal{K}$ was empty, then $\min_{\theta \in \mathcal{K}} W \circ T(\theta) - W(\theta) > 0$, which would contradict (11.26). Hence $\mathcal{L} \cap \mathcal{K}$ is non-empty. For simplicity, we assume in the following that $\mathcal{L} \subseteq \mathcal{K}$, if not, simply replace $\mathcal{L}$ by $\mathcal{L} \cap \mathcal{K}$.

For any $\alpha > 0$, let $[W(\mathcal{L})]_\alpha \stackrel{\text{def}}{=} \{x \in \mathbb{R} :\ \inf_{y \in W(\mathcal{L})} |x - y| < \alpha\}$. Because $W(\mathcal{L})$ is bounded, the set $[W(\mathcal{L})]_\alpha$ is a *finite* union of disjoint bounded open intervals of length at least equal to 2α. Thus there exists an integer $n_\alpha \ge 0$ and real numbers $a_\alpha(1) < b_\alpha(1) < \ldots < a_\alpha(n_\alpha) < b_\alpha(n_\alpha)$ such that

$$[W(\mathcal{L})]_\alpha = \bigcup_{k=1}^{n_\alpha} (a_\alpha(k), b_\alpha(k))\,. \tag{11.29}$$

Note that $W^{-1}([W(\mathcal{L})]_\alpha)$ is an open neighborhood of $\mathcal{L}$, and define

$$\varepsilon \stackrel{\text{def}}{=} \inf_{\{\theta \in \mathcal{K} \setminus W^{-1}([W(\mathcal{L})]_\alpha)\}} \{W \circ T(\theta) - W(\theta)\} > 0\,. \tag{11.30}$$

Write

$$W(\theta^{i+1}) - W(\theta^i) = \left\{W \circ T(\theta^i) - W(\theta^i)\right\} + \left\{W(\theta^{i+1}) - W \circ T(\theta^i)\right\}. \tag{11.31}$$

Because $W(\theta^i) \notin [W(\mathcal{L})]_\alpha$ implies $\theta^i \notin W^{-1}([W(\mathcal{L})]_\alpha)$, we obtain

$$W(\theta^{i+1}) \ge W(\theta^i) + \varepsilon \mathbb{1}_{[W(\mathcal{L})]_\alpha^c}\left(W(\theta^i)\right) + \left\{W(\theta^{i+1}) - W \circ T(\theta^i)\right\}\,. \tag{11.32}$$

By (11.26), $W(\theta^{i+1}) - W \circ T(\theta^i) \to 0$ as $i \to \infty$. Thus by Lemma 11.2.4, the set of limit points of the sequence $\{W(\theta^i)\}$ belongs to one of the intervals $[a_\alpha(k), b_\alpha(k)]$. Because $W(\mathcal{L}) = \bigcap_{\alpha > 0} [W(\mathcal{L})]_\alpha$ and $W(\mathcal{L})$ is a finite set, the sequence $\{W(\theta^i)\}$ must be convergent with a limit that belongs to $W(\mathcal{L})$. Using (11.31) and (11.26) again, this implies that $W \circ T(\theta^i) - W(\theta^i) \to 0$ as $i \to \infty$, showing that all limit points of the sequence $\{\theta^i\}$ belongs to $\mathcal{L}$. The proof of Theorem 11.2.3 follows. □

11.2.2 Convergence of the MCEM Algorithm

Throughout this section, we focus on the case where the complete data likelihood is from an exponential family of distributions. To keep the discussion short, we also consider only the simplest mechanism to draw the missing data, that is conditionally i.i.d. simulations. Many of the assumptions below can be relaxed, but the proof of convergence then becomes more cumbersome and technical (Fort and Moulines, 2003; Kuhn and Lavielle, 2004).

We recall the notations $f(x;\theta)$ for the complete data likelihood, $\mathrm{L}(\theta) = \int f(x;\theta)\,\lambda(dx)$ for the likelihood, and $p(x;\theta) = f(x;\theta)/\mathrm{L}(\theta)$ for the conditional density of the missing data. We will also need the function

$$\bar{S}(\theta) \stackrel{\text{def}}{=} \int S(x)p(x\,;\theta)\lambda(dx)\;, \tag{11.33}$$

where $S(x)$ is the (vector of) sufficient statistic(s) defined below.

Assumption 11.2.5.

(i) Θ *is an open subset of* $\mathbb{R}^{d_\theta}$ *and* $\{f(\cdot;\theta)\}_{\theta\in\Theta}$ *defines an exponential family of positive functions on* X*, that is,*

$$f(x\,;\theta) = \exp[\psi^t(\theta)S(x) - c(\theta)]h(x) \tag{11.34}$$

for some functions $\psi : \mathbb{R}^{d_\theta} \to \mathbb{R}^{d_s}$, $S : \mathsf{X} \to \mathbb{R}^{d_s}$, $c : \Theta \to \mathbb{R}$, *and* $h : \mathsf{X} \to \mathbb{R}^+$.

(ii) The function L *is positive and continuous on* Θ.

(iii) For any $\theta \in \Theta$, $\int |S(x)|p(x\,;\theta)\,\lambda(dx) < \infty$, *and the function* $\bar{S}$ *is continuous on* Θ.

(iv) There exists a closed subset $\mathcal{S} \subseteq \mathbb{R}^{d_s}$ *that contains the convex hull of* $S(\mathsf{X})$ *and is such that for any* $s \in \mathcal{S}$*, the function* $\theta \mapsto \psi^t(\theta)s - c(\theta)$ *has a unique global maximum* $\bar{\theta}(s) \in \Theta$*. In addition, the function* $\bar{\theta}(s)$ *is continuous on* $\mathcal{S}$.

Under the assumptions and definitions given above, the EM and the MCEM recursions may be expressed as

$$\text{EM:}\quad \theta^{i+1} \stackrel{\text{def}}{=} T(\theta^i) = \bar{\theta}\circ\bar{S}(\theta^i)\;, \qquad \text{MCEM:}\quad \hat{\theta}^{i+1} = \bar{\theta}(\hat{S}^{i+1})\;, \tag{11.35}$$

where $\{\hat{S}^i\}$ are the estimates of the complete data sufficient statistics given, for instance, by (11.3) or by an importance sampling estimate of the same quantity.

Assumption 11.2.6. *With*

$$\mathcal{L} \stackrel{\text{def}}{=} \{\theta \in \Theta : \bar{\theta}\circ\bar{S}(\theta) = \theta\} \tag{11.36}$$

being the set of fixed points of the EM algorithm, the image by the function L *of this set* $\mathcal{L}$ *is a finite set of points.*

Recall that if the function L is continuously differentiable, then $\mathcal{L}$ coincides with the set of stationary points of the log-likelihood. That is, $\mathcal{L} = \{\theta \in \Theta : \nabla_\theta \mathrm{L}(\theta) = 0\}$ (see in particular Theorem 10.5.3).

To study the MCEM algorithm, we now state conditions that specify how $\hat{S}^{i+1}$ approximates $\bar{S}(\hat{\theta}^i)$.

Assumption 11.2.7. $\mathrm{L}[\bar{\theta}(\hat{S}^{i+1})] - \mathrm{L}[\bar{\theta} \circ \bar{S}(\hat{\theta}^i)] \to 0$ *a.s. as* $i \to \infty$.

Theorem 11.2.8. *Assume 11.2.5, 11.2.6, and 11.2.7. Assume in addition that, almost surely, the closure of the set* $\{\hat{\theta}^i\}$ *is a compact subset of* Θ. *Then, almost surely, the sequence* $\{\hat{\theta}^i\}$ *converges to the set* $\mathcal{L}$ *and the sequence* $\{\mathrm{L}(\hat{\theta}^i)\}$ *has a limit.*

Proof. From Proposition 10.1.4, each iteration of the EM algorithm increases the log-likelihood, $\mathrm{L}(\bar{\theta} \circ \bar{S}(\theta)) \geq \mathrm{L}(\theta)$, with equality if and only if $\theta \in \mathcal{L}$ (see (11.36)). Thus L is a Lyapunov function for $T = \bar{\theta} \circ \bar{S}$ on Θ. Because T is continuous by assumption, the proof follows from Theorem 11.2.3. □

Assumption 11.2.7 is not a "‘low-level" assumption. It may be expressed differently, using the conditional version of the Borel-Cantelli Lemma.

Lemma 11.2.9 (Conditional Borel-Cantelli Lemma). *Let* $\{\mathcal{G}_k\}$ *be a filtration and let* $\{\zeta_k\}$ *be an* $\{\mathcal{G}_k\}$*-adapted sequence of random variables. Assume that there exists a constant* C *such that for any* k, $0 \leq \zeta_k \leq C$. *Then if* $\sum_{k=1}^{\infty} \mathrm{E}[\zeta_k \,|\, \mathcal{G}_{k-1}] < \infty$ *a.s., it holds that* $\sum_{k=1}^{\infty} \zeta_k < \infty$ *a.s.*

Proof. Set $M_n = \sum_{k=1}^{n} \{\zeta_k - \mathrm{E}[\zeta_k \,|\, \mathcal{G}_{k-1}]\}$. Then $\{M_n\}$ is a square-integrable $\{\mathcal{G}_n\}$-adapted martingale. The angle-bracket process of this martingale (see Dacunha-Castelle and Duflo, 1986, Section 2.6) is bounded by

$$\langle M \rangle_n \stackrel{\text{def}}{=} \sum_{k=1}^{n} \mathrm{E}[M_k^2 \,|\, \mathcal{G}_{k-1}] - M_{k-1}^2 = \sum_{k=1}^{n} \mathrm{E}[(\zeta_k - \mathrm{E}[\zeta_k \,|\, \mathcal{G}_{k-1}])^2 \,|\, \mathcal{G}_{k-1}]$$
$$\leq C \sum_{k=1}^{n} \mathrm{E}[\zeta_k \,|\, \mathcal{G}_{k-1}] < \infty \quad \text{P-a.s.}$$

The proof is concluded by applying Proposition 2.6.29 of Dacunha-Castelle and Duflo (1986), which shows that $\{M_n\}$ converges a.s. to an a.s. finite random variable. □

We may use the conditional Borel-Cantelli lemma to show that Assumption 11.2.7 is implied by the following sufficient condition, which turns out to be more convenient to check.

Lemma 11.2.10. *Assume 11.2.5 and that the following conditions hold.*

(i) The closure of the set $\{\hat{\theta}^i\}$ *is, almost surely, a compact subset of* Θ.

(ii) For any $\epsilon > 0$ and any compact set $\mathcal{K} \subseteq \Theta$,

$$\sum_{i=1}^{\infty} \mathrm{P}\{|\hat{S}^i - \bar{S}(\hat{\theta}^{i-1})| \geq \epsilon \,|\, \mathcal{F}^{i-1}\} \mathbb{1}_{\mathcal{K}}(\hat{\theta}^{i-1}) < \infty \quad a.s. \,, \tag{11.37}$$

where $\mathcal{F}^j \stackrel{\text{def}}{=} \sigma(\hat{\theta}^0, \hat{S}^1, \ldots, \hat{S}^j)$.

Then Assumption 11.2.7 is satisfied.

Note that the indicator random variable is $\mathcal{F}^{i-1}$-measurable, as $\hat{\theta}^{i-1}$ is a deterministic function (the M-step) of the previous estimate $\hat{S}^{i-1}$ of the sufficient statistic.

Proof. We first prove that for any $\epsilon > 0$ and any compact set $\mathcal{K} \subseteq \Theta$,

$$\sum_{i=1}^{\infty} \mathrm{P}\{|\mathrm{L}[\bar{\theta}(\hat{S}^i)] - \mathrm{L}[\bar{\theta} \circ \bar{S}(\hat{\theta}^{i-1})]| \geq \epsilon \,|\, \mathcal{F}^{i-1}\} \mathbb{1}_{\mathcal{K}}(\hat{\theta}^{i-1}) < \infty \quad \text{a.s.} \tag{11.38}$$

In order to do so, note that for any $\delta > 0$ and $\epsilon > 0$,

$$\begin{aligned}\mathrm{P}\{|\mathrm{L}[\bar{\theta}(\hat{S}^i)] - \mathrm{L}[\bar{\theta} \circ \bar{S}(\hat{\theta}^{i-1})]| \geq \epsilon \,|\, \mathcal{F}^{i-1}\} &\leq \mathrm{P}\{|\hat{S}^i - \bar{S}(\hat{\theta}^{i-1})| \geq \delta \,|\, \mathcal{F}^{i-1}\} \\ &+ \mathrm{P}\{|\mathrm{L}[\bar{\theta}(\hat{S}^i)] - \mathrm{L}[\bar{\theta} \circ \bar{S}(\hat{\theta}^{i-1})]| \geq \epsilon,\ |\hat{S}^i - \bar{S}(\hat{\theta}^{i-1})| \leq \delta \,|\, \mathcal{F}^{i-1}\} \,.\end{aligned}$$

In particular, this inequality holds true on the event $\{\hat{\theta}^{i-1} \in \mathcal{K}\}$. Now define the set $\mathcal{T} = \mathcal{S} \cap \{|s| \leq \sup_{\theta \in \mathcal{K}} \|\bar{S}(\theta)\| + \delta\}$. Because $\bar{S}$ is assumed continuous this set is compact, and therefore the function $\mathrm{L} \circ \bar{\theta}$ is uniformly continuous on $\mathcal{T}$. Hence we can find an $\eta > 0$ such that $|\mathrm{L} \circ \bar{\theta}(s) - \mathrm{L} \circ \bar{\theta}(s')| \leq \epsilon$ for any $(s, s') \in \mathcal{T} \times \mathcal{T}$ such that $|s - s'| \leq \eta$. We thus see that on the on the event $\{\hat{\theta}^{i-1} \in \mathcal{K}\}$,

$$\begin{aligned}&\mathrm{P}\{|\mathrm{L}[\bar{\theta}(\hat{S}^i)] - \mathrm{L}[\bar{\theta} \circ \bar{S}(\hat{\theta}^{i-1})]| \geq \epsilon,\ |\hat{S}^i - \bar{S}(\hat{\theta}^{i-1})| \leq \delta \,|\, \mathcal{F}^{i-1}\} \\ &\qquad\qquad \leq \mathrm{P}\{|\hat{S}^i - \bar{S}(\hat{\theta}^{i-1})| \geq \eta \,|\, \mathcal{F}^{i-1}\} \,.\end{aligned}$$

In view of assumption (ii), (11.38) follows.

Combining (11.38) with Lemma 11.2.9 shows that for any compact set $\mathcal{K} \subseteq \Theta$,

$$\lim_{i \to \infty} |\mathrm{L}[\bar{\theta}(\hat{S}^i)] - \mathrm{L}[\bar{\theta} \circ \bar{S}(\hat{\theta}^{i-1})]| \mathbb{1}_{\mathcal{K}}(\hat{\theta}^{i-1}) = 0 \quad \text{a.s.}$$

The proof is concluded by noting that there exists an increasing sequence $\mathcal{K}_1 \subset \mathcal{K}_2 \subset \cdots$ of compact subsets of Θ such that $\Theta = \bigcup_{n=0}^{\infty} \mathcal{K}_n$. □

As discussed previously, there are many different ways to approximate $\bar{S}(\theta)$. To simplify the discussion, we concentrate below on the simple situation of plain Monte Carlo approximation, assuming that

$$\hat{S}^i = m_i^{-1} \sum_{j=1}^{m_i} S(\xi^{i,j}) , \qquad i \geq 1 , \tag{11.39}$$

where m_i is the number of replications in the ith iteration and $\xi^{i,1}, \ldots, \xi^{i,m_i}$ are conditionally i.i.d. given the σ-field $\mathcal{F}^{i-1}$ with common density $p(x; \hat{\theta}^{i-1})$.

Lemma 11.2.11. *Assume 11.2.5 and that the closure of the set $\{\hat{\theta}^i\}$ is, almost surely, a compact subset of Θ. Assume in addition that $\sum_{i=1}^{\infty} m_i^{-r/2} < \infty$ for some $r \geq 2$ and that $\sup_{\theta \in \mathcal{K}} \int |S(x)|^r p(x\,; \theta)\, \lambda(dx) < \infty$ for any compact set $\mathcal{K} \subseteq \Theta$. Then the MCEM sequence $\{\hat{\theta}^i\}$ based on the estimators $\{\hat{S}^i\}$ of the sufficient statistics given by (11.39) satisfies Assumption 11.2.7.*

Proof. The Markov and the Marcinkiewicz-Zygmund (Theorem 9.1.5) inequalities state that for any $r \geq 2$ and any $\epsilon > 0$,

$$\begin{aligned}
&\sum_{i=1}^{\infty} \mathrm{P}\{|\hat{S}^i - \bar{S}(\hat{\theta}^{i-1})| \geq \epsilon \,|\, \mathcal{F}^{i-1}\} \mathbb{1}_{\mathcal{K}}(\hat{\theta}^{i-1}) \\
&\qquad \leq \epsilon^{-r} \sum_{i=1}^{\infty} \mathrm{E}[|\hat{S}^i - \bar{S}(\hat{\theta}^{i-1})|^r \,|\, \mathcal{F}^{i-1}] \mathbb{1}_{\mathcal{K}}(\hat{\theta}^{i-1}) \\
&\qquad \leq C(r)\epsilon^{-r} \sum_{i=1}^{\infty} m_i^{-r/2} \int |S(x)|^r p(x\,; \hat{\theta}^{i-1}) \lambda(dx)\; \mathbb{1}_{\mathcal{K}}(\hat{\theta}^{i-1}) \\
&\qquad \leq C(r)\epsilon^{-r} \sup_{\theta \in \mathcal{K}} \int |S(x)|^r p(x\,; \theta) \lambda(dx) \sum_{i=1}^{\infty} m_i^{-r/2} ,
\end{aligned}$$

where $C(r)$ is a universal constant. The right-hand side is finite by assumption, so that the conditions of Lemma 11.2.10 are satisfied. □

The situation is slightly more complicated when instead of drawing i.i.d. random variables from the density $p(x\,; \hat{\theta}^{i-1})$, we run an ergodic Markov chain with stationary density $p(x\,; \hat{\theta}^{i-1})$. We then need a version of Marcinkiewicz-Zygmund inequality for ergodic Markov chains (see for instance Fort and Moulines, 2003, Section 6). We will not develop further the theory in this direction. All we need to know at this point is that Assumption 11.2.7 still holds true in this case under reasonable conditions.

11.2.3 Rate of Convergence of MCEM

Recall from Section 10.5.2 that the asymptotic behavior of an EM sequence $\{\theta^i\}$ that converges to a local maximum $\theta_\star$ may be (approximately) described by the linear dynamical system

$$(\theta^{i+1} - \theta_\star) = M(\theta^i) - M(\theta_\star) \approx \nabla_\theta M(\theta_\star)(\theta^i - \theta_\star) , \tag{11.40}$$

where the eigenvalues of $\nabla_\theta M(\theta_\star)$ lie in the interval $(0, 1)$ (Proposition 10.5.5). To use this decomposition, we require some additional regularity assumptions.

Assumption 11.2.12.

(i) The functions ψ and c of the exponential family characterization, $\bar{S}$ and ℓ, are twice continuously differentiable on Θ.
(ii) $\bar{\theta}$ is twice continuously differentiable on the interior of $\mathcal{S}$.
(iii) The set $\mathcal{L}$ of stationary points of ℓ is reduced to a single point $\theta_\star$, which is a proper maximizer of ℓ and such that $s_\star = \bar{S}(\theta_\star)$ lies in the interior of $\mathcal{S}$; the matrices $H(\theta_\star)$ and $G(\theta_\star)$ defined by (10.71) and (10.72) are positive definite.

Note that in exponential families, the form taken by $\ell(\theta)$ (see Definition 10.1.5) and the first assumption above imply that the technical condition (b) in Proposition 10.1.6 holds so that Proposition 10.5.5 applies and $\theta_\star$ is a stable stationary point of the EM mapping. The third condition above is overly restrictive and is adopted only to allow for simpler statements. It is possible to obtain similar results assuming only that $\mathcal{L}$ consists of isolated points by properly conditioning on the events $\{|\hat{\theta}^i - \theta_\star| < \epsilon\}$ for $\theta_\star \in \mathcal{L}$ and arbitrary values of $\epsilon > 0$ (see Fort and Moulines, 2003, for details).

It is useful in the following to consider the EM algorithm not directly in the parameter space Θ but in the space $\mathcal{S}$ of the complete data sufficient statistic. In this space, the EM recursion may be written as

$$S^{i+1} \stackrel{\text{def}}{=} \bar{S} \circ \bar{\theta}(S^i) = G(S^i), \qquad \theta^{i+1} = \bar{\theta}(S^{i+1}) . \tag{11.41}$$

If $\theta_\star$ is a fixed point of M, then $s_\star \stackrel{\text{def}}{=} \bar{S}(\theta_\star)$ is a fixed point of G, that is, $s_\star = G(s_\star) = \bar{S} \circ \bar{\theta}(s_\star)$. In addition, $\nabla_\theta M(\theta_\star) = \nabla_s \bar{\theta}(s_\star)\, \nabla_\theta \bar{S}(\theta_\star)$ and $\nabla_s G(s_\star) = \nabla_\theta \bar{S}(\theta_\star)\, \nabla_s \hat{\theta}(s_\star)$, so that $\nabla_s G(s_\star)$ and $\nabla_\theta M(\theta_\star)$ have the same eigenvalues (counting multiplicities).

We now apply this principle to the MCEM algorithm, letting again $\hat{S}^i$ be the estimate of the sufficient statistic at the ith iteration. The difference $\hat{S}^i - s_\star$, where $s_\star = \bar{S}(\theta_\star)$, may be expressed as

$$\begin{aligned} \hat{S}^i - s_\star &= [G(\hat{S}^{i-1}) - G(s_\star)] + [\hat{S}^i - G(\hat{S}^{i-1})] \\ &= \nabla_s G(s_\star)(\hat{S}^{i-1} - s_\star) + (\hat{S}^i - \mathrm{E}[\hat{S}^i \,|\, \mathcal{F}^{i-1}]) + Q^i , \end{aligned}$$

where $\mathcal{F}^{i-1}$ is as in Lemma 11.2.10 and Q^i is a remainder term. For conditionally i.i.d. simulations, $\hat{S}^i$ is given by (11.39) and hence $\mathrm{E}(\hat{S}^i \,|\, \mathcal{F}^{i-1}) = \int S(x) p(x\,; \bar{\theta}(\hat{S}^{i-1}))\lambda(dx) = G(\hat{S}^{i-1})$. Thus the remainder term Q^i is equal to the difference between $G(\hat{S}^{i-1}) - G(s_\star)$ and its first-order approximation $\nabla_s G(s_\star)(\hat{S}^{i-1} - s_\star)$, which we expect to be small for large values of the iteration index i when $\hat{S}^i$ converges to $s_\star$.

For technical reasons, we consider instead the equivalent error decomposition $\hat{S}^i - s_\star = M^i + R^i$, where M^i obeys a linear difference equation driven by the martingale difference,

$$M^0 = 0 \quad \text{and} \quad M^i = \nabla_s G(s_\star) M^{i-1} + (\hat{S}^i - \mathrm{E}[\hat{S}^i \,|\, \mathcal{F}^{i-1}])\mathbb{1}_{\mathcal{C}}(\hat{\theta}^{i-1}) , \tag{11.42}$$

$\mathcal{C} \subset \Theta$ being a compact neighborhood of $\theta_\star = \bar{\theta}(s_\star)$ and R^i is the remainder term. Because the stationary point $s_\star$ is stable, all eigenvalues of $\nabla_s G(s_\star)$ have modulus less than 1, implying that the linear difference equation (11.42) is stable. To go further, we need to strengthen the assumption on the Monte Carlo perturbation.

Assumption 11.2.13. *$\sum m_i^{-r/2} < \infty$ for some $r \geq 2$ and for any compact set $\mathcal{K} \subset \Theta$, $\limsup_{i\to\infty} m_i^{1/2}(\mathrm{E}\,|\hat{S}^i - \mathrm{E}[\hat{S}^i \mid \mathcal{F}^{i-1}]|^r \mathbb{1}_{\mathcal{K}}(\hat{\theta}^{i-1}))^{1/r} < \infty$.*

This condition implies that

$$\sum_{j=1}^{\infty} \mathrm{E}[|\hat{S}^i - \mathrm{E}\{\hat{S}^i \mid \mathcal{F}^{i-1}\}|^r \mid \mathcal{F}^{i-1}]\mathbb{1}_{\mathcal{K}}(\hat{\theta}^{i-1}) < \infty \quad \text{a.s.}$$

Hence by Markov inequality and Lemma 11.2.10, Assumption 11.2.13 implies Assumption 11.2.7.

The following result (adapted from Fort and Moulines, 2003, Theorem 6), which we state without proof, establishes the rate of convergence of M^i and R^i.

Theorem 11.2.14. *Assume 11.2.5, 11.2.7, 11.2.12, 11.2.13, and that $\hat{S}^i \to s_\star$ a.s. Assume in addition that $1 \leq \lim_i m_{i+1}/m_i < |\lambda_{\max}(\nabla_s G(s_\star))|^{-2}$. Then there exists a constant C such that $(\mathrm{E}\,\|M^i\|^r)^{1/r} \leq C m_i^{-1/2}$ and $m_i^{1/2}(\hat{S}^i - s_\star - M^i) \to 0$ a.s., where M^i is as in (11.42).*

To understand the impact of the schedule $\{m_i\}$ on the dispersion of the MCEM estimate, it is appropriate to evaluate the rate of convergence as a function of the total number of simulations. For any sequence $\{a^i\}$, we define the interpolated sequence $\underline{a^i} = a^{\phi(i)}$, where for any integer i, $\phi(i)$ is the largest integer such that

$$\sum_{k=0}^{\phi(i)} m_k < i \leq \sum_{k=0}^{\phi(i)+1} m_k \; .$$

Hence $\underline{a^i}$ is the original sequence reindexed by simulation number rather than by iteration number. In particular, $\underline{\hat{\theta}^i}$ denotes the fit of the parameter after the ith simulation while, as usual, $\hat{\theta}^i$ is the fit of the parameter after the ith iteration. Assume first that the number of simulations increases at a polynomial rate, $m_i \propto i^\alpha$, for some $\alpha > 0$. Then $\phi(i) \propto [(1+\alpha)i]^{1/(1+\alpha)}$ and $\underline{\hat{\theta}^i} = \theta_\star + O_P(i^{-\frac{\alpha}{2(1+\alpha)}})$. Whatever the value of α, the rate of convergence is slower than $i^{-1/2}$. It is worthwhile to note that the rate improves by choosing large values of α; on the simulation scale, the dispersion of the estimator decreases when increasing α. Assume now that the schedule is exponential, $m_i \propto \rho^i$ for some $\rho > 1$. This choice has been advocated by Chan and Ledolter (1995) and in several earlier works on the subject. We obtain similarly that $\underline{\hat{\theta}^i} = \theta_\star + O_P(i^{-1/2})$ whenever $1 < \rho < |\lambda_{\max}[\nabla_s G(s_\star)]|^{-2}$. This analysis suggests that the optimal schedule is exponential, yet the choice of ρ is not obvious as $\lambda_{\max}[\nabla_s G(s_\star)]$ is in general unknown.

We now study the averaged algorithm based on the use of (11.9). Then $\tilde{S}_i - s_\star$ may be decomposed as $\tilde{S}_i - s_\star = \tilde{M}_i + \tilde{R}_i$, where the leading term $\tilde{M}^i$ is given by

$$\tilde{M}_i \stackrel{\text{def}}{=} \left(\sum_{j=0}^{i} m_j\right)^{-1} \sum_{k=0}^{i} \left(\sum_{j=0}^{i-k} m_{j+k} \nabla_s G(s_\star)^j\right) (\hat{S}^k - \mathrm{E}[\hat{S}^k \,|\, \mathcal{F}^{k-1}]) \,.$$

Fort and Moulines (2003, Theorem 8) shows that the following result holds true.

Theorem 11.2.15. *Assume 11.2.5, 11.2.7, 11.2.12, 11.2.13, and that $\hat{S}^i \to s_\star$ a.s. Assume in addition that the following conditions hold true.*

(i) $1 \le \lim_i m_{i+1}/m_i < |\lambda_{\max}[\nabla_s G(s_\star)]|^{-2}$.
(ii) $\lim_{i\to\infty} i(\sum_{j=0}^{i} m_j)^{-1/2} = 0$.

Then there is a constant C such that

$$(\mathrm{E}\,|\tilde{M}_i|^r)^{1/r} \le C \left(\sum_{j=0}^{i} m_j\right)^{-1/2} ,$$

and

$$\left(\sum_{j=0}^{i} m_j\right)^{1/2} (\tilde{S}^i - s_\star - \tilde{M}_i) \to 0 \quad a.s.$$

The L^r-norm of the leading term $\tilde{M}_i$ of the error $\tilde{S}^i - s_\star$ thus decreases as the inverse square root of the *total* number of simulations up to iteration i, both for subexponential and exponential schedules. This implies that the estimator $\tilde{\theta}^i = \bar{\theta}(\tilde{S}^i)$ converges to $\theta_\star$ at a rate inversely proportional to the square root of the total number of simulations up to iteration i. When expressed on the simulation timescale, the previous result shows that the rate of convergence of the interpolated sequence $\underline{\tilde{\theta}^i}$ is proportional to $i^{-1/2}$, the total number of simulations up to time i. Hence the averaging procedure improves the rate of convergence and makes the choice of the sequence $\{m_i\}$ less sensitive.

11.3 Analysis of Stochastic Approximation Algorithms

11.3.1 Basic Results for Stochastic Approximation Algorithms

Since the early work by Kushner and Clark (1978), convergence of stochastic approximation procedures has been thoroughly studied under various sets of assumptions. For a good summary of available results, we recommend in particular the books by Benveniste *et al.* (1990), Duflo (1997), and Kushner and Yin (2003). In the following, we follow the approach recently proposed by Andrieu *et al.* (2005), which is of interest here because it parallels the method adopted in the previous section for the MCEM algorithm. The analysis again consists in decomposing the study of the convergence of stochastic approximation algorithms in two distinct steps.

In the first step, we establish deterministic conditions on a noise sequence $\{\zeta^i\}$ and a step size sequence $\{\gamma_i\}$ under which a deterministic sequence $\{\theta^i\}$ defined as

$$\theta^0 \in \Theta , \quad \theta^{i+1} = \theta^i + \gamma_{i+1}(h(\theta^i) + \zeta^{i+1}) , \qquad i \geq 0 , \tag{11.43}$$

converges to the set of stationary points of h. This first result (Theorem 11.3.2 below) is the analogy of Theorem 11.2.3, which was instrumental in analyzing the convergence of the MCEM algorithm. Because the proof of Theorem 11.3.2 is more technical, however, it is postponed to Section 11.4 and may be omitted in a first reading.

In a second step, which is probabilistic in nature and depends on the distribution of the process $\{\zeta^i\}$, we check that these conditions are satisfied with probability one.

In order to state Theorem 11.3.2, we first need to adopt a strengthened version of Definition (11.2.2).

Definition 11.3.1 (Differential Lyapunov Function). *Let Θ be a subset of $\mathbb{R}^{d_\theta}$, let w be a real function on Θ, and let $h : \Theta \to \mathbb{R}^{d_\theta}$ be a vector-valued function. The function w is said to be a Lyapunov function relative to (h, Θ) if w is continuously differentiable on Θ and $\langle \nabla_\theta w(\theta), h(\theta) \rangle \geq 0$ for any $\theta \in \Theta$, with equality if and only if θ is such that $h(\theta) = 0$.*

In this context, the function h is usually referred to as the *mean field* and the points θ such that $h(\theta) = 0$ are called stationary points (of the mean field). We will denote by $\mathcal{L}$ the set of such points, that is,

$$\mathcal{L} \stackrel{\text{def}}{=} \{\theta \in \Theta : h(\theta) = 0\} . \tag{11.44}$$

To make the connection with Definition (11.2.2), note that if W is a Lyapunov function relative to T in the sense of Definition (11.2.2) and that both functions are continuously differentiable on Θ, then W also is a (differential) Lyapunov function in the sense of Definition 11.3.1 relative to the *gradient field* $h = \nabla_\theta T$. Recall that we adopt in this chapter a definition that is compatible with maximization tasks, whereas the tradition is to consider Lyapunov functions as descent functions (hence replacing $\geq$ by $\leq$ in Definition 11.3.1).

Theorem 11.3.2. *Assume that Θ is an open subset of $\mathbb{R}^{d_\theta}$ and let $h : \Theta \to \mathbb{R}^{d_\theta}$ be continuous. Let $\{\gamma_i\}$ be a positive sequence such that $\gamma_i \to 0$ and $\sum \gamma_i =$*

∞, *and let* $\{\zeta^i\}$ *be a sequence in* $\mathbb{R}^{d_\theta}$ *satisfying* $\lim_{k\to\infty} \sup_{l\geq k} |\sum_{i=k}^{l} \gamma_i \zeta^i| = 0$. *Assume that there exists a Lyapunov function* w *relative to* (h, Θ) *such that* $w(\mathcal{L})$ *is finite, where* $\mathcal{L}$ *is as in* (11.44). *Finally, assume that the sequence* $\{\theta^i\}_{i\geq 0}$ *given by*

$$\theta^i = \theta^{i-1} + \gamma_i h(\theta^{i-1}) + \gamma_i \zeta^i$$

is such that $\{\theta^i\} \subseteq \mathcal{K}$ *for some compact subset* $\mathcal{K}$ *of* Θ *satisfying* $\mathcal{L} \subseteq \mathcal{K}$.

Then the sequence $\{w(\theta^i)\}$ *converges to some* $w_\star$ *in* $w(\mathcal{L})$ *and the sequence* $\{\theta^i\}$ *converges to the set* $\mathcal{L}_{w_\star} = \{\theta \in \mathcal{L} : w(\theta) = w_\star\}$.

11.3.2 Convergence of the Stochastic Gradient Algorithm

We consider the stochastic gradient algorithm defined by (11.17). For simplicity, we set the number of simulations m in each iteration to one, bringing us back to the basic form (11.15). This recursion may be rewritten in Robbins-Monro form $\hat{\theta}^i = \hat{\theta}^{i-1} + \gamma_i h(\hat{\theta}^{i-1}) + \gamma_i \zeta^i$, where

$$h(\theta) = \nabla_\theta \ell(\theta) , \qquad \zeta^i = \nabla_\theta \log f(\xi^i ; \hat{\theta}^{i-1}) - h(\hat{\theta}^{i-1}) . \tag{11.45}$$

Because the mean field h is a gradient, the function $w = \ell$ is a Lyapunov function relative to (Θ, h). To proceed, one needs to specify how the missing data is simulated. We consider the following simple assumption.

Assumption 11.3.3. *For any* $i \geq 1$, *given* $\mathcal{F}^{i-1} = \sigma(\hat{\theta}^0, \xi^1, \ldots, \xi^{i-1})$, *the simulated missing data* ξ^i *is drawn from the density* $p(x ; \hat{\theta}^{i-1})$.

In addition, for some $r > 2$, *the function* $\int |S(x)|^r p(x ; \theta)\, \lambda(dx)$ *is finite and continuous on* Θ.

This assumption can be relaxed to allow for Markovian dependence, a situation that is typical when MCMC methods are used for simulation of the missing data (Andrieu *et al.*, 2005). We may now formulate a general convergence result for the stochastic gradient algorithm under the assumption that the complete data likelihood is from an exponential family of distributions. Note that in the latter case, the representation $f(x ; \theta) = \exp[\psi^t(\theta) S(x) - c(\theta)] h(x)$ implies that the perturbation ζ^i defined in (11.45) may be rewritten as $\zeta^i = [\nabla_\theta \psi(\hat{\theta}^{i-1})]^t (\hat{S}^i - \mathrm{E}[\hat{S}^i \mid \mathcal{F}^{i-1}])$, where $\nabla_\theta \psi(\theta)$ is the Jacobian matrix of ψ and $\hat{S}^i = S(\xi^i)$ is a simulation of the complete data sufficient statistics under the density $p(x ; \hat{\theta}^{i-1})$.

Theorem 11.3.4. *Assume 11.2.5, 11.2.6, and 11.3.3. Assume in addition that* $\ell(\theta)$ *is a continuously differentiable function of* θ, *that*

$$\gamma_k \geq 0 , \sum \gamma_k = \infty \quad \textit{and} \quad \sum \gamma_k^2 < \infty ,$$

and that the closure of the set $\{\hat{\theta}^i\}$ *is a compact subset of* Θ. *Then, almost surely, the sequence* $\hat{\theta}^i$ *given by* (11.15) *satisfies* $\lim_{k\to\infty} \nabla_\theta \ell(\hat{\theta}_k) = 0$.

Proof. Put $M^i = \sum_{j=1}^{i} \gamma_j \zeta^j$. The result will follow from Theorem 11.3.2 provided $\{M^i\}$ has a finite limit a.s., so this is what we will prove.

Using the form of ζ^i given above, we see that the sequence $\{M^i\}$ is an $\{\mathcal{F}^i\}$-martingale satisfying

$$\sum_{i=1}^{\infty} \mathrm{E}[|M^{i+1} - M^i|^2 \,|\, \mathcal{F}^i] \leq \sum_{i=1}^{\infty} \gamma_i^2 \|\nabla_\theta \psi(\hat{\theta}^{i-1})\|^2 \int |S(x)|^2 p(x; \hat{\theta}^{i-1})\, \lambda(dx) \; .$$

Under the stated assumptions the sequence $\{\hat{\theta}^i\}$ a.s. belongs to a compact subset of Θ. Therefore, by Assumption 11.3.3, the right-hand side of the above display is finite a.s., and Dacunha-Castelle and Duflo (1986, Proposition 2.6.29) then shows that M^i has a finite limit almost surely. □

11.3.3 Rate of Convergence of the Stochastic Gradient Algorithm

The results above are of little help in selecting the step size sequence, because they do not tell much about the behavior of the sequence $\{\hat{\theta}^i\}$ when the algorithm approaches convergence. This section is concerned with the rate of convergence, assuming that convergence occurs. To simplify the discussion it is assumed here that, as in Section 11.2.3, $\hat{\theta}^i \to \theta_\star$, which is a *stable stationary point*. That is, a point $\theta_\star$ in Θ satisfying the following conditions: (i) $h(\theta_\star) = 0$, (ii) h is twice differentiable in a neighborhood of $\theta_\star$ and (iii) $J(\theta_\star)$, the Jacobian matrix of h, or, in other words, the Hessian of $\ell(\theta_\star)$, is negative definite. All this is guaranteed by Assumption 11.2.12, under which $\theta_\star$ is a proper maximizer of ℓ.

Write the difference $\hat{\theta}^i - \theta_\star$ as

$$\begin{aligned} \hat{\theta}^i - \theta_\star &= (\hat{\theta}^{i-1} - \theta_\star) + \gamma_i [h(\hat{\theta}^{i-1}) - h(\theta_\star)] + \gamma_i \zeta^i \\ &= (\hat{\theta}^{i-1} - \theta_\star) + \gamma_i J(\theta_\star)(\hat{\theta}^{i-1} - \theta_\star) + \gamma_i \zeta^i + \gamma_i Q^i \; , \end{aligned}$$

where $Q^i = [h(\hat{\theta}^{i-1}) - h(\theta_\star)] - J(\theta_\star)(\hat{\theta}^{i-1} - \theta_\star)$ is the remainder term. This suggests the error decomposition $\hat{\theta}^i - \theta_\star = M^i + R^i$, where M^i obeys a linear difference equation driven (under Assumption 11.3.3) by a martingale difference; $M^0 = 0$ and, for $i \geq 1$,

$$M^i = [I + \gamma_i J(\theta_\star)] M^{i-1} + \gamma_i \zeta^i = \sum_{j=0}^{i} \gamma_j \prod_{l=j+1}^{i} [I + \gamma_l J(\theta_\star)] \zeta^j \; . \tag{11.46}$$

The following result is adapted from Delyon *et al.* (1999, Lemma 6) (see also Kushner and Yin, 2003, Chapter 10).

Theorem 11.3.5. *Assume 11.2.5, 11.2.12, 11.3.3, and that $\hat{\theta}^i \to \theta_\star$ a.s. Assume in addition that $\sum_{i=0}^{\infty} \gamma_i = \infty$, $\sum_{i=0}^{\infty} \gamma_i^2 < \infty$ and that $\gamma_{i+1}^{-1} - \gamma_i^{-1} \to 0$. Then there exists a constant C such that $(\mathrm{E}[\|M^i\|^r])^{1/r} \leq C\gamma_i$ and $\gamma_i^{-1/2}(\hat{\theta}^i - \theta_\star - M^i) \to 0$ a.s., where M^i is as in* (11.46).

Hence M^i is the leading term of the error and A^i is a remainder term. Because the variance of the leading term M^i is proportional to the step size γ_i, this result suggests taking the smallest possible step size compatible with the assumptions. Using "small" step sizes is however clearly not a recommendable practice. Indeed, if the step sizes are not sufficient, it is likely that the algorithm will get stuck at an early stage, failing to come close to the target point. In addition, it is difficult to detect that the step size is converging too quickly to zero, or that it is too small, and therefore there is a substantial ambiguity on how to select an appropriate sequence of step sizes. This difficulty has long been considered as a serious handicap for practical applications of stochastic approximation procedures.

Note that it is possible to carry out a different analysis of stochastic approximation procedures in which the error $\hat{\theta}^i - \theta_\star$ is normalized by the square root of the inverse of the step size γ_i. One may for example prove convergence in distribution of the centered and normalized iterate $\gamma_i^{-1/2}(\hat{\theta}^i - \theta_\star)$, with the variance of the limiting distribution taken as a measure of how fast convergence occurs (Benveniste *et al.*, 1990; Duflo, 1997). It is also possible to analyze scenarios in which the step sizes are essentially constant but assumed sufficiently small (Kushner and Yin, 2003) or to use approaches based on large deviations (Dupuis and Ellis, 1997).

As in the case of MCEM, the averaging procedure partly raises the difficulty discussed above: for the averaged sequence $\{\tilde{\theta}^i\}$ defined in (11.19), the following result, adapted from Delyon *et al.* (1999, Theorem 4), holds.

Theorem 11.3.6. *Under the assumptions of Theorem 11.3.5,*

$$\sqrt{i}(\tilde{\theta}^i - \theta_\star) \xrightarrow{\mathcal{D}} \mathrm{N}(0, H(\theta_\star)^{-1} \Sigma_\star H(\theta_\star)^{-1}) , \tag{11.47}$$

where

$$\Sigma_\star = \psi^t(\theta^*) \int [S(x) - \bar{S}(\theta_\star)][S(x) - \bar{S}(\theta_\star)]^t p(x\,; \theta_\star)\, \lambda(dx)\, \psi(\theta_\star) .$$

As shown by Poznyak and Chikin (1984) and Chikin (1988), the rate $1/\sqrt{i}$ and the asymptotic variance of (11.47) are optimal. This performance may also be achieved using a Gauss-Newton type stochastic approximation algorithm. Such an algorithm would however require knowledge, or estimates of $H(\theta_\star)$, whereas averaging circumvents such difficulties. This result suggests a rather different philosophy for setting the step sizes: because the optimal rate of $1/\sqrt{i}$ can be achieved by averaging, the step sizes $\{\gamma_i\}$ should decrease as slowly as permitted by the assumptions of Theorem 11.3.5 to ensure fast convergence toward the region of interest (hence the choice of a rate $n^{-0.6}$ adopted in Example 11.1.6).

11.3.4 Convergence of the SAEM Algorithm

We consider the stochastic approximation EM (SAEM) algorithm (11.21) and again, for simplicity, with $m = 1$ replication of the missing data in each iteration. In Robbins-Monro form, this algorithm is defined as $\hat{S}^i = S^{i-1} + \gamma_i h(\hat{S}^{i-1}) + \gamma_i \zeta^i$, where the mean field h and the perturbation ζ^i are given by

$$h(s) = \bar{S} \circ \bar{\theta}(s) - s \;, \qquad \zeta^i = S(\xi^i) - \bar{S} \circ \bar{\theta}(\hat{S}^{i-1}) \;. \tag{11.48}$$

The log-likelihood function $\ell(\theta)$ is increased at each iteration of the EM algorithm. We show in the following lemma that this property, in the domain of complete data sufficient statistics, implies that $\ell \circ \bar{\theta}$ is a Lyapunov function for the mean field h.

Lemma 11.3.7. *Assume 11.2.5, items (i) and (ii) of 11.2.12 and set* $w \stackrel{\text{def}}{=} \ell \circ \bar{\theta}$. *Then* $\langle \nabla_s w(s), h(s) \rangle \geq 0$ *for any* $s \in \mathcal{S}$, *where* h *is the mean field of* (11.48). *Moreover,*

$$\{s \in \mathcal{S} : \langle \nabla_s w(s), h(s) \rangle = 0\} = \{s \in \mathcal{S} : \nabla_s w(s) = 0\} \;, \tag{11.49}$$
$$\bar{\theta}(\{s \in \mathcal{S} : \langle \nabla_s w(s), h(s) \rangle = 0\}) = \{\theta \in \Theta : \nabla_\theta \ell(\theta) = 0\} \;. \tag{11.50}$$

Proof. We start by working out an expression for the gradient of w. Under Assumption 11.2.12, the function $\bar{S}$ is continuously differentiable on Θ and the function $\bar{\theta}$ is continuously differentiable on $\mathcal{S}$. Hence h is continuously differentiable on $\mathcal{S}$, so that h is bounded on every compact subset of $\mathcal{S}$. By construction for any $s \in \mathcal{S}$, the function $\bar{\theta}$ satisfies

$$-\nabla_\theta c[\bar{\theta}(s)] + s^t \nabla_\theta \psi[\bar{\theta}(s)] = 0 \;. \tag{11.51}$$

Put $F(s, \theta) = \psi^t(\theta) s - c(\theta)$, so that this relation reads $\nabla_\theta F[s\,; \bar{\theta}(s)] = 0$. Under the assumptions made, we may differentiate this relation with respect to s to obtain

$$\nabla_\theta^2 F[s\,; \bar{\theta}(s)] \, \nabla_s \bar{\theta}(s) = -\nabla_\theta \psi[\bar{\theta}(s)] \;. \tag{11.52}$$

On the other hand, the Fisher identity implies that for any θ,

$$\nabla_\theta \ell(\theta) = -\nabla_\theta c(\theta) + \bar{S}(\theta)^t \nabla_\theta \psi(\theta) \;.$$

Evaluating this equality at $\bar{\theta}(s)$ and using (11.51) yields

$$\begin{aligned} \nabla_\theta \ell[\bar{\theta}(s)] &= \{-s + \bar{S}[\bar{\theta}(s)]\}^t \, \nabla_\theta \psi[\bar{\theta}(s)] \\ &= h(s)^t \, \nabla_\theta \psi[\bar{\theta}(s)] = -h(s)^t \, \nabla_s \bar{\theta}(s)^t \nabla_\theta^2 F[s\,; \bar{\theta}(s)] \;, \end{aligned} \tag{11.53}$$

whence

$$\nabla_s \ell \circ \bar{\theta}(s) = -h(s)^t \, \nabla_s \bar{\theta}(s)^t \, \nabla_\theta^2 F[s\,; \bar{\theta}(s)] \, \nabla_s \bar{\theta}(s) \;. \tag{11.54}$$

Because the $F(s; \theta)$ as a unique proper maximizer in $\theta = \bar{\theta}(s)$, $\nabla_\theta^2 F[s\,; \bar{\theta}(s)]$ is negative definite implying that

$$\langle \nabla_s w(s), h(s) \rangle = -h(s)^t \, \nabla_s \bar{\theta}(s)^t \, \nabla_\theta^2 F[s\,; \bar{\theta}(s)] \, \nabla_s \bar{\theta}(s) \, h(s) \geq 0 \;. \tag{11.55}$$

This is the first claim of the lemma.

Now pick $s_\star \in \mathcal{S}$ to be such that $\langle \nabla w(s_\star), h(s_\star) \rangle = 0$. Under Assumption 11.2.12, the matrix $\nabla^2_\theta F[s_\star\,; \bar{\theta}(s_\star)]$ is negative definite, whence (11.55) shows that $\nabla_s \bar{\theta}(s_\star)\, h(s_\star) = 0$. Inserting this into (11.54) yields $\nabla_s w(s_\star) = 0$, so that

$$\{s \in \mathcal{S} : \langle \nabla_s w(s), h(s) \rangle = 0\} \subseteq \{s \in \mathcal{S} : \nabla_s w(s) = 0\} \; .$$

The reverse inclusion is trivial, and the second claim of the lemma follows. For the final claim, use a similar argument and (11.53) as well as the fact that if $\nabla_\theta \ell(\theta_\star) = 0$ then $h(s_\star) = \bar{S} \circ \bar{\theta}(s_\star) - s_\star = 0$ (for the point $s_\star = \bar{S}(\theta_\star)$). □

We may now formulate a result that is the stochastic counterpart of the general convergence theorem for the EM sequence.

Theorem 11.3.8. *Let $\{\hat{\theta}^i\}$ and $\{\hat{S}^i\}$ be sequences of parameters and complete sufficient statistics, respectively, of the SAEM algorithm* (11.21)*. Assume 11.2.5, 11.2.6, and items (i) and (ii) of 11.2.12 and 11.3.3. Assume in addition that*

$$\gamma_k \geq 0 \; , \; \sum \gamma_k = \infty \quad \text{and} \quad \sum \gamma_k^2 < \infty \; ,$$

and that the closure of the set $\{\hat{S}^i\}$ is a compact subset of $\mathcal{S}$. Then, almost surely, $\lim_{i\to\infty} h(\hat{S}^i) = 0$ and $\lim_{i\to\infty} \nabla_\theta \ell(\hat{\theta}^i) = 0$.

The proof is similar to the one of Theorem 11.3.4 and is omitted.

11.4 Complements

We give below the proof of Theorem 11.3.2, which was omitted in Section 11.3. We first need three lemmas for which the assumptions of Theorem 11.3.2 are assumed to hold.

Lemma 11.4.1. *Let $\mathcal{J} \subset \Theta$ be a compact subset of Θ such that $0 < \inf_{\theta\in\mathcal{J}} \langle \nabla_\theta w(\theta), h(\theta) \rangle$. Then, for any $0 < \delta < \inf_{\theta\in\mathcal{J}} \langle \nabla_\theta w(\theta), h(\theta) \rangle$, there exist constants $\lambda > 0$ and $\beta > 0$, such that, for any γ, $0 \leq \gamma \leq \lambda$, ζ, $|\zeta| \leq \beta$, and $\theta \in \mathcal{J}$,*

$$w[\theta + \gamma h(\theta) + \gamma\zeta] \geq w(\theta) + \gamma\delta \; .$$

Proof. For any $0 < \delta < \inf_{\theta\in\mathcal{J}} \langle \nabla_\theta w, h \rangle$, there exist $\lambda > 0$ and $\beta > 0$ such that for all γ, $0 \leq \gamma \leq \lambda$, ζ, $|\zeta| \leq \beta$ and t, $0 \leq t \leq 1$, we have for all $\theta \in \mathcal{J}$, $\theta + \gamma t h(\theta) + \gamma t \zeta \in \Theta$ and

$$|\langle \nabla_\theta w(\theta), h(\theta) \rangle - \langle \nabla_\theta w[\theta + \gamma t h(\theta) + \gamma t \zeta], h(\theta) + \zeta \rangle| \leq \inf_{\theta \in \mathbb{R}^{d_\theta} \setminus \mathcal{W}} \langle \nabla_\theta w, h \rangle - \delta .$$

Then, for any γ, $0 \leq \gamma \leq \lambda$ and ζ, $|\zeta| \leq \beta$,

$$
\begin{aligned}
w(\theta + \gamma h(\theta) + \gamma\zeta) - w(\theta) &= \gamma \langle \nabla_\theta w(\theta), h(\theta) \rangle \\
&\quad + \gamma \int_0^1 \{ \langle \nabla_\theta w[\theta + t\gamma h(\theta) + t\gamma\zeta], h(\theta) + \zeta \rangle - \langle \nabla_\theta w(\theta), h(\theta) \rangle \} \, dt \\
&\geq \gamma \inf_{\theta \in \mathbb{R}^{d_\theta} \setminus \mathcal{W}} \langle \nabla_\theta w, h \rangle - \gamma \left(\inf_{\theta \in \mathbb{R}^{d_\theta} \setminus \mathcal{W}} |\langle \nabla_\theta w, h \rangle| - \delta \right) \\
&= \gamma\delta \, .
\end{aligned}
$$

□

Lemma 11.4.2. *Let $\mathcal{N} \subset \Theta$ be an open neighborhood of $\mathcal{L}$. There exist positive constants δ, C, ε, and λ (depending only on the sets $\mathcal{N}$ and $\mathcal{K}$), such that for any δ', $0 < \delta' \leq \delta$, λ', $0 < \lambda' \leq \lambda$, one can find an integer N and a sequence $\{\bar{\theta}_j\}_{j \geq N}$ satisfying $\bar{\theta}_j \in \Theta$ for any $j \geq N$ and*

$$
\sup_{j \geq N} |\theta_j - \bar{\theta}_j| \leq \delta' \, , \quad \sup_{j \geq N} \gamma_j \leq \lambda' \, , \quad \text{and} \quad \sup_{j \geq N} |w(\theta_j) - w(\bar{\theta}_j)| \leq \eta \, , \tag{11.56}
$$

$$
w(\bar{\theta}_j) \geq w(\bar{\theta}_{j-1}) + \gamma_j \varepsilon \, \mathbb{1}_{\mathcal{N}^c}(\bar{\theta}_{j-1}) - \gamma_j C \, \mathbb{1}_{\mathcal{N}}(\bar{\theta}_{j-1}) \quad \text{for} \quad j \geq N+1. \tag{11.57}
$$

Proof. Let us choose $\delta_0 > 0$ small enough so that

$$
\mathcal{K}_{\delta_0} \stackrel{\text{def}}{=} \{\theta \in \Theta, \inf_{\theta' \in \mathcal{K}} |\theta - \theta'| \leq \delta_0\} \subset \Theta \, .
$$

The set $\mathcal{K}_{\delta_0} \setminus \mathcal{N}$ is compact and $\inf_{\mathcal{K}_{\delta_0} \setminus \mathcal{N}} \langle \nabla w, h \rangle > 0$. By Lemma 11.4.1, for any ε, $0 < \varepsilon < \inf_{\theta \in \mathcal{K}_{\delta_0} \setminus \mathcal{N}} \langle \nabla w(\theta), h(\theta) \rangle$, one may choose $\lambda > 0$ and $\beta > 0$ small enough so that for any γ, $0 \leq \gamma \leq \lambda$, ζ, $|\zeta| \leq \beta$ and $\theta \in \mathcal{K}_{\delta_0} \setminus \mathcal{N}$, $\theta + \gamma h(\theta) + \gamma\zeta \in \Theta$ and

$$
w[\theta + \gamma h(\theta) + \gamma\zeta] \geq w(\theta) + \gamma\varepsilon \, . \tag{11.58}
$$

Because the function h is continuous on Θ, it is uniformly continuous on each compact subset of Θ, i.e., for any $\eta > 0$ one may choose δ, $0 < \delta \leq \lambda \left\| h \mathbb{1}_{\mathcal{K}} \right\|_\infty \wedge \delta_0$ so that for all $(\theta, \bar{\theta}) \in \mathcal{K}_{\delta_0} \times \mathcal{K}_{\delta_0}$ satisfying $|\theta - \bar{\theta}| \leq \delta$,

$$
|h(\theta) - h(\bar{\theta})| \leq \beta \quad \text{and} \quad |w(\theta) - w(\bar{\theta})| \leq \eta \, . \tag{11.59}
$$

Under the stated conditions for any δ', $0 < \delta' \leq \delta$ and λ', $0 < \lambda' \leq \lambda$ there exists an integer N such that for any $j \geq N+1$, $\gamma_j \leq \lambda'$ and $\left| \sum_{i=N+1}^{j} \gamma_i \zeta^i \right| \leq \delta'$. Define recursively for $j \geq N$ the sequence $\{\bar{\theta}^j\}_{j \geq N}$ as follows: $\bar{\theta}^N = \theta^N$ and for $j \geq N+1$,

$$
\bar{\theta}^j = \bar{\theta}^{j-1} + \gamma_j h(\theta^{j-1}) \, . \tag{11.60}
$$

By construction, for $j \geq N+1$, $\bar{\theta}^j - \theta^j = \sum_{i=N+1}^{j} \gamma_i \zeta^i$, which implies that $\sup_{j \geq N} |\bar{\theta}^j - \theta^j| \leq \delta'$ and thus, for all $j \geq N$, $\bar{\theta}^j \in \mathcal{K}_{\delta_0}$ and $|w(\theta^j) - w(\bar{\theta}^j)| \leq \eta$. On the other hand, for $j \geq N+1$,

$$\bar{\theta}^j = \bar{\theta}^{j-1} + \gamma_j h(\bar{\theta}^{j-1}) + \gamma_j [h(\theta^{j-1}) - h(\bar{\theta}^{j-1})] \,, \tag{11.61}$$

and because $|\bar{\theta}^{j-1} - \theta^{j-1}| \leq \delta' \leq \delta$, (11.59) shows that $|h(\theta^{j-1}) - h(\bar{\theta}^{j-1})| \leq \beta$. Thus, (11.58) implies that, whenever $\bar{\theta}^{j-1} \in \mathcal{K}_{\delta_0} \setminus \mathcal{N}$, $w(\bar{\theta}^j) \geq w(\bar{\theta}^{j-1}) + \gamma_j \varepsilon$. Now (11.59) and (11.60) imply that, for any $j \geq N$,

$$|w(\bar{\theta}^j) - w(\bar{\theta}^{j-1})| \leq \gamma_j \ \|\nabla_\theta w \mathbb{1}_\mathcal{K}\|_\infty \|h \mathbb{1}_\mathcal{K}\|_\infty \ .$$

□

Lemma 11.4.3. *Let ε and C be real constants, n be an integer and let $-\infty < a_1 < b_1 < \cdots < a_n < b_n < \infty$ be real numbers. Let $\{u_j\}$ be a sequence such that* $\limsup u_j < \infty$ *and, for any j,*

$$u_j \geq u_{j-1} + \gamma_j \varepsilon \mathbb{1}_{A^c}(u_{j-1}) - \gamma_j C \mathbb{1}_A(u_{j-1}) \qquad A = \bigcup_{i=1}^n [a_i, b_i] \,. \tag{11.62}$$

Then, the limit points of the sequence $\{u_j\}$ are included in A.

Proof. As $\limsup u_j < \infty$ is bounded, $\{u_j\}$ is infinitely often in the set A and thus in at least one of the intervals $[a_k, b_k]$, $k = 1, \ldots, n$. Choose η, $0 < \eta < \inf_{1 \leq i \leq n-1}(a_{i+1} - b_i)/2$ and let J be sufficiently large so that, for all $j \geq J$, $\gamma_j C \leq \eta$. Assume that $\{u_i\}$ is infinitely often in the interval $[a_k, b_k]$, for some $k = 1, \ldots, n$. Let $p \geq J$ be such that $u_p \in [a_k, b_k]$. We will show by induction that,

$$\text{for any } j \geq p \,, u_j \geq a_k - \eta \,. \tag{11.63}$$

The property is obviously true for $j = p$. Assume now that the property holds true for some $j \geq p$. If $u_j \geq a_k$, then, $u_{j+1} \geq a_k - \eta$. If $a_k - \eta \leq u_j \leq a_k$, then $u_{j+1} \geq u_j + \gamma_j \varepsilon \geq a_k - \eta$, showing (11.63). Because η is arbitrary, $\liminf u_j \geq a_k$, showing that the sequence $\{u_j\}$ is infinitely often in only one of the intervals. Hence, there exists an index j_0 such that, for any $j \geq j_0$, $u_j < a_{k+1}$ (with the convention that $a_{n+1} = \infty$), which is possible only if, for any $j \geq j_0$, $u_j < b_k$. As a consequence, there cannot be an accumulation point in an interval other than $[a_k, b_k]$. □

Proof (Theorem 11.3.2). We first prove that $\lim_{j \to \infty} w(\theta^j)$ exists. For any $\alpha > 0$, define the set $[w(\mathcal{L})]_\alpha = \{x \in \mathbb{R} : \inf_{y \in w(\mathcal{L})} |x - y| < \alpha\}$. Because $\|w \mathbb{1}_\mathcal{L}\|_\infty < \infty$, $[w(\mathcal{L})]_\alpha$ is a finite union of disjoint intervals of length at least equal to 2α. By applying Lemma 11.4.2 with $\mathcal{N} = w^{-1}([w(\mathcal{L})]_\alpha)$, there exist positive constants C, δ, ε, λ, such that for any δ', $0 < \delta' \leq \delta$, λ', $0 < \lambda' \leq \lambda$ and $\eta > 0$, one may find an integer N and a sequence $\{\bar{\theta}^j\}_{j \geq N}$ such that,

$$\sup_{j \geq N} |\theta^j - \bar{\theta}^j| \leq \delta' \,, \sup_{j \geq N} \gamma_j \leq \lambda' \quad \text{and} \quad \sup_{j \geq N} |w(\theta^j) - w(\bar{\theta}^j)| \leq \eta$$

and, for any $j \geq N + 1$,

$$w(\bar{\theta}^j) \geq w(\bar{\theta}^{j-1}) + \gamma_j \varepsilon \mathbb{1}_{[w(\mathcal{L})]_\alpha^c}[w(\bar{\theta}^{j-1})] - \gamma_j C \mathbb{1}_{[w(\mathcal{L})]_\alpha}[w(\bar{\theta}^{j-1})] ,$$

By Lemma 11.4.3, the limit points of the sequence $\{w(\bar{\theta}^j)\}$ are in $[w(\mathcal{L})]_\alpha$ and because $\sup_{j \geq N} |w(\theta^j) - w(\bar{\theta}^j)| \leq \eta$, the limit points of the sequence $\{w(\theta^j)\}$ belong to $[w(\mathcal{L})]_{\alpha+\eta}$. Because α and η are arbitrary, this implies that the limit points of the sequence $\{w(\theta^j)\}$ are included in $\bigcap_{\alpha>0}[w(\mathcal{L})]_\alpha$. Because $w(\mathcal{L})$ is finite, $w(\mathcal{L}) = \bigcap_{\alpha>0}[w(\mathcal{L})]_\alpha$ showing that the limit points of $\{w(\theta^j)\}$ belong to the set $w(\mathcal{L})$.

On the other hand, $\limsup_{j\to\infty} |w(\theta^j) - w(\theta^{j-1})| = 0$, which implies that the set of limit points of $\{w(\theta^j)\}$ is an interval. Because $w(\mathcal{L})$ is finite, the only intervals included in $w(\mathcal{L})$ are isolated points, which shows that the limit $\lim_{j\to\infty} w(\theta^j)$ exists.

We now proceed to proving that all the limit points of the sequence $\{\theta^j\}$ belong to $\mathcal{L}$. Let $\mathcal{N}$ be an arbitrary neighborhood of $\mathcal{L}$. From Lemma 11.4.2, there exist constants C, $\delta > 0$, $\varepsilon > 0$, $\lambda > 0$ such that for any δ', $0 < \delta' \leq \delta$, λ', $0 < \lambda' \leq \lambda$, and $\eta > 0$, one may find an integer N and a sequence $\{\bar{\theta}^j\}_{j \geq N}$ such that

$$\sup_{j\geq N} |\theta^j - \bar{\theta}^j| \leq \delta', \quad \sup_{j\geq N} \gamma_j \leq \lambda' \quad \text{and} \quad \sup_{j\geq N} |w(\theta^j) - w(\bar{\theta}^j)| \leq \eta$$

and, for any $j \geq N+1$,

$$w(\bar{\theta}^j) \geq w(\bar{\theta}^{j-1}) + \gamma_j \varepsilon \mathbb{1}_{\mathcal{N}^c}(\bar{\theta}^{j-1}) - \gamma_j C \mathbb{1}_{\mathcal{N}}(\bar{\theta}^{j-1}) .$$

For $j \geq N$, define $\tau(j) = \inf\left\{k \geq 0, \bar{\theta}^{k+j} \in \mathcal{N}\right\}$. For any integer p, define $\tau^p(j) \stackrel{\text{def}}{=} \tau(j) \wedge p$, where $a \wedge b = \min(a, b)$.

$$w(\bar{\theta}^{j+\tau^p(j)}) - w(\bar{\theta}^j) = \sum_{i=j+1}^{j+\tau^p(j)} [w(\bar{\theta}^i) - w(\bar{\theta}^{i-1})] \geq \varepsilon \sum_{i=j+1}^{j+\tau^p(j)} \gamma_i, \tag{11.64}$$

with the convention that, for any sequence $\{a_i\}$ and any integer l, $\sum_{i=l+1}^{l} a_i = 0$. Therefore,

$$w(\theta^{j+\tau^p(j)}) - w(\theta^j) = w(\theta^{j+\tau^p(j)}) - w(\bar{\theta}^{j+\tau^p(j)}) + \\ w(\bar{\theta}^{j+\tau^p(j)}) - w(\bar{\theta}^j) + w(\bar{\theta}^j) - w(\theta^j) \geq -2\eta + \varepsilon \sum_{i=j+1}^{j+\tau^p(j)} \gamma_i .$$

Because $\{w(\theta^j)\}$ converges, there exists $N' > N$ such that, for all $j \geq N'$,

$$\eta \geq w(\theta^{j+\tau^p(j)}) - w(\theta^j) \geq -2\eta + \varepsilon \sum_{i=j+1}^{j+\tau^p(j)} \gamma_i .$$

This implies that, for all $j \geq N'$ and all integer $p \geq 0$,

$$\sum_{i=j+1}^{j+\tau^p(j)} \gamma_i \leq 3\eta/\varepsilon \,. \tag{11.65}$$

Because $\sum_{i=j+1}^{j+\tau(j)} \gamma_i = \lim_{p\to\infty} \sum_{i=j+1}^{j+\tau^p(j)} \gamma_i$ and $\sum_{i=1}^{\infty} \gamma_i = \infty$, the previous relation implies that, for all $j \geq N'$, $\tau(j) < \infty$ and $\sum_{i=j+1}^{j+\tau(j)} \gamma_i \leq 3\eta/\varepsilon$. For any integer p, $\theta^{j+p} - \theta^j = \sum_{i=j+1}^{j+p} \gamma_i h(\theta^{i-1}) + \sum_{i=j+1}^{j+p} \gamma_i \zeta^i$, which implies that

$$\left|\theta^{j+p} - \theta^j\right| \leq \|h\mathbb{1}_{\mathcal{K}}\|_\infty \sum_{i=j+1}^{j+p} \gamma_i + \left|\sum_{i=j+1}^{j+p} \gamma_i \zeta^i\right| .$$

Applying this inequality for $j \geq N'$ and $p = \tau(j)$ and using that, by definition, $\bar{\theta}^{j+\tau(j)} \in \mathcal{N}$,

$$\begin{aligned}\left|\theta^j - \bar{\theta}^{j+\tau(j)}\right| &\leq |\bar{\theta}^{j+\tau(j)} - \theta^{j+\tau(j)}| + |\theta^{j+\tau(j)} - \theta^j| \\ &\leq \delta' + \|h\mathbb{1}_{\mathcal{K}}\|_\infty \, 3\eta/\varepsilon + \left|\sum_{i=j+1}^{j+\tau(j)} \gamma_i \zeta^i\right| .\end{aligned}$$

Because η, δ', and ε' can be arbitrarily small, and $\sup_{l\geq k} |\sum_{i=k}^{l} \gamma_i \zeta^i|$ tends to zero, the latter inequality shows that all limit points of the sequence $\{\theta^j\}$ belong to $\mathcal{N}$. Because $\mathcal{N}$ is arbitrary, all limit points of $\{\theta^j\}$ belong to $\mathcal{L}$. □

12

Statistical Properties of the Maximum Likelihood Estimator

The maximum likelihood estimator (MLE) is one of the backbones of statistics, and as we have seen in previous chapters, it is very much appropriate also for HMMs, even though numerical approximations are required when the state space is not finite. A standard result in statistics says that, except for "atypical cases", the MLE is consistent, asymptotically normal with asymptotic (scaled) variance equal to the inverse Fisher information matrix, and efficient. The purpose of the current chapter is to show that these properties are indeed true for HMMs as well, provided some conditions of rather standard nature hold. We will also employ the asymptotic results obtained to verify the validity of certain likelihood-based tests.

Recall that the distribution (law) P of $\{Y_k\}_{k\geq 0}$ depends on a parameter θ that lies in a parameter space Θ, which we assume is a subset of $\mathbb{R}^{d_\theta}$ for some d_θ. Commonly, θ is a vector containing some components that parameterize the transition kernel of the hidden Markov chain—such as the transition probabilities if the state space X is finite—and other components that parameterize the conditional distributions of the observations given the states. Throughout the chapter, it is assumed that the HMM model is, for all θ, fully dominated in the sense of Definition 2.2.3 and that the underlying Markov chain is positive (see Definition 14.2.26).

Assumption 12.0.1.

(i) *There exists a probability measure* λ *on* $(\mathsf{X}, \mathcal{X})$ *such that for any* $x \in \mathsf{X}$ *and any* $\theta \in \Theta$, $Q_\theta(x, \cdot) \ll \lambda$ *with transition density* q_θ. *That is,* $Q_\theta(x, A) = \int q_\theta(x, x')\,\lambda(dx')$ *for* $A \in \mathcal{X}$.

(ii) *There exists a probability measure* μ *on* $(\mathsf{Y}, \mathcal{Y})$ *such that for any* $x \in \mathsf{X}$ *and any* $\theta \in \Theta$, $G_\theta(x, \cdot) \ll \mu$ *with transition density function* g_θ. *That is,* $G_\theta(x, A) = \int g_\theta(x, y)\,\mu(dy)$ *for* $A \in \mathcal{Y}$.

(iii) *For any* $\theta \in \Theta$, Q_θ *is positive, that is,* Q_θ *is phi-irreducible and admits a (necessarily unique) invariant distribution denoted by* π_θ.

In this chapter, we will generally assume that Θ is compact. Furthermore, $\theta_\star$ is used to denote the true parameter, that is, the parameter corresponding to the data that we actually observe.

12.1 A Primer on MLE Asymptotics

The standard asymptotic properties of the MLE hinge on three basic results: a law of large numbers for the log-likelihood, a central limit theorem for the score function, and a law of large of numbers for the observed information. More precisely,

(i) for all $\theta \in \Theta$, $n^{-1}\ell_n(\theta) \to \ell(\theta)$ $\mathrm{P}_{\theta_\star}$-a.s. uniformly over compact subsets of Θ, where $\ell_n(\theta)$ is the log-likelihood of the parameter θ given the first n observations and $\ell(\theta)$ is a continuous deterministic function with a unique global maximum at $\theta_\star$;
(ii) $n^{-1/2}\nabla_\theta \ell_n(\theta_\star) \to \mathrm{N}(0, \mathcal{J}(\theta_\star))$ $\mathrm{P}_{\theta_\star}$-weakly, where $\mathcal{J}(\theta)$ is the Fisher information matrix at θ (we do not provide a more detailed definition at the moment);
(iii) $\lim_{\delta \to 0} \lim_{n \to \infty} \sup_{|\theta - \theta_\star| \le \delta} \| -n^{-1}\nabla_\theta^2 \ell_n(\theta) - \mathcal{J}(\theta_\star)\| = 0$ $\mathrm{P}_{\theta_\star}$-a.s.

The function ℓ in (i) is sometimes referred to as the *contrast function*. We note that $-n^{-1}\nabla_\theta^2 \ell_n(\theta)$ in (iii) is the observed information matrix, so that (iii) says that the observed information should converge to the Fisher information in a certain uniform sense. This uniformity may be replaced by conditions on the third derivatives of the log-likelihood, which is common in statistical textbooks, but as we shall see, it is cumbersome enough even to deal with second derivatives of the log-likelihood for HMMs, whence avoiding third derivatives is preferable.

Condition (i) assures strong consistency of the MLE, which can be shown using an argument that goes back to Wald (1949). The idea of the argument is as follows. Denote by $\widehat{\theta}_n$ the maximum the ML estimator; $\ell_n(\widehat{\theta}_n) \ge \ell_n(\theta)$ for any $\theta \in \Theta$. Because ℓ has a unique global maximum at $\theta_\star$, $\ell(\theta_\star) - \ell(\theta) \ge 0$ for any $\theta \in \Theta$ and, in particular, $\ell(\theta_\star) - \ell(\widehat{\theta}_n) \ge 0$. We now combine these two inequalities to obtain

$$
\begin{aligned}
0 &\le \ell(\theta_\star) - \ell(\widehat{\theta}_n) \\
&\le \ell(\theta_\star) - n^{-1}\ell_n(\theta_\star) + n^{-1}\ell_n(\theta_\star) - n^{-1}\ell_n(\widehat{\theta}_n) + n^{-1}\ell_n(\widehat{\theta}_n) - \ell(\widehat{\theta}_n) \\
&\le 2 \sup_{\theta \in \Theta} |\ell(\theta) - n^{-1}\ell_n(\theta)| .
\end{aligned}
$$

Therefore, by taking the compact subset in (i) above as Θ itself, $\ell(\widehat{\theta}_n) \to \ell(\theta_\star)$ $\mathrm{P}_{\theta_\star}$-a.s. as $n \to \infty$, which in turn implies, as ℓ is continuous with a unique global maximum at $\theta_\star$, that the MLE converges to $\theta_\star$ $\mathrm{P}_{\theta_\star}$-a.s.. In other words, the MLE is strongly consistent.

Provided strong consistency holds, properties (ii) and (iii) above yield asymptotic normality of the MLE. In fact, we must also assume that $\theta_\star$ is an interior point of Θ and that the Fisher information matrix $\mathcal{J}(\theta_\star)$ is non-singular. Then we can for sufficiently large n make a Taylor expansion around $\theta_\star$, noting that the gradient of ℓ_n vanishes at the MLE $\widehat{\theta}_n$ because $\theta_\star$ is maximal there,

$$0 = \nabla_\theta \ell_n(\widehat{\theta}_n) = \nabla_\theta \ell_n(\theta_\star) + \left\{ \int_0^1 \nabla_\theta^2 \ell_n[\theta_\star + t(\widehat{\theta}_n - \theta_\star)]\, dt \right\} (\widehat{\theta}_n - \theta_\star) \; .$$

From this expansion we obtain

$$n^{1/2}(\widehat{\theta}_n - \theta_\star) = \left\{ -n^{-1} \int_0^1 \nabla_\theta^2 \ell_n[\theta_\star + t(\widehat{\theta}_n - \theta_\star)]\, dt \right\}^{-1} n^{-1/2} \nabla_\theta \ell_n(\theta_\star) \; .$$

Now $\widehat{\theta}_n$ converges to $\theta_\star$ $\mathrm{P}_{\theta_\star}$-a.s. and so, using (iii), the first factor on the right-hand side tends to $\mathcal{J}(\theta_\star)^{-1}$ $\mathrm{P}_{\theta_\star}$-a.s. The second factor converges weakly to $\mathrm{N}(0, \mathcal{J}(\theta_\star))$; this is (ii). Cramér-Slutsky's theorem hence tells us that $n^{1/2}(\widehat{\theta}_n - \theta_\star)$ tends $\mathrm{P}_{\theta_\star}$-weakly to $\mathrm{N}(0, \mathcal{J}^{-1}(\theta_\star))$, and this is the standard result on asymptotic normality of the MLE.

In an entirely similar way properties (ii) and (iii) also show that for any $u \in \mathbb{R}^{d_\theta}$ (recall that Θ is a subset of $\mathbb{R}^{d_\theta}$),

$$\ell_n(\theta_\star + n^{-1/2} u) - \ell_n(\theta_\star) = n^{-1/2} u^T \nabla_\theta \ell_n(\theta_\star) + \frac{1}{2} u^T [-n^{-1} \nabla_\theta^2 \ell_n(\theta_\star)] u + R_n(u) \; ,$$

where $n^{-1/2} \nabla_\theta \ell_n(\theta_\star)$ and $-n^{-1} \nabla_\theta^2 \ell_n(\theta_\star)$ converge as described above, and where $R_n(u)$ tends to zero $\mathrm{P}_{\theta_\star}$-a.s. Such an expansion is known as *local asymptotic normality (LAN)* of the model, cf. Ibragimov and Hasminskii (1981, Definition II.2.1). Under this condition, it is known that so-called *regular* estimators (a property possessed by all "sensible" estimators) cannot have an asymptotic covariance matrix smaller than $\mathcal{J}^{-1}(\theta_\star)$ (Ibragimov and Hasminskii, 1981, p. 161). Because this limit is obtained by the MLE, this estimator is efficient.

Later on in this chapter, we will also exploit properties (i)–(iii) to derive asymptotic properties of likelihood ratio and other tests for lower dimensional hypotheses regarding θ.

12.2 Stationary Approximations

In this section, we will introduce a way of obtaining properties (i)–(iii) for HMMs; more detailed descriptions are given in subsequent sections.

Before proceeding, we will be precise on the likelihood we shall analyze. In this chapter, we generally make the assumption that the sequence $\{X_k\}_{k \geq 0}$ is stationary; then $\{X_k, Y_k\}_{k \geq 0}$ is stationary as well. Then there is obviously a corresponding likelihood. However, it is sometimes convenient to work with a likelihood $\mathrm{L}_{x_0,n}(\theta)$ that is conditional on an initial state x_0,

$$\mathrm{L}_{x_0,n}(\theta) = \int g_\theta(x_0, Y_0) \prod_{i=1}^{n} q_\theta(x_{i-1}, x_i) g_\theta(x_i, Y_i)\, \lambda(dx_i)\;. \tag{12.1}$$

We could also want to replace the fixed initial state by an initial distribution ν on $(\mathsf{X}, \mathcal{X})$, giving

$$\mathrm{L}_{\nu,n}(\theta) = \int_{\mathsf{X}} \mathrm{L}_{x_0,n}(\theta)\, \nu(dx_0)\;.$$

The stationary likelihood is then $\mathrm{L}_{\pi_\theta,n}(\theta)$, which we will simply denote by $\mathrm{L}_n(\theta)$. The advantage of working with the stationary likelihood is of course that it is the correct likelihood for the model and may hence be expected to provide better finite-sample performance. The advantage of assuming a fixed initial state x_0—and hence adopting the likelihood $\mathrm{L}_{x_0,n}(\theta)$—is that the stationary distribution π_θ is not always available in closed form when X is not finite. It is however important that $g_\theta(x_0, Y_0)$ is positive $\mathrm{P}_{\theta_\star}$-a.s.; otherwise the log-likelihood may not be well-defined. In fact, we shall require that $g_\theta(x_0, Y_0)$ is, $\mathrm{P}_{\theta_\star}$-a.s., bounded away from zero. In the following, we always assume that this condition is fulfilled. A further advantage of $\mathrm{L}_{x_0,n}(\theta)$ is that the methods described in the current chapter may be extended to Markov-switching autoregressions (Douc *et al.*, 2004), and then the stationary likelihood is almost never computable, not even when X is finite. Throughout the rest of this chapter, we will work with $\mathrm{L}_{x_0,n}(\theta)$ unless noticed, where $x_0 \in \mathsf{X}$ is chosen to satisfy the above positivity assumption but otherwise arbitrarily. The MLE arising from this likelihood has the same asymptotic properties as has the MLE arising from $\mathrm{L}_n(\theta)$, provided the initial stationary distribution π_θ has smooth second-order derivatives (cf. Bickel *et al.*, 1998), whence from an asymptotic point of view there is no loss in using the incorrect likelihood $\mathrm{L}_{x_0,n}(\theta)$.

We now return to the analysis of log-likelihood and items (i)–(iii) above. In the setting of i.i.d. observations, the log-likelihood $\ell_n(\theta)$ is a sum of i.i.d. terms, and so (i) and (iii) follow from uniform versions of the strong law of large numbers and (ii) is a consequence of the simplest central limit theorem. In the case of HMMs, we can write $\ell_{x_0,n}(\theta)$ as a sum as well:

$$\ell_{x_0,n}(\theta) = \sum_{k=0}^{n} \log \left[\int g_\theta(x_k, Y_k)\, \phi_{x_0,k|k-1}[Y_{0:k-1}](dx_k\,;\theta) \right] \tag{12.2}$$

$$= \sum_{k=0}^{n} \log \left[\int g_\theta(x_k, Y_k)\, \mathrm{P}_\theta(X_k \in dx_k \mid Y_{0:k-1}, X_0 = x_0) \right]\;, \tag{12.3}$$

where $\phi_{x_0,k|k-1}[Y_{0:k-1}](\cdot\,;\theta)$ is the predictive distribution of the state X_k given the observations $Y_{0:k-1}$ and $X_0 = x_0$. These terms do not form a stationary sequence however, so the law of large numbers—or rather the ergodic theorem—does not apply directly. Instead we must first approximate $\ell_{x_0,n}(\theta)$ by the partial sum of a stationary sequence.

When the joint Markov chain $\{X_k, Y_k\}$ has an invariant distribution, this chain is stationary provided it is started from its invariant distribution. In this case, we can (and will!) extend it to a stationary sequence $\{X_k, Y_k\}_{-\infty<k<\infty}$ with doubly infinite time, as we can do with any stationary sequence. Having done this extension, we can imagine a predictive distribution of the state X_k given the infinite past $Y_{-\infty:k-1}$ of observations. A key feature of these variables is that they now form a stationary sequence, whence the ergodic theorem applies. Furthermore we can approximate $\ell_{x_0,n}(\theta)$ by

$$\ell_n^s(\theta) = \sum_{k=0}^{n} \log \left[\int g_\theta(x_k, Y_k) \, \mathrm{P}_\theta(X_k \in dx_k \,|\, Y_{-\infty:k-1}) \right] , \tag{12.4}$$

where superindex s stands for "stationary". Heuristically, one would expect this approximation to be good, as observations far in the past do not provide much information about the current one, at least not if the hidden Markov chain enjoys good mixing properties. What we must do is thus to give a precise definition of the predictive distribution $\mathrm{P}_\theta(X_k \in \cdot \,|\, Y_{-\infty:k-1})$ given the infinite past, and then show that it approximates the predictive distribution $\phi_{x_0,k|k-1}(\cdot\,;\theta)$ well enough that the two sums (12.2) and (12.4), after normalization by n, have the same asymptotic behavior. We can treat the score function similarly by defining a sequence that forms a stationary martingale increment sequence; for sums of such sequences there is a central limit theorem.

The cornerstone in this analysis is the result on conditional mixing stated in Section 4.3. We will rephrase it here, but before doing so we state a first assumption. It is really a variation of Assumption 4.3.24, adapted to the dominated setting and uniform in θ.

Assumption 12.2.1.

(i) The transition density $q_\theta(x, x')$ of $\{X_k\}$ satisfies $0 < \sigma^- \le q_\theta(x, x') \le \sigma^+ < \infty$ for all $x, x' \in \mathsf{X}$ and all $\theta \in \Theta$, and the measure λ is a probability measure.

(ii) For all $y \in \mathsf{Y}$, the integral $\int_{\mathsf{X}} g_\theta(x, y)\, \lambda(dx)$ is bounded away from 0 and ∞ on Θ.

Part (i) of this assumption often, but not always holds when the state space X is finite or compact. Note that Assumption 12.2.1 says that for all $\theta \in \Theta$, the whole state space X is a 1-small set for the transition kernel Q_θ, which implies that for all $\theta \in \Theta$, the chain is phi-irreducible and strongly aperiodic (see Section 14.2 for definitions). It also ensures that there exists a stationary distribution π_θ for Q_θ. In addition, the chain is uniformly geometrically ergodic in the sense that for any $x \in \mathsf{X}$ and $n \ge 0$, $\|Q_\theta^n(x, \cdot) - \pi_\theta\|_{\mathrm{TV}} \le (1 - \sigma^-)^n$. Under Assumption 12.0.1, it holds that $\pi_\theta \ll \lambda$, and we use the same notation for this distribution and its density with respect to the dominating measure λ.

Using the results of Section 14.3, we conclude that the state space $\mathsf{X} \times \mathsf{Y}$ is 1-small for the joint chain $\{X_k, Y_k\}$. Thus the joint chain is also phi-irreducible and strongly aperiodic, and it admits a stationary distribution with density $\pi_\theta(x) g_\theta(x, y)$ with respect to the product measure $\lambda \otimes \mu$ on $(\mathsf{X} \times \mathcal{Y}, \mathcal{X} \otimes \mathcal{Y})$ The joint chain also is uniformly geometrically ergodic.

Put $\rho = 1 - \sigma^-/\sigma^+$; then $0 \leq \rho < 1$. The important consequence of Assumption 12.2.1 that we need in the current chapter is Proposition 4.3.26. It says that if Assumption 12.2.1 holds true, then for all $k \geq 1$, all $y_{0:n}$ and all initial distributions ν and ν' on $(\mathsf{X}, \mathcal{X})$,

$$\left\| \int_{\mathsf{X}} \mathrm{P}_\theta(X_k \in \cdot \mid X_0 = x, Y_{0:n} = y_{0:n}) \left[\nu(dx) - \nu'(dx)\right] \right\|_{\mathrm{TV}} \leq \rho^k . \tag{12.5}$$

12.3 Consistency

12.3.1 Construction of the Stationary Conditional Log-likelihood

We shall now construct $\mathrm{P}_\theta(X_k \in dx_k \mid Y_{-\infty:k-1})$ and $\int g_\theta(x_k, Y_k) \, \mathrm{P}_\theta(X_k \in dx_k \mid Y_{-\infty:k-1})$. The latter variable will be defined as the limit of

$$\mathrm{H}_{k,m,x}(\theta) \stackrel{\text{def}}{=} \int g_\theta(x_k, Y_k) \, \mathrm{P}_\theta(X_k \in dx_k \mid Y_{-m+1:k-1}, X_{-m} = x) \tag{12.6}$$

as $m \to \infty$. Note that $\mathrm{H}_{k,m,x}(\theta)$ is the conditional density of Y_k given $Y_{-m+1:k-1}$ and $X_{-m} = x$, under the law P_θ. Put

$$h_{k,m,x}(\theta) \stackrel{\text{def}}{=} \log \mathrm{H}_{k,m,x}(\theta) \tag{12.7}$$

and consider the following assumption.

Assumption 12.3.1. $b^+ = \sup_\theta \sup_{x,y} g_\theta(x, y) < \infty$ *and* $\mathrm{E}_{\theta_\star} |\log b^-(Y_0)| < \infty$, *where* $b^-(y) = \inf_\theta \int_{\mathsf{X}} g_\theta(x, y) \, \lambda(dx)$.

Lemma 12.3.2. *The following assertions hold true* $\mathrm{P}_{\theta_\star}$*-a.s. for all indices* k, m *and* m' *such that* $k > -(m \wedge m')$:

$$\sup_{\theta \in \Theta} \sup_{x, x' \in \mathsf{X}} |h_{k,m,x}(\theta) - h_{k,m',x'}(\theta)| \leq \frac{\rho^{k+(m \wedge m')-1}}{1 - \rho} , \tag{12.8}$$

$$\sup_{\theta \in \Theta} \sup_{m \geq -(k-1)} \sup_{x \in \mathsf{X}} |h_{k,m,x}(\theta)| \leq |\log b^+| \vee |\log(\sigma^- b^-(Y_k))| . \tag{12.9}$$

Proof. Assume that $m' \geq m$ and write

$$\mathrm{H}_{k,m,x}(\theta) = \iint \left[\int g_\theta(x_k, Y_k) q_\theta(x_{k-1}, x_k)\, \lambda(dx_k) \right] \\ \times \mathrm{P}_\theta(X_{k-1} \in dx_{k-1} \mid Y_{-m+1:k-1}, X_{-m} = x_{-m})\, \delta_x(dx_{-m}) \,, \quad (12.10)$$

$$\mathrm{H}_{k,m',x'}(\theta) = \iint \left[\int g_\theta(x_k, Y_k) q_\theta(x_{k-1}, x_k)\, \lambda(dx_k) \right] \\ \times \mathrm{P}_\theta(X_{k-1} \in dx_{k-1} \mid Y_{-m+1:k-1}, X_{-m} = x_{-m}) \\ \times \mathrm{P}_\theta(X_{-m} \in dx_{-m} \mid Y_{-m'+1:k-1}, X_{-m'} = x') \,, \quad (12.11)$$

and invoke (12.5) to see that

$$|\mathrm{H}_{k,m,x}(\theta) - \mathrm{H}_{k,m',x'}(\theta)| \leq \rho^{k+m-1} \sup_{x_{k-1}} \int g_\theta(x_k, Y_k) q_\theta(x_{k-1}, x_k)\, \lambda(dx_k) \\ \leq \rho^{k+m-1} \sigma^+ \int g_\theta(x_k, Y_k)\, \lambda(dx_k) \,. \quad (12.12)$$

Note that the step from the total variation bound to the bound on the difference between the integrals does not need a factor "2", because the integrands are non-negative. Also note that (12.5) is stated for $m = m' = 0$, but its initial time index is of course arbitrary. The integral in (12.10) can be bounded from below as

$$\mathrm{H}_{k,m,x}(\theta) \geq \sigma^- \int g_\theta(x_k, Y_k)\, \lambda(dx_k) \,, \quad (12.13)$$

and the same lower bound holds for (12.11). Combining (12.12) with these lower bounds and the inequality $|\log x - \log y| \leq |x - y|/(x \wedge y)$ shows that

$$|h_{k,m,x}(\theta) - h_{k,m',x'}(\theta)| \leq \frac{\sigma^+}{\sigma^-} \rho^{k+m-1} = \frac{\rho^{k+m-1}}{1-\rho} \,,$$

which is the first assertion of the lemma. Furthermore note that (12.10) and (12.13) yield

$$\sigma^- b^-(Y_k) \leq \mathrm{H}_{k,m,x}(\theta) \leq b^+ \,, \quad (12.14)$$

which implies the second assertion. □

Equation (12.8) shows that for any given k and x, $\{h_{k,m,x}(\theta)\}_{m \geq -(k-1)}$ is a uniform (in θ) Cauchy sequence as $m \to \infty$, $\mathrm{P}_{\theta_\star}$-a.s., whence there is a $\mathrm{P}_{\theta_\star}$-a.s. limit. Moreover, again by (12.8), this limit does not depend on x, so we denote it by $h_{k,\infty}(\theta)$. Our interpretation of this limit is as $\log \mathrm{E}_\theta\left[g_\theta(X_k, Y_k) \mid Y_{-\infty:k-1} \right]$. Furthermore (12.9) shows that provided Assumption 12.3.1 holds, $\{h_{k,m,x}(\theta)\}_{m \geq -(k-1)}$ is uniformly bounded in $L^1(\mathrm{P}_{\theta_\star})$, so that $h_{k,\infty}(\theta)$ is in $L^1(\mathrm{P}_{\theta_\star})$ and, by the dominated convergence theorem, the limit holds in this mode as well. Finally, by its definition $\{h_{k,\infty}(\theta)\}_{k \geq 0}$ is a stationary process, and it is ergodic because $\{Y_k\}_{-\infty < k < \infty}$ is. We summarize these findings.

Proposition 12.3.3. *Assume 12.0.1, 12.2.1, and 12.3.1 hold. Then for each* $\theta \in \Theta$ *and* $x \in \mathsf{X}$*, the sequence* $\{h_{k,m,x}(\theta)\}_{m \geq -(k-1)}$ *has,* $\mathrm{P}_{\theta_\star}$*-a.s., a limit* $h_{k,\infty}(\theta)$ *as* $m \to \infty$*. This limit does not depend on* x*. In addition, for any* $\theta \in \Theta$*,* $h_{k,\infty}(\theta)$ *belongs to* $L^1(\mathrm{P}_{\theta_\star})$*, and* $\{h_{k,m,x}(\theta)\}_{m \geq -(k-1)}$ *also converges to* $h_{k,\infty}(\theta)$ *in* $L^1(\mathrm{P}_{\theta_\star})$ *uniformly over* $\theta \in \Theta$ *and* $x \in \mathsf{X}$*.*

Having come thus far, we can quantify the approximation of the log-likelihood $\ell_{x_0,n}(\theta)$ by $\ell_n^s(\theta)$.

Proposition 12.3.4. *For all* $n \geq 0$ *and* $\theta \in \Theta$*,*

$$|\ell_{x_0,n}(\theta) - \ell_n^s(\theta)| \leq |\log g_\theta(x_0, Y_0)| + h_{0,\infty}(\theta) + \frac{1}{(1-\rho)^2} \quad \mathrm{P}_{\theta_\star}\text{-a.s.}$$

Proof. Letting $m' \to \infty$ in (12.8) we obtain $|h_{k,0,x_0}(\theta) - h_{k,\infty}(\theta)| \leq \rho^{k-1}/(1-\rho)$ for $k \geq 1$. Therefore, $\mathrm{P}_{\theta_\star}$-a.s.,

$$\begin{aligned}
|\ell_{x_0,n}(\theta) - \ell_n^s(\theta)| &= \left| \sum_{k=0}^{n} h_{k,0,x_0}(\theta) - \sum_{k=0}^{n} h_{k,\infty}(\theta) \right| \\
&\leq |\log g_\theta(x_0, Y_0)| + h_{0,\infty}(\theta) + \sum_{k=1}^{n} \frac{\rho^{k-1}}{1-\rho} .
\end{aligned}$$

□

12.3.2 The Contrast Function and Its Properties

Because $h_{k,\infty}(\theta)$ is in $L^1(\mathrm{P}_{\theta_\star})$ under the assumptions made above, we can define the real-valued function $\ell(\theta) \stackrel{\text{def}}{=} \mathrm{E}_{\theta_\star}[h_{k,\infty}(\theta)]$. It does not depend on k, by stationarity. This is the contrast function $\ell(\theta)$ referred to above. By the ergodic theorem $n^{-1}\ell_n^s(\theta) \to \ell(\theta)$ $\mathrm{P}_{\theta_\star}$-a.s., and by Proposition 12.3.4, $n^{-1}\ell_{x_0,n}(\theta) \to \ell(\theta)$ $\mathrm{P}_{\theta_\star}$-a.s. as well. As noted above, however, we require this convergence to be uniform in θ, which is not guaranteed so far. In addition, we require $\ell(\theta)$ to be continuous and possess a unique global maximum at $\theta_\star$; the latter is an identifiability condition. In the rest of this section, we address continuity and convergence; identifiability is addressed in the next one.

To ensure continuity we need a natural assumption on continuity of the building blocks of the likelihood.

Assumption 12.3.5. *For all* $(x, x') \in \mathsf{X} \times \mathsf{X}$ *and* $y \in \mathsf{Y}$*, the functions* $\theta \mapsto q_\theta(x, x')$ *and* $\theta \mapsto g_\theta(x, y)$ *are continuous.*

The following result shows that $h_{k,\infty}(\theta)$ is then continuous in $L^1(\mathrm{P}_{\theta_\star})$.

Proposition 12.3.6. *Assume 12.0.1, 12.2.1, 12.3.1, and 12.3.5. Then for any* $\theta \in \Theta$*,*

$$\mathrm{E}_{\theta_\star}\left[\sup_{\theta'\in\Theta:\,|\theta'-\theta|\le\delta}|h_{0,\infty}(\theta')-h_{0,\infty}(\theta)|\right]\to 0 \quad \textit{as } \delta\to 0\,,$$

and $\theta\mapsto\ell(\theta)$ *is continuous on* Θ.

Proof. Recall that $h_{0,\infty}(\theta)$ is the limit of $h_{0,m,x}(\theta)$ as $m\to\infty$. We first prove that for any $x\in\mathsf{X}$ and any $m>0$, the latter quantity is continuous in θ and then use this to show continuity of the limit. Recall the interpretation of $\mathrm{H}_{0,m,x}(\theta)$ as a conditional density and write

$$\mathrm{H}_{0,m,x}(\theta) = \frac{\int\cdots\int\prod_{i=-m+1}^{0}q_\theta(x_{i-1},x_i)g_\theta(x_i,Y_i)\,\lambda(dx_{-m+1})\cdots\lambda(dx_0)}{\int\cdots\int\prod_{i=-m+1}^{-1}q_\theta(x_{i-1},x_i)g_\theta(x_i,Y_i)\,\lambda(dx_{-m+1})\cdots\lambda(dx_{-1})} \tag{12.15}$$

The integrand in the numerator is, by assumption, continuous and bounded by $(\sigma^+b^+)^m$, whence dominated convergence shows that the numerator is continuous with respect to θ (recall that λ is assumed finite). Likewise the denominator is continuous, and it is bounded from below by $(\sigma^-)^{m-1}\prod_{-m+1}^{-1}b^-(Y_i)>0$ $\mathrm{P}_{\theta_\star}$-a.s. Thus $\mathrm{H}_{0,m,x}(\theta)$ and $h_{0,m,x}(\theta)$ are continuous as well. Because $h_{0,m,x}(\theta)$ converges to $h_{0,\infty}(\theta)$ uniformly in θ as $m\to\infty$, $\mathrm{P}_{\theta_\star}$-a.s., $h_{0,\infty}(\theta)$ is continuous $\mathrm{P}_{\theta_\star}$-a.s. The uniform bound (12.9) assures that we can invoke dominated convergence to obtain the first part of the proposition.

The second part is a corollary of the first one, as

$$\begin{aligned}\sup_{\theta':\,|\theta'-\theta|\le\delta}|\ell(\theta')-\ell(\theta)| &= \sup_{\theta':\,|\theta'-\theta|\le\delta}|\,\mathrm{E}_{\theta_\star}[h_{0,\infty}(\theta')-h_{0,\infty}(\theta)]| \\ &\le \mathrm{E}_{\theta_\star}\left[\sup_{\theta':\,|\theta'-\theta|\le\delta}|h_{0,\infty}(\theta')-h_{0,\infty}(\theta)|\right].\end{aligned}$$

□

We can now proceed to show uniform convergence of $n^{-1}\ell_{x_0,n}(\theta)$ to $\ell(\theta)$.

Proposition 12.3.7. *Assume 12.0.1, 12.2.1, 12.3.1, and 12.3.5. Then*

$$\sup_{\theta\in\Theta}|n^{-1}\ell_{x_0,n}(\theta)-\ell(\theta)|\to 0 \quad \mathrm{P}_{\theta_\star}\textit{-a.s. as } n\to\infty.$$

Proof. First note that because Θ is compact, it is sufficient to prove that for all $\theta\in\Theta$,

$$\limsup_{\delta\to 0}\limsup_{n\to\infty}\sup_{\theta':\,|\theta'-\theta|\le\delta}|n^{-1}\ell_{x_0,n}(\theta')-\ell(\theta)| = 0 \quad \mathrm{P}_{\theta_\star}\text{-a.s.}$$

Now write

$$\begin{aligned}
&\limsup_{\delta\to 0}\limsup_{n\to\infty}\sup_{\theta'\colon|\theta'-\theta|\le\delta}|n^{-1}\ell_{x_0,n}(\theta')-\ell(\theta)| \\
&\quad=\limsup_{\delta\to 0}\limsup_{n\to\infty}\sup_{\theta'\colon|\theta'-\theta|\le\delta}|n^{-1}\ell_{x_0,n}(\theta')-n^{-1}\ell^s_n(\theta)| \\
&\quad\le\limsup_{\delta\to 0}\limsup_{n\to\infty}\sup_{\theta'\colon|\theta'-\theta|\le\delta}n^{-1}|\ell_{x_0,n}(\theta')-\ell^s_n(\theta')| \\
&\qquad+\limsup_{\delta\to 0}\limsup_{n\to\infty}\sup_{\theta'\colon|\theta'-\theta|\le\delta}n^{-1}|\ell^s_n(\theta')-\ell^s_n(\theta)| \, .
\end{aligned}$$

The first term on the right-hand side vanishes by Proposition 12.3.4 (note that Lemma 12.3.2 shows that $\sup_{\theta'}|h_{0,\infty}(\theta')|$ is in $L^1(\mathrm{P}_{\theta_\star})$ and hence finite $\mathrm{P}_{\theta_\star}$-a.s.). The second term is bounded by

$$\begin{aligned}
&\limsup_{\delta\to 0}\limsup_{n\to\infty}\sup_{\theta'\colon|\theta'-\theta|\le\delta}n^{-1}\left|\sum_{k=0}^{n}(h_{k,\infty}(\theta')-h_{k,\infty}(\theta))\right| \\
&\quad\le\limsup_{\delta\to 0}\limsup_{n\to\infty}n^{-1}\sum_{k=0}^{n}\sup_{\theta'\colon|\theta'-\theta|\le\delta}|h_{k,\infty}(\theta')-h_{k,\infty}(\theta)| \\
&\quad=\limsup_{\delta\to 0}\mathrm{E}_{\theta_\star}\left[\sup_{\theta'\colon|\theta'-\theta|\le\delta}|h_{0,\infty}(\theta')-h_{0,\infty}(\theta)|\right]=0\,,
\end{aligned}$$

with convergence $\mathrm{P}_{\theta_\star}$-a.s. The two final steps follow by the ergodic theorem and Proposition 12.3.6 respectively. The proof is complete. □

At this point, we thus know that $n^{-1}\ell_{x_0,n}$ converges uniformly to ℓ. The same conclusion holds when other initial distributions ν are put on X_0, provided $\sup_\theta|\log\int g_\theta(x,Y_0)\,\nu(dx)|$ is finite $\mathrm{P}_{\theta_\star}$-a.s. When ν is the stationary distribution π_θ, uniform convergence can in fact be proved without this extra regularity assumption by conditioning on the previous state X_{-1} to get rid of the first two terms in the bound of Proposition 12.3.4; cf. Douc *et al.* (2004).

The uniform convergence of $n^{-1}\ell_{x_0,n}(\theta)$ to $\ell(\theta)$ can be used—with an argument entirely similar to the one of Wald outlined in Section 12.1—to show that the MLE converges a.s. to the set, $\Theta_\star$ say, of global maxima of ℓ. Because ℓ is continuous, we know that $\Theta_\star$ is closed and hence also compact. More precisely, for any (open) neighborhood of $\Theta_\star$, the MLE will be in that neighborhood for large n, $\mathrm{P}_{\theta_\star}$-a.s. We say that the MLE converges to $\Theta_\star$ *in the quotient topology*. This way of describing convergence was used, in the context of HMMs, by Leroux (1992). The purpose of the identifiability constraint, that $\ell(\theta)$ has a *unique* global maximum at $\theta_\star$, is thus to ensure that $\Theta_\star$ consists of the single point $\theta_\star$ so that the MLE indeed converges to the point $\theta_\star$.

12.4 Identifiability

As became obvious in the previous section, the set of global maxima of ℓ is of intrinsic importance, as this set constitutes the possible limit points of the

MLE. The definition of $\ell(\theta)$ as a limit is however usually not suitable for extracting relevant information about the set of maxima, and the purpose of this section is to derive a different characterization of the set of global maxima of ℓ.

12.4.1 Equivalence of Parameters

We now introduce the notion of *equivalence of parameters.*

Definition 12.4.1. *Two points $\theta, \theta' \in \Theta$ are said to be* equivalent *if they govern identical laws for the process $\{Y_k\}_{k\geq 0}$, that is, if $\mathrm{P}_\theta = \mathrm{P}_{\theta'}$.*

We note that, by virtue of Kolmogorov's extension theorem, θ and θ' are equivalent if and only if the finite-dimensional distributions $\mathrm{P}_\theta(Y_1 \in \cdot, Y_2 \in \cdot, \ldots, Y_n \in \cdot)$ and $\mathrm{P}_{\theta'}(Y_1 \in \cdot, Y_2 \in \cdot, \ldots, Y_n \in \cdot)$ agree for all $n \geq 1$.

We will show that a parameter $\theta \in \Theta$ is a global maximum point of ℓ if and only if θ is equivalent to $\theta_\star$. This implies that the limit points of the MLE are those points θ that govern the same law for $\{Y_k\}_{k\geq 0}$ as does $\theta_\star$. This is the best we can hope for because there is no way—even with an infinitely large sample of Ys!—to distinguish between the true parameter $\theta_\star$ and a different but equivalent parameter θ. Naturally we would like to conclude that no parameter other than $\theta_\star$ itself is equivalent to $\theta_\star$. This is not always the case however, in particular when X is finite and we can number the states arbitrarily. We will discuss this matter further after proving the following result.

Theorem 12.4.2. *Assume 12.0.1, 12.2.1, and 12.3.1. Then a parameter $\theta \in \Theta$ is a global maximum of ℓ if and only if θ is equivalent to $\theta_\star$.*

An immediate implication of this result is that $\theta_\star$ is a global maximum of ℓ.

Proof. By the definition of $\ell(\theta)$ and Proposition 12.3.3,

$$\begin{aligned}\ell(\theta_\star) - \ell(\theta) &= \mathrm{E}_{\theta_\star}\left[\lim_{m\to\infty} h_{1,m,x}(\theta_\star)\right] - \mathrm{E}_{\theta_\star}\left[\lim_{m\to\infty} h_{1,m,x}(\theta)\right] \\ &= \lim_{m\to\infty} \mathrm{E}_{\theta_\star}\left[h_{1,m,x}(\theta_\star)\right] - \lim_{m\to\infty} \mathrm{E}_{\theta_\star}\left[h_{1,m,x}(\theta)\right] \\ &= \lim_{m\to\infty} \mathrm{E}_{\theta_\star}\left[h_{1,m,x}(\theta_\star) - h_{1,m,x}(\theta)\right] ,\end{aligned}$$

where $h_{k,m,x}(\theta)$ is given in (12.7). Next, write

$$\begin{aligned}&\mathrm{E}_{\theta_\star}\left[h_{1,m,x}(\theta_\star) - h_{1,m,x}(\theta)\right] \\ &\qquad = \mathrm{E}_{\theta_\star}\left\{\mathrm{E}_{\theta_\star}\left[\log \frac{\mathrm{H}_{1,m,x}(\theta_\star)}{\mathrm{H}_{1,m,x}(\theta)} \,\middle|\, Y_{-m+1:0}, X_{-m} = x\right]\right\} ,\end{aligned}$$

where $\mathrm{H}_{k,m,x}(\theta)$ is given in (12.6). Recalling that $\mathrm{H}_{1,m,x}(\theta)$ is the conditional density of Y_1 given $Y_{-m+1:0}$ and $X_{-m} = x$, we see that the inner (conditional) expectation on the right-hand side is a Kullback-Leibler divergence and hence non-negative. Thus the outer expectation and the limit $\ell(\theta_\star) - \ell(\theta)$ are non-negative as well, so that $\theta_\star$ is a global mode of ℓ.

Now pick $\theta \in \Theta$ such that $\ell(\theta) = \ell(\theta_\star)$. Throughout the remainder of the proof, we will use the letter p to denote (possibly conditional) densities of random variables, with the arguments of the density indicating which random variables are referred to. For any $k \geq 1$,

$$\begin{aligned}
\mathrm{E}_{\theta_\star} &[\log p_\theta(Y_{1:k} | Y_{-m+1:0}, X_{-m} = x)] \\
&= \sum_{i=1}^{k} \mathrm{E}_{\theta_\star} [\log p_\theta(Y_i | Y_{-m+1:i-1}, X_{-m} = x)] \\
&= \sum_{i=1}^{k} \mathrm{E}_{\theta_\star} [h_{i,m,x}(\theta)]
\end{aligned}$$

so that, employing stationarity,

$$\lim_{m \to \infty} \mathrm{E}_{\theta_\star} [\log p_\theta(Y_{1:k} | Y_{-m+1:0}, X_{-m} = x)] = k\ell(\theta) .$$

Thus for any positive integer $n < k$,

$$\begin{aligned}
0 &= k(\ell(\theta_\star) - \ell(\theta)) \\
&= \lim_{m \to \infty} \mathrm{E}_{\theta_\star} \left[\log \frac{p_{\theta_\star}(Y_{1:k} | Y_{-m+1:0}, X_{-m} = x)}{p_\theta(Y_{1:k} | Y_{-m+1:0}, X_{-m} = x)} \right] \\
&= \lim_{m \to \infty} \left\{ \mathrm{E}_{\theta_\star} \left[\log \frac{p_{\theta_\star}(Y_{k-n+1:k} | Y_{-m+1:0}, X_{-m} = x)}{p_\theta(Y_{k-n+1:k} | Y_{-m+1:0}, X_{-m} = x)} \right] \right. \\
&\qquad \left. + \mathrm{E}_{\theta_\star} \left[\log \frac{p_{\theta_\star}(Y_{1:k-n} | Y_{k-n+1:k}, Y_{-m+1:0}, X_{-m} = x)}{p_\theta(Y_{1:k-n} | Y_{k-n+1:k}, Y_{-m+1:0}, X_{-m} = x)} \right] \right\} \\
&\geq \limsup_{m \to \infty} \mathrm{E}_{\theta_\star} \left[\log \frac{p_{\theta_\star}(Y_{1:n} | Y_{n-k-m+1:n-k}, X_{n-k-m} = x)}{p_\theta(Y_{1:n} | Y_{n-k-m+1:n-k}, X_{n-k-m} = x)} \right] ,
\end{aligned}$$

where the inequality follows by using stationarity for the first term and noting that the second term is non-negative as an expectation of a (conditional) Kullback-Leibler divergence as above. Hence we have inserted a gap between the variables $Y_{1:n}$ whose density we examine and the variables $Y_{n-k-m+1:n-k}$ and X_{n-k-m} that appear as a condition. The idea is now to let this gap tend to infinity and to show that in the limit the condition has no effect. Next we shall thus show that

$$\begin{aligned}
\lim_{k \to \infty} \sup_{m \geq k} \Bigg| \mathrm{E}_{\theta_\star} &\left[\log \frac{p_{\theta_\star}(Y_{1:n} | Y_{-m+1:-k}, X_{-m} = x)}{p_\theta(Y_{1:n} | Y_{-m+1:-k}, X_{-m} = x)} \right] \\
&- \mathrm{E}_{\theta_\star} \left[\log \frac{p_{\theta_\star}(Y_{1:n})}{p_\theta(Y_{1:n})} \right] \Bigg| = 0 . \qquad (12.16)
\end{aligned}$$

Combining (12.16) with the previous inequality, it is clear that if $\ell(\theta) = \ell(\theta_\star)$, then $\mathrm{E}_{\theta_\star} \{\log[p_{\theta_\star}(Y_{1:n}) / p_\theta(Y_{1:n})]\} = 0$, that is, the Kullback-Leibler divergence between the n-dimensional densities $p_{\theta_\star}(y_{1:n})$ and $p_\theta(y_{1:n})$ vanishes. This implies, by the information inequality, that these densities coincide except on a set with $\mu^{\otimes n}$-measure zero, so that the n-dimensional laws of $\mathrm{P}_{\theta_\star}$ and P_θ agree. Because n was arbitrary, we find that $\theta_\star$ and θ are equivalent.

What remains to do is thus to prove (12.16). To that end, put $U_{k,m}(\theta) = \log p_\theta(Y_{1:n}|Y_{-m+1:-k}, X_{-m} = x)$ and $U(\theta) = \log p_\theta(Y_{1:n})$. Obviously, it is enough to prove that for all $\theta \in \Theta$,

$$\lim_{k\to\infty} \mathrm{E}_{\theta_\star}\left[\sup_{m\geq k} |U_{k,m}(\theta) - U(\theta)|\right] = 0 \,. \tag{12.17}$$

To do that we write

$$p_\theta(Y_{1:n}|Y_{-m+1:-k}, X_{-m} = x) = \iint p_\theta(Y_{1:n}|X_0 = x_0)\, Q_\theta^k(x_{-k}, dx_0) \\ \times \mathrm{P}_\theta(X_{-k} \in dx_{-k} \,|\, Y_{-m+1:-k}, X_{-m} = x)$$

and

$$p_\theta(Y_{1:n}) = \iint p_\theta(Y_{1:n}|X_0 = x_0)\, Q_\theta^k(x_{-k}, dx_0)\, \pi_\theta(dx_{-k}) \,,$$

where π_θ is the stationary distribution of $\{X_k\}$. Realizing that $p_\theta(Y_{1:n}|X_0 = x_0)$ is bounded from above by $(b^+)^n$ (condition on $X_{1:n}$!) and that the transition kernel Q_θ satisfies the Doeblin condition (see Definition 4.3.12) and is thus uniformly geometrically ergodic (see Definition 4.3.15 and Lemma 4.3.13), we obtain

$$\sup_{m\geq k} |p_\theta(Y_{1:n}|Y_{-m+1:-k}, X_{-m} = x) - p_\theta(Y_{1:n})| \leq (b^+)^n (1-\sigma^-)^k \tag{12.18}$$

$\mathrm{P}_{\theta_\star}$-a.s.. Moreover, the bound

$$p_\theta(Y_{1:n}|X_0 = x_0) = \int \cdots \int \prod_{i=1}^n q_\theta(x_{i-1}, x_i) g_\theta(x_i, Y_i)\, \lambda(dx_i) \\ \geq (\sigma^-)^n \prod_{i=1}^n b^-(Y_i)$$

implies that $p_\theta(Y_{1:n}|Y_{-m+1:-k}, X_{-m} = x)$ and $p_\theta(Y_{1:n})$ both obey the same lower bound. Combined with the observation $b^-(Y_i) > 0$ $\mathrm{P}_{\theta_\star}$-a.s., which follows from Assumption 12.3.1, and the bound $|\log(x) - \log(y)| \leq |x-y|/x \wedge y$, (12.18) shows that

$$\lim_{k\to\infty} \sup_{m\geq k} |U_{k,m}(\theta) - U(\theta)| \to 0 \quad \mathrm{P}_{\theta_\star}\text{-a.s.}$$

Now (12.17) follows from dominated convergence provided

$$\mathrm{E}_\theta\left[\sup_k \sup_{m\geq k} U_{k,m}(\theta)\right] < \infty \,.$$

Using the aforementioned bounds, we conclude that this expectation is indeed finite. □

We remark that the basic structure of the proof is potentially applicable also to models other than HMMs. Indeed, using the notation of the proof, we may define ℓ as $\ell(\theta) = \lim_{m\to\infty} \mathrm{E}_{\theta_\star}[\log p_\theta(Y_1|Y_{-m:1})]$, a definition that does not exploit the HMM structure. Then the first part of the proof, up to (12.16), does not use the HMM structure either, so that all that is needed, in a more general framework, is to verify (12.16) (or, more precisely, a version thereof not containing X_{-m}). For particular other processes, this could presumably be carried out using, for instance, suitable mixing properties.

The above theorem shows that the points of global maxima of ℓ—forming the set of possible limit points of the MLE—are those that are statistically equivalent to $\theta_\star$. This result, although natural and important (but not trivial!), is however yet of a somewhat "high level" character, that is, not verifiable in terms of "low level" conditions. We would like to provide some conditions, expressed directly in terms of the Markov chain and the conditional distributions $g_\theta(x, y)$, that give information about parameters that are equivalent to $\theta_\star$ and, in particular, when there is no other such parameter than $\theta_\star$. We will do this using the framework of mixtures of distributions.

12.4.2 Identifiability of Mixture Densities

We first define what is meant by a mixture density.

Definition 12.4.3. *Let $f_\phi(y)$ be a parametric family of densities on* Y *with respect to a common dominating measure μ and parameter ϕ in some set Φ. If π is a probability measure on Φ, then the density*

$$f_\pi(y) = \int_\Phi f_\phi(y)\,\pi(d\phi)$$

is called a mixture density*; the distribution π is called the* mixing distribution.

We say that the class of (all) mixtures of (f_ϕ) is identifiable *if $f_\pi = f_{\pi'}$ μ-a.e. if and only if $\pi = \pi'$.*

Furthermore we say that the class of finite mixtures of (f_ϕ) is identifiable *if for all probability measures π and π' with finite support, $f_\pi = f_{\pi'}$ μ-a.e. if and only if $\pi = \pi'$.*

In other words, the class of all mixtures of (f_ϕ) is identifiable if the two distributions with densities f_π and $f_{\pi'}$ respectively agree only when $\pi = \pi'$. Yet another way to put this property is to say that identifiability means that the mapping $\pi \mapsto f_\pi$ is one-to-one (injective). A way, slightly Bayesian, of thinking of a mixture distribution that is often intuitive and fruitful is the following. Draw $\phi \in \Phi$ with distribution π and then Y from the density f_ϕ. Then, Y has density f_π.

Many important and commonly used parametric classes of densities are identifiable. We mention the following examples.

(i) The Poisson family (Feller, 1943). In this case, $\mathsf{Y} = \mathbb{Z}_+$, $\Phi = \mathbb{R}_+$, ϕ is the mean of the Poisson distribution, μ is counting measure, and $f_\phi(y) = \phi^y \mathrm{e}^{-\phi}/y!$.
(ii) The Gamma family (Teicher, 1961), with the mixture being either on the scale parameter (with a fixed form parameter) or on the form parameter (with a fixed scale parameter). The class of joint mixtures over both parameters is not identifiable however, but the class of joint *finite* mixtures is identifiable.
(iii) The normal family (Teicher, 1960), with the mixture being either on the mean (with fixed variance) or on the variance (with fixed mean). The class of joint mixtures over both mean and variance is not identifiable however, but the class of joint *finite* mixtures is identifiable.
(iv) The Binomial family $\mathrm{Bin}(N, p)$ (Teicher, 1963), with the mixture being on the probability p. The class of finite mixtures is identifiable, provided the number of components k of the mixture satisfies $2k - 1 \le N$.

Further reading on identifiability of mixtures is found, for instance, in Titterington *et al.* (1985, Section 3.1).

A very useful result on mixtures, taking identifiability in one dimension into several dimensions, is the following.

Theorem 12.4.4 (Teicher, 1967). *Assume that the class of all mixtures of the family (f_ϕ) of densities on Y with parameter $\phi \in \Phi$ is identifiable. Then the class of all mixtures of the n-fold product densities $f_\phi^{(n)}(y) = f_{\phi_1}(y_1) \cdots f_{\phi_n}(y_n)$ on $y \in \mathsf{Y}^n$ with parameter $\phi \in \Phi^n$ is identifiable. The same conclusion holds true when "all mixtures" is replaced by "finite mixtures".*

12.4.3 Application of Mixture Identifiability to Hidden Markov Models

Let us now explain how identifiability of mixture densities applies to HMMs. Assume that $\{X_k, Y_k\}$ is an HMM such that the conditional densities $g_\theta(x, y)$ all belong to a single parametric family. Then given $X_k = x$, Y_k has conditional density $g_{\phi(x)}$ say, where $\phi(x)$ is a function mapping the current state x into the parameter space Φ of the parametric family of densities. Now assume that the class of all mixtures of this family of densities is identifiable, and that we are given a true parameter $\theta_\star$ of the model as well as an equivalent other parameter θ. Associated with these two parameters are two mappings $\phi_\star(x)$ and $\phi(x)$, respectively, as above. As $\theta_\star$ and θ are equivalent, the n-dimensional restrictions of $\mathrm{P}_{\theta_\star}$ and P_θ coincide; that is, $\mathrm{P}_{\theta_\star}(Y_{1:n} \in \cdot)$ and $\mathrm{P}_\theta(Y_{1:n} \in \cdot)$ agree. Because the class of all mixtures of (g_ϕ) is identifiable, Theorem 12.4.4 tells us that the n-dimensional distributions of the processes $\{\phi_\star(X_k)\}$ and $\{\phi(X_k)\}$ agree. That is, for all subsets $A \subseteq \Phi^n$,

$$\begin{aligned} &\mathrm{P}_{\theta_\star}\{(\phi_\star(X_1), \phi_\star(X_2), \ldots, \phi_\star(X_n)) \in A\} \\ &\quad = \mathrm{P}_\theta\{(\phi(X_1), \phi(X_2), \ldots, \phi(X_n)) \in A\} . \end{aligned}$$

This condition is often informative for concluding $\theta = \theta_\star$.

Example 12.4.5 (Normal HMM). Assume that X is finite, say $\mathsf{X} = \{1, 2, \ldots, r\}$, and that $Y_k | X_k = i \sim \mathrm{N}(\mu_i, \sigma^2)$. The parameters of the model are the transition probabilities q_{ij} of $\{X_k\}$, the μ_i and σ^2. We thus identify $\phi(x) = \mu_x$. If $\theta_\star$ and θ are two equivalent parameters, the laws of the processes $\{\mu_{\star X_k}\}$ and $\{\mu_{X_k}\}$ are thus the same, and in addition $\sigma_\star^2 = \sigma^2$. Here $\mu_{\star i}$ denotes the μ_i-component of $\theta_\star$, etc. Assuming the $\mu_{\star i}$ to be distinct, this can only happen if the *sets* $\{\mu_{\star 1}, \ldots, \mu_{\star r}\}$ and $\{\mu_1, \ldots, \mu_r\}$ are identical. We may thus conclude that the sets of means must be the same for both parameters, but they need not be enumerated in the same order. Thus there is a permutation $\{c(1), c(2), \ldots, c(r)\}$ of $\{1, 2, \ldots, r\}$ such that $\mu_{c(i)} = \mu_{\star i}$ for all $i \in \mathsf{X}$. Now because the laws of $\{\mu_{\star X_k}\}$ under $\mathrm{P}_{\theta_\star}$ and $\{\mu_{c(X_k)}\}$ under P_θ coincide with the μ_is being distinct, we conclude that the laws of $\{X_k\}$ under $\mathrm{P}_{\theta_\star}$ and of $\{c(X_k)\}$ under P_θ also agree, which in turn implies $q_{\star ij} = q_{c(i),c(j)}$ for all $i, j \in \mathsf{X}$.

Hence any parameter θ that is equivalent to $\theta_\star$ is in fact identical, up to a permutation of state indices. Sometimes the parameter space is restricted by, for instance, requiring the means μ_i to be sorted: $\mu_1 < \mu_2 < \ldots < \mu_r$, which removes the ambiguity. Such a restriction is not always desirable though; for example, in a Bayesian framework, it destroys exchangeability of the parameter in the posterior distribution (see Chapter 13).

In the current example, we could also have allowed the variance σ^2 to depend on the state, $Y_k | X_k = i \sim \mathrm{N}(\mu_i, \sigma_i^2)$, reaching the same conclusion. The assumption of conditional normality is of course not crucial either; any family of distributions for which finite mixtures are identifiable would do. ■

Example 12.4.6 (General Stochastic Volatility). In this example, we consider a stochastic volatility model of the form $Y_k | X_k = x \in \mathrm{N}(0, \sigma^2(x))$, where $\sigma^2(x)$ is a mapping from X to $\mathbb{R}_+$. Thus, we identify $\phi(x) = \sigma^2(x)$. Again assume that we are given a true parameter $\theta_\star$ as well as another parameter θ, which is equivalent to $\theta_\star$. Because all variance mixtures of normal distributions are identifiable, the laws of $\{\sigma_\star^2(X_k)\}$ under $\mathrm{P}_{\theta_\star}$ and of $\{\sigma^2(X_k)\}$ under P_θ agree. Assuming for instance that $\sigma_\star^2(x) = \sigma^2(x) = x$ (and hence also $\mathsf{X} \subseteq \mathbb{R}_+$), we conclude that the laws of $\{X_k\}$ under $\mathrm{P}_{\theta_\star}$ and P_θ, respectively, agree. For particular models of the transition kernel Q of $\{X_k\}$, such as the finite case of the previous example, we may then be able to show that $\theta = \theta_\star$, possibly up to a permutation of state indices. ■

Example 12.4.7. Sometimes a model with finite state space is identifiable even though the conditional densities $g(x, \cdot)$ are identical for several x. For instance, consider a model on the state space $\mathsf{X} = \{0, 1, 2\}$ with $Y_k | X_k = i \sim \mathrm{N}(\mu_i, \sigma^2)$, the constraints $\mu_0 = \mu_1 < \mu_2$, and transition probability matrix

$$Q = \begin{pmatrix} q_{00} & q_{01} & 0 \\ q_{10} & q_{11} & q_{12} \\ 0 & q_{21} & q_{22} \end{pmatrix} .$$

The Markov chain $\{X_k\}$ is thus a (discrete-time) birth-and-death process in the sense that it can change its state index by at most one in each step. This model is similar to models used in modeling ion channel dynamics (cf. Fredkin and Rice, 1992). Because $\mu_1 < \mu_2$, we could then think of states 0 and 1 as "closed" and of state 2 as "open".

Now assume that θ is equivalent to $\theta_\star$. Just as in Example 12.4.5, we may then conclude that the law of $\{\mu_{\star X_k}\}$ under $\mathrm{P}_{\theta_\star}$ and that of $\{\mu_{X_k}\}$ under P_θ agree, and hence, because of the constraints on the μs, that the laws of $\{\mathbb{1}(X_k \in \{0,1\}) + \mathbb{1}(X_k = 2)\}$ under $\mathrm{P}_{\theta_\star}$ and P_θ agree. In other words, after lumping states 0 and 1 of the Markov chain we obtain processes with identical laws. This in particular implies that the distributions under $\mathrm{P}_{\theta_\star}$ and P_θ of the sojourn times in the state aggregate $\{0,1\}$ coincide. The probability of such a sojourn having length 1 is q_{12}, whence $q_{12} = q_{\star 12}$ must hold. For length 2, the corresponding probability is $q_{11}q_{12}$, whence $q_{11} = q_{\star 11}$ follows and then also $q_{10} = q_{\star 10}$ as rows of Q sum up to unity. For length 3, the probability is $q_{11}^2 q_{12} + q_{10}q_{01}q_{12}$, so that finally $q_{01} = q_{\star 01}$ and $q_{00} = q_{\star 00}$. We may thus conclude that $\theta = \theta_\star$, that is, the model is identifiable. The reason that identifiability holds despite the means μ_i being non-distinct is the special structure of Q. For further reading on identifiability of lumped Markov chains, see Ito *et al.* (1992). ∎

12.5 Asymptotic Normality of the Score and Convergence of the Observed Information

We now turn to asymptotic properties of the score function and the observed information. The score function will be discussed in some detail, whereas for the information matrix we will just state the results.

12.5.1 The Score Function and Invoking the Fisher Identity

Define the score function

$$\nabla_\theta \ell_{x_0,n}(\theta) = \sum_{k=0}^{n} \nabla_\theta \log \left[\int g_\theta(x_k, Y_k)\, \mathrm{P}_\theta(X_k \in dx_k \mid Y_{0:k-1}, X_0 = x_0) \right] . \tag{12.19}$$

To make sure that this gradient indeed exists and is well-behaved enough for our purposes, we make the following assumptions.

Assumption 12.5.1. *There exists an open neighborhood $\mathcal{U} = \{\theta : |\theta - \theta_\star| < \delta\}$ of $\theta_\star$ such that the following hold.*

(i) For all $(x, x') \in \mathsf{X} \times \mathsf{X}$ and all $y \in \mathsf{Y}$, the functions $\theta \mapsto q_\theta(x, x')$ and $\theta \mapsto g_\theta(x, y)$ are twice continuously differentiable on $\mathcal{U}$.

(ii)

$$\sup_{\theta \in \mathcal{U}} \sup_{x,x'} \|\nabla_\theta \log q_\theta(x,x')\| < \infty$$

and

$$\sup_{\theta \in \mathcal{U}} \sup_{x,x'} \|\nabla_\theta^2 \log q_\theta(x,x')\| < \infty \, .$$

(iii)

$$\mathrm{E}_{\theta_\star}\left[\sup_{\theta \in \mathcal{U}} \sup_{x} \|\nabla_\theta \log g_\theta(x,Y_1)\|^2\right] < \infty$$

and

$$\mathrm{E}_{\theta_\star}\left[\sup_{\theta \in \mathcal{U}} \sup_{x} \|\nabla_\theta^2 \log g_\theta(x,Y_1)\|\right] < \infty \, .$$

(iv) For μ-almost all $y \in \mathsf{Y}$, there exists a function $f_y : \mathsf{X} \to \mathbb{R}_+$ in $L^1(\lambda)$ such that $\sup_{\theta \in \mathcal{U}} g_\theta(x,y) \le f_y(x)$.

(v) For λ-almost all $x \in \mathsf{X}$, there exist functions $f_x^1 : \mathsf{Y} \to \mathbb{R}_+$ and $f_x^2 : \mathsf{Y} \to \mathbb{R}_+$ in $L^1(\mu)$ such that $\|\nabla_\theta g_\theta(x,y)\| \le f_x^1(y)$ and $\|\nabla_\theta^2 g_\theta(x,y)\| \le f_x^2(y)$ for all $\theta \in \mathcal{U}$.

These assumptions assure that the log-likelihood is twice continuously differentiable, and also that the score function and observed information have finite moments of order two and one, respectively, under $\mathrm{P}_{\theta_\star}$. The assumptions are natural extensions of standard assumptions that are used to prove asymptotic normality of the MLE for i.i.d. observations. The asymptotic results to be derived below are valid also for likelihoods obtained using a distribution ν_θ for X_0 (such as the stationary one), provided this distribution satisfies conditions similar to the above ones: for all $x \in \mathsf{X}$, $\theta \mapsto \nu_\theta(x)$ is twice continuously differentiable on $\mathcal{U}$, and the first and second derivatives of $\theta \mapsto \log \nu_\theta(x)$ are bounded uniformly over $\theta \in \mathcal{U}$ and $x \in \mathsf{X}$.

We shall now study the score function and its asymptotics in detail. Even though the log-likelihood is differentiable, one must take some care to arrive at an expression for the score function that is useful. A tool that is often useful in the context of models with incompletely observed data is the so-called *Fisher identity*, which we encountered in Section 10.1.3. Invoking this identity, which holds in a neighborhood of $\theta_\star$ under Assumption 12.5.1, we find that (cf. (10.29))

$$\nabla_\theta \ell_{x_0,n}(\theta) = \nabla_\theta \log g_\theta(x_0,Y_0) + \mathrm{E}_\theta\left[\sum_{k=1}^n \phi_\theta(X_{k-1},X_k,Y_k) \,\middle|\, Y_{0:n}, X_0 = x_0\right] , \tag{12.20}$$

where $\phi_\theta(x,x',y') = \nabla_\theta \log[q_\theta(x,x')g_\theta(x',y')]$. However, just as when we obtained a law of large numbers for the normalized log-likelihood, we want to express the score function as a sum of increments, conditional scores. For that purpose we write

$$\nabla_\theta \ell_{x_0,n}(\theta) = \nabla_\theta \ell_{x_0,0}(\theta) + \sum_{k=1}^{n} \{\nabla_\theta \ell_{x_0,k}(\theta) - \nabla_\theta \ell_{x_0,k-1}(\theta)\} = \sum_{k=0}^{n} \dot{h}_{k,0,x_0}(\theta) , \tag{12.21}$$

where $\dot{h}_{0,0,x_0} = \nabla_\theta \log g_\theta(x_0, Y_0)$ and, for $k \geq 1$,

$$\dot{h}_{k,0,x}(\theta) = \mathrm{E}_\theta \left[\sum_{i=1}^{k} \phi_\theta(X_{i-1}, X_i, Y_i) \,\middle|\, Y_{0:k}, X_0 = x \right] - \mathrm{E}_\theta \left[\sum_{i=1}^{k-1} \phi_\theta(X_{i-1}, X_i, Y_i) \,\middle|\, Y_{0:k-1}, X_0 = x \right] .$$

Note that $\dot{h}_{k,0,x}(\theta)$ is the gradient with respect to θ of the conditional log-likelihood $h_{k,0,x}(\theta)$ as defined in (12.7). It is a matter of straightforward algebra to check that (12.20) and (12.21) agree.

12.5.2 Construction of the Stationary Conditional Score

We can extend, for any integers $k \geq 1$ and $m \geq 0$, the definition of $\dot{h}_{k,0,x}(\theta)$ to

$$\dot{h}_{k,m,x}(\theta) = \mathrm{E}_\theta \left[\sum_{i=-m+1}^{k} \phi_\theta(X_{i-1}, X_i, Y_i) \,\middle|\, Y_{-m+1:k}, X_{-m} = x \right] - \mathrm{E}_\theta \left[\sum_{i=-m+1}^{k-1} \phi_\theta(X_{i-1}, X_i, Y_i) \,\middle|\, Y_{-m+1:k-1}, X_{-m} = x \right]$$

with the aim, just as before, to let $m \to \infty$. This will yield a definition of $\dot{h}_{k,\infty}(\theta)$; the dependence on x will vanish in the limit. Note however that the construction below does not show that this quantity is in fact the gradient of $h_{k,\infty}(\theta)$, although one can indeed prove that this is the case.

As noted in Section 12.1, we want to prove a central limit theorem (CLT) for the score function evaluated at the true parameter. A quite general way to do that is to recognize that the corresponding score increments form, under reasonable assumptions, a martingale increment sequence with respect to the filtration generated by the observations. This sequence is not stationary though, so one must either use a general martingale CLT or first approximate the sequence by a stationary martingale increment sequence. We will take the latter approach, and our approximating sequence is nothing but $\{\dot{h}_{k,\infty}(\theta_\star)\}$.

We now proceed to the construction of $\dot{h}_{k,\infty}(\theta)$. First write $\dot{h}_{k,m,x}(\theta)$ as

$$\dot{h}_{k,m,x}(\theta) = \mathrm{E}_\theta[\phi_\theta(X_{k-1}, X_k, Y_k) \,|\, Y_{-m+1:k}, X_{-m} = x] + \sum_{i=-m+1}^{k-1} (\mathrm{E}_\theta[\phi_\theta(X_{i-1}, X_i, Y_i) \,|\, Y_{-m+1:k}, X_{-m} = x] - \mathrm{E}_\theta[\phi_\theta(X_{i-1}, X_i, Y_i) \,|\, Y_{-m+1:k-1}, X_{-m} = x]) . \tag{12.22}$$

The following result shows that it makes sense to take the limit as $m \to \infty$ in the previous display.

Proposition 12.5.2. *Assume 12.0.1, 12.2.1, and 12.5.1 hold. Then for any integers* $1 \le i \le k$, *the sequence* $\{\mathrm{E}_\theta[\phi_\theta(X_{i-1}, X_i, Y_i) \,|\, Y_{-m+1:k}, X_{-m} = x]\}_{m \ge 0}$ *converges* $\mathrm{P}_{\theta_\star}$*-a.s. and in* $L^2(\mathrm{P}_{\theta_\star})$, *uniformly with respect to* $\theta \in \mathcal{U}$ *and* $x \in \mathsf{X}$, *as* $m \to \infty$. *The limit does not depend on* x.

We interpret and write this limit as $\mathrm{E}_\theta[\phi_\theta(X_{i-1}, X_i, Y_i) \,|\, Y_{-\infty:k}]$.

Proof. The proof is entirely similar to that of Proposition 12.3.3. For any $(x, x') \in \mathsf{X} \times \mathsf{X}$ and non-negative integers $m' \ge m$,

$$
\begin{aligned}
&\big|\mathrm{E}_\theta[\phi_\theta(X_{i-1}, X_i, Y_i) \,|\, Y_{-m+1:k}, X_{-m} = x] \\
&\qquad\qquad - \mathrm{E}_\theta[\phi_\theta(X_{i-1}, X_i, Y_i) \,|\, Y_{-m'+1:k}, X_{-m'} = x']\big| \\
&\qquad = \bigg| \iiint \phi_\theta(x_{i-1}, x_i, Y_i)\, Q_\theta(x_{i-1}, dx_i) \\
&\qquad\qquad \times \mathrm{P}_\theta(X_{i-1} \in dx_{i-1} \,|\, Y_{-m+1:k}, X_{-m} = x_{-m}) \\
&\qquad\qquad \times [\delta_x(dx_{-m}) - \mathrm{P}_\theta(X_{-m} \in dx_{-m} \,|\, Y_{-m'+1:k}, X_{-m'} = x')] \bigg| \\
&\qquad \le 2 \sup_{x, x'} \|\phi_\theta(x, x', Y_i)\| \rho^{(i-1)+m} ,
\end{aligned}
\tag{12.23}
$$

where the inequality stems from (12.5). Setting $x = x'$ in this display shows that $\{\mathrm{E}_\theta[\phi_\theta(X_{i-1}, X_i, Y_i) \,|\, Y_{-m+1:k}, X_{-m} = x]\}_{m \ge 0}$ is a Cauchy sequence, thus converging $\mathrm{P}_{\theta_\star}$-a.s. The inequality also shows that the limit does not depend on x. Moreover, because for any non-negative integer m, $x \in \mathsf{X}$ and $\theta \in \mathcal{U}$,

$$
\|\mathrm{E}_\theta[\phi_\theta(X_{i-1}, X_i, Y_i) \,|\, Y_{-m+1:k}, X_{-m} = x]\| \le \sup_{x, x'} \|\phi_\theta(x, x', Y_i)\|
$$

with the right-hand side belonging to $L^2(\mathrm{P}_{\theta_\star})$. The inequality (12.23) thus also shows that $\{\mathrm{E}_\theta[\phi_\theta(X_{i-1}, X_i, Y_i) \,|\, Y_{-m+1:k}, X_{-m} = x]\}_{m \ge 0}$ is a Cauchy sequence in $L^2(\mathrm{P}_{\theta_\star})$ and hence converges in $L^2(\mathrm{P}_{\theta_\star})$. □

With the sums arranged as in (12.22), we can let $m \to \infty$ and define, for $k \ge 1$,

$$
\begin{aligned}
&\dot{h}_{k,\infty}(\theta) = \mathrm{E}_\theta[\phi_\theta(X_{k-1}, X_k, Y_k) \,|\, Y_{-\infty:k}] \\
&\quad + \sum_{i=-\infty}^{k-1} \left(\mathrm{E}_\theta[\phi_\theta(X_{i-1}, X_i, Y_i) \,|\, Y_{-\infty:k}] - \mathrm{E}_\theta[\phi_\theta(X_{i-1}, X_i, Y_i) \,|\, Y_{-\infty:k-1}] \right) .
\end{aligned}
$$

The following result gives an L^2-bound on the difference between $\dot{h}_{k,m,x}(\theta)$ and $\dot{h}_{k,\infty}(\theta)$.

Lemma 12.5.3. *Assume 12.0.1, 12.2.1, 12.3.1, and 12.5.1 hold. Then for* $k \geq 1$,

$$(\mathrm{E}_\theta \|\dot{h}_{k,m,x}(\theta) - \dot{h}_{k,\infty}(\theta)\|^2)^{1/2} \\ \leq 12 \left(\mathrm{E}_\theta \left[\sup_{x,x' \in \mathsf{X}} \|\phi_\theta(x, x', Y_1)\|^2 \right] \right)^{1/2} \frac{\rho^{(k+m)/2-1}}{1-\rho} .$$

Proof. The idea of the proof is to match, for each index i of the sums expressing $\dot{h}_{k,m,x}(\theta)$ and $\dot{h}_{k,\infty}(\theta)$, pairs of terms that are close. To be more precise, we match

1. The first terms of $\dot{h}_{k,m,x}(\theta)$ and $\dot{h}_{k,\infty}(\theta)$;
2. For i close to k,

   $$\mathrm{E}_\theta[\phi_\theta(X_{i-1}, X_i, Y_i) \,|\, Y_{-m+1:k}, X_{-m} = x]$$

   and

   $$\mathrm{E}_\theta[\phi_\theta(X_{i-1}, X_i, Y_i) \,|\, Y_{-\infty:k}] ,$$

   and similarly for the corresponding terms conditioned on $Y_{-m+1:k-1}$ and $Y_{-\infty:k-1}$, respectively;
3. For i far from k,

   $$\mathrm{E}_\theta[\phi_\theta(X_{i-1}, X_i, Y_i) \,|\, Y_{-m+1:k}, X_{-m} = x]$$

   and

   $$\mathrm{E}_\theta[\phi_\theta(X_{i-1}, X_i, Y_i) \,|\, Y_{-m+1:k-1}, X_{-m} = x] ,$$

   and similarly for the corresponding terms conditioned on $Y_{-\infty:k}$ and $Y_{-\infty:k-1}$, respectively.

We start with the second kind of matches (of which the first terms are a special case). Taking the limit in $m' \to \infty$ in (12.23), we see that

$$\| \mathrm{E}_\theta[\phi_\theta(X_{i-1}, X_i, Y_i) \,|\, Y_{-m+1:k}, X_{-m} = x] - \mathrm{E}_\theta[\phi_\theta(X_{i-1}, X_i, Y_i) \,|\, Y_{-\infty:k}]\| \\ \leq 2 \sup_{x,x' \in \mathsf{X}} \|\phi_\theta(x, x', Y_i)\| \rho^{(i-1)+m} .$$

This bound remains the same if k is replaced by $k-1$. Obviously, it is small if i is far away from m, that is, close to k.

For the third kind of matches, we need a total variation bound that works "backwards in time". Such a bound reads

$$\|\mathrm{P}_\theta(X_i \in \cdot \,|\, Y_{-m+1:k}, X_{-m} = x) \\ - \mathrm{P}_\theta(X_i \in \cdot \,|\, Y_{-m+1:k-1}, X_{-m} = x)\|_{\mathrm{TV}} \leq \rho^{k-1-i} .$$

The proof of this bound is similar to that of Proposition 4.3.23 and uses the time-reversed process. We postpone the proof to the end of this section. We may also let $m \to \infty$ and omit the condition on X_{-m} without affecting the bound. As a result of these bounds, we have

$$\begin{aligned}\| \mathrm{E}_\theta[\phi_\theta(X_{i-1}, X_i, Y_i) \mid Y_{-m+1:k}, X_{-m} = x] \\ - \mathrm{E}_\theta[\phi_\theta(X_{i-1}, X_i, Y_i) \mid Y_{-m+1:k-1}, X_{-m} = x]\| \\ \leq 2 \sup_{x,x' \in \mathsf{X}} \|\phi_\theta(x, x', Y_i)\| \rho^{k-1-i} ,\end{aligned}$$

with the same bound being valid if the conditioning is on $Y_{-\infty:k}$ and $Y_{-\infty:k-1}$, respectively. This bound is small if i is far away from k.

Combining these two kinds of bounds and using Minkowski's inequality for the L^2-norm, we find that $(\mathrm{E}_\theta \|\dot{h}_{k,m,x}(\theta) - \dot{h}_{k,\infty}(\theta)\|^2)^{1/2}$ is bounded by

$$\begin{aligned}&2\rho^{k+m-1} + 2 \times 2 \sum_{i=-m+1}^{k-1} (\rho^{k-i-1} \wedge \rho^{i+m-1}) + 2 \sum_{i=-\infty}^{-m} \rho^{k-i-1} \\ &\leq 4 \frac{\rho^{k+m-1}}{1-\rho} + 4 \sum_{-\infty < i \leq (k-m)/2} \rho^{k-i-1} + 4 \sum_{(k-m)/2 \leq i < \infty} \rho^{i+m-1} \\ &\leq 12 \frac{\rho^{(k+m)/2-1}}{1-\rho}\end{aligned}$$

up to the factor $(\mathrm{E}_\theta \sup_{x,x' \in \mathsf{X}} \|\phi_\theta(x, x', Y_i)\|^2)^{1/2}$. The proof is complete. □

We now establish the "backwards in time" uniform forgetting property, which played a key role in the above proof.

Proposition 12.5.4. *Assume 12.0.1, 12.2.1, and 12.3.1 hold. Then for any integers i, k, and m such that $m \geq 0$ and $-m < i < k$, any $x_{-m} \in \mathsf{X}$, $y_{-m+1:k} \in \mathsf{Y}^{k+m}$, and $\theta \in \mathcal{U}$,*

$$\begin{aligned}\|\mathrm{P}_\theta(X_i \in \cdot \mid Y_{-m+1:k} = y_{-m+1:k}, X_{-m} = x_{-m}) \\ - \mathrm{P}_\theta(X_i \in \cdot \mid Y_{-m+1:k-1} = y_{-m+1:k-1}, X_{-m} = x_{-m})\|_{\mathrm{TV}} \leq \rho^{k-1-i} .\end{aligned}$$

Proof. The cornerstone of the proof is the observation that conditional on $Y_{-m+1:k}$ and X_{-m}, the time-reversed process X with indices from k down to $-m$ is a non-homogeneous Markov chain satisfying a uniform mixing condition. We shall indeed use a slight variant of the backward decomposition developed in Section 3.3.2. For any $j = -m+1, \ldots, k-1$, we thus define the backward kernel (cf. (3.39)) by

$$\mathrm{B}_{x_{-m},j}[y_{-m+1:j}](x, f) = \frac{\int \cdots \int \prod_{u=-m+1}^{j} q(x_{u-1}, x_u) g(x_u, y_u)\, \lambda(dx_u)\, f(x_j) q(x_j, x)}{\int \cdots \int \prod_{u=-m+1}^{j} q(x_{u-1}, x_u) g(x_u, y_u)\, \lambda(dx_u)\, q(x_j, x)} \tag{12.24}$$

for any $f \in \mathcal{F}_b(\mathsf{X})$. For brevity, we do not indicate the dependence of the quantities involved on θ. We note that the integral of the denominator of this display is bounded from below by $(\sigma^-)^{m+j} \prod_{-m+1}^{j} \int g_\theta(x_u, y_u)\, \lambda(dx_u)$, and is hence positive $\mathrm{P}_{\theta_\star}$-a.s. under Assumption 12.3.1.

It is trivial that for any $x \in \mathsf{X}$,

$$\int \cdots \int \prod_{u=-m+1}^{j} q(x_{u-1}, x_u) g(x_u, y_u)\, \lambda(dx_u)\, f(x_j) q(x_j, x) =$$

$$\int \cdots \int \prod_{u=-m+1}^{j} q(x_{u-1}, x_u) g(x_u, y_u)\, \lambda(dx_u) q(x_j, x) \mathrm{B}_{x_{-m},j}[y_{-m+1:j}](x, f) ,$$

which implies that

$$\begin{aligned}\mathrm{E}_\theta[f(X_j) \,|\, X_{j+1:k}, Y_{-m+1:k} = y_{-m+1:k}, X_{-m} = x] \\ = \mathrm{B}_{x_{-m},j}[y_{-m+1:j}](X_{j+1}, f) .\end{aligned}$$

This is the desired Markov property referred to above.

Along the same lines as in the proof of Proposition 4.3.26, we can show that the backward kernels satisfy a Doeblin condition,

$$\frac{\sigma^-}{\sigma^+} \nu_{x_{-m},j}[y_{-m+1:j}] \le \mathrm{B}_{x_{-m},j}[y_{-m+1:j}](x, \cdot) \le \frac{\sigma^+}{\sigma^-} \nu_{x_{-m},j}[y_{-m+1:j}] ,$$

where for any $f \in \mathcal{F}_b(\mathsf{X})$,

$$\nu_{x_{-m},j}[y_{-m+1:j}](f) = \frac{\int \cdots \int \prod_{u=-m+1}^{j} q_\theta(x_{u-1}, x_u) g_\theta(x_u, y_u)\, \lambda(dx_u)\, f(x_j)}{\int \cdots \int \prod_{u=-m+1}^{j} q_\theta(x_{u-1}, x_u) g_\theta(x_u, y_u)\, \lambda(dx_u)} .$$

Thus Lemma 4.3.13 shows that the Dobrushin coefficient of each backward kernel is bounded by $\rho = 1 - \sigma^-/\sigma^+$.

Finally

$$\begin{aligned}&\mathrm{P}_\theta(X_i \in \cdot \,|\, Y_{-m+1:k-1} = y_{-m+1:k-1}, X_{-m} = x_{-m}) \\ &= \int \mathrm{P}_\theta(X_i \in \cdot \,|\, Y_{-m+1:k-1} = y_{-m+1:k-1}, X_{-m} = x_{-m}, X_{k-1} = x_{k-1}) \\ &\qquad \times \mathrm{P}_\theta(X_{k-1} \in dx_{k-1} \,|\, Y_{-m+1:k-1} = y_{-m+1:k-1}, X_{-m} = x_{-m})\end{aligned}$$

and

$$\begin{aligned}&\mathrm{P}_\theta(X_i \in \cdot \,|\, Y_{-m+1:k} = y_{-m+1:k}, X_{-m} = x_{-m}) \\ &= \int \mathrm{P}_\theta(X_i \in \cdot \,|\, Y_{-m+1:k-1} = y_{-m+1:k-1}, X_{-m} = x_{-m}, X_{k-1} = x_{k-1}) \\ &\qquad \times \mathrm{P}_\theta(X_{k-1} \in dx_{k-1} \,|\, Y_{-m+1:k} = y_{-m+1:k}, X_{-m} = x_{-m}) ,\end{aligned}$$

so that the two distributions on the left-hand sides can be considered as the result of running the above-described reversed conditional Markov chain from index $k-1$ down to index i, using two different initial conditions. Therefore, by Proposition 4.3.10, they differ by at most ρ^{k-1-i} in total variation distance. The proof is complete. □

12.5.3 Weak Convergence of the Normalized Score

We now return to the question of a weak limit of the normalized score $n^{-1/2}\sum_{k=0}^{n}\dot{h}_{k,0,x_0}(\theta_\star)$. Using Lemma 12.5.3 and Minkowski's inequality, we see that

$$\left[\mathrm{E}_{\theta_\star}\left\|n^{-1/2}\sum_{k=0}^{n}(\dot{h}_{k,0,x_0}(\theta_\star)-\dot{h}_{k,\infty}(\theta_\star))\right\|^2\right]^{1/2}$$
$$\le n^{-1/2}\sum_{k=0}^{n}\left[\mathrm{E}_{\theta_\star}\|\dot{h}_{k,0,x_0}(\theta_\star)-\dot{h}_{k,\infty}(\theta_\star)\|^2\right]^{1/2}\to 0 \quad \text{as } n\to\infty\,,$$

whence the limiting behavior of the normalized score agrees with that of $n^{-1/2}\sum_{k=0}^{n}\dot{h}_{k,\infty}(\theta_\star)$. Now define the filtration $\mathcal{F}$ by $\mathcal{F}_k=\sigma(Y_i,\ -\infty<i\le k)$ for all integer k. By conditional dominated convergence,

$$\mathrm{E}_{\theta_\star}\left[\sum_{i=-\infty}^{k-1}(\mathrm{E}_{\theta_\star}[\phi_{\theta_\star}(X_{i-1},X_i,Y_i)\,|\,Y_{-\infty:k}]\right.$$
$$\left.-\ \mathrm{E}_{\theta_\star}[\phi_{\theta_\star}(X_{i-1},X_i,Y_i)\,|\,Y_{-\infty:k-1}])\,|\,\mathcal{F}_{k-1}\right]=0\,,$$

and Assumption 12.5.1 implies that

$$\mathrm{E}_{\theta_\star}[\phi_{\theta_\star}(X_{k-1},X_k,Y_k)\,|\,Y_{-\infty:k-1}]$$
$$=\mathrm{E}_{\theta_\star}[\mathrm{E}_{\theta_\star}[\phi_{\theta_\star}(X_{k-1},X_k,Y_k)\,|\,Y_{-\infty:k-1},X_{k-1}]\,|\,\mathcal{F}_{k-1}]=0\,.$$

It is also immediate that $\dot{h}_{k,\infty}(\theta_\star)$ is $\mathcal{F}_k$-measurable. Hence the sequence $\{\dot{h}_{k,\infty}(\theta_\star)\}_{k\ge 0}$ is a $\mathrm{P}_{\theta_\star}$-martingale increment sequence with respect to the filtration $\{\mathcal{F}_k\}_{k\ge 0}$ in $L^2(\mathrm{P}_{\theta_\star})$. Moreover, this sequence is stationary because $\{Y_k\}_{-\infty<k<\infty}$ is. Any stationary martingale increment sequence in $L^2(\mathrm{P}_{\theta_\star})$ satisfies a CLT (Durrett, 1996, p. 418), that is, $n^{-1/2}\sum_0^n\dot{h}_{k,\infty}(\theta_\star)\to \mathrm{N}(0,\mathcal{J}(\theta_\star))$ $\mathrm{P}_{\theta_\star}$-weakly, where

$$\mathcal{J}(\theta_\star)\stackrel{\text{def}}{=}\mathrm{E}_{\theta_\star}[\dot{h}_{1,\infty}(\theta_\star)\dot{h}^t_{1,\infty}(\theta_\star)] \tag{12.25}$$

is the limiting Fisher information.

Because the normalized score function has the same limiting behavior, the following result is immediate.

Theorem 12.5.5. *Under Assumptions 12.0.1, 12.2.1, 12.3.1, and 12.5.1,*

$$n^{-1/2}\nabla_\theta\ell_{x_0,n}(\theta_\star)\to\mathrm{N}(0,\mathcal{J}(\theta_\star))\quad \mathrm{P}_{\theta_\star}\text{-weakly}$$

for all $x_0\in\mathsf{X}$, where $\mathcal{J}(\theta_\star)$ is the limiting Fisher information as defined above.

We remark that above, we have normalized sums with indices from 0 to n, that is, with $n+1$ terms, by $n^{1/2}$ rather than by $(n+1)^{1/2}$. This of course does not affect the asymptotics. However, if $\mathcal{J}(\theta_\star)$ is estimated for the purpose of making a confidence interval for instance, then one may well normalize it using the number $n+1$ of observed data.

12.5.4 Convergence of the Normalized Observed Information

We shall now very briefly discuss the asymptotics of the observed information matrix, $-\nabla^2_\theta \ell_{x_0,n}(\theta)$. To handle this matrix, one can employ the so-called *missing information principle* (see Section 10.1.3 and (10.30)). Because the complete information matrix, just as the complete score, has a relatively simple form, this principle allows us to study the asymptotics of the observed information in a fashion similar to what was done above for the score function. The analysis becomes more difficult however, as covariance terms, arising from the conditional variance of the complete score, also need to be accounted for. In addition, we need the convergence to be uniform in a certain sense. We state the following theorem, whose proof can be found in Douc *et al.* (2004).

Theorem 12.5.6. *Under Assumptions 12.0.1, 12.2.1, 12.3.1, and 12.5.1,*

$$\lim_{\delta \to 0} \lim_{n \to \infty} \sup_{|\theta - \theta_\star| \le \delta} \|(-n^{-1} \nabla^2_\theta \ell_{x_0,n}(\theta)) - \mathcal{J}(\theta_\star)\| = 0 \quad \mathrm{P}_{\theta_\star}\text{-a.s.}$$

for all $x_0 \in \mathsf{X}$.

12.5.5 Asymptotics of the Maximum Likelihood Estimator

The general arguments in Section 12.1 and the theorems above prove the following result.

Theorem 12.5.7. *Assume 12.0.1, 12.2.1, 12.3.1, 12.3.5, and 12.5.1, and that* $\theta_\star$ *is identifiable, that is,* θ *is equivalent to* $\theta_\star$ *only if* $\theta = \theta_\star$ *(possibly up to a permutation of states if* X *is finite). Then the following hold true.*

(i) The MLE $\widehat{\theta}_n = \widehat{\theta}_{x_0,n}$ *is strongly consistent:* $\widehat{\theta}_n \to \theta_\star$ $\mathrm{P}_{\theta_\star}$*-a.s. as* $n \to \infty$.

(ii) If the Fisher information matrix $\mathcal{J}(\theta_\star)$ *defined above is non-singular and* $\theta_\star$ *is an interior point of* Θ*, then the MLE is asymptotically normal:*

$$n^{1/2}(\widehat{\theta}_n - \theta_\star) \to \mathrm{N}(0, \mathcal{J}(\theta_\star)^{-1}) \quad \mathrm{P}_{\theta_\star}\text{-weakly as } n \to \infty$$

for all $x_0 \in \mathsf{X}$.

(iii) The normalized observed information at the MLE is a strongly consistent estimator of $\mathcal{J}(\theta_\star)$:

$$-n^{-1} \nabla^2_\theta \ell_{x_0,n}(\widehat{\theta}_n) \to \mathcal{J}(\theta_\star) \quad \mathrm{P}_{\theta_\star}\text{-a.s. as } n \to \infty.$$

As indicated above, the MLE $\widehat{\theta}_n$ depends on the initial state x_0, but that dependence will generally not be included in the notation.

The last part of the result is important, as is says that confidence intervals or regions and hypothesis tests based on the estimate $-(n+1)^{-1} \nabla^2_\theta \ell_{x_0,n}(\widehat{\theta}_n)$ of $\mathcal{J}(\theta_\star)$ will asymptotically be of correct size. In general, there is no closed-form expression for $\mathcal{J}(\theta_\star)$, so that it needs to be estimated in one way or another. The observed information is obviously one way to do that, while another one is to simulate data $Y_{1:N}^*$ from the HMM, using the MLE, and then computing $-(N+1)^{-1}\nabla_\theta^2 \ell_{x_0,N}(\widehat{\theta}_n)$ for this set of simulated data and some x_0. An advantage of this approach is that N can be chosen arbitrarily large. Yet another approach, motivated by (12.25), is to estimate the Fisher information by the empirical covariance matrix of the conditional scores of (12.19) at the MLE, that is, by $(n+1)^{-1}\sum_0^n [S_{k|k-1}(\widehat{\theta}_n) - \bar{S}(\widehat{\theta}_n)][S_{k|k-1}(\widehat{\theta}_n) - \bar{S}(\widehat{\theta}_n)]^t$ with $S_{k|k-1}(\theta) = \nabla_\theta \log \int g_\theta(x, Y_k)\, \phi_{x_0,k|k-1}[Y_{0:k-1}](dx\,;\theta)$ and $\bar{S}(\theta) = (n+1)^{-1}\sum_0^n S_{k|k-1}(\theta)$. This estimate can of course also be computed from estimated data, then using an arbitrary sample size. The conditional scores may be computed as $S_{k|k-1}(\theta) = \nabla_\theta \ell_{x_0,k}(\theta) - \nabla_\theta \ell_{x_0,k-1}(\theta)$, where the scores are computed using any of the methods of Section 10.2.3.

12.6 Applications to Likelihood-based Tests

The asymptotic properties of the score function and observed information have immediate implications for the asymptotics of the MLE, as has been described in previous sections. However, there are also other conclusions that can be drawn from these convergence results.

One such application is the validity of some classical procedures for testing whether $\theta_\star$ lies in some subset, Θ_0 say, of the parameter space Θ. Suppose that Θ_0 is an $(d_\theta - s)$-dimensional subset that may be expressed in terms of constraints $R_i(\theta) = 0$, $i = 1, 2, \ldots, s$, and that there is an equivalent formulation $\theta_i = b_i(\gamma)$, $i = 1, 2, \ldots, d_\theta$, where γ is the "constrained parameter" lying in a subset Γ of $\mathbb{R}^{d_\theta - s}$. We also let $\gamma_\star$ be a point such that $\theta_\star = b(\gamma_\star)$. Each function R_i and b_i is assumed to be continuously differentiable and such that the matrices

$$C_\theta = \left(\frac{\partial R_i}{\partial \theta_j}\right)_{s \times d_\theta} \qquad \text{and} \qquad D_\gamma = \left(\frac{\partial b_i}{\partial \gamma_j}\right)_{d_\theta \times (d_\theta - s)}$$

have full rank (s and $d_\theta - s$ respectively) in a neighborhood of $\theta_\star$ and $\gamma_\star$, respectively.

Perhaps the simplest example is when we want to test a simple (point) null hypothesis $\theta_\star = \theta_0$ versus the alternative $\theta_\star \neq \theta_0$. Then, we take $R_i(\theta) = \theta_i - \theta_{0i}$ and $b_i(\gamma) = \theta_{i0}$ for $i = 1, 2, \ldots, d_\theta$. In this case, γ is void as $s = d_\theta$ and hence $d_\theta - s = 0$. Furthermore, C is the identity matrix and D is void.

Now suppose that we want to test the equality $\theta_i = \theta_{i0}$ only for i in a subset K of the d_θ coordinates of θ, where K has cardinality s. The constraints we employ are then $R_i(\theta) = \theta_i - \theta_{0i}$ for $i \in K$; furthermore, γ comprises θ_i for $i \notin K$ and, using the $d_\theta - s$ indices not in K for γ, $b_i(\gamma) = \theta_{0i}$ for $i \in K$ and $b_i(\gamma) = \gamma_i$ otherwise. Again it is easy to check that C and D are constant and of full rank.

Example 12.6.1 (Normal HMM). A slightly more involved example concerns the Gaussian hidden Markov model with finite state space $\{1, 2, \dots, r\}$ and conditional distributions $Y_k | X_k = i \sim \mathrm{N}(\mu_i, \sigma_i^2)$. Suppose that we want to test for equality of all of the r component-wise conditional variances σ_i^2: $\sigma_1^2 = \sigma_2^2 = \dots = \sigma_r^2$. Then, the R-functions are for instance $\sigma_i^2 - \sigma_r^2$ for $i = 1, 2, \dots, r-1$. The parameter γ is obtained by removing from θ all σ_i^2 and then adding a common conditional variance σ^2; those b-functions referring to any of the σ_i^2 evaluate to σ^2. The matrices C and D are again constant and of full rank. ■

A further application, to test the structure of conditional covariance matrices in a conditionally Gaussian HMM with multivariate output, can be found in Giudici *et al.* (2000).

There are many different tests available for testing the null hypothesis $\theta_\star \in \Theta_0$ versus the alternative $\theta_\star \in \Theta \setminus \Theta_0$. One is the generalized likelihood ratio test, which uses the test statistic

$$\lambda_n = 2 \left\{ \sup_{\theta \in \Theta} \ell_{x_0,n}(\theta) - \sup_{\theta \in \Theta_0} \ell_{x_0,n}(\theta) \right\} .$$

Another one is the Wald test, which uses the test statistic

$$W_n = nR(\widehat{\theta}_n)^t [C_{\widehat{\theta}_n} \mathcal{J}_n(\widehat{\theta}_n)^{-1} C^t_{\widehat{\theta}_n}]^{-1} R(\widehat{\theta}_n) ,$$

where $R(\theta)$ is the $s \times 1$ vector of R-functions evaluated at θ, and $\mathcal{J}_n(\theta) = -n^{-1} \nabla_\theta^2 \ell_{x_0,n}(\theta)$ is the observed information evaluated at θ. Yet another test is based on the Rao statistic, defined as

$$V_n = n^{-1} S_n(\widehat{\theta}_n^0) \mathcal{J}_n(\widehat{\theta}_n^0)^{-1} S_n(\widehat{\theta}_n^0)^t ,$$

where $\widehat{\theta}_n^0$ is the MLE over Θ_0, that is, the point where $\ell_{x_0,n}(\theta)$ is maximized subject to the constraint $R_i(\theta) = 0$, $1 \le i \le s$, and $S_n(\theta) = \nabla_\theta \ell_{x_0,n}(\theta)$ is the score function at θ. This test is also known under the names *efficient score test* and *Lagrange multiplier test*. The Wald and Rao test statistics are usually defined using the true Fisher information $\mathcal{J}(\theta)$ rather than the observed one, but as $\mathcal{J}(\theta)$ is generally infeasible to compute for HMMs, we replace it by the observed counterpart.

Statistical theory for i.i.d. data suggests that the likelihood ratio, Wald and Rao test statistics should all converge weakly to a χ^2 distribution with s degrees of freedom provided $\theta_\star \in \Theta_0$ holds true, so that an approximate p-value of the test of this null hypothesis can be computed by evaluating the complementary distribution function of the χ_s^2 distribution at the point λ_n, W_n, or V_n, whichever is preferred. We now state formally that this procedure is indeed correct.

Theorem 12.6.2. *Assume 12.0.1, 12.2.1, 12.3.1, 12.3.5, and 12.5.1 as well as the conditions stated on the functions R_i and b_i above. Also assume that*

$\theta_\star$ is identifiable, that is, θ is equivalent to $\theta_\star$ only if $\theta = \theta_\star$ (possibly up to a permutation of states if X is finite), that $\mathcal{J}(\theta_\star)$ is non-singular, and that $\theta_\star$ and $\gamma_\star$ are interior points of Θ and Γ, respectively. Then if $\theta_\star \in \Theta_0$ holds true, each of the test statistics λ_n, W_n, and V_n converges $\mathrm{P}_{\theta_\star}$-weakly to the χ^2_s distribution as $n \to \infty$.

The proof of this result follows, for instance, Serfling (1980, Section 4.4). The important observation is that the validity of the proof does not hinge on independence of the data but on asymptotic properties of the score function and the observed information, properties that have been established for HMMs in this chapter.

It is important to realize that a key assumption for Theorem 12.6.2 to hold is that $\theta_\star$ is identifiable, so that $\widehat{\theta}_n$ converges to a unique point $\theta_\star$. As a result, the theorem does not apply to the problem of testing the number of components of a finite state HMM. In the normal HMM for instance, with $Y_k|X_k = i \sim \mathrm{N}(\mu_i, \sigma_i^2)$, one can indeed effectively remove one component by invoking the constraints $\mu_1 - \mu_2 = 0$ and $\sigma_1^2 - \sigma_2^2 = 0$, say. In this way, within Θ_0, components 1 and 2 collapse into a single one. However, any $\theta \in \Theta_0$ is then non-identifiable as the transition probabilities q_{12} and q_{21}, among others, can be chosen arbitrarily without changing the dynamics of the model. Hence Theorem 12.6.2 does not apply, and in fact we know from Chapter 15 that the limiting distribution of the likelihood ratio test statistic for selecting the number of components in a finite state HMM is much more complex than a χ^2 distribution. The reason that Theorem 12.6.2 fails is that its proof crucially depends on a unique point $\theta_\star$ to which $\widehat{\theta}_n$ converges and around which log-likelihoods can be Taylor-expanded.

12.7 Complements

The theoretical statistical aspects of HMMs and related models have essentially been developed since 1990. The exception is the seminal paper Baum and Petrie (1966) and the follow-up Petrie (1969), which both consider HMMs for which X and Y are finite. Such HMMs can be viewed as a process obtained by lumping states of a Markov chain living on a larger set $\mathsf{X} \times \mathsf{Y}$, and this idea lies behind much of the analysis in these early papers. Yet Baum and Petrie (1966) contains the basic idea used in the current chapter, namely that of defining log-likelihoods, score functions, etc., conditional on the "infinite past", and bounds that quantify how far these variables are from their counterparts conditional on a finite past. Baum and Petrie (1966) established consistency and asymptotic normality of the MLE, while Petrie (1969) took a closer look at identifiability, and in fact a lot more, which was not studied in detail in the first paper.

Leroux (1992) was the first to carry out some analysis on more general HMMs, with finite X but general Y. He proved consistency of the MLE by an approach based on Kingman's subadditive ergodic theorem and did also provide a very useful discussion on identifiability on which much of the above one (Section 12.4) is based. Leroux's approach was thus not based on conditioning on the "infinite past"; the subadditive ergodic approach however has the drawback that it applies to analysis of the log-likelihood only and not to the score function or observed information. A few years later, Bickel and Ritov (1996) took the first steps toward an analysis of the MLE for models of the kind studied by Leroux. Their results imply so-called local asymptotic normality (LAN) of the log-likelihood, but not asymptotic normality of the MLE without some extra assumptions. This result was instead obtained by Bickel *et al.* (1998), who based their analysis on the "infinite past" approach almost entirely, employing bounds on conditional mixing rates similar to those of Baum and Petrie (1966). This analysis was generalized to models with compact X by Jensen and Petersen (1999). Finally, as mentioned above, Douc *et al.* (2004) took this approach to the point where autoregression is also allowed, using the mixing rate bound of Proposition 4.3.23. Neither Bickel *et al.* (1998) nor Jensen and Petersen (1999) used uniform forgetting to derive their bounds, but both of them can easily be stated in such terms. Higher order derivatives of the log-likelihood are studied in Bickel *et al.* (2002).

A quite different approach to studying likelihood asymptotics is to express the log-likelihood through the predictor,

$$\ell_{x_0,n}(\theta) = \sum_{k=1}^{n} \log \int_{\mathsf{X}} g_\theta(x, Y_k)\, \phi_{x_0,k|k-1}(dx;\theta) ,$$

cf. Chapter 3, and then differentiating the recursive formula (3.27) for $\phi_{x_0,k|k-1}$ with respect to θ to obtain recursive expressions for the score function and observed information. This approach is technically more involved than that using the "infinite past" but does allow for analysis of recursive estimators such as recursive maximum likelihood. Le Gland and Mevel (2000) studied the recursive approach for HMMs with finite state space, and Douc and Matias (2002) extended the results to HMMs on compact state spaces.

As good as all of the results above can be extended to Markov-switching autoregressions; see Douc *et al.* (2004). Under Assumption 12.2.1, the conditional chain then still satisfies the same favorable mixing properties as in Section 4.3. The log-likelihood, score function, and observed observation can be analyzed using the ideas exposed in this chapter; we just need to replace some of the assumptions by analogs including regressors (lagged Ys). Other papers that examine asymptotics of estimators in Markov-switching autoregressions include Francq and Roussignol (1997), Krishnamurthy and Rydén (1998), and Francq and Roussignol (1998). Markov-switching GARCH models were studied by Francq *et al.* (2001).

13

Fully Bayesian Approaches

Some previous chapters have already mentioned MCMC and conditional (or posterior) distributions, especially in the set-up of posterior state estimation and simulation. The spirit of this chapter is obviously different in that it covers the fully Bayesian processing of HMMs, which means that, besides the hidden states and their conditional (or parameterized) distributions, the model parameters are assigned probability distributions, called *prior distributions*, and the inference on these parameters is of Bayesian nature, that is, conditional on the observations (or the *data*). Because more advanced Markov chain Monte Carlo methodology is also needed for this fully Bayesian processing, additional covering of MCMC methods, like reversible jump techniques, will be given in this chapter (Section 13.2). The emphasis is put on HMMs with finite state space (X is finite), but some facts are general and the case of continuous state space is addressed at some points.

13.1 Parameter Estimation

13.1.1 Bayesian Inference

Although the whole apparatus of modern Bayesian inference cannot be discussed here (we refer the reader to, e.g., Robert, 2001, or Gelman *et al.*, 1995), we briefly recall the basics of a Bayesian analysis of a statistical model, and we also introduce some notation not used in earlier chapters.

Given a general parameterized model

$$Y \sim p(y|\theta), \quad \theta \in \Theta ,$$

where $p(y|\theta)$ thus denotes a parameterized density, the idea at the core of Bayesian analysis is to provide an inferential assessment (on θ) *conditional on the realized value of* Y, which we denote (as usual) by y. Obviously, to give a proper probabilistic meaning to this conditioning, θ itself must be embedded with a probability distribution called the *prior distribution*, which is denoted by $\pi(d\theta)$. The choice of this prior distribution is often decided on practicality grounds rather than strong subjective belief or overwhelming prior information, but there also exist less disturbing (or subjective) choices called *non-informative priors*, as we will discuss below.

Definition 13.1.1 (Bayesian Model). *A* Bayesian model *is given by the completion of a statistical model*

$$Y \sim p(y|\theta), \quad \theta \in \Theta ,$$

with a probability distribution $\pi(d\theta)$, *called the* prior distribution, *on the parameter space* Θ.

The associated posterior distribution *is given by Bayes' theorem as the conditional distribution of* θ *given the observation* y,

$$\pi(d\theta|y) = \frac{p(y|\theta)\pi(d\theta)}{\int_\Theta p(y|\xi)\,\pi(d\xi)} . \tag{13.1}$$

The density $p(y|\theta)$ is the likelihood of the model and will also be denoted by $\mathrm{L}(y|\theta)$ as in previous chapters. Note that in this chapter, we always assume that both the prior and the posterior distributions admits densities that we denote by $\pi(\theta)$ and $\pi(\theta|y)$, respectively. For the sake of notational simplicity, the dominating measure for both of these densities, whose exact specification is not important here, is denoted by $d\theta$.

Once the prior distribution is selected, Bayesian inference is, in principle, "over", that is, completely determined, as the estimation, testing, and evaluation procedures are provided by the prior and the associated loss function. For instance, if the loss function for the evaluation of estimators is the quadratic loss function

$$\mathrm{loss}(\theta, \hat{\theta}) = \|\theta - \hat{\theta}\|^2 ,$$

the corresponding Bayes procedure is the *expected* value of θ, either under the prior distribution (when no observation is available), or under the posterior distribution,

$$\hat{\theta} = \int \theta\,\pi(d\theta|y) = \frac{\int \theta\,p(y|\theta)\,\pi(d\theta)}{\int p(y|\theta)\,\pi(d\theta)} .$$

When no specific loss function is available, this estimator is often used as the default estimator, although alternatives also are available.

A specific alternative is the *maximum marginal posterior estimator*, defined as

$$\hat{\theta}_i = \arg\max_{\theta_i} \pi_i(\theta_i|y)$$

for each component θ_i of the vector θ. A difficulty with this estimator is that the *marginal posteriors*

$$\pi_i(\theta_i|y) = \int \pi(\theta|y)\,d\theta_{-i} ,$$

where $\theta_{-i} = \{\theta_j,\ j \neq i\}$, are often intractable, especially in the setting of latent variable models like HMMs.

Another alternative, not to be confused with the previous one, is the *maximum a posteriori estimator* (MAP),

$$\hat{\theta} = \arg\max_\theta \pi(\theta|y) = \arg\max_\theta \pi(\theta)p(y|\theta)\ , \tag{13.2}$$

which is thus in principle easier to compute because the function to maximize is usually provided in closed form. However, numerical problems make the optimization involved in finding the MAP far from trivial. Note also here the similarity of (13.2) with the maximum likelihood estimator: the influence of the prior distribution $\pi(\theta)$ progressively disappears with the number of observations and the MAP estimator recovers the asymptotic properties of the MLE. This is, of course, only true if the support of the distribution π contains the true value, and if latent variables like the hidden states of the HMM—the number of which grows linearly with n—are not adjoined to the parameter vector θ. See Schervish (1995) for more details on the asymptotics of Bayesian estimators.

We will discuss in more detail the important issue of selection of the prior distribution for HMMs in Section 13.1.2, but at this point we note that when the model is from an exponential family of distributions, in so-called *natural parameterization* (which corresponds to the case $\psi(\theta) = \theta$ in Definition 10.1.5),

$$p(y|\theta) = \exp\left\{\theta^t S(y) - c(\theta)\right\} h(y)\ ,$$

there exists a generic class of priors called the *class of conjugate priors*,

$$\pi(\theta|\xi,\lambda) \propto \exp\left\{\theta^t \xi - \lambda c(\theta)\right\}\ ,$$

which are parameterized by a positive real value λ and a vector ξ of the same dimension as the sufficient statistic $S(y)$. These parameterized prior distributions on θ are thus such that the posterior distribution can be written as

$$\pi(\theta|\xi,\lambda,y) = \pi[\theta|\xi'(y),\lambda'(y)]\ . \tag{13.3}$$

Equation (13.3) simply says that the conjugate prior is such that the prior and posterior densities belong to the same parametric family of densities, but with different parameters. Indeed, the parameters of the posterior density are "updated", using the observations, relative to the prior parameters. To avoid confusion, the parameters involved in the prior distribution on the model parameter are usually called *hyperparameters*.

Example 13.1.2 (Normal Distribution). Consider a normal $\mathrm{N}(\mu,\sigma^2)$ distribution for Y and assume we have i.i.d. observations $y_0, y_1, \ldots, y_n$. Assuming μ is to be estimated, the conjugate prior associated with this distribution is, again, normal $\mathrm{N}(\alpha,\beta)$, as then

$$\pi(\mu|y_{0:n}) \propto \exp\{-(\mu-\alpha)^2/2\beta\} \prod_{k=0}^{n} \exp\{-(y_k-\mu)^2/2\sigma^2\}$$
$$\propto \exp\left\{-\frac{1}{2}\left[\mu^2\left(\frac{1}{\beta}+\frac{n+1}{\sigma^2}\right)-2\mu\left(\frac{\alpha}{\beta}+\frac{S}{\sigma^2}\right)\right]\right\} ,$$

where S is the sum of the y_k. Inspecting the right-hand side shows that it is proportional (in μ) to the density of a normal distribution with mean $(S + \alpha\sigma^2/\beta)/[(n+1)+\sigma^2/\beta]$ and variance $\sigma^2/[(n+1)+\sigma^2/\beta]$.

In the case where σ^2 is to be estimated and μ is known, the conjugate prior is instead the inverse Gamma distribution $\mathrm{IG}(\kappa,\gamma)$, with density

$$\pi(\sigma^2|\gamma,\kappa) = \frac{\gamma^\kappa}{\Gamma(\kappa)(\sigma^2)^{\kappa+1}} \mathrm{e}^{-\gamma/\sigma^2} .$$

Indeed, with this prior,

$$\pi(\sigma^2|y_{1:n}) \propto (\sigma^2)^{-(\kappa+1)} \mathrm{e}^{-\gamma/\sigma^2} \prod_{k=0}^{n} \frac{1}{\sqrt{\sigma^2}} \exp\{-(y_k-\mu)^2/2\sigma^2\}$$
$$= (\sigma^2)^{-[(n+1)/2+\kappa+1]} \exp\{-(S^{(2)}/2+\gamma)/\sigma^2\} ,$$

where $S^{(2)} = \sum_{k=0}^{n}(y_k-\mu)^2$. Hence, the posterior distribution of σ^2 is the inverse gamma distribution $\mathrm{IG}((n+1)/2+\kappa, S^{(2)}/2+\gamma)$. ■

As argued in Robert (2001), there is no compelling reason to choose *these* priors, except for their simplicity, but the restrictive aspect of conjugate priors can be attenuated by using *hyperpriors* on the hyperparameters. Those hyperpriors can be chosen amongst so-called non-informative (or vague) priors to attenuate the impact on the resulting inference. As an aside related to this point, let us recall that the introduction of vague priors within the Bayesian framework allows for a "closure" of this framework, in the sense that limits of Bayes procedures are also Bayes procedures for non-informative priors.

Example 13.1.3 (Normal Distribution, Continued). A limiting case of the conjugate $\mathrm{N}(\alpha,\beta)$ prior is obtained when letting β go to infinity. In this case, the posterior $\pi(\theta|y)$ is the same as the posterior obtained with the "flat" prior $\pi(\theta) = 1$, which is not the density of a probability distribution but simply the density of Lebesgue measure! ■

Although this sounds like an invalid extension of the probabilistic framework, it is quite correct to define posterior distributions associated with positive σ-finite measures π, then viewing (13.1) as a formal expression valid as long as the integral in the denominator is finite (almost surely). More detailed accounts are provided in Hartigan (1983), Berger (1985), or Robert (2001, Section 1.5) about this possibility of using σ-finite measures (sometimes called *improper* priors) in settings where true probability prior distributions are too difficult to come up with or too subjective to be accepted by all. Let us conclude this aside with the remark that *location models*

$$y \sim p(y - \theta)$$

are usually associated with flat priors $\pi(\theta) = 1$, whereas *scale models*

$$y \sim \frac{1}{\theta} f\left(\frac{1}{\theta}\right)$$

are usually associated with the log-transform of a flat prior, that is,

$$\pi(\theta) = \frac{1}{\theta} .$$

13.1.2 Prior Distributions for HMMs

In the specific set-up of HMMs, there are typically two separate entities of the parameter vector θ. That is, θ can be decomposed as

$$\theta = (\eta, \zeta) ,$$

where η parameterizes the transition pdf $q(\cdot, \cdot) = q_\eta(\cdot, \cdot)$ and ζ parameterizes the conditional distribution of $Y_{0:n}$ given $X_{0:n}$, with marginal conditional pdf $g(\cdot, \cdot) = g_\zeta(\cdot, \cdot)$. The reason for this decomposition should be clear from Chapter 10 on the EM framework: when conditioned on the (latent) chain $X_{0:n}$, the parameter ζ is estimated as in a regular (non-latent) model, whereas the parameter η only depends on the chain $X_{0:n}$. A particular issue is the distribution ν of the initial state X_0. In general, it is assumed either that X_0 is fixed and known (ν is then degenerate); or that X_0 is random, unknown, and ν is parameterized by a separate parameter; or that X_0 is random, unknown, and with ν being parameterized by η. In the latter case, a standard setting is that $\{X_k\}_{k \geq 0}$ is assumed stationary—so that the HMM as a whole is stationary—and ν is then the stationary distribution of the transition kernel $Q = Q_\eta$. A particular instance of the second case is to assume that ν is fixed, for example uniform on X. We remark that if ν is parameterized by a separate parameter, for instance the probabilities $(\nu_1, \ldots, \nu_r)$ themselves, there is of course no hope of being able to estimate this parameter consistently, as there is only one variable X_0—that we do not even observe!—whose distribution is given by ν.

The above is formalized in the following separation lemma about θ.

Lemma 13.1.4. *Assume that the prior distribution $\pi(\theta)$ is such that*

$$\pi(\theta) = \pi_\eta(\eta)\pi_\zeta(\zeta) \tag{13.4}$$

and that the distribution of X_0 depends on η or on another separate parameter. Then, given $x_{0:n}$ and $y_{0:n}$, η and ζ are conditionally independent, and the conditional posterior distribution of η does not depend on the observations $y_{0:n}$.

Proof. The proof is straightforward: given that the posterior distribution $\pi(\theta|x_{0:n}, y_{0:n})$ factorizes as

$$\begin{aligned}
&\pi_\eta(\eta)\,\pi_\zeta(\zeta)\nu_\eta(x_0)\prod_{k=1}^{n} q_\eta(x_{k-1}, x_k)\prod_{k=0}^{n} g_\zeta(x_k, y_k) \\
&= \pi_\eta(\eta)\nu_\eta(x_0)\prod_{k=1}^{n} q_\eta(x_{k-1}, x_k) \times \pi_\zeta(\zeta)\prod_{k=0}^{n} g_\zeta(x_k, y_k) \qquad (13.5)
\end{aligned}$$

up to a normalizing constant, the two subvectors η and ζ are indeed conditionally independent. Independence of the conditional distribution of η from $y_{0:n}$ is obvious from (13.5). □

A practical consequence of Lemma 13.1.4 is therefore that we can conduct Bayesian inference about η and ζ separately, conditional on the (latent) chain $X_{0:n}$ (and of course on the observables $Y_{0:n}$). Conditional inference is of interest because of its relation with the *Gibbs sampler* (see Chapter 6) associated with this model, as will be made clearer in Section 13.1.4.

In the case where the latent variables are finite, that is, when X is finite, a reparameterization of X into $\{1, \ldots, r\}$ allows for use of the "classical" conjugate *Dirichlet* prior on the transition probability matrix $q = (q_{ij})$, $\text{Dir}_r(\delta_1, \ldots, \delta_r)$. These priors generalize the Beta (of type one) distribution as priors on the simplex of $\mathbb{R}^r$.

Definition 13.1.5 (Dirichlet Distribution). *A Dirichlet* $\text{Dir}_r(\delta_1, \ldots, \delta_r)$ *distribution is a distribution on the subset* $q_1 + \ldots + q_r = 1$ *of* $\mathbb{R}^r$, *given by the density*

$$\pi(q_1, \ldots, q_r) = \frac{\Gamma(\delta_1 + \ldots + \delta_r)}{\Gamma(\delta_1)\cdots\Gamma(\delta_r)}\, q_1^{\delta_1 - 1}\cdots q_r^{\delta_r - 1}\,\mathbb{1}\{q_1 + \ldots + q_r = 1\}\ ,$$

where all $\delta_i > 0$.

We remark that the above density is with respect to Lebesgue measure on the subset that supports the distribution. Of particular interest is the choice $\delta_i = 1$ for all i, in which case the density is constant and hence the distribution uniform.

Under the assumption that ν is known or with a distribution parameterized by a separate parameter, we then have the following conjugacy result.

Lemma 13.1.6. *The Dirichlet prior is a conjugate distribution for the transition probability matrix* Q *of the Markov chain* $X_{1:n}$ *in the following sense. Assume that each row of* Q *has a prior distribution that is Dirichlet,*

$$(q_{i1}, \ldots, q_{ir}) \sim \text{Dir}_r(\delta_1, \ldots, \delta_r)\ ,$$

with the rows being a priori independent, and that the distribution ν *of* X_0 *is either fixed or parameterized by a separate parameter. Then, given the Markov chain, the rows of* Q *are conditionally independent and*

$$(q_{i1}, \ldots, q_{ir})|x_{1:n} \sim \mathrm{Dir}_r(\delta_1 + n_{i1}, \ldots, \delta_r + n_{ir}) ,$$

where n_{ij} denotes the number of transitions from i to j in the sequence $x_{0:n}$.

Proof. Given that the parameters of Q only depend on $X_{0:n}$, we have

$$\pi(Q|x_{0:n}) \propto \pi(Q) \prod_{k=1}^{n} q_{x_{k-1}x_k} \propto \prod_{i,j} q_{ij}^{\delta_j + n_{ij} - 1} .$$

□

We remark that in the case where the distribution ν of X_0 is the stationary distribution of Q, there is no conjugate distribution because of the non-exponential relation between this stationary distribution and Q. This does not mean that Bayesian inference is not possible, but simulation from the posterior distribution of Q is less straightforward in this case.

Simulation from a Dirichlet distribution is easy: if $\xi_1, \ldots, \xi_r$ are independent with ξ_i having a $\mathrm{Ga}(\delta_i, 1)$ distribution, then the r-tuple

$$\left(\frac{\xi_1}{\sum_{i=1}^r \xi_i}, \frac{\xi_2}{\sum_{i=1}^r \xi_i}, \ldots, \frac{\xi_r}{\sum_{i=1}^r \xi_i} \right)$$

has a $\mathrm{Dir}_r(\delta_1, \ldots, \delta_r)$ distribution.

Example 13.1.7 (Normal HMM). Assume that $\{X_k\}_{k \geq 0}$ is a finite Markov chain on $\mathsf{X} = \{1, \ldots, r\}$ and that, conditional on $X_k = i$, Y_k has a $\mathrm{N}(\mu_i, \sigma_i^2)$ distribution.

A typical prior for this model may look as follows. On the transition probability matrix Q we put a $\mathrm{Dir}_r(\delta_1, \ldots, \delta_r)$ distribution on each row, with independence between rows. A standard choice is to set the δ_j equal; often $\delta_j = 1$. The means and variances of the normal distributions are assumed *a priori* independent and with conjugate priors, that is, a $\mathrm{N}(\alpha, \beta)$ prior for each mean μ_i and a $\mathrm{IG}(\kappa, \gamma)$ prior for each variance σ_i^2 (cf. Example 13.1.2).

The joint prior thus becomes

$$\begin{aligned}
\pi(\theta) &= \pi(Q, \mu_1, \ldots, \mu_r, \sigma_1^2, \ldots, \sigma_r^2) \\
&= \prod_{i=1}^{r} \frac{\Gamma(\delta_1 + \ldots + \delta_r)}{\Gamma(\delta_1) \cdots \Gamma(\delta_r)} \prod_{j=1}^{r} q_{ij}^{\delta_j - 1} \\
&\quad \times \prod_{i=1}^{r} \frac{1}{\sqrt{2\pi\beta}} \mathrm{e}^{-(\mu_i - \alpha)^2/2\beta} \\
&\quad \times \prod_{i=1}^{r} \frac{\gamma^\kappa (\sigma^2)^{-(\kappa+1)}}{\Gamma(\kappa)} \mathrm{e}^{-\gamma/\sigma^2} .
\end{aligned}$$

It is often appropriate to consider one or several of α, β, κ, and γ as unknown random quantities themselves, and hence put hyperpriors on them. These quantities are then adjoined to θ, and their prior densities are adjoined to the above prior. Richardson and Green (1997) and Robert *et al.* (2000), for instance, contain such examples. ■

In the above example, the initial distribution ν was not mentioned. Indeed, it was tacitly assumed that the initial distribution ν is given by Q, for example as the stationary distribution. From a simulation point of view this is inconvenient however, as the posterior distributions of the rows of Q are then no longer Dirichlet; cf. the remark below Lemma 13.1.6. A different assumption, more appealing from this simulation point of view, is to assume that ν is fixed, typically uniform on $\{1, \ldots, r\}$. We may also assume that ν is unknown and equip it with a $\mathrm{Dir}(\delta'_1, \ldots, \delta'_r)$ prior, usually with all δ'_i equal. Then ν is adjoined to θ and the Dirichlet density goes into the prior. Finally, we may also assume that X_0 is fixed and known, equal to 1, say. This implies that the prior is not exchangeable though, and the structure of the implied non-exchangeability is difficult to describe (see below). Therefore, in practice the two alternatives of setting ν as the uniform distribution or assigning it a Dirichlet prior are the most appealing. In the latter case, as remarked above Lemma 13.1.4, ν cannot be estimated consistently.

13.1.3 Non-identifiability and Label Switching

An issue of particular interest for the choice of the loss function or, correspondingly, of the Bayes estimator, is *non-identifiability*. This is a problem that primarily arises in the case of finite state space X. Hence, assume $\mathsf{X} = \{1, \ldots, r\}$.

To start with, we will make assumptions about the parameterization of the HMM. We assume that θ decomposes into (η, ζ) as in (13.4), that η simply comprises the transition probabilities q_{ij} themselves, and that ζ further decomposes as $\zeta = (\zeta_1, \ldots, \zeta_r)$, where ζ_i parameterizes the conditional density $g(i, \cdot)$ in a way that is identical for each i. Hence, all $g(i, \cdot)$ belong to the same parametric family. A typical example is to take, as in the above example, the $g(i, \cdot)$ as normal distributions $\mathrm{N}(\mu_i, \sigma_i^2)$, in which case $\zeta_i = (\mu_i, \sigma_i^2)$. The initial distribution ν is assumed to be the stationary distribution of Q, or to be fixed and uniform on X, or to be given by a separate set $(\nu_1, \ldots, \nu_r)$ of probabilities. Under these conditions, the likelihood $\mathrm{L}(y_{0:n}|\theta)$ is *invariant* under permutation of state indices. More precisely, if $(s_1, \ldots, s_r)$ is a permutation of $\{1, \ldots, r\}$, then

$$\mathrm{L}[y_{0:n}|(\nu_i), (q_{ij}), (\zeta_i)] = \mathrm{L}[y_{0:n}|(\nu_{s_i}), (q_{s_i,s_j}), (\zeta_{s_i})] .$$

This equality simply says that if we renumber the states in X and permute the parameter indices accordingly, the likelihood remains unchanged.

We now turn to a second set of assumptions. A density on $\mathbb{R}^r$ is said to be *exchangeable* if it is invariant under permutations of the components. We will assume that the joint prior for $(q(i,j))$, (ζ_i), and (ν_i) is exchangeable,

$$\pi[(\nu_i), (q_{ij}), (\zeta_i)] = \pi[(\nu_{s_i}), (q_{s_i,s_j}), (\zeta_{s_i})] .$$

This exchangeability condition is very often occurring in practice. It holds, for instance, if the three entities involved are *a priori* independent with an independent Dirichlet $\text{Dir}_r(\delta, \ldots, \delta)$ prior on each row of the transition probability matrix, independent identical priors on the ζ_i and, when applicable, a Dirichlet $\text{Dir}_r(\delta', \ldots, \delta')$ prior on (ν_i).

Under the above two sets of assumptions, because $\pi(\theta|y_{0:n})$ is proportional to $\pi(\theta)L(y_{0:n}|\theta)$ in θ, the posterior will also be exchangeable,

$$\pi[(\nu_i), (q(i,j)), (\zeta_i)|y_{0:n}] = \pi[(\nu_{s_i}), (q(s_i, s_j)), (\zeta_{s_i})|y_{0:n}] .$$

This non-identifiability feature has the serious consequence that, from a Bayesian point of view, within each block of parameters *all marginals are the same*! Indeed, for example,

$$\pi(\zeta_1, \ldots, \zeta_r|y_{0:n}) = \pi(\zeta_{s_1}, \ldots, \zeta_{s_r}|y_{0:n}) . \tag{13.6}$$

Thus, for $1 \le i \le r$, the density π_{ζ_i} defined as

$$\pi_{\zeta_i}(\zeta_i|y_{0:n}) = \int \pi(\zeta_1, \ldots, \zeta_r|y_{0:n}) \, d\zeta_{-i} ,$$

is independent of i. Therefore, both the posterior mean and the maximum marginal posterior estimators are ruled out in exchangeable settings, as they only depend on the marginals.

A practical consequence of this lack of identifiability is so-called *label switching*, illustrated in Figure 13.1. This figure provides an MCMC sequence for both the standard deviations σ_i and the stationary probabilities of Q for an HMM with three Gaussian components $\text{N}(0, \sigma_i^2)$. The details will be discussed below, but the essential feature of this graph is the continuous shift between the three levels of each component σ_i, which translates the equivalence between $(\sigma_1, \sigma_2, \sigma_3)$ and any of its permutations for the posterior distribution. As discussed by Celeux *et al.* (2000), this behavior does not always occur in a regular MCMC implementation. In the current case, it is induced by the underlying reversible jump algorithm (see Section 13.2.3). We stress that label switching as such is not a result of exploring the posterior surface by simulation but is rather an intrinsic property of the model and its prior.

Lack of identifiability also creates a difficulty with the maximum *a posteriori* estimator in that the exchangeability property implies that there are a multiple of $r!$ (local and global) modes of the posterior surface, given (13.6). It is therefore difficult to devise efficient algorithms that can escape a particular mode to provide a fair picture of the overall, multimodal posterior surface. For instance, Celeux *et al.* (2000) had to resort to *simulated tempering*, a sort of inverted *simulated annealing*, to achieve a proper exploration.

A common approach to combat problems caused by lack of identifiability is to put constraints on the prior, in that certain parameters are required to appear in ascending or descending order. For instance, in the above

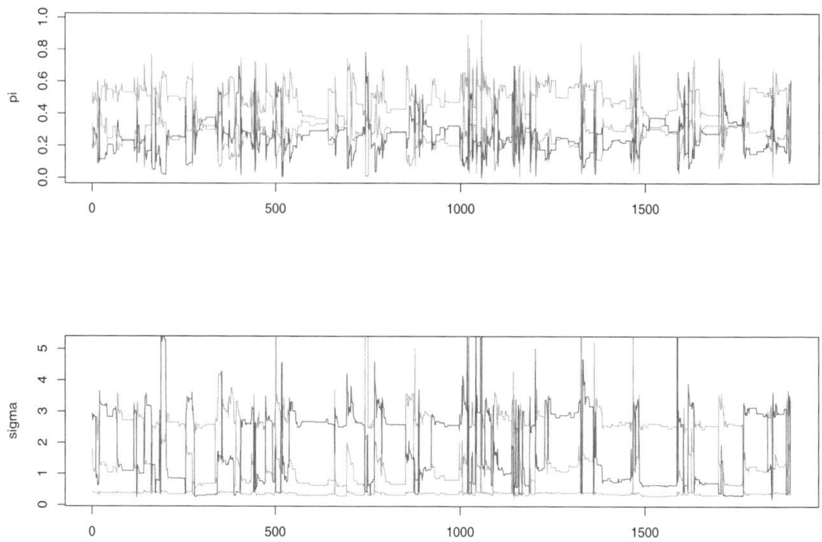

Fig. 13.1. Representation of an MCMC sequence simulated from the posterior distribution associated with a Gaussian HMM with three hidden states, Gaussian components $\mathrm{N}(0, \sigma_i^2)$, and a data set made of a sequence of wind intensities in Athens (Greece). The top graph plots the sequence of stationary probabilities of the transition probability matrix Q and the bottom graph the sequence of σ_i. *Source: Cappé et al. (2003).*

example, we could set the prior density to zero outside the region where $\mu_1 < \mu_2 < \ldots < \mu_r$. That is, we require the normal means to appear in ascending order. Such a constraint does not affect the MAP, but it does affect the marginal posterior distributions—obviously, the marginal posterior distribution functions of the μ_i become stochastically ordered—and hence, for instance, the posterior means of individual parameters. It is important to realize that marginal posterior distributions of parameters not directly involved in the constraint, for instance the σ_i^2 in the current example, are also affected. Even more importantly, if an ordering constraint is put on a different set of parameters, $\sigma_1^2 < \sigma_2^2 < \ldots < \sigma_r^2$ for example, then the marginal posterior distributions will be affected in a different way. Hence, ordering constraints are not a tool that is unambiguous in the sense that any constraint leads to the same marginal posterior distributions. This is illustrated in Richardson and Green (1997). From a practical point of view, in an MCMC simulation, ordering can be imposed at each step of the sampler, but we could also design a sampler without such constraints and do the sorting as a part of post-processing of the sampler output. This approach obviously greatly simplifies investigations of how constraints on different sets of parameters affect the results. Stephens (2000b) discusses the label switching problem in a general decision theoretic framework. In particular, he demonstrates that sorting means, variances, etc., sometimes gives results that are difficult to interpret, and he suggests, in the contexts of i.i.d. observations from a finite mixture, a relabeling algorithm based on probabilities of the each observation belonging to a certain mixture component.

If we put a sorting constraint on the parameters, we implicitly construct a new prior that is zero in regions where the constraint does not hold. Moreover, because a parameter can be permuted in $r!$ different ways, the new prior is equal to the original prior multiplied by $r!$ in the region where the constraint does hold, in order to make it integrate to unity (over the constrained space). A similar but slightly different view, suggested by Stephens (2000a), is to think of the $r!$ permutations of a given parameter as a single element of an *equivalence class* of parameters; the effective parameter space is then the space of such equivalence classes. Again, because a parameter of order r can be permuted in $r!$ different ways, each element of the equivalence class $[\theta]$ has a prior that is $r!$ times the prior $\pi(\theta)$ of any of its particular representations θ. This distinction between a parameter and its corresponding equivalence class and the factor $r!$ are not important when r is fixed, but it becomes important when r is variable, and we attempt to estimate it, as discussed in Section 13.2.

Lack of identifiability can also be circumvented by using a loss function that is impervious to label switching, that is, invariant under permutation of the label indices. For instance, in the case of mixtures, Celeux *et al.* (2000) employed a loss function for the estimation of the parameter θ may based on the Kullback-Leibler divergence,

$$\text{loss}(\theta, \hat{\theta}) = \int \log \frac{p(y_{0:n}|\theta)}{p(y_{0:n}|\hat{\theta})} p(y_{0:n}|\theta)\, dy_{0:n} \, .$$

13.1.4 MCMC Methods for Bayesian Inference

Analytic computation of Bayesian estimates like the posterior mean or posterior mode is most generally infeasible for HMMs, except for the simplest models. We now review *simulation-based* methods that follow the general MCMC scheme introduced in Chapter 6 and provide Monte Carlo approximations of the posterior distribution of the parameters θ given the observable $Y_{0:n}$. As noted in Chapter 6, the distribution of $X_{0:n}$ given both $Y_{0:n}$ and θ is often manageable (when X is finite notably). Likewise, the conditional distribution of the parameters given $Y_{0:n}$ *and* $X_{0:n}$ is usually simple enough in HMMs, especially when conjugate priors are used (as in Example 13.1.7). What remains to be exposed here is how to bridge the gap between these two conditionals.

The realization that for HMMs, the distribution of interest involves two separate entities, θ and $X_{0:n}$, for which the two conditional distributions $\pi(\theta|x_{0:n}, y_{0:n})$ and $\pi(x_{0:n}|\theta, y_{0:n})$ are available or may be sampled from, suggests the use of a two stage Gibbs sampling strategy as defined in Chapter 6 (see Algorithm 6.2.13). The simplest version of the Gibbs sampler, which will be referred to as *global updating of the hidden chain*, goes as follows.

Algorithm 13.1.8. Iterate:

1. Simulate θ from $\pi(\theta|x_{1:n}, y_{0:n})$.
2. Simulate $X_{0:n}$ from $\pi(x_{0:n}|\theta, y_{0:n})$.

This means that, if we can simulate the parameters based on the completed model (and this is usually the case, see Example 13.1.10 for instance) and the missing states $X_{0:n}$ conditionally on the parameters and $Y_{0:n}$ (see Chapter 6), we can implement this two-stage Gibbs sampler, also called *data augmentation* by Tanner and Wong (1987). We note that θ typically is multivariate, and it is then often broken down into several components; accordingly, the first step above then breaks down into several sub-steps. Similar comments apply if there are hyperparameters with their own priors in the model; we can view them as part of θ even though they are often updated separately.

By global updating we mean that the trajectory of the hidden chain is updated as a whole from its joint conditional distribution given the parameter θ and the data $Y_{0:n}$. This corresponds to the partitioning $(\theta, X_{0:n})$ of the state space of the Gibbs sampler. Another possible partitioning is $(\theta, X_0, X_1, \ldots, X_n)$, which leads to an earlier and more "rudimentary" version of the Gibbs sampler (Robert *et al.*, 1993). In this algorithm, only one hidden variable X_k is updated at a time, and we refer to this scheme as *local updating of the hidden chain.* The algorithm thus looks as follows.

Algorithm 13.1.9. Iterate:

1. Simulate θ from $\pi(\theta|x_{1:n}, y_{1:n})$.
2. For $k = 0, 1, \ldots, n$, simulate X_k from $\pi(x_k|\theta, y_{1:n}, x_{1:k-1}, x_{k+1:n})$.

This algorithm only updates one state at a time, and, because

$$\pi(x_k|\theta, y_{0:n}, x_{0:k-1}, x_{k+1:n})$$

reduces to

$$\pi(x_k|\theta, y_k, x_{k-1}, x_{k+1}) \propto q_\theta(x_{k-1}, x_k)q_\theta(x_k, x_{k+1})g_\theta(x_k, y_k)$$

where the first factor on the right-hand side is replaced by $\nu_\theta(x_0)$ for $k = 0$ and the second factor is replaced by unity for $k = n$; this means that each X_k is updated conditional upon its neighbors, as seen in Chapter 6.

In the above algorithm, the X_k are updated in a fixed linear order, but there is nothing that prevents us from using a different order or from picking the variable X_k to be updated at random. Of course there are schemes intermediate between the extremes global and local updating. We might, for example, update blocks of X_k; like for local updating, these blocks may be of fixed size and updated in a specific order, but size and order may also be chosen at random as in (Shephard and Pitt, 1997).

Example 13.1.10 (Normal HMM, Continued). Let us return to the HMM and prior given in Example 13.1.7. To compute the respective full conditionals in the Gibbs sampler, we note again that each such distribution, or density, is proportional (in the component to be updated) to the product of the prior and the likelihood. For example,

$$\begin{aligned}
&\pi(\mu_1, \ldots, \mu_r | Q, \sigma_1^2, \ldots, \sigma_r^2, x_{1:n}, y_{0:n}) \\
&\quad \propto \pi(Q, \mu_1, \ldots, \mu_r, \sigma_1^2, \ldots, \sigma_r^2) \\
&\quad\quad \times p(x_{0:n}|Q) \mathrm{L}(y_{0:n}|x_{0:n}, \mu_1, \ldots, \mu_r, \sigma_1^2, \ldots, \sigma_r^2) \\
&\quad = \pi(Q)\pi(\mu_1) \cdots \pi(\mu_r)\pi(\sigma_1^2) \cdots \pi(\sigma_r^2) p(x_{0:n}|Q) \prod_{k=0}^{n} g_{(\mu,\sigma)}(x_k, y_k) \ .
\end{aligned}$$

By picking out the factors on the right-hand side that contain the appropriate variables, we can find their full conditional. We now detail this process for each of the variables involved.

The conditional pdf of $\mu_1, \ldots, \mu_r$ is proportional to

$$\begin{aligned}
&\prod_{i=1}^{r} \exp\left\{-(\mu_i - \alpha)^2/2\beta\right\} \prod_{k=0}^{n} \exp\left\{-(y_k - \mu_{x_k})/2\sigma_{x_k}^2\right\} \\
&\qquad \propto \prod_{i=1}^{r} \exp\left\{-\frac{1}{2}[\mu_i^2(\beta^{-1} + n_i\sigma_i^{-2}) - 2\mu_i(\alpha\beta^{-1} + S_i\sigma_i^{-2})]\right\} ,
\end{aligned}$$

where n_i is the number of x_k with $x_k = i$ and S_i is the sum of the corresponding y_k; $S_i = \sum_{\{k:\, x_k=i\}} y_k$. We can conclude that the full conditional distribution of $\mu_1, \ldots, \mu_r$ is such that these variables are conditionally independent and

$$\mu_i \,|\, Q, \sigma_1^2, \ldots, \sigma_r^2, x_{0:n}, y_{0:n} \sim \mathrm{N}\left(\frac{\alpha\sigma_i^2/\beta + S_i}{\sigma_i^2/\beta + n_i}, \frac{1}{1/\beta + n_i/\sigma_i^2}\right) . \tag{13.7}$$

This can also be understood in the following way: given $X_{0:n}$ all the observations are independent, and to obtain the posterior for μ_i we only need to consider observations governed by this regime. As the μ_i are *a priori* independent, they will be so *a posteriori* as well. The above formula is then a standard result of Bayesian statistics (cf. Example 13.1.2).

In a similar fashion, one finds that

$$\begin{aligned}
&\pi(\sigma_1^2, \ldots, \sigma_r^2 | Q, \mu_1, \ldots, \mu_r, x_{0:n}, y_{0:n}) \\
&\qquad\qquad \propto \prod_{i=1}^{r} (\sigma_i^2)^{-(\kappa + n_i/2 + 1)} \exp\left\{-(\gamma + S_i^{(2)}/2)/\sigma_i^2\right\} ,
\end{aligned}$$

where $S_i^{(2)} = \sum_{\{k:\, x_k=i\}} (y_k - \mu_i)^2$. Hence, the full conditional distribution of $\sigma_1^2, \ldots, \sigma_1^2$ is such that these variables are conditionally independent, and

$$\sigma_i^2 \mid Q, \mu_1, \ldots, \mu_r, x_{0:n}, y_{0:n} \sim \mathrm{IG}(\kappa + n_i/2, (\gamma + S_i^{(2)}/2)) . \qquad (13.8)$$

This result is indeed also an immediate consequence of Example 13.1.2.

The full conditional distribution of the transition matrix Q was essentially derived in Lemma 13.1.6; the rows are conditionally independent with the ith row following a Dirichlet distribution $\mathrm{Dir}_r(\delta_1 + n_{ij}, \ldots, \delta_r + n_{ir})$. Here n_{ij} is the number of transitions from state i to j, that is, $n_{ij} = \#\{0 \le k \le n-1 : x_k = i, x_{k+1} - j\}$.

Several types of MCMC moves are typically put together in what is often called a *sweep* of the algorithm. Thus, one sweep of the Gibbs sampler with local updating for the present model looks as follows.

Algorithm 13.1.11.

1. Simulate the μ_i independently according to (13.7).
2. Simulate the σ_i^2 independently according to (13.8).
3. Simulate the rows of Q independently, with the ith row from $\mathrm{Dir}_r(\delta_1 + n_{i1}, \ldots, \delta_r + n_{ir})$.
4. For $k = 0, 1, \ldots, n$, simulate X_k with unnormalized probabilities

$$\mathrm{P}(X_k = i \mid \theta, y_k, x_{k-1}, x_{k+1}) \propto q(x_{k-1}, i) q(i, x_{k+1}) \frac{1}{\sigma_i} \mathrm{e}^{-(y_k - \mu_i)^2/2\sigma_i^2} ;$$

for $k = 0$ the first factor is replaced by $\nu(x_0)$, and for $k = n$ the factor $q(i, x_{k+1})$ is replaced by unity.

If ν is the stationary distribution of Q, simulation of Q requires a Metropolis-Hastings step; a sensible proposal is then the same Dirichlet as above. If ν is rather a separate parameter, Q is updated as above and, provided the prior on $(\nu_1, \ldots, \nu_r)$ is a Dirichlet as in Example 13.1.7, this vector is updated with full conditional distribution $\mathrm{Dir}_r(\delta_1' + t_1, \ldots, \delta_r' + t_r)$ with $t_i = \mathbb{1}\{x_0 = i\}$. Of course, global updating of $X_{0:n}$ could have been used as well, which would modify step 4 of the algorithm only. ■

The Gibbs sampler with local updating should mix and explore the posterior surface much more slowly than when global updating is used. It must be considered, however, that the simulation of the whole vector of states, $X_{1:n}$, is more time-consuming in that it requires the use of the forward or backward formulas (Section 6.1.2). A numerical comparison of the two approaches by Robert *et al.* (1999), using several specially designed convergence monitoring tools, did not exhibit an overwhelming advantage in favor of global updating, even without taking into account the additional $\mathrm{O}(n^2)$ computational time required by this approach. On the other hand, Scott (2002) provided an example showing a significant advantage for global updating in terms of autocovariance decay. It is thus difficult to make a firm recommendation on which updating scheme to use. One may start by running local updating, and if its mixing behavior is poor, try global updating as well. We do remark, however, that when the state space X is continuous, there is seldom any alternative to local updating. In addition, with continuous X, local updating must in general be carried out by a Metropolis-Hastings step, as the full conditional distribution seldom lends itself to direct simulation (see Section 6.3). The next example demonstrates a somewhat more complicated use of the single site Gibbs sampling strategy.

Example 13.1.12 (Capture-Recapture, Continued). Let us now consider Gibbs simulation from the posterior distribution of the parameters in the capture-recapture model of Example 1.3.4. The parameters are divided into (a) the capture probabilities $p_k(i)$, indexed by the capture zone i $(i = 1, 2, 3)$, and (b) the movement probabilities $q_k(i, j)$ $(i, j = 1, 2, 3, \dagger)$, which are the probabilities that the lizard is in zone j at time $k + 1$ given that it is in zone i at time k. For instance, the probability $q_k(\dagger, \dagger)$ is equal to 1, because of the absorbing nature of $\dagger$. We also denote by $\varphi_k(i)$ the *survival probability* at time k in zone i, that is,

$$\varphi_k(i) = 1 - q_k(i, \dagger) ,$$

and by $\psi_k(i, j)$ the *effective probability of movement* for the animals remaining in the system, that is,

$$\psi_k(i, j) = q_k(i, j)/\varphi_k(i) .$$

If we denote $\psi_k(i) = (\psi_k(i, 1), \psi_k(i, 2), \psi_k(i, 3))$, the prior distributions are chosen to be

$$p_k(i) \sim \text{Be}(a, b), \quad \varphi_k(i) \sim \text{Be}(\alpha, \beta), \quad \psi_k(i) \sim \text{Dir}_3(\gamma_1, \gamma_2, \gamma_3) ,$$

where the hyperparameters $a, b, \gamma_1, \gamma_2, \gamma_3$ are known.

The probabilities of capture $p_k(i)$ depend on the zone of capture i and the missing data structure of the model, which must be taken into account. Slightly modifying the notations of Example 1.3.4, we let y^*_{km} be the position of animal m at time k and x_{km} its capture indicator, the observations can be written in the form $y_{km} = x_{km} y^*_{km}$, where $y_{km} = 0$ corresponds to a missing observation. The sequence of y^*_{km} for a given m then corresponds to a *non-homogeneous Markov chain*, with transition matrix $Q_k = (q_k(i, j))$. *Conditionally on* y^*_{km}, the X_{km} then are Bernoulli variables with probability of success $p_k(y^*_{km})$.

The Gibbs sampler associated with this model has the following steps.

Algorithm 13.1.13.

1. Simulate y^*_{km} for sites such that $x_{km} = 0$.
2. Generate $(0 \le k \le n)$

$$\begin{aligned} p_k(i) &\sim \text{Be}(a + u_k(i), b + v_k(i)) , \\ \varphi_k(m) &\sim \text{Be}(\alpha + w_k(i), \beta + w_k(i, \dagger)) , \\ \psi_k(i) &\sim \text{Dir}_3(\gamma_1 + w_k(i, 1), \gamma_2 + w_k(i, 2), \gamma_3 + w_k(i, 3)) , \end{aligned}$$

where $u_k(i)$ denotes the number of captures in i at time k, $v_k(i)$ the number of animals unobserved at time i for which the simulated y_{km} is equal to i, $w_k(i,j)$ the number of passages (observed or simulated) from i to j, $w_k(i,\dagger)$ the number of (simulated) passages from i to $\dagger$, and

$$w_k(i) = w_k(i,1) + w_k(i,2) + w_k(i,3) .$$

Step `1.` must be decomposed into conditional sub-steps to account for the Markovian nature of the observations; in a full Gibbs strategy, y^*_{km} can be simulated conditionally on $y^*_{(k-1)m}$ and $y^*_{(k+1)k}$ when $x_{km} = 0$. If $k \neq n$, the missing data are simulated according to

$$\mathrm{P}(y^*_{km} = j \mid y^*_{(k-1)m} = i, y^*_{(k+1)m} = \ell, x_{km} = 0) \propto q_{k-1}(i,j)(1 - p_k(j))q_k(j,\ell)$$

and

$$\mathrm{P}(y^*_{nm} = j \mid y^*_{(n-1)m} = i, x_{nm} = 0) \propto q_{n-1}(i,j)(1 - p_n(j)) .$$

■

So far, we have dealt with MCMC algorithms for which the state space of the sampler consists of the parameter θ and the hidden chain $X_{0:n}$; both are random, unobserved quantities—θ because we are in a Bayesian framework and $X_{0:n}$ because of its role in the model as a latent variable. However, it is quite possible to devise MCMC algorithms for which the sampler state space comprises θ alone and not the hidden chain. In particular, when the state space X of the hidden chain is finite, we know that the likelihood may be computed exactly. In such a case the completion step, that is, the simulation of $X_{0:n}$, does not appear as a necessity any longer, and alternative Metropolis-Hastings steps can be used instead.

Example 13.1.14 (Normal HMM, Continued). In Cappé *et al.* (2003), the simulation of the parameters of the normal components, as well as of the parameters of the transition probability matrix, was done through simple random walk proposals: for the means μ_j the proposed move is

$$\mu'_j = \mu_j + \varepsilon_i ,$$

where $\varepsilon_i \sim \mathrm{N}(0, \tau_\mu)$ and τ_μ is a parameter that may be adjusted to optimize performance of the sampler. Because the proposal is symmetric, the acceptance ratio is simple; it is

$$\frac{\pi(\theta')\mathrm{L}(y_{0:n}|\theta')}{\pi(\theta)\mathrm{L}(y_{0:n}|\theta)} ,$$

where L is the likelihood computed via the forward algorithm (Section 5.1.1).

For the variances σ^2_j, the proposed move is a multiplicative random walk

$$\log \sigma'_j = \log \sigma_j + \varepsilon_j ,$$

where $\varepsilon_j \sim \mathrm{N}(0, \tau_\sigma)$, with acceptance ratio

$$\frac{\pi(\theta')\mathrm{L}(y_{0:n}|\theta')}{\pi(\theta)\mathrm{L}(y_{0:n}|\theta)} \prod_j \frac{\sigma'_j}{\sigma_j} ,$$

the last term being the ratio of the Jacobians incurred by working on the log-scale. To describe the above proposal, we also sometimes say that σ'_j follows a log-normal $\mathrm{LN}(\log \sigma_j, \tau_\sigma)$ distribution.

In the case of the transition probability matrix, Q, the move is slightly more involved due to the constraint on the sums of the rows, Q being a stochastic matrix. Cappé *et al.* (2003) solved this difficulty by reparameterizing each row $(q_{i1}, \dots, q_{ir})$ as

$$q_{ij} = \frac{\omega_{ij}}{\sum_\ell \omega_{i\ell}}, \qquad \omega_{ij} > 0 ,$$

so that the summation constraint on the q_{ij} does not hinder the random walk. Obviously the ω_{ij} are not identifiable, but as we are only interested in the q_{ij}, this is not a true difficulty. On the opposite, using overparameterized representations often helps with the mixing of the corresponding MCMC algorithms, as they are less constrained by the data set or the likelihood. The proposed move on the ω_{ij} is

$$\log \omega'_{ij} = \log \omega_{ij} + \varepsilon_{ij} ,$$

where $\varepsilon_{ij} \sim \mathrm{N}(0, \tau_\omega)$, with acceptance ratio

$$\frac{\pi(\theta')\mathrm{L}(y_{0:n}|\theta')}{\pi(\theta)\mathrm{L}(y_{0:n}|\theta)} \prod_{i,j} \frac{\omega'_{ij}}{\omega_{ij}} .$$

Note that this reparameterization of the model forces us to select a prior distribution on the ω_{ij} rather than on the q_{ij}. The choice $\omega_{ij} \sim \mathrm{Ga}(\delta_j, 1)$ is natural in that it gives a $\mathrm{Dir}_r(\delta_1, \dots, \delta_r)$ distribution on the corresponding $(q_{i1}, \dots, q_{ir})$. We also note that it is not difficult to show that if $(\omega_{i1}, \dots, \omega_{ir})$ is reparameterized into $S_i = \sum_1^r \omega_{ij}$ and $(q_{i1}, \dots, q_{ir})$, then, given $x_{0:n}$, S_i and $(q_{i1}, \dots, q_{ir})$ are conditionally independent and distributed as $\mathrm{Ga}(\sum_1^r \delta_j, 1)$ and $\mathrm{Dir}_r(\delta_1 + n_{i1}, \dots, \delta_r + n_{ir})$ respectively. This proves that the ω-parameterization does nothing but introduce a new parameter for each row, the sum S_i, that is independent of everything else and hence totally irrelevant for the inference. The point of introducing this extra variable is only to simplify the design of Metropolis-Hastings moves. If the initial distribution ν is also a parameter of the model, it can be recast in a similar fashion.

Figure 13.1 provides an illustration of this simulation scheme in the special case of a Gaussian HMM with zero means. Over the 2,000 MCMC iterations represented on both graphs, there are periods where the value of the σ_i or of the stationary probabilities of Q do not change: these periods correspond to sequences of proposed values that are rejected at the Metropolis-Hastings stage. Note that the rejection periods are not the same for the σ_i and the stationary probabilities. This is due to the fact that there is a Metropolis-Hastings stage for each group of parameters. ■

Another alternative stands at the opposite end of the range of possibilities: the parameters of the model can be integrated out when conjugate priors are used, as demonstrated by Liu (1994), Chen and Liu (1996), and Casella *et al.* (2000) in the case of mixture and switching regression models. In such schemes, each site X_k is typically sampled conditionally on all the other sites, with the model parameters fully integrated out.

13.2 Reversible Jump Methods

So far we have not touched upon the topic of the unknown number of states in an HMM and of the estimation of this number via Bayesian procedures. After a short presentation of variable dimension models and of their meaning, we introduce the adequate MCMC methodology to deal with this additional level of complexity.

13.2.1 Variable Dimension Models

In general, a variable dimension model is, to quote Peter Green, a "model where one of the things you do not know is the number of things you do not know". In other words, this pertains to a statistical model where the dimension of the parameter space is not "known". This is not a formal enough definition, obviously, and we need to provide a more rigorous perspective.

Definition 13.2.1 (Variable Dimension Model). *A variable dimension model is defined as a collection of models (or parameter spaces),*

$$\Theta_r, \quad r = 1, \ldots, R\,,$$

associated with a collection of priors on these spaces,

$$\pi_r(\theta_r), \quad r = 1, \ldots, R\,,$$

and a prior distribution on (the indices of) these spaces,

$$\varrho(r), \quad r = 1, \ldots, R\,.$$

In the following, we shall consider that a variable dimension model is associated with a probability distribution on the space

$$\Theta = \bigcup_{r=1}^{R} \{r\} \times \Theta_r\,, \tag{13.9}$$

where the union is of course one of disjoint sets. An element θ of Θ may thus always be written as $\theta = (r, \theta_r)$, where θ_r is an element of Θ_r. Obviously, this convention is somewhat redundant, as we generally know by looking at the second component of θ to which of the sets in (13.9) θ belongs, but it will greatly simplify matters from a notational point of view. The target density will be denoted by

$$\pi(\theta) = \pi(r, \theta_r) = \varrho(r)\pi_r(\theta_r) \ .$$

In order to avoid tedious (but straightforward) constructions, we do not fully specify the dominating measure used for defining the above density, and we will also, when needed and unambiguous from the context, use the notation $\pi(d\theta)$ to refer to the probability measure itself. On the individual parameter spaces Θ_r, we denote the dominating measure by $d\theta_r$ as previously.

For HMMs, the space Θ_r is in general that of parameters for HMMs with r states for the hidden Markov chain. We remark that strictly speaking, a model is not identical to a parameter space, as the parameter space alone does not tell anything about the model structure. Two completely different models could well have identical parameter spaces. In the development below, this distinction between model and parameter space is not important however, and we will work with the parameter spaces only.

In the Bayesian framework exposed above, the dimension r of the model is treated as a usual parameter. The aim is to address the two problems of testing—deciding which model is best—and estimation—determining the parameters of the best fitting model—simultaneously. Conceptually, a variable dimension model is more complicated only because the prior and posterior distributions live in the space Θ defined in (13.9), whose structure is quite complex. Interestingly, by integrating out the index part of the model, we simply end up with mixture representations both for the distribution of the data,

$$\sum_{r=1}^{R} \varrho(r)p(y) \ ,$$

and for the *predictive distribution* (given observations y_{obs})

$$\sum_{r=1}^{R} \varrho(r|y_{\text{obs}}) \int p(y|\theta_r)\pi_r(\theta_r|y_{\text{obs}}) \, d\theta_r \ .$$

This mixture representation, called *model averaging* in the Bayesian literature, is interesting because it suggests the use of predictors that are *not* obtained by selecting a particular model from the R possible ones but rather consist in taking all the options into account simultaneously, weighting them by their posterior odds $\varrho(r|y_{\text{obs}})$. The variability due to the selection of the model is thus accounted for.

Note also that in defining the variable dimension model, we have chosen a completely new set of parameters for each model Θ_r and set the parameter space as the union of the model parameter spaces Θ_r, even though some parameters may have a similar meaning in two different models. For instance, when comparing an AR(p) and an AR($p+1$) model, it could be posited that the first p autoregressive coefficients would remain the same for the AR(p) and AR($p+1$) models, i.e., that an AR(p) model is simply an AR($p+1$) model with an extra zero coefficient. We argue on the opposite that they should be distinguished as *entities* because the models are different and also because, for instance, the best fitting AR($p+1$) model is not necessarily a straight modification of the best fitting AR(p) model by adding an extra term while keeping the other ones fixed. Similarly, even though the variance σ^2 has the same formal meaning for all values of p in the autoregressive case, we insist on using a different variance parameter for each value of p.

This is not the only possible perspective on this problem however, and many prefer to use some parameters common to all models in order to reduce model and computational complexity. In some sense, the reversible jump technique to be discussed in Section 13.2.3 is based on this assumption of exchangeable parameters between models, using proposal distributions that modify only a part of the parameter vector to move between models.

Given a variable dimension model, there is an additional computational difficulty in representing, or simulating from, the posterior distribution in that a sampler must move both *within* and *between* models Θ_r. Although the former pertains to previous developments (Section 13.1.4), the latter requires a sound measure-theoretic basis to lead to correct MCMC moves, that is, to moves that validate $\pi(\theta|y_{0:n})$ as the stationary distribution of the simulated Markov chain. There have been several earlier approaches in the literature, using for instance birth-and-death processes (Geyer and Møller, 1994) or pseudo-priors (Carlin and Chib, 1995), but the general formalization of this problem has been realized by Green (1995).

13.2.2 Green's Reversible Jump Algorithm

Green's (1995) algorithm is basically of Metropolis-Hastings type with specific trans-dimensional proposals carefully designed to move between different models in a way that is consistent with the desired stationary distribution of the MCMC algorithm. We discuss here only the simplest, and more common, application of Green's ideas in which the moves from higher to lower dimensional models are deterministic and refer to Green (1995) or Richardson and Green (1997) for more involved proposals.

We describe below the structure of moves between two different models Θ_s and Θ_l, where Θ_l say is of larger dimension than is Θ_s ("s" is for small and "l" for large). If the Markov chain is currently in state $\theta_s \in \Theta_s$, Green's algorithm uses an auxiliary random variable, which we denote by v, and a function m that maps the pair (θ_s, v) into a proposed new state $\theta_l \in \Theta_l$. The only requirement is that m be differentiable with an inverse mapping m^{-1} that is also differentiable. If (θ_s, v) is the point that corresponds to θ_l trough m^{-1}, we will use the notations

$$\theta_s = m^{-1}_{\text{param}}(\theta_l) \qquad \text{and} \qquad v = m^{-1}_{\text{aux}}(\theta_l)$$

for the associated projections of $m^{-1}(\theta_l)$. The reverse move from Θ_l to Θ_s is deterministic and simply consists in jumping back to the point $\theta_s = m^{-1}_{\text{param}}(\theta_l)$. Obviously, this dimension-changing move alone may fail to explore the whole space, and it is necessary to propose usual fixed dimension moves as well as these trans-dimensional moves. For the moment we can ignore this fact however, as we are going to show that the trans-dimensional move alone is π reversible. We shall assume that when in state $\theta_s \in \Theta_s$, the move to Θ_l is attempted with probability $P_{s,l}$ and that the auxiliary variable v has a density p. Conversely, when in Θ_l, the move to Θ_s is attempted with probability $P_{l,s}$. The moves are then accepted with probability $\alpha(\theta_s, \theta_l)$ in the first case and $\alpha(\theta_l, \theta_s)$ in the second one, where it is understood that the chain stays in its current state in case of rejection.

To determine the correct form of the acceptance probability α, we will check that the transition kernel corresponding to the mechanism described above does satisfy the detailed balance condition (2.12) for the target π. A first remark is that given the structure of the state space Θ, which is a union of disjoint sets, one can fully specify probability distributions on Θ by their operation on test functions $\bar{f}_q$ of the form

$$\bar{f}_q(\theta) = \bar{f}_q(r, \theta_r) = \begin{cases} 0 & \text{if } r \neq q \,, \\ f_r(\theta_r) & \text{otherwise} \,, \end{cases} \tag{13.10}$$

for some $q = 1, \ldots, R$ and $f_q \in \mathcal{F}_{\mathrm{b}}(\Theta_q)$. For such a test function,

$$\mathrm{E}_\pi(\bar{f}_q) = \varrho(q) \int_{\Theta_q} f_q(\theta_q) \pi_q(\theta_q) \, d\theta_q \,.$$

The second important remark is that when examining the proof of the reversibility of the usual Metropolis-Hastings algorithm (Proposition 6.2.6), it is seen that the form of the acceptance probability α is entirely determined by what happens when the chain really moves. The part that concerns rejection is fully determined by the fact that the transition kernel must be a probability kernel, that is, integrate to unity. Hence, in the case under consideration, we may check the detailed balance condition for test functions of the form given in (13.10) only, with $q = s$ and $q = l$. We will denote these functions by $\bar{f}_s$ and $\bar{f}_l$ respectively (with associated functions $f_s \in \mathcal{F}_{\mathrm{b}}(\Theta_s)$ and $f_l \in \mathcal{F}_{\mathrm{b}}(\Theta_l)$).

Denoting by K the transition kernel associated with the move between Θ_s and Θ_l described above, we have

$$\iint \bar{f}_s(\theta) \, \pi(d\theta) \times K(\theta, d\theta') \bar{f}_l(\theta') =$$
$$\int \varrho(s) \pi_s(\theta_s) f_s(\theta_s) \left\{ \int P_{s,l} \, \alpha[\theta_s, m(\theta_s, v)] p(v) f_l[m(\theta_s, v)] \, dv \right\} d\theta_s \,.$$

Now apply the change of variables formula to replace the pair (θ_s, v) by θ_l. This yields

$$\iint \bar{f}_s(\theta)\,\pi(d\theta) \times K(\theta, d\theta')\bar{f}_l(\theta') =$$
$$\int f_s[m_{\text{param}}^{-1}(\theta_l)] f_l(\theta_l) \varrho(s) \pi_s(m_{\text{param}}^{-1}(\theta_l)) \frac{P_{s,l}\alpha(\theta_s,\theta_l)p[m_{\text{aux}}^{-1}(\theta_l)]}{J_{s,l}(\theta_l)}\, d\theta_l \; , \tag{13.11}$$

where $J_{s,l}(\theta_l)$ is the absolute value of the determinant of the Jacobian matrix associated with the mapping m. It may be evaluated either as

$$J_{s,l}(\theta_l) = \left| \frac{\partial m(\theta_s, v)}{\partial(\theta_s, v)} \right|_{(\theta_s,v)=m^{-1}(\theta_l)}$$

or

$$J_{s,l}(\theta_l) = \left| \frac{\partial m^{-1}(\theta_l)}{\partial \theta_l} \right|^{-1} .$$

Because the reverse move is deterministic, the opposite case is much simpler and

$$\iint \bar{f}_l(\theta)\,\pi(d\theta) \times K(\theta, d\theta')\bar{f}_s(\theta') =$$
$$\int \varrho(l)\pi_l(\theta_l) f_l(\theta_l) \left\{ P_{l,s}\alpha[\theta_l, m_{\text{param}}^{-1}(\theta_l)] f_s[m_{\text{param}}^{-1}(\theta_l)] \right\} d\theta_l \; . \tag{13.12}$$

To ensure that (13.11) and (13.12) coincide for all choices of the functions f_s and f_l, the acceptance probability must satisfy

$$\frac{\varrho(s)\pi_s(\theta_s)P_{s,l}p(v)}{J_{s,l}(\theta_l)}\alpha(\theta_s,\theta_l) = \varrho(l)\pi_l(\theta_l)P_{l,s}\alpha(\theta_l,\theta_s) \; , \tag{13.13}$$

where it is understood that θ_s, θ_l and v satisfy $\theta_l = m(\theta_s, v)$. By analogy with the case of the usual Metropolis-Hastings algorithm, it is possible to find a solution to the above equation of the form

$$\alpha(\theta_s,\theta_l) = A(\theta_s,\theta_l) \wedge 1 \qquad \text{and} \qquad \alpha(\theta_l,\theta_s) = A^{-1}(\theta_s,\theta_l) \wedge 1$$

by setting

$$A(\theta_s,\theta_l) = \frac{\varrho(l)\pi_l(\theta_l)P_{l,s}}{\varrho(s)\pi_s(\theta_s)P_{s,l}p(v)} J_{s,l}(\theta_l) \; . \tag{13.14}$$

Indeed, with this choice both sides of (13.13) evaluate to

$$\varrho(l)\pi_l(\theta_l)P_{l,s} \wedge \frac{\varrho(s)\pi_s(\theta_s)P_{s,l}p(v)}{J_{s,l}(\theta_l)} .$$

Thus (13.14) defines the applicable acceptance ratio to be used with Green's reversible jump move. At this level the formulation of Green's algorithm is rather abstract, but we hope it will be more clear after studying the following example.

Example 13.2.2 (Normal HMM, Continued). We shall extend Example 13.1.14 to allow for moving between HMMs of different orders using reversible jump MCMC. We will discuss two different kinds of dimension-changing moves, or, rather, pair of moves: birth/death and split/combine. In a *birth move*, the order of the Markov chain is increased by one by adding a new state, and the *death move* works in the reverse way by deleting an existing state. The *split move* takes an existing state and splits it in two, whereas the *combine* (also called *merge*) *move* takes a pair of states and tries to combine them into one. We will now in detail describe these moves and how their acceptance ratios are computed.

We start with the birth move. Suppose that the current MCMC state is (r, θ_r), and that we attempt to add a new state, that we denote by i_0, to the HMM. We first draw the random variables

$$\begin{aligned} &\mu_{i_0} \sim \mathrm{N}(\alpha, \beta), && \sigma^2_{i_0} \sim \mathrm{IG}(\kappa, \gamma), \\ &\omega_{i_0,j} \sim \mathrm{Ga}(\delta_j, 1) \text{ for } j = 1, \ldots, r, && \omega_{i,i_0} \sim \mathrm{Ga}(\delta_{i_0}, 1) \text{ for } i = 1, \ldots, r, \\ &\omega_{i_0,i_0} \sim \mathrm{Ga}(\delta_{i_0}, 1), \end{aligned}$$

all independently. In other words, the parameters that go with the new state are drawn from their respective priors. These parameters correspond to the auxiliary variable v_{birth} for the birth move. The remaining parameters, that is, the components of θ_r, are simply copied to the proposed new state θ_{r+1}. Therefore, the corresponding mapping m_{birth} is simply the identity; no particular transformation is required to obtain the proposed new state in Θ_{r+1}. In the death move, the attempted move is to delete a state, denoted by i_0, that is chosen at random. The auxiliary variables μ_{i_0}, etc., of the associated birth move are trivially recovered; they are just components of the state i_0 that is proposed to be deleted!

Next in turn is the computation of the acceptance ratio, which is in fact quite simple in this particular case. Because the mapping m_{birth} is the identity mapping, its Jacobian is the identity matrix, with determinant one. The remaining factors of (13.14) become

$$\begin{aligned} &\frac{\varrho(r+1)\pi_{r+1}(\theta_{r+1})\mathrm{L}(y_{0:n}|\theta_{r+1})(r+1)!}{\varrho(r)\pi_r(\theta_r)\mathrm{L}(y_{0:n}|\theta_r)r!} \times \frac{P_d(r+1)/(r+1)}{P_b(r)} \\ &\times \frac{1}{p_\mu(\mu_{i_0})p_{\sigma^2}(\sigma^2_{i_0})\prod_{i=1}^r p_\omega(\omega_{i,i_0})\prod_{j=1}^r p_\omega(\omega_{i_0,j})p_\omega(\omega_{i_0,i_0})} \,. \end{aligned} \tag{13.15}$$

This ratio deserves some further comments. The first factor is the ratio of posterior densities. The factorials arise from the fact that, as the prior is exchangeable—the prior as well as the posterior are invariant under permutations of states—we cannot distinguish between parameters that are identical up to such permutations. Thus our effective parameter space for r-order HMMs is that of equivalence classes of parameters that are identical up to permutations, and the prior of such an equivalence class is $r!$ times the original prior of one of its representations (cf. Section 13.1.3). When r stays put, this distinction between a parameter and its equivalence class is unimportant, but it becomes important when r is allowed to vary as ignoring it would lead to incorrect acceptance ratios.

The remaining factors in (13.15) are as follows: $P_b(r)$ is the probability of proposing a birth move when the current state is of order r, $P_d(r+1)$ is the probability of proposing a death move when the current state is of order $r+1$, so that $P_d(r+1)/(r+1)$ is the probability of proposing to kill the specific state i_0 of θ_{r+1}, and the product of densities p_μ, p_{σ^2} and p_ω forms the joint proposal density p_{birth} of the birth move.

Now, because the proposal densities p_μ, etc., are identical to the priors of the corresponding parameters, and because the components in θ_r remain the same in θ_{r+1}, there will be cancellations in (13.15), leading to the simplified expression

$$\frac{\varrho(r+1)\mathrm{L}(y_{0:n}|\theta_{r+1})}{\varrho(r)\mathrm{L}(y_{0:n}|\theta_r)} \times \frac{P_d(r+1)}{P_b(r)} \,. \tag{13.16}$$

The acceptance ratio for the death move is the inverse of the above, which completes the description of the birth/death move.

We now turn to the split/combine move. Starting with the split move, suppose that the current MCMC state is θ_r, of order r. The split move selects a state, i_0 say, and attempts to split it into two new ones, i_1 and i_2. The parameters of the corresponding normal distribution must be "split" as well. This can be done as follows.

(i) Split μ_{i_0} as

$$\mu_{i_1} = \mu_{i_0} - \sigma_{i_0}\varepsilon_\mu, \quad \mu_{i_2} = \mu_{i_0} + \sigma_{i_0}\varepsilon_\mu, \quad \text{with } \varepsilon_\mu \sim \mathrm{N}(0, \tau'_\mu),$$

and split $\sigma^2_{i_0}$ as

$$\sigma^2_{i_1} = \sigma^2_{i_0}\xi_\sigma, \quad \sigma^2_{i_2} = \sigma^2_{i_0}/\xi_\sigma, \quad \text{with } \xi_\sigma \sim \mathrm{LN}(0, \tau'_\sigma).$$

(ii) Split column i_0 as

$$\omega_{i,i_1} = \omega_{i,i_0}u_i, \quad \omega_{i,i_2} = \omega_{i,i_0}(1-u_i), \quad \text{with } u_i \sim \mathrm{U}(0,1) \text{ for } i \neq i_0.$$

(iii) Split row i_0 as

$$\omega_{i_1,j} = \omega_{i_0,j}\xi_j, \quad \omega_{i_2,j} = \omega_{i_0,j}/\xi_j, \quad \text{with } \xi_j \sim \mathrm{LN}(0, \tau'_\omega) \text{ for } j \neq i_0.$$

(iv) Split ω_{i_0,i_0} as

$$\begin{aligned}
\omega_{i_1,i_1} &= \omega_{i_0,i_0}u_{i_0}\xi_{i_1}, & \omega_{i_1,i_2} &= \omega_{i_0,i_0}(1-u_{i_0})\xi_{i_2}, \\
\omega_{i_2,i_1} &= \omega_{i_0,i_0}u_{i_0}/\xi_{i_1}, & \omega_{i_2,i_2} &= \omega_{i_0,i_0}(1-u_{i_0})/\xi_{i_2},
\end{aligned}$$

where $u_{i_0} \sim \mathrm{U}(0,1)$ and $\xi_{i_1}, \xi_{i_2} \sim \mathrm{LN}(0, \tau'_\xi)$.

These formulas deserve some comments. Step (ii) is sensible in the way that the transition probability of moving from state i to i_0 is distributed between the probabilities of moving to the new states i_1 and i_2, respectively. We note that state i_0 can be split into states (i_1, i_2) with corresponding normal parameters $(\mu_{i_1}, \sigma^2_{i_1})$ and $(\mu_{i_2}, \sigma^2_{i_2})$, but also into the same pair but in reverse order (the corresponding ω are then also reversed). This gives an identical parameter in terms of equivalence classes as defined above. In fact, the densities of these two proposals are identical, as u and $1-u$ have the same distribution, and likewise for ε and $-\varepsilon$, and ξ and $1/\xi$, respectively (here subscripts on these variables are omitted).

The move that reverses the above operations, that is, the combine move, goes as follows. Select two distinct states i_1 and i_2 at random, and attempt to combine them into a single state i_0 as follows.

(i') Let $\mu_{i_0} = (\mu_{i_1} + \mu_{i_2})/2$ and let $\sigma^2_{i_0} = (\sigma^2_{i_1}\sigma^2_{i_2})^{1/2}$.
(ii') Let $\omega_{i,i_0} = \omega_{i,i_1} + \omega_{i,i_2}$ for $i \neq i_0$.
(iii') Let $\omega_{i_0,j} = (\omega_{i_1,j}\omega_{i_2,j})^{1/2}$ for $j \neq i_0$.
(iv') Let $\omega_{i_0,i_0} = (\omega_{i_1,i_1}\omega_{i_2,i_1})^{1/2} + (\omega_{i_1,i_2}\omega_{i_2,i_2})^{1/2}$.

Along the way, we recover the values of the auxiliary variables of the split move.

The auxiliary variables ε_μ, ξ_σ, etc., constitute the vector v_{split} of the split move. The mapping m_{split} is not the identity, as for the birth move, but rather given by steps (i)–(iv) above. We will now detail the computation of the corresponding Jacobian and its determinant. The transformation we need to examine is thus the one taking the components of an rth order parameter θ_r and the auxiliary variables into an $(r+1)$-th order parameter θ_{r+1} by a split move. In this transformation most components, namely all that are not associated with state i_0 that is split, are simply copied to the new parameter θ_{r+1}, and they do not affect any of the other components of θ_{r+1}. Thus the Jacobian will be block diagonal with respect to these components, and the block corresponding to the copied components is an identity matrix. In effect, this means that the Jacobian determinant equals the Jacobian determinant associated with the components actually involved in the split only. Analyzing this part closer, we find further structure implying diagonal blocks, namely the structure found in steps (i)–(iv) above. The sets of parameters and auxiliary variables involved in each of these steps are disjoint, meaning that the Jacobian will be block diagonal with respect to the structure of the steps and its determinant will be the product of the determinants given by each of the steps.

(i) For this step, taking $(\mu_{i_0}, \varepsilon_\mu, \sigma^2_{i_0}, \xi_\sigma)$ into $(\mu_{i_1}, \mu_{i_2}, \sigma^2_{i_1}, \sigma^2_{i_1})$, the Jacobian is

$$\begin{pmatrix} 1 & \sigma_{i_0} & \varepsilon_\mu/2\sigma_{i_0} & 0 \\ 1 & -\sigma_{i_0} & -\varepsilon_\mu/2\sigma_{i_0} & 0 \\ 0 & 0 & \xi_\sigma & \sigma^2_{i_0} \\ 0 & 0 & 1/\xi_\sigma & -\sigma^2_{i_0}/\xi^2_\sigma \end{pmatrix},$$

given that we differentiate with respect to $\sigma_{i_0}^2$, not σ_{i_0}. The (modulus of the) determinant of this matrix is $4\sigma_{i_0}^3/\xi_\sigma$.

(ii) For this step, the Jacobian is further block diagonal with respect to each $i \neq i_0$. For each such i, the step takes (ω_{i,i_0}, u_i) into $(\omega_{i,i_1}, \omega_{i,i_2})$, with Jacobian

$$\begin{pmatrix} u_i & 1-u_i \\ \omega_{i,i_0} & -\omega_{i,i_0} \end{pmatrix}$$

and (modulus of the) determinant ω_{i,i_0}. The overall Jacobian determinant of this step is thus $\prod_{i\neq i_0} \omega_{i,i_0}$.

(iii) For this step, the Jacobian is also further block diagonal with respect to $j \neq i_0$. For a specific j, the step takes $(\omega_{i_0,j}, \xi_j)$ into $(\omega_{i_1,j}, \omega_{i_2,j})$, with Jacobian

$$\begin{pmatrix} \xi_j & 1/\xi_j \\ \omega_{i_0,j} & -\omega_{i_0,j}/\xi_j^2 \end{pmatrix}$$

and (modulus of the) determinant $2\omega_{i_0,j}/\xi_j$. The overall Jacobian determinant of this step is thus $2^{r-1}\prod_{j\neq i_0} \omega_{i_0,j}/\xi_j$.

(iv) For this step, taking $(\omega_{i_0,i_0}, u_{i_0}, \xi_{i_1}, \xi_{i_2})$ into $(\omega_{i_1,i_1}, \omega_{i_1,i_2}, \omega_{i_2,i_1}, \omega_{i_2,i_2})$, the Jacobian is

$$\begin{pmatrix} u_{i_0}\xi_{i_1} & (1-u_{i_0})\xi_{i_2} & u_{i_0}/\xi_{i_1} & (1-u_{i_0})/\xi_{i_1} \\ \omega_{i_0,i_0}\xi_{i_1} & -\omega_{i_0,i_0}\xi_{i_2} & \omega_{i_0,i_0}/\xi_{i_1} & -\omega_{i_0,i_0}/\xi_{i_2} \\ \omega_{i_0,i_0}u_{i_0} & 0 & -\omega_{i_0,i_0}u_{i_0}/\xi_{i_1}^2 & 0 \\ 0 & \omega_{i_0,i_0}(1-u_{i_0}) & 0 & -\omega_{i_0,i_0}(1-u_{i_0})/\xi_{i_2}^2 \end{pmatrix}.$$

Some algebra shows that the (modulus of the) determinant of this matrix is $4\omega_{i_0,i_0}^3 u_{i_0}(1-u_{i_0})/\xi_{i_1}\xi_{i_2}$.

Finally we arrive at the overall Jacobian determinant (in absolute value) of the split move,

$$J_{\text{split}} = \left| 2^{r+3} \frac{\sigma_{i_0}^3 \omega_{i_0,i_0}^3 u_{i_0}(1-u_{i_0})}{\xi_\sigma \xi_{i_1} \xi_{i_2}} \prod_{i\neq i_0} \omega_{i,i_0} \prod_{j\neq i_0} \frac{\omega_{i_0,j}}{\xi_j} \right| .$$

The acceptance ratio for the split/combine move is thus

$$\begin{aligned} &\frac{\varrho(r+1)\pi_{r+1}(\theta_{r+1})\mathrm{L}(y_{0:n}|\theta_{r+1})(r+1)!}{\varrho(r)\pi_r(\theta_r)\mathrm{L}(y_{0:n}|\theta_r)r!} \times \frac{P_c(r+1)/[(r+1)r/2]}{P_s(r)/r} \\ &\quad \times \frac{1}{2p_{\varepsilon_\mu}(\varepsilon_\mu)p_{\xi_\sigma}(\xi_\sigma)p_{\xi_{i_1}}(\xi_{i_1})p_{\xi_{i_2}}(\xi_{i_2})\prod_{j\neq i_0} p_{\xi_j}(\xi_j)} \times J_{\text{split}} \\ &= \frac{\varrho(r+1)\pi_{r+1}(\theta_{r+1})\mathrm{L}(y_{0:n}|\theta_{r+1})}{\varrho(r)\pi_r(\theta_r)\mathrm{L}(y_{0:n}|\theta_r)} \times \frac{P_c(r+1)}{P_s(r)} \\ &\quad \times \frac{1}{p_{\varepsilon_\mu}(\varepsilon_\mu)p_{\xi_\sigma}(\xi_\sigma)p_{\xi_{i_1}}(\xi_{i_1})p_{\xi_{i_2}}(\xi_{i_2})\prod_{j\neq i_0} p_{\xi_j}(\xi_j)} \times J_{\text{split}} . \end{aligned}$$

Here $P_s(r)/r$ and $P_c(r+1)/[(r+1)r/2]$ are the probabilities to propose to split a specific component out of r and to propose to combine a specific pair out of $(r+1)r/2$ (the number of pairs selected from $r+1$ items) possible ones, respectively. For the auxiliary variable densities, we note that the uniform variables involved have densities equal to unity, and that the factor 2 arises from the above observation that there are two different combinations of auxiliary variables that have equal density and that result in identical parameters after the split. The acceptance rate for the combine move is the inverse of the above.

Just as for MCMC algorithms with fixed r, several types of moves are typically put together into a sweep. For the current algorithm, a sweep may look as follows.

(a) Update the means μ_i while letting r stay fixed.
(b) Update the variances σ_i^2 while letting r stay fixed.
(c) Update the ω_{ij} while letting r stay fixed.
(d) Propose a birth move or a death move, with probabilities $P_b(r)$ and $P_d(r)$, respectively.
(e) Propose a split move or a combine move, with probabilities $P_s(r)$ and $P_c(r)$, respectively.

Obviously, $P_b(r) + P_d(r) = 1$ and $P_s(r) + P_c(r) = 1$ must hold for all r. Typically, all these probabilities are set to 1/2, except for $P_b(1) = P_s(1) = 1$, $P_d(1) = P_c(1) = 0$, $P_b(R) = P_s(R) = 0$, and $P_d(R) = P_c(R) = 1$, where R is the maximum number of states allowed by the prior. Steps (a)–(c) above may be accomplished by Metropolis-Hastings steps as in Example 13.1.14 but may also be done by completing the data through simulation of the hidden chain $X_{0:n}$ followed by a Gibbs step for updating μ_i and σ_i^2 conditional on both the data and the hidden chain. The ω_{ij} may also be updated this way, by simulating the row sums and the q_{ij} separately and then computing the corresponding ω_{ij}. ■

The above reversible jump MCMC algorithm was implemented and run on a data set consisting of 600 monthly returns (in percent) from the Japanese stock index Nikkei over the time period 1950–1999; Graflund and Nilsson (2003) contains a fuller description of this time series as well as an ML-based statistical analysis using the normal HMMs. The mean of the data was 1.14, and its minimal and maximal values were −29.8 and 24.6, respectively. In our implementation, we put a uniform prior on r over the range $1, 2, \ldots, R$ with $R = 10$, and took $\alpha = 0$, $\beta = 40$, $\kappa = 1$, $\gamma = 2$, and $\delta_j = 1$ for all j. Updating of the μ_i and the σ_i^2 for fixed r was done through imputation of the hidden chain followed by Gibbs sampling, whereas the ω_{ij} were updated through a $\mathrm{N}(0, 0.1^2)$ increment random walk Metropolis-Hastings proposal on each $\log \omega_{ij}$. The birth, death, split, and combine proposal probabilities $P_b(r)$, etc., were all set to 1/2 with the aforementioned modifications at the boundaries $r = 1$ and $r = R$. In the split move, we used $\tau'_\mu = \tau'_\sigma = \tau'_\omega = 0.5$.

The algorithm was run for 100,000 burn-in sweeps and then for another 2,000,000 sweeps during which its output was monitored. The acceptance rate for the update-ω_{ij} move, the split/combine move, and the birth/death move was about 34%, 1.8%, and 1.4%, respectively. A higher rate for the dimension-changing moves would indeed be desirable, and this could perhaps be achieved with modified moves. We did some experimentation with other values for κ, γ, and the τ', but without obtaining much variation in the acceptance rates.

The estimated posterior probabilities for r were 0.000, 0.307, 0.500, 0.156, 0.029, 0.006, and 0.001 for $r = 1, 2, \ldots, 7$ and below 0.001 for larger r. Graflund and Nilsson (2003) estimated the same kind of HMM from the data but using ML implemented through simulated annealing, arriving at the estimated p-value 0.60 for testing $r = 2$ vs. $r = 3$. They thus adopted $r = 2$ as their order estimate, whereas the reversible jump MCMC analysis above gives the largest posterior probability for $r = 3$. However, our particular choice of prior may have a substantial effect on the posterior for r, and a Bayes factor analysis, which we did not carry out, may also give a different conclusion. Indeed, hierarchical priors are often used to attenuate the effect of the prior on the posterior (Richardson and Green, 1997; Robert *et al.*, 2000). We stress that the algorithm outlined above should be viewed as an example of a reversible jump MCMC algorithm that may be modified and tuned for different applications, rather than as a "ready-to-use" algorithm that suits every need. As another example of posterior analysis, we extracted the MCMC samples with $r = 2$ components, permuted the component indices for each such sample to make the means μ_i sorted (there was label switching in the MCMC output), and computed the posterior means: $\overline{\mu}_1 = 0.755$ and $\overline{\mu}_2 = 1.568$. This is to be compared to the MLEs $\hat{\mu}_1 = 0.847$ and $\hat{\mu}_2 = 1.531$ reported by Graflund and Nilsson (2003). The credibility intervals we obtained were quite wide; the 95% intervals for μ_1 and μ_2 (after sorting) read $(-0.213, 1.460)$ and $(1.102, 2.074)$ respectively, both covering the respective MLE.

13.2.3 Alternative Sampler Designs

Reversible jump MCMC algorithms have in common with more conventional Metropolis-Hastings algorithms that they generally contain some parameters that need to be "fine tuned" in order to optimize their performance. In the example above, these parameters are τ'_μ, τ'_σ and τ'_ω. Often the only way to do this fine tuning is through a set of pilot runs during which acceptance probabilities and other statistics related to the mixing of the algorithm are monitored.

For any particular variable-dimension statistical model, there is an infinite number of ways of designing reversible jump algorithms. The above example is only one of them for the normal HMM. Other structures of the split/combine move, for instance, may prove more efficient with certain combinations of priors and/or data. Designing a reversible jump algorithm is by no means an automated procedure but needs to be guided by experimentation and, when available, experience. The recent paper by Brooks *et al.* (2003) does outline, however, some general ideas about how to construct efficient reversible jump algorithms by setting up rules to calibrate the jump proposals.

Above, we motivated the factorial $r!$ that is adjoined to the posterior density by an argument based on equivalence classes of parameters. Richardson and Green (1997) motivated them by saying that the actual parameter space is the one only containing parameters such that the normal means, for instance, appear in ascending order: $\mu_1 < \mu_2 < \ldots < \mu_r$, cf. Section 13.1.3. We note that sorting of this kind may become necessary even without restrictions on the prior, as we have seen that with an exchangeable prior, the marginal posterior densities of the means, for example, are generally identical. We prefer to view such sorting as a part of the post-processing of the MCMC sampler output, however, rather than as an intrinsic property of the algorithm itself. Sorting afterwards simplifies, for example, examination of how sorting with respect to different sets of parameters (means or variances, for example) affect the inference.

As a consequence of the assumption of sorted means, Richardson and Green (1997) also restrict the split move, disallowing it to separate the normal means so far apart that the ordering is violated, and the combine move is restricted accordingly in that it may only attempt to combine states with adjacent normal means. We make some comments on this approach. The first is that this restriction on the split/combine move is by no means necessary; if a split move violates the ordering, we can view that parameter as the equivalent one obtained upon sorting the means followed by a corresponding permutation of the remaining coordinates. The combine move is then allowed to attempt merging any pair of states. A second comment is that the above restriction on the split and combine moves may prove useful, even when we do not make any restrictions on the prior. With r states, there are $r(r-1)/2$ different pairs to combine, and one can imagine that pairs with means (or variances) far apart are less likely to generate a successful combine move. Therefore, restricting the combine move to consider states with adjacent means (or variances) only may lead to an increased acceptance probability for this move. If this strategy is adopted, the split move must be restricted accordingly, as the split/combine pair (as all other pairs) must be reversible: what one move may do the other one must be able to undo.

We also mention the option to include the hidden chain $\{X_k\}_{k\geq 0}$ in the MCMC state space, that is, adjoining it to the parameter θ. This choice was made by Richardson and Green (1997) in the setting of mixtures, and followed up for HMMs by Robert *et al.* (2000). These papers also provide suggestions for other designs of split/combine moves. In addition, the latter paper contains a lot of fine tuning done in the process of increasing acceptance rates. Including the hidden chain in the MCMC sampler simplifies the computation of the posterior density, as the likelihood involved is then $\mathrm{L}(y_{0:n}|x_{0:n}, \theta_r)$ rather than $\mathrm{L}(y_{0:n}|\theta_r)$, and the former is simply a product of scalars. On the other hand, in the birth move the new state i_0 must be assigned to some X_k and, similarly, in the split move each X_k equal to i_0 must be relabeled either i_1 or i_2. The simulation mechanisms for doing so may be quite complex, cf. (Robert *et al.*, 2000), and computationally demanding.

13.2.4 Alternatives to Reversible Jump MCMC

Reversible jump MCMC has had a vast impact on variable-dimension Bayesian inference, but there certainly are some other approaches that deserve to be discussed.

Brooks *et al.* (2003) reassess the reversible jump methodology through a global *saturation scheme*. They consider a series of models Θ_r $(r = 1, \ldots, R)$ such that $\max_r \dim(\Theta_r) = r_{\max} < \infty$. The parameter $\theta_r \in \Theta_r$ is then completed with an auxiliary variable U_r such that

$$\dim(\theta_r, u_r) = r_{\max}$$

and $U_r \sim q_r(u_r)$. Brooks *et al.* (2003) define in addition a vector ω_r of dimension $r_{\max}$ with i.i.d. components, distributed from $\psi(\omega_r)$, and assign the following joint prior to a parameter in Θ_r,

$$\pi(r, \theta_r)\, q_r(u_r) \prod_{i=1}^{r_{\max}} \psi(\omega_i)\ .$$

Within this augmented (or *saturated*) framework, there is no varying dimension anymore because, for all models, the whole vector (θ_r, u_r, ω) is of fixed dimension. Therefore, moves between models can be defined just as freely as moves between points of each model—see also (Godsill, 2001) for a similar development. Brooks *et al.* (2003) propose a three stage MCMC update.

Algorithm 13.2.3.

1. Update the current value of the parameter, θ_r.
2. Update u_r and ω conditional on θ_r.
3. Update the model index r into r' using the bijection.

$$(\theta_{r'}, u_{r'}) = m(\theta_r, u_r)\ .$$

Note that, for specific models, saturation schemes appear rather naturally. For instance, the case of a noisily observed time series with abrupt changes corresponds to a variable dimension model, when considered in continuous time (Green, 1995; Hodgson, 1998). Its discrete time counterpart however may be reparameterized by using indicators X_k that a change occurs at index k (for all indices) rather than the indices of change points (Chib, 1998; Lavielle and Lebarbier, 2001). The resulting model is then a fixed dimension model, whatever the number of change points in the series.

Petris and Tardella (2003) devised an approach that is close to a saturation scheme in the sense that it constructs a density on the subspace of largest dimension. However, it does not construct the extra variables u_k explicitly but rather embeds the densities on lower dimensional subspaces into a function on the subspace of largest dimension that effectively incorporates all densities. This approach has not yet been tested on HMMs.

Reversible jump algorithms operate in discrete time, but similar algorithms may be formulated in continuous time. Stephens (2000a) suggested such an algorithm, built on birth/death moves only, for mixture distribution, and Cappé *et al.* (2003) extended the framework to allow for other kinds of dimension-changing moves like split/combine. In this continuous time approach, there are no acceptance probabilities and birth moves are always accepted, but model parameters that are unlikely, in the sense of having low posterior density, are assigned large death rates and are hence abandoned quickly. Similar remarks apply to split/combine moves. Moves that update model parameters without changing its dimension may also be incorporated. Cappé *et al.* (2003) also compared the discrete and continuous time approaches and concluded that the differences between them are very minor, but with the continuous time approach generally requiring more computing time.

13.3 Multiple Imputations Methods and Maximum *a Posteriori*

We consider in this last section a class of methods, which methods are arguably less directly connected with the Bayesian framework and which may also be envisioned as extensions or variants of the approaches discussed in Chapter 11. Rather than simulating from the posterior distribution of the parameters, we now consider maximizing it to determine the so-called maximum *a posteriori* (or MAP) point estimate. In contrast to the methods of Chapters 10–11, which could also be used in this context (Remark 10.2.1), the techniques to be discussed below explicitly use parameter simulation in addition to hidden state simulation. The primary objective of these techniques is not (only) to compensate for the lack of exact smoothing computations in many models of interest, but also to perform some form of *random search optimization*—see discussion in the introduction of Chapter 11—which is (hopefully) more robust to the presence of local maxima in the function to be optimized.

We already mentioned, in conjunction with identifiability issues, the difficulties in using, in a Bayesian context, marginal posterior means as parameter estimates in HHMs. Identifiability can be forced upon the parameter θ by imposing some artificial identifying constraint such as ascending means, as mentioned above, or as in Robert and Titterington (1998) for instance. Even in that case, the posterior mean is a poor candidate for Bayesian inference, given that it heavily depends on the identifying constraints (see Celeux *et al.*, 2000, for an illustration in the setting of mixtures). Therefore in many cases, the remaining candidate is the MAP estimate,

$$\begin{aligned}\widehat{\theta}^{\text{MAP}} &= \arg\max_\theta \int \pi(\theta, x_{0:n}|y_{0:n})\pi(\theta, x_{0:n})\, dx_{0:n} \\ &= \arg\max_\theta \pi(\theta|y) \;. \end{aligned} \tag{13.17}$$

As previously discussed, the methods of either Chapter 10 or 11 may be used to determine the MAP estimator, depending on whether or not the marginalization in (13.17) can be performed exactly. The structure of (13.17) also suggests a specific class of optimization algorithms which implement the *simulated annealing* principle originally proposed by Metropolis *et al.* (1953).

13.3.1 Simulated Annealing

Simulated annealing methods are a non-homogeneous variant of MCMC algorithms used to perform global optimization. The word "global" is used to emphasize that the ultimate goal is convergence to the actual maxima of the function of interest—the so-called *global maxima*—whether or not the function does possess local maxima. The terminology is borrowed from metallurgy where a slow decrease of the temperature of a metal—the annealing process—is used to obtain a minimum energy crystalline structure. By analogy, simulated annealing is a random search technique that explores the parameter space Θ, using a non-homogeneous Markov chain $\{\theta^i\}_{i\geq 0}$ whose transition kernels K_i are tailored to have invariant probability density functions

$$\pi_{M_i}(\theta|y_{0:n}) \propto \pi^{M_i}(\theta|y_{0:n}) \;, \tag{13.18}$$

$\{M_i\}_{i\geq 1}$ being a positive increasing sequence tending to infinity. The intuition behind simulated annealing is that as M_i tends to infinity, $\pi^{M_i}(\theta|y)$ concentrates itself upon the set of global modes of the posterior distribution. It has been shown under various assumptions that convergence to the set of global maxima is indeed ensured for sequences $\{M_i\}_{i\geq 1}$ growing at a logarithmic rate (Laarhoven and Arts, 1987). Using the metallurgic analogy again, the sequence $\{M_i\}_{i\geq 1}$ is often called a *cooling schedule*, and the reciprocal of M_i is known as the *temperature*.

In simple situations where the posterior $\pi(\theta|Y_{0:n})$ is known (up to a constant), sampling from a kernel K_i that has (13.18) as invariant density may be done using the Metropolis-Hastings algorithm (see Section 6.2.3). For HMMs however, this situation is the exception rather than the rule, and the posterior is only available in closed form in models where exact smoothing is feasible, such as normal HMMs with finite state space. To overcome this difficulty, Doucet *et al.* (2002) developed a novel approach named *SAME* (for *state augmentation for marginal estimation*), also studied by Gaetan and Yao (2003) under the name MEM (described as multiple-inputed Metropolis version of the EM algorithm). We adopt here the terminology proposed by Doucet *et al.* (2002).

13.3.2 The SAME Algorithm

The key argument behind SAME is that upon restricting the M_i to be integers, the probability density function π_{M_i} in (13.18) may be viewed as the marginal posterior in an artificially augmented probability model. Hence one may use standard MCMC techniques to draw from this augmented probability model, and therefore the simulated annealing strategy is feasible for general missing data models. The concentrated distribution π_{M_i} is obtained by artificially replicating the latent variables in the model, in our case the hidden states $X_{0:n}$.

To make the argument more precise, denote by M the current value of M_i and consider M artificial copies of the hidden state sequence, denoted by $X_{0:n}(1), \ldots, X_{0:n}(M)$. The fictitious probability model postulates that these sequences are *a priori* independent with common parameter θ and observed sequence $Y_{0:n}$, leading to a posterior joint density defined by

$$
\begin{aligned}
\pi_M[\theta, x_{0:n}(1), \ldots, x_{0:n}(M) | y_{0:n}] &\propto \prod_{m=1}^{M} \pi[\theta, x_{0:n}(m) | y_{0:n}] \\
&\propto \left\{ \prod_{m=1}^{M} p[x_{0:n}(m) | y_{0:n}, \theta] \right\} \pi(\theta)^M ,
\end{aligned}
\tag{13.19}
$$

where $\pi(\cdot|y_{0:n})$ is the joint posterior distribution corresponding to the model, $p(\cdot|y_{0:n}, \theta)$ the likelihood, and π is the prior. This distribution does not correspond to a real phenomenon but it is a properly defined density in that it is positive, and the right-hand side can be normalized so that (13.19) integrates to unity.

Now the marginal distribution of θ in (13.19), obtained by integration over all replications of $X_{0:n}$, is

$$
\begin{aligned}
&\pi_M(\theta|y_{0:n}) \\
&\quad = \int \cdots \int \pi_M[\theta, x_{0:n}(1), \ldots, x_{0:n}(M) | y_{0:n}] \, dx_{0:n}(1) \cdots dx_{0:n}(M) \\
&\quad \propto \int \cdots \int \prod_{m=1}^{M} \pi[\theta, x_{0:n}(m) | y_{0:n}] \, dx_{0:n}(1) \cdots dx_{0:n}(M) \\
&\quad = \pi^M(\theta|y_{0:n}) \; .
\end{aligned}
$$

Hence an MCMC algorithm in the augmented space, with invariant distribution $\pi_M[\theta, x_{0:n}(1), \ldots, x_{0:n}(M)|y_{0:n}]$, is such that the simulated sequence of parameter $\{\theta^i\}_{i \geq 0}$ marginally admits π_M in (13.18) as invariant distribution.

An important point here is that when an MCMC sampler is available for the density $\pi(\theta, x_{0:n}|y_{0:n})$, it is usually easy to construct an MCMC sampler with target density (13.19) as the replications of $X_{0:n}$ are statistically independent conditional on θ in this fictitious model, that is,

$$\pi_M[x_{0:n}(1), \ldots, x_{0:n}(M)|y_{0:n}, \theta] = \prod_{m=1}^{M} \pi[x_{0:n}(m)|y_{0:n}, \theta] , \qquad (13.20)$$

and for θ, the full conditional distribution satisfies

$$\pi_M[\theta|y_{0:n}, x_{0:n}(1), \ldots, x_{0:n}(M)] \propto \prod_{m=1}^{M} \pi[\theta|y_{0:n}, x_{0:n}(m)] . \qquad (13.21)$$

According to (13.20), the sampling step for $x_{0:n}(k)$ is identical to its counterpart in a standard data augmentation sampler with target distribution $\pi[\theta, x_{0:n}(k)|y_{0:n}]$, whereas the sampling step for θ involves a draw from (13.21). If $\pi(\theta|y_{0:n}, x_{0:n})$ belongs to an exponential family of densities, then sampling from (13.21) is straightforward, as the product of conditionals in (13.21) is also a member of this exponential family. In other cases, (13.21) can be simulated using a Metropolis-Hastings step—Gaetan and Yao (2003) for instance used random walk Metropolis-Hastings proposals. For normal HMMs, the SAME algorithm may be implemented as follows.

Example 13.3.1 (SAME for Normal HMMs). Assume that the state space X is $\{1, \ldots, r\}$ and that the conditional distributions are normal, $Y_k|X_k = j \sim \mathrm{N}(\mu_j, \sigma_j^2)$. Conjugate priors are assumed, that is, $\mu_j \sim \mathrm{N}(\alpha, \beta)$, $\sigma_j^2 \sim \mathrm{IG}(\kappa, \gamma)$ and $q_{j,\cdot} \sim \mathrm{Dir}_r(\delta, \ldots, \delta)$ with independence between the μ_j, the σ_j^2, and the rows of Q. We assume (for simplicity) that the initial distribution ν is fixed and known. To avoid confusion with simulation indices (which are indicated by superscripts), we will use the notation υ_j rather than σ_j^2 for the components' variances.

Examining Example 13.1.10, we find that the full conditional distribution of the means μ_j is such that they are conditionally independent with

$$\mu_j \mid \upsilon_j, x_{0:n}(1), \ldots, x_{0:n}(M), y_{0:n} \qquad (13.22)$$
$$\sim \mathrm{N}\left(\frac{M\alpha\upsilon_j/\beta + \sum_{m=1}^{M} S_j(m)}{M\upsilon_j/\beta + \sum_{m=1}^{M} n_j(m)}, \frac{1}{M/\beta + \sum_{m=1}^{M} n_j(m)/\upsilon_j} \right) ,$$

where $S_j(m) = \sum_{0 \le k \le n:\, x_k(m)=j} y_k$ is the sum statistic associated with the mth replication of $X_{0:n}$ and state j and, similarly, $n_j(m) = \#\{0 \le k \le n : x_k(m) = j\}$ is the number of $x_k(m)$ with $x_k(m) = j$.

In an analogous way, we find that the full conditional distribution of the variances υ_j is such that they are conditionally independent with

$$\upsilon_j \mid \mu_j, x_{0:n}(1), \ldots, x_{0:n}(M), y_{0:n} \qquad (13.23)$$
$$\sim \mathrm{IG}\left(M(\kappa+1) - 1 + \frac{1}{2}\sum_{m=1}^{M} n_j(m), M\gamma + \frac{1}{2}\sum_{m=1}^{M} S_j^{(2)}(m) \right) ,$$

where $S_j^{(2)}(m) = \sum_{0 \le k \le n:\, x_k(m)=j} (y_k - \mu_j)^2$, and that the full conditional distribution of Q is such that the rows are conditionally independent with

$$(q_{j1}, \ldots, q_{jr}) \mid x_{0:n}(1), \ldots, x_{0:n}(M) \tag{13.24}$$
$$\sim \mathrm{Dir}_r \left(M(\delta - 1) + 1 + \sum_{m=1}^{M} n_{j1}(m), \ldots, M(\delta - 1) + 1 + \sum_{m=1}^{M} n_{jr}(m) \right) ,$$

where $n_{jl}(m) = \#\{0 \leq k \leq n-1 : x_k(m) = j, x_{k+1}(m) = l\}$ is the number of transitions from state j to l in the mth replication. Hence the SAME algorithm looks as follows.

Algorithm 13.3.2. Initialize the algorithm with $\theta^0 = \{\{\mu_j^0, \upsilon_j^0\}_{j=1,\ldots,r}, Q^0\}$ and select a schedule $\{M_i\}_{i\geq 0}$. Then for $i \geq 1$,

- Simulate the M_i missing data replications $X_{0:n}^i(1), \ldots, X_{0:n}^i(M_i)$ independently under the common distribution $\pi(x_{0:n}|\theta^{i-1})$;
- Simulate $\mu_1^i, \ldots, \mu_r^i$ independently from the normal distributions (13.22);
- Simulate $\upsilon_1^i, \ldots, \upsilon_r^i$ independently from the inverse Gamma distributions (13.23), using the newly simulated μ_j^i to evaluate $S_j^{(2)}(m)$ for $j = 1, \ldots, r$ and $m = 1, \ldots, M$;
- Simulate the rows of Q^i independently from the Dirichlet distributions (13.24).

The simulation of the replications $X_{0:n}^i(m)$ can be carried out using the forward filtering-backward sampling recursion developed in Section 6.1.2. ■

It should be clear from the above example that the SAME approach is strikingly close to the SEM and MCEM methods discussed in Sections 11.1.7 and 11.1.1, respectively. Indeed, taking the log, (13.19) may be rewritten as

$$\log \pi_M[\theta, x_{0:n}(1), \ldots, x_{0:n}(M)|y_{0:n}] = C^{st}$$
$$+ M \left\{ \left[\frac{1}{M} \sum_{m=1}^{M} \log p(x_{0:n}(m)|y_{0:n}, \theta) \right] + \log \pi(\theta) \right\} , \tag{13.25}$$

where the constant does not depend on the parameter θ. The term in braces in (13.25) is recognized as a Monte Carlo approximation of the intermediate quantity of EM for this problem, with the addition of the prior term (see Remark 10.2.1). Hence replacing the parameter simulation step in the SAME algorithm by a maximization step lead us back to the MCEM approach. In the example of Algorithm 13.3.2, the MCEM update can be obtained by setting the new values of the parameter to the modes of (13.22)–(13.24), that is,

$$\mu_j^* = \frac{\alpha \upsilon_j/\beta + M^{-1} \sum_{m=1}^{M} S_j(m)}{\upsilon_j/\beta + M^{-1} \sum_{m=1}^{M} n_j(m)} ,$$
$$\upsilon_j^* = \frac{\gamma + (1/2) M^{-1} \sum_{m=1}^{M} S_j^{(2)}(m)}{(\kappa + 1) + (1/2) M^{-1} \sum_{m=1}^{M} n_j(m)} ,$$
$$q_{jl}^* = \frac{(\delta - 1) + M^{-1} \sum_{m=1}^{M} n_{jl}(m)}{r(\delta - 1) + M^{-1} \sum_{l=1}^{r} \sum_{m=1}^{M} n_{jl}(m)} .$$

These equations can also be obtained from the M-step update equations (10.41)–(10.43) of the EM algorithm for the normal HMM, taking into account the prior terms and replacing the posterior expectations by their Monte Carlo approximation. It is also of interest that the distributions (13.22)–(13.24), from which simulation is done in the SAME approach, have variances that decrease proportionally to $1/M$; hence the distributions get more and more concentrated around the modes given above as the number of replications increases.

The interest of SAME, however, is that it exactly implements the simulated annealing principle for which a number of convergence results have been obtained in the literature. In particular, both Doucet and Robert (2002) and Gaetan and Yao (2003) provide some conditions under which the distribution of the ith parameter estimate θ^i converges to a measure that is concentrated on the set of global maxima of the marginal posterior. Although very appealing, these results do imply restrictive conditions on the model, requiring in particular that the likelihood be bounded from above and below. In addition, those results apply only for very slow logarithmic rates of increase of $\{M_i\}_{i\geq 1}$, with appropriate choice of multiplicative constants. Many authors, among which are Doucet *et al.* (2002), recommend using faster schedules in practice, reporting for instance good results with sequences $\{M_i\}_{i\geq 1}$ that grow linearly. We conclude this brief exposition with an example that illustrates the importance of the choice of a proper schedule—see Doucet *et al.* (2002), Gaetan and Yao (2003), and Jacquier *et al.* (2004) for further applications of the method.

Example 13.3.3 (Binary Deconvolution Model, Continued). We consider again the noisy binary deconvolution model of Example 10.3.2, which served for illustrating the EM and quasi-Newton methods. Recall that this model is a four-state normal HMM for which the transition parameters are known, the variances υ_j are constrained to equal a common value that we denote by υ, the means are given by $\mu_j = s_j^t h$ where h is a two-dimensional vector of unknown filter coefficients, and s_1 to s_4 are fixed two-dimensional vectors.

For easier comparison with the results discussed in Example 10.3.2, we select improper priors for the parameters, which amounts to setting $\alpha = 0$ and $\beta = \infty$ in (13.22) and $\kappa = -1$ and $\gamma = 0$ in (13.23). Hence the SAME algorithm will directly maximize the likelihood. Taking into account the constraints mentioned above, the posteriors in (13.22) and (13.23) should then be replaced by

$$h \mid \upsilon, x_{0:n}(1), \ldots, x_{0:n}(M), y_{0:n} \\ \sim \mathrm{N}\left(\Pi[x_{0:n}(1\!:\!M)] \sum_{m=1}^{M} \sum_{k=0}^{n} y_k x_k(m),\ \Pi[x_{0:n}(1\!:\!M)] \right) ,$$

where

$$\Pi[x_{0:n}(1:M)] = \left[\sum_{m=1}^{M}\sum_{k=0}^{n} x_k(m)x_k(m)^t\right]^{-1} ,$$

and

$$v \mid h, x_{0:n}(1), \ldots, x_{0:n}(M), y_{0:n} \sim \mathrm{IG}\left(\frac{M(n+1)}{2} - 1, \frac{1}{2}\sum_{m=1}^{M}\sum_{k=0}^{n}[y_k - x_k(m)x_k(m)^t]^2\right) .$$

Note that for this discrete-state space model, the likelihood is indeed computable exactly for all values of the parameters h and v. Hence we could also imagine implementing the simulated annealing approach directly, without resorting to the SAME completion mechanism. This example nonetheless constitutes a realistic testbed for the SAME algorithm with the advantage that the likelihood can be plotted exactly and its maximum determined with high precision by the deterministic methods discussed in Example 10.3.2.

The data is the same as in Example 10.3.2, leading to the profile likelihood surface shown in Figure 10.1. Recall that for the sake of clarity, we only consider the estimated values of h although the variance v is also treated as a parameter. For this problem, we fixed the total number of simulations of the missing state trajectories $X_{0:n}$ to 10,000 and then evaluated different schedules of the form $M_i = 1 + \lfloor ai \rfloor$ for various values of a and such that the overall number of simulations, $\sum_{i=1}^{i_{\max}} M_i$, equals 10,000. Hence $i_{\max}$ is not fixed and varies depending on the cooling schedule. These choices will be discussed below, but we can already note that 10,000 is a rather large number of simulations for this problem. Recall for instance from Figure 10.1 that the convergence of EM is quite fast in this problem (compared with the model of Example 11.1.2 for instance), although it sometimes converges to a local mode that, as we will see below, is very unlikely compared to the MLE.

Table 13.1 summarizes the results obtained over 100 independent replications of the SAME trajectories started from the first two starting points considered in Figure 10.1. The first column shows that the simple MCMC simulation without cooling schedule ($M_i = 1$) is indeed very efficient at finding the global mode of the likelihood. Indeed, once in its steady-state, the MCMC simulations spend about 640 times more time in the vicinity of the global mode than in the local mode. This finding is coherent with the log-likelihood difference between the two points (labeled "MLE" and "LOC", respectively) in Figure 10.1, which corresponds to a factor 937 once converted back to a linear scale. Hence the likelihood indeed has a local mode but one that is very unlikely compared to the MLE. Letting a simple MCMC chain run long enough is thus sufficient to end up in the vicinity of the global mode with high probability (640/641). Because of the correlation between successive values of the parameters however, this phenomenon does not manifest itself as fast as expected and 210 iterations are necessary to ensure that 95% out of the

a	0	1/72	1/12	1/2	1
$i_{\max}$	10000	1163	483	198	140
$M_{i_{\max}}$	1	17	41	100	141
Starting from point 1 in Figure 10.1					
# converged	99	92	78	79	95
std. error	0.122	0.028	0.017	0.014	0.010
Starting from point 2 in Figure 10.1					
# converged	100	87	61	52	36
std. error	0.121	0.029	0.018	0.013	0.009

Table 13.1. Summary of results of the SAME algorithm for 100 runs and different rates of increase a. The upper part of the table pertains to trajectories started from the point labeled "1" in Figure 10.1 and the lower part to those started from the point labeled "2" in Figure 10.1. "# converged" is the number of sequences that converged to the MLE and not to the local mode, and "std. error" is the average L^2-norm of the distance to the MLE for those trajectories (for comparison purposes, the L^2-norm of the MLE itself is 1.372). The random seeds used for the simulations were the same for all values of a.

200 trajectories started from either of the two starting points indeed visit the neighborhood of the global mode. Likewise, although some of the trajectories do visit the mirror modes that have identical likelihood for negative values of h_0 (see Example 10.3.2), none of the trajectories was found to switch between positive and negative values of h_0 once converged[1]. The Gibbs sampler is thus unable to connect these two regions of the posterior, which are however equally probable. This phenomenon has been observed in various other missing data settings by Celeux *et al.* (2000). In this example these mixing problems rapidly get more severe as M_i increases. Accordingly, the number of trajectories in Table 13.1 that do eventually reach the MLE drops down as the linear factor a is set to higher values. The picture is somewhat more complicated in the case of the first starting point, as the number of trajectories that reach the MLE first decreases ($a = 1/72, 1/12$) before increasing again. The explanation for this behavior is to be found in Figure 10.1, which shows that the trajectory of the EM algorithm started from this point does converge to the MLE, in contrast with what happens for the second starting point. Hence for this first starting point, when M_i increases sufficiently rapidly, the SAME algorithm mimics the EM trajectory (with some random fluctuations) and eventually converges to the MLE. This behavior is illustrated in Figure 13.2.

In this example, it turns out that in order to guarantee that the SAME algorithm effectively reaches the MLE, it is very important that M_i stays exactly equal to one for a large number of iterations, preferably a few hundreds, but fifty is really a minimum. The logarithmic rates of increase of M_i that

[1]In Table 13.1, the trajectories that converge to minus the MLE are counted as having converged, as we know that it corresponds to an identifiability issue inherent to the model.

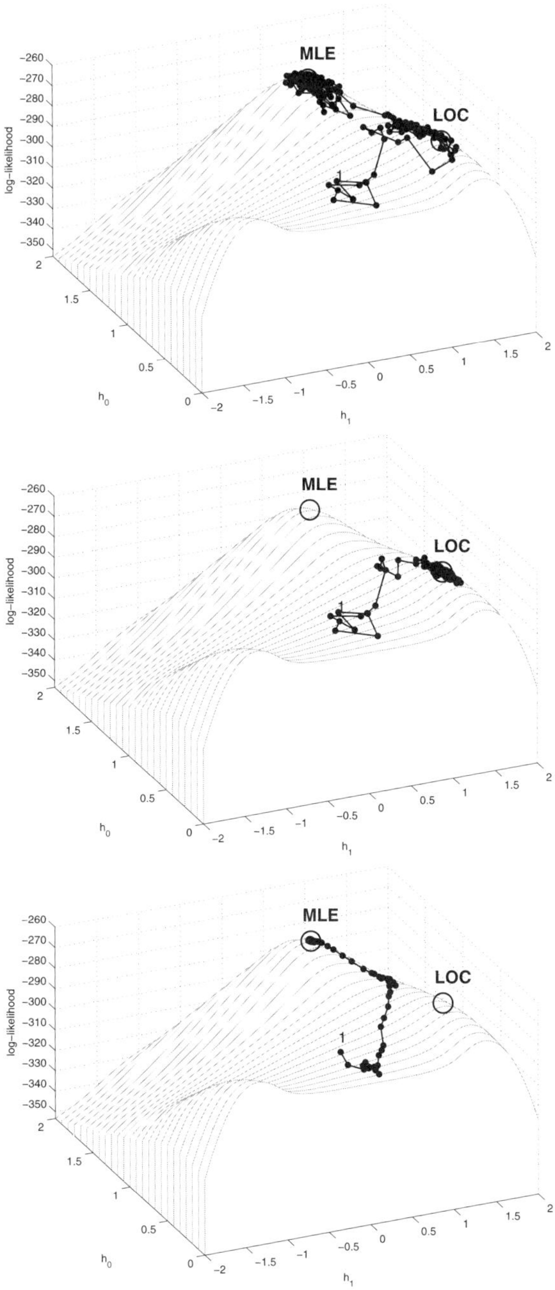

Fig. 13.2. Same profile log-likelihood surface as in Figure 10.1. The trajectories show the first 200 SAME estimates for, from top to bottom, $a = 0$, $a = 1/12$, and $a = 1$, started at the point labeled "1" in Figure 10.1. The same random seed was used for all three cases.

are compatible with this constraint and with the objective of using an overall number of simulations equal to 10,000 typically end up with $M_{i_{\max}}$ being of the order three and are thus roughly equivalent to the MCMC run ($a = 0$) in Table 13.1. Note that the error obtained with this simple scheme is not that bad, being about ten times smaller than the L^2 norm of the MLE. The factor $a = 1/72$, which gives a reasonable probability of convergence to the MLE from both points, provides an error that is further reduced by a factor of ten. ■

We would like to point out that—especially when the answer is known as in this toy example!—it is usually possible to find out by trial-and-error cooling schedules that are efficient for the problem (data and model) at hand. In the case of Example 13.3.3, setting $M_i = 1$ for the first 280 iterations and letting $M_i = 4, 16, 36, 64, 100$ for the last five iterations (500 simulations in total) is very successful with 98 (resp. 96) trajectories converging to the MLE and an average error of 0.018 (resp. 0.020) when started from the two initial points under consideration. The last five iterations in this cooling schedule follow a square progression that was used for the MCEM algorithm in Example 11.1.3. Note that rather than freezing the parameter by abruptly increasing M_i, one could use instead the averaging strategy (see Section 11.1.2) advocated by Gaetan and Yao (2003). Clearly, one-size-fits-all cooling schedules such as linear or logarithmic rates of increase may be hard to adjust to a particular problem, at least when the overall number of simulations is limited to a reasonable amount. This observation contrasts with the behavior observed for the MCEM and SAEM algorithms in Chapter 11, which are more robust in this respect, particularly for the latter. Remember however that we are here tackling a much harder problem in trying not only to avoid all local maxima but also to ensure that the parameter estimate eventually gets reasonably close to the actual global maximum.

There is no doubt that simulated annealing strategies in general, and SAME in particular, are very powerful tools for global maximization of the likelihood or marginal posterior in HMMs. Their usefulness in practical situations however depends crucially on the ability to select proper finite-effort cooling schedules, which may itself be a difficult issue.

Part III

Background and Complements

14

Elements of Markov Chain Theory

14.1 Chains on Countable State Spaces

We review the key elements of the mathematical theory developed for studying the limiting behavior of Markov chains. In this first section, we restrict ourselves to the case where the state space X is countable, which is conceptually simpler. On our way, we will also meet a number of important concepts to be used in the next section when dealing with Markov chains on general state spaces.

14.1.1 Irreducibility

Let $\{X_k\}_{k\geq 0}$ be a Markov chain on a countable state space X with transition matrix Q. For any $x \in X$, we define the first hitting time σ_x on x and the return time τ_x to x respectively as

$$\sigma_x = \inf\{n \geq 0 : X_n = x\} , \tag{14.1}$$

$$\tau_x = \inf\{n \geq 1 : X_n = x\} , \tag{14.2}$$

where, by convention, $\inf \emptyset = +\infty$. The successive hitting times $\sigma_x^{(n)}$ and return times $\tau_x^{(n)}$, $n \geq 0$, are defined inductively by

$$\sigma_x^{(0)} = 0, \; \sigma_x^{(1)} = \sigma_x, \; \sigma_x^{(n+1)} = \inf\{k > \sigma_x^{(n)} : \; X_k = x\} ,$$

$$\tau_x^{(0)} = 0, \; \tau_x^{(1)} = \tau_x, \; \tau_x^{(n+1)} = \inf\{k > \tau_x^{(n)} : \; X_k = x\} .$$

For two states x and y, we say that state x *leads to* state y, which we write $x \to y$, if $\mathrm{P}_x(\sigma_y < \infty) > 0$. In words, x leads to y if the state y can be reached from x. An alternative, equivalent definition is that there exists some integer $n \geq 0$ such that the n-step transition probability $Q^n(x, y) > 0$. If both x leads to y and y leads to x, then we say that the x and y *communicate*, which we write $x \leftrightarrow y$.

Theorem 14.1.1. *The relation "$\leftrightarrow$" is an equivalence relation on* X.

Proof. We need to prove that the relation $\leftrightarrow$ is reflexive, symmetric, and transitive. The first two properties are immediate because, by definition, for all $x, y \in \mathsf{X}$, $x \leftrightarrow x$ (reflexivity), and $x \leftrightarrow y$ if and only if $y \leftrightarrow x$ (symmetry).

For any pairwise distinct $x, y, z \in \mathsf{X}$, $\{\sigma_y + \sigma_z \circ \theta^{\sigma_y} < \infty\} \subset \{\sigma_z < \infty\}$ (if the chain reaches y at some time and later z, it certainly reaches z). The strong Markov property (Theorem 2.1.6) implies that

$$\begin{aligned}\mathrm{P}_x(\sigma_z < \infty) \geq \mathrm{P}_x(\sigma_y + \sigma_z \circ \theta^{\sigma_y} < \infty) &= \mathrm{E}_x[\mathbb{1}_{\{\sigma_y<\infty\}} \mathbb{1}_{\{\sigma_z<\infty\}} \circ \theta^{\sigma_y}] \\ &= \mathrm{E}_x[\mathbb{1}_{\{\sigma_y<\infty\}} \, \mathrm{P}_{X_{\sigma_y}}(\sigma_z < \infty)] = \mathrm{P}_x(\sigma_y < \infty) \, \mathrm{P}_y(\sigma_z < \infty) \,.\end{aligned}$$

In words, if the chain can reach y from x and z from y, it can reach z from x by going through y. Hence if $x \to y$ and $y \to z$, then $x \to z$ (transitivity). □

For $x \in \mathsf{X}$, we denote the equivalence class of x with respect to the relation "$\leftrightarrow$" by $C(x)$. Because "$\leftrightarrow$" is an equivalence relation, there exists a collection $\{x_i\}$ of states, which may be finite or infinite, such that the classes $\{C(x_i)\}$ form a partition of the state space X.

Definition 14.1.2 (Irreducibility). *If* $C(x) = \mathsf{X}$ *for some* $x \in \mathsf{X}$ *(and then for all* $x \in \mathsf{X}$*), the Markov chain is called* irreducible.

14.1.2 Recurrence and Transience

When a state is visited by the Markov chain, it is natural to ask how often the state is visited in the long-run. Define the *occupation time* of the state x as

$$\eta_x \stackrel{\text{def}}{=} \sum_{n=0}^{\infty} \mathbb{1}_x(X_n) = \sum_{j=1}^{\infty} \mathbb{1}_{\{\sigma_x^{(j)}<\infty\}} \,.$$

If the expected number of visits to x starting from x is finite, that is, if $\mathrm{E}_x[\eta_x] < \infty$, then the state x is called *transient*. Otherwise, if $\mathrm{E}_x[\eta_x] = \infty$, x is said to be *recurrent*. When X is countable, the recurrence or transience of a state x can be expressed in terms of the probability $\mathrm{P}_x(\tau_x < \infty)$ that the chain started in x ever returns to x.

Proposition 14.1.3. *For any* $x \in \mathsf{X}$ *the following hold true,*

(i) If x *is recurrent, then* $\mathrm{P}_x(\eta_x = \infty) = 1$ *and* $\mathrm{P}_x(\tau_x < \infty) = 1$.
(ii) If x *is transient, then* $\mathrm{P}_x(\eta_x < \infty) = 1$ *and* $\mathrm{P}_x(\tau_x < \infty) < 1$.
(iii) $\mathrm{E}_x[\eta_x] = 1/[1 - \mathrm{P}_x(\tau_x < \infty)]$, *with* $1/0 = \infty$.

Proof. By construction,

$$\mathrm{E}_x[\eta_x] = \sum_{k=1}^{\infty} \mathrm{P}_x(\eta_x \geq k) = \sum_{k=1}^{\infty} \mathrm{P}_x(\sigma_x^{(k)} < \infty) \,.$$

Applying strong Markov property (Theorem 2.1.6) for $n > 1$, we obtain

$$\begin{aligned} \mathrm{P}_x(\sigma_x^{(n)} < \infty) &= \mathrm{P}_x(\sigma_x^{(n-1)} < \infty, \tau_x \circ \theta^{\sigma_x^{(n-1)}} < \infty) \\ &= \mathrm{E}_x[\mathbb{1}_{\{\sigma_x^{(n-1)}<\infty\}} \mathrm{P}_{X_{\sigma_x^{(n-1)}}}(\tau_x < \infty)] . \end{aligned}$$

If $\sigma_x^{(n-1)} < \infty$, then $X_{\sigma_x^{(n-1)}} = x$ P_x-a.s., so that

$$\mathrm{P}_x(\sigma_x^{(n)} < \infty) = \mathrm{P}_x(\tau_x < \infty)\, \mathrm{P}_x(\sigma_x^{(n-1)} < \infty) .$$

By definition $\mathrm{P}_x(\sigma_x < \infty) = 1$, whence $\mathrm{P}_x(\sigma_x^{(n)} < \infty) = [\mathrm{P}_x(\tau_x < \infty)]^{n-1}$ and

$$\mathrm{E}_x[\eta_x] = \sum_{n=1}^{\infty} [\mathrm{P}_x(\tau_x < \infty)]^{n-1} .$$

This proves part (iii).

Now assume x is recurrent. Then by definition $\mathrm{E}_x[\eta_x] = \infty$, and hence $\mathrm{P}_x(\tau_x < \infty) = 1$ and $\mathrm{P}_x(\tau_x^{(n)} < \infty) = 1$ for all $n \geq 1$. Thus $\eta_x = \infty$ P_x-a.s.

If x is transient then $\mathrm{E}_x[\eta_x] < \infty$, which implies $\mathrm{P}_x(\tau_x < \infty) < 1$. □

For a recurrent state x, the occupation time of x is infinite with probability one under P_x; essentially, once the chain started from x returns to x with probability one, it returns a second time with probability one, and so on. Thus the occupation time of a state has a remarkable property, not shared by all random variables: if the expectation of the occupation time is infinite, then the actual number of returns is infinite with probability one. The mean of the occupation time of a state obeys the so-called maximum principle.

Proposition 14.1.4. *For all x and y in X,*

$$\mathrm{E}_x[\eta_y] = \mathrm{P}_x(\sigma_y < \infty)\, \mathrm{E}_y[\eta_y] , \tag{14.3}$$

with the convention $0 \times \infty = 0$.

Proof. It follows from the definition that $\eta_y \mathbb{1}_{\{\sigma_y=\infty\}} = 0$ and $\eta_y \mathbb{1}_{\{\sigma_y<\infty\}} = \eta_y \circ \theta^{\sigma_y} \mathbb{1}_{\{\sigma_y<\infty\}}$. Thus, applying the strong Markov property,

$$\begin{aligned} \mathrm{E}_x[\eta_y] &= \mathrm{E}_x[\mathbb{1}_{\{\sigma_y<\infty\}}\eta_y] = \mathrm{E}_x[\mathbb{1}_{\{\sigma_y<\infty\}}\, \eta_y \circ \theta^{\sigma_y}] \\ &= \mathrm{E}_x[\mathbb{1}_{\{\sigma_y<\infty\}}\, \mathrm{E}_{X_{\sigma_y}}[\eta_y]] = \mathrm{P}_x(\sigma_y < \infty)\, \mathrm{E}_y[\eta_y] . \end{aligned}$$

□

Corollary 14.1.5. *If $\mathrm{E}_x[\eta_y] = \infty$ for some x, then y is recurrent. If X is finite, then there exists at least one recurrent state.*

Proof. By Proposition 14.1.4, $\mathrm{E}_y[\eta_y] \geq \mathrm{E}_x[\eta_y]$, so that $\mathrm{E}_x[\eta_y] = \infty$ implies that $\mathrm{E}_y[\eta_y] = \infty$, that is, y is recurrent.

Next, obviously $\sum_{y\in\mathsf{X}} \eta_y = \infty$ and thus for all $x \in \mathsf{X}$, $\sum_{y\in\mathsf{X}} \mathrm{E}_x[\eta_y] = \infty$. Hence if X is finite, given $x \in \mathsf{X}$ there necessarily exists at least one $y \in \mathsf{X}$ such that $\mathrm{E}_x[\eta_y] = \infty$, which implies that y is recurrent. □

Our next result shows that a recurrent state can only lead to another recurrent state.

Proposition 14.1.6. *Let x be a recurrent state. Then for $y \in \mathsf{X}$, either of the following two statements holds true.*

(i) x leads to y, $\mathrm{E}_x[\eta_y] = \infty$, y is recurrent and leads to x, and $\mathrm{P}_x(\tau_y < \infty) = \mathrm{P}_y(\tau_x < \infty) = 1$;
(ii) x does not lead to y and $\mathrm{E}_x[\eta_y] = 0$.

Proof. Assume that x leads to y. Then there exists an integer k such that $Q^k(x,y) > 0$. Applying the Chapman-Kolmogorov equations, we obtain $Q^{n+k}(x,y) \geq Q^n(x,x)Q^k(x,y)$ for all n. Hence

$$\mathrm{E}_x[\eta_y] \geq \sum_{n=1}^{\infty} Q^{n+k}(x,y) \geq \sum_{n=1}^{\infty} Q^n(x,x)Q^k(x,y) = \mathrm{E}_x[\eta_x]Q^k(x,y) = \infty \, .$$

Thus y is also recurrent by Corollary 14.1.5. Because x is recurrent, the strong Markov property implies that

$$\begin{aligned} 0 = \mathrm{P}_x(\tau_x = \infty) &\geq \mathrm{P}_x(\tau_y < \infty,\, \tau_x = \infty) \\ &= \mathrm{P}_x(\tau_y < \infty,\, \tau_x \circ \theta^{\tau_y} = \infty) = \mathrm{P}_x(\tau_y < \infty)\,\mathrm{P}_y(\tau_x = \infty) \, . \end{aligned}$$

Because x leads to y, $\mathrm{P}_x(\tau_y < \infty) > 0$, whence $\mathrm{P}_y(\tau_x = \infty) = 0$. Thus y leads to x and moreover $\mathrm{P}_y(\tau_x < \infty) = 1$. By symmetry, $\mathrm{P}_x(\tau_y < \infty) = 1$.

If x does not lead to y then Proposition 14.1.4 shows that $\mathrm{E}_x[\eta_y] = 0$. □

For a recurrent state x, the equivalence class $C(x)$ (with respect to the relation of communication defined in Section 14.1.1) may thus be equivalently defined as

$$C(x) = \{y \in \mathsf{X} : \mathrm{E}_x[\eta_y] = \infty\} = \{y \in \mathsf{X} : \mathrm{P}_x(\tau_y < \infty) = 1\} \, . \tag{14.4}$$

If $y \notin C(x)$, then $\mathrm{P}_x(\eta_y = 0) = 1$, which implies that $\mathrm{P}_x(X_n \in C(x)$ for all $n \geq 0) = 1$. In words, the chain started from the recurrent state x forever stays in $C(x)$ and visits each state of $C(x)$ infinitely many times.

The behavior of a Markov chain can thus be described as follows. If a chain is not irreducible, there may exist several equivalence classes of communication. Some of them contain only transient states, and some contain only recurrent states. The latter are then called recurrence classes. If a chain starts from a recurrent state, then it remains in its recurrence class forever. If it starts from a transient state, then either it stays in the class of transient states forever, which implies that there exist infinitely many transient states, or it reaches a recurrent state and then remains in its recurrence class forever.

In contrast, if the chain is irreducible, then all the states are either transient or recurrent. This is called the *solidarity property* of an irreducible chain. We now summarize the previous results.

Theorem 14.1.7. *Consider an irreducible Markov chain on a countable state space* X. *Then every state is either transient, and the chain is called* transient, *or every state is recurrent, and the chain is called* recurrent. *Moreover, either of the following two statements holds true for all* x *and* y *in* X.

(i) $\mathrm{P}_x(\tau_y < \infty) = 1$, $\mathrm{E}_x[\eta_y] = \infty$ *and the chain is recurrent.*
(ii) $\mathrm{P}_x(\tau_x < \infty) < 1$, $\mathrm{E}_x[\eta_y] < \infty$ *and the chain is transient.*

Remark 14.1.8. Note that in the transient case, we do not necessarily have $\mathrm{P}_x(\tau_y < \infty) < 1$ for all x and y in X. For instance, if Q is a transition matrix on $\mathbb{N}$ such that $Q(n, n+1) = 1$ for all n, then $\mathrm{P}_k(\tau_n < \infty) = 1$ for all $k < n$. Nevertheless all states are obviously transient because $X_n = X_0 + n$. ■

14.1.3 Invariant Measures and Stationarity

For many purposes, we might want the marginal distribution of $\{X_k\}$ not to depend on k. If this is the case, then by the Markov property it follows that the finite-dimensional distributions of $\{X_k\}$ are invariant under translation in time, and $\{X_k\}$ is thus a stationary process. Such considerations lead us to invariant distributions. A non-negative vector $\{\pi(x)\}_{x\in\mathsf{X}}$ with the property

$$\pi(y) = \sum_{x\in\mathsf{X}} \pi(x) Q(x, y) , \qquad y \in \mathsf{X} ,$$

will be called *invariant.* If the invariant vector π is summable, then we assume it is a probability distribution, that is, it sums to one. Such distributions are also called *stationary distributions* or *stationary probability measures.* The key result concerning the existence of invariant vectors is the following.

Theorem 14.1.9. *Consider an irreducible recurrent Markov chain* $\{X_k\}_{k\geq 0}$ *on a countable state space* X. *Then there exists a unique (up to a scaling factor) invariant measure* π. *Moreover* $0 < \pi(x) < \infty$ *for all* $x \in \mathsf{X}$. *This measure is summable if and only if there exists a state* x *such that*

$$\mathrm{E}_x[\tau_x] < \infty . \tag{14.5}$$

In this case, $\mathrm{E}_y[\tau_y] < \infty$ *for all* $y \in \mathsf{X}$ *and the unique invariant probability measure is given by*

$$\pi(x) = 1/\mathrm{E}_x[\tau_x] , \qquad x \in \mathsf{X} . \tag{14.6}$$

Proof. Let Q be the transition matrix of the chain. Pick an arbitrary state $x \in \mathsf{X}$ and define the measure λ_x by

$$\lambda_x(y) = \mathrm{E}_x\left[\sum_{k=0}^{\tau_x - 1} \mathbb{1}_y(X_k)\right] = \mathrm{E}_x\left[\sum_{k=1}^{\tau_x} \mathbb{1}_y(X_k)\right] . \tag{14.7}$$

That is, $\lambda_x(y)$ is the expected number of visits to the state y before the first return to x, given that the chain starts in x. Let f be a non-negative function on X. Then

$$\lambda_x(f) = \mathrm{E}_x\left[\sum_{k=0}^{\tau_x - 1} f(X_k)\right] = \sum_{k=0}^{\infty} \mathrm{E}_x\left[\mathbb{1}_{\{\tau_x > k\}} f(X_k)\right] .$$

Using this identity and the fact that $Qf(X_k) = \mathrm{E}_x[f(X_{k+1}) \,|\, \mathcal{F}_k^X]$ P_x-a.s. for all $k \geq 1$, we find that

$$\begin{aligned}
\lambda_x(Qf) &= \sum_{k=0}^{\infty} \mathrm{E}_x[\mathbb{1}_{\{\tau_x > k\}}\, Qf(X_k)] = \sum_{k=0}^{\infty} \mathrm{E}_x\{\mathbb{1}_{\{\tau_x > k\}}\, \mathrm{E}_x[f(X_{k+1}) \,|\, \mathcal{F}_k^X]\} \\
&= \sum_{k=0}^{\infty} \mathrm{E}_x[\mathbb{1}_{\{\tau_x > k\}} f(X_{k+1})] = \mathrm{E}_x\left[\sum_{k=1}^{\tau_x} f(X_k)\right] ,
\end{aligned}$$

showing that $\lambda_x(Qf) = \lambda_x(f) - f(x) + \mathrm{E}_x[f(X_{\tau_x})] = \lambda_x(f)$. Because f was arbitrary, we see that $\lambda_x Q = \lambda_x$; the measure λ_x is invariant. For any other state y, the chain may reach y before returning to x when starting in x, as it is irreducible. This proves that $\lambda_x(y) > 0$. Moreover, again by irreducibility, we can pick an $m > 0$ such that $Q^m(y,x) > 0$. By invariance $\lambda_x(x) = \sum_{z \in \mathsf{X}} \lambda_x(z) Q^m(z,x) \geq \lambda_x(y) Q^m(y,x)$, and as $\lambda_x(x) = 1$, we see that $\lambda_x(y) < \infty$

We now prove that the invariant measure is unique up to a scaling factor. The first step consists in proving that if π is an invariant measure such that $\pi(x) = 1$, then $\pi \geq \lambda_x$. It suffices to show that, for any $y \in \mathsf{X}$ and any integer n,

$$\pi(y) \geq \sum_{k=1}^{n} \mathrm{E}_x[\mathbb{1}_y(X_k) \mathbb{1}_{\{\tau_x \geq k\}}] . \tag{14.8}$$

The proof is by induction. The inequality is immediate for $n = 1$. Assume that (14.8) holds for some $n \geq 1$. Then

$$\begin{aligned}
\pi(y) &= Q(x,y) + \sum_{z \neq x} \pi(z) Q(z,y) \\
&\geq Q(x,y) + \sum_{k=1}^{n} \mathrm{E}_x[Q(X_k,y) \mathbb{1}_{\{x\}^c}(X_k) \mathbb{1}_{\{\tau_x \geq k\}}] \\
&\geq Q(x,y) + \sum_{k=1}^{n} \mathrm{E}_x[\mathbb{1}_y(X_{k+1}) \mathbb{1}_{\{\tau_x \geq k+1\}}] \\
&= \sum_{k=1}^{n+1} \mathrm{E}_x[\mathbb{1}_{\{y\}}(X_k) \mathbb{1}_{\{\tau_x \geq k\}}] ,
\end{aligned}$$

showing the induction. We will now show that $\pi = \lambda_x$. The proof is by contradiction. Assume that $\pi(z) > \lambda_x(z)$ for some $z \in \mathsf{X}$. Then

$$1 = \pi(x) = \pi Q(x) = \sum_{z \in \mathsf{X}} \pi(z) Q(z,x) > \sum_{z \in \mathsf{X}} \lambda_x(z) Q(z,x) = \lambda_x(x) = 1 ,$$

which cannot be true.

The measure λ_x is summable if and only if

$$\infty > \sum_{y \in \mathsf{X}} \lambda_x(y) = \sum_{y \in \mathsf{X}} \mathrm{E}_x \left[\sum_{k=0}^{\tau_x - 1} \mathbb{1}_{\{X_k = y\}} \right] = \mathrm{E}_x[\tau_x] .$$

Thus the unique invariant measure is summable if and only if a state x satisfying this relation exists. On the other hand, if such a state x exists then, by uniqueness of the invariant measure, $\mathrm{E}_y[\tau_y] < \infty$ must hold for all states y. In this case, the invariant probability measure, π say, satisfies $\pi(x) = \lambda_x(x)/\lambda_x(\mathsf{X}) = 1/\mathrm{E}_x[\tau_x]$. Because the reference state x was in fact arbitrary, we find that $\pi(y) = 1/\mathrm{E}_x[\tau_y]$ for all states y. □

It is natural to ask what can be inferred from the knowledge that a chain possesses an invariant probability measure. The next proposition gives a partial answer.

Proposition 14.1.10. *Let Q be a transition matrix and π an invariant probability measure. Then every state x such that $\pi(x) > 0$ is recurrent. If Q is irreducible, then it is recurrent.*

Proof. Let $y \in \mathsf{X}$. If $\pi(y) > 0$ then $\sum_{n=0}^{\infty} \pi Q^n(y) = \sum_{n=0}^{\infty} \pi(y) = \infty$. On the other hand, by Proposition 14.1.4,

$$\begin{aligned} \sum_{n=0}^{\infty} \pi Q^n(y) &= \sum_{x \in \mathsf{X}} \pi(x) \sum_{n=0}^{\infty} Q^n(x,y) \\ &= \sum_{x \in \mathsf{X}} \pi(x) \mathrm{E}_x[\eta_y] \leq \mathrm{E}_y[\eta_y] \sum_{x \in \mathsf{X}} \pi(x) = \mathrm{E}_y[\eta_y] . \end{aligned} \tag{14.9}$$

Thus $\pi(y) > 0$ implies $\mathrm{E}_y[\eta_y] = \infty$, that is, y is recurrent. □

Let $\{X_k\}$ be an irreducible Markov chain. If there exists an invariant probability measure, the chain is called *positivc recurrent*; otherwise it is called *null.* Note that null chains can be either null recurrent or transient. Transient chains are always null, though they may admit an invariant measure.

14.1.4 Ergodicity

A key result for positive recurrent irreducible chains is that the transition laws converge, in a suitable sense, to the invariant vector π. The classical result is the following.

Proposition 14.1.11. *Consider an irreducible and positive recurrent Markov chain on a countable state space. Then for any states x and y,*

$$n^{-1} \sum_{i=1}^{n} Q^n(x, y) \to \pi(y) \quad \text{as } n \to \infty \,. \tag{14.10}$$

The use of the Césaro limit can be avoided if the chain is *aperiodic*. The simplest definition of aperiodicity is that a state x is aperiodic if $Q^k(x, x) > 0$ for all k sufficiently large or, equivalently, that the *period* of the state x is one. The *period* of x is defined as the greatest common divisor of the set $I(x) = \{n > 0 : Q^n(x, x) > 0\}$. For irreducible chains, the following result holds true.

Proposition 14.1.12. *If the chain is irreducible, then all states have the same period. If the transition matrix Q is irreducible and aperiodic, then for all x and y in X, there exists $n(x, y) \in \mathbb{N}$ such that $Q^k(x, y) > 0$ for all $k \geq n(x, y)$.*

Thus, an irreducible chain can be said to be aperiodic if the common period of all states is one.

The traditional pointwise convergence (14.10) of transition probabilities has been replaced in more recent research by convergence in *total variation* (see Definition 4.3.1). The convergence result may then be formulated as follows.

Theorem 14.1.13. *Consider an irreducible and aperiodic positive recurrent Markov chain on a countable state space X with transition matrix Q and invariant probability distribution π. Then for all initial distributions ξ and ξ' on X,*

$$\|\xi Q^n - \xi' Q^n\|_{\mathrm{TV}} \to 0 \quad \text{as } n \to \infty \,. \tag{14.11}$$

In particular, for any $x \in \mathsf{X}$ we may set $\xi = \delta_x$ and $\xi' = \pi$ to obtain

$$\|Q^n(x, \cdot) - \pi\|_{\mathrm{TV}} \to 0 \quad \text{as } n \to \infty \,. \tag{14.12}$$

The proof of this result, and indeed the focus on convergence in total variation, follows using of the coupling technique. We postpone the presentation of this technique to Section 14.2.4 because essentially the same ideas can be applied to Markov chains on general state spaces.

14.2 Chains on General State Spaces

In this section, we extend the concepts and results pertaining to countable state spaces to general ones. In the following, X is an arbitrary set, and we just require that it is equipped with a countably generated σ-field $\mathcal{X}$. By $\{X_k\}_{k \geq 0}$ we denote an X-valued Markov chain with transition kernel Q. It is defined on a probability space $(\Omega, \mathcal{F}, \mathrm{P})$, and $\mathbb{F}^X = \{\mathcal{F}_k^X\}_{k\geq 0}$ denotes the natural filtration of $\{X_k\}$.

For any set $A \in \mathcal{X}$, we define the first *hitting time* σ_A and *return time* τ_A respectively by

$$\sigma_A = \inf\{n \geq 0 : X_n \in A\} , \tag{14.13}$$

$$\tau_A = \inf\{n \geq 1 : X_n \in A\} , \tag{14.14}$$

where, by convention, $\inf \emptyset = +\infty$. The successive hitting times $\sigma_A^{(n)}$ and return times $\tau_A^{(n)}$, $n \geq 0$, are defined inductively by

$$\sigma_A^{(0)} = 0,\ \sigma_A^{(1)} = \sigma_A,\ \sigma_A^{(n+1)} = \inf\{k > \sigma_A^{(n)} : X_k \in A\} ,$$

$$\tau_A^{(0)} = 0,\ \tau_A^{(1)} = \tau_A,\ \tau_A^{(n+1)} = \inf\{k > \tau_A^{(n)} : X_k \in A\} .$$

We again define the *occupation time* η_A as the number of visits by $\{X_k\}$ to A,

$$\eta_A \stackrel{\text{def}}{=} \sum_{k=0}^{\infty} \mathbb{1}_A(X_k) . \tag{14.15}$$

14.2.1 Irreducibility

The first step to develop a theory on general state spaces is to define a suitable concept of irreducibility. The definition of irreducibility adopted for countable state spaces does not extend to general ones, as the probability of reaching single point x in the state space is typically zero.

Definition 14.2.1 (Phi-irreducibility). *The transition kernel Q, or the Markov chain $\{X_k\}_{k\geq 0}$ with transition kernel Q, is said to be* phi-irreducible *if there exists a measure ϕ on $(\mathsf{X}, \mathcal{X})$ such that for any $A \in \mathcal{X}$ with $\phi(A) > 0$, $\mathrm{P}_x(\tau_A < \infty) > 0$ for all $x \in \mathsf{X}$. Such a measure is called an* irreducibility measure *for Q.*

Phi-irreducibility is a weaker property than irreducibility of a transition kernel on a countable state space. If a transition kernel on a countable state space is irreducible, then it is phi-irreducible, and any measure is an irreducibility measure. The converse is not true. For instance, the transition kernel

$$Q = \begin{pmatrix} 0 & 1 \\ 0 & 1 \end{pmatrix}$$

on $\{0, 1\}$ is phi-irreducible (δ_1 is an irreducibility measure for Q) but not irreducible.

In general, there are infinitely many irreducibility measures, and two irreducibility measures are not necessarily equivalent. For instance, if ϕ is an irreducibility measure and $\hat{\phi}$ is absolutely continuous with respect to ϕ, then

$\hat{\phi}$ is also an irreducibility measure. Nevertheless, as shown in the next result, there exist *maximal irreducibility measures* ψ, which are such that any irreducibility measure ϕ is absolutely continuous with respect to ψ.

Theorem 14.2.2. *Let Q be a phi-irreducible transition kernel on $(\mathsf{X}, \mathcal{X})$. Then there exists an irreducibility measure ψ such that all irreducibility measures are absolutely continuous with respect to ψ and for all $A \in \mathcal{X}$,*

$$\psi(A) > 0 \;\Leftrightarrow\; \mathrm{P}_x(\tau_A < \infty) > 0 \text{ for all } x \in \mathsf{X} . \tag{14.16}$$

Proof. Let ϕ be an irreducibility measure and $\epsilon \in (0,1)$. Let ϕ_ϵ be the measure defined by $\phi_\epsilon = \phi K_\epsilon$, where K_ϵ is the *resolvent kernel* defined by

$$K_\epsilon(x, A) \stackrel{\text{def}}{=} (1-\epsilon) \sum_{k \geq 0} \epsilon^k Q^k(x, A) , \qquad x \in \mathsf{X}, A \in \mathcal{X} . \tag{14.17}$$

We will first show that ϕ_ϵ is an irreducibility measure. Let $A \in \mathcal{X}$ be such that $\phi_\epsilon(A) > 0$ and define

$$\bar{A} = \{x \in \mathsf{X} : \mathrm{P}_x(\sigma_A < \infty) > 0\} = \{x \in \mathsf{X} : K_\epsilon(x, A) > 0\} . \tag{14.18}$$

By definition, $\phi_\epsilon(A) > 0$ implies that $\phi(\bar{A}) > 0$. Define $\bar{A}_m = \{x \in \mathsf{X} : \mathrm{P}_x(\sigma_A < \infty) \geq 1/m\}$. By construction, $\bar{A} = \bigcup_{m>0} \bar{A}_m$, and because $\phi(\bar{A}) > 0$, there exists m such that $\phi(\bar{A}_m) > 0$. Because ϕ is an irreducibility measure, $\mathrm{P}_x(\tau_{\bar{A}_m} < \infty) > 0$ for all $x \in \mathsf{X}$. Hence by the strong Markov property, for all $x \in \mathsf{X}$,

$$\begin{aligned} \mathrm{P}_x(\tau_A < \infty) &\geq \mathrm{P}_x(\tau_{\bar{A}_m} + \sigma_A \circ \theta^{\tau_{\bar{A}_m}} < \infty \,, \tau_{\bar{A}_m} < \infty) \\ &= \mathrm{E}_x[\mathbb{1}_{\{\tau_{\bar{A}_m} < \infty\}} \mathrm{P}_{X_{\tau_{\bar{A}_m}}}(\sigma_A < \infty)] \geq \frac{1}{m} \mathrm{P}_x(\tau_{\bar{A}_m} < \infty) > 0 , \end{aligned}$$

showing that ϕ_ϵ is an irreducibility measure.

Now for $m \geq 0$ the Chapman-Kolmogorov equations imply

$$\int_{\mathsf{X}} \phi_\epsilon(dx)\, \epsilon^m Q^m(x, A) = (1-\epsilon) \int_{\mathsf{X}} \sum_{n=m}^{\infty} \epsilon^n Q^n(x, A)\, \phi(dx) \leq \phi_\epsilon(A) .$$

Therefore, if $\phi_\epsilon(A) = 0$ then $\phi_\epsilon K_\epsilon(A) = 0$, which in turn implies $\phi_\epsilon(\bar{A}) = 0$. Summarizing the results above, for any $A \in \mathcal{X}$,

$$\phi_\epsilon(A) > 0 \Leftrightarrow \phi_\epsilon\left(\{x \in \mathsf{X} : \mathrm{P}_x(\sigma_A < \infty) > 0\}\right) > 0 . \tag{14.19}$$

This proves (14.16)

To conclude we must show that all irreducibility measures are absolutely continuous with respect to ϕ_ϵ. Let $\hat{\phi}$ be an irreducibility measure and let $C \in \mathcal{X}$ be such that $\hat{\phi}(C) > 0$. Then $\phi_\epsilon\left(\{x \in \mathsf{X} : \mathrm{P}_x(\sigma_C < \infty) > 0\}\right) = \phi_\epsilon(\mathsf{X}) > 0$, which, by (14.19), implies that $\phi_\epsilon(C) > 0$. This exactly says that $\hat{\phi}$ is absolutely continuous with respect to ϕ_ϵ. □

A set $A \in \mathcal{X}$ is said to be *accessible for the kernel* Q (or Q-*accessible*, or simply *accessible* if there is no risk of confusion) if $\mathrm{P}_x(\tau_A < \infty) > 0$ for all $x \in \mathsf{X}$. The family of accessible sets is denoted $\mathcal{X}^+$. If ψ is a maximal irreducibility measure the set A is accessible if and only if $\psi(A) > 0$.

Example 14.2.3 (Autoregressive Model). The first-order autoregressive model on $\mathbb{R}$ is defined iteratively by $X_n = \phi X_{n-1} + U_n$, where ϕ is a real number and $\{U_n\}$ is an i.i.d. sequence. If Γ is the probability distribution of the noise sequence $\{U_n\}$, the transition kernel of this chain is given by $Q(x, A) = \Gamma(A - \phi x)$. The autoregressive model is phi-irreducible provided that the noise distribution has an everywhere positive density with respect to Lebesgue measure λ^{Leb}. If we take $\phi = \lambda^{\mathrm{Leb}}$, it is easy to see that whenever $\lambda^{\mathrm{Leb}}(A) > 0$, we have $\Gamma(A - \phi x) > 0$ for any x, and so $Q(x, A) > 0$ in just one step. ∎

Example 14.2.4. The Metropolis-Hastings algorithm, introduced in Chapter 6, provides another typical example of a general state-space Markov chain. For simplicity, we assume here that $\mathsf{X} = \mathbb{R}^d$, which we equip with the Borel σ-field $\mathcal{X} = \mathcal{B}(\mathbb{R}^d)$. Assume that we are given a probability density function π on with respect to Lebesgue measure λ^{Leb}. Let r be a transition density kernel. Starting from $X_n = x$, a candidate transition x' is generated from $r(x, \cdot)$ and accepted with probability

$$\alpha(x, x') = \frac{\pi(x')\, r(x', x)}{\pi(x)\, r(x, x')} \wedge 1 \,. \tag{14.20}$$

The transition kernel of the Metropolis-Hastings chain is given by

$$Q(x, A) = \int_A \alpha(x, x') r(x, x')\, \lambda^{\mathrm{Leb}}(dx') \\ + \mathbb{1}_x(A) \int [1 - \alpha(x, x')] r(x, x')\, \lambda^{\mathrm{Leb}}(dx') \,. \tag{14.21}$$

There are various sufficient conditions for the Metropolis-Hastings algorithm to be phi-irreducible (Roberts and Tweedie, 1996; Mengersen and Tweedie, 1996). For the Metropolis-Hastings chain, it is simple to check that the chain is phi-irreducible if for λ^{Leb}-almost all $x' \in \mathsf{X}$, the condition $\pi(x') > 0$ implies that $r(x, x') > 0$ for any $x \in \mathsf{X}$. ∎

14.2.2 Recurrence and Transience

In view of the discussion above, it is not sensible to define recurrence and transience in terms of the expectation of the occupation measure of a state, but for phi-irreducible chains it makes sense to consider the occupation measure of accessible sets.

Definition 14.2.5 (Uniform Transience and Recurrence). *A set $A \in \mathcal{X}$ is called* uniformly transient *if* $\sup_{x \in A} \mathrm{E}_x[\eta_A] < \infty$. *A set $A \in \mathcal{X}$ is called* recurrent *if* $\mathrm{E}_x[\eta_A] = +\infty$ *for all $x \in A$.*

Obviously, if $\sup_{x \in \mathsf{X}} \mathrm{E}_x[\eta_A] < \infty$, then A is uniformly transient. In fact the reverse implication holds true too, because if the chain is started outside A it cannot hit A more times, on average, than if it is started at "the most favorable location" in A. Thus an alternative definition of a uniformly transient set is $\sup_{x \in \mathsf{X}} \mathrm{E}_x[\eta_A] < \infty$.

The main result on phi-irreducible transition kernels is the following recurrence/transience dichotomy, which parallels Theorem 14.1.7 for countable state-space Markov chains.

Theorem 14.2.6. *Let Q be a phi-irreducible transition kernel (or Markov chain). Then either of the following two statements holds true.*

(i) Every accessible set is recurrent, in which case we call Q recurrent.
(ii) There is a countable cover of X with uniformly transient sets, in which case we call Q transient.

In the next section, we will prove Theorem 14.2.6 in the particular case where the chain possesses an *accessible atom* (see Definition 14.2.7); the proof is then very similar to that for countable state space. In the general case, the proof is more involved. It is necessary to introduce *small sets* and the so-called *splitting construction*, which relates the chain to one that does possess an accessible atom.

14.2.2.1 Transience and Recurrence for Chains Possessing an Accessible Atom

Definition 14.2.7 (Atom). *A set $\alpha \in \mathcal{X}$ is called an* atom *if there exists a probability measure ν on $(\mathsf{X}, \mathcal{X})$ such that $Q(x, A) = \nu(A)$ for all $x \in \alpha$ and $A \in \mathcal{X}$.*

Atoms behave the same way as do individual states in the countable state space case. Although any singleton $\{x\}$ is an atom, it is not necessarily accessible, so that Markov chain theory on general state spaces differs from the theory of countable state space chains.

If α is an atom for Q, then for any $m \geq 1$ it is an atom for Q^m. Therefore we denote by $Q^m(\alpha, \cdot)$ the common value of $Q^m(x, \cdot)$ for all $x \in \alpha$. This implies that if the chain starts from within the atom, the distribution of the whole chain does not depend on the precise starting point. Therefore we will also use the notation P_α instead of P_x for any $x \in \alpha$.

Example 14.2.8 (Random Walk on the Half-Line). The random walk on the half-line (RWHL) is defined by an initial condition $X_0 \geq 0$ and the recursion

$$X_{k+1} = (X_k + W_{k+1})^+ , \qquad k \geq 0 , \tag{14.22}$$

where $\{W_k\}_{k\geq 1}$ is an i.i.d. sequence of random variables, independent of X_0, with distribution function Γ on $\mathbb{R}$. This process is a Markov chain with transition kernel Q defined by

$$Q(x, A) = \Gamma(A - x) + \Gamma((-\infty\,, -x])\mathbb{1}_A(0) , \qquad x \in \mathbb{R}_+,\ A \in \mathcal{B}(\mathbb{R}_+) ,$$

where $A - x = \{y - x : y \in A\}$. The set $\{0\}$ is an atom, and it is accessible if and only if $\Gamma((-\infty\,, 0]) > 0$. ■

We now prove Theorem 14.2.6 when there exists an accessible atom.

Proposition 14.2.9. *Let $\{X_k\}_{k\geq 0}$ be a Markov chain that possesses an accessible atom α, with associated probability measure ν. Then the chain is phi-irreducible, ν is an irreducibility measure, and a set $A \in \mathcal{X}$ is accessible if and only if $\mathrm{P}_\alpha(\tau_A < \infty) > 0$.*

Moreover, α is recurrent if and only if $\mathrm{P}_\alpha(\tau_\alpha < \infty) = 1$ and (uniformly) transient otherwise, and the chain is recurrent if α is recurrent and transient otherwise.

Proof. For all $A \in \mathcal{X}$ and $x \in \mathsf{X}$, the strong Markov property yields

$$\begin{aligned} \mathrm{P}_x(\tau_A < \infty) &\geq \mathrm{P}_x(\tau_\alpha + \tau_A \circ \theta^{\tau_\alpha} < \infty,\ \tau_\alpha < \infty) \\ &= \mathrm{E}_x[\mathrm{P}_{X_{\tau_\alpha}}(\tau_A < \infty)\mathbb{1}_{\{\tau_\alpha < \infty\}}] \\ &= \mathrm{P}_\alpha(\tau_A < \infty)\,\mathrm{P}_x(\tau_\alpha < \infty) \\ &\geq \nu(A)\,\mathrm{P}_x(\tau_\alpha < \infty) . \end{aligned}$$

Because α is accessible, $\mathrm{P}_x(\tau_\alpha < \infty) > 0$ for all $x \in \mathsf{X}$. Thus for any $A \in \mathcal{X}$ satisfying $\nu(A) > 0$, it holds that $\mathrm{P}_x(\tau_A < \infty) > 0$ for all $x \in \mathsf{X}$, showing that ν is an irreducibility measure. The above display also shows that A is accessible if and only if $\mathrm{P}_\alpha(\tau_A < \infty)$.

Now let $\sigma_\alpha^{(n)}$ be the successive hitting times of α (see (14.13)). The strong Markov property implies that for any $n > 1$,

$$\mathrm{P}_\alpha(\sigma_\alpha^{(n)} < \infty) = \mathrm{P}_\alpha(\tau_\alpha < \infty)\,\mathrm{P}_\alpha(\sigma_\alpha^{(n-1)} < \infty) .$$

Hence, as for discrete state spaces, $\mathrm{P}_\alpha(\sigma_\alpha^{(n)} < \infty) = [\mathrm{P}_\alpha(\tau_\alpha < \infty)]^{n-1}$ and $\mathrm{E}_\alpha[\eta_\alpha] = 1/[1 - \mathrm{P}_\alpha(\tau_\alpha < \infty)]$. This proves that α is recurrent if and only if $\mathrm{P}_\alpha(\tau_\alpha < \infty) = 1$.

Assume that α is recurrent. Because the atom α is accessible, for any $x \in \mathsf{X}$, there exists r such that $Q^r(x, \alpha) > 0$. If $A \in \mathcal{X}^+$ there exists s such that $Q^s(\alpha, A) > 0$. By the Chapman-Kolmogorov equations,

$$\sum_{n\geq 1} Q^{r+s+n}(x, A) \geq Q^r(x, \alpha)\left[\sum_{n\geq 1} Q^n(\alpha, \alpha)\right] Q^s(\alpha, A) = \infty .$$

Hence $\mathrm{E}_x[\eta_A] = \infty$ for all $x \in \mathsf{X}$ and A is recurrent. Because A was an arbitrary accessible set, the chain is recurrent.

Assume now that α is transient, in which case $\mathrm{E}_\alpha(\eta_\alpha) < \infty$. Then, following the same line of reasoning as in the discrete state space case (proof of Proposition 14.1.4), we obtain that for all $x \in \mathsf{X}$,

$$\mathrm{E}_x[\eta_\alpha] = \mathrm{P}_x(\tau_\alpha < \infty)\,\mathrm{E}_\alpha[\eta_\alpha] \le \mathrm{E}_\alpha[\eta_\alpha] \,. \tag{14.23}$$

Define $B_j = \{x : \sum_{n=1}^{j} Q^n(x, \alpha) \ge 1/j\}$. Then $\cup_{j=1}^{\infty} B_j = \mathsf{X}$ because α is accessible. Applying the definition of the sets B_j and the Chapman Kolmogorov equations, we find that

$$\sum_{k=1}^{\infty} Q^k(x, B_j) \le \sum_{k=1}^{\infty} Q^k(x, B_j) \inf_{y \in B_j} j \sum_{\ell=1}^{j} Q^\ell(y, \alpha)$$
$$\le j \sum_{k=1}^{\infty} \sum_{\ell=1}^{j} \int_{B_j} Q^k(x, dy)\, Q^\ell(y, \alpha) \le j^2 \sum_{k=1}^{\infty} Q^k(x, \alpha) = j^2\, \mathrm{E}_x[\eta_\alpha] < \infty \,.$$

The sets B_j are thus uniformly transient. The proof is complete. □

14.2.2.2 Small Sets and the Splitting Construction

We now return to the general phi-irreducible case. In order to prove Theorem 14.2.6, we need to introduce the splitting technique. To do so, we need to define a class of sets (containing accessible sets) that behave the same way in many respects as do atoms. We shall see this in many of the results below, which exactly mimic the atomic case results they generalize. These sets are called *small sets*.

Definition 14.2.10 (Small Set). *Let Q and ν be a transition kernel and a probability measure, respectively, on $(\mathsf{X}, \mathcal{X})$, let m be a positive integer and $\epsilon \in (0, 1]$. A set $C \in \mathcal{X}$ is called a (m, ϵ, ν)-*small set *for Q, or simply a* small set, *if $\nu(C) > 0$ and for all $x \in C$ and $A \in \mathcal{X}$,*

$$Q^m(x, A) \ge \epsilon \nu(A) \,.$$

If $\epsilon = 1$ then C is an atom for the kernel Q^m.

Trivially, any individual point is a small set, but small sets that are not accessible are of limited interest. If the state space is countable and Q is irreducible, then every finite set is small. The minorization measure associated to an accessible small set provides an irreducibility measure.

Proposition 14.2.11. *Let C be an accessible (m, ϵ, ν)-small set for the transition kernel Q on $(\mathsf{X}, \mathcal{X})$. Then ν is an irreducibility measure.*

Proof. Let $A \in \mathcal{X}$ be such that $\nu(A) > 0$. The strong Markov property yields

$$\mathrm{P}_x(\tau_A < \infty) \geq \mathrm{P}_x(\tau_C < \infty,\, \tau_A \circ \theta^{\tau_C} < \infty) = \mathrm{E}_x[\mathbb{1}_{\{\tau_C < \infty\}} \mathrm{P}_{X_{\tau_C}}(\tau_A < \infty)] \,.$$

Because C is a small set, for all $y \in C$ it holds that

$$\mathrm{P}_y(\tau_A < \infty) \geq \mathrm{P}_y(X_m \in A) = Q^m(y, A) \geq \epsilon\nu(A) \,.$$

Because C is accessible and $\nu(A) > 0$, for all $x \in \mathsf{X}$ it holds that

$$\mathrm{P}_x(\tau_A < \infty) \geq \epsilon\nu(A)\, \mathrm{P}_x(\tau_C < \infty) > 0 \,.$$

Thus A is accessible, whence ν is an irreducibility measure. □

An important result due to Jain and Jamison (1967) states that if the transition kernel is phi-irreducible, then small sets do exist. For a proof see Nummelin (1984, p. 16) or Meyn and Tweedie (1993, Theorem 5.2.2).

Proposition 14.2.12. *If the transition kernel Q on $(\mathsf{X}, \mathcal{X})$ is phi-irreducible, then every accessible set contains an accessible small set.*

Given the existence of just one small set from Proposition 14.2.12, we may show that it is possible to cover X with a countable number of small sets in the phi-irreducible case.

Proposition 14.2.13. *Let Q be a phi-irreducible transition kernel on $(\mathsf{X}, \mathcal{X})$.*

(i) If $C \in \mathcal{X}$ is an (m, ϵ, ν)-small set and for any $x \in D$ we have $Q^n(x, C) \geq \delta$, then D is $(m + n, \delta\epsilon, \nu)$-small set.
(ii) If Q is phi-irreducible then there exists a countable collection of small sets C_i such that $\mathsf{X} = \bigcup_i C_i$.

Proof. Using the Chapman-Kolmogorov equations, we find that for any $x \in D$,

$$Q^{n+m}(x, A) \geq \int_C Q^n(x, dy)\, Q^m(y, A) \geq \epsilon Q^n(x, C)\nu(A) \geq \epsilon\delta\nu(A) \,,$$

showing part (i). Because Q is phi-irreducible, by Proposition 14.2.12 there exists an accessible (m, ϵ, ν)-small set C. Moreover, by the definition of phi-irreducibility, the sets $C(n, m) = \{x : Q^n(x, C) \geq 1/m\}$ cover X and, by part (i), each $C(n, m)$ is small. □

Proposition 14.2.14. *If Q is phi-irreducible and transient, then every accessible small set is uniformly transient.*

Proof. Let C be an accessible (m, ϵ, ν)-small set. If Q is transient, there exists at least one $A \in \mathcal{X}^+$ that is uniformly transient. For $\delta \in (0, 1)$, by the Chapman-Kolmogorov equations,

$$
\begin{aligned}
\mathrm{E}_x[\eta_A] &= \sum_{k=0}^{\infty} Q^k(x,A) \geq (1-\delta) \sum_{p=0}^{\infty} \delta^p \sum_{k=0}^{\infty} Q^{k+m+p}(x,A) \\
&\geq (1-\delta) \sum_{p=0}^{\infty} \delta^p \sum_{k=0}^{\infty} \int_C Q^k(x,dx') \int Q^m(x',dx'')\, Q^p(x'',A) \\
&\geq \epsilon \sum_{k=0}^{\infty} Q^k(x,C) \times (1-\delta) \sum_{p=0}^{\infty} \delta^p \nu Q^p(A) = \epsilon\, \mathrm{E}_x[\eta_C]\, \nu K_\delta(A) \;,
\end{aligned}
$$

where K_δ is the resolvent kernel (14.17). Because C is an accessible small set, Proposition 14.2.11 shows that ν is an irreducibility measure. By Theorem 14.2.2, νK_δ is a maximal irreducibility measure, so that $\nu K_\delta(A) > 0$. Thus $\sup_{x \in \mathsf{X}} \mathrm{E}_x[\eta_C] < \infty$ and we conclude that C is uniformly transient (see the remark following Definition 14.2.5). □

Example 14.2.15 (Autoregressive Process, Continued). Suppose that the noise distribution in Example 14.2.3 has an everywhere positive continuous density γ with respect to Lebesgue measure λ^{Leb}. If $C = [-M, M]$ and $\epsilon = \inf_{|x| \leq (1+\phi)M} \gamma(u)$, then for $A \subseteq C$,

$$
Q(x,A) = \int_A \gamma(x' - \phi x)\, dx' \geq \epsilon \lambda^{\mathrm{Leb}}(A) \;.
$$

Hence the compact set C is small. Obviously $\mathbb{R}$ is covered by a countable collection of small sets and every accessible set (here sets with non-zero Lebesgue measure) contains a small set. ■

Example 14.2.16 (Metropolis-Hastings Algorithm, Continued). Similar results hold for the Metropolis-Hastings algorithm of Example 14.2.4 if $\pi(x)$ and $r(x,x')$ are positive and continuous for all $(x,x') \in \mathsf{X} \times \mathsf{X}$. Suppose that C is compact with $\lambda^{\mathrm{Leb}}(C) > 0$. By positivity and continuity, we then have $d = \sup_{x \in C} \pi(x) < \infty$ and $\varepsilon = \inf_{(x,x') \in C \times C} q(x,x') > 0$. For any $A \subseteq C$, define

$$
R_x(A) \stackrel{\mathrm{def}}{=} \left\{ x' \in A : \frac{\pi(x') q(x',x)}{\pi(x) q(x,x')} < 1 \right\} ,
$$

the region of possible rejection. Then for any $x \in C$,

$$
\begin{aligned}
Q(x,A) &\geq \int_A q(x,x') \alpha(x,x')\, dx' \\
&\geq \int_{R_x(A)} \frac{q(x',x)}{\pi(x)} \pi(x')\, dx' + \int_{A \setminus R_x(A)} q(x,x')\, dx' \\
&\geq \frac{\varepsilon}{d} \int_{R_x(A)} \pi(x')\, dx' + \frac{\varepsilon}{d} \int_{A \setminus R_x(A)} \pi(x')\, dx' \\
&= \frac{\varepsilon}{d} \int_A \pi(x')\, dx' \;.
\end{aligned}
$$

Thus C is small and, again, X can be covered by a countable collection of small sets. ∎

We now show that it is possible to define a Markov chain with an atom, the so-called *split chain*, whose properties are directly related to those of the original chain. This technique was introduced by Nummelin (1978) (Athreya and Ney, 1978, introduced, independently, a virtually identical concept) and allows extending results valid for Markov chain possessing an accessible atom to irreducible Markov chains that only possess small sets. The basic idea is as follows. Suppose the chain admits a $(1, \epsilon, \nu)$-small set C. Then as long as the chain does not enter C, the transition kernel Q is used to generate the trajectory. However, as soon as the chain hits C, say $X_n \in C$, a zero-one random variable d_n is drawn, independent of everything else. The probability that $d_n = 1$ is ϵ, and hence $d_n = 0$ with probability $1 - \epsilon$. Then if $d_n = 1$, the next value X_{n+1} is drawn from ν; otherwise X_{n+1} is drawn from the kernel

$$R(x, A) = [1 - \epsilon \mathbb{1}_C(x)]^{-1}[Q(x, A) - \epsilon \mathbb{1}_C(x)\nu(A)] ,$$

with $x = X_n$. It is immediate that $\epsilon\nu(A) + (1 - \epsilon)R(x, A) = Q(x, A)$ for all $x \in C$, so X_{n+1} is indeed drawn from the correct (conditional) distribution. Note also that $R(x, \cdot) = Q(x, \cdot)$ for $x \notin C$. So, what is gained by this approach? What is gained is that whenever $X_n \in C$ and $d_n = 1$, the next value of the chain will be independent of X_n (because it is drawn from ν). This is often called a *regeneration time*, as the joint chain $\{(X_k, d_k)\}$ in a sense "restarts" and forgets its history. In technical terms, the state $C \times \{1\}$ in the extended state space is as atom, and it will be accessible provided C is.

We now make this formal. Thus we define the so-called *extended state space* as $\check{\mathsf{X}} = \mathsf{X} \times \{0, 1\}$ and let $\check{\mathcal{X}}$ be the associated product σ-field. We associate to every measure μ on $(\mathsf{X}, \mathcal{X})$ the split measure $\mu^\star$ on $(\check{\mathsf{X}}, \check{\mathcal{X}})$ as the unique measure satisfying, for $A \in \mathcal{X}$,

$$\begin{aligned} \mu^\star(A \times \{0\}) &= (1 - \epsilon)\mu(A \cap C) + \mu(A \cap C^c) , \\ \mu^\star(A \times \{1\}) &= \epsilon\mu(A \cap C) . \end{aligned}$$

If Q is a transition kernel on $(\mathsf{X}, \mathcal{X})$, we define the kernel $Q^\star$ on $\mathsf{X} \times \check{\mathcal{X}}$ by $Q^\star(x, \check{A}) = [Q(x, \cdot)]^\star(\check{A})$ for $x \in \mathsf{X}$ and $\check{A} \in \check{\mathcal{X}}$.

Assume now that Q is a phi-irreducible transition kernel and let C be a $(1, \epsilon, \nu)$-small set. We define the split transition kernel $\check{Q}$ on $\check{\mathsf{X}} \times \check{\mathcal{X}}$ as follows. For any $x \in \mathsf{X}$ and $\check{A} \in \check{\mathcal{X}}$,

$$\check{Q}((x, 0), \check{A}) = R^\star(x, \check{A}) , \tag{14.24}$$

$$\check{Q}((x, 1), \check{A}) = \nu^\star(\check{A}) . \tag{14.25}$$

Examining the above technicalities, we find that transitions into $C^c \times \{1\}$ have zero probability from everywhere, so that $d_n = 1$ can only occur if $X_n \in C$. Because $d_n = 1$ indicates a regeneration time, from within C, this is logical. Likewise we find that given a transition to some $y \in C$, the conditional probability that $d_n = 1$ is ϵ, wherever the transition took place from. Thus the above split transition kernel corresponds to the following simulation scheme for $\{(X_k, d_k)\}$. Assume (X_k, d_k) are given. If $X_k \notin C$, then draw X_{k+1} from $Q(X_k, \cdot)$. If $X_k \in C$ and $d_n = 1$, then draw X_{k+1} from ν, otherwise from $R(X_k, \cdot)$. If the realized X_{k+1} is not in C, then set $d_{k+1} = 0$; if X_{k+1} is in C, then set $d_{k+1} = 1$ with probability ϵ, and otherwise set $d_{k+1} = 0$.

Split measures operate on the split kernel in the following way. For any measure μ on $(\mathsf{X}, \mathcal{X})$,

$$\mu^\star \check{Q} = (\mu Q)^\star . \tag{14.26}$$

For any probability measure $\check{\mu}$ on $\check{\mathcal{X}}$, we denote by $\check{\mathrm{P}}_{\check{\mu}}$ and $\check{\mathrm{E}}_{\check{\mu}}$, respectively, the probability distribution and the expectation on the canonical space $(\check{\mathsf{X}}^{\mathbb{N}}, \check{\mathcal{X}}^{\otimes \mathbb{N}})$ such that the coordinate process, denoted $\{(X_k, d_k)\}_{k \geq 0}$, is a Markov chain with initial probability measure $\check{\mu}$ and transition kernel $\check{Q}$. We also denote by $\{\check{\mathcal{F}}_k\}_{k \geq 0}$ the natural filtration of this chain and, as usual, by $\{\mathcal{F}_k^X\}_{k \geq 0}$ the natural filtration of $\{X_k\}_{k \geq 0}$.

Proposition 14.2.17. *Let Q be a phi-irreducible transition kernel on $(\mathsf{X}, \mathcal{X})$, let C be an accessible $(1, \epsilon, \nu)$-small set for Q and let μ be a probability measure on $(\mathsf{X}, \mathcal{X})$. Then for any bounded $\mathcal{X}$-measurable function f and any $k \geq 1$,*

$$\check{\mathrm{E}}_{\mu^\star}[f(X_k) \,|\, \mathcal{F}_{k-1}^X] = Qf(X_{k-1}) \quad \check{\mathrm{P}}_{\mu^\star}\text{-a.s.} \tag{14.27}$$

Before giving the proof, we discuss the implications of this result. It implies that under $\check{\mathrm{P}}_{\mu^\star}$, $\{X_k\}_{k \geq 0}$ is a Markov chain (with respect to its natural filtration) with transition kernel Q and initial distribution μ. By abuse of notation, we can identify $\{X_k\}$ with the coordinate process associated to the canonical space $\mathsf{X}^{\mathbb{N}}$. Denote by P_μ the probability measure on $(\mathsf{X}^{\mathbb{N}}, \mathcal{X}^{\otimes \mathbb{N}})$ such that $\{X_k\}_{k \geq 0}$ is a Markov chain with transition kernel Q and initial distribution μ (see Section 2.1.2.1) and denote by E_μ the associated expectation operator. Then Proposition 14.2.17 yields the following identity. For any bounded $\mathcal{F}_\infty^X$-measurable random variable Y,

$$\check{\mathrm{E}}_{\mu^\star}[Y] = \mathrm{E}_\mu[Y] . \tag{14.28}$$

Proof (of Proposition 14.2.17). We have, $\mu^\star$-a.s.,

$$\check{\mathrm{E}}_{\mu^\star}[f(X_k) \,|\, \check{\mathcal{F}}_{k-1}] = \mathbb{1}_{\{d_{k-1}=1\}} \nu(f) + \mathbb{1}_{\{d_{k-1}=0\}} Rf(X_{k-1}) .$$

Because $\check{\mathrm{P}}_{\check{\mu}}(d_{k-1} = 1 \,|\, \mathcal{F}_{k-1}^X) = \epsilon \mathbb{1}_C(X_{k-1})$ $\check{\mathrm{P}}_{\mu^\star}$-a.s., it holds that

$$\begin{aligned}
\check{\mathrm{E}}_{\mu^\star}[f(X_k) \,|\, \mathcal{F}_{k-1}^X] &= \check{\mathrm{E}}_{\mu^\star}\{\check{\mathrm{E}}[f(X_k) \,|\, \check{\mathcal{F}}_{k-1}] \,|\, \mathcal{F}_{k-1}^X\} \\
&= \epsilon \mathbb{1}_C(X_{k-1}) \nu(f) + [1 - \epsilon \mathbb{1}_C(X_{k-1})] Rf(X_{k-1}) \\
&= Qf(X_{k-1}) .
\end{aligned}$$

□

Corollary 14.2.18. *Under the assumptions of Proposition 14.2.17,* $\mathsf{X} \times \{1\}$ *is an accessible atom and* $\nu^\star$ *is an irreducibility measure for the split kernel* $\check{Q}$. *More generally, if* $B \in \mathcal{X}$ *is accessible for* Q, *then* $B \times \{0,1\}$ *is accessible for the split kernel.*

Proof. Because $\check{\alpha} = \mathsf{X} \times \{1\}$ is an atom for the split kernel $\check{Q}$, Proposition 14.2.9 shows that $\nu^\star$ is an irreducibility measure if $\check{\alpha}$ is accessible. Applying (14.28) we obtain for $x \in \mathsf{X}$,

$$\begin{aligned}
\check{\mathrm{P}}_{(x,1)}(\tau_{\check{\alpha}} < \infty) &= \check{\mathrm{P}}_{(x,1)}(d_n = 1 \text{ for some } n \geq 1) \\
&\geq \check{\mathrm{P}}_{(x,1)}(d_1 = 1) = \epsilon\nu(C) > 0\,, \\
\check{\mathrm{P}}_{(x,0)}(\tau_{\check{\alpha}} < \infty) &= \check{\mathrm{P}}_{(x,0)}((X_n, d_n) \in C \times \{1\} \text{ for some } n \geq 1) \\
&\geq \check{\mathrm{P}}_{(x,0)}(\tau_{C\times\{0,1\}} < \infty\,, d_{\tau_{C\times\{0,1\}}} = 1) = \epsilon\,\mathrm{P}_x(\tau_C < \infty) > 0\,.
\end{aligned}$$

Thus $\check{\alpha}$ is accessible and $\nu^\star$ is an irreducibility measure for $\check{Q}$. This implies, by Theorem 14.2.2, that for all $\eta \in (0,1)$, $\nu^\star \check{K}_\eta$ is a maximal irreducibility measure for the split kernel $\check{Q}$; here K_η is the resolvent kernel (14.17) associated to $\check{Q}$. By straightforward applications of the definitions, it is easy to check that $\nu^\star \check{K}_\eta = (\nu K_\eta)^\star$. Moreover, ν is an irreducibility measure for Q, and νK_η is a maximal irreducibility measure for Q (still by Proposition 14.2.11 and Theorem 14.2.2). If B is accessible, then $\nu K_\eta(B) > 0$ and

$$\nu^\star \check{K}_\eta(B \times \{0,1\}) = (\nu K_\eta)^\star(B \times \{0,1\}) = \nu K_\eta(B) > 0.$$

Thus $B \times \{0,1\}$ is accessible for $\check{Q}$. □

14.2.2.3 Transience/Recurrence Dichotomy for General Phi-irreducible Chains

Using the splitting construction, we are now able to prove Theorem 14.2.6 for chains not possessing accessible atoms. We first consider the simple case in which the chain possesses a 1-small set.

Proposition 14.2.19. *Let* Q *be a phi-irreducible transition kernel that admits an accessible* $(1, \epsilon, \nu)$*-small set* C. *Then* Q *is either recurrent or transient. It is recurrent if and only if the small set* C *is recurrent.*

Proof. Because the split chain possesses an accessible atom, by Proposition 14.2.9 the split chain is phi-irreducible and either recurrent or transient. Applying (14.28) we can write

$$\check{\mathrm{E}}_{\delta_x^\star}[\eta_{B\times\{0,1\}}] = \mathrm{E}_x[\eta_B]\,. \tag{14.29}$$

Assume first that the split chain is recurrent. Let B be an accessible set for Q. By Proposition 14.2.17, $B \times \{0,1\}$ is accessible for the split chain. Hence

$\check{\mathrm{E}}_{\delta_x^\star}[\eta_{B\times\{0,1\}}] = \infty$ for all $x \in B$, so that, by (14.29), $\mathrm{E}_x[\eta_B] = \infty$ for all $x \in B$.

Conversely, if the split chain is transient, then by Proposition 14.2.9 the atom $\check{\alpha}$ is transient. For $j \geq 1$, define $B_j = \{x : \sum_{l=1}^{j} \check{Q}^l((x,0),\check{\alpha}) \geq 1/j\}$. Because $\check{\alpha}$ is accessible, $\cup_{j=1}^{\infty} B_j = \mathsf{X}$. By the same argument as in the proof of Proposition 14.2.9, the sets $B_j \times \{0,1\}$ are uniformly transient for the split chain. Hence, by (14.29), the sets B_j are uniformly transient for Q.

It remains to prove that if the small set C is recurrent, then the chain is recurrent. We have just proved that Q is recurrent if and only if $\check{Q}$ is recurrent and, by Proposition 14.2.9, this is true if and only if the atom $\check{\alpha}$ is recurrent. Thus we only need to prove that if C is recurrent then $\check{\alpha}$ is recurrent. If C is recurrent, then (14.29) yields for all $x \in C$,

$$\check{\mathrm{E}}_{\delta_x^\star}[\eta_{\check{\alpha}}] \geq \epsilon \check{\mathrm{E}}_{\delta_x^\star}[\eta_{C\times\{0,1\}}] = \epsilon\, \mathrm{E}_x[\eta_C] = \infty\ .$$

Using the definition of $\delta_x^\star$, this implies that there exists $\check{x} \in \check{\mathsf{X}}$ such that $\check{\mathrm{E}}_{\check{x}}[\eta_{\check{\alpha}}] = \infty$. This observation and (14.23) imply that $\check{\mathrm{E}}_{\check{\alpha}}[\eta_{\check{\alpha}}] = \infty$, that is, the atom is recurrent. □

Using the resolvent kernel, the previous results can be extended to the general case where an accessible small set exists, but not necessarily a 1-small one.

Proposition 14.2.20. *Let Q be transition kernel.*

(i) If Q is phi-irreducible and admits an accessible (m,ϵ,ν)-small set C, then for any $\eta \in (0,1)$, C is an accessible $(1,\epsilon',\nu)$-small set for the resolvent kernel $K_\eta = (1-\eta)\sum_{k=0}^{\infty} \eta^k Q^k$ with $\epsilon' = (1-\eta)\eta^m\epsilon$.
(ii) A set is recurrent (resp. uniformly transient) for Q if and only if it is recurrent (resp. uniformly transient) for K_η for some (hence for all) $\eta \in (0,1)$.
(iii) Q is recurrent (resp. transient) if and only if K_η is recurrent (resp. transient) for some (hence for all) $\eta \in (0,1)$.

Proof. For any $\eta > 0$, $x \in C$, and $A \in \mathcal{X}$,

$$K_\eta(x,A) \geq (1-\eta)\eta^m Q^m(x,A) \geq (1-\eta)\eta^m \epsilon\nu(A) = \epsilon'\nu(A)\ .$$

Thus C is a $(1,\epsilon',\nu)$-small set for K_η, showing part (i). The remaining claims follow from the identity

$$\sum_{n\geq 1} K_\eta^n = \frac{1-\eta}{\eta}\sum_{n\geq 0} Q^n\ .$$

□

14.2.2.4 Harris Recurrence

As for countable state spaces, it is sometimes useful to consider stronger recurrence properties, expressed in terms of return probabilities rather than mean occupation times.

Definition 14.2.21 (Harris Recurrence). *A set $A \in \mathcal{X}$ is said to be* Harris recurrent *if* $\mathrm{P}_x(\tau_A < \infty) = 1$ *for any* $x \in \mathsf{X}$. *A phi-irreducible Markov chain is said to be* Harris (recurrent) *if any accessible set is Harris recurrent.*

It is intuitively obvious that, as for countable state spaces, Harris recurrence implies recurrence.

Proposition 14.2.22. *A Harris recurrent set is recurrent.*

Proof. Let A be a Harris recurrent set. Because for $j \geq 1$, $\sigma_A^{(j+1)} = \tau_A \circ \theta^{\sigma_A^{(j)}}$ on the set $\{\sigma_A^{(j)} < \infty\}$, the strong Markov property implies that for any $x \in A$,

$$\mathrm{P}_x(\sigma_A^{(j+1)} < \infty) = \mathrm{E}_x\left[\mathrm{P}_{X_{\sigma_A^{(j)}}}(\tau_A < \infty)\mathbb{1}_{\{\sigma_A^{(j)} < \infty\}}\right] = \mathrm{P}_x(\sigma_A^{(j)} < \infty) .$$

Because $\mathrm{P}_x(\sigma_A^{(1)} < \infty) = 1$ for $x \in A$, we obtain that for all $x \in A$ and all $j \geq 1$, $\mathrm{P}_x(\sigma_A^{(j)} = 1)$ and $\mathrm{E}_x[\eta_A] = \sum_{j=1}^{\infty} \mathrm{P}_x(\sigma_A^{(j)} < \infty) = \infty$. □

Even though all transition kernels may not be Harris recurrent, the following theorem provides a very useful decomposition of the state space of a recurrent phi-irreducible transition kernel. For a proof of this result, see Meyn and Tweedie (1993, Theorem 9.1.5)

Theorem 14.2.23. *Let Q be a phi-irreducible recurrent transition kernel on a state space* X *and let ψ be a maximal irreducibility measure. Then* $\mathsf{X} = \mathsf{N} \cup \mathsf{H}$, *where* N *is covered by a countable family of uniformly transient sets,* $\psi(\mathsf{N}) = 0$ *and every accessible subset of* H *is Harris recurrent.*

As a consequence, if A is an accessible set of a recurrent phi-irreducible chain, then there exists a set $A' \subseteq A$ such that $\psi(A \setminus A') = 0$ for any maximal irreducibility measure ψ, and $\mathrm{P}_x(\tau_{A'} < \infty) = 1$ for all $x \in A'$.

Example 14.2.24. To understand why a recurrent Markov chain can fail to be Harris, consider the following elementary example of a chain on $\mathsf{X} = \mathbb{N}$. Let the transition kernel Q be given by $Q(0,0) = 1$ and for $x \geq 1$, $Q(x, x+1) = 1 - 1/x^2$ and $Q(x,0) = 1/x^2$. Thus the state 0 is absorbing. Because $Q(x,0) > 0$ for any $x \in \mathsf{X}$, δ_0 is an irreducibility measure. In fact, by application of Theorem 14.2.2, this measure is maximal. The set $\{0\}$ is an atom and because $\mathrm{P}_0(\tau_{\{0\}} < \infty) = 1$, the chain is recurrent by Proposition 14.2.9.

The chain is not Harris recurrent, however. Indeed, for any $x \geq 1$ we have

$$\mathrm{P}_x(\tau_0 \geq k) = \mathrm{P}_x(X_1 \neq 0, \dots, X_{k-1} \neq 0) = \prod_{j=x}^{x+k-1} (1 - 1/j^2) .$$

Because $\prod_{j=2}^{\infty}(1 - 1/j^2) > 0$, we obtain that $\mathrm{P}_x(\tau_0 = \infty) = \lim_{k\to\infty} \mathrm{P}_x(\tau_0 \geq k) > 0$ for any $x \geq 2$, so that the accessible state 0 is not certainly reached from such an initial state. Comparing to Theorem 14.2.23, we see that the decomposition of the state space is given by $\mathsf{H} = \{0\}$ and $\mathsf{N} = \{1, 2, \dots\}$. ■

14.2.3 Invariant Measures and Stationarity

On general state spaces, we again further classify chains using *invariant measures*. A σ-finite measure μ is called *Q-sub-invariant* if $\mu \geq \mu Q$ and *Q-invariant* if $\mu = \mu Q$.

Theorem 14.2.25. *A phi-irreducible recurrent transition kernel (or Markov chain) admits a unique (up to a multiplicative constant) invariant measure which is also a maximal irreducibility measure.*

This result leads us to define the following classes of chains.

Definition 14.2.26 (Positive and Null Chains). *A phi-irreducible transition kernel (or Markov chain) is called* positive *if it admits an invariant probability measure; otherwise it is called* null.

We now prove the existence of an invariant measure when the chain admits an accessible atom. The invariant measure is defined as for countable state spaces, by replacing an individual state by the atom. Thus define the measure μ_α on $\mathcal{X}$ by

$$\mu_\alpha(A) = \mathrm{E}_\alpha\left[\sum_{n=1}^{\tau_\alpha} \mathbb{1}_A(X_n)\right], \qquad A \in \mathcal{X} . \tag{14.30}$$

Proposition 14.2.27. *Let α be an accessible atom for the transition kernel Q. Then μ_α is Q-sub-invariant. It is invariant if and only if the atom α is recurrent. In that case, any Q-invariant measure μ is proportional to μ_α, and μ_α is a maximal irreducibility measure.*

Proof. By the definition of μ_α and the strong Markov property,

$$\begin{aligned}\mu_\alpha Q(A) &= \mathrm{E}_\alpha\left[\sum_{k=1}^{\tau_\alpha} Q(X_k, A)\right] = \mathrm{E}_\alpha\left[\sum_{k=2}^{\tau_\alpha+1} \mathbb{1}_A(X_k)\right] \\ &= \mu_\alpha(A) - \mathrm{P}_\alpha(X_1 \in A) + \mathrm{E}_\alpha[\mathbb{1}_A(X_{\tau_\alpha+1})\mathbb{1}_{\{\tau_\alpha<\infty\}}] .\end{aligned}$$

Applying the strong Markov property once again yields

$$
\begin{aligned}
\mathrm{E}_\alpha[\mathbb{1}_A(X_{\tau_\alpha+1})\mathbb{1}_{\{\tau_\alpha<\infty\}}] &= \mathrm{E}_\alpha\{\mathrm{E}_\alpha[\mathbb{1}_A(X_1)\circ\theta^{\tau_\alpha}\mid\mathcal{F}^X_{\tau_\alpha}]\mathbb{1}_{\{\tau_\alpha<\infty\}}\} \\
&= \mathrm{E}_\alpha[\mathrm{P}_{X_{\tau_\alpha}}(X_1\in A)\mathbb{1}_{\{\tau_\alpha<\infty\}}] = \mathrm{P}_\alpha(X_1\in A)\,\mathrm{P}_\alpha(\tau_\alpha<\infty)\,.
\end{aligned}
$$

Thus $\mu_\alpha Q(A) = \mu_\alpha(A) - \mathrm{P}_\alpha(X_1\in A)[1-\mathrm{P}_\alpha(\tau_\alpha<\infty)]$. This proves that μ_α is sub-invariant, and invariant if and only if $\mathrm{P}_\alpha(\tau_\alpha<\infty)=1$.

Now let μ be an invariant non-trivial measure and let A be an accessible set such that $\mu(A)<\infty$. Then there exists an integer n such that $Q^n(\alpha,A)>0$. Because μ is invariant, it holds that $\mu=\mu Q^n$, so that

$$
\infty > \mu(A) = \mu Q^n(A) \geq \mu(\alpha)Q^n(\alpha,A)\,.
$$

This implies that $\mu(\alpha)<\infty$. Without loss of generality, we can assume $\mu(\alpha)>0$; otherwise we replace μ by $\mu+\mu_\alpha$. Assuming $\mu(\alpha)>0$, there is then no loss of generality in assuming $\mu(\alpha)=1$.

The next step is to prove that if μ is an invariant measure such that $\mu(\alpha)=1$, then $\mu\geq\mu_\alpha$. To prove this it suffices to prove that for all $n\geq1$,

$$
\mu(A) \geq \sum_{k=1}^{n} \mathrm{P}_\alpha(X_k\in A,\,\tau_\alpha\geq k)\,.
$$

We prove this inequality by induction. For $n=1$ we can write

$$
\mu(A) = \mu Q(A) \geq \mu(\alpha)Q(\alpha,A) = Q(\alpha,A) = \mathrm{P}_\alpha(X_1\in A)\,.
$$

Now assume now that the inequality holds for some $n\geq1$. Then

$$
\begin{aligned}
\mu(A) &= Q(\alpha,A) + \int_{\alpha^c}\mu(dy)\,Q(y,A) \\
&\geq Q(\alpha,A) + \sum_{k=1}^{n}\mathrm{E}_\alpha[Q(X_k,A)\mathbb{1}_{\{\tau_\alpha\geq k\}}\mathbb{1}_{\{X_k\notin\alpha\}}] \\
&\geq Q(\alpha,A) + \sum_{k=1}^{n}\mathrm{E}_\alpha[Q(X_k,A)\mathbb{1}_{\{\tau_\alpha\geq k+1\}}]\,.
\end{aligned}
$$

Because $\{\tau_\alpha\geq k+1\}\in\mathcal{F}^X_k$, the Markov property yields

$$
\mathrm{E}_\alpha[Q(X_k,A)\mathbb{1}_{\{\tau_\alpha\geq k+1\}}] = \mathrm{P}_\alpha(X_{k+1}\in A,\,\tau_\alpha\geq k+1)\,,
$$

whence

$$
\mu(A) \geq Q(\alpha,A) + \sum_{k=2}^{n+1}\mathrm{P}_\alpha(X_k\in A,\,\tau_\alpha\geq k) = \sum_{k=1}^{n+1}\mathrm{P}_\alpha(X_k\in A,\,\tau_\alpha\geq k)\,.
$$

This completes the induction, and we conclude that $\mu\geq\mu_\alpha$.

Assume that there exists a set A such that $\mu(A)>\mu_\alpha(A)$. It is straightforward that μ and μ_α are both invariant for the resolvent kernel K_δ (see

(14.17)), for any $\delta \in (0,1)$. Because α is accessible, $K_\delta(x,\alpha) > 0$ for all $x \in \mathsf{X}$. Hence $\int_A \mu(dx)\, Q(x,\alpha) > \int_A \mu_\alpha(dx)\, Q(x,\alpha)$, which implies that

$$\begin{aligned} 1 = \mu(\alpha) = \mu K_\delta(\alpha) &= \int_A \mu(dx)\, K_\delta(x,\alpha) + \int_{A^c} \mu(dx)\, K_\delta(x,\alpha) \\ &> \int_A \mu_\alpha(dx)\, K_\delta(x,\alpha) + \int_{A^c} \mu_\alpha(dx)\, K_\delta(x,\alpha) = \mu_\alpha K_\delta(\alpha) = \mu_\alpha(\alpha) = 1. \end{aligned}$$

This contradiction shows that $\mu = \mu_\alpha$.

We finally prove that μ_α is a maximal irreducibility measure. Let ψ be a maximal irreducibility measure and assume that $\psi(A) = 0$. Then $\mathrm{P}_x(\tau_A < \infty) = 0$ for ψ-almost all $x \in \mathsf{X}$. This obviously implies that $\mathrm{P}_x(\tau_A < \infty) = 0$ for ψ-almost all $x \in \alpha$. Because $\mathrm{P}_x(\tau_A < \infty)$ is constant over α, we find that $\mathrm{P}_x(\tau_A < \infty) = 0$ for all $x \in \alpha$, and this yields $\mu_\alpha(A) = 0$. Thus μ_α is absolutely continuous with respect to ψ, hence an irreducibility measure. Let again K_δ be the resolvent kernel. By Theorem 14.2.2, $\mu_\alpha K_\delta$ is a maximal irreducibility measure. But, as noted above, $\mu_\alpha K_\epsilon = \mu_\alpha$, and therefore μ_α is a maximal irreducibility measure. □

Proposition 14.2.28. *Let Q be a recurrent phi-irreducible transition kernel that admits an accessible $(1,\epsilon,\nu)$-small set C. Then it admits a non-trivial invariant measure, unique up to multiplication by a constant and such that $0 < \pi(C) < \infty$, and any invariant measure is a maximal irreducibility measure.*

Proof. By (14.26), $(\mu Q)^\star = \mu^\star \check{Q}$, so that μ is Q-invariant if and only if $\mu^\star$ is $\check{Q}$-invariant. Let $\check{\mu}$ be a $\check{Q}$-invariant measure and define

$$\mu = \int_{C\times\{0\}} \check{\mu}(d\check{x})\, R(x,\cdot) + \int_{C^c\times\{0\}} \check{\mu}(d\check{x})\, Q(x,\cdot) + \check{\mu}(\mathsf{X}\times\{1\})\nu\,.$$

By application of the definition of the split kernel and measures, it can be checked that $\check{\mu}\check{Q} = \mu^\star$. Hence $\mu^\star = \check{\mu}\check{Q} = \check{\mu}$. We thus see that $\mu^\star$ is $\check{Q}$-invariant, which, as noted above, implies that μ is Q-invariant. Hence we have shown that there exists a Q-invariant measure if and only if there exists a $\check{Q}$-invariant one.

If Q is recurrent then C is recurrent, and as appears in the proof of Proposition 14.2.28 this implies that the atom $\check{\alpha}$ is recurrent for the split chain $\check{Q}$. Thus, by Proposition 14.2.9 the kernel $\check{Q}$ is recurrent, and by Proposition 14.2.27 it admits an invariant measure that is unique up to a scaling factor. Hence Q also admits an invariant measure, unique up to a scaling factor and such that $0 < \pi(C) < \infty$.

Let μ be Q-invariant. Then $\mu^\star$ is $\check{Q}$-invariant and hence, by Proposition 14.2.27, a maximal irreducibility measure. If $\mu(A) > 0$, then $\mu^\star(A \times \{0,1\}) = \mu(A) > 0$. Thus $A \times \{0,1\}$ is accessible, and this implies that A is accessible. We conclude that μ is an irreducibility measure, and it is maximal because it is K_η-invariant. □

If the kernel Q is phi-irreducible and admits an accessible (m, ϵ, ν)-small set C, then, by Proposition 14.2.20, for any $\eta \in (0,1)$ the set C is an accessible $(1, \epsilon', \nu)$-small set for the resolvent kernel K_η. If C is recurrent for Q, it is also recurrent for K_η and therefore, by Proposition 14.2.19, K_η has a unique invariant probability measure. The following result shows that this probability measure is invariant also for Q.

Lemma 14.2.29. *A measure μ on $(\mathsf{X}, \mathcal{X})$ is Q-invariant if and only if μ is K_η-invariant for some (hence for all) $\eta \in (0,1)$.*

Proof. If $\mu Q = \mu$, then obviously $\mu Q^n = \mu$ for all $n \geq 0$, so that $\mu K_\eta = \mu$. Conversely, assume that $\mu K_\eta = \mu$. Because $K_\eta = \eta Q K_\eta + (1-\eta) Q^0$ and $Q K_\eta = K_\eta Q$, it holds that

$$\mu = \mu K_\eta = \eta \mu Q K_\eta + (1-\eta)\mu = \eta \mu K_\eta Q + (1-\eta)\mu = \eta \mu Q + (1-\eta)\mu \,.$$

Hence $\eta \mu Q = \eta \mu$, which concludes the proof. □

14.2.3.1 Drift Conditions

We first give a sufficient condition for a chain to be positive, based on the expectation of the return time to an accessible small set.

Proposition 14.2.30. *Let Q be a transition kernel that admits an accessible small set C such that*

$$\sup_{x \in C} \mathrm{E}_x[\tau_C] < \infty \,. \tag{14.31}$$

Then the chain is positive and the invariant probability measure π satisfies, for all $A \in \mathcal{X}$,

$$\pi(A) = \int_C \pi(dy)\, \mathrm{E}_y\left[\sum_{k=0}^{\tau_C - 1} \mathbb{1}_A(X_k)\right] = \int_C \pi(dy)\, \mathrm{E}_y\left[\sum_{k=1}^{\tau_C} \mathbb{1}_A(X_k)\right] \,. \tag{14.32}$$

If f is a non-negative measurable function such that

$$\sup_{x \in C} \mathrm{E}_x\left[\sum_{k=0}^{\tau_C - 1} f(X_k)\right] < \infty \,, \tag{14.33}$$

then f is integrable with respect to π and

$$\pi(f) = \int_C \pi(dy)\, \mathrm{E}_y\left[\sum_{k=0}^{\tau_C - 1} f(X_k)\right] = \int_C \pi(dy)\, \mathrm{E}_y\left[\sum_{k=1}^{\tau_C} f(X_k)\right] \,.$$

Proof. First note that by Proposition 14.2.11, Q is phi-irreducible. Equation (14.31) implies that for all $\mathrm{P}_x(\tau_C < \infty) = 1$ $x \in C$, that is, C is Harris recurrent. By Proposition 14.2.22, C is recurrent, and so, by Proposition 14.2.19, Q is recurrent. Let π be an invariant measure such that $0 < \pi(C) < \infty$, the existence of which is given by Proposition 14.2.28. Then define a measure μ_C on $\mathcal{X}$ by

$$\mu_C(A) \stackrel{\text{def}}{=} \int_C \pi(dy)\, \mathrm{E}_y \left[\sum_{k=1}^{\tau_C} \mathbb{1}_A(X_k) \right] .$$

Because $\tau_C < \infty$ P_y-a.s. for all $y \in C$, it holds that $\mu_C(C) = \pi(C)$. Then we can show that $\mu_C(A) = \pi(A)$ for all $A \in \mathcal{X}$. The proof is along the same lines as the proof of Proposition 14.2.27 and is therefore omitted. Thus, μ_C is invariant. In addition, we obtain that for any measurable set A,

$$\int_C \pi(dy)\, \mathrm{E}_y \left[\mathbb{1}_A(X_0)\right] = \pi(A \cap C) = \mu_C(A \cap C) = \int_C \pi(dy)\, \mathrm{E}_y \left[\mathbb{1}_A(X_{\tau_C})\right] ,$$

and this yields

$$\mu_C(A) = \int_C \pi(dy)\, \mathrm{E}_y \left[\sum_{k=1}^{\tau_C} \mathbb{1}_A(X_k) \right] = \int_C \pi(dy)\, \mathrm{E}_y \left[\sum_{k=0}^{\tau_C - 1} \mathbb{1}_A(X_k) \right] .$$

We thus obtain the following equivalent expressions for μ_C:

$$\begin{aligned} \mu_C(A) &= \int_C \pi(dy)\, \mathrm{E}_y \left[\sum_{k=0}^{\tau_C - 1} \mathbb{1}_A(X_k) \right] = \int_C \mu_C(dy)\, \mathrm{E}_y \left[\sum_{k=0}^{\tau_C - 1} \mathbb{1}_A(X_k) \right] \\ &= \int_C \mu_C(dy)\, \mathrm{E}_y \left[\sum_{k=1}^{\tau_C} \mathbb{1}_A(X_k) \right] = \int_C \pi(dy)\, \mathrm{E}_y \left[\sum_{k=1}^{\tau_C} \mathbb{1}_A(X_k) \right] \\ &= \pi(A) . \end{aligned}$$

Hence

$$\pi(\mathsf{X}) = \int_C \pi(dy)\, \mathrm{E}_y \left[\sum_{k=0}^{\tau_C - 1} \mathbb{1}_{\mathsf{X}}(X_k) \right] \le \pi(C) \sup_{y \in C} \mathrm{E}_y[\tau_C] < \infty ,$$

so that any invariant measure is finite and the chain is positive. Finally, under (14.33) we obtain that

$$\pi(f) = \int_C \pi(dy)\, \mathrm{E}_y \left[\sum_{k=0}^{\tau_C - 1} f(X_k) \right] \le \pi(C) \sup_{y \in C} \mathrm{E}_y \left[\sum_{k=1}^{\tau_C - 1} f(X_k) \right] < \infty .$$

□

Except in specific examples (where, for example, the invariant distribution is known in advance), it may be difficult to decide if a chain is positive or null. To check such properties, it is convenient to use *drift conditions*.

Proposition 14.2.31. *Assume that there exists a set $C \in \mathcal{X}$, two measurable functions $1 \le f \le V$, and a constant $b > 0$ such that*

$$QV \le V - f + b\mathbb{1}_C \; . \tag{14.34}$$

Then

$$\mathrm{E}_x[\tau_C] \le V(x) + b\mathbb{1}_C(x) \; , \tag{14.35}$$

$$\mathrm{E}_x\left[V(X_{\tau_C})\right] + \mathrm{E}_x\left[\sum_{k=0}^{\tau_C - 1} f(X_k)\right] \le V(x) + b\mathbb{1}_C(x) \; . \tag{14.36}$$

If C is an accessible small set and V is bounded on C, then the chain is positive recurrent and $\pi(f) < \infty$.

Proof. Set for $n \ge 1$,

$$M_n = \left[V(X_n) + \sum_{k=0}^{n-1} f(X_k)\right] \mathbb{1}_{\{\tau_C \ge n\}} \; .$$

Then

$$\begin{aligned}
\mathrm{E}[M_{n+1} \,|\, \mathcal{F}_n] &= \left[QV(X_n) + \sum_{k=0}^{n} f(X_k)\right] \mathbb{1}_{\{\tau_C \ge n+1\}} \\
&\le \left[V(X_n) - f(X_n) + b\mathbb{1}_C(X_n) + \sum_{k=0}^{n} f(X_k)\right] \mathbb{1}_{\{\tau_C \ge n+1\}} \\
&= \left[V(X_n) + \sum_{k=0}^{n-1} f(X_k)\right] \mathbb{1}_{\{\tau_C \ge n+1\}} \le M_n \; ,
\end{aligned}$$

as $\mathbb{1}_C(X_n)\mathbb{1}_{\{\tau_C \ge n+1\}} = 0$. Hence $\{M_n\}_{n \ge 1}$ is a non-negative super-martingale. For any integer n, $\tau_C \wedge n$ is a bounded stopping time, and Doob's optional stopping theorem shows that for any $x \in \mathsf{X}$,

$$\mathrm{E}_x\left[M_{\tau_C \wedge n}\right] \le \mathrm{E}_x\left[M_1\right] \le V(x) + b\mathbb{1}_C(x) \; . \tag{14.37}$$

Applying this relation with $f \equiv 1$ yields for any $x \in \mathsf{X}$ and $n \ge 0$,

$$\mathrm{E}_x\left[\tau_C \wedge n\right] \le V(x) + b\mathbb{1}_C(x) \; ,$$

and (14.35) follows using monotone convergence. This implies in particular that $\mathrm{P}_x(\tau_C < \infty) = 1$ for any $x \in \mathsf{X}$. The proof of (14.36) follows similarly from (14.37) by the letting $n \to \infty$ and $\pi(f)$ is finite by (14.33). □

Example 14.2.32 (Random Walk on the Half-Line, Continued). Consider again the model of Example 14.2.8. Previously we have seen that sets of the form $[0, c]$ are small. If $\Gamma((-\infty\ , -c]) > 0$, then for $x \in [0, c]$,

$$Q(x, A) \geq \Gamma((-\infty\,, -c])\mathbb{1}_A(0)\;;$$

otherwise there exists an integer m such that $\Gamma^{*m}((-\infty\,, -c]) > 0$, whence

$$Q^m(x, A) \geq \Gamma^{*m}((-\infty\,, -c])\mathbb{1}_A(0)\;.$$

To prove recurrence for $\mu < 0$, we apply Proposition 14.2.31. Because $\mu < 0$, there exists $c > 0$ such that $\int_{-c}^{\infty} w\,\Gamma(dw) \leq \mu/2 < 0$. Thus taking $V(x) = x$ for $x > c$,

$$\begin{aligned} QV(x) - V(x) &= \int_{-\infty}^{\infty} [(x+w)_+ - x]\,\Gamma(dw) \\ &= -x\Gamma((-\infty\,, -x]) + \int_{-x}^{\infty} w\,\Gamma(dw) \leq \mu/2\;. \end{aligned}$$

Hence the chain is positive recurrent.

Consider now the case $\mu > 0$. In view of Proposition 14.2.9, we have to show that the atom $\{0\}$ is transient. For any n, $X_n \geq X_0 + \sum_{i=1}^n W_i$. Define $C_n = \left\{\left|n^{-1}\sum_{i=1}^n W_i - \mu\right| \geq \mu/2\right\}$ and write D_n for $\{X_n = 0\}$. The strong law of large numbers implies that $\mathrm{P}_0(D_n \text{ i.o.}) \leq \mathrm{P}_0(C_n \text{ i.o.}) = 0$. Hence the atom $\{0\}$ is transient, and so is the chain.

When $\mu = 0$, additional assumptions on Γ are needed to prove the recurrence of the RWHL (see for instance Meyn and Tweedie, 1993, Lemma 8.5.2). ■

Example 14.2.33 (Autoregressive Model, Continued). Consider again the model of Example 14.2.3 and assume that the noise process has zero mean and finite variance. Choosing $V(x) = x^2$ we have

$$PV(x) = \mathrm{E}[(\phi x + U_1)^2] = \phi^2 V(x) + \mathrm{E}[U_1^2]\;,$$

so that (14.34) holds when $C = [-M, M]$ for some large enough M, provided $|\phi| < 1$. Because we know that every compact set is small if the noise process has an everywhere continuous positive density, Proposition 14.2.31 shows that the chain is positive recurrent. Note that this approach provides an existence result but does not help us to determine π. If $\{U_k\}$ are Gaussian with zero mean and variance σ^2, then one can check that the invariant distribution also is Gaussian with zero mean and variance $\sigma^2/(1 - \phi^2)$. ■

Theorem 14.2.25 shows that if a chain is phi-irreducible and recurrent then the chain is positive, that is, it admits a unique invariant probability measure π. In certain situations, and in particular when dealing with MCMC procedures, it is known that Q admits an invariant probability measure, but it is not known, *a priori*, that the chain is recurrent. The following result shows that positivity implies recurrence.

Proposition 14.2.34. *If the Markov kernel Q is positive, then it is recurrent.*

Proof. Suppose that the chain is positive and let π be an invariant probability measure. If Q is transient, the state space X is covered by a countable family $\{A_j\}$ of uniformly transient subsets (see Theorem 14.2.6). For any j and k,

$$k\pi(A_j) = \sum_{n=1}^{k} \pi Q^n(A_j) \leq \int \pi(dx)\, \mathrm{E}_x[\eta_{A_j}] \leq \sup_{x\in\mathsf{X}} \mathrm{E}_x[\eta_{A_j}] \,. \tag{14.38}$$

The strong Markov property implies that

$$\begin{aligned} \mathrm{E}_x[\eta_{A_j}] &= \mathrm{E}_x[\eta_{A_j} \mathbb{1}_{\{\sigma_{A_j}<\infty\}}] \\ &\leq \mathrm{E}_x\{\mathbb{1}_{\{\sigma_{A_j}<\infty\}}\, \mathrm{E}_{X_{\sigma_{A_j}}}[\eta_{A_j}]\} \leq \sup_{x\in A_j} \mathrm{E}_x[\eta_{A_j}]\, \mathrm{P}_x(\sigma_{A_j} < \infty) \,. \end{aligned}$$

Thus, the left-hand side of (14.38) is bounded as $k \to \infty$. This implies that $\pi(A_j) = 0$, and hence $\pi(\mathsf{X}) = 0$. This is a contradiction so the chain cannot be transient. □

14.2.4 Ergodicity

In this section, we study the convergence of iterates Q^n of the transition kernel to the invariant distribution. As for discrete state spaces case, we first need to avoid periodic behavior that prevents the iterates to converge. In the discrete case, the period of a state x is defined as the greatest common divisor of the set of time points $\{n \geq 0 : Q^n(x,x) > 0\}$. Of course this notion does not extend to general state spaces, but for phi-irreducible chains we may define the period of accessible small sets. More precisely, let Q be a phi-irreducible transition kernel with maximal irreducibility measure ψ. By Theorem 14.2.11, there exists an accessible (m, ϵ, ψ)-small set C. Because ψ is a maximal irreducibility measure, $\psi(C) > 0$, so that when the chain starts from C there is a positive probability that the it will return to C at time m. Let

$$E_C \stackrel{\text{def}}{=} \{n \geq 1 : \text{the set } C \text{ is } (n, \epsilon_n, \psi)\text{-small for some } \epsilon_n > 0\} \tag{14.39}$$

be the set of time points for which C is small with minorizing measure ψ. Note that for n and m in E_C, $B \in \mathcal{X}^+$ and $x \in C$,

$$Q^{n+m}(x, B) \geq \int_C Q^m(x, dx')\, Q^n(x', B) \geq \epsilon_m \epsilon_n \psi(C)\psi(B) > 0 \,,$$

showing that E_C is closed under addition. There is thus a natural period for E_C, given by the greatest common divisor. Similar to the discrete case (see Proposition 14.1.12), this period d may be shown to be independent of the particular choice of the small set C (see for instance Meyn and Tweedie, 1993, Theorem 5.4.4).

Proposition 14.2.35. *Suppose that Q is phi-irreducible with maximal irreducibility measure ψ. Let C be an accessible (m, ϵ, ψ)-small set and let d be the greatest common divisor of the set E_C, defined in (14.39). Then there exist disjoint sets $D_1, \ldots, D_d$ (a d-*cycle*) such that*

(i) for $x \in D_i$, $Q(x, D_{i+1}) = 1$, $i = 0, \ldots, d-1$ (mod d);
(ii) the set $N = (\cup_{i=1}^d D_i)^c$ is ψ-null.

The d-cycle is maximal in the sense if $D'_1, \ldots, D'_{d'}$ is a d'-cycle, then d' divides d, and if $d = d'$, then up to a permutation of indices D'_i and D_i are ψ-almost equal.

It is obvious from the this theorem that the period d does not depend on the choice of the small set C and that any small set must be contained (up to ψ-null sets) inside one specific member of a d-cycle. This in particular implies that if there exists an accessible $(1, \epsilon, \psi)$-small set C, then $d = 1$. This suggests the following definition

Definition 14.2.36 (Aperiodicity). *Suppose that Q is a phi-irreducible transition kernel with maximal irreducibility measure ψ. The largest d for which a d-cycle exists is called the* period *of Q. When $d = 1$, the chain is called* aperiodic. *When there exists a $(1, \epsilon, \psi)$-small set C, the chain is called* strongly aperiodic.

In all the examples considered above, we have shown the existence of a 1-small set; therefore all these Markov chains are strongly aperiodic.

Now we can state the main convergence result, formulated and proved by Athreya *et al.* (1996). It parallels Theorem 14.1.13.

Theorem 14.2.37. *Let Q be a phi-irreducible positive aperiodic transition kernel. Then for π-almost all x,*

$$\lim_{n \to \infty} \|Q^n(x, \cdot) - \pi\|_{\mathrm{TV}} = 0 . \tag{14.40}$$

If Q is Harris recurrent, the convergence occurs for all $x \in \mathsf{X}$.

Although this result does not provide information on the rate of convergence to the invariant distribution, its assumptions are quite minimal. In fact, it may be shown that these assumptions are essentially necessary and sufficient. If $\|Q^n(x, \cdot) - \pi\|_{\mathrm{TV}} \to 0$ for any $x \in \mathsf{X}$, then by Nummelin (1984, Proposition 6.3), the chain is π-irreducible, aperiodic, positive Harris, and π is an invariant distribution. This form of the ergodicity theorem is of particular interest in cases where the invariant distribution is explicitly known, as in Markov chain Monte Carlo. It provides conditions that are simple and easy to verify, and under which an MCMC algorithm converges to its stationary distribution.

Of course the exceptional null set for non-Harris recurrent chain is a nuisance. The example below however shows that there is no way of getting rid of it.

Example 14.2.38. In the model of Example 14.2.24, $\pi = \delta_0$ is an invariant probability measure. Because $Q^n(x,0) = \mathrm{P}_x(\tau_{\{0\}} \leq n)$ for any $n \geq 0$, $\lim_{n\to\infty} Q^n(x,0) = \mathrm{P}_x(\tau_{\{0\}} < \infty)$. We have previously shown that $\mathrm{P}_x(\tau_{\{0\}} < \infty) = 1 - \mathrm{P}_x(\tau_{\{0\}} = \infty) < 1$ for $x \geq 2$, whence $\limsup \|Q^n(x,\cdot) - \pi\|_{\mathrm{TV}} \neq 0$ for such x. ■

Fortunately, in many cases it is not hard to show that a chain is Harris.

A proof of Theorem 14.2.37 from first principles is given by Athreya *et al.* (1996). We give here a proof due to Rosenthal (1995), based on pathwise coupling (see Rosenthal, 2001; Roberts and Rosenthal, 2004). The same construction is used to compute bounds on $\|Q^n(x,\cdot) - \pi\|_{\mathrm{TV}}$. Before proving the theorem, we briefly introduce the pathwise coupling construction for phi-irreducible Markov chains and present the associated Lindvall inequalities.

14.2.4.1 Pathwise Coupling and Coupling Inequalities

Suppose that we have two probability measures ξ and ξ' on $(\mathsf{X}, \mathcal{X})$ that are such that $\frac{1}{2}\|\xi - \xi'\|_{\mathrm{TV}} \leq 1 - \epsilon$ for some $\epsilon \in (0,1]$ or, equivalently (see (4.19)), that there exists a probability measure ν such that $\epsilon\nu \leq \xi \wedge \xi'$. Because ξ and ξ' are probability measures, we may construct a probability space $(\Omega, \mathcal{F}, \mathrm{P})$ and X-valued random variables X and X' such that $\mathrm{P}(X \in \cdot) = \xi(\cdot)$ and $\mathrm{P}(X' \in \cdot) = \xi'$, respectively. By definition, for any $A \in \mathcal{X}$,

$$|\xi(A) - \xi'(A)| = |\,\mathrm{P}(X \in A) - \mathrm{P}(X' \in A)| = |\,\mathrm{E}[\mathbb{1}_A(X) - \mathbb{1}_A(X')]| \quad (14.41)$$

$$= |\,\mathrm{E}[(\mathbb{1}_A(X) - \mathbb{1}_A(X'))\mathbb{1}_{\{X \neq X'\}}]| \leq \mathrm{P}(X \neq X')\;, \quad (14.42)$$

so that the total variation distance between the laws of two random elements is bounded by the probability that they are unequal. Of course, this inequality is not in general sharp, but we can construct on an appropriately defined probability space $(\tilde{\Omega}, \tilde{\mathcal{F}}, \tilde{\mathrm{P}})$ two X-valued random variables X and X' with laws ξ and ξ' such that $\tilde{\mathrm{P}}(X = X') \geq 1 - \epsilon$. The construction goes as follows. We draw a Bernoulli random variable d with probability of success ϵ. If $d = 0$, we then draw X and X' independently from the distributions $(1-\epsilon)^{-1}(\xi - \epsilon\nu)$ and $(1-\epsilon)^{-1}(\xi' - \epsilon\nu)$, respectively. If $d = 1$, we draw X from ν and set $X = X'$. Note that for any $A \in \mathcal{X}$,

$$\tilde{\mathrm{P}}(X \in A) = \tilde{\mathrm{P}}(X \in A \,|\, d = 0)\tilde{\mathrm{P}}(d = 0) + \tilde{\mathrm{P}}(X \in A \,|\, d = 1)\tilde{\mathrm{P}}(d = 1)$$
$$= (1-\epsilon)\{(1-\epsilon)^{-1}[\xi(A) - \epsilon\nu(A)]\} = \xi(A)$$

and, similarly, $\tilde{\mathrm{P}}(X' \in A) = \xi'(A)$. Thus, marginally the random variables X and X' are distributed according to ξ and ξ'. By construction, $\tilde{\mathrm{P}}(X = X') \geq \mathrm{P}(d = 1) \geq \epsilon$, showing that X and X' are equal with probability at least ϵ. Therefore the *coupling bound* (14.41) can be made sharp by using an appropriate construction. Note that this construction may be used to derive bounds on distances between probability measures that generalize the total variation; we will consider in the sequel the V-total variation.

Definition 14.2.39 (V-Total Variation). *Let $V : \mathsf{X} \to [1, \infty)$ be a measurable function. The* V-total variation distance *between two probability measures ξ and ξ' on $(\mathsf{X}, \mathcal{X})$ is*

$$\|\xi - \xi'\|_V \stackrel{\text{def}}{=} \sup_{|f| \leq V} |\xi(f) - \xi'(f)| .$$

If $V \equiv 1$, $\| \cdot \|_1$ is the total variation distance.

When applied to Markov chains, the whole idea of coupling is to construct on an appropriately defined probability space two Markov chains $\{X_k\}$ and $\{X'_k\}$ with transition kernel Q and initial distributions ξ and ξ', respectively, in such a way that $X_n = X'_n$ for all indices n after a random time T, referred to as the *coupling time*. The coupling procedure attempts to *couple* the two Markov chains when they simultaneously enter a coupling set.

Definition 14.2.40 (Coupling Set). *Let $\bar{C} \subseteq \mathsf{X} \times \mathsf{X}$, $\epsilon \in (0, 1]$ and let $\boldsymbol{\nu} = \{\nu_{x,x'}, x, x' \in \mathsf{X}\}$ be transition kernels from $\bar{C}$ (endowed with the trace σ-field) to $(\mathsf{X}, \mathcal{X})$. The set $\bar{C}$ is a $(1, \epsilon, \boldsymbol{\nu})$-*coupling set *if for all $(x, x') \in \bar{C}$ and all $A \in \mathcal{X}$,*

$$Q(x, A) \wedge Q(x', A) \geq \epsilon \, \nu_{x,x'}(A) . \tag{14.43}$$

By applying Lemma 4.3.5, this condition can be stated equivalently as: there exists $\epsilon \in (0, 1]$ such that for all $(x, x') \in \bar{C}$,

$$\frac{1}{2} \|Q(x, \cdot) - Q(x', \cdot)\|_{\mathrm{TV}} \leq 1 - \epsilon . \tag{14.44}$$

For simplicity, only one-step minorization is considered in this chapter. Adaptations to m-step minorization (replacing Q by Q^m in (14.43)) can be carried out as in Rosenthal (1995). Condition (14.43) is often satisfied by setting $\bar{C} = C \times C$ for a $(1, \epsilon, \nu)$-small set C. Indeed, in that case, for all $(x, x') \in C \times C$ and $A \in \mathcal{X}$,

$$Q(x, A) \wedge Q(x', A) \geq \epsilon \nu(A) .$$

The case $\epsilon = 1$ needs some consideration. If there exists an atom, say α, i.e., there exists a probability measure ν such that for all $x \in \alpha$ and $A \in \mathcal{X}$, $Q(x, A) = \nu(A)$, then $\bar{C} = \alpha \times \alpha$ is a $(1, 1, \boldsymbol{\nu})$-coupling set with $\nu_{x,x'} = \nu$ for all $(x, x') \in \bar{C}$. Conversely, assume that $\bar{C}$ is a $(1, 1, \boldsymbol{\nu})$-coupling set. The alternative characterization (14.44) shows that $Q(x, \cdot) = Q(x', \cdot)$ for all $(x, x') \in \bar{C}$, that is, $\bar{C}$ is an atom. This also implies that the set $\bar{C}$ contains a set $\alpha_1 \times \alpha_2$, where α_1 and α_2 are atoms for Q.

We now introduce the coupling construction. Let $\bar{C}$ be a $(1, \epsilon, \boldsymbol{\nu})$-coupling set. Define $\bar{\mathsf{X}} = \mathsf{X} \times \mathsf{X}$ and $\bar{\mathcal{X}} = \mathcal{X} \otimes \mathcal{X}$. Let $\bar{Q}$ be a transition kernel on $(\bar{\mathsf{X}}, \bar{\mathcal{X}})$ given for all A and A' in $\mathcal{X}$ by

$$\bar{Q}(x, x'; A \times A') = Q(x, A)Q(x', A')\mathbb{1}_{\bar{C}^c}(x, x') + \\ (1-\epsilon)^{-2}[Q(x, A) - \epsilon\nu_{x,x'}(A)][Q(x', A') - \epsilon\nu_{x,x'}(A')]\mathbb{1}_{\bar{C}}(x, x') \quad (14.45)$$

if $\epsilon < 1$ and $\bar{Q} = Q \otimes Q$ if $\epsilon = 1$. For any probability measure $\bar{\mu}$ on $(\bar{\mathsf{X}}, \bar{\mathcal{X}})$, let $\bar{\mathrm{P}}_{\bar{\mu}}$ be the probability measure on the canonical space $(\bar{\mathsf{X}}^{\mathbb{N}}, \bar{\mathcal{X}}^{\mathbb{N}})$ such that the coordinate process $\{\bar{X}_k\}$ is a Markov chain with respect to its natural filtration and with initial distribution $\bar{\mu}$ and transition kernel $\bar{Q}$. As usual, denote the associated expectation operator by $\bar{\mathrm{E}}_{\bar{\mu}}$.

We now define a transition kernel $\tilde{Q}$ on the space $\tilde{X} \stackrel{\text{def}}{=} \mathsf{X} \times \mathsf{X} \times \{0, 1\}$ endowed with the product σ-field $\tilde{\mathcal{X}}$ by, for any $x, x' \in \mathsf{X}$ and $A, A' \in \mathcal{X}$,

$$\tilde{Q}\left((x, x', 0), A \times A' \times \{0\}\right) = [1 - \epsilon\mathbb{1}_{\bar{C}}(x, x')]\bar{Q}((x, x'), A \times A') , \quad (14.46)$$

$$\tilde{Q}\left((x, x', 0), A \times A' \times \{1\}\right) = \epsilon\mathbb{1}_{\bar{C}}(x, x')\nu_{x,x'}(A \cap A') , \quad (14.47)$$

$$\tilde{Q}\left((x, x', 1), A \times A' \times \{1\}\right) = Q(x, A \cap A') . \quad (14.48)$$

For any probability measure $\tilde{\mu}$ on $(\tilde{\mathsf{X}}, \tilde{\mathcal{X}})$, let $\tilde{\mathrm{P}}_{\tilde{\mu}}$ be the probability measure on the canonical space $(\tilde{\mathsf{X}}^{\mathbb{N}}, \tilde{\mathcal{X}}^{\otimes\mathbb{N}})$ such that the coordinate process $\{\tilde{X}_k\}$ is a Markov chain with transition kernel $\tilde{Q}$ and initial distribution $\tilde{\mu}$. The corresponding expectation operator is denoted by $\tilde{\mathrm{E}}_{\tilde{\mu}}$.

The transition kernel $\tilde{Q}$ can be described algorithmically. Given $\tilde{X}_0 = (X_0, X'_0, d_0) = (x, x', d)$, $\tilde{X}_1 = (X_1, X'_1, d_1)$ is obtained as follows.

- If $d = 1$, then draw X_1 from $Q(x, \cdot)$ and set $X'_1 = X_1$, $d_1 = 1$.
- If $d = 0$ and $(x, x') \in \bar{C}$, flip a coin with probability of heads ϵ.
  - If the coin comes up heads, draw X_1 from $\nu_{x,x'}$ and set $X'_1 = X_1$ and $d_1 = 1$.
  - If the coin comes up tails, draw (X_1, X'_1) from $\bar{Q}(x, x'; \cdot)$ and set $d_1 = 0$.
- If $d = 0$ and $(x, x') \notin \bar{C}$, draw (X_1, X'_1) from $\bar{Q}(x, x'; \cdot)$ and set $d_1 = 0$.

The variable d_n is called the *bell variable*; it indicates whether coupling has occurred by time n ($d_n = 1$) or not ($d_n = 0$). The first index n at which $d_n = 1$ is the coupling time;

$$T = \inf\{k \geq 1 : d_k = 1\}.$$

If $d_n = 1$, then $X_k = X'_k$ for all $k \geq n$. The coupling construction is carried out in such a way that under $\tilde{\mathrm{P}}_{\xi\otimes\xi'\otimes\delta_0}$, $\{X_k\}$ and $\{X'_k\}$ are Markov chains with transition kernel Q with initial distributions ξ and ξ', respectively.

The coupling construction allows deriving quantitative bounds on the (V-)total variation distance in terms of the tail probability of the coupling time.

Proposition 14.2.41. *Assume that the transition kernel Q admits a $(1, \epsilon, \boldsymbol{\nu})$-coupling set. Then for any probability measures ξ and ξ' on $(\mathsf{X}, \mathcal{X})$ and any measurable function $V : \mathsf{X} \to [1, \infty)$,*

$$\|\xi Q^n - \xi' Q^n\|_{\mathrm{TV}} \leq 2\tilde{\mathrm{P}}_{\xi\otimes\xi'\otimes\delta_0}(T > n) , \tag{14.49}$$

$$\|\xi Q^n - \xi' Q^n\|_V \leq 2\tilde{\mathrm{E}}_{\xi\otimes\xi'\otimes\delta_0}[\bar{V}(X_n, X'_n)\mathbb{1}_{\{T>n\}}] , \tag{14.50}$$

where $\bar{V} : \mathsf{X} \times \mathsf{X} \to [1, \infty)$ *is defined by* $\bar{V}(x, x') = \{V(x) + V(x')\}/2$.

Proof. We only need to prove (14.50) because (14.49) is obtained by setting $V \equiv 1$. Pick a function f such that $|f| \leq V$ and note that $[f(X_n) - f(X'_n)]\mathbb{1}_{\{d_n=1\}} = 0$. Hence

$$\begin{aligned}|\xi Q^n f - \xi' Q^n f| &= |\tilde{\mathrm{E}}_{\xi\otimes\xi'\otimes\delta_0}[f(X_n) - f(X'_n)]| \\ &= |\tilde{\mathrm{E}}_{\xi\otimes\xi'\otimes\delta_0}[(f(X_n) - f(X'_n))\mathbb{1}_{\{d_n=0\}}]| \\ &\leq 2\tilde{\mathrm{E}}_{\xi\otimes\xi'\otimes\delta_0}[\bar{V}(X_n, X'_n)\mathbb{1}_{\{d_n=0\}}] .\end{aligned}$$

□

We now provide an alternative expression of the coupling inequality that only involves the process $\{\bar{X}_k\}$. Let $\sigma_{\bar{C}}$ be the hitting time on the coupling set $\bar{C}$ by this process, define $K_0(\epsilon) = 1$, and for all $n \geq 1$,

$$K_n(\epsilon) = \begin{cases} \mathbb{1}_{\{\sigma_{\bar{C}} \geq n\}} & \text{if } \epsilon = 1 ; \\ \prod_{j=0}^{n-1}[1 - \epsilon \mathbb{1}_{\bar{C}}(\bar{X}_j)] & \text{if } \epsilon \in (0, 1) . \end{cases} \tag{14.51}$$

Proposition 14.2.42. *Assume that the transition kernel* Q *admits a* $(1, \epsilon, \boldsymbol{\nu})$*-coupling set. Let* ξ *and* ξ' *be probability measures on* $(\mathsf{X}, \mathcal{X})$ *and let* $V : \mathsf{X} \to [1, \infty)$ *be a measurable function. Then*

$$\|\xi Q^n - \xi' Q^n\|_V \leq 2\bar{\mathrm{E}}_{\xi\otimes\xi'}[\bar{V}(X_n, X'_n)K_n(\epsilon)] , \tag{14.52}$$

with $\bar{V}(x, x') = [V(x) + V(x')]/2$.

Proof. We show that for any probability measure $\bar{\mu}$ on $(\bar{\mathsf{X}}, \bar{\mathcal{X}})$,

$$\tilde{\mathrm{E}}_{\bar{\mu}\otimes\delta_0}[\bar{V}(X_n, X'_n)\mathbb{1}_{\{T>n\}}] = \bar{\mathrm{E}}_{\bar{\mu}}[\bar{V}(X_n, X'_n)K_n(\epsilon)] .$$

To do this, we shall prove by induction that for any $n \geq 0$ and any bounded $\bar{\mathcal{X}}$-measurable functions $\{f_j\}_{j\geq 0}$,

$$\tilde{\mathrm{E}}_{\bar{\mu}\otimes\delta_0}\left[\prod_{j=0}^{n} f_j(X_j, X'_j)\,\mathbb{1}_{\{T>n\}}\right] = \bar{\mathrm{E}}_{\bar{\mu}}\left[\prod_{j=0}^{n} f_j(X_j, \bar{X}_j)\,K_n(\epsilon)\right] . \tag{14.53}$$

This is obviously true for $n = 0$. For $n \geq 0$, put $\chi_n = \prod_{j=0}^{n} f_j(X_j, X'_j)$. The induction assumption and the identity $\{T > n+1\} = \{d_{n+1} = 0\}$ yield

$$\begin{aligned}\tilde{\mathrm{E}}_{\bar{\mu}\otimes\delta_0}[\chi_{n+1}\mathbb{1}_{\{T>n+1\}}] &= \tilde{\mathrm{E}}_{\bar{\mu}\otimes\delta_0}[\chi_n\; f_{n+1}(X_{n+1}, X'_{n+1})\mathbb{1}_{\{d_{n+1}=0\}}] \\ &= \tilde{\mathrm{E}}_{\bar{\mu}\otimes\delta_0}\{\chi_n\; \tilde{\mathrm{E}}[f_{n+1}(X_{n+1}, X'_{n+1})\mathbb{1}_{\{d_{n+1}=0\}} \,|\, \tilde{\mathcal{F}}_n]\mathbb{1}_{\{d_n=0\}}\} \\ &= \tilde{\mathrm{E}}_{\bar{\mu}\otimes\delta_0}\{\chi_n[1 - \epsilon\mathbb{1}_{\bar{C}}(X_n, X'_n)]\bar{Q}f_{n+1}(X_n, X'_n)\mathbb{1}_{\{d_n=0\}}\} \\ &= \bar{\mathrm{E}}_{\bar{\mu}}[\chi_n\bar{Q}f_{n+1}(\bar{X}_n)K_{n+1}(\epsilon)] = \bar{\mathrm{E}}_{\bar{\mu}}[\chi_{n+1}K_{n+1}(\epsilon)] .\end{aligned}$$

This concludes the induction and the proof. □

14.2.4.2 Proof of Theorem 14.2.37

We preface the proof of Theorem 14.2.37 by two technical lemmas that establish some elementary properties of a chain on the product space with transition kernel $Q \otimes Q$.

Lemma 14.2.43. *Suppose that Q is a phi-irreducible aperiodic transition kernel. Then for any n, Q^n is phi-irreducible and aperiodic.*

Proof. Propositions 14.2.11 and 14.2.12 show that there exists an accessible (m, ϵ, ν)-small set C and that ν is an irreducibility measure. Because Q is aperiodic, there exists a sequence $\{\epsilon_k\}$ of positive numbers and an integer n_C such that for all $n \geq n_C$, $x \in C$, and $A \in \mathcal{X}$, $Q^n(x, A) \geq \epsilon_n \nu(A)$. In addition, because C is accessible, there exists p such that $Q^p(x, C) > 0$ for any $x \in \mathsf{X}$. Therefore for any $n \geq n_C$ and any $A \in \mathcal{X}$ such that $\nu(A) > 0$,

$$Q^{n+p}(x, A) \geq \int_C Q^p(x, dx')\, Q^n(x', A) \geq \epsilon_n \nu(A) Q^p(x, C) > 0 \,. \tag{14.54}$$

□

Lemma 14.2.44. *Let Q be an aperiodic positive transition kernel with invariant probability measure π. Then $Q \otimes Q$ is phi-irreducible, $\pi \otimes \pi$ is $Q \otimes Q$-invariant, and $Q \otimes Q$ is positive. If C is an accessible (m, ϵ, ν)-small set for Q, then $C \times C$ is an accessible $(m, \epsilon^2, \nu \otimes \nu)$-small set for $Q \otimes Q$.*

Proof. Because Q is phi-irreducible and admits π as an invariant probability measure, π is a maximal irreducibility measure for Q. Let C be an accessible (m, ϵ, ν)-small set for Q. Then for $(x, x') \in C \times C$ and $A \in \mathcal{X} \otimes \mathcal{X}$,

$$(Q \otimes Q)^m(x, x'; A) = \iint_A Q^m(x, dy)\, Q^m(x', dy') \geq \epsilon^2 \nu \otimes \nu(A) \,.$$

Because $\nu \otimes \nu(C \times C) = [\nu(C)]^2 > 0$, this shows that $C \times C$ is a $(1, \epsilon^2, \nu \otimes \nu)$-small set for $Q \otimes Q$. By (14.54) there exists an integer n_x such that for any $n \geq n_x$, $Q^n(x, C) > 0$. This implies that for any $(x, x') \in \mathsf{X} \times \mathsf{X}$ and any $n \geq n_x \vee n_{x'}$,

$$(Q \otimes Q)^n(x, x'; C \times C) = Q^n(x, C) Q^n(x', C) > 0 \,,$$

showing that $C \times C$ is accessible. Because $C \times C$ is a small set, Proposition 14.2.11 shows that $Q \otimes Q$ is phi-irreducible. In addition, $\pi \otimes \pi$ is invariant for $Q \otimes Q$, so that $\pi \otimes \pi$ is a maximal irreducibility measure and $Q \otimes Q$ is positive. □

We have now all the necessary ingredients to prove Theorem 14.2.37.

Proof (of Theorem 14.2.37). By Lemma 14.2.43, Q^m is phi-irreducible for any integer m. By Proposition 14.2.12, there exists an accessible (m, ϵ, ν)-small set C with $\nu(C) > 0$. Lemma 4.3.8 shows that for all integers n,

$$\|Q^n(x, \cdot) - Q^n(x', \cdot)\|_{\mathrm{TV}} \leq \|Q^{m[n/m]}(x, \cdot) - Q^{m[n/m]}(x', \cdot)\|_{\mathrm{TV}} \,.$$

Hence it suffices to prove that (14.40) holds for Q^m and we may thus without loss of generality assume that $m = 1$.

For any probability measure μ on $(\mathsf{X} \times \mathsf{X}, \mathcal{X} \otimes \mathcal{X})$, let $\mathrm{P}^\star_\mu$ denote the probability measure on the canonical space $((\mathsf{X} \times \mathsf{X})^{\mathbb{N}}, (\mathcal{X} \otimes \mathcal{X})^{\otimes \mathbb{N}})$ such that the canonical process $\{(X_k, X'_k)\}_{k \geq 0}$ is a Markov chain with transition kernel $Q \otimes Q$ and initial distribution μ. By Lemma 14.2.44, $Q \otimes Q$ is positive, and it is recurrent by Proposition 14.2.34.

Because $\pi \otimes \pi(C \times C) = \pi^2(C) > 0$, by Theorem 14.2.23 there exist two measurable sets $\bar{C} \subseteq C \times C$ and $\bar{H} \subseteq \mathsf{X} \times \mathsf{X}$ such that $\pi \otimes \pi(C \times C \setminus \bar{C}) = 0$, $\pi \times \pi(H) = 1$, and for all $(x, x') \in \bar{H}$, $\mathrm{P}^\star_{x,x'}(\tau_{\bar{C}} < \infty) = 1$. Moreover, the set $\bar{C}$ is a $(1, \epsilon, \boldsymbol{\nu})$-coupling set with $\nu_{x,x'} = \nu$ for all $(x, x') \in \bar{C}$.

Let the transition kernel $\bar{Q}$ be defined by (14.45) if $\epsilon < 1$ and by $\bar{Q} = Q \otimes Q$ if $\epsilon = 1$. For $\epsilon = 1$, $\bar{\mathrm{P}}_{x,x'} = \mathrm{P}^\star_{x,x'}$. Now assume that $\epsilon \in (0, 1)$. For $(x, x') \notin \bar{C}$, $\bar{\mathrm{P}}_{x,x'}(\tau_{\bar{C}} = \infty) = \mathrm{P}^\star_{x,x'}(\tau_{\bar{C}} = \infty)$. For $(x, x') \in \bar{C}$, noting that $\bar{Q}(x, x', A) \leq (1 - \epsilon)^{-2} Q \otimes Q(x, x', A)$ we obtain

$$\begin{aligned}
\bar{\mathrm{P}}_{x,x'}(\tau_{\bar{C}} = \infty) &= \bar{\mathrm{P}}_{x,x'}(\tau_{\bar{C}} = \infty \,|\, (X_1, X'_1) \notin C \times C)\, \bar{Q}(x, x', \bar{C}^c) \\
&\leq (1 - \epsilon)^{-2} Q \otimes Q(x, x', \bar{C}^c)\, \mathrm{P}^\star_{x,x'}(\tau_{\bar{C}} = \infty \,|\, \bar{X}_1 \notin \bar{C}) \\
&= (1 - \epsilon)^{-2}\, \mathrm{P}^\star_{x,x'}(\tau_{\bar{C}} = \infty) = 0 \,.
\end{aligned}$$

Thus, for all $\epsilon \in (0, 1]$ the set $\bar{C}$ is Harris-recurrent for the kernel $\bar{Q}$. This implies that $\lim_{n\to\infty} \bar{\mathrm{E}}_{x,x'}[K_n(\epsilon)] = 0$ for all $(x, x') \in \bar{H}$ and, using Proposition 14.2.42, we conclude that (14.40) is true. □

14.2.5 Geometric Ergodicity and Foster-Lyapunov Conditions

Theorem 14.2.37 implies forgetting of the initial distribution and convergence to stationarity but does not provide us with rates of convergence. In this section, we show how to adapt the construction above to derive explicit bounds on $\|\xi Q^n - \xi' Q^n\|_V$. We focus on conditions that imply geometric convergence.

Definition 14.2.45 (Geometric Ergodicity). *A positive aperiodic transition kernel Q with invariant probability measure π is said to be V-*geometrically ergodic *if there exist constants $\rho \in (0, 1)$ and $M < \infty$ such that*

$$\|Q^n(x, \cdot) - \pi\|_V \leq M V(x) \rho^n \quad \textit{for } \pi\textit{-almost all } x. \tag{14.55}$$

We now present conditions that ensure geometric ergodicity.

Definition 14.2.46 (Foster-Lyapunov Drift Condition). *A transition kernel Q is said to satisfy a Foster-Lyapunov drift condition outside a set $C \in \mathcal{X}$ if there exists a measurable function $V : \mathsf{X} \to [1, \infty]$, bounded on C, and non-negative constants $\lambda < 1$ and $b < \infty$ such that*

$$QV \leq \lambda V + b\mathbb{1}_C . \tag{14.56}$$

If Q is phi-irreducible and satisfies a Foster-Lyapunov condition outside a small set C, then C is accessible and, writing $QV \leq V - (1 - \lambda)V + b\mathbb{1}_C$, Proposition 14.2.31 shows that Q is positive and $\pi(V) < \infty$.

Example 14.2.47 (Random Walk on the Half-Line, Continued). Assume that for the model of Example 14.2.8 there exists $z > 0$ such that $\mathrm{E}[\mathrm{e}^{zW_1}] < \infty$. Then because $\mu < 0$, there exists $z > 0$ such that $\mathrm{E}[\mathrm{e}^{zW_1}] < 1$. Define $z_0 = \arg\min_{z>0} \mathrm{E}[\mathrm{e}^{zW_1}]$ and $V(x) = \mathrm{e}^{z_0 x}$, and choose $x_0 > 0$ such that $\lambda = \mathrm{E}[\mathrm{e}^{z_0 W_1}] + \mathrm{P}(W_1 < -x_0) < 1$. Then for $x > x_0$,

$$QV(x) = \mathrm{E}[\mathrm{e}^{z_0(x+W_1)_+}] = \mathrm{P}(W_1 \leq -x) + \mathrm{e}^{z_0 x}\, \mathrm{E}[\mathrm{e}^{z_0 W_1} \mathbb{1}_{\{W_1 > -x\}}] \leq \lambda V(x) .$$

Hence the Foster-Lyapunov drift condition holds outside the small set $[0, x_0]$, and the RWHL is geometrically ergodic. For a sharper choice of the constants z_0 and λ, see Scott and Tweedie (1996, Theorem 4.1). ■

Example 14.2.48 (Metropolis-Hastings Algorithm, Continued). Consider the Metropolis-Hastings algorithm of Example 14.2.4 with random walk proposal kernel $r(x, x') = r(|x - x'|)$. Geometric ergodicity of the Metropolis-Hastings algorithm on $\mathbb{R}^d$ is largely a property of the tails of the stationary distribution π. Conditions for geometric ergodicity can be shown to be, essentially, that the tails are exponential or lighter (Mengersen and Tweedie, 1996) and that in higher dimensions the contours of π are regular near ∞ (see for instance Jarner and Hansen, 2000). To understand how the tail conditions come into play, consider the case where π is a probability density on $\mathsf{X} = \mathbb{R}^+$. We suppose that π is log-concave in the upper tail, that is, that there exists $\alpha > 0$ and M such that for all $x' \geq x \geq M$,

$$\log \pi(x) - \log \pi(x') \geq \alpha(x' - x) . \tag{14.57}$$

To simplify the proof, we assume that π is non-increasing, but this assumption is unnecessary. Define $A_x = \{x' \in \mathbb{R}^+ : \pi(x') \leq \pi(x)\}$ and $R_x = \{x' \in \mathbb{R}^+, \pi(x) > \pi(x')\}$, the acceptance and (possible) rejection regions for the chain started from x. Because π is non-increasing, these sets are simple: $A_x = [0, x]$ and $R_x = (x, \infty) \cup (-\infty, 0)$. If we relax the monotonicity conditions, the acceptance and rejection regions become more involved, but because π is log-concave and thus in particular monotone in the upper tail, A_x and R_x are essentially intervals when x is sufficiently large.

For any function $V : \mathbb{R}^+ \to [1, +\infty)$ and $x \in \mathbb{R}^+$,

$$\frac{QV(x)}{V(x)} = 1 + \int_{A_x} r(x'-x)\left[\frac{V(x')}{V(x)} - 1\right] dx' \\ + \int_{R_x} r(x'-x)\frac{\pi(x')}{\pi(x)}\left[\frac{V(x')}{V(x)} - 1\right] dx' .$$

We set $V(x) = \mathrm{e}^{sx}$ for some $s \in (0, \alpha)$. Because π is log-concave, $\pi(x')/\pi(x) \leq \mathrm{e}^{-\alpha(x'-x)}$ when $x' \geq x \geq M$. For $x \geq M$, it follows from elementary calculations that

$$\limsup_{x\to\infty} \frac{QV(x)}{V(x)} \leq 1 - \int_0^\infty r(u)(1-\mathrm{e}^{-su})[1-\mathrm{e}^{-(\alpha-s)u}]\, du < 1 ,$$

showing that the random walk Metropolis-Hastings algorithm on the positive real line satisfies the Foster-Lyapunov condition when π is monotone and log-concave in the upper tail. ■

The main result guaranteeing geometric ergodicity is the following.

Theorem 14.2.49. *Let Q be a phi-irreducible aperiodic positive transition kernel with invariant distribution π. Also assume that Q satisfies a Foster-Lyapunov drift condition outside a small set C with drift function V. Then $\pi(V)$ is finite and Q is V-geometrically ergodic.*

In fact, it follows from Meyn and Tweedie (1993, Theorems 15.0.1 and 16.0.1) that the converse is also true: if a phi-irreducible aperiodic kernel is V-geometrically ergodic, then there exists an accessible small set C such that V is a drift function outside C.

For the sake of brevity and simplicity, we now prove Theorem 14.2.49 under the additional assumption that the level sets of V are all $(1, \epsilon, \nu)$-small. In that case, it is possible to define a coupling set $\bar{C}$ and a transition kernel $\bar{Q}$ that satisfies a (bivariate) Foster-Lyapunov drift condition outside $\bar{C}$. The geometric ergodicity of the transition kernel Q is then proved under this assumption. This is the purpose of the following propositions.

Proposition 14.2.50. *Let Q be a kernel that satisfies the Foster-Lyapunov drift condition (14.56) with respect to a $(1, \epsilon, \nu)$-small set C and a function V whose level sets are $(1, \epsilon, \nu)$-small. Then for any $d > 1$, the set $C' = C \cup \{x \in \mathsf{X} : V(x) \leq d\}$ is small, $C' \times C'$ is a $(1, \epsilon, \nu)$-coupling set, and the kernel $\bar{Q}$, defined as in (14.45), satisfies the drift condition (14.58) with $\bar{C} = C' \times C'$, $\bar{V}(x, x') = (1/2)[V(x) + V(x')]$, and $\bar{\lambda} = \lambda + b/(1+d)$ provided $\bar{\lambda} < 1$.*

Proof. For $(x, x') \notin \bar{C}$ we have $(1+d)/2 \leq \bar{V}(x, x')$. Therefore

$$\bar{Q}\bar{V}(x, x') \leq \lambda\bar{V}(x, x') + \frac{b}{2} \leq \left(\lambda + \frac{b}{1+d}\right)\bar{V}(x, x') ,$$

and for $(x, x') \in \bar{C}$ it holds that

$$\bar{Q}\bar{V}(x, x') = \frac{1}{2(1-\epsilon)}[QV(x) + QV(x') - 2\epsilon\nu(V)]$$
$$\leq \frac{\lambda}{(1-\epsilon)}\bar{V}(x, x') + \frac{b - \epsilon\nu(V)}{1-\epsilon} .$$

□

Proposition 14.2.51. *Assume that Q admits a $(1, \epsilon, \boldsymbol{\nu})$-coupling set $\bar{C}$ and that there exists a choice of the kernel $\bar{Q}$ for which there is a measurable function $\bar{V} : \bar{\mathsf{X}} \to [1, \infty)$, $\bar{\lambda} \in (0, 1)$ and $\bar{b} > 0$ such that*

$$\bar{Q}\bar{V} \leq \bar{\lambda}\bar{V} + \bar{b}\mathbb{1}_{\bar{C}} . \tag{14.58}$$

Let $W : \mathsf{X} \to [1, \infty)$ be a measurable function such that $W(x) + W(x') \leq 2\bar{V}(x, x')$ for all $(x, x') \in \mathsf{X} \times \mathsf{X}$. Then there exist $\rho \in (0, 1)$ and $c > 0$ such that for all $(x, x') \in \mathsf{X} \times \mathsf{X}$,

$$\|Q^n(x, \cdot) - Q^n(x', \cdot)\|_W \leq c\bar{V}(x, x')\rho^n . \tag{14.59}$$

Proof. By Proposition 14.2.41, proving (14.59) amounts to proving the requested bound for $\bar{\mathrm{E}}_{x,x'}[\bar{V}(\bar{X}_n)K_n(\epsilon)]$. We only consider the case $\epsilon \in (0, 1)$, the case $\epsilon = 1$ being easier. Write $\bar{x} = (x, x')$. By induction, the drift condition (14.58) implies that

$$\bar{\mathrm{E}}_{\bar{x}}[\bar{V}(\bar{X}_n)] = \bar{Q}^n\bar{V}(\bar{x}) \leq \bar{\lambda}^n\bar{V}(\bar{x}) + \bar{b}\sum_{j=0}^{n-1}\bar{\lambda}^j \leq \bar{V}(\bar{x}) + \bar{b}/(1 - \bar{\lambda}) . \tag{14.60}$$

Recall that $K_n(\epsilon) = (1 - \epsilon)^{\eta_n(\bar{C})}$ for $\epsilon \in (0, 1)$, where $\eta_n(\bar{C}) = \sum_0^{n-1} \mathbb{1}_{\bar{C}}(X_j)$ is the number of visits to the coupling set $\bar{C}$ before time n. Hence $K_n(\epsilon)$ is $\bar{\mathcal{F}}_{n-1}$-measurable. Let $j \leq n + 1$ be an arbitrary positive integer to be chosen later. Then (14.60) yields

$$\bar{\mathrm{E}}_{\bar{x}}[\bar{V}(\bar{X}_n)K_n(\epsilon)\mathbb{1}_{\{\eta_n(\bar{C})\geq j\}}] \leq (1 - \epsilon)^j\bar{\mathrm{E}}_{\bar{x}}[\bar{V}(\bar{X}_n)]\mathbb{1}_{\{j\leq n\}}$$
$$\leq [\bar{V}(\bar{x}) + \bar{b}/(1 - \bar{\lambda})](1 - \epsilon)^j\mathbb{1}_{\{j\leq n\}} . \tag{14.61}$$

Put $M = \sup_{\bar{x}\in\bar{C}} \bar{Q}\bar{V}(\bar{x})/V(\bar{x})$ and $B = 1 \vee [M(1 - \epsilon)/\bar{\lambda}]$. For $k = 0, \ldots, n$, define $Z_k = \bar{\lambda}^{-k}[(1 - \epsilon)/B]^{\eta_k(\bar{C})}\bar{V}(\bar{X}_k)$. Because $\eta_n(\bar{C})$ is $\bar{\mathcal{F}}_{n-1}$-measurable, we obtain

$$\bar{\mathrm{E}}_{\bar{x}}[Z_n \,|\, \bar{\mathcal{F}}_{n-1}] = \bar{\lambda}^{-n}\bar{Q}\bar{V}(\bar{X}_{n-1})[(1 - \epsilon)/B]^{\eta_n(\bar{C})}$$
$$\leq \bar{\lambda}^{-n+1}\bar{V}(\bar{X}_{n-1})[(1 - \epsilon)/B]^{\eta_n(\bar{C})}\mathbb{1}_{\bar{C}^c}(\bar{X}_{n-1})$$
$$+ \bar{\lambda}^{-n}M\bar{V}(\bar{X}_{n-1})[(1 - \epsilon)/B]^{\eta_n(\bar{C})}\mathbb{1}_{\bar{C}}(\bar{X}_{n-1}) .$$

Using the relations $\eta_n(\bar{C}) = \eta_{n-1}(\bar{C}) + 1_{\bar{C}}(\bar{X}_{n-1})$ and $M(1 - \epsilon) \leq B\bar{\lambda}$, we find that $\bar{\mathrm{E}}_{\bar{x}}[Z_n \,|\, \bar{\mathcal{F}}_{n-1}] \leq Z_{n-1}$ and, by induction, $\bar{\mathrm{E}}_{\bar{x}}[Z_n] \leq \bar{\mathrm{E}}_{\bar{x}}[Z_0] = \bar{V}(\bar{x})$. Hence, as $B \geq 1$,

$$\bar{\mathrm{E}}_{\bar{x}}[\bar{V}(\bar{X}_n)K_n(\epsilon)\mathbb{1}_{\{\eta_n(\bar{C})<j\}}] \leq \bar{\lambda}^n B^j \bar{\mathrm{E}}_{\bar{x}}[Z_n] \leq \bar{\lambda}^n B^j \, \bar{V}(\bar{x}) \,. \tag{14.62}$$

Gathering (14.61) and (14.62) yields

$$\bar{\mathrm{E}}_{\bar{x}}[\bar{V}(\bar{X}_n)K_n(\epsilon)] \leq [\bar{V}(\bar{x}) + \bar{b}/(1-\bar{\lambda})]\,[(1-\epsilon)^j \mathbb{1}_{\{j\leq n\}} + \bar{\lambda}^n B^j] \,.$$

If $B = 1$, choosing $j = n+1$ yields (14.59) with $\rho = \bar{\lambda}$, and if $B > 1$ then set $j = [\alpha n]$ with $\alpha \in (0,1)$ such that $\log(\bar{\lambda}) + \alpha \log(B) < 0$; this choice yields (14.59) with $\rho = (1-\epsilon)^\alpha \vee (\bar{\lambda} B^\alpha) < 1$. □

Example 14.2.52 (Autoregressive Model, Continued). In the model of Example 14.2.3, we have verified that $V(x) = 1 + x^2$ satisfies (14.56) when the noise variance is finite. We can deduce from Theorem 14.2.49 a variety of results: the stationary distribution has finite variance and the iterates $Q^n(x,\cdot)$ of the transition kernel converge to the stationary distribution π geometrically fast in V-total variation distance. Thus there exist constants C and $\rho < 1$ such that for any $x \in \mathsf{X}$, $\|Q^n(x,\cdot) - \pi\|_V \leq C(1+x^2)\rho^n$. This implies in particular that for any $x \in \mathsf{X}$ and any function f such that $\sup_{x\in\mathsf{X}}(1+x^2)^{-1}|f(x)| < \infty$, $\mathrm{E}_x[f(X_n)]$ converges to the limiting value

$$\mathrm{E}_\pi[f(X_n)] = \sqrt{\frac{1-\phi^2}{2\pi\sigma^2}} \int \exp\left[-\frac{(1-\phi^2)x^2}{2\sigma^2}\right] f(x)\,dx$$

geometrically fast. This applies for the mean, $f(x) = x$, and the second moment, $f(x) = x^2$ (though in this case convergence can be derived directly from the autoregression). ■

14.2.6 Limit Theorems

One of the most important problems in probability theory is the investigation of limit theorems for appropriately normalized sums of random variables. The case of independent random variables is fairly well understood, but less is known about dependent random variables such as Markov chains. The purpose of this section is to study several basic limit theorems for additive functionals of Markov chains.

14.2.6.1 Law of Large Numbers

Suppose that $\{X_k\}$ is a Markov chain with transition kernel Q and initial distribution ν. Assume that Q is phi-irreducible and aperiodic and has a stationary distribution π. Let f be a π-integrable function; $\pi(|f|) < \infty$. We say that the sequence $\{f(X_k)\}$ satisfies a law of large numbers (LLN) if for any initial distribution ν on $(\mathsf{X}, \mathcal{X})$, the sample mean $n^{-1}\sum_{k=1}^n f(X_k)$ converges to $\pi(f)$ P_ν-a.s.

For i.i.d. samples, classical theory shows that the LLN holds provided $\pi(|f|) < \infty$. The following theorem shows that the LLN holds for ergodic Markov chains; it does not require any conditions on the rate of convergence to the stationary distribution.

Theorem 14.2.53. *Let Q be a positive Harris recurrent transition kernel with invariant distribution π. Then for any real π-integrable function f on X and any initial distribution ν on $(\mathsf{X}, \mathcal{X})$,*

$$\lim_{n\to\infty} n^{-1} \sum_{k=1}^{n} f(X_k) = \pi(f) \quad \mathrm{P}_\nu\text{-a.s.} \tag{14.63}$$

The LLN can be obtained from general ergodic theorems for stationary processes. An elementary proof can be given when the chain possesses an accessible atom. The basic technique is then the regeneration method, which consists in dividing the chain into blocks between the chain's successive returns to the atom. These blocks are independent (see Lemma 14.2.54 below) and standard limit theorems for i.i.d. random variables yield the desired result. When the chain has no atom, one may still employ this technique by replacing the atom by a suitably chosen small set and using the splitting technique (see for instance Meyn and Tweedie, 1993, Chapter 17).

Lemma 14.2.54. *Let Q be a positive Harris recurrent transition kernel that admits an accessible atom α. Define for any measurable function f,*

$$s_j(f) = \left(\sum_{k=1}^{\tau_\alpha} f(X_k)\right) \circ \theta^{\tau_\alpha^{(j-1)}}, \qquad j \geq 1\,. \tag{14.64}$$

Then for any initial distribution ν on $(\mathsf{X}, \mathcal{X})$, $k \geq 0$ and functions $\{\Psi_j\}$ in $\mathcal{F}_\mathrm{b}(\mathbb{R})$,

$$\mathrm{E}_\nu\left[\prod_{j=1}^{k} \Psi_j(s_j(f))\right] = \mathrm{E}_\nu\left[\Psi_1(s_1(f))\right] \prod_{j=2}^{k} \mathrm{E}_\alpha\left[\Psi_j(s_j(f))\right]\,.$$

Proof. Because the atom α is accessible and the chain is Harris recurrent, $\mathrm{P}_x(\tau_\alpha^{(k)} < \infty) = 1$ for any $x \in \mathsf{X}$. By the strong Markov property, for any integer k,

$$\begin{aligned}
&\mathrm{E}_\nu[\Psi_1(s_1(f)) \cdots \Psi_k(s_k(f))] \\
&\quad = \mathrm{E}_\nu[\Psi_1(s_1(f)) \cdots \Psi_{k-1}(s_{k-1}(f))\, \mathrm{E}_\alpha[\Psi_k(s_k(f)) \mid \mathcal{F}_{\tau_\alpha^{(k-1)}}] \mathbb{1}_{\{\tau_\alpha^{(k-1)} < \infty\}}] \\
&\quad = \mathrm{E}_\nu[\Psi_1(s_1(f)) \cdots \Psi_{k-1}(s_{k-1}(f))]\, \mathrm{E}_\alpha[\Psi_k(s_1(f))]\,.
\end{aligned}$$

The desired result in then obtained by induction. □

Proof (of Theorem 14.2.53 when there is an accessible atom). First assume that f is non-negative. Denote the accessible atom by α and define

$$\eta_n = \sum_{k=1}^{n} \mathbb{1}_\alpha(X_k)\,, \tag{14.65}$$

the occupation time of the atom α up to time n. We now split the sum $\sum_{k=1}^{n} f(X_k)$ into sums over the excursions between successive visits to α,

$$\sum_{k=1}^{n} f(X_k) = \sum_{j=1}^{\eta_n} s_j(f) + \sum_{k=\tau_\alpha^{(\eta_n)}+1}^{n} f(X_k) .$$

This decomposition shows that

$$\sum_{j=1}^{\eta_n} s_j(f) \leq \sum_{k=1}^{n} f(X_k) \leq \sum_{j=1}^{\eta_n+1} s_j(f) . \tag{14.66}$$

Because Q is Harris recurrent and α is accessible, $\eta_n \to \infty$ P_ν-a.s. as $n \to \infty$. Hence $s_1(f)/\eta_n \to 0$ and $(\eta_n - 1)/\eta_n \to 1$ P_ν-a.s. By Lemma 14.2.54 the variables $\{s_j(f)\}_{j\geq 2}$ are i.i.d. under P_ν. In addition $\mathrm{E}_\nu[s_j(f)] = \mu_\alpha(f)$ for $j \geq 2$ with μ_α, defined in (14.30), being an invariant measure. Because all invariant measures are constant multiples of μ_α and $\pi(|f|) < \infty$, $\mathrm{E}_\alpha[s_j(f)]$ is finite. Writing

$$\frac{1}{\eta_n} \sum_{j=1}^{\eta_n} s_j(f) = \frac{s_1(f)}{\eta_n} + \frac{\eta_n - 1}{\eta_n} \frac{1}{\eta_n - 1} \sum_{j=2}^{\eta_n} s_j(f) ,$$

the LLN for i.i.d. random variables shows that

$$\lim_{n\to\infty} \frac{1}{\eta_n} \sum_{j=1}^{\eta_n} s_j(f) = \mu_\alpha(f) \quad \mathrm{P}_\nu\text{-a.s.} ,$$

whence, by (14.66), the same limit holds for $\eta_n^{-1} \sum_1^n f(X_k)$. Because $\pi(1) = 1$, $\mu_\alpha(1)$ is finite too. Applying the above result with $f \equiv 1$ yields $n/\eta_n \to \mu_\alpha(1)$, so that $n^{-1} \sum_1^n f(X_k) \to \mu_\alpha(f)/\mu_\alpha(1) = \pi(f)$ P_ν-a.s. This is the desired result when $f \geq 0$. The general case is is handled by splitting f into its positive and negative parts. □

14.2.6.2 Central Limit Theorems

We say that $\{f(X_k)\}$ satisfies a central limit theorem (CLT) if there is a constant $\sigma^2(f) \geq 0$ such that the normalized sum $n^{-1/2} \sum_{k=1}^{n} \{f(X_k) - \pi(f)\}$ converges P_ν-weakly to a Gaussian distribution with zero mean and variance $\sigma^2(f)$ (we allow for the special case $\sigma^2(f) = 0$ corresponding to weak convergence to the constant 0). CLTs are essential for understanding the error occurring when approximating $\pi(f)$ by the sample mean $n^{-1} \sum_{k=1}^{n} f(X_k)$ and are thus a topic of considerable importance.

For i.i.d. samples, classical theory guarantees a CLT as soon as $\pi(|f|^2) < \infty$. This is not true in general for Markov chains; the CLTs that are available do require some additional assumptions on the rate of convergence and/or the existence of higher order moments of f under the stationary distribution.

Theorem 14.2.55. *Let Q be a phi-irreducible aperiodic positive Harris recurrent transition kernel with invariant distribution π. Let f be a measurable function and assume that there exists an accessible small set C satisfying*

$$\int_{x\in C}\pi(dx)\,\mathrm{E}_x\left[\left(\sum_{k=1}^{\tau_C}|f|(X_k)\right)^2\right]<\infty \quad \textit{and} \quad \int_C \pi(dx)\,\mathrm{E}_x[\tau_C^2]<\infty\,. \tag{14.67}$$

Then $\pi(f^2)<\infty$ and $\{f(X_k)\}$ satisfies a CLT.

Proof. To start with, it follows from the expression (14.32) for the stationary distribution that

$$\pi(f^2)=\int_C \pi(dx)\,\mathrm{E}_x\left[\sum_{k=1}^{\tau_C} f^2(X_k)\right]\le\int_C \pi(dx)\,\mathrm{E}_x\left[\left(\sum_{k=1}^{\tau_C}|f(X_k)|\right)^2\right]<\infty\,.$$

We now prove the CLT under the additional assumption that the chain admits an accessible atom α. The proof in the general phi-irreducible case can be obtained using the splitting construction. The proof is along the same lines as for the LLN. Put $\bar{f}=f-\pi(f)$. By decomposing the sum $\sum_{k=1}^n \bar{f}(X_k)$ into excursions between successive visits to the atom α, we obtain

$$n^{-1/2}\left|\sum_{k=1}^n \bar{f}(X_k)-\sum_{j=2}^{\eta_n} s_j(\bar{f})\right|\le n^{-1/2}s_1(|\bar{f}|)+n^{-1/2}s_{\eta_n+1}(|\bar{f}|)\,, \tag{14.68}$$

where η_n and $s_j(f)$ are defined in (14.65) and (14.64). It is clear that the first term on the right-hand side of this display vanishes (in P_ν-probability) as $n\to\infty$. For the second one, the strong LLN (Theorem 14.2.53) shows that $n^{-1}\sum_1^n s_j^2(|\bar{f}|)$ has an P_ν-a.s. finite limit, whence, P_ν-a.s.,

$$\limsup_{n\to\infty}\frac{s_n^2(|\bar{f}|)}{n}=\limsup_{n\to\infty}\left[\frac{1}{n}\sum_{j=1}^n s_j^2(|\bar{f}|)-\frac{n+1}{n}\frac{1}{n+1}\sum_{j=1}^{n+1}s_j^2(|\bar{f}|)\right]=0\,.$$

The strong LLN with $f=\mathbb{1}_\alpha$ also shows that $\eta_n/n\to\pi(\alpha)$ P_ν-a.s., so that $s_{\eta_n}^2(|\bar{f}|)/\eta_n\to 0$ and $n^{-1/2}s_{\eta_n+1}(|\bar{f}|)\to 0$ P_ν-a.s.

Thus $n^{-1/2}\sum_1^n \bar{f}(X_k)$ and $n^{-1/2}\sum_2^{\eta_n} s_j(\bar{f})$ have the same limiting behavior. By Lemma 14.2.54, the blocks $\{s_j^2(|\bar{f}|)\}_{j\ge 2}$ are i.i.d. under P_ν. Thus, by the CLT for i.i.d. random variables, $n^{-1/2}\sum_2^n s_j(\bar{f})$ converges P_ν-weakly to a Gaussian law with zero mean and some variance $\sigma^2<\infty$; that the variance is indeed finite follows as above with the small set C being the accessible atom α. The so-called Ascombe's theorem (see for instance Gut, 1988, Theorem 3.1) then implies that $\eta_n^{-1/2}\sum_2^{\eta_n}\bar{f}(X_k)$ converges P_ν-weakly to the same Gaussian law. Thus we may conclude that $n^{-1/2}\sum_2^{\eta_n}\bar{f}(X_k)=(\eta_n/n)^{1/2}\eta_n^{-1/2}\sum_2^{\eta_n}\bar{f}(X_k)$ converges P_ν-weakly to a Gaussian law with zero mean and variance $\pi(\alpha)\sigma^2$. By (14.68), so does $n^{-1/2}\sum_1^n \bar{f}(X_k)$. □

The condition (14.67) is stated in terms of the second moment of the excursion between two successive visits to a small set and appears rather difficult to verify directly. More explicit conditions can be obtained, in particular if we assume that the chain is V-geometrically ergodic.

Proposition 14.2.56. *Let Q be a phi-irreducible, aperiodic, positive Harris reccurrent kernel that Q satisfies a Foster-Lyapunov drift condition (see Definition 14.2.46) outside an accessible small set C, with drift function V. Then any measurable function f such that $|f|^2 \leq V$ satisfies a CLT.*

Proof. Minkovski's inequality implies that

$$\mathrm{E}_x \left[\left(\sum_{k=0}^{\tau_C - 1} |f(X_k)| \right)^2 \right] \leq \left\{ \sum_{k=0}^{\infty} \sqrt{\mathrm{E}_x[f^2(X_k) \mathbb{1}_{\{\tau_C > k\}}]} \right\}^{1/2} \leq \left\{ \sum_{k=0}^{\infty} \sqrt{\mathrm{E}_x[V(X_k) \mathbb{1}_{\{\tau_C > k\}}]} \right\}^{1/2} .$$

Put $M_k = \lambda^{-k} V(X_k) \mathbb{1}_{\{\tau_C \geq k\}}$, where λ is as in (14.56). Then for $k \geq 1$,

$$\begin{aligned} \mathrm{E}[M_{k+1} \,|\, \mathcal{F}_k] &\leq \lambda^{-(k+1)} \, \mathrm{E}[V(X_{k+1}) \,|\, \mathcal{F}_k] \mathbb{1}_{\{\tau_C \geq k+1\}} \\ &\leq \lambda^{-k} V(X_k) \mathbb{1}_{\{\tau_C \geq k+1\}} \leq M_k \,, \end{aligned}$$

showing that $\{M_k\}$ is a super-martingale. Thus $\mathrm{E}_x[M_k] \leq \mathrm{E}_x[M_1]$ for any $x \in C$, which implies that for $k \geq 1$,

$$\sup_{x \in C} \mathrm{E}_x[V(X_k) \mathbb{1}_{\{\tau_C \geq k\}}] \leq \lambda^k \left[\sup_{x \in C} V(x) + b \right] .$$

□

14.3 Applications to Hidden Markov Models

As discussed in Section 2.2, an HMM is best defined as a Markov chain $\{X_k, Y_k\}_{k \geq 0}$ on the product space $(\mathsf{X} \times \mathsf{Y}, \mathcal{X} \otimes \mathcal{Y})$. The transition kernel of this joint chain has a simple structure reflecting the conditional independence assumptions that are imposed. Let Q and G denote, respectively, a Markov transition kernel on $(\mathsf{X}, \mathcal{X})$ and a transition kernel from $(\mathsf{X}, \mathcal{X})$ to $(\mathsf{Y}, \mathcal{Y})$. The transition kernel of the joint chain $\{X_k, Y_k\}_{k \geq 0}$ is given by, for any $(x, y) \in \mathsf{X} \times \mathsf{Y}$,

$$T[(x,y), C] = \iint_C Q(x, dx') \, G(x', dy) \,, \qquad (x,y) \in \mathsf{X} \times \mathsf{Y}, \; C \in \mathcal{X} \otimes \mathcal{Y} \,. \tag{14.69}$$

This chain is said to be hidden because only a component (here $\{Y_k\}_{k\geq 0}$) is observed. Of course, the process $\{Y_k\}$ is not a Markov chain, but nevertheless most of the properties of this process are inherited from stability properties of the hidden chain. In this section, we establish stability properties of the kernel T of the joint chain.

14.3.1 Phi-irreducibility

Phi-irreducibility of the joint chain T is inherited from irreducibility of the hidden chain, and the maximal irreducibility measures of the joint and hidden chains are related in a simple way. Before stating the precise result, we recall (see Section 2.1.1) that if ϕ is a measure on $(\mathsf{X}, \mathcal{X})$, we define the measure $\phi \otimes G$ on $(\mathsf{X} \times \mathsf{Y}, \mathcal{X} \otimes \mathcal{Y})$ by

$$\phi \otimes G(A) \stackrel{\text{def}}{=} \iint_A \mu(dx)\, G(x, dy) \;, \qquad A \in \mathcal{X} \otimes \mathcal{Y} \;.$$

Proposition 14.3.1. *Assume that Q is phi-irreducible, and let ϕ be an irreducibility measure for Q. Then $\phi \otimes G$ is an irreducibility measure for T. If ψ is a maximal irreducibility measure for Q, then $\psi \otimes G$ is a maximal irreducibility measure for T.*

Proof. Let $A \in \mathcal{X} \otimes \mathcal{Y}$ be a set such that $\phi \otimes G(A) > 0$. Denote by Ψ_A the function $\Psi_A(x) = \int_{\mathsf{Y}} G(x, dy)\, \mathbb{1}_A(x, y)$ for $x \in \mathsf{X}$. By Fubini's theorem,

$$\phi \otimes G(A) = \iint \phi(dx)\, G(x, dy)\, \mathbb{1}_A(x, y) = \int \phi(dx)\, \Psi_A(x) \;,$$

and the condition $\phi \otimes G(A) > 0$ implies that $\phi(\{\Psi_A > 0\}) > 0$. Because $\{\Psi_A > 0\} = \bigcup_{m=0}^{\infty} \{\Psi_A \geq 1/m\}$, we have $\phi(\{\Psi_A \geq 1/m\}) > 0$ for some integer m. Because ϕ is an irreducibility measure, for any $x \in \mathsf{X}$ there exists an integer $k \geq 0$ such that $Q^k(x, \{\Psi_A \geq 1/m\}) > 0$. Therefore for any $y \in \mathsf{Y}$,

$$\begin{aligned} T^k[(x, y), A] &= \iint Q^k(x, dx')\, G(x', dy')\, \mathbb{1}_A(x', y') = \int Q^k(x, dx')\, \Psi_A(x') \\ &\geq \int_{\{\Psi_A \geq 1/m\}} Q^k(x, dx')\, \Psi_A(x') \geq \frac{1}{m} Q^k(x, \{\Psi_A \geq 1/m\}) > 0 \;, \end{aligned}$$

showing that $\phi \otimes G$ is an irreducibility measure for T.

Morever, using Theorem 14.2.2, we see that a maximal irreducibility measure ψ_T for T is given by, for any $\delta \in (0, 1)$ and $A \in \mathcal{X} \otimes \mathcal{Y}$,

$$\begin{aligned} \psi_T(A) &= \iint \phi(dx)\, G(x, dy)\, (1 - \delta) \sum_{m=0}^{\infty} \delta^m T^m[(x, y), A] \\ &= \iint (1 - \delta) \sum_{m=0}^{\infty} \delta^m \int \phi(dx)\, Q^m(x, dx')\, G(x', dy')\, \mathbb{1}_A(x', y') \\ &= \iint \psi(dx')\, G(x', dy')\, \mathbb{1}_A(x', y') = \psi \otimes G(A) \;, \end{aligned}$$

where

$$\psi(B) = \int \phi(dx)\,(1-\delta) \sum_{m=0}^{\infty} \delta^m Q^m(x,B)\;, \qquad B \in \mathcal{X}\;.$$

By Theorem 14.2.2, ψ is a maximal irreducibility measure for Q. In addition, if $\hat{\psi}$ is a maximal irreducibility measure for Q, then $\hat{\psi}$ is equivalent to ψ. Because for any $A \in \mathcal{X} \otimes \mathcal{Y}$,

$$\hat{\psi} \otimes G(A) = \iint \hat{\psi}(dx)\, G(x,dy)\, \mathbb{1}_A(x,y) = \iint \psi \otimes G(dx,dy)\, \frac{d\hat{\psi}}{d\psi}(x) \mathbb{1}_A(x,y)\;,$$

$\hat{\psi} \otimes G(A) = 0$ whenever $\psi \otimes G(A) = 0$. Thus $\hat{\psi} \otimes G \ll \psi \otimes G$. Exchanging ψ and $\hat{\psi}$ shows that $\psi \otimes G$ and $\hat{\psi} \otimes G$ are indeed equivalent, which concludes the proof. □

Example 14.3.2 (Normal HMM). Consider a normal HMM (see Section 1.3.2). In this case, the state space X of the hidden chain is finite, $\mathsf{X} = \{1, 2, \ldots, r\}$ and $\mathsf{Y} = \mathbb{R}$. The hidden chain is governed by a transition matrix $Q = [Q(x,y)]_{1 \le x,y \le r}$. Conditionally on the state $x \in \mathsf{X}$, the distribution of the observation is Gaussian with mean μ_x and variance σ_x^2. Hence the transition kernel T for the joint Markov chain is given by, for any $(x,y) \in \mathsf{X} \times \mathsf{Y}$ and $A \in \mathcal{B}(\mathbb{R})$,

$$T[(x,y), \{x'\} \times A] = Q(x,x') \int_A \frac{1}{\sqrt{2\pi\sigma_{x'}^2}} \exp\left[-\frac{1}{2} \frac{(y' - \mu_{x'})^2}{\sigma_{x'}^2}\right] dy'\;.$$

If the transition matrix Q is irreducible (all states in X communicate), then Q is positive. For any $x \in \mathsf{X}$, δ_x is an irreducibility measure for Q and T is phi-irreducible with irreducibility measure $\delta_x \otimes G$. Denote by π the unique invariant probability measure for Q. Then π is also a maximal irreducibility measure, whence $\pi \otimes G$ is a maximal irreducibility measure for T. ■

Example 14.3.3 (Stochastic Volatility Model). The canonical stochastic volatility model (see Example 1.3.13) is given by

$$\begin{aligned} X_{k+1} &= \phi X_k + \sigma U_k\;, \qquad && U_k \sim \mathrm{N}(0,1)\;, \\ Y_k &= \beta \exp(X_k/2) V_k\;, && V_k \sim \mathrm{N}(0,1)\;, \end{aligned}$$

We have established (see Example 14.2.3) that because $\{U_k\}$ has a positive density on $\mathbb{R}^+$, the chain $\{X_k\}$ is phi-irreducible and λ^{Leb} is an irreducibility measure. Therefore $\{X_k, Y_k\}$ is also phi-irreducible and $\lambda^{\text{Leb}} \otimes \lambda^{\text{Leb}}$ is a maximal irreducibility measure. ■

14.3.2 Atoms and Small Sets

It is possible to relate atoms and small sets of the joint chain to those of the hidden chain. Examples of HMMs possessing accessible atoms are numerous, even when the state space of the joint chain is general. They include in particular the Markov chains whose hidden state space X is finite.

Example 14.3.4 (Normal HMM, Continued). For the normal HMM (see Example 14.3.2), it holds that $T[(x, y), \cdot] = T[(x, y'), \cdot]$ for any $(y, y') \in \mathbb{R} \times \mathbb{R}$. Hence $\{x\} \times \mathbb{R}$ is an atom for T. ■

When accessible atoms do not exist, it is important to determine small sets. Here again the small sets of the joint chain can easily be related to those of the hidden chain.

Lemma 14.3.5. *Let m be a positive integer, $\epsilon > 0$ and let η be a probability measure on $(\mathsf{X}, \mathcal{X})$. Let $C \in \mathcal{X}$ be an (m, ϵ, η)-small set for the transition kernel Q, that is, $Q^m(x, A) \geq \epsilon \mathbb{1}_C(x)\eta(A)$ for all $x \in \mathsf{X}$ and $A \in \mathcal{X}$. Then $C \times \mathsf{Y}$ is an $(m, \epsilon, \eta \otimes G)$-small set for the transition kernel T defined in* (2.14), *that is,*

$$T^m[(x, y), A] \geq \epsilon \mathbb{1}_C(x)\, \eta \otimes G(A) , \qquad (x, y) \in \mathsf{X} \times \mathsf{Y},\ A \in \mathcal{X} \otimes \mathcal{Y} .$$

Proof. Pick $(x, y) \in C \times \mathsf{Y}$. Then

$$\begin{aligned} T^m[(x, y), A] &= \iint Q^m(x, dx')\, G(x', dy')\, \mathbb{1}_A(x', y') \\ &\geq \epsilon \iint \eta(dx')\, G(x', dy')\, \mathbb{1}_A(x', y') . \end{aligned}$$

□

If the Markov transition kernel Q on $(\mathsf{X}, \mathcal{X})$ is phi-irreducible (with maximal irreducibility measure ψ), then we know from Proposition 14.2.12 that there exists an accessible small set C. That is, there exists a set $C \in \mathcal{X}$ with $\mathrm{P}_x(\tau_C < \infty) > 0$ for all $x \in \mathsf{X}$ and such that C is (m, ϵ, η)-small for some triple (m, ϵ, η) with $\eta(C) > 0$. Then Lemma 14.3.5 shows that $C \times \mathsf{Y}$ is an $(m, \epsilon, \eta \otimes G)$-small set for the transition kernel T.

Example 14.3.6 (Stochastic Volatility Model, Continued). We have shown in Example 14.2.3 that any compact set $K \subset \mathbb{R}$ is small for the first-order autoregression constituting the hidden chain of the stochastic volatility model of Example 14.3.3. Therefore any set $K \times \mathbb{R}$, where K a compact subset of $\mathbb{R}$, is small for the joint chain $\{X_k, Y_k\}$. ■

The simple relations between the small sets of the joint chain and those of the hidden chain immediately imply that T and Q have the same period.

Proposition 14.3.7. *Suppose that Q is phi-irreducible and has period d. Then T is phi-irreducible and has the same period d. In particular, if Q is aperiodic, then so is T.*

Proof. Let C be an accessible (m, ϵ, η)-small set for Q with $\eta(C) > 0$. Define E_C as the set of time indices for which C is a small set with minorizing probability measure η,

$$E_C \stackrel{\text{def}}{=} \{n \geq 0 : C \text{ is } (n, \epsilon, \eta)\text{-small for some } \epsilon > 0\} .$$

The period of the set C is given by the greatest common divisor of E_C. Proposition 14.2.35 shows that this value is in fact common to the chain as such and does not depend on the particular small set chosen. Lemma 14.3.5 shows that $C \times \mathsf{Y}$ is an $(m, \epsilon, \eta \otimes G)$-small set for the joint Markov chain with transition kernel T, and that $\eta \otimes G(C \times \mathsf{Y}) = \eta(C) > 0$. The set $E_{C \times \mathsf{Y}}$ of time indices for which $C \times \mathsf{Y}$ is a small set for T with minorizing measure $\eta \otimes G$ is thus, using Lemma 14.3.5 again, equal to E_C. Thus the period of the set C is also the period of the set $C \times \mathsf{Y}$. Because the period of T does not depend on the choice of the small set $C \times \mathsf{Y}$, it follows that the periods of Q and T coincide. □

14.3.3 Recurrence and Positive Recurrence

As the following result shows, recurrence and transience of the joint chain follows directly from the corresponding properties of the hidden chain.

Proposition 14.3.8. *Assume that the hidden chain is phi-irreducible. Then the following statements hold true.*

(i) The joint chain is transient (recurrent) if and only if the hidden chain is transient (recurrent).
(ii) The joint chain is positive if and only if the hidden chain is positive. In addition, if the hidden chain is positive with stationary distribution π, then $\pi \otimes G$ is the stationary distribution of the joint chain.

Proof. First assume that the transition kernel Q is transient, that is, that there is a countable cover $\mathsf{X} = \cup_i A_i$ of X with uniformly transient sets,

$$\sup_{x \in A_i} \mathrm{E}_x \left[\sum_{n=1}^{\infty} \mathbb{1}_{A_i}(X_n) \right] < \infty .$$

Then the sets $\{A_i \times \mathsf{Y}\}_{i \geq 1}$ form a countable cover of $\mathsf{X} \times \mathsf{Y}$, and these sets are uniformly transient because

$$\mathrm{E}_x \left[\sum_{n=1}^{\infty} \mathbb{1}_{A_i \times \mathsf{Y}}(X_n, Y_n) \right] = \mathrm{E}_x \left[\sum_{n=1}^{\infty} \mathbb{1}_{A_i}(X_n) \right] . \tag{14.70}$$

Thus the joint chain is transient.

Conversely, assume that the joint chain is transient. Because the hidden chain is phi-irreducible, Proposition 14.2.13 shows that there is a countable cover $\mathsf{X} = \cup_i A_i$ of X with sets that are small for Q. At least one of these, say A_1, is accessible for Q. By Lemma 14.3.5, the sets $A_i \times \mathsf{Y}$ are small. By Proposition 14.3.1, $A_1 \times \mathsf{Y}$ is accessible and, because T is transient, Proposition 14.2.14 shows that $A_1 \times \mathsf{Y}$ is uniformly transient. Equation (14.70) then shows that A_1 is uniformly transient, and because A_1 is accessible, we conclude that Q is transient.

Thus the hidden chain is transient if and only if the joint chain is so. The transience/recurrence dichotomy (Theorem 14.2.6) then implies that the hidden chain is recurrent if and only if the joint chain is so, which completes the proof of (i).

We now turn to (ii). First assume that the hidden chain is positive recurrent, that is, that there exists a unique stationary probability measure π satisfying $\pi Q = \pi$. Then the probability measure $\pi \otimes G$ is stationary for the transition kernel T of the joint chain, because

$$\begin{aligned}(\pi \otimes G)T(A) &= \int \cdots \int \pi(dx)\, G(x, dy)\, Q(x, dx')\, G(x', dy')\, \mathbb{1}_A(x', y') \\ &= \iiint \pi(dx)\, Q(x, dx')\, G(x', dy')\, \mathbb{1}_A(x', y') \\ &= \iint \pi(dx')\, G(x', dy')\, \mathbb{1}_A(x', y') = \pi \otimes G(A) .\end{aligned}$$

Because the joint chain admits a stationary distribution it is positive, and by Proposition 14.2.34 it is recurrent.

Conversely, assume that the joint chain is positive. Denote by $\bar{\pi}$ the (unique) stationary probability measure of T. Thus for any $\bar{A} \in \mathcal{X} \otimes \mathcal{Y}$, we have

$$\begin{aligned}&\iint \bar{\pi}(dx, dy)\, Q(x, dx')\, G(x', dy')\, \mathbb{1}_{\bar{A}}(x', y') \\ &\qquad = \iint \bar{\pi}(dx, \mathsf{Y})\, Q(x, dx')\, G(x', dy')\, \mathbb{1}_{\bar{A}}(x', y') = \bar{\pi}(\bar{A}) .\end{aligned}$$

Setting $\bar{A} = A \times \mathsf{Y}$ for $A \in \mathcal{X}$, this display implies that

$$\int \bar{\pi}(dx, \mathsf{Y})\, Q(x, A) = \bar{\pi}(A \times \mathsf{Y}) .$$

This shows that $\pi(A) = \bar{\pi}(A \times \mathsf{Y})$ is a stationary distribution for the hidden chain. Hence the hidden chain is positive and recurrent. □

When the joint (or hidden) chain is positive, it is natural to study the rate at which it converges to stationarity.

Proposition 14.3.9. *Assume that the hidden chain satisfies a uniform Doeblin condition, that is, there exists a positive integer m, $\epsilon > 0$ and a family $\{\eta_{x,x'}, (x, x') \in \mathsf{X} \times \mathsf{X}\}$ of probability measures such that*

$$Q^m(x, A) \wedge Q^m(x', A) \geq \epsilon \eta_{x,x'}(A), \quad A \in \mathcal{X},\ (x, x') \in \mathsf{X} \times \mathsf{X} .$$

Then the joint chain also satisfies a uniform Doeblin condition. Indeed, for all (x, y) and (x', y') in $\mathsf{X} \times \mathsf{Y}$ and all $\bar{A} \in \mathcal{X} \otimes \mathcal{Y}$,

$$T^m[(x,y),\bar{A}] \wedge T^m[(x',y'),\bar{A}] \geq \epsilon\bar{\eta}_{x,x'}(\bar{A}) ,$$

where

$$\bar{\eta}_{x,x'}(\bar{A}) = \int \eta_{x,x'}(dx)\, G(x,dy)\, \mathbb{1}_{\bar{A}}(x,y) .$$

The proof is along the same lines as the proof of Lemma 14.3.5 and is omitted. This proposition in particular implies that the ergodicity coefficients for the kernels T^m and Q^m coincide; $\delta(T^m) = \delta(Q^m)$. A straightforward but useful application of this result is when the hidden Markov chain is defined on a finite state space. If the transition matrix Q of this chain is primitive, that is, there exists a positive integer m such that $Q^m(x,x') > 0$ for all $(x,x') \in \mathsf{X} \times \mathsf{X}$ (or, equivalently, if the chain Q is irreducible and aperiodic), then the joint Markov chain satisfies a uniform Doeblin condition and the ergodicity coefficient of the joint chain is bounded as $\delta(T^m) \leq 1 - \epsilon$ with

$$\epsilon = \inf_{(x,x') \in \mathsf{X} \times \mathsf{X}} \sup_{x'' \in \mathsf{X}} [Q^m(x,x'') \wedge Q^m(x',x'')] .$$

A similar result holds when the hidden chain satisfies a Foster-Lyapunov drift condition instead of a uniform Doeblin condition. This result is of particular interest when dealing with hidden Markov models on state spaces that are not finite or bounded.

Proposition 14.3.10. *Assume that Q is phi-irreducible, aperiodic, and satisfies a Foster-Lyapunov drift condition (Definition 14.2.46) with drift function V outside a set C. Then the transition kernel T also satisfies a Foster-Lyapunov drift condition with drift function V outside the set $C \times \mathsf{Y}$,*

$$T[(x,y),V] \leq \lambda V(x) + b\mathbb{1}_{C \times \mathsf{Y}}(x,y) .$$

Here on the left-hand side, we wrote V also for a function on $\mathsf{X} \times \mathsf{Y}$ defined by $V(x,y) = V(x)$.

The proof is straightforward. Proposition 14.2.50 yields an explicit bound on the rate of convergence of the iterates of the Markov chain to the stationary distribution. This result has a lot of interesting consequences.

Proposition 14.3.11. *Suppose that Q is phi-irreducible, aperiodic, and satisfies a Foster-Lyapunov drift condition with drift function V outside a small set C. Then the transition kernel T is positive and aperiodic with invariant distribution $\pi \otimes G$, where π is the invariant distribution of Q. In addition, for any measurable function $f : \mathsf{X} \times \mathsf{Y} \to \mathbb{R}$, the following statements hold true.*

(i) If $\sup_{x \in \mathsf{X}}[V(x)]^{-1} \int G(x,dy)\,|f(x,y)| < \infty$, then there exist $\rho \in (0,1)$ and $K < \infty$ (not depending on f) such that for any $n \geq 0$ and $(x,y) \in \mathsf{X} \times \mathsf{Y}$,

$$|T^n f(x,y) - \pi \otimes G(f)| \leq K\rho^n V(x) \sup_{x' \in \mathsf{X}} [V(x')]^{-1}\; intG(x',dy)\,|f(x',y)| .$$

(ii) If $\sup_{x \in \mathsf{X}} [V(x)]^{-1} \int G(x, dy)\, f^2(x, y) < \infty$*, then* $\mathrm{E}_{\pi \otimes G}[f^2(X_0, Y_0)] < \infty$ *and there exist* $\rho \in (0, 1)$ *and* $K < \infty$ *(not depending on* f*) such that for any* $n \geq 0$,

$$|\mathrm{Cov}_\pi[f(X_n, Y_n), f(X_0, Y_0)]| \\ \leq K\rho^n \pi(V) \left\{ \sup_{x \in \mathsf{X}} [V(x)]^{-1/2} \int G(x, dy)\, |f(x, y)| \right\}^2 .$$

Proof. First note that

$$|T^n f(x, y) - \pi \otimes G(f)| = \left| \iint [Q^n(x, dx') - \pi(dx')] G(x', dy')\, f(x', y') \right| \\ \leq \|Q^n(x, \cdot) - \pi\|_V \sup_{x' \in \mathsf{X}} [V(x')]^{-1} \int G(x', dy)\, |f(x', y)| .$$

Now part (i) follows from the geometric ergodicity of Q (Theorem 14.2.49). Next, because $\pi(V) < \infty$,

$$\mathrm{E}_{\pi \otimes G}[f^2(X_0, Y_0)] = \iint \pi(dx)\, G(x, dy)\, f^2(x, y) \\ \leq \pi(V) \sup_{x \in \mathsf{X}} [V(x)]^{-1} \int G(x, dy)\, f^2(x, y) < \infty ,$$

implying that $|\mathrm{Cov}_\pi[|f(X_n, Y_n)|, |f(X_0, Y_0)|]| \leq \mathrm{Var}_\pi[f(X_0, Y_0)] < \infty$. In addition

$$\begin{aligned} &\mathrm{Cov}_\pi[f(X_n, Y_n), f(X_0, Y_0)] \\ &\quad = \mathrm{E}_\pi\{\mathrm{E}[f(X_n, Y_n) - \pi \otimes G(f) \mid \mathcal{F}_0] f(X_0, Y_0)\} \\ &\quad = \iint \pi \otimes G(dx, dy)\, f(x, y) \iint [Q^n(x, dx') - \pi(dx')]\, G(x', dy')\, f(x', y') . \end{aligned} \tag{14.71}$$

By Jensen's inequality $\int G(x, dy)\, |f(x, y)| \leq [\int G(x, dy)\, f^2(x, y)]^{1/2}$ and

$$QV^{1/2}(x) \leq [QV(x)]^{1/2} \leq [\lambda V(x) + b\mathbb{1}_C(x)]^{1/2} \leq \lambda^{1/2} V^{1/2}(x) + b^{1/2} \mathbb{1}_C(x) ,$$

showing that Q also satisfies a Foster-Lyapunov condition outside C with drift function $V^{1/2}$. By Theorem 14.2.49, there exists $\rho \in (0, 1)$ and a constant K such that

$$\left| \iint [Q^n(x, dx') - \pi(dx)]\, G(x', dy')\, f(x', y') \right| \\ \leq \|Q^n(x, \cdot) - \pi\|_{V^{1/2}} \sup_{x' \in \mathsf{X}} V^{-1/2}(x) \int G(x', dy)\, |f(x', y)| \\ \leq K\rho^n V^{1/2}(x) \sup_{x' \in \mathsf{X}} V^{-1/2}(x') \int G(x', dy)\, |f(x', y)| .$$

Part (ii) follows by plugging this bound into (14.71). □

Example 14.3.12 (Stochastic Volatility Model, Continued). In the model of Example 14.3.3, we set $V(x) = e^{x^2/2\delta^2}$ for $\delta > \sigma_U$. It is easily shown that

$$QV(x) = \frac{\rho}{\sigma_U} \exp\left[\frac{x^2}{2\delta^2} \frac{\phi^2(\rho^2 + \delta^2)}{\delta^2}\right] ,$$

where $\rho^2 = \sigma_U^2\delta^2/(\delta^2 - \sigma_U^2)$. We may choose δ large enough that $\phi^2(\rho^2 + \delta^2)/\delta^2 < 1$. Then $\limsup_{|x|\to\infty} QV(x)/V(x) = 0$ so that Q satisfies a Foster-Lyapunov condition with drift function $V(x) = e^{x^2/2\delta^2}$ outside a compact set $[-M, +M]$. Because every compact set is small, the assumptions of Proposition 14.3.11 are satisfied, showing that the joint chain is positive. Set $f(x, y) = |y|$. Then $\int G(x, dy)\,|y| = \beta e^{x/2}\sqrt{2/\pi}$. Proposition 14.3.11(ii) shows that $\mathrm{Var}_\pi(Y_0) < \infty$ and that the autocovariance function $\mathrm{Cov}(|Y_n|, |Y_0|)$ decreases to zero exponentially fast. ■

15

An Information-Theoretic Perspective on Order Estimation

Statistical inference in hidden Markov models with finite state space X has to face a serious problem: order identification. The order of an HMM $\{Y_k\}_{k\geq 1}$ over Y (in this chapter, we let indices start at 1) is the minimum size of the hidden state space X of an HMM over (X, Y) that can generate $\{Y_k\}_{k\geq 1}$. In many real-life applications of HMM modeling, no hints about this order are available. As order misspecification is an impediment to parameter estimation, consistent order identification is a prerequisite to HMM parameter estimation.

Furthermore, HMM order identification is a distinguished representative of a family of related problems that includes Markov order identification. In all those problems, a nested family of models is given, and the goal is to identify the smallest model that contains the distribution that has generated the data. Those problems differ in an essential way according to whether identifiability does or does not depend on correct order specification.

Order identification problems are related to composite hypothesis testing. As the performance of generalized likelihood ratio testing in this framework is still a matter of debate, order identification problems constitute benchmarks for which the performance of generalized likelihood ratio testing can be investigated (see Zeitouni *et al.*, 1992). As a matter of fact, analyzing order identification issues boils down to understanding the simultaneous behavior of (possibly infinitely) many maximum likelihood estimators. When identifiability depends on correct order specification, universal coding arguments have proved to provide very valuable insights into the behavior of likelihood ratios. This is the main reason why source coding concepts and techniques have become a standard tool in the area.

This chapter presents four kinds of results: first, in a Bayesian setting, a general consistency result provides hints about the ideal penalties that could be used in penalized maximum likelihood order estimation. Then universal coding arguments are shown to provide a general construction of strongly consistent order estimators. Afterwards, a general framework for analyzing the Bahadur efficiency of order estimation procedures is presented, following the lines of Gassiat and Boucheron (2003). Consistency and efficiency results hold for HMMs. As explained below, refining those consistency and efficiency results requires a precise understanding of the behavior of likelihood ratios. As of writing this text, in the HMM setting, this precise picture is beyond our understanding. But such a work has been carried recently out for Markov order estimation. In order to give a flavor of what remains to be done concerning HMMs, this chapter reports in detail the recent *tour de force* by Csiszár and Shields (2000) who show that the Bayesian information criterion provides a strongly consistent Markov order estimator.

15.1 Model Order Identification: What Is It About?

In preceding chapters, we have been concerned with inference problems in HMMs for which the hidden state space is known in advance: it might be either finite with known cardinality or compact under restrictive conditions; see the assumptions on the transition kernel of the hidden chain to ensure consistency of the MLE in Chapter 12. In this chapter, we focus on HMMs with finite state space of unknown cardinality. Moreover, the set Y in which the observations $\{Y_k\}_{k\geq 1}$ take values is assumed to be *finite* and fixed. Let $\mathcal{M}^r$ denote the set of distributions of Y-valued processes $\{Y_k\}_{k\geq 1}$ that can be generated by an HMM with hidden state space X of cardinality r.

The parameter space associated with $\mathcal{M}^r$ is Θ^r. Note that even if all finite-dimensional distributions of $\{Y_k\}_{k\geq 1}$ are known, deciding whether the distribution of $\{Y_k\}_{k\geq 1}$ belongs to $\mathcal{M}^r$ or even to $\cup_r \mathcal{M}^r$ is not trivial (Finesso, 1991, Chapter 1). Elementary arguments show that $\mathcal{M}^r \subseteq \mathcal{M}^{r+1}$; further reflection verifies that this inclusion is strict. Hence for a fixed observation set Y, the sequence $(\mathcal{M}^r)_{r\geq 1}$ defines a nested sequence of models. We may now define the main topic of this chapter: the order of an HMM.

Definition 15.1.1. *The order of an HMM $\{Y_k\}_{k\geq 1}$ over Y is the smallest integer r such that the distribution of $\{Y_k\}_{k\geq 1}$ belongs to $\mathcal{M}^r$.*

Henceforth, when dealing with an HMM $\{Y_k\}_{k\geq 1}$, its order will be denoted by $r_\star$, and $\theta_\star$ will denote a parameterization of this distribution in $\Theta^{r_\star}$. The distribution of the process will be denoted by $\mathrm{P}_\star$.

Assume for a moment that we are given an infinite sequence of observations of an HMM $\{Y_k\}_{k\geq 1}$: $y_1, \ldots, y_k, \ldots$, that we are told that the order of $\{Y_k\}_{k\geq 1}$ is at most some r_0, and that we are asked to estimate a parameterization of the distribution of $\{Y_k\}_{k\geq 1}$. It might seem that the MLE in Θ^{r_0} would perform well in such a situation. Unfortunately, if the order of $\{Y_k\}_{k\geq 1}$ is strictly smaller than r_0, maximum likelihood estimation will run into trouble. As a matter of fact, if $r_\star < r_0$, then $\theta_\star$ is not identifiable in Θ^{r_0}. Hence, when confronted with such an estimation problem, it is highly reasonable to first estimate $r_\star$ and then to proceed to maximum likelihood estimation of $\theta_\star$.

The *order estimation* question is then the following: given an outcome $y_{1:n}$ of the process $\{Y_k\}_{k\geq 1}$ with distribution in $\cup_r \mathcal{M}^r$, can we identify $r_\star$?

Definition 15.1.2. *An order estimation procedure is a sequence of estimators* $\widehat{r}_1, \ldots, \widehat{r}_n, \ldots$ *that, given input sequences of length* $1, \ldots n, \ldots$*, outputs estimates* $\widehat{r}_n(y_{1:n})$ *of* $r_\star$*.*

A sequence of estimators is strongly consistent if the sequence $\widehat{r}_1, \ldots \widehat{r}_n, \ldots$ converges to $r_\star$ $\mathrm{P}_\star$-a.s.

15.2 Order Estimation in Perspective

The ambition of this chapter is not only to provide a state-of-the-art exposition of order estimation in HMMs but also to provide a perspective. There are actually many other order estimation problems in the statistical or the information-theoretical literature. All pertain to the estimation of the dimension of a model. We may quote for example the following.

- Estimating the order of a Markov process. In that case, the order should be understood as the Markov order of the process (Finesso, 1991; Finesso *et al.*, 1996; Csiszár and Shields, 2000; Csiszár, 2002). See Section 15.8 for precise definitions and recent advances on this topic.
- Estimating the order of semi-Markov models, which have proved to be valuable tools in telecommunication engineering.
- Estimating the order in stochastic context-free grammars, which are currently considered in genomics (Durbin *et al.*, 1998).
- Estimating the number of populations in a mixture (Dacunha-Castelle and Gassiat, 1997a,b, 1999; Gassiat, 2002).
- Estimating the number of change points in detection problems.
- Estimating the order of ARMA models (Azencott and Dacunha-Castelle, 1984; Dacunha-Castelle and Gassiat, 1999; Boucheron and Gassiat, 2004).

Hence, HMM order estimation is both interesting *per se* and as a paradigm of a rich family of statistical problems for which the general setting is the following. Let $\{\mathcal{M}^r\}_{r \geq 1}$ be a nested sequence of models (sets of probability distributions) for sequences $\{Y_k\}_{k \geq 1}$ on a set Y. For any P in $\cup_r \mathcal{M}^r$, the order is the smallest integer r such that P belongs to $\mathcal{M}^r$. Our two technical questions will be the following.

(i) Does there exist (strongly) consistent order estimators? Is it possible to design generic order estimation procedures?
(ii) How efficient are the putative consistent order estimators?

The analysis of order estimation problems is currently influenced by the theory of *universal coding* from information theory and by the theory of *composite hypothesis testing* from plain old statistics. The first perspective provides a convenient framework for designing consistent order estimators, and the second provides guidelines in the analysis of the performance of order estimators. As a matter of fact, code-based order estimators turn out to be analyzed as penalized maximum likelihood estimators.

Definition 15.2.1. *Let* $\{\text{pen}(n,r)\}_{n,r}$ *denote a family of non-negative numbers. A* penalized maximum likelihood *(PML) order estimator is defined by*

$$\widehat{r}_n \stackrel{\text{def}}{=} \arg\max_r \left[\sup_{\text{P}\in\mathcal{M}^r} \log \text{P}(y_{1:n}) - \text{pen}(n,r) \right] .$$

The main point now becomes the choice of the penalty $\text{pen}(n,r)$. To ensure consistency and/or efficiency,

$$\sup_{\text{P}\in\mathcal{M}^r} \log \text{P}(y_{1:n}) - \sup_{\text{P}\in\mathcal{M}^{r_\star}} \log \text{P}(y_{1:n}) \tag{15.1}$$

has to be compared with

$$\text{pen}(n,r) - \text{pen}(n,r_\star) .$$

In case $r < r_\star$, this is related to Shannon-McMillan-Breiman theorems (see Section 15.4.2), and if the penalty grows slower than n, PML order estimators do not underestimate the order (see Lemma 15.6.2). Moreover the probability of underestimating the order decreases exponentially with rate proportional to n, and the better the constant is, the more efficient is the estimation. Asymptotic behavior of this error thus comes from a large deviations analysis of the likelihood process (see Theorem 15.7.2 and 15.7.7).

The analysis of the overestimation error follows different considerations. A first simple remark is that it depends on whether the parameter describing the distribution of the observations *is or is not* identifiable as an element of a model of larger order. When the parameter is still identifiable in larger models, stochastic behavior of the maximum likelihood statistic is well understood and can be cast into the old framework created by Wilks, Wald, and Chernoff. In this case, weak consistency of PML order estimators is achieved as soon as the penalties go to infinity with n and the set of possible orders is bounded. When the parameter is no longer identifiable in larger models, stochastic description of the maximum likelihood statistic has to be investigated on an ad hoc basis. Indeed, for general HMMs, the likelihood ratio statistic is stochastically unbounded even for bounded parameters (see Kéribin and Gassiat, 2000), and we are not even aware of a candidate for penalties warranting weak consistency of PML order estimators. Note that one can however use marginal likelihoods to build weakly consistent order estimators (see Gassiat, 2002).

From now on, we will mainly focus on finite sets Y. In this case, ideas and results from information theory may be used to build consistent order estimators, without assuming any *a priori* upper bound on the order (see Lemma 15.6.3). Though the likelihood ratio (15.1) may be unbounded for $r > r_\star$, its rate of growth is smaller than n. The asymptotic characterization of the decay of the overestimation error should thus resort to a moderate deviations analysis of the likelihood process.

Consistency and efficiency theorems are stated in Sections 15.6 and 15.7. Although they apply to HMMs, in order to outline the key ingredients of the proofs, those theorems are stated and derived in a general setting, Though the results might seem satisfactory, they fall short of closing the story. Indeed, for example, lower bounds on penalties warranting strongly consistent order identification for HMMs has only received very partial (and far too conservative) answers . In practice, the question is important when underestimation has to be avoided at (almost) any price. The theoretical counterpart is also fascinating, as it is connected to non-asymptotic evaluation of stochastic deviations of likelihoods (in the range of large and moderate deviations). This is why we shall also consider in more detail the problem of Markov order estimation. A process $\{Y_k\}_{k\geq 1}$ with distribution $\mathrm{P}_\star$ is said to be Markov of order r if for every $y_{1:n+1} \in \mathsf{Y}^{n+1}$,

$$\mathrm{P}_\star(y_{n+1} \mid y_{1:n}) = \mathrm{P}_\star(y_{n+1} \mid y_{n-r+1:n}) \; .$$

For Markov models, whatever the value of r, the maximum likelihood estimator is uniquely defined and it can be computed easily from a (r-dependent) finite-dimensional sufficient statistic. Martingale tools may be used to obtain non-asymptotic tail inequalities for maximum likelihoods. Section 15.8 reports a recent *tour de force* by Csiszár and Shields (2000), who show that the Bayesian information criterion provides a strongly consistent Markov order estimator. Of course, though this order estimation problem is apparently very similar to the HMM order estimation problem, this similarity should be taken cautiously. Indeed, maximum likelihood estimators in an HMM may not be computed directly using finite-dimensional statistics. However, we believe that our current understanding of Markov order estimation will provide insights into the HMM order estimation problem. Moreover, designing the right non-asymptotic deviation inequalities has become a standard approach in the analysis of model selection procedures (see Barron *et al.*, 1999). This work still has to be done for HMMs.

We will start the technical exposition by describing the relationship between order estimation and hypothesis testing.

15.3 Order Estimation and Composite Hypothesis Testing

If we have a consistent order estimation procedure, we should be able to manufacture a sequence of consistent tests for the following questions: is the true order larger than $1, \ldots, r, \ldots$? We may indeed phrase the following composite hypothesis testing problem:

H0: The source belongs to $\mathcal{M}^{r_0}$;
H1: The source belongs to $(\cup_r \mathcal{M}^r) \setminus \mathcal{M}^{r_0}$.

To put things in perspective, in this paragraph we will focus on testing whether some probability distribution P belongs to some subset $\mathcal{M}^0$ (H0) of some set $\mathcal{M}$ of distributions over Y^∞. Hypothesis H1 corresponds to $\mathrm{P} \in \mathcal{M}^1 = \mathcal{M} \setminus \mathcal{M}^0$.

A test on samples of length n is a function T_n that maps Y^n on $\{0, 1\}$. If $T_n(y_{1:n}) = 1$, the test rejects H0 in favor of H1, otherwise the test does not reject. The region K_n on which the test rejects H0 is called the critical region. The power function π_n of the test maps distributions P to the probability of the critical region,

$$\pi_n(\mathrm{P}) \stackrel{\text{def}}{=} \mathrm{P}(Y_{1:n} \in K_n) .$$

If $\pi_n(\mathrm{P}) \leq \alpha$ for all $\mathrm{P} \in \mathcal{M}^0$, the test T_n is said to be of *level* α. The goal of test design is to achieve high power at low level. In many settings of interest, the determination of the highest achievable power at a given level for a given sample size n is beyond our capabilities. This motivates asymptotic analysis. A sequence of tests T_n is asymptotically of level α if for all $\mathrm{P} \in \mathcal{M}^0$,

$$\limsup_{n \to \infty} \mathrm{P}(K_n) \leq \alpha .$$

A sequence of tests T_n with power functions π_n is consistent at level α if all but finitely many T_n have level α, and if $\pi_n(\mathrm{P}) \to 1$ for all $\mathrm{P} \in \mathcal{M}^1$.

When comparing two simple hypotheses, the question is solved by the Neyman-Pearson lemma. This result asserts that it is enough to compare the ratio of likelihoods of observations according to the two hypotheses with a threshold. When dealing with composite hypotheses, things turn out to be more difficult. In the context of nested models, the generalized likelihood ratio test is defined in the following way.

Definition 15.3.1. *Let $\mathcal{M}^0$ and $\mathcal{M}$ denote two sets of distributions on Y^∞, with $\mathcal{M}^0 \subseteq \mathcal{M}$. Then the nth likelihood ratio test between $\mathcal{M}^0$ and $\mathcal{M} \setminus \mathcal{M}^0$ has critical region*

$$K_n \stackrel{\text{def}}{=} \left\{ y_{1:n} : \sup_{\mathrm{P} \in \mathcal{M}^0} \log \mathrm{P}(y_{1:n}) \leq \sup_{\mathrm{P} \in \mathcal{M}} \log \mathrm{P}(y_{1:n}) - \mathrm{pen}(n) \right\} ,$$

where the penalty $\mathrm{pen}(n)$ *defines an n-dependent threshold.*

Increasing the penalty shrinks the critical region and tends to diminish the level of the test. As a matter of fact, in order to get a non-trivial level, $\mathrm{pen}(n)$ should be positive. The definition of the generalized likelihood ratio test raises two questions.

1. How should $\mathrm{pen}(n)$ be chosen to warrant strong consistency?
2. Is generalized likelihood ratio testing the best way to design a consistent test?

It turns out that the answers to these two questions depend on the properties of maximum likelihood in the models $\mathcal{M}^0$ and $\mathcal{M}$. Moreover, the way to get the answers depends on the models under consideration. In order to answer the first question, we need to understand the behavior of

$$\sup_{\mathrm{P}\in\mathcal{M}} \log \mathrm{P}(Y_{1:n}) - \sup_{\mathrm{P}\in\mathcal{M}^0} \log \mathrm{P}(Y_{1:n})$$

under the two hypotheses.

Let $\mathcal{M}^0$ denote Markov chains of order r and let $\mathcal{M}$ denote Markov chains of order $r+1$. If $\mathrm{P}_\star$ defines a Markov chain of order r, then as n tends to infinity, $2[\sup_{\mathrm{P}\in\mathcal{M}} \log \mathrm{P}(Y_{1:n}) - \sup_{\mathrm{P}\in\mathcal{M}^0} \log \mathrm{P}(Y_{1:n})]$ converges in distribution to a χ^2 random variable with $|\mathsf{Y}|^r(|\mathsf{Y}|-1)^2$ degrees of freedom. As a consequence of the law of the iterated logarithm, $\mathrm{P}_\star$-a.s., it should be of order $\log\log n$ as n tends to infinity (see Finesso, 1991, and Section 15.8). Hence in such a case, a good understanding of the behavior of maximum likelihood estimates provides hints for designing consistent testing procedures. As already pointed out such a knowledge is not available for HMMs. As of this writing the best and most useful insights into the behavior of $\sup_{\mathrm{P}\in\mathcal{M}} \log \mathrm{P}(Y_{1:n}) - \sup_{\mathrm{P}\in\mathcal{M}^0} \log \mathrm{P}(Y_{1:n})$ when $\mathcal{M}$ denotes HMMs of order r and $\mathcal{M}^0$ denotes HMMs of order $r' < r$, can be found in the universal coding literature.

15.4 Code-based Identification

15.4.1 Definitions

The pervasive influence of concepts originating from universal coding theory in the literature dedicated to Markov order or HMM order estimation should not be a surprise. Recall that by the Kraft-McMillan inequality (Cover and Thomas, 1991), a uniquely decodable code on Y^n defines a (sub)-probability on Y^n, and conversely, for any probability distribution P on Y^n, there exists a uniquely decodable code for Y^n such that the length of the codeword associated with $y_{1:n}$ is upper-bounded by $\lceil \log \mathrm{P}\{y_{1:n}\} \rceil + 1$. Henceforth, the probability associated with a code will be called the *coding probability*, and the logarithm of the coding probability will represent the *ideal codeword length* associated with the coding probability.

For each n, let R^n denote a coding probability for Y^n. The family (R^n) is not necessarily compatible—in other words it is not necessarily the nth dimensional marginal of a distribution on Y^∞. We shall denote by subscripts the marginals: for a probability P on Y^∞, P_n is the marginal distribution of $Y_{1:n}$.

The redundancy of R^n with respect to $\mathrm{P} \in \mathcal{M}$ is defined as the Kullback divergence between P_n and R^n, denoted by

$$D(\mathrm{P}_n \,|\, \mathsf{R}^n) \,.$$

The family (R^n) of coding probabilities is a universal coding probability for model $\mathcal{M}$ if and only if

$$\sup_{\mathrm{P}\in\mathcal{M}} \lim_n n^{-1} D(\mathrm{P}_n \,|\, \mathsf{R}^n) = 0 \,.$$

The quantity $\sup_{\mathrm{P}\in\mathcal{M}} D(\mathrm{P}_n \mid \mathsf{R}^n)$ is called the redundancy rate of the family (R^n) with respect to $\mathcal{M}$.

The following coding probability has played a distinguished role in the areas of universal coding and prediction of individual sequences.

Definition 15.4.1. *Given a model $\mathcal{M}$ of probability distributions over Y^n, the normalized maximum likelihood (NML) coding probability induced by $\mathcal{M}$ on Y^n is defined by*

$$NML^n(y_{1:n}) = \frac{\sup_{\mathrm{P}\in\mathcal{M}} \mathrm{P}(y_{1:n})}{\mathcal{C}_n} ,$$

where

$$\mathcal{C}_n \stackrel{\text{def}}{=} \sum_{y_{1:n}\in\mathsf{Y}^n} \sup_{\mathrm{P}\in\mathcal{M}} \mathrm{P}(y_{1:n}) .$$

The maximum point-wise regret of a coding probability R^n with respect to the model $\mathcal{M}$ is defined as

$$\max_{y_{1:n}\in\mathsf{Y}^n} \sup_{\mathrm{P}\in\mathcal{M}} \log \frac{\mathrm{P}(y_{1:n})}{\mathsf{R}^n(y_{1:n})} .$$

Note that NML^n achieves the same regret $\log \mathcal{C}_n$ over all strings from Y^n. No coding probability can achieve a smaller maximum point-wise regret. This is why NML coders are said to achieve minimax point-wise regret over $\mathcal{M}$.

During the last two decades, precise bounds on $\mathcal{C}_n$ have been determined for different kinds of models, notably for the class of product distributions (memoryless sources), for the class of Markov chains of order r (Markov sources), and for the class of hidden Markov sources of order r.

The relevance of bounds on $\mathcal{C}_n$ to our problem is immediate. Let $\mathcal{C}_n$ be defined with respect to $\mathcal{M}$ and let $\mathrm{P}_\star$ denote the true distribution, which is assumed to belong to $\mathcal{M}$. Then

$$\sup_{\mathrm{P}\in\mathcal{M}} \log \mathrm{P}(y_{1:n}) - \log \mathrm{P}_\star(y_{1:n}) = \log \mathrm{NML}^n(y_{1:n}) - \log \mathrm{P}_\star(y_{1:n}) + \log \mathcal{C}_n .$$

On the right-hand side of this inequality, the two quantities that show up refer to two fixed probabilities. After exponentiation, those two quantities may take part into summations over $y_{1:n}$ as will be seen for example when proving consistency of penalized maximum likelihood order estimators (see Lemma 15.6.3). One possible (conservative) choice of the penalty term will be made by comparison with normalizing constants $\mathcal{C}_n$

The NML coding probability is one among many universal coding probabilities that have been investigated in the literature. For models like HMMs with fixed order r, the parameter space Θ^r can be endowed with a probability space structure. A prior probability ω can be defined on Θ^r, and under mild measurability assumptions this in turn defines a probability distribution P on Y^∞,

$$\mathrm{P} = \int_{\Theta^r} \mathrm{P}_\theta \; \omega(d\theta) \; , \tag{15.2}$$

where P_θ is the probability distribution on Y^∞ of the HMM with parameter θ. Such coding probabilities are called mixture coders. Historically, several prior probabilities on Θ have been considered. Uniform (or Laplace) priors were considered first, but Dirichlet distributions soon gained much attention.

Definition 15.4.2. *A Dirichlet-*$(\alpha_1, \ldots, \alpha_r)$ *distribution is a distribution on the simplex of* $\mathbb{R}^r$ *given by the density*

$$\omega(q_1, \ldots, q_r | \alpha_1, \ldots, \alpha_r) = \frac{\Gamma(\alpha_1 + \ldots + \alpha_r)}{\Gamma(\alpha_1) \cdots \Gamma(\alpha_r)} \, q_1^{\alpha_1 - 1} \cdots q_r^{\alpha_r - 1} \, \mathbb{1}_{q_1 + \ldots + q_r = 1} \; ,$$

where the α_i *are all positive.*

Though the Dirichlet prior has a venerable history in Bayesian inference, in this chapter we will stick to the information-theoretical tradition and call the resulting coding probability the Krichevsky-Trofimov mixture.

Definition 15.4.3. *The Krichevsky-Trofimov mixture (KT) is defined by providing* Θ^r *with a product of Dirichlet-*$(1/2, \ldots 1/2)$ *distributions. More precisely, such a distribution is assigned to* $\nu_\theta(\cdot)$ *in the simplex of* $\mathbb{R}^r$*, to each row* $G_\theta(i, \cdot)$*, in the simplex of* $\mathbb{R}^s$ *where* $s = |\mathsf{Y}|$*, and to each row* $Q_\theta(i, \cdot)$ *in the simplex of* $\mathbb{R}^r$*,*

$$\begin{aligned} \omega_{KT}(d\theta) &\stackrel{\text{def}}{=} \left[\frac{\Gamma\left(\frac{r}{2}\right)}{\Gamma\left(\frac{1}{2}\right)^r} \prod_{i=1}^r \nu_\theta(i)^{-1/2} \right] \\ &\times \prod_{i=1}^r \left[\frac{\Gamma\left(\frac{r}{2}\right)}{\Gamma\left(\frac{1}{2}\right)^r} \prod_{j=1}^r Q_\theta(i,j)^{-1/2} \right] \times \left[\frac{\Gamma\left(\frac{d}{2}\right)}{\Gamma\left(\frac{1}{2}\right)^d} \prod_{j=1}^d G_\theta(i,j)^{-1/2} \right] . \end{aligned} \tag{15.3}$$

Krichevsky-Trofimov mixtures define a compatible family of probability distributions over Y^n for $n \geq 1$. This is in sharp contrast with NML distributions and is part of the reason why KT mixtures became so popular in source coding theory.

Resorting to coding-theoretical concepts provides a framework for defining an order estimation procedure known as minimum description length (MDL) order estimation. MDL was introduced and popularized by J. Rissanen in the late 1970s. Although MDL has often been promoted by borrowing material from medieval philosophy, we will see later that it can be justified using some non-trivial mathematics for Markov order estimation.

Definition 15.4.4. *Assume that* μ *is a probability distribution on the set of possible orders and that for each order* r *and* $n \geq 1$, R_r^n *defines a coding probability for* Y^n *with respect to* $\mathcal{M}^r$*. Then the MDL order estimator is defined by*

$$\widehat{r} \stackrel{\text{def}}{=} \arg\max\nolimits_r [\log R_r^n(y_{1:n}) + \log \mu(r)] \; .$$

Note that if the coding probability BBr_r^n turns out to be the normalized maximum likelihood distribution, the MDL order estimator is a special kind of penalized maximum likelihood (PML) order estimator.

The Bayesian information criterion (BIC) order estimator is nothing but another distinguished member of the family of penalized maximum likelihood order estimators. It is closely related to but different from the MDL order estimator derived from the NML coding probability.

Definition 15.4.5. *Let* $\dim(r)$ *be the dimension of the parameter space* Θ^r *in* $\mathcal{M}^r$. *Then the* BIC order estimator *is defined by*

$$\widehat{r} \stackrel{\text{def}}{=} \arg\max_r \left[\sup_{\mathrm{P}\in\mathcal{M}^r} \log \mathrm{P}(y_{1:n}) - \frac{\dim(r)}{2} \log n \right] .$$

Schwarz introduced the BIC in the late 1970s using Bayesian reasoning, and using Laplace's trick to simplify high-dimensional integrals. The validity of this trick and the relevance of Bayesian reasoning to the minimax framework has to be checked on an ad hoc basis.

15.4.2 Information Divergence Rates

The order estimators we have in mind (MDL, BIC, PML) are related to generalized likelihood ratio testing. In order to prove their consistency, we need strong laws of large numbers concerning logarithms of likelihood ratios. In the stationary independent case, those laws of large numbers reduce to the classical laws of large numbers for sums of independent random variables. Such strong laws have proved to be fundamental tools both in statistics and in information theory. In general (that is, not necessarily i.i.d. settings), the laws of large numbers we are looking for have been called asymptotic equipartition principles for information in information theory or Shannon-McMillan-Breiman (SMB) theorems in ergodic theory (Barron, 1985).

Before formulating SMB theorems in a convenient form, let us recall some basic facts about likelihood ratios. Let P and P$'$ denote two probabilities over Y^∞ such that for every n, P'_n is absolutely continuous with respect to P_n. Then under P, the ratio $\mathrm{P}'_n \,/\, \mathrm{P}_n$ is a martingale with expectation less than or equal than 1. By monotonicity and concavity of the logarithm, $\log \mathrm{P}'_n \,/\, \mathrm{P}_n$ is a super-martingale with non-positive expectation. It follows from a theorem due to Doob that this super-martingale converges a.s. to an integrable random variable. If the expectation of the latter random variable is infinite, P is singular with respect to P$'$. In such a setting, the rate of growth of $\log \mathrm{P}'_n \,/\, \mathrm{P}_n$ is a matter of concern. If the two distributions are product probabilities, the log-likelihood ratio is a sum of independent random variables and grows linearly with n if the factors are identical. Moreover, the strong law of large numbers tells us that $n^{-1} \log \mathrm{P}'_n \,/\, \mathrm{P}_n$ converges a.s. to a fixed value, which is called the information divergence rate between the two distributions.

How robust is this observation? This is precisely the topic of SMB theorems.

Definition 15.4.6. *A set $\mathcal{M}$ of process laws over* Y *is said to satisfy a* generalized AEP *if the following holds.*

(i) For every pair of laws P *and* P′ *from $\mathcal{M}$, the relative entropy rate (information divergence rate) between* P *and* P′,

$$\lim_{n\to\infty} \frac{1}{n} D(\mathrm{P}_n \mid \mathrm{P}'_n) ,$$

exists. It is denoted by $D_\infty(\mathrm{P} \mid \mathrm{P}')$.

(ii) Furthermore, if P *and* P′ *are stationary ergodic, then*

$$\lim_{n\to\infty} \frac{1}{n} \log \frac{\mathrm{P}(Y_{1:n})}{\mathrm{P}'(Y_{1:n})} = D_\infty(\mathrm{P} \mid \mathrm{P}') \qquad \mathrm{P}\text{-}a.s.$$

Remark 15.4.7. In the i.i.d. setting, the AEP boils down to the usual strong law of large numbers.

■

The cases of Markov models and hidden Markov models can be dealt with using Barron's generalized Shannon-McMillan-Breiman theorem, which we state here.

Theorem 15.4.8. *Let* Y *be a standard Borel space and let $\{Y_k\}_{k\geq 1}$ be a* Y*-valued stochastic ergodic process distributed according to* P. *Let* P′ *denote a distribution over* Y^∞*, which is assumed to be Markovian of order r, and such that for each n,* P_n *has a density with respect to* P'_n*. Then*

$$n^{-1} \log \frac{d\,\mathrm{P}}{d\,\mathrm{P}'}(Y_{1:n})$$

converges P*-a.s. to the relative entropy rate between the two distributions,*

$$D_\infty(\mathrm{P} \mid \mathrm{P}') = \lim_n n^{-1} D(\mathrm{P}_n \mid \mathrm{P}'_n) = \sup_n n^{-1} D(\mathrm{P}_n \mid \mathrm{P}'_n) .$$

From Barron's theorem, it is immediate that the collection of Markov models satisfies the generalized AEP. The status of HMMs is less straightforward. There are actually several proofs that HMMs satisfy the generalized AEP (see Finesso, 1991). The argument we present here simply resorts to the extended chain device.

Theorem 15.4.9. *The collection of HMMs over some finite observation alphabet* Y *satisfies the generalized AEP.*

Proof. Let P and P′ denote two HMMs over some finite observation alphabet Y. Let ϕ_n and ϕ'_n denote the associated prediction filters. Then under P and P′ the sequence $\{Y_n, \phi_n, \phi'_n\}$ is a Markov chain over $\mathsf{Y} \times \mathbb{R}^r \times \mathbb{R}^{r'}$, which may be regarded as a standard Borel space. Moreover

$$\log \mathrm{P}(y_{1:n}) = \log \mathrm{P}(y_{1:n}, \phi_{1:n}, \phi'_{1:n}) .$$

Applying Theorem 15.4.8 to the sequence $\{Y_n, \phi_n, \phi'_n\}$ finishes the proof. □

Knowing that some collection of models satisfies the generalized AEP allows us to test between two elements picked from the collection. When performing order estimation, we need more than that. If ML estimation is consistent, we need to have for every $\mathrm{P}_\star \in \mathcal{M}^{r_\star} \setminus \mathcal{M}^{r_\star - 1}$,

$$\limsup_n \sup_{\mathrm{P} \in \mathcal{M}^{r_\star - 1}} n^{-1} \log \frac{\mathrm{P}(Y_{1:n})}{\mathrm{P}_\star(Y_{1:n})} < 0 \qquad \mathrm{P}_\star\text{-a.s.}$$

If the collection of models satisfies the generalized AEP, this should at least imply that

$$\inf_{\mathrm{P} \in \mathcal{M}^{r_\star - 1}} D_\infty(\mathrm{P}_\star \mid \mathrm{P}) > 0 \,.$$

We recall here some results concerning divergence rates of stationary HMMs that may be found in Gassiat and Boucheron (2003). Here $\mathcal{M}^r$ is the set of stationary HMMs of order at most r.

Lemma 15.4.10. *$D_\infty(\cdot \mid \cdot)$ is lower semi-continuous on $\cup_r \mathcal{M}^r \times \cup_r \mathcal{M}^r$.*

Lemma 15.4.11. *If P is a stationary but not necessarily ergodic HMM of order r, it can be represented as a mixture of ergodic HMMs $(\mathrm{P}_i)_{i \le i(r)}$ having disjoint supports on $\mathsf{X} \times \mathsf{Y}$,*

$$\mathrm{P} = \sum_{i=1}^{d} \lambda_i \mathrm{P}_i \,,$$

where $\sum_i \lambda_i = 1$, $\lambda_i \ge 0$ and $i(r)$ depends on r only. If P' is a stationary ergodic HMM then

$$D_\infty(\mathrm{P} \mid \mathrm{P}') = \sum_i \lambda_i D_\infty(\mathrm{P}_i \mid \mathrm{P}') \,,$$

$$D_\infty(\mathrm{P}' \mid \mathrm{P}) = \inf_i D_\infty(\mathrm{P}' \mid \mathrm{P}_i) \,.$$

Lemma 15.4.12. *If $\mathrm{P}_\star$ is a stationary ergodic HMM of order $r_\star$ and $r < r_\star$, then*

$$\inf_{\mathrm{P} \in \mathcal{M}^r} D_\infty(\mathrm{P} \mid \mathrm{P}_\star) > 0 \quad \text{and} \quad \inf_{\mathrm{P} \in \mathcal{M}^r} D_\infty(\mathrm{P}_\star \mid \mathrm{P}) > 0 \,.$$

15.5 MDL Order Estimators in Bayesian Settings

Under mild but non-trivial conditions on universal redundancy rates, the above-described order estimators are strongly consistent in a minimax setting. In this section, we will present a result that might seem to be a definitive one.

Recall that two probability distributions Q and Q' are orthogonal or mutually singular if there exists a set A such that $\mathsf{Q}(A) = 1 = \mathsf{Q}'(A^c)$.

Theorem 15.5.1. *Let $\{\Theta^r\}_{r\geq 1}$ denote a collection of models and let Q_r denote coding probabilities defined by* (15.2) *with prior probabilities ω_r. Let $L(r)$ denote the length of a prefix binary encoding of the integer r. Assume that the probabilities Q_r are mutually singular on the asymptotic σ-field. If the order estimator is defined as*

$$\widehat{r}_n \stackrel{\text{def}}{=} \arg\min_r \left[-\log_2 Q_r(y_{1:n}) + L(r) \right] ,$$

then for all $r_\star$ and $\omega_{r_\star}$-almost all θ, $\widehat{r}_n$ converges to $r_\star$ a.s.

Proof. Define $\mathrm{Q}^\star$ as the double mixture

$$\mathrm{Q}^\star = C \sum_{r\neq r_\star} 2^{-L(r)} \mathrm{Q}_r ,$$

where $C \geq 1$ is a normalization factor. Under the assumptions of the theorem $\mathrm{Q}^\star$ and $\mathrm{Q}_{r_\star}$ are mutually singular on the asymptotic σ-field. Moreover for all $y_{1:n}$,

$$\mathrm{Q}^\star(y_{1:n}) \geq C \sup_{r\neq r_\star} \left[2^{-L(r)} \mathrm{Q}_r(y_{1:n}) \right] ,$$

which is equivalent to

$$-\log_2 \mathrm{Q}^\star(y_{1:n}) \leq -\log_2 C + \inf_{r\neq r_\star} \left[L(r) - \log_2 \mathrm{Q}_r(y_{1:n}) \right] .$$

On the other hand, a standard martingale argument tells us that $\mathrm{Q}_{r_\star}$-a.s.,

$$\log_2 \frac{\mathrm{Q}_{r_\star}(y_{1:n})}{\mathrm{Q}^\star(y_{1:n})}$$

converges to a limit, and the fact that $\mathrm{Q}_{r_\star}$ and $\mathrm{Q}^\star$ are mutually singular entails that this limit is infinite $\mathrm{Q}_{r_\star}$-a.s. Hence $\mathrm{Q}_{r_\star}$-a.s., for all sufficiently large n

$$-\log_2 \mathrm{Q}_{r_\star}(y_{1:n}) + L(r_\star) < \inf_{r\neq r_\star} \left[L(r) - \log_2 \mathrm{Q}_r(y_{1:n}) \right] .$$

This implies that $\mathrm{Q}_{r_\star}$-a.s., for all sufficiently large n, $\widehat{r}_n = r_\star$, which is the desired result. □

Remark 15.5.2. Theorem 15.5.1 should not be misinterpreted. It does not prevent the fact that for some θ in a set with null $\omega_{r_\star}$ probability, the order estimator might be inconsistent. Neither does the theorem give a way to identify those θ for which the order estimator is consistent. ■

15.6 Strongly Consistent Penalized Maximum Likelihood Estimators for HMM Order Estimation

In this section, we give general results concerning order estimation in the framework of nested sequences of models, and we then state their application to stationary HMMs. We shall consider penalized ML estimators $\widehat{r}_n$.

Assumption 15.6.1.

(i) The sequence of models satisfies the generalized AEP (Definition 15.4.6).

(ii) Whenever $\mathrm{P}_\star$ *is stationary ergodic of order* $r_\star$ *and* $r < r_\star$*,*

$$\inf_{\mathrm{P}\in\mathcal{M}^r} D_\infty(\mathrm{P}_\star \mid \mathrm{P}) > 0 \,.$$

(iii) For any $\epsilon > 0$ *and any* r*, there exists a sieve* $(\mathrm{P}_i)_{i\in I_\epsilon^r}$*, that is, a finite set* I_ϵ^r *such that* $P_i \in \mathcal{M}^r$ *with all* P_i *being stationary ergodic, and a* n_ϵ^r *such that for all* $\mathrm{P} \in \mathcal{M}^r$ *there is an* $i \in I_\epsilon^r$ *such that*

$$n^{-1}|\log \mathrm{P}(y_{1:n}) - \log \mathrm{P}_i(y_{1:n})| \leq \epsilon$$

for all $n \geq n_\epsilon^r$ *and all* $y_{1:n}$.

Non-trivial upper bounds on point-wise minimax regret for the different models at hand will enable us to build strongly consistent code-based order estimators.

Lemma 15.6.2. *Let the penalty function* $\mathrm{pen}(n,r)$ *be non-decreasing in* r *and such that* $\mathrm{pen}(n,r)/n \to 0$*. Let* $\{\widehat{r}_n\}$ *denote the sequence of penalized maximum likelihood order estimators defined by* pen()*. Then under Assumption 15.6.1,* $\mathrm{P}_\star$*-a.s.,* $\widehat{r}_n \geq r_\star$ *eventually.*

Proof. Throughout "infinitely often" will be abbreviated "i.o." Write

$$\{\widehat{r}_n < r_\star \text{ i.o.}\} = \bigcup_{r<r_\star} \{\widehat{r}_n = r \text{ i.o.}\}$$

and note that

$$\begin{aligned}
\{\widehat{r}_n = r \text{ i.o.}\} &\subseteq \left\{ \sup_{\mathrm{P}\in\mathcal{M}^r} \log \mathrm{P}(y_{1:n}) \geq \log \mathrm{P}_\star(y_{1:n}) - \mathrm{pen}(n, r_\star) \text{ i.o.}\right\} \\
&\subseteq \left\{ \max_{i\in I_\epsilon^r} \log \mathrm{P}_i(y_{1:n}) \geq \log \mathrm{P}_\star(y_{1:n}) - n\epsilon - \mathrm{pen}(n, r_\star) \text{ i.o.}\right\} \\
&\subseteq \bigcup_{i\in I_\epsilon^r} \left\{ \limsup n^{-1}[\log \mathrm{P}_i(y_{1:n}) - \log \mathrm{P}_\star(y_{1:n})] \geq -\epsilon \right\} ,
\end{aligned}$$

where $(\mathrm{P}_i)_{i\in I_\epsilon^r}$ is the sieve for $\mathcal{M}^r$ given by Assumption 15.6.1(iii). Now, by Assumption 15.6.1(i), $n^{-1}[\log \mathrm{P}_i(y_{1:n}) - \log \mathrm{P}_\star(y_{1:n})]$ converges $\mathrm{P}_\star$-a.s. to $-D_\infty(\mathrm{P}_\star \mid \mathrm{P}_i)$, and by Assumption 15.6.1(ii), as soon as

$$\epsilon < \min_{r<r_\star} \inf_{\mathrm{P}\in\mathcal{M}^r} D_\infty(\mathrm{P}_\star \mid \mathrm{P}) \,,$$

one obtains $\mathrm{P}_\star(\widehat{r} < r \text{ i.o.}) = 0$. □

A possibly very conservative way of choosing penalties may be justified in a straightforward way by universal coding arguments. Let C_n^r denote the normalizing constant in the definition of the NML coding probability induced by $\mathcal{M}^r$ on Y^n.

Lemma 15.6.3. *Let the penalty function be* $\mathrm{pen}(n,r) = \sum_{r'=0}^{r}(\log C_n^{r'} + 2\log n)$ *and let* $\{\widehat{r}_n\}$ *denote the sequence of penalized maximum likelihood order estimators defined by* pen(). *Then* $\mathrm{P}_\star$*-a.s.,* $\widehat{r}_n \le r_\star$ *eventually.*

Proof. Let r denote an integer larger than $r_\star$. Then

$$
\begin{aligned}
&\mathrm{P}_\star(\widehat{r}_n = r) \\
&\quad \le \mathrm{P}_\star\left\{\log \mathrm{P}_\star(Y_{1:n}) \le \sup_{\mathrm{P}\in\mathcal{M}^r} \log \mathrm{P}(Y_{1:n}) - \mathrm{pen}(n,r) + \mathrm{pen}(n,r_\star)\right\} \\
&\quad \le \mathrm{P}_\star\left\{\log \mathrm{P}_\star(Y_{1:n}) \le \log \mathrm{NML}_n^r(Y_{1:n}) - \sum_{r'=r_\star+1}^{r-1} \log C_n^{r'} - 2(r-r_\star)\log n\right\} \\
&\quad \le \sum_{y_{1:n}} \exp[\log \mathrm{P}_\star(y_{1:n})] \\
&\qquad\qquad \times \mathbb{1}_{\{\log \mathrm{P}_\star(y_{1:n}) \le \log \mathrm{NML}_n^r(y_{1:n}) - \sum_{r'=r_\star+1}^{r-1} \log C_n^{r'} - 2(r-r_\star)\log n\}} \\
&\quad \le \sum_{y_{1:n}} \mathrm{NML}_n^r(y_{1:n}) \exp\left[-\sum_{r'=r_\star+1}^{r-1} \log C_n^{r'} - 2(r-r_\star)\log n\right] \\
&\quad \le \exp\left[-\sum_{r'=r_\star+1}^{r-1} \log C_n^{r'} - 2(r-r_\star)\log n\right] \\
&\quad \le n^{-2(r-r_\star)} ,
\end{aligned}
$$

because $\sum_{r'=r_\star+1}^{r-1} \log C_n^{r'} = 0$ for $r = r_\star + 1$.

By the union bound,

$$
\mathrm{P}_\star(\widehat{r}_n > r_\star) = \sum_{r > r_\star} \mathrm{P}_\star(\widehat{r}_n = r) \le \frac{n^{-2}}{1-n^{-2}} ,
$$

whence

$$
\sum_n \mathrm{P}_\star(\widehat{r}_n > r_\star) \le \sum_n 1 \wedge \frac{n^{-2}}{1-n^{-2}} < \infty .
$$

Applying the Borel-Cantelli lemma, we may now conclude that $\mathrm{P}_\star$-a.s., order over-estimation occurs only finitely many times. □

In order to show the existence of strongly consistent order estimators for HMMs, it remains to check that Assumption 15.6.1 holds and that the penalties used in the statement of Lemma 15.6.3 satisfy the conditions stated in Lemma 15.6.2, that is, for all $r \ge 1$,

$$
\lim_n \frac{1}{n} \sum_{r' \le r} \left(\log C_n^{r'} + 2\log n\right) = 0 .
$$

This last point follows immediately from the following result from universal coding theory.

Lemma 15.6.4. *For all r, all $n > r$ and all $y_{1:n}$,*

$$\log C_n^r = \log \frac{\sup_{\mathrm{P} \in \mathcal{M}^r} \mathrm{P}(y_{1:n})}{\mathrm{NML}_n^r(y_{1:n})} \leq \frac{r(r+d-2)}{2} \log n + c_{r,d}(n),$$

where for $n \geq 4$, $c_{r,d}(n)$ may be chosen as

$$c_{r,d}(n) = \log r + r \left(-\log \frac{\Gamma\left(\frac{r}{2}\right)\Gamma\left(\frac{d}{2}\right)}{\Gamma\left(\frac{1}{2}\right)\Gamma\left(\frac{1}{2}\right)} + \frac{r^2+d^2}{4n} + \frac{1}{6n} \right).$$

Concerning Assumption 15.6.1, part (i) is Theorem 15.4.9 and part(ii) is Lemma 15.4.12. Now for any positive δ, let us denote by Θ_δ^r the set of HMM parameters in Θ^r such that each coordinate is lower-bounded by δ.

For any $\theta \in \Theta^r$, there exists $\theta_\delta \in \Theta_\delta^r$ such that for any n and any $y_{1:n}$,

$$n^{-1} |\log \mathrm{P}_\theta(y_{1:n}) - \log \mathrm{P}_{\theta_\delta}(y_{1:n})| \leq \frac{r^2+d^2}{2} \delta .$$

A glimpse at the proof of this fact in Liu and Narayan (1994) reveals that this statement still holds when θ_δ is constrained to lie in a sieve for Θ_δ^r, defined as a finite subset $(\theta_i)_{i \in I}$ such that for all $\theta \in \Theta^r$, at least one θ_i in the sieve is within L_∞-distance smaller than δ away from θ.

This may be summarized in the following way.

Corollary 15.6.5. *Let $\mathrm{P}_\star$ be an HMM of order $r_\star$ and let $\{\widehat{r}_n\}$ be the sequence of penalized ML order estimators defined in Lemma 15.6.3. Then $\mathrm{P}_\star$-a.s., $\widehat{r}_n = r_\star$ eventually.*

Remark 15.6.6. Resorting to universal coding arguments to cope with our poor understanding of the maximum likelihood in misspecified HMMs provides us with a Janus-faced result: on one hand it allows us to describe a family of strongly consistent order estimators that will prove to be optimal as far as under-estimation is concerned; on the other hand the question raised by Kieffer (1993) about the consistency of BIC and MDL for HMM order estimation remains open. ■

15.7 Efficiency Issues

How efficient are the aforementioned order estimation procedures? The notions of efficiency that have been considered in the order estimation literature have been shaped on the testing theory setting. As a matter of fact, the classical efficiency notions have emerged from the analysis of the simple hypotheses testing problem. Determining how those notions could be tailored to the nested composite hypothesis testing problem is still a subject of debate.

Among the various notions of efficiency, or even of asymptotic relative efficiency that are regarded as relevant in testing theory, Pitman's efficiency focuses on the minimal sample size that is required to achieve simultaneously a given level and a given power at alternatives. Up to our knowledge, Pitman's efficiency for Markov order or HMM order estimation related problems has not been investigated. This is due to the lack of non-asymptotic results concerning estimation procedures for HMM and Markov chains.

The notion of efficiency that has been assessed in the order estimation literature is rather called Bahadur relative efficiency in the statistical literature and error exponents in the information-theoretical literature. When testing a simple hypothesis against another simple hypothesis in the memoryless setting, a classical result by Chernoff tells us that comparing likelihood ratios to a fixed threshold, both level and power may decay exponentially fast with respect to the number of observations. In that setting, Bahadur-efficient testing procedures are those that achieve the largest exponents. Viewing that set of circumstances, there have been several attempts to generalize those results to the composite hypothesis setting. Part of the difficulty lies in stating the proper questions.

Although consistency issues concerning the BIC and MDL criteria for HMM order estimation have not yet been clarified, our understanding of efficiency issues concerning HMM order identification recently underwent significant progress. In this section, we give general results concerning efficiency of order estimation in the framework of nested sequences of models; these results apply to stationary HMMs.

15.7.1 Variations on Stein's Lemma

The next theorems are extensions of Stein's lemma to the order estimation problem. Theorem 15.7.2 aims at determining the best underestimation exponent for a class of order estimators that ultimately overestimate the order with a probability bounded away from 1. Theorem 15.7.4 aims at proving that the best overestimation exponent should be trivial in most cases of interest.

Assumption 15.7.1.

(i) The sequence of models satisfies the general AEP (Definition 15.4.6).

(ii) For any r, there exists $\mathcal{M}_0^r \subseteq \mathcal{M}^r$ such that any P in $\mathcal{M}_0^r$ is stationary ergodic and has true order at most r, and such that for any $\mathrm{P}_\star \in \mathcal{M}_0^{r_\star}$,

$$\inf_{\mathrm{P}\in\mathcal{M}^r} D_\infty(\mathrm{P} \mid \mathrm{P}_\star) = \inf_{\mathrm{P}\in\mathcal{M}_0^r} D_\infty(\mathrm{P} \mid \mathrm{P}_\star) .$$

Versions of the following theorem have been proved iby Finesso *et al.* (1996) for Markov chains and by Gassiat and Boucheron (2003) for HMMs.

Theorem 15.7.2. *Let the sequence $\{\mathcal{M}^r\}_{r\geq 1}$ of nested models satisfy Assumption 15.7.1. Let $\{\widehat{r}_n\}_{n\geq 1}$ denote a sequence of order estimators such that for some $\alpha < 1$, all $r_\star$ and all $\mathrm{P}_\star \in \mathcal{M}_0^{r_\star}$,*

$$\mathrm{P}_\star(\widehat{r}_n(Y_{1:n}) > r_\star) \leq \alpha$$

for $n \geq T_1(\mathrm{P}_\star, \alpha, r_\star)$. *Then for all* $r_\star$ *and all* $\mathrm{P}_\star \in \mathcal{M}_0^{r_\star}$,

$$\liminf_{n\to\infty} n^{-1} \log \mathrm{P}_\star(\widehat{r}_n(Y_{1:n}) < r_\star) \geq -\min_{r' < r_\star} \inf_{\mathrm{P}\in\mathcal{M}^{r'}} D_\infty(\mathrm{P} \mid \mathrm{P}_\star) .$$

Proof. Fix $\mathrm{P}_\star \in \mathcal{M}_0^{r_\star}$. Let $\mathrm{P}' \in \mathcal{M}_0^{r'}$ with $r' < r_\star$ and define

$$\begin{aligned} A_n(\mathrm{P}') &\stackrel{\text{def}}{=} \{y_{1:n} \ : \ \widehat{r}_n(y_{1:n}) \leq r'\} , \\ B_n(\mathrm{P}') &\stackrel{\text{def}}{=} \{y_{1:n} \ : \ n^{-1} \log \frac{\mathrm{P}'(y_{1:n})}{\mathrm{P}^\star(y_{1:n})} \leq D_\infty(\mathrm{P}' \mid \mathrm{P}_\star) + \epsilon\} . \end{aligned}$$

For $n > T_1(\mathrm{P}', \alpha, r')$,

$$\mathrm{P}'(A_n(\mathrm{P}')) > 1 - \alpha ,$$

and as $\cup_r \mathcal{M}^r$ is assumed to satisfy the generalized AEP, for all $n > T_3(\epsilon, \mathrm{P}', \mathrm{P}_\star)$ it holds that

$$\mathrm{P}'(B_n(\mathrm{P}')) > 1 - \epsilon . \tag{15.4}$$

If $n > T_2(\alpha, \epsilon, \mathrm{P}') = \max[T_1(\alpha, r'), T_3(\epsilon, \mathrm{P}', \mathrm{P}_\star)]$, then

$$\begin{aligned} \mathrm{P}_\star(\widehat{r}_n(Y_{1:n}) < r_\star) &= \mathrm{E}_{\mathrm{P}_\star}[\mathbb{1}_{\{\widehat{r}_n < r_\star\}}] \\ &\quad \text{is an equality if } \mathrm{P}_\star \text{ and } \mathrm{P}' \text{ have the same} \\ &\quad \text{support set for finite marginals} \\ &\geq \mathrm{E}_{\mathrm{P}'}\left[\frac{\mathrm{P}_\star(Y_{1:n})}{\mathrm{P}'(Y_{1:n})} \mathbb{1}_{\{\widehat{r}_n < r_\star\}}\right] \\ &\quad \text{as } r' < r_\star \\ &\geq \mathrm{E}_{\mathrm{P}'}\left[\frac{\mathrm{P}_\star(Y_{1:n})}{\mathrm{P}'(Y_{1:n})} \mathbb{1}_{A_n(\mathrm{P}')}\right] \\ &\quad \text{from the definition of } B_n(\mathrm{P}') \\ &\geq \mathrm{E}_{\mathrm{P}'}\left[\mathbb{1}_{A_n(\mathrm{P}')} \mathbb{1}_{B_n(\mathrm{P}')} \mathrm{e}^{-n[D(\mathrm{P}' \mid \mathrm{P}_\star)+\epsilon]}\right] \\ &\geq \mathrm{E}_{\mathrm{P}'}\left[\mathbb{1}_{A_n(\mathrm{P}')} \mathbb{1}_{B_n(\mathrm{P}')}\right] \mathrm{e}^{-n[D(\mathrm{P}' \mid \mathrm{P}^\star)+\epsilon]} \\ &\quad \text{from the union bound, and by the AEP} \\ &\geq (1 - \alpha - \epsilon)\mathrm{e}^{-n[D(\mathrm{P}' \mid \mathrm{P}^\star)+\epsilon]} . \end{aligned}$$

Now optimizing with respect to θ' and r' and letting ϵ tend to zero, the theorem follows. □

Remark 15.7.3. Assessing that the upper bound on underestimation exponent is positive amounts to checking properties of relative entropy rates. ■

Theorem 15.7.2 holds for stationary HMMs. Assumption 15.7.1(i) is Theorem 15.4.9, and part (ii) is verified by taking $\mathcal{M}_0^r$ as the distributions of stationary ergodic HMMs with order at most r. Then Theorem 15.7.2 follows using Lemmas 15.4.10 and 15.4.11.

Another Stein-like argument provides an even more clear-cut statement concerning possible overestimation exponents. Such a statement seems to be a hallmark of a family of embedded composite testing problems. It shows that in many circumstances of interest, we cannot hope to achieve both non-trivial under- and overestimation exponents. Versions of this theorem have been proved by Finesso *et al.* (1996) for Markov chains and by Gassiat and Boucheron (2003) for HMMs.

Theorem 15.7.4. *Let the sequence* $\{\mathcal{M}^r\}_{r\geq 1}$ *of nested models satisfy Assumption 15.7.1. Assume also that for* $\mathrm{P} \in \mathcal{M}_0^r \subseteq \mathcal{M}^r$ *there exists a sequence* $\{\mathrm{P}^m\}_m$ *of elements in* $\mathcal{M}_0^{r+1} \setminus \mathcal{M}^r$ *such that*

$$\lim_{m\to\infty} D_\infty(\mathrm{P}^m \mid \mathrm{P}) = 0 \; .$$

Assume that $\{\widehat{r}_n\}_n$ *is a consistent order estimation procedure. Then for all* $\mathrm{P} \in \mathcal{M}_0^{r_\star}$ *having order* $r_\star$*,*

$$\liminf_{n\to\infty} \frac{1}{n} \log \mathrm{P}(\widehat{r}_n > r_\star) = 0 \; .$$

The change of measure argument that proved effective in the proof of Theorem 15.7.2 can now be applied for each $\mathrm{P} \in \mathcal{M}_0^r$.

Proof. Let P denote a distribution in $\mathcal{M}_0^{r_\star}$ having order $r_\star$ and let $\{\mathrm{P}^m\}$ denote a sequence as above. Let ϵ denote a small positive real. Fix m sufficiently large that $D_\infty(\mathrm{P}^m \mid \mathrm{P}) \leq \epsilon$ and then n sufficiently large that

$$\mathrm{P}^m \left(n^{-1} \log \frac{d\,\mathrm{P}^m}{d\,\mathrm{P}}(Y_{1:n}) \geq D_\infty(\mathrm{P}_m \mid \mathrm{P}) + \epsilon \right) \leq \epsilon$$

while

$$\mathrm{P}_n^m(\widehat{r}_n = r_\star + 1) \geq 1 - \epsilon \; .$$

We may now lower bound the overestimation probability as

$$\begin{aligned}
\mathrm{P}(\widehat{r}_n > r_\star) &\geq \mathrm{P}(\widehat{r}_n = r_\star + 1) \\
&\geq \mathrm{E}_{\mathrm{P}^m}\left[\frac{d\,\mathrm{P}}{d\,\mathrm{P}^n} \mathbb{1}_{\{\widehat{r}_n = r_\star + 1\}} \right] \\
&\geq \mathrm{E}_{\mathrm{P}^m}\left[\frac{d\,\mathrm{P}_n}{d\,\mathrm{P}_n^m} \mathbb{1}_{\{\widehat{r}_n = r_\star + 1\}} \right] \\
&\geq \mathrm{E}_{\mathrm{P}^m}\left[\exp\left(-\log \frac{d\,\mathrm{P}_n^m}{d\,\mathrm{P}_n} \right) \mathbb{1}_{\{\widehat{r}_n = r_\star + 1\}} \right] \\
&\geq \mathrm{e}^{-2n\epsilon}(1 - 2\epsilon) \; .
\end{aligned}$$

Hence $\liminf_n n^{-1} \log \mathrm{P}_n(\widehat{r}_n > r_\star) \geq -2\epsilon$. As ϵ may be arbitrarily small, this finishes the proof. □

This theorem holds for stationary HMMs; see Gassiat and Boucheron (2003).

The message of this section is rather straightforward: in order estimation problems like HMM order estimation, underestimation corresponds to large deviations of the likelihood process, whereas overestimation corresponds to moderate deviations of the likelihood process. In the Markov order estimation problem, the large-scale typicality theorem of Csiszár and Shields allows us to assign a quantitative meaning to this statement.

15.7.2 Achieving Optimal Error Exponents

Stein-like theorems (Theorems 15.7.2 and 15.7.4) provide a strong incentive to investigate underestimation exponents of the consistent order estimators that have been described in Section 15.6. As those estimators turn out to be penalized maximum likelihood estimators, what is at stake here is the (asymptotic) optimality of generalized likelihood ratio testing. In some situations, generalized likelihood ratio testing fails to be optimal. We will show that this is not the case in the order estimation problems we have in mind.

As will become clear from the proof, as soon as the NML normalizing constant $\log \mathcal{C}_n^r/n$ tends to 0 as n tends to infinity, NML code-based order estimators exhibit the same property.

Assumption 15.7.5.

(i) The sequence of models satisfies the AEP.

(ii) Each model $\mathcal{M}^r$ can be endowed with a topology under which it is sequentially compact.

(iii) Relative entropy rates satisfy the semi-continuity property: if P^m and $\mathrm{P}'^{,m}$ are stationary ergodic and converge respectively to P and P', then $D_\infty(\mathrm{P} \mid \mathrm{P}') \leq \liminf_m D_\infty(\mathrm{P}^m \mid \mathrm{P}'^{,m})$.

(iv) For any $\epsilon > 0$ and any r, there exists a sieve $(\mathrm{P}_i)_{i \in I_\epsilon^r}$, that is, a finite set I_ϵ^r such that $\mathrm{P}_i \in \mathcal{M}^r$ with all P_i ergodic and such that the following hold true.

(a) Assumption 15.6.1(iii) is satisfied.

(b) For each stationary ergodic distribution $\mathrm{P}_\star \in \cup_r \mathcal{M}^r$ with order $r_\star$ and for every finite subset $\mathcal{P}$ of the union $\cup_\epsilon \{\mathrm{P}_i : i \in I_\epsilon^r\} \subseteq \mathcal{M}^{r_\star}$ of all sieves, the log-likelihood process $\{\log \mathrm{P}(Y_{1:n})\}_{\mathrm{P} \in \mathcal{P}}$ satisfies a large deviation principle with good rate function $J_\mathcal{P}$ and rate n.

Moreover, any sample path $\{u(\mathrm{P})\}_{\mathrm{P} \in \mathcal{P}}$ of the log-likelihood process indexed by $\mathcal{P}$ that satisfies $J_\mathcal{P}(u) < \infty$ enjoys the representation property that there exists a distribution $\mathrm{P}_u \in \mathcal{M}^{r_\star}$ such that

$$
\begin{aligned}
u(\mathrm{P}) &= \lim_n n^{-1} \mathrm{E}_{\mathrm{P}_u}[\log \mathrm{P}(Y_{1:n})] , \qquad \mathrm{P} \in \mathcal{P} , \\
J_\mathcal{P}(u) &\geq D_\infty(\mathrm{P}_u \mid \mathrm{P}_\star) .
\end{aligned}
$$

(v) For any $r_1 < r_2$, if $\mathrm{P}_1 \in \mathcal{M}^{r_1}$ and $\mathrm{P}_2 \in \mathcal{M}^{r_2}$ satisfy $D_\infty(\mathrm{P}_2 \mid \mathrm{P}_1) = 0$, then $\mathrm{P}_2 = \mathrm{P}_1 \in \mathcal{M}^{r_1}$.

(vi) If $\mathrm{P} \in \mathcal{M}^{r_\star}$ *is not stationary ergodic, it can be represented as a finite mixture of ergodic components* $(\mathrm{P}_i)_{i \le i(r_\star)}$ *(where* $i(r_\star)$ *depends only on* $r_\star$*) in* $\mathcal{M}^{r_\star}$, $\sum_i \lambda_i \mathrm{P}_i = \mathrm{P}$, *and for all ergodic* P' *in* $\mathcal{M}$,

$$D_\infty(\mathrm{P} \mid \mathrm{P}') = \sum_{i \le i(r_\star)} \lambda_i D_\infty(\mathrm{P}_i \mid \mathrm{P}') .$$

Remark 15.7.6. Assumption 15.7.5 holds for HMMs. This is not obvious at all and follows from available LDPs for additive functionals of Markov chains, the extended chain device, and ad hoc considerations. The interested reader may find complete proofs and relevant information in Gassiat and Boucheron (2003). ∎

Theorem 15.7.7. *Assume that the sequence of nested models* $(\mathcal{M}^r)$ *satisfies Assumptions 15.7.1 and 15.7.5. If* $\mathrm{pen}(n,r)$ *is non-negative and for each* r, $\mathrm{pen}(n,r)/n \to 0$ *as* $n \to \infty$, *the penalized maximum likelihood order estimators achieve the optimal underestimation exponent,*

$$\min_{r<r_\star} \inf_{\mathrm{P} \in \mathcal{M}^r} D_\infty(\mathrm{P} \mid \mathrm{P}_\star) .$$

The optimality of this exponent comes from Theorem 15.7.2, which holds under Assumption 15.7.1. Hence the proof of Theorem 15.7.7 consists in proving that the exponent is achievable.

Proof. An application of the union bound entails that

$$\limsup n^{-1} \log \mathrm{P}_\star(\widehat{r}_n < r_\star) \le \max_{r<r_\star} \limsup n^{-1} \log \mathrm{P}_\star(\widehat{r}_n = r) .$$

Hence the problem reduces to checking that for each $r < r_\star$,

$$\limsup \frac{1}{n} \log \mathrm{P}_\star(\widehat{r}_n = r) \le - \inf_{\mathrm{P} \in \mathcal{M}^r} D_\infty(\mathrm{P} \mid \mathrm{P}_\star) .$$

Fix $r < r_\star$. The proof will be organized in two steps. First, we will check that for each $\epsilon > 0$ we can find some $\widehat{\mathrm{P}}_\epsilon \in I^r_\epsilon$ and some P_ϵ such that

$$\begin{aligned} D_\infty(\mathrm{P}_\epsilon \mid \widehat{\mathrm{P}}_\epsilon) &\le 3\epsilon , \\ \limsup_n n^{-1} \log \mathrm{P}_\star(\widehat{r}_n = r) &\le -D_\infty(\mathrm{P}_\epsilon \mid \mathrm{P}_\star) . \end{aligned}$$

In the second step, we let ϵ tend to 0 to check that there exists some $\bar{\mathrm{P}}$ in $\mathcal{M}^r$ such that

$$\lim_n n^{-1} \log \mathrm{P}_\star(\widehat{r}_n = r) \le -D_\infty(\bar{\mathrm{P}} \mid \mathrm{P}_\star) .$$

Let us choose $\epsilon > 0$ and n_ϵ large enough that $\mathrm{pen}(n, r_\star) \le \epsilon n$ for $n \ge n_\epsilon$. Under Assumption 15.7.5(iv)(a), we get for $n \ge n_\epsilon \vee n^r_\epsilon$,

$$\begin{aligned}
&\log \mathrm{P}_\star(\widehat{r}_n = r) \\
&\leq \log \mathrm{P}_\star\left(\sup_{\mathrm{P}\in\mathcal{M}^r} \log \mathrm{P}(Y_{1:n}) - \sup_{\mathrm{P}\in\mathcal{M}^{r_\star}} \log \mathrm{P}(Y_{1:n}) \geq \mathrm{pen}(n,r) - \mathrm{pen}(n,r_\star)\right) \\
&\leq \log \mathrm{P}_\star\left(\max_{i\in I_\epsilon^r} n^{-1} \log \mathrm{P}_i(Y_{1:n}) - \max_{i\in I_\epsilon^{r_\star}} n^{-1} \log \mathrm{P}(Y_{1:n}) \geq -2\epsilon \right) .
\end{aligned}$$

We may divide by n, take the lim sup of the two expressions as n tends to infinity, and use Assumption 15.7.5(iv)(b) to obtain

$$\limsup n^{-1} \log \mathrm{P}_\star(\widehat{r}_n = r) \leq -\inf \left\{ J_{\mathcal{P}}(u) \ : \ \sup_{i\in I_\epsilon^r} u(P_i) - \sup_{i\in I_\epsilon^{r_\star}} u(P_i) \geq -2\epsilon \right\}$$

with

$$\mathcal{P} = \{\mathrm{P}_i : i \in I_\epsilon^r\} \cup \{\mathrm{P}_i : i \in I_\epsilon^{r_\star}\} .$$

The infimum on the right-hand side of the inequality is attained at some path u_ϵ. Hence, using again Assumption 15.7.5(iv)(b),

$$\limsup n^{-1} \log \mathrm{P}_\star(\widehat{r}_n = r) \leq -D_\infty(\mathrm{P}_\epsilon \mid \mathrm{P}_\star) , \tag{15.5}$$

where $\mathrm{P}_\epsilon \in \mathcal{M}^{r_\star}$,

$$u_\epsilon(\mathrm{P}) = \lim n^{-1} \mathrm{E}_{\mathrm{P}_\epsilon}[\log \mathrm{P}(Y_{1:n})] , \qquad \mathrm{P} \in \mathcal{P} , \tag{15.6}$$

and

$$\sup_{i\in I_\epsilon^r} u_\epsilon(\mathrm{P}_i) - \sup_{i\in I_\epsilon^{r_\star}} u_\epsilon(\mathrm{P}_i) \geq -2\epsilon . \tag{15.7}$$

Pick $\tilde{\mathrm{P}}_\epsilon \in \{\mathrm{P}_i\}_{i\in I_\epsilon^{r_\star}}$ such that for $n \geq n_\epsilon^{r_\star}$,

$$n^{-1} |\log \tilde{\mathrm{P}}_\epsilon(y_{1:n}) - \log \mathrm{P}_\epsilon(y_{1:n})| \leq \epsilon$$

and $\widehat{\mathrm{P}}_\epsilon$ such that

$$\sup_{i\in I_\epsilon^r} u_\epsilon(\mathrm{P}_i^r) = u_\epsilon(\widehat{\mathrm{P}}_\epsilon) . \tag{15.8}$$

Then

$$\begin{aligned}
\limsup n^{-1} \mathrm{E}_{\mathrm{P}_\epsilon}[\log \mathrm{P}_\epsilon(Y_{1:n})] &\leq \limsup n^{-1} \mathrm{E}_{\mathrm{P}_\epsilon}[\log \tilde{\mathrm{P}}_\epsilon(Y_{1:n})] + \epsilon \\
&= u_\epsilon(\tilde{\mathrm{P}}_\epsilon) + \epsilon \\
&\leq u_\epsilon(\widehat{\mathrm{P}}_\epsilon) + 3\epsilon \\
&= \lim n^{-1} \mathrm{E}_{\mathrm{P}_\epsilon}[\log \widehat{\mathrm{P}}_\epsilon(Y_{1:n})] + 3\epsilon .
\end{aligned}$$

Here we used (15.6) for the second step, then (15.8) and (15.7), and finally (15.6) again. Using Assumption 15.7.5(i) we thus finally obtain

$$D_\infty(\mathrm{P}_\epsilon \,|\, \widehat{\mathrm{P}}_\epsilon) \leq 3\epsilon .$$

Let us now proceed to the second step. It remains to check that if we let ϵ tend to 0, the sequence $(\mathrm{P}_\epsilon)_\epsilon$ obtained in (15.5) has an accumulation point in $\mathcal{M}^r$.

Note that $\widehat{\mathrm{P}}_\epsilon$ is ergodic and let $\sum_i \lambda_{i,\epsilon} \mathrm{P}_{i,\epsilon}$ denote the ergodic decomposition of P_ϵ. Then

$$D_\infty(\mathrm{P}_\epsilon \,|\, \widehat{\mathrm{P}}_\epsilon) = \sum_i \lambda_{i,\epsilon} D_\infty(\mathrm{P}_{i,\epsilon} \,|\, \widehat{\mathrm{P}}_\epsilon) \,.$$

Extract a subsequence of $(\lambda_{i,\epsilon})$ and $(\mathrm{P}_{i,\epsilon})$ converging to λ_i and P_i, respectively, and such that $\bar{\mathrm{P}} = \sum_i \lambda_i \mathrm{P}_i$, while $\widehat{\mathrm{P}}$ is the corresponding accumulation point of the sequence $\widehat{\mathrm{P}}_\epsilon$. We may then apply the semi-continuity property to obtain

$$\sum_i \lambda_i D_\infty(\mathrm{P}_i \,|\, \widehat{\mathrm{P}}) = 0 \,.$$

This leads, using Assumption 15.7.5(v) and (vi), to $\sum_i \lambda_i \mathrm{P}_i = \widehat{\mathrm{P}}$, that is, $\bar{\mathrm{P}} = \widehat{\mathrm{P}} \in \mathcal{M}^r$. Using the semi-continuity property again we find that

$$\lim_\epsilon D_\infty(\mathrm{P}_\epsilon \mid \mathrm{P}_\star) = \lim_\epsilon \sum_i \lambda_{i,\epsilon} D_\infty(\mathrm{P}_{i,\epsilon} \mid \mathrm{P}_\star) \geq D_\infty(\bar{\mathrm{P}} \mid \mathrm{P}_\star) \,,$$

whence

$$\limsup n^{-1} \mathrm{P}_\star(\widehat{r}_n = r) \leq - \inf_{\mathrm{P} \in \mathcal{M}^r} D_\infty(\mathrm{P} \mid \mathrm{P}_\star) \,.$$

□

15.8 Consistency of the BIC Estimator in the Markov Order Estimation Problem

Though consistency of the BIC estimator for HMM order is still far from being established, recent progress concerning the Markov order estimation problem raises great expectations. As a matter of fact, the following was established by Csiszár and Shields and recently refined by Csiszár (Csiszár and Shields, 2000; Csiszár, 2002).

Theorem 15.8.1. *For any stationary irreducible Markov process with distribution* $\mathrm{P}_\star$ *over the finite set* Y *and of order* $r_\star$*, the BIC order estimator converges to* $r_\star$ $\mathrm{P}_\star$*-a.s.*

The proof of this remarkable theorem follows from a series of technical lemmas concerning the behavior of maximum likelihood estimators in models $\mathcal{M}^r$ for $r \geq r_\star$. In the Markov order estimation problem, such precise results can be obtained at a reasonable price, thanks to the fact that maximum likelihood estimates coincide with simple functions of empirical measures. Here we follow the argument presented by Csiszár (2002).

First note that underestimation issues are dealt with using Lemma 15.6.2. Theorem 15.8.1 actually follows almost directly from the following result. Let $\widehat{\mathrm{P}}^r$ denote the MLE of the probability distribution in $\mathcal{M}^r$ on the sample $y_{1:n}$.

Theorem 15.8.2. *For any stationary irreducible Markov process with distribution* $\mathrm{P}_\star$ *of order* $r_\star$ *over the finite set* Y,

$$\sup_{r \geq r_\star} \frac{1}{|\mathcal{S}_r|} \frac{1}{\log n} \left[\log \widehat{\mathrm{P}}^r(y_{1:n}) - \log \mathrm{P}_\star(y_{1:n}) \right] \to 0 \qquad \mathrm{P}_\star\text{-}a.s.$$

Here $\mathcal{S}_r$ denotes the subset of patterns from $|\mathsf{Y}|^r$ that have non-zero stationary probability. To emphasize the power of this theorem, let us first use it to derive Theorem 15.8.1.

Proof (of Theorem 15.8.1). The event $\{\widehat{r}_n > r_\star \text{ i.o.}\}$ equals the event

$$\{\exists r > r_\star : \log \widehat{\mathrm{P}}^r(y_{1:n}) - \log \widehat{\mathrm{P}}^{r_\star}(y_{1:n}) \geq \mathrm{pen}(n, r) - \mathrm{pen}(n, r_\star) \text{ i.o.}\} ,$$

which is included in

$$\{\exists r > r_\star : \log \widehat{\mathrm{P}}^r(y_{1:n}) - \log \mathrm{P}_\star(y_{1:n}) \geq \mathrm{pen}(n, r) - \mathrm{pen}(n, r_\star) \text{ i.o.}\} .$$

By Theorem 15.8.2, it follows that for any $\eta > 0$, $\mathrm{P}_\star$-a.s.,

$$\sup_{r \geq r_\star} \frac{1}{|\mathcal{S}_r|} \frac{1}{\log n} \left\{ \log \widehat{\mathrm{P}}^r(y_{1:n}) - \log \mathrm{P}_\star(y_{1:n}) \right\} < \eta .$$

Finally, for large n, for the BIC criterion, $\mathrm{pen}(n, r) \geq (1/2)|\mathcal{S}_r| \times (|\mathsf{Y}| - 1) \log n$. □

Remark 15.8.3. Viewing the proof of strong consistency of the BIC Markov order estimator, one may wonder whether an analogous result holds for MDL order estimators derived from NML coding probabilities or KT coding probabilities. If no *a priori* restriction on the order is enforced, the answer is negative: there exists at least one stationary ergodic Markov chain (the uniform memoryless source) for which unrestricted MDL order estimators overestimate the order infinitely often with probability one.

But if the search for r in $\max_r \{-\log \mathsf{Q}^{n,r}(y_{1:n}) - \log \mu(r)\}$ is restricted to some finite range $\{0, \ldots, \alpha \log n\}$ where α is small enough (depending on the unknown $\mathrm{P}_\star$) and does not depend on n, then the MDL order estimator derived by taking $\mathrm{NML}^{n,r}$ as the rth coding probability turns out to be strongly consistent. The reason why this holds is that in order to prove strong consistency, we need to control

$$\log \mathcal{C}_n^r - \frac{|\mathcal{S}_{r+1}| - |\mathcal{S}_r|}{2} \log n$$

over a large range of values of r for all sufficiently large n. Sharp estimates of the minimax point-wise regret of NML for Markov sources of order r have recently been obtained. It is not clear whether such precise estimates can be obtained for models like HMMs where maximum likelihood is not as well-behaved as in the Markov chain setting. ■

Throughout this section, $\mathrm{P}_\star$ denotes the distribution of a stationary irreducible Markov chain of order $r_\star$ over Y. For all r and all $a_{1:r} \in \mathsf{Y}^r$,

$$N_n(a_{1:r}) \stackrel{\text{def}}{=} \sum_{i=1}^{n+1-r} \mathbb{1}_{\cap_{j=1}^r \{Y_{i+j-1}=a_j\}}$$

is the number of times the pattern $a_{1:r}$ occurs in the sequence $y_{1:n}$. The MLE of the conditional distribution in $\mathcal{M}^r$ (r-transitions) is

$$\widehat{\mathrm{P}}^r(a_{r+1} \,|\, a_{1:r}) = \frac{N_n(a_{1:r+1})}{N_{n-1}(a_{1:r})}$$

for all $a_{1:r+1} \in \mathsf{Y}^{r+1}$, whenever $N_{n-1}(a_{1:r}) > 0$.

The proof of Theorem 15.8.2 is decomposed into two main parts. The easiest part relates $\log \widehat{\mathrm{P}}^r(y_{1:n}) - \log \mathrm{P}_\star(y_{1:n})$ and a χ^2 distance between the empirical transition kernel $\widehat{\mathrm{P}}^r_n$ and $\mathrm{P}_\star$, under conditions that aver to be almost surely satisfied by sample paths of irreducible Markov chains. This relationship (Lemma 15.8.4) is a quantitative version of the asymptotic equivalence between relative entropy and χ^2 distance (see Csiszár, 1990, for more information on this topic). The most original part actually proves that the almost sure convergence of $\widehat{\mathrm{P}}^r$ to $\mathrm{P}_\star$ is uniform over $r \geq r_\star$.

Lemma 15.8.4. *Let P and P' be two probability distributions on $\{1, \ldots, m\}$. If $P'(i)/2 \leq P(i) \leq 2P'(i)$ for all i then $D(P \,|\, P') \leq \chi^2(P, P')$, where $\chi^2(P, P') = \sum_{i=1}^m \{P(i) - P'(i)\}^2 / P'(i)$.*

A simple corollary of this lemma is the following.

Corollary 15.8.5. *Let r be an integer such that $r \geq r_\star$. If $y_{1:n}$ is such that for all $a_{1:r+1} \in \mathcal{S}_{r+1}$,*

$$\frac{1}{2}\,\mathrm{P}_\star(a_{r+1} \,|\, a_{1:r}) \leq \frac{N_n(a_{1:r+1})}{N_{n-1}(a_{1:r})} \leq 2\,\mathrm{P}_\star(a_{r+1} \,|\, a_{1:r}) \;,$$

then

$$\log \widehat{\mathrm{P}}^r(y_{1:n}) - \log \mathrm{P}_\star(y_{1:n}) \leq \sum_{a_{1:r} \in \mathcal{S}_r} N_n(a_{1:r}) \chi^2(\widehat{\mathrm{P}}^r(\cdot \,|\, a_{1:r}), \mathrm{P}_\star(\cdot \,|\, a_{1:r})) \;.$$

15.8.1 Some Martingale Tools

The proof of Theorem 15.8.2 relies on martingale arguments. The basic tools of martingale theory we need are gathered here.

In the sequel, ϕ denotes the convex function $\phi(x) \stackrel{\text{def}}{=} \exp(x) - x - 1$ and $\phi^\star$ its convex dual, $\phi^\star(y) = \sup_x (yx - \phi(x)) = (y+1)\log(y+1) - y$ for $y \geq -1$ and ∞ otherwise. We will use repeatedly the classical inequality

$$\phi^\star(x) \geq \frac{x^2}{1+x/3}\,, \qquad x \geq 0\,.$$

The following lemma is usually considered as an extension of the Bennett inequality to martingales with bounded increments. Various proofs may be found in textbooks on probability theory such as Neveu (1975) or Dacunha-Castelle and Duflo (1986).

Lemma 15.8.6. *Let $\{\mathcal{F}_n\}_{n\geq 1}$ denote a filtration and let $\{Z_n\}_{n\geq 1}$ denote a centered square-integrable martingale with respect to this filtration, with increments bounded by* 1. *Let $\langle Z\rangle_n \stackrel{\text{def}}{=} \sum_{s=1}^n \mathrm{E}[(Z_s - Z_{s-1})^2 \mid \mathcal{F}_{s-1}]$ be its bracket. Then for all λ, the random variables*

$$\exp[\lambda Z_n - \phi(\lambda)\langle Z\rangle_n]$$

form an $\{\mathcal{F}_n\}$-adapted super-martingale.

Let us now recall Doob's maximal inequality and the optional sampling principle. Doob's maximal inequality asserts that if $\{Z_n\}$ is a a super-martingale, then for all n_0 and all $x > 0$,

$$\mathrm{P}\left(\sup_{n\geq n_0} Z_n \geq x\right) \leq \frac{\mathrm{E}[(Z_{n_0})_+]}{x}\,. \tag{15.9}$$

Recall that a random variable T is a stopping time with respect to a filtration $\{\mathcal{F}_n\}$ if for each n the event $\{T \leq n\}$ is $\mathcal{F}_n$-measurable.

The optional sampling theorem asserts that if $T_1, T_2, \ldots, T_k, \ldots$ form an increasing sequence of stopping times with respect to $\{\mathcal{F}_n\}$, then the sequence $\{Z_{T_i}\}$ is a $\{\mathcal{F}_{T_i}\}$-adapted super-martingale.

Considering a stopping time T and the increasing sequence $\{T \vee n\}$ of stopping times, it follows from Lemma 15.8.6, Doob's maximal inequality, and the optional sampling theorem that if $\{Z_n\}$ is a martingale with increments bounded by 1, then for any stopping time T,

$$\mathrm{P}\left(\exists n \geq T:\ |Z_n| > \frac{\phi(\lambda)}{\lambda}\langle Z_n\rangle + \alpha\right) \leq 2\exp(-\alpha\lambda)\,. \tag{15.10}$$

Let $B_1 \leq B_2$ be two numbers. If the stopping times T_1 and T_2 are defined by $T_1 = \inf\{n : \langle Z\rangle_n \geq B_1\}$ and $T_2 = \inf\{n : \langle Z\rangle_n \geq B_2\}$, (15.10) entails that for any $x > 0$,

$$\begin{aligned}
\mathrm{P}\left(\exists n \in \{T_1, \ldots T_2\} :\ |Z_n| > x\right) &\leq 2\exp\left\{-B_2 \sup_\lambda \left[\lambda\frac{x}{B_2} - \phi(\lambda)\right]\right\} \\
&= 2\exp\left\{-B_2\phi^\star\left(\frac{x}{B_2}\right)\right\} \\
&\leq 2\exp\left\{-\frac{x^2}{2\big(B_2 + x/3\big)}\right\}\,.
\end{aligned} \tag{15.11}$$

This inequality will aver to be the workhorse in the proof of Theorem 15.8.2.

15.8.2 The Martingale Approach

The following observation has proved to be crucial in the developments that started with Finesso (1991) and culminated in Csiszár (2002). For each $r > r_\star$ and $a_{1:r} \in \mathsf{Y}^r$, the random variables $Z_n(a_{1:r})$ defined by

$$Z_n(a_{1:r}) \stackrel{\text{def}}{=} N_n(a_{1:r}) - N_{n-1}(a_{1:r-1}) \times \mathrm{P}_\star(a_r \mid a_{1:r-1})$$

form an $\{\mathcal{F}_n\}$-adapted martingale. Moreover, this martingale has increments bounded by 1, and the associated bracket has the form

$$\langle Z(a_{1:r})\rangle_n = N_{n-1}(a_{1:r-1})\,\mathrm{P}_\star(a_r \mid a_{1:r-1})[1 - \mathrm{P}_\star(a_r \mid a_{1:r-1})] \,. \tag{15.12}$$

Note that $|Z_n(a_{1:r})| < x$ implies that

$$|\widehat{\mathrm{P}}^{r-1}(a_r \mid a_{1:r-1}) - \mathrm{P}_\star(a_r \mid a_{1:r-1})| < \frac{x}{N_{n-1}(a_{1:r-1})} \,.$$

Hence bounds on the deviations of the martingales $Z_n(a_{1:r})$ for $a_{1:r} \in \mathcal{S}_r \subseteq \mathsf{Y}^r$ are of immediate relevance to the characterization of $\widehat{\mathrm{P}}^{r-1}$.

The following lemma will be the fundamental bridging block in the proof of the large scale typicality Theorem 15.8.1.

Lemma 15.8.7. *Let ξ and η be two positive reals, $r > r_\star$, $a_{1:r} \in \mathcal{S}_r$ and let Z_n denote the martingale associated with $a_{1:r}$. Then for any $\theta > 1$ and any integer $m \geq 0$,*

$$\mathrm{P}_\star\left\{\exists n : \theta^m \leq \langle Z\rangle_n \leq \theta^{m+1},\ \ |Z_n| \geq \sqrt{\langle Z\rangle_n \max[\xi r, \eta \log\log(\langle Z\rangle_n)]}\right\}$$
$$\leq 2\exp\left(-\frac{\max[\xi r, \eta\log\log(\theta^m)]}{2\theta\{1 + (1/3)\sqrt{\max[\xi r, \eta\log\log(\theta^m)]/\theta^{m+2}}\}}\right) \,. \tag{15.13}$$

Proof. Let the stopping time T_m be defined as the first instant n such that $\langle Z\rangle_n \geq \theta^m$. Note that $\langle Z\rangle_n \geq \theta_m$ for n between T_m and T_{m+1}, and we may take $x = \sqrt{\theta^m \max[\xi r, \eta\log\log\theta^m]}$ and $B_2 = \theta^{m+1}$ in (15.11). □

Remark 15.8.8. If $a_{1:r} \in \mathcal{S}_r$, ergodicity implies that $\mathrm{P}_\star$-a.s., $\langle Z(a_{1:r})\rangle_n$ converges to infinity. Choosing $\xi = 0$ and taking $\eta = 2\theta(1+\alpha)$ with $\alpha > 0$, the previous lemma asserts that

$$\mathrm{P}_\star\left\{\exists n :\ \theta^m \leq \langle Z\rangle_n \leq \theta^{m+1},\ \ |Z_n| \geq \sqrt{2\theta(1+\alpha)\langle Z\rangle_n \log\log(\langle Z\rangle_n)}\right\}$$
$$\leq 2\exp\left[\frac{(1+\alpha)\log\log\theta^m}{1 + \frac{1}{3}\sqrt{\frac{2(1+\alpha)\log\log\theta^m}{\theta^{m+1}}}}\right] \,.$$

The sum over m of the right-hand side is finite. Thus by the Borel-Cantelli lemma, $\mathrm{P}_\star$-a.s., the event on the left-hand side only occurs for finitely many m. Combining these two observations and letting θ tend to 1 and α tend to 0 completes the proof that $\mathrm{P}_\star$-a.s.,

$$\limsup_n \frac{|Z_n|}{\sqrt{2\langle Z\rangle_n \log\log\langle Z\rangle_n}} \leq 1 . \tag{15.14}$$

Note that by Corollary 15.8.5 this entails that for some fixed $r > r_\star$, $\mathrm{P}_\star$-a.s., eventually for all $a_{1:r} \in \mathcal{S}_r$,

$$\frac{N_{n-1}(a_{1:r})}{|\mathsf{Y}|}\chi^2[\widehat{\mathrm{P}}^r(\cdot\,|\,a_{1:r}), \mathrm{P}_\star(\cdot\,|\,a_{1:r})] \leq 2\log\log N_{n-1}(a_{1:r})$$

and

$$\frac{1}{|\mathsf{Y}||\mathcal{S}_r|}[\log\widehat{\mathrm{P}}^r(a_{1:r}) - \log\mathrm{P}_\star(a_{1:r})] \leq 2\log\log n .$$

If we were ready to assume that $r_\star$ is smaller than some given upper bound on the true order, this would be enough to ensure almost sure consistency of penalized maximum likelihood order estimators by taking

$$\mathrm{pen}(n,r) = 2|\mathsf{Y}|^{r+1}\log\log n .$$

■

15.8.3 The Union Bound Meets Martingale Inequalities

The following lemma will allow us to control $\sup_{r:r_\star\leq r\leq\alpha\log n}\{\log\widehat{\mathrm{P}}^r - \log\mathrm{P}_\star\}$.

Lemma 15.8.9. *For every $\delta > 0$ there exists $\alpha > 0$ (depending on $\mathrm{P}_\star$) such that eventually almost surely as $n \to \infty$, for all $a_{1:r}$ in $\mathcal{S}_r$ with $r_\star < r \leq \alpha\log n$,*

$$|Z_n(a_{1:r})| \leq \sqrt{\delta\langle Z(a_{1:r})\rangle_n \log\langle Z(a_{1:r})\rangle_n} .$$

Let the event $D_n^{\xi,c,\eta}(a_{1:r})$ be defined by

$$D_n^{\xi,c,\eta}(a_{1:r}) \stackrel{\text{def}}{=} \Big\{y_{1:n} :\ \langle Z(a_{1:r})\rangle_n > cr,$$
$$|Z_n(a_{1:r})| \geq \sqrt{\langle Z(a_{1:r})\rangle_n \max[\xi r, \eta\log\log(\langle Z(a_{1:r})\rangle_n)]}\Big\} .$$

Lemma 15.8.10. *Let ξ, η and c be chosen in a way that there exists $\theta > 1$ such that*

$$\xi > 2\log|\mathsf{Y}|\left[\theta + \frac{\sqrt{\xi}}{3}\max(c^{-1/2},1)\right] \tag{15.15}$$

and

$$\eta > \frac{\xi}{\frac{\xi}{2[\theta+\sqrt{\xi}/3\max(c^{-1/2},1)]} - \log|\mathsf{Y}|} . \tag{15.16}$$

Then

$$\limsup_n \sum_{r\geq r_\star} \sum_{a_{1:r}\in\mathcal{S}_r} \mathbb{1}_{D_n^{\xi,c,\eta}(a_{1:r})} = 0 \qquad \mathrm{P}_\star\text{-}a.s.$$

Proof. Fix $\theta > 1$ in such a way that (15.15) and (15.16) are satisfied. For each integer m, let the event $E_m^{\xi,c,\eta}(a_{1:r})$ be defined by

$$E_m^{\xi,c,\eta}(a_{1:r}) \stackrel{\text{def}}{=} \Big\{ y_{1:\infty} : \theta^m > cr,\ \exists a_{1:r},\ \exists n \in \{T_m(a_{1:r}), \ldots, T_{m+1}(a_{1:r})\}\ ,$$
$$|Z_n(a_{1:r})| \geq \sqrt{\langle Z(a_{1:r})\rangle_n \max[\xi r, \eta \log\log(\langle Z(a_{1:r})\rangle_n)]} \Big\}\ .$$

The lemma will be proved in two steps. We will first check that $\mathrm{P}_\star$-a.s., only finitely many events $E_m^{\xi,c,\eta}(a_{1:r})$ occur. Then we will check that on a set of sample paths that has probability 1, this entails that only finitely many events $D_n^{\xi,c,\eta}(a_{1:r})$ occur.

Note that

$$\max[\xi r, \eta\log\log(\theta^m)] = \begin{cases} \xi r & \text{if } r \geq \frac{\eta}{\xi}\log\log\theta^m\ , \\ \eta\log\log(\theta^m) & \text{otherwise}\ . \end{cases}$$

To alleviate notations, let μ be defined as

$$\mu = \frac{\xi}{2\Big[\theta + \frac{\sqrt{\xi}}{3}\max(c^{-1/2}, 1)\Big]} - \log|\mathsf{Y}|\ .$$

Then

$$\mathrm{E}\Big[\sum_m \sum_r \sum_{a_{1:r}} \mathbb{1}_{E_m^{\xi,c,\eta}(a_{1:r})}\Big]$$
$$\leq \sum_m \sum_{\frac{\eta}{\xi}\log\log\theta^m \leq r \leq \theta^m/c} |\mathsf{Y}|^r \exp\left[-\frac{\xi r}{2(\theta + \frac{1}{3}\sqrt{\frac{\xi r}{\theta^m}})}\right]$$
$$+ \sum_{r_\star < r \leq \frac{\eta}{\xi}\log\log\theta^m} |\mathsf{Y}|^r \exp\left[-\frac{\eta\log\log\theta^m \xi r}{2(1 + \frac{1}{3}\sqrt{\frac{\eta\log\log\theta^m}{\theta^m}})}\right]$$
$$\leq \sum_m \exp\Big(-\frac{\mu\eta}{\xi}\log\log\theta^m\Big) \times \left[\frac{1}{|\mathsf{Y}|-1} + \frac{1}{1-\exp(-\mu)}\right]\ .$$

Note that as $\mu\eta > \xi$, by (15.15), the last sum is finite. This shows that our first goal is attained.

Now as $\mathrm{P}_\star$ is assumed to be ergodic, $\mathrm{P}_\star$-a.s., for all $r > r_\star$ and all $a_{1:r} \in \mathcal{S}_r$, $\langle Z(a_{1:r})\rangle_n$ tends to infinity. Let us consider such a sample path. Then if infinitely many events of the form $D_n^{\xi,c,\eta}(a_{1:r})$ occur for a fixed pattern $a_{1:r}$, also infinitely many events of the form $E_m^{\xi,c,\eta}(a_{1:r})$ occur for the same fixed pattern.

If there exists an infinite sequence $\{a_{1:r_n}\}$ of patterns such that the events $D_n^{\xi,c,\eta}(a_{1:r_n})$ occur for infinitely many n, then infinitely many events of the form $E_{m_n}^{\xi,c,\eta}(a_{1:r_n})$ also occur. □

In order to prove Lemma 15.8.9, we will need lower bounds on $\mathrm{P}_\star\{a_{1:r}\}$ for $r \le r_n$ and $a_{1:r} \in \mathcal{S}_r$. As P has Markov order $r_\star$ we have

$$\mathrm{P}_\star(a_{1:r}) = \mathrm{P}_\star(a_{1:r_\star}) \prod_{j=r_\star+1}^{r} \mathrm{P}_\star(a_j \mid a_{j-1:j-r_\star}) .$$

Now let $\gamma = \min_{a_{1:r_\star} \in \mathcal{S}_{r_\star}} \mathrm{P}_\star(a_{1:r_\star})$ and $\kappa = \min_{a_{1:r_\star+1} \in \mathcal{S}_{r_\star+1}} \mathrm{P}_\star(a_{r_\star+1} \mid a_{1:r_\star})$. Then

$$\min_{a_{1:r} \in \mathcal{S}_r} \mathrm{P}_\star(a_{1:r}) \ge \gamma \kappa^{r-r_\star} . \tag{15.17}$$

Proof (of Lemma 15.8.9). We will rely on Lemma 15.8.10 and we thus fix η, ξ and c to satisfy the conditions of this lemma. The challenge will consist in checking that for every $\delta > 0$ we can find some $\alpha > 0$ such that

(i) $\mathrm{P}_\star$-a.s. all the “clocks” associated with patterns in $\cup_{r\in\{r_\star,\dots,\alpha \log n\}} \mathcal{S}_r$ move sufficiently fast, that is, for all sufficiently large n,

$$\langle Z(a_{1:r}) \rangle_n > r \quad \text{for all } \; a_{1:r} \in \cup_{r\in\{r_\star,\dots,\alpha \log n\}} \mathcal{S}_r \; ;$$

(ii) For all sufficiently large n,

$$\max[\xi r, \eta \log\log\langle Z(a_{1:r}) \rangle_n] \le \delta \log n \quad \text{for all } \; a_{1:r} \in \cup_{r\in\{r_\star,\dots,\alpha \log n\}} \mathcal{S}_r \, .$$

Let us first make a few observations. If $1 - \epsilon_{r-1} < |N_{n-1}(a_{1:r-1})/(n-r+1)\,\mathrm{P}_\star(a_{1:r-1})| < 1 + \epsilon_{r-1}$ and

$$|Z_n(a_{1:r})| < \sqrt{\langle Z(a_{1:r}) \rangle_n \max[\xi r, \eta \log\log\langle Z(a_{1:r}) \rangle_n]} \, ,$$

then

$$\begin{aligned}
&N_n(a_{1:r}) \\
&\quad > N_{n-1}(a_{1:r-1})\,\mathrm{P}_\star(a_r \mid a_{1:r-1}) \\
&\qquad - \sqrt{\langle Z(a_{1:r}) \rangle_n \max[\xi r, \eta \log\log\langle Z(a_{1:r}) \rangle_n]} \\
&\quad > (n-r+1)\,\mathrm{P}_\star(a_{1:r}) \times \\
&\qquad \left\{ 1 - \epsilon_{r-1} - \frac{\sqrt{(1+\epsilon_{r-1}) \max[\xi r, \eta \log\log(2(n-r+1))\,\mathrm{P}_\star(a_{1:r-1})]}}{\sqrt{(n-r+1)\,\mathrm{P}_\star(a_{1:r})}} \right\} \\
&\quad > (n-r+1)\,\mathrm{P}_\star(a_{1:r}) \left\{ 1 - \epsilon_{r-1} - \frac{2\sqrt{\max[\xi r, \eta \log\log(2n)]}}{\sqrt{n \gamma \kappa^{r-r_\star}}} \right\}
\end{aligned}$$

and

$$
\begin{aligned}
N_n(a_{1:r}) &< N_{n-1}(a_{1:r-1}) \, \mathrm{P}_\star(a_{1:r-1}) \\
&\quad + \sqrt{\langle Z(a_{1:r})\rangle_n \max[\xi r, \eta \log\log\langle Z(a_{1:r})\rangle_n]} \\
&< (n-r+1) \, \mathrm{P}_\star(a_{1:r}) \left\{ 1 + \epsilon_{r-1} + \frac{2\sqrt{\max[\xi r, \eta \log\log(2n)]}}{\sqrt{n\gamma\kappa^{r-r_\star}}} \right\} .
\end{aligned}
$$

Now $\mathrm{P}_\star$-a.s., for n large enough and all $a_{1:r_\star} \in \mathcal{S}_{r_\star}$,

$$
1 - \epsilon_{r_\star} < |N_n(a_{1:r_\star})/(n-r+1) \ \mathrm{P}_\star(a_{1:r_\star})| < 1 + \epsilon_{r_\star} .
$$

Let α be such that $\alpha < 1/\log(1/\kappa)$. Then for $r < (\eta/\xi)\log\log n$, we may choose $\epsilon_r(n)$ in such a way that

$$
\epsilon_r(n) \le \epsilon_{r_\star}(n) + \frac{\eta}{\xi}\log\log(2n) \frac{2\sqrt{\eta\log\log 2n}}{\sqrt{n\gamma\kappa^{(\eta/\xi)\log\log(2n)}}} + 2\alpha\log n \frac{\sqrt{\xi\alpha\log n}}{n^{1/4}\sqrt{\gamma}}
$$

for all $r \le \alpha\log n$. Hence for sufficiently large n, we have $\epsilon_r(n) \le 1/2$ for all $r \le \alpha \log n$.

This however implies that $\mathrm{P}_\star$-a.s. for all sufficiently large n, all $r \le \alpha\log n$ and all $a_{1:r} \in \mathcal{S}_r$,

$$
\langle Z(a_{1:r})\rangle_n \ge \frac{1}{2}(n-r+1)\gamma\kappa^r > cr .
$$

By Lemma 15.8.10, this renders that $\mathrm{P}_\star$-a.s., for all sufficiently large n, all $r \le \alpha\log n$ and all $a_{1:r} \in \mathcal{S}_r$,

$$
|Z_n(a_{1:r})| \le \sqrt{\langle Z(a_{1:r})\rangle_n \max[\xi r, \eta\log\log\langle Z(a_{1:r})\rangle_n]} .
$$

If α is sufficiently small, the right-hand side of this display is smaller than $\sqrt{\delta\langle Z(a_{1:r})\rangle_n \log\langle Z(a_{1:r})\rangle_n}$ in the range of r considered. □

The next lemma will prove crucial when checking the most delicate part of the BIC consistency theorem. It will allow us to rule out (almost surely) the possibility that the BIC order estimator jitters around $\log n$ for infinitely many values of n.

For any $\xi > 0$, any $c > 0$ and any $a_{1:r}$, define the event $B_n^{\xi,c}(a_{1:r})$ by

$$
\begin{aligned}
B_n^{\xi,c}(a_{1:r}) \stackrel{\text{def}}{=} \Big\{ & y_{1:n} : \langle Z(a_{1:r})\rangle_n > cr \text{ and} \\
& |Z_n(a_{1:r})| \ge \sqrt{\langle Z(a_{1:r})\rangle_n \max[\xi r, 4\log\log\langle Z(a_{1:r})\rangle_n]} \Big\} .
\end{aligned}
$$

Lemma 15.8.11. *Let $\xi > 0$ and $c > 0$ be such that $\sqrt{\xi} < 3/2$. Then*

$$
\limsup_n \sup_{r > r_\star} \frac{1}{|\mathcal{S}_r|} \sum_{a_{1:r}\in\mathcal{S}_r} \mathbb{1}_{B_n^{\xi,c}(a_{1:r})} = 0 \qquad \mathrm{P}_\star\text{-}a.s.
$$

Proof. Choose $\theta > 1$ such that $\theta(1 + \frac{1}{3}\sqrt{\xi}) \leq 3/2$. In the sequel, we only consider those m such that $\theta\big(1 + \frac{1}{3}\sqrt{\frac{4\log\log\theta^m}{\theta^{m+2}}}\big) \leq 3/2$. Put

$$C_m^{\xi,c}(a_{1:r}) \stackrel{\text{def}}{=} \Big\{y_{1:\infty} : \exists n : \theta^m \leq \langle Z(a_{1:r})\rangle_n \leq \theta^{m+1},\ \ \theta^m > cr \\ \text{and } |Z_n(a_{1:r})| \geq \sqrt{\langle Z(a_{1:r})\rangle_n \max[\xi r, 4\log\log\langle Z(a_{1:r})\rangle_n]}\Big\} .$$

The proof is carried in two steps:

(i) Proving that $\mathrm{P}_\star$-a.s.,

$$\limsup_M \sum_{m>M} \sum_{r>r_\star} \frac{1}{|\mathcal{S}_r|} \sum_{a_{1:r}\in\mathcal{S}_r} \mathbb{1}_{C_m^{\xi,c}(a_{1:r})} = 0 ; \tag{15.18}$$

(ii) Proving that this entails

$$\limsup_n \sup_{r>r_\star} \frac{1}{|\mathcal{S}_r|} \sum_{a_{1:r}\in\mathcal{S}_r} \mathbb{1}_{B_n^{\xi,c}(a_{1:r})} = 0 . \tag{15.19}$$

Note that when dealing with $|\mathcal{S}_r|^{-1}\sum_{a_{1:r}\in\mathcal{S}_r} \mathbb{1}_{C_m^{\xi,c}(a_{1:r})}$, we adapt the time-scale at which we analyze $Z_n(a_{1:r})$ to the pattern. This allows us to formulate a rather strong statement: not only does

$$u_m = \sum_{r>r_\star} \frac{1}{|\mathcal{S}_r|} \sum_{a_{1:r}\in\mathcal{S}_r} \mathbb{1}_{C_m^{\xi,c}(a_{1:r})}$$

tend to 0 as m tends to infinity, but the series $\sum_m u_m$ is convergent.

Let us start with the first step. Thanks to our assumptions on the values of ξ and m,

$$\begin{aligned}
&\mathrm{E}\Big[\sum_{r>r_\star} \frac{1}{|\mathcal{S}_r|} \sum_{a_{1:r}\in\mathcal{S}_r} \mathbb{1}_{C_m^{\xi,c}(a_{1:r})}\Big] \\
&\qquad \leq \sum_{\frac{4}{\xi}\log\log\theta^m < r < \frac{\theta^m}{c}} \exp\left[-\frac{\xi r}{2\theta(1+\frac{\sqrt{\xi}}{3})}\right] \\
&\qquad\quad + \sum_{r<\frac{4}{\xi}\log\log\theta^m} \exp\left[-\frac{4\log\log\theta^m}{2\theta\big(1+\frac{1}{3}\sqrt{\frac{4\log\log\theta^m}{\theta^{m+2}}}\big)}\right] \\
&\qquad \leq \exp\Big(-\frac{4}{3}\log\log\theta^m\Big)\Big[\frac{1}{1-\exp(-1/3)} + \frac{4}{\xi}\log\log\theta^m\Big] .
\end{aligned}$$

Hence

$$\sum_{m>M} \mathrm{E}\Big[\sum_{r>r_\star} \frac{1}{|\mathcal{S}_r|} \sum_{a_{1:r}\in\mathcal{S}_r} \mathbb{1}_{C_m^{\xi,c}(a_{1:r})}\Big] < \infty ,$$

which shows that (15.18) holds $\mathrm{P}_\star$-a.s.

Let us now proceed to the second step. As $\mathrm{P}_\star$ is assumed ergodic, it is enough to consider sequences $y_{1:\infty}$ such that $\langle Z(a_{1:r})\rangle_n$ tends to infinity for all $a_{1:r}$.

Assume that there exists a sequence $\{r_n\}$ such that for some $\alpha > 0$, for infinitely many n,

$$\frac{1}{|\mathcal{S}_{r_n}|} \sum_{a_{1:r_n} \in \mathcal{S}_{r_n}} \mathbb{1}_{B_n^{\xi,c}(a_{1:r_n})} > \alpha \;.$$

If the sequence r_n has an accumulation point r, then there exists some $a_{1:r}$ such that $B_n^{\xi,c}(a_{1:r_n})$ occurs for infinitely many n. This however implies that infinitely many events $C_m^{\xi,c}(a_{1:r})$ occur, which means that whatever M,

$$\sum_{m>M} \frac{1}{|\mathcal{S}_r|} \mathbb{1}_{C_m^{\xi,r}(a_{1:r})} = \infty \;.$$

If the sequence r_n is increasing then for each n such that

$$\frac{1}{|\mathcal{S}_{r_n}|} \sum_{a_{1:r_n} \in \mathcal{S}_{r_n}} \mathbb{1}_{B_n^{\xi,c}(a_{1:r_n})} > \alpha$$

holds, also

$$\frac{1}{|\mathcal{S}_{r_n}|} \sum_{a_{1:r_n}} \sum_{m>\log_\theta(cr_n)} \mathbb{1}_{C_m^{\xi,r}(a_{1:r_n})} > \alpha \;.$$

Hence, whatever M,

$$\sum_{m>M} \sum_{r>\theta^m/c} \frac{1}{|\mathcal{S}_r|} \sum_{a_{1:r} \in \mathcal{S}_r} \mathbb{1}_{C_m^{\xi,c}(a_{1:r})} > \alpha \;.$$

□

Remark 15.8.12. Lemmas 15.8.10 and 15.8.11 are proved in a very similar way, they have a similar form, but convey a different message. In Lemma 15.8.10, the constant η may be taken rather close to 2 and the constants in the lemma may be considered as trade-offs between the constants that show up in the law of the iterated logarithm and the constants that may be obtained if the union bound has to be used repeatedly. Note that if the conditions of Lemma 15.8.10 are to be met, for a given ξ we cannot look for arbitrarily small c.

This is sharp contrast with the setting of Lemma 15.8.11. There the constant η was deliberately set to 4, and the freedom allowed by this convention, as well as by the normalizing factors $1/|\mathcal{S}_r|$, allows us to consider arbitrarily small c. ■

Proof (of Theorem 15.8.2). First note that if $|\mathcal{S}_r|$ does not grow exponentially fast in r, then the Markov chain has zero entropy rate, it is a deterministic process and the likelihood ratios of interest are equal to 1. Thus there is nothing to do.

Let us hence thus assume that there exists some $h > 0$ such that for all sufficiently large r, $\log |\mathcal{S}_r| \geq hr$. Then

$$\frac{1}{|\mathcal{S}_r|}\frac{1}{\log n}[\log \widehat{\mathrm{P}}^r(y_{1:n}) - \log \mathrm{P}_\star(y_{1:n})] \leq \mathrm{e}^{-hr} \log \frac{1}{\gamma \kappa^n} .$$

Hence for $r \geq (C/h)\log n$ with $C > -\log\kappa$, the quantity tends to 0 as n tends to infinity. It thus remains to prove that for every $\delta > 0$,

$$\sup_{r_\star \leq r \leq \frac{C}{h}\log n} \frac{1}{|\mathcal{S}_r|}\frac{1}{\log n}[\log \widehat{\mathrm{P}}^r(y_{1:n}) - \log \mathrm{P}_\star(y_{1:n})] \geq \delta$$

occurs only finitely many times.

Assume $\delta < 1/4$. Then by Lemma 15.8.9 there exists some $\alpha > 0$ depending on $\mathrm{P}_\star$ and δ such that for all sufficiently large n, all r such that $r_\star < r < \alpha \log n$ and all $a_{1:r} \in \mathcal{S}_r$,

$$|Z_n(a_{1:r})| < \sqrt{\delta \langle Z(a_{1:r})\rangle_n \log \langle Z(a_{1:r})\rangle_n} . \tag{15.20}$$

But this inequality shows that

$$|\widehat{\mathrm{P}}^r(a_r \mid a_{1:r-1}) - \mathrm{P}_\star(a_r \mid a_{1:r-1})| \leq \sqrt{\delta \frac{\mathrm{P}_\star(a_r \mid a_{1:r-1}) \log N_{n-1}(a_{1:r-1})}{N_{n-1}(a_{1:r-1})}} .$$

Hence $\mathrm{P}_\star$-a.s., for all sufficiently large n and all $r_\star < r < \alpha \log n$,

$$\frac{N_{n-1}(a_{1:r-1})}{|\mathsf{Y}|}\chi^2[\widehat{\mathrm{P}}^r(\cdot \mid a_{1:r-1}), \mathrm{P}_\star(\cdot \mid a_{1:r-1})] \leq \delta \log n . \tag{15.21}$$

On the other hand, notice that if

$$|Z_n(a_{1:r})| \leq \frac{1}{2}\langle Z(a_{1:r})\rangle_n ,$$

then

$$|\widehat{\mathrm{P}}^r_n(a_r \mid a_{1:r-1}) - \mathrm{P}_\star(a_r \mid a_{1:r-1})| \leq \frac{1}{2}\mathrm{P}_\star(a_r \mid a_{1:r-1}) .$$

Hence by Corollary 15.8.5, as $\delta \log u < u/4$, $\mathrm{P}_\star$-a.s., for all sufficiently large n and all $r_\star < r < \alpha \log n$,

$$\frac{1}{|\mathcal{S}_r| \log n}[\log \widehat{\mathrm{P}}^r_n(y_{1:n}) - \log \mathrm{P}_\star(y_{1:n})] \leq \delta .$$

Thus $\mathrm{P}_\star$-a.s., for sufficiently large n,

$$\sup_{r < r_\star < \alpha \log n} \frac{1}{|\mathcal{S}_r| \log n}[\log \widehat{\mathrm{P}}^r_n(y_{1:n}) - \log \mathrm{P}_\star(y_{1:n})] \leq \delta .$$

Let us now consider those r such that $\alpha \log n \le r \le (C/h)\log n$. Choose ξ_2 and c_2 such that for some (irrelevant) $\eta > 2$, the conditions of Lemma 15.8.10 are satisfied. Note that for n sufficiently large, for all r such that $\alpha \log n \le r \le (C/h)\log n$, $\max(\xi_2 r, \eta \log\log n) = \xi_2 r$.

Let $\xi_1 > 0$ and $c_1 > 0$ be chosen in such a way that $c_1 + \xi_1 < h\delta/C$. We will use Lemma 15.8.11 with those constants. Recall that c_1 and ξ_1 may be chosen arbitrarily close to 0 (see Remark 15.8.12).

Let $G_1^{r,n}$, $G_2^{r,n}$, $G_3^{r,n}$ and $G_4^{r,n}$ be defined by

$$
\begin{aligned}
G_1^{r,n} &= \{a_{1:r-1} : N_{n-1}(a_{1:r-1}) < c_1 r\} \cap \mathcal{S}_{r-1} \;, \\
G_2^{r,n} &= \{a_{1:r-1} : c_1 r \ge N_{n-1}(a_{1:r-1}) \\
&\qquad \text{and for all } a \in \mathsf{Y},\ |Z_n(a_{1:r-1}, a)| < \sqrt{\xi_1 r \langle Z(a_{1:r-1}, a)\rangle_n}\} \;, \\
G_3^{r,n} &= \{a_{1:r-1} : c_1 r \le N_{n-1}(a_{1:r-1}) < c_2 r \\
&\qquad \text{and for some } a \in \mathsf{Y},\ |Z_n(a_{1:r-1}, a)| < \sqrt{\xi_1 r \langle Z(a_{1:r-1}, a)\rangle_n}\} \;, \\
G_4^{r,n} &= \{a_{1:r-1} \;:\; c_2 r < N_{n-1}(a_{1:r-1}) \\
&\qquad \text{and for all } a \in \mathsf{Y},\ |Z_n(a_{1:r-1}, a)| < \sqrt{\xi_2 r \langle Z(a_{1:r-1}, a)\rangle_n}\Big\} \setminus G_2^{r,n} \;.
\end{aligned}
$$

By Lemma 15.8.10, $\mathrm{P}_\star$-a.s., for sufficiently large n and all r such that $\alpha \log n \le r \le (C/h)\log n$,

$$
G_1^{r,n} \cup G_2^{r,n} \cup G_3^{r,n} \cup G_4^{r,n} = \mathcal{S}_{r-1} \;.
$$

Moreover by Lemma 15.8.11, $\mathrm{P}_\star$-a.s., for sufficiently large n and the same r,

$$
\frac{|G_3^{r,n}| + |G_4^{r,n}|}{|\mathcal{S}_{r-1}|} < \delta \;.
$$

By the definition of $G_2^{r,n}$ and $G_4^{r,n}$, we are in a position to use Corollary 15.8.5 to obtain

$$
N_{n-1}(a_{1:r-1}) D(\widehat{\mathrm{P}}_n(\cdot \,|\, a_{1:r-1}) \,|\, \mathrm{P}_\star(\cdot \,|\, a_{1:r-1})) \;\le\; \begin{cases} \xi_1 r \text{ if } a_{1:r-1} \in G_2^{r,n} \;, \\ \xi_2 r \text{ if } a_{1:r-1} \in G_4^{r,n} \;. \end{cases} \tag{15.22}
$$

Thus $\mathrm{P}_\star$-a.s., for sufficiently large n and all r such that $\alpha \log n \le r \le (C/h)\log n$,

$$
\begin{aligned}
&\log \widehat{\mathrm{P}}^r(y_{1:n}) - \log \mathrm{P}_\star(y_{1:n}) \\
&\quad \le \sum_{i \in G_i^{r,n}} \sum_{a_{1:r-1} \in G_i^{r,n}} N_{n-1}(a_{1:r-1}) D(\widehat{\mathrm{P}}_n(\cdot \,|\, a_{1:r-1}) \,|\, \mathrm{P}_\star(\cdot \,|\, a_{1:r-1})) \\
&\qquad\qquad \le |G_1^{r,n}| c_1 r \log \frac{1}{\kappa} + |G_2^{r,n}| \xi_1 r + |G_3^{r,n}| c_2 r \log \frac{1}{\kappa} + |G_4^{r,n}| \xi_2 r \;.
\end{aligned}
$$

Dividing both sides by $|\mathcal{S}_r| \log n$, we find for the range of r of interest that

$$\frac{1}{|\mathcal{S}_r| \log n} [\log \widehat{\mathrm{P}}^r(y_{1:n}) - \log \mathrm{P}_\star(y_{1:n})] \\ \leq \frac{C}{h} \left[c_1 + \xi_1 + c_2 \frac{|G_3^{r,n}|}{|\mathcal{S}_r|} + \frac{|G_4^{r,n}|}{|\mathcal{S}_r|} \xi_2 \right] .$$

As we may choose $c_1 + \xi_1 \leq h\delta/C$, $\mathrm{P}_\star$-a.s., for sufficiently large n,

$$\sup_{r:\, \alpha \log n \leq r \leq \frac{C}{h} \log n} \frac{1}{|\mathcal{S}_r| \log n} [\log \widehat{\mathrm{P}}^r(y_{1:n}) - \log \mathrm{P}_\star(y_{1:n})] \leq \delta .$$

□

15.9 Complements

The order estimation problem for HMMs and Markov processes became an active topic in the information theory literature in the late 1980s. Early references can be found in Finesso (1991) and Ziv and Merhav (1992). Other versions of the order estimation problem had been tackled even earlier, see Haughton (1988). We refer to Chambaz (2003, Chapter 7) for a brief history of order identification.

The definition of HMM order used in this chapter is classical. A general discussion concerning HMM order and related notions like rank can be found in Finesso (1991).

An early discussion of order estimation issues in ARMA modeling is presented in Azencott and Dacunha-Castelle (1984). Finesso (1991) credits the latter reference for major influence on his work on Markov order estimation. The connections between the performance of generalized likelihood ratio testing and the behavior of maximum likelihood ratios was outlined in Finesso (1991). Using the law of iterated logarithms for the empirical measure of Markov chains in order to identify small penalties warranting consistency in Markov order estimation also goes back to Finesso (1991)

The connections between order estimation and hypothesis testing has been emphasized in the work of Merhav and collaborators (Zeitouni and Gutman, 1991; Zeitouni *et al.*, 1992; Ziv and Merhav, 1992; Feder and Merhav, 2002). Those papers present various settings for composite hypothesis testing in which generalized likelihood ratio testing may or may not be asymptotically optimal.

Though the use of universal coding arguments in order identification is already present in Finesso (1991), Zeitouni and Gutman (1991), and Ziv and Merhav (1992), the paper by Kieffer (1993) provides the most striking exposition of the connections between order identification and universal coding. Versions of Lemmas 15.6.2 and 15.6.3 are at least serendipitous in Kieffer (1993). Results of Section 15.6 can be regarded as elaboration of ideas exposed by Kieffer.

The proof of the first inequality in Lemma 15.6.4 goes back to Shtarkov (1987). The proof of the second inequality for HMMs is due to Csiszár (1990). Variants of the result have been used by Finesso (1991) and Liu and Narayan (1994).

Section 15.8 is mainly borrowed from Csiszár (2002), although the results presented here were already contained in Csiszár and Shields (2000) but justified with different proofs. The use of non-asymptotic tail inequalities (concentration inequalities) for the analysis of model selection procedure has become a standard approach in modern statistics (see Bartlett *et al.*, 2002, and references therein for more examples on this topic).

Section 15.7 is largely inspired by Gassiat and Boucheron (2003), and further results in this direction can be found in Chambaz (2003) and Boucheron and Gassiat (2004).

Part IV

Appendices

A

Conditioning

A.1 Probability and Topology Terminology and Notation

By a *measurable space* is meant a pair $(\mathsf{X}, \mathcal{X})$ with X being a set and $\mathcal{X}$ being a σ-field of subsets of X. The sets in the σ-field are called *measurable sets.* We will always assume that for any $x \in \mathsf{X}$, the singleton set $\{x\}$ is measurable. Typically, if X is a topological space, then $\mathcal{X}$ is the Borel σ-field, that is, the σ-field generated by the open subsets of X. If X is a discrete set (that is, finite or countable), then $\mathcal{X}$ is the power set $\mathcal{P}(\mathsf{X})$, the collection of all subsets of X.

A *positive measure* on a measurable space $(\mathsf{X}, \mathcal{X})$[1] is a measure such that $\mu(A) \geq 0$, for all $A \in \mathcal{X}$, and $\mu(\mathsf{X}) > 0$. A *probability measure* is a positive measure with unit total mass, $\mu(\mathsf{X}) = 1$. All measures will be assumed to be σ-finite.

Let $(\Omega, \mathcal{F})$ and $(\mathsf{X}, \mathcal{X})$ be two measurable spaces. A function $X : \Omega \to \mathsf{X}$ is said to be *measurable* if the set $X^{-1}(A) \in \mathcal{F}$ for all $A \in \mathcal{X}$. If $(\mathsf{X}, \mathcal{X}) = (\mathbb{R}, \mathcal{B}(\mathbb{R}))$ where $\mathcal{B}(\mathbb{R})$ is the Borel σ-field, X is said to be real-valued random variable. By abuse of notation, but in accordance with well-established traditions, the phrase "random variable" usually refers to a real-valued random variable. If X is not the real numbers $\mathbb{R}$, we often write "X-valued random variable".

A σ-field $\mathcal{G}$ on Ω such that $\mathcal{G} \subseteq \mathcal{F}$ is called a *sub-σ-field of $\mathcal{F}$*. If X is a random variable (real-valued or not) such that $X^{-1}(A) \in \mathcal{G}$ for all $A \in \mathcal{X}$ for such a sub-σ-field $\mathcal{G}$, then X is said to be *$\mathcal{G}$-measurable.* If X denotes an X-valued mapping on Ω, then the *σ-field generated by* X, denoted by $\sigma(X)$, is the smallest σ-field on Ω that makes X measurable. It can be expressed as $\sigma(X) = X^{-1}(\mathcal{X}) = \{X^{-1}(B) : B \in \mathcal{X}\}$. Typically it is assumed that X is a random variable, that is, X is $\mathcal{F}$-measurable, and then $\sigma(X)$ is a sub-σ-field of

[1]In some situations, such as when X is a countable set, the σ-field under consideration is unambiguous and essentially unique and we may omit $\mathcal{X}$ for notational simplicity.

$\mathcal{F}$. If Z is a real-valued random variable that is $\sigma(X)$-measurable, then there exists a measurable function $g : \mathsf{X} \to \mathbb{R}$ such that $Z = g \circ X = g(X)$.

If $(\Omega, \mathcal{F})$ is a measurable space and P is a probability measure on $\mathcal{F}$, the triplet $(\Omega, \mathcal{F}, \mathrm{P})$ is called a *probability space.* We then write $\mathrm{E}[X]$ for the expectation of a random variable X on $(\Omega, \mathcal{F})$, meaning the (Lebesgue) integral $\int_\Omega X \, d\mathrm{P}$. The *image of* P *by* X, denoted by P^X, is the probability measure defined by $\mathrm{P}^X(B) = \mathrm{P}(X^{-1}(B))$. As good as all random variables (real-valued or not) in this book are assumed to be defined on a probability space denoted by $(\Omega, \mathcal{F}, \mathrm{P})$, and in most cases this probability space is not mentioned explicitly. The space Ω is sometimes called the *sample space.*

Finally, a few words on topological spaces. A topological space is a set Y equipped with a topology $\mathcal{T}$. A topological space $(\mathsf{Y}, \mathcal{T})$ is called *metrizable* if there exists a metric $d : \mathsf{Y} \times \mathsf{Y} \to [0, \infty]$ such that the topology induced by d is $\mathcal{T}$. If (Y, d) is a metric space, a *Cauchy sequence* in this space is a sequence $\{y_n\}_{n \geq 0}$ in Y such that $d(y_n, y_m) \to 0$ as $n, m \to \infty$. A metric space (Y, d) is called *complete* if every Cauchy sequence in Y has a limit in Y. A topological space $(\mathsf{Y}, \mathcal{T})$ is called a *Polish space* if $(\mathsf{Y}, \mathcal{T})$ is *separable* (i.e., it admits a countable dense subset) and metrizable for some metric d such that the metric space (Y, d) is complete. As a trivial example, $\mathbb{R}^n$ equipped with the Euclidean distance is the most elementary example of a Polish space.

A.2 Conditional Expectation

Let $(\Omega, \mathcal{F}, \mathrm{P})$ be a probability space. For $p > 0$ we denote by $\mathcal{L}^p(\Omega, \mathcal{F}, \mathrm{P})$ the space of random variables X such that $\mathrm{E}\,|X|^p < \infty$, and by $\mathcal{L}^+(\Omega, \mathcal{F}, \mathrm{P})$ the space of random variables X such that $X \geq 0$ P-a.s. If we identify random variables that are equal P-a.s., we get respectively the spaces $L^p(\Omega, \mathcal{F}, \mathrm{P})$ and $L^+(\Omega, \mathcal{F}, \mathrm{P})$. We allow random variables to assume the values $\pm\infty$.

Lemma A.2.1. *Let $(\Omega, \mathcal{F}, \mathrm{P})$ be a probability space, let $X \in \mathcal{L}^+(\Omega, \mathcal{F}, \mathrm{P})$, and let $\mathcal{G}$ be a sub-σ-field of $\mathcal{F}$. Then there exists $Y \in \mathcal{L}^+(\Omega, \mathcal{G}, \mathrm{P})$ such that*

$$\mathrm{E}[XZ] = \mathrm{E}[YZ] \tag{A.1}$$

for all $Z \in \mathcal{L}^+(\Omega, \mathcal{G}, \mathrm{P})$. If $Y' \in \mathcal{L}^+(\Omega, \mathcal{G}, \mathrm{P})$ also satisfies (A.1)*, then $Y = Y'$ P-a.s.*

A random variable with the above properties is called a *version of the conditional expectation* of X given $\mathcal{G}$, and we write $Y = \mathrm{E}[X \mid \mathcal{G}]$. Conditional expectations are thus defined up to P-almost sure equality. Hence, when writing $\mathrm{E}[X \mid \mathcal{G}] = Y$ for instance, we always mean that this relations holds P-a.s., that is, Y is a version of the conditional expectation.

One can indeed extend the definition of the conditional expectation to random variables that do not belong to $\mathcal{L}^+(\Omega, \mathcal{F}, \mathrm{P})$. We follow here the approach outlined in Shiryaev (1996, Section II.7).

Definition A.2.2 (Conditional Expectation). *Let $(\Omega, \mathcal{F}, \mathrm{P})$ be a probability space, let X be a random variable and let $\mathcal{G}$ be a a sub-σ field of $\mathcal{F}$. Define $X^+ \stackrel{\text{def}}{=} \max(X,0)$ and $X^- \stackrel{\text{def}}{=} -\min(X,0)$. If*

$$\min\{\mathrm{E}[X^+ \,|\, \mathcal{G}], \mathrm{E}[X^- \,|\, \mathcal{G}]\} < \infty \quad \mathrm{P}\text{-}a.s.\,,$$

then (a version of) the conditional expectation of X given $\mathcal{G}$ is defined by

$$\mathrm{E}[X \,|\, \mathcal{G}] = \mathrm{E}[X^+ \,|\, \mathcal{G}] - \mathrm{E}[X^- \,|\, \mathcal{G}]\ ;$$

on the set of probability 0 of sample points where $\mathrm{E}[X^+ \,|\, \mathcal{G}]$ and $\mathrm{E}[X^- \,|\, \mathcal{G}]$ are both infinite, the above difference is assigned an arbitrary value, for instance, zero.

In particular, if $\mathrm{E}[|X| \,|\, \mathcal{G}] < \infty$ P-a.s., then $\mathrm{E}[X^+ \,|\, \mathcal{G}] < \infty$ and $\mathrm{E}[X^- \,|\, \mathcal{G}] < \infty$ P-a.s., and we may always define the conditional expectation in this context. Note that for $X \in \mathcal{L}^1(\Omega, \mathcal{F}, \mathrm{P})$, $\mathrm{E}[X^+] < \infty$ and $\mathrm{E}[X^-] < \infty$. By applying (A.1) with $Z \equiv 1$, $\mathrm{E}[\mathrm{E}(X^+ \,|\, \mathcal{G})] = \mathrm{E}[X^+] < \infty$ and $\mathrm{E}[\mathrm{E}(X^- \,|\, \mathcal{G})] = \mathrm{E}[X^-] < \infty$. Therefore, $\mathrm{E}[X^+ \,|\, \mathcal{G}] < \infty$ and $\mathrm{E}[X^- \,|\, \mathcal{G}] < \infty$, and thus the conditional expectation is always defined for $X \in \mathcal{L}^1(\Omega, \mathcal{F}, \mathrm{P})$.

Let Y be a random variable and let $\sigma(X)$ be the sub-σ-field generated by a random variable X. If $\mathrm{E}[Y \,|\, \sigma(X)]$ is well-defined, we write $\mathrm{E}[Y \,|\, X]$ rather than $\mathrm{E}[Y \,|\, \sigma(X)]$. This is called the *conditional expectation of Y given X*. By construction, $\mathrm{E}[Y \,|\, X]$ is a $\sigma(X)$-measurable random variable. Thus (cf. Section A.1), there exists a real measurable function g on X such that $\mathrm{E}[Y \,|\, X] = g(X)$. The choice of g is unambiguous in the sense that any two functions g and $\tilde{g}$ satisfying this equality must be equal P^X-a.s. We sometimes write $\mathrm{E}[Y \,|\, X = x]$ for such a $g(x)$.

Many of the useful properties of expectations extend to conditional expectations. We state below some these useful properties. In the following statements, all equalities and inequalities between random variables, and convergence of such, should be understood to hold P-a.s.

Proposition A.2.3 (Elementary Properties of Conditional Expectation).

(a) If $X \leq Y$ and, either, $X \geq 0$ and $Y \geq 0$, or $\mathrm{E}[|X| \,|\, \mathcal{G}] < \infty$ and $\mathrm{E}[|Y| \,|\, \mathcal{G}] < \infty$, then $\mathrm{E}[X \,|\, \mathcal{G}] \leq \mathrm{E}[Y \,|\, \mathcal{G}]$.

(b) If $\mathrm{E}[|X| \,|\, \mathcal{G}] < \infty$, then $|\mathrm{E}[X \,|\, \mathcal{G}]| \leq \mathrm{E}[|X| \,|\, \mathcal{G}]$.

(c) If $X \geq 0$ and $Y \geq 0$, then for any non-negative real numbers a and b,

$$\mathrm{E}[aX + bY \,|\, \mathcal{G}] = a\,\mathrm{E}[X \,|\, \mathcal{G}] + b\,\mathrm{E}[Y \,|\, \mathcal{G}]\ .$$

If $\mathrm{E}[|X| \,|\, \mathcal{G}] < \infty$ and $\mathrm{E}[|Y| \,|\, \mathcal{G}] < \infty$, the same equality holds for arbitrary real numbers a and b.

(d) If $\mathcal{G} = \{\emptyset, \Omega\}$ is the trivial σ-field and $X \geq 0$ or $\mathrm{E}\,|X| < \infty$, then $\mathrm{E}[X \,|\, \mathcal{G}] = \mathrm{E}[X]$.

(e) If $\mathcal{H}$ is a sub-σ-field of $\mathcal{F}$ such that $\mathcal{G} \subseteq \mathcal{H}$ and $X \geq 0$, then

$$\mathrm{E}[\mathrm{E}(X \mid \mathcal{H}) \mid \mathcal{G}] = \mathrm{E}[X \mid \mathcal{G}] . \tag{A.2}$$

If $\mathrm{E}[|X| \mid \mathcal{G}] < \infty$, then $\mathrm{E}[|X| \mid \mathcal{H}] < \infty$ and (A.2) holds.

(f) Assume that X is independent of $\mathcal{G}$, in the sense that $\mathrm{E}[XY] = \mathrm{E}[X]\,\mathrm{E}[Y]$ for all $\mathcal{G}$-measurable random variables Y. If, in addition, either $X \geq 0$ or $\mathrm{E}\,|X| < \infty$, then

$$\mathrm{E}[X \mid \mathcal{G}] = \mathrm{E}[X] . \tag{A.3}$$

(g) If X is $\mathcal{G}$-measurable, $X \geq 0$, and $Y \geq 0$, then

$$\mathrm{E}[XY \mid \mathcal{G}] = X\,\mathrm{E}[Y \mid \mathcal{G}] . \tag{A.4}$$

The same conclusion holds if $\mathrm{E}[|XY| \mid \mathcal{G}]$, $|X|$, and $\mathrm{E}[|Y| \mid \mathcal{G}]$ are all finite.

Proof. (a): Assume that X and Y are non-negative. By (A.1), for any $A \in \mathcal{G}$,

$$\mathrm{E}[\mathrm{E}(X \mid \mathcal{G})\mathbb{1}_A] = \mathrm{E}[X\mathbb{1}_A] \leq \mathrm{E}[Y\mathbb{1}_A] = \mathrm{E}[\mathrm{E}(Y \mid \mathcal{G})\mathbb{1}_A] .$$

Setting, for any $M > 0$, $A_M = \{\mathrm{E}[X \mid \mathcal{G}] - \mathrm{E}[Y \mid \mathcal{G}] \geq 1/M\}$, the above relation implies that $P(A_M) = 0$. Therefore, $\mathrm{P}\{\mathrm{E}[X \mid \mathcal{G}] - \mathrm{E}[Y \mid \mathcal{G}] > 0) = 0$. For general X and Y, the condition $X \leq Y$ implies that $X^+ \leq Y^+$ and $Y^- \leq X^-$; therefore $\mathrm{E}[X^+ \mid \mathcal{G}] \leq \mathrm{E}[Y^+ \mid \mathcal{G}]$ and $\mathrm{E}[Y^- \mid \mathcal{G}] \leq \mathrm{E}[X^- \mid \mathcal{G}]$, which proves the desired result.

(b): This part follows from the preceding property, on observing that $-|X| \leq X \leq |X|$.

(c): Assume first that X, Y, a, and b are all non-negative, Then, for any $A \in \mathcal{G}$,

$$\begin{aligned}
\mathrm{E}[\mathrm{E}(aX + bY \mid \mathcal{G})\mathbb{1}_A] &= \mathrm{E}[(aX + bY)\mathbb{1}_A] = a\,\mathrm{E}[X\mathbb{1}_A] + b\,\mathrm{E}[Y\mathbb{1}_A] \\
&= a\,\mathrm{E}[\mathrm{E}(X \mid \mathcal{G})\mathbb{1}_A] + b\,\mathrm{E}[\mathrm{E}(Y \mid \mathcal{G})\mathbb{1}_A] \\
&= \mathrm{E}\left\{[a\,\mathrm{E}(X \mid \mathcal{G}) + b\,\mathrm{E}(Y \mid \mathcal{G})]\,\mathbb{1}_A\right\} ,
\end{aligned}$$

which establishes the first part of (c). For arbitrary reals a and b, and X and Y such that $\mathrm{E}[|X| \mid \mathcal{G}] < \infty$ and $\mathrm{E}[|Y| \mid \mathcal{G}] < \infty$, (b) and the first part of (c) shows that

$$\mathrm{E}[|aX + bY| \mid \mathcal{G}] \leq |a|\,\mathrm{E}[|X| \mid \mathcal{G}] + |b|\,\mathrm{E}[|Y| \mid \mathcal{G}] < \infty ,$$

whence $\mathrm{E}[(aX + bY) \mid \mathcal{G}]$ is well-defined. We will now show that, for two non-negative random variables U and V satisfying $\mathrm{E}[U \mid \mathcal{G}] < \infty$ and $\mathrm{E}[V \mid \mathcal{G}] < \infty$,

$$\mathrm{E}[U - V \mid \mathcal{G}] = \mathrm{E}[U \mid \mathcal{G}] - \mathrm{E}[V \mid \mathcal{G}] . \tag{A.5}$$

Applying again the first part of (c) and noting that $(U-V)^+ = (U-V)\mathbb{1}_{\{U \geq V\}}$ and $(U - V)^- = (V - U)\mathbb{1}_{\{V \geq U\}}$, we find that

$$\begin{aligned}
\mathrm{E}[U - V \,|\, \mathcal{G}] + \mathrm{E}[V \mathbb{1}_{\{U \geq V\}} \,|\, \mathcal{G}] - \mathrm{E}[U \mathbb{1}_{\{V > U\}} \,|\, \mathcal{G}] \\
&= \mathrm{E}[(U - V)\mathbb{1}_{\{U \geq V\}} \,|\, \mathcal{G}] + \mathrm{E}[V \mathbb{1}_{\{U \geq V\}} \,|\, \mathcal{G}] \\
&\quad - \{\mathrm{E}[(V - U)\mathbb{1}_{\{V \geq U\}} \,|\, \mathcal{G}] + \mathrm{E}[U \mathbb{1}_{\{V > U\}} \,|\, \mathcal{G}]\} \\
&= \mathrm{E}[U \mathbb{1}_{\{U \geq V\}} \,|\, \mathcal{G}] - \mathrm{E}[V \mathbb{1}_{\{V > U\}} \,|\, \mathcal{G}] \,.
\end{aligned}$$

Moving the two last terms on the left-hand side to the right-hand side establishes (A.5). Finally, the second part of (c) follows by splitting aX and bY into their positive and negative parts $(aX)^+$ and $(aX)^-$ etc., and using the above linearity.

(e): Suppose first that $X \geq 0$, and pick $A \in \mathcal{G}$. Then A is in $\mathcal{H}$ as well, so that, using (A.1) repeatedly,

$$\mathrm{E}\left(\mathbb{1}_A \mathrm{E}\left[\mathrm{E}(X \,|\, \mathcal{H}) | \mathcal{G}\right]\right) = \mathrm{E}\left[\mathbb{1}_A \mathrm{E}\left(X \,|\, \mathcal{H}\right)\right] = \mathrm{E}[\mathbb{1}_A X] = \mathrm{E}\left[\mathbb{1}_A \mathrm{E}\left(X \,|\, \mathcal{G}\right)\right].$$

This establishes (e) for non-negative random variables. Suppose now that $\mathrm{E}[|X| \,|\, \mathcal{G}] < \infty$. For any integer $M \geq 0$, put $A_M = \{\mathrm{E}[X \,|\, \mathcal{H}] > M\}$, and put $A = \{\mathrm{E}[X \,|\, \mathcal{H}] = \infty\}$. Then A_M is in $\mathcal{H}$, and so is $A = \cap_M A_M$. Moreover,

$$\begin{aligned}
M \,\mathrm{E}[\mathbb{1}_A \,|\, \mathcal{G}] &\leq \mathrm{E}[M \mathbb{1}_{A_M} \,|\, \mathcal{G}] \leq \mathrm{E}\left[\mathrm{E}\left(|X| \,|\, \mathcal{H}\right) \mathbb{1}_{A_M} \,|\, \mathcal{G}\right] \\
&\leq \mathrm{E}\left[\mathrm{E}\left(|X| \,|\, \mathcal{H}\right) \,|\, \mathcal{G}\right] = \mathrm{E}[|X| \,|\, \mathcal{G}] < \infty \,.
\end{aligned}$$

Because M is arbitrary in this display, $\mathrm{E}[\mathbb{1}_A \,|\, \mathcal{G}] = 0$, implying that $E[\mathbb{1}_A] = 0$. Hence, $P(A) = 0$, that is, $\mathrm{E}[X \,|\, \mathcal{H}] < \infty$. The second part of (e) now follows from (c) applied to $\mathrm{E}[X^+ \,|\, \mathcal{H}]$ and $-\mathrm{E}[X^- \,|\, \mathcal{H}]$.

(f): If $X \geq 0$, then (A.1) implies that for any $A \in \mathcal{G}$,

$$\mathrm{E}[\mathbb{1}_A \mathrm{E}(X \,|\, \mathcal{G})] = \mathrm{E}[\mathbb{1}_A X] = \mathrm{E}[\mathbb{1}_A \mathrm{E}(X)] \,.$$

This proves the first part of (f). If $\mathrm{E}\,|X| < \infty$, then $\mathrm{E}[X^+] < \infty$ and $\mathrm{E}[X^-] < \infty$, and the proof follows by linearity.

(g): For $X \geq 0$ and $Y \geq 0$, (A.1) shows that, for any $A \in \mathcal{G}$,

$$\mathrm{E}[\mathbb{1}_A \mathrm{E}(XY \,|\, \mathcal{G})] = \mathrm{E}[\mathbb{1}_A XY] = \mathrm{E}[\mathbb{1}_A X \,\mathrm{E}(Y \,|\, \mathcal{G})] \,.$$

Thus, the first part of (g) follows. For X and Y such that $|X|$, $\mathrm{E}[|Y| \,|\, \mathcal{G}]$, and $\mathrm{E}[|XY| \,|\, \mathcal{G}]$ are all finite, the random variables $\mathrm{E}[X^+Y^+ \,|\, \mathcal{G}]$, $\mathrm{E}[X^+Y^- \,|\, \mathcal{G}]$, $\mathrm{E}[X^-Y^+ \,|\, \mathcal{G}]$, and $\mathrm{E}[X^-Y^- \,|\, \mathcal{G}]$ are finite too. Therefore, applying (c),

$$\mathrm{E}[XY \,|\, \mathcal{G}] = \mathrm{E}[X^+Y^+ \,|\, \mathcal{G}] + \mathrm{E}[X^-Y^- \,|\, \mathcal{G}] - \mathrm{E}[X^+Y^- \,|\, \mathcal{G}] - \mathrm{E}[X^-Y^+ \,|\, \mathcal{G}] \,.$$

The preceding result shows that the four terms on the right-hand side equal $X^+ \mathrm{E}[Y^+ \,|\, \mathcal{G}]$, $X^- \mathrm{E}[Y^- \,|\, \mathcal{G}]$, $X^+ \mathrm{E}[Y^- \,|\, \mathcal{G}]$, and $X^- \mathrm{E}[Y^+ \,|\, \mathcal{G}]$, respectively. Because these four random variables are finite, the result follows. □

Proposition A.2.4. *Let $\{X_n\}_{n \geq 0}$ be a sequence of random variables.*

(i) If $X_n \geq 0$ and $X_n \uparrow X$, then $\mathrm{E}[X_n \,|\, \mathcal{G}] \uparrow \mathrm{E}[X \,|\, \mathcal{G}]$.

(ii) *If* $X_n \leq Y$, $\mathrm{E}[|Y| \,|\, \mathcal{G}] < \infty$, *and* $X_n \downarrow X$ *with* $\mathrm{E}[|X| \,|\, \mathcal{G}] < \infty$, *then* $\mathrm{E}[X_n \,|\, \mathcal{G}] \downarrow \mathrm{E}[X \,|\, \mathcal{G}]$.

(iii) *If* $|X_n| \leq Z$, $\mathrm{E}[Z \,|\, \mathcal{G}] < \infty$, *and* $X_n \to X$, *then* $\mathrm{E}[X_n \,|\, \mathcal{G}] \to \mathrm{E}[X \,|\, \mathcal{G}]$ *and* $\mathrm{E}[|X_n - X| \,|\, \mathcal{G}] \to 0$.

Proof. (i): Proposition A.2.3(a) shows that $\mathrm{E}[X_n \,|\, \mathcal{G}] \leq \mathrm{E}[X_{n+1} \,|\, \mathcal{G}]$; hence, $\lim_{n\to\infty} \mathrm{E}[X_n \,|\, \mathcal{G}]$ exists P-a.s. Because $\lim_{n\to\infty} \mathrm{E}[X_n \,|\, \mathcal{G}]$ is a limit of $\mathcal{G}$-measurable random variables, it is $\mathcal{G}$-measurable. By (A.1) and the monotone convergence theorem, for any $A \in \mathcal{G}$,

$$\mathrm{E}[\mathbb{1}_A \lim \mathrm{E}(X_n \,|\, \mathcal{G})] = \lim \mathrm{E}[\mathbb{1}_A \mathrm{E}(X_n \,|\, \mathcal{G})] = \lim \mathrm{E}[\mathbb{1}_A X_n] = \mathrm{E}[\mathbb{1}_A X] \,.$$

Because the latter relation holds for all $A \in \mathcal{G}$, Lemma A.2.1 shows that $\lim \mathrm{E}(X_n \,|\, \mathcal{G}) = \mathrm{E}(X \,|\, \mathcal{G})$.

(ii): First note that, as $\{X_n\}$ decreases to X, we have $X \leq X_n \leq Y$ for all n. This implies $|X_n| \leq |X| + |Y|$, and we conclude that $\mathrm{E}[|X_n| \,|\, \mathcal{G}] < \infty$ for all n. Now set $Z_n = Y - X_n$. Then, $Z_n \geq 0$ and $Z_n \uparrow Y - X$. Therefore, using (i) and Proposition A.2.3(c),

$$\begin{aligned} \mathrm{E}[Y \,|\, \mathcal{G}] - \mathrm{E}[X_n \,|\, \mathcal{G}] &= \mathrm{E}[Z_n \,|\, \mathcal{G}] \uparrow \mathrm{E}[\lim Z_n \,|\, \mathcal{G}] \\ &= \mathrm{E}[Y - X \,|\, \mathcal{G}] = \mathrm{E}[Y \,|\, \mathcal{G}] - \mathrm{E}[X \,|\, \mathcal{G}] \,. \end{aligned}$$

(iii): Set $Z_n = \sup_{m \geq n} |X_m - X|$. Because $X_n \to X$, $Z_n \downarrow 0$. By Proposition A.2.3(b) and (c),

$$|\mathrm{E}(X_n \,|\, \mathcal{G}) - \mathrm{E}(X \,|\, \mathcal{G})| \leq \mathrm{E}[|X_n - X| \,|\, \mathcal{G}] \leq \mathrm{E}[Z_n \,|\, \mathcal{G}] \,.$$

Because $Z_n \downarrow 0$ and $Z_n \leq 2Z$, (ii) shows that $\mathrm{E}[Z_n \,|\, \mathcal{G}] \downarrow 0$. □

The following equality plays a key role in several parts of the book, and we thus provide a simple proof of this result.

Proposition A.2.5 (Rao-Blackwell Inequality). *Let* $(\Omega, \mathcal{F}, \mathrm{P})$ *be a probability space, let* X *be a random variable such that* $\mathrm{E}[X^2] < \infty$, *and let* $\mathcal{G}$ *be a sub-*σ*-field of* $\mathcal{F}$. *Then*

$$\mathrm{Var}[X] = \mathrm{Var}[\mathrm{E}(X \,|\, \mathcal{G})] + \mathrm{E}[\mathrm{Var}(X \,|\, \mathcal{G})] \,, \tag{A.6}$$

where the conditional variance $\mathrm{Var}(X \,|\, \mathcal{G})$ *is defined as*

$$\mathrm{Var}(X \,|\, \mathcal{G}) \stackrel{\text{def}}{=} \mathrm{E}[(X - \mathrm{E}[X \,|\, \mathcal{G}])^2 \,|\, \mathcal{G}] \,. \tag{A.7}$$

This implies in particular that $\mathrm{Var}[\mathrm{E}(X \,|\, \mathcal{G})] \leq \mathrm{Var}[X]$, *where the inequality is strict unless* X *is* $\mathcal{G}$*-measurable.*

Proof. Without loss of generality, we may assume that $\mathrm{E}[X] = 0$. Write

$$\mathrm{E}[(X - \mathrm{E}[X \,|\, \mathcal{G}])^2 \,|\, \mathcal{G}] = \mathrm{E}[X^2 \,|\, \mathcal{G}] - (\mathrm{E}[X \,|\, \mathcal{G}])^2 \,.$$

Taking expectation on both sides and noting that $\mathrm{E}[\mathrm{E}(X \,|\, \mathcal{G})] = \mathrm{E}[X] = 0$ yields (A.6). □

A.3 Conditional Distribution

Definition A.3.1 (Version of Conditional Probability). *Let* $(\Omega, \mathcal{F}, \mathrm{P})$ *be a probability space and let* $\mathcal{G}$ *be a sub-*σ*-field of* $\mathcal{F}$*. For any event* $F \in \mathcal{F}$*,* $\mathrm{P}(F \mid \mathcal{G}) = \mathrm{E}[\mathbb{1}_F \mid \mathcal{G}]$ *is called* a version of the conditional probability of F with respect to $\mathcal{G}$.

We might expect a version of the conditional probability $F \to \mathrm{P}(F \mid \mathcal{G})$ to be a probability measure on $\mathcal{F}$. If $\{F_n\}_{n \geq 0}$ is a sequence of disjoint subsets of $\mathcal{F}$, then Propositions A.2.3–(c) and A.2.4–(i) show that

$$\mathrm{P}\left(\bigcup_{n=0}^{\infty} F_n \,\middle|\, \mathcal{G} \right) = \sum_{n=0}^{\infty} \mathrm{P}(F_n \mid \mathcal{G}),$$

or, more precisely, that $\sum_{n=0}^{\infty} \mathrm{P}(F_n \mid \mathcal{G})$ is a version of the conditional expectation of $\bigcup_{n=0}^{\infty} F_n$ given $\mathcal{G}$. This version is defined up to a P-null set. However, this null set may depend on the sequence $\{F_n\}_{n \geq 0}$. Because unless in very specific cases the σ-field $\mathcal{F}$ is not countable, there is no guarantee that it is possible to choose versions of the conditional distribution for each set F that are such that the sub-additive property holds for *all* sequences $\{F_n\}_{n \geq 0}$ except on a P-null set. This leads to the need for and definition of regular conditional probabilities

Definition A.3.2 (Regular Conditional Probability). *Let* $(\Omega, \mathcal{F}, \mathrm{P})$ *be a probability space and let* $\mathcal{G}$ *be a sub-*σ*-field of* $\mathcal{F}$*. A* regular version of the conditional probability of P given $\mathcal{G}$ *is a function*

$$\mathrm{P}^{\mathcal{G}} : \Omega \times \mathcal{F} \to [0, 1]$$

such that

(i) For all $F \in \mathcal{F}$*,* $\omega \mapsto \mathrm{P}^{\mathcal{G}}(\omega, F)$ *is* $\mathcal{G}$*-measurable and is a version of the conditional probability of* F *given* $\mathcal{G}$*,* $\mathrm{P}^{\mathcal{G}}(\cdot, F) = \mathrm{P}[F \mid \mathcal{G}]$*;*
(ii) For P*-almost every* ω*, the mapping* $F \mapsto \mathrm{P}^{\mathcal{G}}(\omega, F)$ *is a probability measure on* $\mathcal{F}$*.*

Closely related to regular conditional probabilities is the notion of regular conditional distribution.

Definition A.3.3 (Regular Conditional Distribution of Y **Given** $\mathcal{G}$**).** *Let* $(\Omega, \mathcal{F}, \mathrm{P})$ *be a probability space and let* $\mathcal{G}$ *be a sub-*σ*-field of* $\mathcal{F}$*. Let* $(\mathsf{Y}, \mathcal{Y})$ *be a measurable space and let* Y *be an* Y*-valued random variable. A* regular version of the conditional distribution of Y given $\mathcal{G}$ *is a function*

$$\mathrm{P}^{Y|\mathcal{G}} : \Omega \times \mathcal{Y} \to [0, 1]$$

such that

(i) *For all* $E \in \mathcal{Y}$, $\omega \mapsto \mathrm{P}^{Y|\mathcal{G}}(\omega, E)$ *is* $\mathcal{G}$*-measurable and is a version of the conditional probability of* P^Y *given* $\mathcal{G}$, $\mathrm{P}^{Y|\mathcal{G}}(\cdot, E) = \mathrm{E}[\mathbb{1}_E(Y) \,|\, \mathcal{G}]$;
(ii) *For* P*-almost every* ω, $E \mapsto \mathrm{P}^{Y|\mathcal{G}}(\omega, E)$ *is a probability measure on* $\mathcal{Y}$.

In the sequel, we will focus exclusively on regular conditional distributions. When a regular version of a conditional distribution of Y given $\mathcal{G}$ exists, conditional expectations can be written as integrals for each ω.

Theorem A.3.4. *Let* $(\Omega, \mathcal{F}, \mathrm{P})$ *be a probability space and let* $\mathcal{G}$ *be a sub-*σ*-field of* $\mathcal{F}$*. Let* $(\mathsf{Y}, \mathcal{Y})$ *be a measurable space, let* Y *be an* Y*-valued random variable and let* $\mathrm{P}^{Y|\mathcal{G}}$ *be a regular version of the conditional expectation of* Y *given* $\mathcal{G}$*. Then for any real-valued measurable function* g *on* Y *such that* $\mathrm{E}\,|g(Y)| < \infty$, g *is integrable with respect to* $\mathrm{P}^{Y|\mathcal{G}}(\omega, \cdot)$*, that is,* $\int_{\mathsf{Y}} |g(y)|\, \mathrm{P}^{Y|\mathcal{G}}(\omega, dy) < \infty$*, for* P*-almost every* ω*, and*

$$\mathrm{E}[g(Y) \,|\, \mathcal{G}] = \int g(y)\, \mathrm{P}^{Y|\mathcal{G}}(\cdot, dy) . \tag{A.8}$$

That is, $\int g(y)\, \mathrm{P}^{Y|\mathcal{G}}(\cdot, dy)$ *is a version of the conditional expectation of* $g(Y)$ *given* $\mathcal{G}$.

The key question is now the existence of regular conditional probabilities. It is known that *regular conditional probabilities exist under most conditions encountered in practice*, but we should keep in mind that they *do not always exist.* This topic requires some care, because the existence of these regular versions requires some additional assumptions on the topology of the probability space (see Dudley, 2002, Chapter 10).

Here is a main theorem on existence and uniqueness of regular conditional probabilities. It is not stated under the weakest possible topological assumptions, but nevertheless the assumptions of this theorem are mild and are verified in all situations considered in this book.

Theorem A.3.5. *Let* $(\Omega, \mathcal{F}, \mathrm{P})$ *be a probability space and let* $\mathcal{G}$ *be a sub-*σ*-field of* $\mathcal{F}$*. Let* Y *be a Polish space, let* $\mathcal{Y}$ *be its Borel* σ*-field, and let* Y *be an* Y*-valued random variable. Then there exists a regular version of the conditional distribution of* Y *given* $\mathcal{G}$, $\mathrm{P}^{Y|\mathcal{G}}$*, and this version is unique in the sense that for any other regular version* $\bar{\mathrm{P}}^{Y|\mathcal{G}}$ *of this distribution, for* P*-almost every* ω *it holds that*

$$\mathrm{P}^{Y|\mathcal{G}}(\omega, F) = \bar{\mathrm{P}}^{Y|\mathcal{G}}(\omega, F) \quad \textit{for all } F \in \mathcal{F} .$$

For a proof, see Dudley (2002, Theorem 10.2.2).

Finally ,it is of interest to define the regular conditional distribution of a random variable Y given another random variable X.

Definition A.3.6 (Regular Conditional Distribution of Y Given X). *Let* $(\Omega, \mathcal{F}, \mathrm{P})$ *be a probability space and let* X *and* Y *be random variables with* *values in the measurable spaces* $(\mathsf{X}, \mathcal{X})$ *and* $(\mathsf{Y}, \mathcal{Y})$*, respectively. Then a* regular version of the conditional distribution of Y given $\sigma(X)$ *is a function*

$$\mathrm{P}^{Y|X} : \mathsf{X} \times \mathcal{Y} \to [0, 1]$$

such that

(i) For all $E \in \mathcal{Y}$, $x \to \mathrm{P}^{Y|X}(x, E)$ *is* $\mathcal{X}$*-measurable and*

$$\mathrm{P}^{Y|X}(x, E) = \mathrm{E}[\mathbb{1}_E(Y) \,|\, X = x] \; ; \tag{A.9}$$

(ii) For P^X*-almost every* $x \in \mathsf{X}$, $E \mapsto \mathrm{P}^{Y|X}(x, E)$ *is a probability measure on* $\mathcal{Y}$.

When a regular version of a conditional distribution of Y given X exists, conditional expectations can be written as integrals for each x.

Theorem A.3.7. *Let* $(\Omega, \mathcal{F}, \mathrm{P})$ *be a probability space, let* X *and* Y *be random variables with values in the measurable spaces* $(\mathsf{Y}, \mathcal{Y})$ *and* $(\mathsf{X}, \mathcal{X})$*, respectively, and let* $\mathrm{P}^{Y|X}$ *be a regular version of the conditional expectation of* Y *given* X*.*

Then if for any real-valued measurable function g *on* Y *such that* $\mathrm{E}\,|g(Y)| < \infty$, g *is integrable with respect to* $\mathrm{P}^{Y|X}(x, \cdot)$ *for* P^X*-almost every* x *and*

$$\mathrm{E}[g(Y)|X = x] = \int g(y) \; \mathrm{P}^{Y|X}(x, dy) \; . \tag{A.10}$$

Moreover, for any a real-valued measurable function g *on the measurable space* $(\mathsf{X} \times \mathsf{Y}, \mathcal{X} \otimes \mathcal{Y})$ *such that* $\mathrm{E}\,|g(X, Y)| < \infty$, $g(x, \cdot)$ *is integrable with respect to* $\mathrm{P}^{Y|X}(x, \cdot)$ *for* P^x*-almost every* x *and*

$$\mathrm{E}[g(X, Y)] = \int \left\{ \int g(x, y) \; \mathrm{P}^{Y|X}(x, dy) \right\} \mathrm{P}^X(dx) \; , \tag{A.11}$$

$$\mathrm{E}[g(X, Y)|X = x] = \int g(x, y) \; \mathrm{P}^{Y|X}(x, dy) \; . \tag{A.12}$$

We conclude this section by stating conditions upon which there exists a regular conditional probability of Y given X.

Theorem A.3.8. *Let* $(\Omega, \mathcal{F}, \mathrm{P})$ *be a probability space and let* X *and* Y *be random variables with values in the measurable spaces* $(X, \mathcal{X})$ *and* $(\mathsf{Y}, \mathcal{Y})$*, respectively, with* Y *being Polish space and* $\mathcal{Y}$ *being its Borel* σ*-field. Then there exists a regular version* $\mathrm{P}^{Y|X}$ *of the conditional distribution of* Y *given* X *and this version is unique.*

A.4 Conditional Independence

Concepts of conditional independence play an important role in hidden Markov models and, more generally, in all models involving complex dependence structures among sets of random variables. This section covers the general definition of conditional independence as well as some basic properties. Further readings on this topic include the seminal paper by Dawid (1980) as well as more condensed expositions such as (Cowell *et al.*, 1999, Chapter 5).

Definition A.4.1 (Conditional Independence). *Let $(\Omega, \mathcal{F}, \mathrm{P})$ be a probability space and let $\mathcal{G}$ and $\mathcal{G}_1, \ldots, \mathcal{G}_n$ be sub-σ-fields of $\mathcal{F}$. Then $\mathcal{G}_1, \ldots, \mathcal{G}_n$ are said to be* P-conditionally independent *given $\mathcal{G}$ if for any bounded random variables $X_1, \ldots, X_n$ measurable with respect to $\mathcal{G}_1, \ldots, \mathcal{G}_n$, respectively,*

$$\mathrm{E}[X_1 \cdots X_n \,|\, \mathcal{G}] = \mathrm{E}[X_1 \,|\, \mathcal{G}] \cdots \mathrm{E}[X_n \,|\, \mathcal{G}] .$$

If $Y_1, \ldots, Y_n$ and Z are random variables, then $Y_1, \ldots, Y_n$ are said to be conditionally independent given Z *if the sub-σ-fields $\sigma(Y_1), \ldots, \sigma(Y_n)$ are P-conditionally independent given $\sigma(Z)$.*

Intuition suggests that if two random variables X and Y are independent given a third one, Z say, then the conditional distribution of X given Y *and* Z should be governed by the value of Z alone, further information about the value of Y being irrelevant. The following result shows that this intuition is not only correct but could in fact serve as an alternative definition of conditional independence of two variables given a third one.

Proposition A.4.2. *Let $(\Omega, \mathcal{F}, \mathrm{P})$ be a probability space and let $\mathcal{A}$, $\mathcal{B}$, and $\mathcal{C}$ be sub-σ-fields of $\mathcal{F}$. Then $\mathcal{A}$ and $\mathcal{B}$ are P-conditionally independent given $\mathcal{C}$ if and only if for any bounded $\mathcal{A}$-measurable random variable X,*

$$\mathrm{E}[X \,|\, \mathcal{B} \vee \mathcal{C}) = \mathrm{E}[X \,|\, \mathcal{C}] , \tag{A.13}$$

where $\mathcal{B} \vee \mathcal{C}$ denotes the σ-field generated by $\mathcal{B} \cup \mathcal{C}$.

Proposition A.4.2 is sometimes used as an alternative definition of conditional independence: it is said that $\mathcal{A}$ and $\mathcal{B}$ are P-conditionally independent given $\mathcal{C}$ if for all $\mathcal{A}$-measurable non-negative random variables X there exists a version of the conditional expectation $\mathrm{E}[X \,|\, \mathcal{B} \vee \mathcal{C}]$ that is $\mathcal{C}$-measurable (Dawid, 1980, Definition 5.1).

Following the suggestion of Dawid (1980), the notation

$$\mathcal{A} \perp\!\!\!\perp \mathcal{B} \,|\, \mathcal{C} \;\; [\mathrm{P}]$$

is used to denote that the sub-σ-fields $\mathcal{A}$ and $\mathcal{B}$ are conditionally independent given $\mathcal{C}$, under the probability P. In the case where $\mathcal{A} = \sigma(X)$, $\mathcal{B} = \sigma(Y)$, and $\mathcal{C} = \sigma(Z)$ with X, Y, and Z being random variables, the simplified notation

$X \perp\!\!\!\perp Y \,|\, Z$ [P] will be used. In accordance with Definition A.4.1, we shall then say that X and Y are conditionally independent given Z under P.

The following proposition states a number of useful properties of conditional independence.

Proposition A.4.3. *Let $(\Omega, \mathcal{F}, \mathrm{P})$ be a probability space and let $\mathcal{A}$, $\mathcal{B}$, $\mathcal{C}$ and $\mathcal{D}$ be sub-σ-fields of $\mathcal{F}$. Then the following properties hold true.*

1. *(Symmetry) If $\mathcal{A} \perp\!\!\!\perp \mathcal{B} \,|\, \mathcal{C}$ [P], then $\mathcal{B} \perp\!\!\!\perp \mathcal{A} \,|\, \mathcal{C}$ [P].*
2. *(Decomposition) If $\mathcal{A} \perp\!\!\!\perp (\mathcal{B} \vee \mathcal{C}) \,|\, \mathcal{D}$ [P], then $\mathcal{A} \perp\!\!\!\perp \mathcal{B} \,|\, \mathcal{D}$ [P] and $\mathcal{A} \perp\!\!\!\perp \mathcal{C} \,|\, \mathcal{D}$ [P].*
3. *(Weak Union) If $\mathcal{A} \perp\!\!\!\perp (\mathcal{B} \vee \mathcal{D}) \,|\, \mathcal{C}$ [P], then $\mathcal{A} \perp\!\!\!\perp \mathcal{D} \,|\, \mathcal{B} \vee \mathcal{C}$ [P].*
4. *(Contraction) If $\mathcal{A} \perp\!\!\!\perp \mathcal{D} \,|\, \mathcal{B} \vee \mathcal{C}$ [P] and $\mathcal{A} \perp\!\!\!\perp \mathcal{B} \,|\, \mathcal{C}$ [P], then $\mathcal{A} \perp\!\!\!\perp \mathcal{B} \vee \mathcal{D} \,|\, \mathcal{C}$ [P].*

In the theory of Bayesian networks (also called graphical models), as introduced by Pearl (1988), these four properties are referred to as the *semi-graphoid inference axioms* (Cowell *et al.*, 1999).

B

Linear Prediction

This appendix provides a brief introduction to the theory of linear prediction of random variables. Further reading includes Brockwell and Davis (1991, Chapter 2), which provides a proof of the projection theorem (Theorem B.2.4 below), as well as Williams (1991) or Jacod and Protter (2000, Chapter 22). The results below are used in Chapter 5 to derive the particular form taken by the filtering and smoothing recursions in linear state-space models.

B.1 Hilbert Spaces

Definition B.1.1 (Inner Product Space). *A real linear space $\mathcal{H}$ is said to be an* inner product space *if for each pair of elements x and y in $\mathcal{H}$ there is a real number $\langle x, y\rangle$, called the* inner product *(or,* scalar product*) of x and y, such that*

(a) $\langle x, y\rangle = \langle y, x\rangle$,
(b) $\langle \alpha x + \beta y, z\rangle = \alpha\langle x, z\rangle + \beta\langle y, z\rangle$ *for z in $\mathcal{H}$ and real α and β,*
(c) $\langle x, x\rangle \geq 0$ *and* $\langle x, x\rangle = 0$ *if and only if* $x = 0$.

Two elements x and y such that $\langle x, y\rangle = 0$ are said to be *orthogonal.*

The *norm* $\|x\|$ of an element x of an inner product space is defined as

$$\|x\| = \sqrt{\langle x, x\rangle}. \tag{B.1}$$

The norm satisfies

(a) $\|x + y\| \leq \|x\| + \|y\|$ (triangle inequality),
(b) $\|\alpha x\| = |\alpha|\|x\|$ for real α,
(c) $\|x\| \geq 0$ and $\|x\| = 0$ if and only if $x = 0$.

These properties justify the use of the terminology "norm" for $\|\cdot\|$. In addition, the Cauchy-Schwarz inequality $|\langle x, y\rangle| \leq \|x\|\,\|y\|$ holds, with equality if and only if $y = \alpha x$ for some real α.

Definition B.1.2 (Convergence in Norm). *A sequence* $\{x_k\}_{k\geq 0}$ *of elements of an inner product space* $\mathcal{H}$ *is said to* converge in norm *to* $x \in \mathcal{H}$ *if* $\|x_n - x\| \to 0$ *as* $n \to \infty$.

It is readily verified that a sequence $\{x_k\}_{k\geq 0}$ that converges in norm to some element x satisfies $\limsup_{n\geq 0} \sup_{m\geq n} \|x_m - x_n\| = 0$. Any sequence, convergent or not, with this property is said to be a *Cauchy sequence*. Thus any convergent sequence is a Cauchy sequence. If the reverse implication holds true as well, that any Cauchy sequence is convergent (in norm), then the space is said to be *complete*. A complete inner product space is called a *Hilbert space*.

Definition B.1.3 (Hilbert Space). *A* Hilbert space $\mathcal{H}$ *is an inner product space that is complete, that is, an inner product space in which every Cauchy sequence converges in norm to some element in* $\mathcal{H}$.

It is well-known that $\mathbb{R}^k$ equipped with the inner product $\langle x, y\rangle = \sum_{i=1}^{k} x_i y_i$, where $x = (x_1, \ldots, x_k)$ and $y = (y_1, \ldots, y_k)$, is a Hilbert space. A more sophisticated example is the space of square integrable random variables. Let $(\Omega, \mathcal{F}, \mathrm{P})$ be a probability space and let $\mathcal{L}^2(\Omega, \mathcal{F}, \mathrm{P})$ be the space of square integrable random variables on $(\Omega, \mathcal{F}, \mathrm{P})$. For any two elements X and Y in $\mathcal{L}^2(\Omega, \mathcal{F}, \mathrm{P})$ we define

$$\langle X, Y\rangle = \mathrm{E}(XY) . \tag{B.2}$$

It is easy to check that $\langle X, Y\rangle$ satisfies all the properties of an inner product except for the last one: if $\langle X, Y\rangle = 0$, then it does not follow that $X(\omega) = 0$ for all $\omega \in \Omega$, but only that $\mathrm{P}\{\omega \in \Omega : X(\omega) = 0\} = 1$. This difficulty is circumvented by saying that the random variables X and Y are *equivalent* if $\mathrm{P}(X = Y) = 1$. This equivalence relation partitions $\mathcal{L}^2(\Omega, \mathcal{F}, \mathrm{P})$ into classes of random variables such that any two random variables in the same class are equal with probability one. The space $L^2(\Omega, \mathcal{F}, \mathrm{P})$ is the set of these equivalence classes with inner product still defined by (B.2). Because each class is uniquely determined by specifying any one of the random variables in it, we shall continue to use the notation X and Y for the elements in L^2 and to call them random variables, although it is sometimes important that X stands for an equivalence class of random variables. A well-known result in functional analysis is the following.

Proposition B.1.4. *The space* $\mathcal{H} = L^2(\Omega, \mathcal{F}, \mathrm{P})$ *equipped with the inner product* (B.2) *is a Hilbert space.*

Norm convergence of a sequence $\{X_n\}$ in $L^2(\Omega, \mathcal{F}, \mathrm{P})$ to a limit X means that

$$\|X_n - X\|^2 = \mathrm{E}\,|X_n - X|^2 \to 0 \quad \text{as} \quad n \to \infty.$$

Norm convergence of X_n to X in an L^2-space is often called *mean square convergence*.

B.2 The Projection Theorem

Before introducing the notion of projection in Hilbert spaces in general and in L^2-spaces in particular, some definitions are needed.

Definition B.2.1 (Closed Subspace). *A linear subspace $\mathcal{M}$ of a Hilbert space $\mathcal{H}$ is said to be closed if $\mathcal{M}$ contains all its limit points. That is, if $\{x_n\}$ is a sequence in $\mathcal{M}$ converging to some element $x \in \mathcal{H}$, then $x \in \mathcal{M}$.*

The lemma below is a direct consequence of the fact that the inner product is continuous mapping from $\mathcal{H}$ to $\mathbb{R}$.

Lemma B.2.2 (Closedness of Finite Spans). *If $y_1, \ldots, y_n$ is a finite family of elements of $\mathcal{H}$, then the linear subspace spanned by $y_1, \ldots, y_n$,*

$$\operatorname{span}(y_1, \ldots, y_n) \stackrel{\text{def}}{=} \left\{ x \in \mathcal{H} : x = \sum_{i=1}^{n} \alpha_i y_i, \text{ for some } \alpha_1, \ldots, \alpha_n \in \mathbb{R} \right\} ,$$

is a closed subspace of $\mathcal{H}$.

Definition B.2.3 (Orthogonal Complement). *The* orthogonal complement *$\mathcal{M}^\perp$ of a subset $\mathcal{M}$ of $\mathcal{H}$ is the set of all elements of $\mathcal{H}$ that are orthogonal to every element of $\mathcal{M}$: $x \in \mathcal{M}^\perp$ if and only if $\langle x, y \rangle = 0$ for every $y \in \mathcal{M}$.*

Theorem B.2.4 (The Projection Theorem). *Let $\mathcal{M}$ be a closed linear subspace of a Hilbert space $\mathcal{H}$ and let $x \in \mathcal{H}$. Then the following hold true.*

(i) There exists a unique element $\hat{x} \in \mathcal{M}$ such that

$$\|x - \hat{x}\| = \inf_{y \in \mathcal{M}} \|x - y\| .$$

(ii) $\hat{x}$ is the unique element of $\mathcal{M}$ such that

$$(x - \hat{x}) \in \mathcal{M}^\perp .$$

The element $\hat{x}$ is referred to as the projection *of x onto $\mathcal{M}$.*

Corollary B.2.5 (The Projection Mapping). *If $\mathcal{M}$ is a closed linear subspace of the Hilbert space $\mathcal{H}$ and I is the identity mapping on $\mathcal{H}$, then there is a unique mapping from $\mathcal{H}$ onto $\mathcal{M}$, denoted* $\operatorname{proj}(\cdot|\mathcal{M})$, *such that $I - \operatorname{proj}(\cdot|\mathcal{M})$ maps $\mathcal{H}$ onto $\mathcal{M}^\perp$.* $\operatorname{proj}(\cdot|\mathcal{M})$ *is called the* projection mapping *onto $\mathcal{M}$.*

The following properties of the projection mapping can be readily obtained from Theorem B.2.4.

Proposition B.2.6 (Properties of the Projection Mapping). *Let $\mathcal{H}$ be a Hilbert space and let* $\operatorname{proj}(\cdot|\mathcal{M})$ *denote the projection mapping onto a closed linear subspace $\mathcal{M}$. Then the following properties hold true.*

(i) For all x, y in $\mathcal{H}$ and real α, β,

$$\mathrm{proj}(\alpha x + \beta y|\mathcal{M}) = \alpha\,\mathrm{proj}(x|\mathcal{M}) + \beta\,\mathrm{proj}(y|\mathcal{M}) .$$

(ii) $x = \mathrm{proj}(x|\mathcal{M}) + \mathrm{proj}(x|\mathcal{M}^{\perp})$.
(iii) $\|x\|^2 = \|\mathrm{proj}(x|\mathcal{M})\|^2 + \|\mathrm{proj}(x|\mathcal{M}^{\perp})\|^2$.
(iv) $x \mapsto \mathrm{proj}(x|\mathcal{M})$ *is continuous.*
(v) $x \in \mathcal{M}$ *if and only if* $\mathrm{proj}(x|\mathcal{M}) = x$ *and* $x \in \mathcal{M}^{\perp}$ *if and only if* $\mathrm{proj}(x|\mathcal{M}^{\perp}) = 0$.
(vi) If $\mathcal{M}_1$ *and* $\mathcal{M}_2$ *are two closed linear subspaces of* $\mathcal{H}$*, then* $\mathcal{M}_1 \subseteq \mathcal{M}_2$ *if and only if for all* $x \in \mathcal{H}$,

$$\mathrm{proj}(\mathrm{proj}(x|\mathcal{M}_2)\,|\,\mathcal{M}_1) = \mathrm{proj}(x|\mathcal{M}_1) .$$

When the space $\mathcal{H}$ is an L^2-space, the following terminology is often preferred.

Definition B.2.7 (Best Linear Prediction). *If $\mathcal{M}$ is a closed subspace of $L^2(\Omega, \mathcal{F}, \mathrm{P})$ and $X \in L^2(\Omega, \mathcal{F}, \mathrm{P})$, then the* best linear predictor *(also called* minimum mean square error linear predictor*) of X in $\mathcal{M}$ is the element $\hat{X} \in \mathcal{M}$ such that*

$$\|X - \hat{X}\|^2 \stackrel{\text{def}}{=} \mathrm{E}(X - \hat{X})^2 \leq \mathrm{E}(X - Y)^2 \quad \textit{for all } Y \in \mathcal{M} .$$

The "best linear predictor" is clearly just an alternative denomination for $\mathrm{proj}(X|\mathcal{M})$, taking the probabilistic context into account. Interestingly, the projection theorem implies that $\hat{X}$ is also the unique element in $\mathcal{M}$ such that

$$\langle X - \hat{X}, Y\rangle \stackrel{\text{def}}{=} \mathrm{E}[(X - \hat{X})Y] = 0 \quad \text{for all } Y \in \mathcal{M} .$$

An immediate consequence of Proposition B.2.6(iii) is that the *mean square prediction error* $\|X - \hat{X}\|^2$ may be written in two other equivalent and often useful ways, namely

$$\|X - \hat{X}\|^2 \stackrel{\text{def}}{=} \mathrm{E}[(X - \hat{X})^2] = \mathrm{E}[X(X - \hat{X})] = \mathrm{E}[X^2] - \mathrm{E}[\hat{X}^2] .$$

C

Notations

C.1 Mathematical

i	imaginary unit, $\mathrm{i}^2 = -1$
e	base of natural logarithm, $\mathrm{e} = 2.7182818\ldots$
$\lfloor x \rfloor$	largest integer less than or equal to x (integer part)
$\lceil x \rceil$	smallest integer larger than or equal to x
$x \wedge y$	minimum of x and y
$x \vee y$	maximum of x and y
$\langle u, v \rangle$	scalar product of vectors u and v
$z_{k:l}$	collection $z_k, z_{k+1}, \ldots, z_l$
A^t	transpose of matrix A
$\lvert \mathsf{S} \rvert$	cardinality of (finite) set S
$\mathbb{1}_A$	indicator function of set A
$\lVert f \rVert_\infty$	supremum of function f
$\operatorname{osc}(f)$	oscillation (global modulus of continuity) of f
$\dot{f}$	derivative of (real-valued) f
$\nabla_\theta f(\theta')$ or $\nabla_\theta\, f(\theta)\rvert_{\theta=\theta'}$	gradient of f at θ'
$\nabla^2_\theta f(\theta')$ or $\nabla^2_\theta\, f(\theta)\rvert_{\theta=\theta'}$	Hessian of f at θ'
$(\mathsf{Z}, \mathcal{Z})$	measurable space
$\mathcal{F}_{\mathrm{b}}(\mathsf{Z})$	bounded measurable functions on $(\mathsf{Z}, \mathcal{Z})$
$\mathcal{G} \vee \mathcal{F}$	minimal σ-field generated by σ-fields $\mathcal{G}$ and $\mathcal{F}$
$\mu \otimes \nu$, $\mu^{\otimes 2}$	product measures
$\mathcal{G}^{\otimes n}$	product σ-field
$\lVert \xi \rVert_{\mathrm{TV}}$	total variation norm of signed measure ξ
$\lVert f \rVert_{\nu,\infty}$	essential supremum of a measurable function f (with respect to the measure ν)
$\operatorname{osc}_\nu(f)$	essential oscillation semi-norm

C.2 Probability

P, E	probability, expectation
$\xrightarrow{\mathcal{D}}$	convergence in distribution
$\xrightarrow{\mathrm{P}}$	convergence in probability
$\xrightarrow{\text{a.s.}}$	almost sure convergence
L^1, L^2	integrable and square integrable functions
$\|X\|_p$	L^p norm of X $([\mathrm{E}\,\|X\|^p]^{1/p})$
$\mathrm{span}(X_1, X_2)$	linear span in Hilbert space, usually $L^2(\Omega, \mathcal{F}, \mathrm{P})$
$\mathrm{proj}(X\|\mathcal{M})$	projection onto a linear subspace
$X \perp\!\!\!\perp Y \mid Z$ [P]	X and Y are conditionally independent given Z (with respect to the probability P)
N	Gaussian distribution, $\mathrm{N}(\mu, \sigma^2)$
LN	log-normal distribution, $\mathrm{LN}(\log(\mu), \sigma^2)$
Dir	Dirichlet distribution, $\mathrm{Dir}_r(\alpha_1, \ldots, \alpha_r)$
Ga	gamma distribution, $\mathrm{Ga}(\alpha, \beta)$
IG	inverse gamma distribution
U	uniform distribution, $\mathrm{U}([a, b])$
Bin	binomial distribution, $\mathrm{Bin}(n, p)$
Be	beta distribution, $\mathrm{Be}(\alpha, \beta)$
Mult	multinomial distribution, $\mathrm{Mult}(n, (\omega_1, \ldots, \omega_N))$

C.3 Hidden Markov Models

$\{X_k\}_{k\geq 0}$	hidden states
$(\mathsf{X}, \mathcal{X})$	state space of the hidden states
$Q(x, dx')$	transition kernel of the hidden chain
$q(x, x')\,\lambda(dx')$	*idem*, in fully dominated models
ν	initial distribution (probability density function with respect to λ in fully dominated models)
π	stationary distribution of $\{X_k\}_{k\geq 0}$ (if any)
r	$\|\mathsf{X}\|$ in finite HMMs
$\{Y_k\}_{k\geq 0}$	observations
$(\mathsf{Y}, \mathcal{Y})$	observation space
$G(x, dy)$	conditional likelihood kernel
$g(x, y)\,\mu(dy)$	*idem*, in partially dominated models
$g_k(x)$	$g(x, Y_k)$—"implicit conditioning convention"
$\mathrm{P}_\nu, \mathrm{E}_\nu$	probability, expectation under the model, assuming initial distribution ν

Smoothing

$\phi_{\nu,k}$ or $\phi_{\nu,k\|k}$	filtering distribution
$\phi_{\nu,k\|k-1}$	predictive distribution
$c_{\nu,k}$	normalization constant for the filter
$\mathrm{L}_{\nu,n}$	likelihood
$\ell_{\nu,n}$	log-likelihood
$\phi_{\nu,k\|n}$, $\phi_{\nu,k:l\|n}$	marginal of joint smoothing distribution
$\alpha_{\nu,k}$	forward measure
$\beta_{k\|n}$	backward function
$\bar{\alpha}_{\nu,k}$	normalized forward measure
$\bar{\beta}_{k\|n}$	normalized backward function
$\mathrm{F}_{k\|n}$	forward smoothing kernel
$\mathrm{B}_{\nu,n}$	backward smoothing kernel
$\tau_{\nu,n}$	recursive smoother

In several chapters, explicit dependence with respect to the initial distribution ν is omitted; in a few others, the above notations are followed by an expression of the form $[Y_{k:l}]$ to highlight dependence with respect to the relevant observations.

Parametric HMMs

θ	parameter vector
d_θ	dimension of the parameter
$\theta_\star$	actual (true) value of parameter
$\mathcal{J}(\theta)$	Fisher information matrix
$\ell_n^s(\theta)$	stationary version of the log-likelihood
$\ell(\theta)$	limiting contrast [of $n^{-1}\ell_{\nu,n}(\theta)$]
$\mathcal{Q}(\theta\,;\theta')$	intermediate quantity of EM
S	complete-data sufficient statistic in exponential family
d_s	dimension of S

State-Space Models

$X_{k+1} = A_k X_k + R_k U_k$	state (dynamic) equation
$Y_k = B_k X_k + S_k V_k$	observation equation
d_x, d_u, d_y, d_v	dimensions of X_k, U_k, Y_k and V_k
$\hat{X}_{k\|k}, \Sigma_{k\|k}$	filtered moments
$\hat{X}_{k\|k-1}, \Sigma_{k\|k-1}$	predicted moments
$\hat{X}_{k\|n}, \Sigma_{k\|n}$	smoothed moments
$\kappa_{k\|n}, \Pi_{k\|n}$	idem in information parameterization
ϵ_k, Γ_k	innovation and associated covariance matrix
H_k	Kalman gain (prediction)
K_k	Kalman gain (filtering)

Hierarchical HMMs

$\{C_k\}_{k\geq 0}$	hierarchic component of the states (usually indicator variables)
$(\mathsf{C}, \mathcal{C})$	space of hierarchic component
Q_C	transition kernel of $\{C_k\}_{k\geq 0}$
ν_C	distribution of C_0
$\{W_k\}_{k\geq 0}$	intermediate component of the states
$(\mathsf{W}, \mathcal{W})$	space of intermediate component
$Q_W[(w,c),w')]$	conditional transition kernel of $\{W_k\}_{k\geq 0}$ given $\{C_k\}_{k\geq 0}$
$\psi_{\nu,k:l\|n}$	distribution of $C_{k:l}$ given $Y_{0:n}$
$\varphi_{k+1\|k}$	predictive distribution of W_{k+1} given $Y_{0:n}$ and $C_{0:k+1}$

C.4 Sequential Monte Carlo

$\hat{\mu}_N^{\mathrm{MC}}(f)$	Monte Carlo estimate of $\mu(f)$ (from N i.i.d. draws)
$\tilde{\mu}_{\nu,N}^{\mathrm{IS}}(f)$	unnormalized importance sampling estimate (using ν as instrumental distribution)
$\widehat{\mu}_{\nu,N}^{\mathrm{IS}}(f)$	importance sampling estimate
$\hat{\mu}_{\nu,N}^{\mathrm{SIR}}(f)$	sampling importance resampling estimate
$T_k^{\mathrm{u}}(x,dx')$	$(\mathrm{L}_{k+1}/\mathrm{L}_k)^{-1}Q(x,dx')\,g_{k+1}(x') \;\propto\; Q(x,dx')\,g_{k+1}(x')$
T_k	"optimal" instrumental kernel (T_k^{u} normalized)
γ_k	normalization function of T_k^{u}
$\{\xi_k^i\}_{i=1,\dots,N}$	population of particles at time index k
$\{\omega_k^i\}_{i=1,\dots,N}$	associated importance weights (usually unnormalized)
$\xi_{0:k}^i, \xi_{0:k}^i(l)$	path particle and lth element in the trajectory [by convention $\xi_k^i = \xi_{0:k}^i(k)$]

References

Akashi, H. and Kumamoto, H. (1977) Random sampling approach to state estimation in switching environment. *Automatica*, **13**, 429–434.

Anderson, B. D. O. and Moore, J. B. (1979) *Optimal Filtering*. Prentice-Hall.

Andrews, D. F. and Mallows, C. L. (1974) Scale mixtures of normal distributions. *J. Roy. Statist. Soc. Ser. B*, **36**, 99–102.

Andrieu, C., Davy, M. and Doucet, A. (2003) Efficient particle filtering for jump Markov systems. Application to time-varying autoregressions. *IEEE Trans. Signal Process.*, **51**, 1762–1770.

Andrieu, C., Moulines, E. and Priouret, P. (2005) Stability of stochastic approximation under verifiable conditions. *SIAM J. Control Optim.* To appear.

Askar, M. and Derin, H. (1981) A recursive algorithm for the Bayes solution of the smoothing problem. *IEEE Trans. Automat. Control*, **26**, 558–561.

Atar, R. and Zeitouni, O. (1997) Exponential stability for nonlinear filtering. *Ann. Inst. H. Poincaré Probab. Statist.*, **33**, 697–725.

Athreya, K. B., Doss, H. and Sethuraman, J. (1996) On the convergence of the Markov chain simulation method. *Ann. Statist.*, **24**, 69–100.

Athreya, K. B. and Ney, P. (1978) A new approach to the limit theory of recurrent Markov chains. *Trans. Am. Math. Soc.*, **245**, 493–501.

Azencott, R. and Dacunha-Castelle, D. (1984) *Séries d'observations irrégulières.* Masson.

Bahl, L., Cocke, J., Jelinek, F. and Raviv, J. (1974) Optimal decoding of linear codes for minimizing symbol error rate. *IEEE Trans. Inform. Theory*, **20**, 284–287.

Baldi, P. and Brunak, S. (2001) *Bioinformatics. The Machine Learning Approach.* MIT Press.

Ball, F. G., Cai, Y., Kadane, J. B. and O'Hagan, A. (1999) Bayesian inference for ion channel gating mechanisms directly from single channel recordings, using Markov chain Monte Carlo. *Proc. Roy. Soc. London A*, **455**, 2879–2932.

Ball, F. G. and Rice, J. H. (1992) Stochastic models for ion channels: Introduction and bibliography. *Math. Biosci.*, **112**, 189–206.

Barron, A. (1985) The strong ergodic theorem for densities; generalized Shannon-McMillan-Breiman theorem. *Ann. Probab.*, **13**, 1292–1303.

Barron, A., Birgé, L. and Massart, P. (1999) Risk bounds for model selection via penalization. *Probab. Theory Related Fields*, **113**, 301–413.

Bartlett, P., Boucheron, S. and Lugosi, G. (2002) Model selection and error estimation. *Machine Learning*, **48**, 85–113.

Baum, L. E. and Eagon, J. A. (1967) An inequality with applications to statistical estimation for probalistic functions of Markov processes and to a model for ecology. *Bull. Am. Math. Soc.*, **73**, 360–363.

Baum, L. E. and Petrie, T. P. (1966) Statistical inference for probabilistic functions of finite state Markov chains. *Ann. Math. Statist.*, **37**, 1554–1563.

Baum, L. E., Petrie, T. P., Soules, G. and Weiss, N. (1970) A maximization technique occurring in the statistical analysis of probabilistic functions of Markov chains. *Ann. Math. Statist.*, **41**, 164–171.

Benveniste, A., Métivier, M. and Priouret, P. (1990) *Adaptive Algorithms and Stochastic Approximations*, vol. 22. Springer. Translated from the French by Stephen S. S. Wilson.

Berger, J. O. (1985) *Statistical Decision Theory and Bayesian Analysis.* Springer, 2nd ed.

Bertozzi, T., Le Ruyet, D., Rigal, G. and Han, V.-T. (2003) Trellis-based search of the maximum a posteriori sequence using particle filtering. In *Proc. IEEE Int. Conf. Acoust., Speech, Signal Process.*, vol. 6, 693–696.

Berzuini, C., Best, N., Gilks, W. R. and Larizza, C. (1997) Dynamic conditional independence models and Markov Chain Monte Carlo methods. *J. Am. Statist. Assoc.*, **92**, 1403–1412.

Berzuini, C. and Gilks, W. R. (2001) Resample-move filtering with cross-model jumps. In *Sequential Monte Carlo Methods in Practice* (eds. A. Doucet, N. De Freitas and N. Gordon). Springer.

Besag, J. (1989) Towards Bayesian image analysis. *J. Applied Statistics*, **16**, 395–407.

Bickel, P. J. and Doksum, K. A. (1977) *Mathematical Statistics.* Prentice-Hall.

Bickel, P. J. and Ritov, Y. (1996) Inference in hidden Markov models I. Local asymptotic normality in the stationary case. *Bernoulli*, **2**, 199–228.

Bickel, P. J., Ritov, Y. and Rydén, T. (1998) Asymptotic normality of the maximum likelihood estimator for general hidden Markov models. *Ann. Statist.*, **26**, 1614–1635.

— (2002) Hidden Markov model likelihoods and their derivatives behave like i.i.d. ones. *Ann. Inst. H. Poincaré Probab. Statist.*, **38**, 825–846.

Billingsley, P. (1995) *Probability and Measure.* Wiley, 3rd ed.

Bollerslev, T., Engle, R. F. and Nelson, D. (1994) ARCH models. In *Handbook of Econometrics* (eds. R. F. Engle and D. McFadden). North-Holland.

Bonnans, J. F. and Shapiro, A. (1998) Optimization problems with perturbations: a guided tour. *SIAM Rev.*, **40**, 228–264.

Booth, J. and Hobert, J. (1999) Maximizing generalized linear mixed model likelihoods with an automated monte carlo EM algorithm. *J. Roy. Statist. Soc. Ser. B*, **61**, 265–285.

Borovkov, A. A. (1998) *Ergodicity and Stability of Stochastic Systems.* Wiley.

Boucheron, S. and Gassiat, E. (2004) Error exponents in AR order testing. Preprint.

Boyd, S. and Vandenberghe, L. (2004) *Convex Optimization.* Cambridge University Press.

Boyles, R. (1983) On the convergence of the EM algorithm. *J. Roy. Statist. Soc. Ser. B*, **45**, 47–50.

Brandière, O. (1998) The dynamic system method and the traps. *Adv. Appl. Probab.*, **30**, 137–151.

Briers, M., Doucet, A. and Maskell, S. (2004) Smoothing algorithms for state-space models. *Tech. Rep.*, University of Cambridge, Department of Engineering.

Brockwell, P. J. and Davis, R. A. (1991) *Time Series: Theory and Methods.* Springer, 2nd ed.

Brooks, S. P., Giudici, P. and Roberts, G. O. (2003) Efficient construction of reversible jump Markov chain Monte Carlo proposal distributions. *J. Roy. Statist. Soc. Ser. B*, **65**, 1–37.

Bryson, A. and Frazier, M. (1963) Smoothing for linear and nonlinear dynamic systems. *Tech. Rep.*, Aero. Sys. Div. Wrigth-Patterson Air Force Base.

Budhiraja, A. and Ocone, D. (1997) Exponential stability of discrete-time filters for bounded observation noise. *Systems Control Lett.*, **30**, 185–193.

Bunke, H. and Caelli, T. (eds.) (2001) *Hidden Markov Models: Applications in Computer Vision.* World Scientific.

Burges, C. and Karlin, S. (1997) Prediction of complete gene structures in human genomic dna. *J. Mol. Biol.*, **268**, 78–94.

Caines, P. E. (1988) *Linear Stochastic Systems.* Wiley.

Campillo, F. and Le Gland, F. (1989) MLE for patially observed diffusions: Direct maximization vs. the EM algorithm. *Stoch. Proc. App.*, **33**, 245–274.

Cappé, O. (2001a) Recursive computation of smoothed functionals of hidden Markovian processes using a particle approximation. *Monte Carlo Methods Appl.*, **7**, 81–92.

— (2001b) Ten years of hmms (online bibliography 1989–2000). URL `http://www.tsi.enst.fr/~cappe/docs/hmmbib.html`.

Cappé, O., Buchoux, V. and Moulines, E. (1998) Quasi-Newton method for maximum likelihood estimation of hidden Markov models. In *Proc. IEEE Int. Conf. Acoust., Speech, Signal Process.*, vol. 4, 2265–2268.

Cappé, O., Doucet, A., Lavielle, M. and Moulines, E. (1999) Simulation-based methods for blind maximum-likelihood filter identification. *Signal Process.*, **73**, 3–25.

Cappé, O., Robert, C. P. and Rydén, T. (2003) Reversible jump, birth-and-death and more general continuous time Markov chain Monte Carlo samplers. *J. Roy. Statist. Soc. Ser. B*, **65**, 679–700.

Cardoso, J.-F., Lavielle, M. and Moulines, E. (1995) Un algorithme d'identification par maximum de vraisemblance pour des données incomplètes. *C. R. Acad. Sci. Paris Série I Statistique*, **320**, 363–368.

Carlin, B. P. and Chib, S. (1995) Bayesian model choice via Markov chain Monte Carlo. *J. Roy. Statist. Soc. Ser. B*, **57**, 473–484.

Carpenter, J., Clifford, P. and Fearnhead, P. (1999) An improved particle filter for non-linear problems. *IEE Proc., Radar Sonar Navigation*, **146**, 2–7.

Carter, C. K. and Kohn, R. (1994) On Gibbs sampling for state space models. *Biometrika*, **81**, 541–553.

— (1996) Markov chain Monte Carlo in conditionnaly Gaussian state space models. *Biometrika*, **83**, 589–601.

Casella, G., Robert, C. P. and Wells, M. T. (2000) Mixture models, latent variables and partitioned importance sampling. *Tech. Rep.*, CREST, INSEE, Paris.

Castledine, B. (1981) A Bayesian analysis of multiple-recapture sampling for a closed population. *Biometrika*, **67**, 197–210.

Celeux, G. and Diebolt, J. (1985) The SEM algorithm: a probabilistic teacher algorithm derived from the EM algorithm for the mixture problem. *Comput. Statist.*, **2**, 73–82.

— (1990) Une version de type recuit simulé de l'algorithme EM. *C. R. Acad. Sci. Paris Sér. I Math.*, **310**, 119–124.

Celeux, G., Hurn, M. and Robert, C. P. (2000) Computational and inferential difficulties with mixture posterior distributions. *J. Am. Statist. Assoc.*, **95**, 957–979.

Cérou, F., Le Gland, F. and Newton, N. (2001) Stochastic particle methods for linear tangent filtering equations. In *Optimal Control and PDE's - Innovations and Applications, in Honor of Alain Bensoussan's 60th Anniversary* (eds. J.-L. Menaldi, E. Rofman and A. Sulem), 231–240. IOS Press.

Chambaz, A. (2003) *Segmentation spatiale et sélection de modèle.* Ph.D. thesis, Université Paris-Sud.

Chan, K. S. and Ledolter, J. (1995) Monte carlo EM estimation for time series models involving counts. *J. Am. Statist. Assoc.*, **90**, 242–252.

Chang, R. and Hancock, J. (1966) On receiver structures for channels having memory. *IEEE Trans. Inform. Theory*, **12**, 463–468.

Chen, M. H. and Shao, Q. M. (2000) *Monte Carlo Methods in Bayesian Computation.* Springer.

Chen, R. and Liu, J. S. (1996) Predictive updating method and Bayesian classification. *J. Roy. Statist. Soc. Ser. B*, **58**, 397–415.

— (2000) Mixture Kalman filter. *J. Roy. Statist. Soc. Ser. B*, **62**, 493–508.

Chib, S. (1998) Estimation and comparison of multiple change point models. *J. Econometrics*, **86**, 221–241.

Chigansky, P. and Lipster, R. (2004) Stability of nonlinear filters in nonmixing case. *Ann. Appl. Probab.*, **14**, 2038–2056.

Chikin, D. O. (1988) Convergence of stochastic approximation procedures in the presence of dependent noise. *Autom. Remote Control*, **1**, 50–61.

Churchill, G. (1992) Hidden Markov chains and the analysis of genome structure. *Computers & Chemistry*, **16**, 107–115.

Collings, I. B. and Rydén, T. (1998) A new maximum likelihood gradient algorithm for on-line hidden Markov model identification. In *Proc. IEEE Int. Conf. Acoust., Speech, Signal Process.*, vol. 4, 2261–2264.

Cover, T. M. and Thomas, J. A. (1991) *Elements of Information Theory.* Wiley.

Cowell, R. G., Dawid, A. P., Lauritzen, S. L. and Spiegelhalter, D. J. (1999) *Probabilistic Networks and Expert Systems.* Springer.

Crisan, D., Del Moral, P. and Lyons, T. (1999) Discrete filtering using branching and interacting particle systems. *Markov Process. Related Fields*, **5**, 293–318.

Crisan, D. and Doucet, A. (2002) A survey of convergence results on particle filtering methods for practitioners. *IEEE Trans. Signal Process.*, **50**, 736–746.

Csiszár, I. (1990) Class notes on information theory and statistics. University of Maryland.

— (2002) Large-scale typicality of Markov sample paths and consistency of MDL order estimators. *IEEE Trans. Inform. Theory*, **48**, 1616–1628.

Csiszár, I. and Shields, P. (2000) The consistency of the BIC Markov order estimator. *Ann. Statist.*, **28**, 1601–1619.

Dacunha-Castelle, D. and Duflo, M. (1986) *Probability and Statistics. Vol. II.* Springer. Translated from the French by D. McHale.

Dacunha-Castelle, D. and Gassiat, E. (1997a) The estimation of the order of a mixture model. *Bernoulli*, **3**, 279–299.

— (1997b) Testing in locally conic models and application to mixture models. *ESAIM Probab. Statist.*, **1**, 285–317.

— (1999) Testing the order of a model using locally conic parametrization: population mixtures and stationary arma processes. *Ann. Statist.*, **27**, 1178–1209.

Damien, P., Wakefield, J. and Walker, S. (1999) Gibbs sampling for Bayesian non-conjugate and hierarchical models by using auxiliary variables. *J. Roy. Statist. Soc. Ser. B*, **61**, 331–344.

Damien, P. and Walker, S. (1996) Sampling probability densities via uniform random variables and a Gibbs sampler. *Tech. Rep.*, Business School, University of Michigan.

Dawid, A. P. (1980) Conditional independence for statistical operations. *Ann. Statist.*, **8**, 598–617.

Del Moral, P. (1996) Nonlinear filtering: interacting particle solution. *Markov Process. Related Fields*, **2**, 555–579.

— (1998) Measure-valued processes and interacting particle systems. Application to nonlinear filtering problems. *Ann. Appl. Probab.*, **8**, 69–95.

— (2004) *Feynman-Kac Formulae. Genealogical and Interacting Particle Systems with Applications.* Springer.

Del Moral, P. and Guionnet, A. (1998) Large deviations for interacting particle systems: applications to non-linear filtering. *Stoch. Proc. App.*, **78**, 69–95.

Del Moral, P. and Jacod, J. (2001) Interacting particle filtering with discrete-time observations: Asymptotic behaviour in the Gaussian case. In *Stochastics in Finite and Infinite Dimensions: In Honor of Gopinath Kallianpur* (eds. T. Hida, R. L. Karandikar, H. Kunita, B. S. Rajput, S. Watanabe and J. Xiong), 101–122. Birkhäuser.

Del Moral, P. and Ledoux, M. (2000) Convergence of empirical processes for interacting particle systems with applications to nonlinear filtering. *J. Theoret. Probab.*, **13**, 225–257.

Del Moral, P., Ledoux, M. and Miclo, L. (2003) On contraction properties of Markov kernels. *Probab. Theory Related Fields*, **126**, 395–420.

Del Moral, P. and Miclo, L. (2001) Genealogies and increasing propagation of chaos for feynman-kac and genetic models. *Ann. Appl. Probab.*, **11**, 1166–1198.

Delyon, B., Lavielle, M. and Moulines, E. (1999) On a stochastic approximation version of the EM algorithm. *Ann. Statist.*, **27**.

Dempster, A. P., Laird, N. M. and Rubin, D. B. (1977) Maximum likelihood from incomplete data via the EM algorithm. *J. Roy. Statist. Soc. Ser. B*, **39**, 1–38 (with discussion).

Devroye, L. (1986) *Non-Uniform Random Variate Generation.* Springer. URL `http://cgm.cs.mcgill.ca/~luc/rnbookindex.html`.

Devroye, L. and Klincsek, T. (1981) Average time behavior of distributive sorting algorithms. *Computing*, **26**, 1–7.

Diaconis, P. and Freedman, D. (1999) Iterated random functions. *SIAM Rev.*, **47**, 45–76.

Diebolt, J. and Ip, E. H. S. (1996) Stochastic EM: method and application. In *Markov Chain Monte Carlo in Practice* (eds. W. R. Gilks, S. Richardson and D. J. Spiegelhalter), 259–273. Chapman.

Dobrushin, R. (1956) Central limit theorem for non-stationary Markov chains. I. *Teor. Veroyatnost. i Primenen.*, **1**, 72–89.

Doob, J. L. (1953) *Stochastic Processes.* Wiley.

Douc, R. and Matias, C. (2002) Asymptotics of the maximum likelihood estimator for general hidden Markov models. *Bernoulli.*

Douc, R., Moulines, E. and Rydén, T. (2004) Asymptotic properties of the maximum likelihood estimator in autoregressive models with Markov regime. *Ann. Statist.*, **32**, 2254–2304.

Doucet, A. and Andrieu, C. (2001) Iterative algorithms for state estimation of jump Markov linear systems. *IEEE Trans. Signal Process.*, **49**, 1216–1227.

Doucet, A., De Freitas, N. and Gordon, N. (eds.) (2001a) *Sequential Monte Carlo Methods in Practice.* Springer.

Doucet, A., Godsill, S. and Andrieu, C. (2000a) On sequential Monte-Carlo sampling methods for Bayesian filtering. *Stat. Comput.*, **10**, 197–208.

Doucet, A., Godsill, S. and Robert, C. P. (2002) Marginal maximum a posteriori estimation using Markov chain Monte Carlo. *Stat. Comput.*, **12**, 77–84.

Doucet, A., Gordon, N. and Krishnamurthy, V. (2001b) Particle filters for state estimation of jump Markov linear systems. *IEEE Trans. Signal Process.*, **49**, 613–624.

Doucet, A., Logothetis, A. and Krishnamurthy, V. (2000b) Stochastic sampling algorithms for state estimation of jump Markov linear systems. *IEEE Trans. Automat. Control*, **45**, 188–202.

Doucet, A. and Robert, C. P. (2002) Marginal maximum a posteriori estimation for hidden Markov models. *Tech. Rep.*, CEREMADE, Université Paris Dauphine.

Doucet, A. and Tadić, V. B. (2003) Parameter estimation in general state-space models using particle methods. *Ann. Inst. Statist. Math.*, **55**, 409–422.

Dudley, R. M. (2002) *Real Analysis and Probability*. Cambridge University Press.

Duflo, M. (1997) *Random Iterative Models*, vol. 34. Springer. Translated from the 1990 French original by S. S. Wilson and revised by the author.

Dupuis, J. A. (1995) Bayesian estimation of movement probabilities in open populations using hidden Markov chains. *Biometrika*, **82**, 761–772.

Dupuis, P. and Ellis, R. S. (1997) *A Weak Convergence Approach to the Theory of Large Deviations*. Wiley.

Dupuis, P. and Simha, R. (1991) On sampling controlled stochastic approximation. *IEEE Trans. Automat. Control*, **36**, 915–924.

Durbin, J. and Koopman, S. J. (2000) Time series analysis of non-Gaussian observations based on state space models from both classical and Bayesian perspectives. *J. Roy. Statist. Soc. Ser. B*, **62**, 3–29.

— (2002) A simple and efficient simulation smoother for state space time series analysis. *Biometrika*, **89**, 603–616.

Durbin, R., Eddy, S., Krogh, A. and Mitchison, G. (1998) *Biological Sequence Analysis: Probabilistic Models of Proteins and Nucleic Acids*. Cambridge University Press.

Durrett, R. (1996) *Probability: Theory and Examples*. Duxbury Press, 2nd ed.

Elliott, E. O. (1963) Estimates of error rates for codes on burst-noise channels. *Bell System Tech. J.*, 1977–1997.

Elliott, R. J. (1993) New finite dimensional filters and smoothers for Markov chains observed in Gaussian noise. *IEEE Trans. Signal Process.*, **39**, 265–271.

Elliott, R. J., Aggoun, L. and Moore, J. B. (1995) *Hidden Markov Models: Estimation and Control*. Springer.

Elliott, R. J. and Krishnamurthy, V. (1999) New finite-dimensional filters for parameter estimation of discrete-time linear Gaussian models. *IEEE Trans. Automat. Control*, **44**.

Engle, R. F. (1982) Autoregressive conditional heteroscedasticity with estimates of the variance of united kingdom inflation. *Econometrica*, **50**, 987–1007.

Ephraim, Y. and Merhav, N. (2002) Hidden Markov processes. *IEEE Trans. Inform. Theory*, **48**, 1518–1569.

Evans, M. and Swartz, T. (1995) Methods for approximating integrals in Statistics with special emphasis on Bayesian integration problems. *Statist. Sci.*, **10**, 254–272.

— (2000) *Approximating Integrals via Monte Carlo and Deterministic Methods*. Oxford University Press.

Fearnhead, P. (1998) *Sequential Monte Carlo methods in filter theory*. Ph.D. thesis, University of Oxford.

Fearnhead, P. and Clifford, P. (2003) On-line inference for hidden Markov models via particle filters. *J. Roy. Statist. Soc. Ser. B*, **65**, 887–899.

Feder, M. and Merhav, N. (2002) Universal composite hypothesis testing: a competitive minimax and its applications. *IEEE Trans. Inform. Theory*, **48**, 1504–1517.

Feller, W. (1943) On a general class of "contagious" distributions. *Ann. Math. Statist.*, **14**, 389–399.

— (1971) *An Introduction to Probability Theory and its Applications*. Wiley.

Fessler, J. A. and Hero, A. O. (1995) Penalized maximum-likelihood image reconstruction using space-alternating generalized em algorithms. *IEEE Trans. Image Process.*, **4**, 1417–29.

Fichou, J., Le Gland, F. and Mevel, L. (2004) Particle based methods for parameter estimation and tracking: Numerical experiments. *Tech. Rep.*, INRIA.

Finesso, L. (1991) *Consistent estimation of the order for Markov and hidden Markov Chains*. Ph.D. thesis, Maryland University.

Finesso, L., Liu, C. and Narayan, P. (1996) The optimal error exponent for Markov order estimation. *IEEE Trans. Inform. Theory*, **42**, 1488–1497.

Fletcher, R. (1987) *Practical Methods of Optimization*. Wiley.

Fong, W., Godsill, S., Doucet, A. and West, M. (2002) Monte carlo smoothing with application to audio signal enhancement. *IEEE Trans. Signal Process.*, **50**, 438–449.

Fonollosa, J. A. R., Anton-Haro, C. and Fonollosa, J. R. (1997) Blind channel estimation and data detection using hidden Markov models. *IEEE Trans. Signal Process.*, **45**, 241–246.

Fort, G. and Moulines, E. (2003) Convergence of the Monte Carlo expectation maximization for curved exponential families. *Ann. Statist.*, **31**, 1220–1259.

Francq, C. and Roussignol, M. (1997) On white noises driven by hidden Markov chains. *J. Time Ser. Anal.*, **18**, 553–578.

— (1998) Ergodicity of autoregressive processes with Markov-switching and consistency of the maximum-likelihood estimator. *Statistics*, **32**, 151–173.

Francq, C., Roussignol, M. and Zakoian, J.-M. (2001) Conditional heteroskedasticity driven by hidden Markov chains. *J. Time Ser. Anal.*, **2**, 197–220.

Fraser, D. and Potter, J. (1969) The optimum linear smoother as a combination of two optimum linear filters. *IEEE Trans. Automat. Control*, **4**, 387–390.

Fredkin, D. R. and Rice, J. A. (1992) Maximum-likelihood-estimation and identification directly from single-channel recordings. *Proc. Roy. Soc. London Ser. B*, **249**, 125–132.

Frey, B. J. (1998) *Graphical Models for Machine Learning and Digital Communication*. MIT Press.

Frühwirth-Schnatter, S. (1994) Data augmentation and dynamic linear models. *J. Time Ser. Anal.*, **15**.

Gaetan, C. and Yao, J.-F. (2003) A multiple-imputation Metropolis version of the EM algorithm. *Biometrika*, **90**, 643–654.

Gassiat, E. (2002) Likelihood ratio inequalities with applications to various mixtures. *Ann. Inst. H. Poincaré Probab. Statist.*, **38**, 887–906.

Gassiat, E. and Boucheron, S. (2003) Optimal error exponents in hidden Markov models order estimation. *IEEE Trans. Inform. Theory*, **49**, 964–980.

Gauvain, J.-L. and Lee, C.-H. (1994) Maximum a posteriori estimation for multivariate Gaussian mixture observations of Markov chains. *IEEE Trans. Speech Audio Process.*, **2**, 291–298.

Gelfand, A. E. and Carlin, B. P. (1993) Maximum-likelihood estimation for constrained or missing-data models. *Can. J. Statist.*, **21**, 303–311.

Gelfand, A. E. and Smith, A. F. M. (1990) Sampling based approaches to calculating marginal densities. *J. Am. Statist. Assoc.*, **85**, 398–409.

Gelman, A. (1995) Methods of moments using monte-carlo simulation. *J. Comput. Graph. Statist.*, **4**, 36–54.

Gelman, A., Carlin, J. B., Stern, H. S. and Rubin, D. B. (1995) *Bayesian Data Analysis*. Chapman.

Geman, S. and Geman, D. (1984) Stochastic relaxation, Gibbs distributions and the Bayesian restoration of images. *IEEE Trans. Pattern Anal. Mach. Intell.*, **6**, 721–741.

Gentle, J. E. (1998) *Random Number Generation and Monte Carlo Methods*. Springer.

Geweke, J. (1989) Bayesian inference in econometric models using Monte-Carlo integration. *Econometrica*, **57**, 1317–1339.

Geyer, C. J. (1996) Estimation and optimization of functions. In *Markov Chain Monte Carlo in Practice* (eds. W. R. Gilks, S. Richardson and D. J. Spiegelhalter). Chapman.

Geyer, C. J. and Møller, J. (1994) Simulation procedures and likelihood inference for spatial point processes. *Scand. J. Statist.*, **21**, 359–373.

Geyer, C. J. and Thompson, E. A. (1992) Constrained Monte Carlo maximum likelihood for dependent data. *J. Roy. Statist. Soc. Ser. B*, **54**, 657–699.

Ghosh, D. (1989) Maximum likelihood estimation of the dynamic shock-error model. *J. Econometrics*, **41**.

Gilbert, E. N. (1960) Capacity of a burst-noise channel. *Bell System Tech. J.*, 1253–1265.

Giudici, P., Rydén, T. and Vandekerkhove, P. (2000) Likelihood-ratio tests for hidden Markov models. *Biometrics*, **56**, 742–747.

Glynn, P. W. and Iglehart, D. (1989) Importance sampling for stochastic simulations. *Management Science*, **35**, 1367–1392.

Godsill, S. J. (2001) On the relationship between MCMC methods for model uncertainty. *J. Comput. Graph. Statist.*, **10**, 230–248.

Godsill, S. J. and Rayner, P. J. W. (1998) *Digital Audio Restoration: A Statistical Model-Based Approach.* Springer.

Gordon, N., Salmond, D. and Smith, A. F. (1993) Novel approach to nonlinear/non-Gaussian Bayesian state estimation. *IEE Proc. F, Radar Signal Process.*, **140**, 107–113.

Graflund, A. and Nilsson, B. (2003) Dynamic portfolio selection: the relevance of switching regimes and investment horizon. *Eur. Financial Management*, **9**, 47–68.

Green, P. J. (1990) On use of the EM algorithm for penalized likelihood estimation. *J. Roy. Statist. Soc. Ser. B*, **52**, 443–452.

— (1995) Reversible jump Markov chain Monte Carlo computation and Bayesian model determination. *Biometrika*, **82**, 711–732.

Gu, M. G. and Kong, F. H. (1998) A stochastic approximation algorithm with Markov chain Monte-Carlo method for incomplete data estimation problems. *Proc. Natl. Acad. Sci. USA*, **95**, 7270–7274.

Gu, M. G. and Li, S. (1998) A stochastic approximation algorithm for maximum-likelihood estimation with incomplete data. *Can. J. Statist.*, **26**, 567–582.

Gu, M. G. and Zhu, H.-T. (2001) Maximum likelihood estimation for spatial models by Markov chain Monte Carlo stochastic approximation. *J. Roy. Statist. Soc. Ser. B*, **63**, 339–355.

Gupta, N. and Mehra, R. (1974) Computational aspects of maximum likelihood estimation and reduction in sensitivity function calculations. *IEEE Trans. Automat. Control*, **19**, 774–783.

Gut, A. (1988) *Stopped Random Walks.* Springer.

Hamilton, J. and Susmel, R. (1994) Autoregressive conditional heteroskedasticity and changes of regime. *J. Econometrics*, **64**, 307–333.

Hamilton, J. D. (1989) A new approach to the economic analysis of nonstationary time series and the business cycle. *Econometrica*, **57**, 357–384.

— (1994) *Time Series Analysis.* Princeton University Press.

Hamilton, J. D. and Raj, B. (eds.) (2003) *Advances in Markov-Switching Models: Applications in Business Cycle Research and Finance (Studies in Empirical Economics).* Springer.

Hammersley, J. M. and Handscomb, D. C. (1965) *Monte Carlo Methods.* Methuen & Co.

Handschin, J. (1970) Monte Carlo techniques for prediction and filtering of non-linear stochastic processes. *Automatica*, **6**, 555–563.

Handschin, J. and Mayne, D. (1969) Monte Carlo techniques to estimate the conditionnal expectation in multi-stage non-linear filtering. In *Int. J. Control*, vol. 9, 547–559.

Hartigan, J. A. (1983) *Bayes Theory.* Springer.

Hastings, W. K. (1970) Monte Carlo sampling methods using Markov chains and their application. *Biometrika*, **57**, 97–109.

Haughton, D. M. (1988) On the choice of a model to fit data from an exponential family. *Ann. Statist.*, **16**, 342–355.

Ho, Y. C. and Lee, R. C. K. (1964) A Bayesian approach to problems in stochastic estimation and control. *IEEE Trans. Automat. Control*, **9**, 333–339.

Hobert, J. P., Jones, G. L., Presnell, B. and Rosenthal, J. S. (2002) On the applicability of regenerative simulation in Markov chain Monte Carlo. *Biometrika*, **89**, 731–743.

Hodgson, M. E. A. (1998) *Reversible jump Markov chain Monte Carlo and inference for ion channel data.* Ph.D. thesis, University of Bristol.

Horn, R. A. and Johnson, C. R. (1985) *Matrix Analysis.* Cambridge University Press.

Hull, J. and White, A. (1987) The pricing of options on assets with stochastic volatilities. *J. Finance*, **42**, 281–300.

Hürzeler, M. and Künsch, H. R. (1998) Monte Carlo approximations for general state-space models. *J. Comput. Graph. Statist.*, **7**, 175–193.

Ibragimov, I. A. and Hasminskii, R. Z. (1981) *Statistical Estimation. Asymptotic Theory.* Springer.

Ito, H., Amari, S. I. and Kobayashi, K. (1992) Identifiability of hidden Markov information sources and their minimum degrees of freedom. *IEEE Trans. Inform. Theory*, **38**, 324–333.

Jacod, J. and Protter, P. (2000) *Probability Essentials.* Springer.

Jacquier, E., Johannes, M. and Polson, N. G. (2004) MCMC maximum likelihood for latent state models. *Tech. Rep.*, Columbia University.

Jacquier, E., Polson, N. G. and Rossi, P. E. (1994) Bayesian analysis of stochastic volatility models (with discussion). *J. Bus. Econom. Statist.*, **12**, 371–417.

Jain, N. and Jamison, B. (1967) Contributions to Doeblin's theory of Markov processes. *Z. Wahrsch. Verw. Geb.*, **8**, 19–40.

Jamshidian, M. and Jennrich, R. J. (1997) Acceleration of the EM algorithm using quasi-Newton methods. *J. Roy. Statist. Soc. Ser. B*, **59**, 569–587.

Jarner, H., larsen, T. S., Krogh, A., Saxild, H. H., Brunak, S. and Knudsen, S. (2001) Sigma A recognition sites in the Bacilius subtilis genome. *Microbiology*, **147**, 2417–2424.

Jarner, S. and Hansen, E. (2000) Geometric ergodicity of Metropolis algorithms. *Stoch. Proc. App.*, **85**, 341–361.

Jelinek, F. (1997) *Statistical Methods for Speech Recognition.* MIT Press.

Jensen, F. V. (1996) *An Introduction to Bayesian Networks.* UCL Press.

Jensen, J. L. and Petersen, N. V. (1999) Asymptotic normality of the maximum likelihood estimator in state space models. *Ann. Statist.*, **27**, 514–535.

De Jong, P. (1988) A cross validation filter for time series models. *Biometrika*, **75**, 594–600.

De Jong, P. and Shephard, N. (1995) The simulation smoother for time series models. *Biometrika*, **82**, 339–350.

Jordan, M. I. (ed.) (1999) *Learning in Graphical Models.* MIT Press.

Jordan, M. I. (2004) Graphical models. *Statist. Sci.*, **19**, 140–155.

Julier, S. J. and Uhlmann, J. K. (1997) A new extension of the Kalman filter to nonlinear systems. In *AeroSense: The 11th International Symposium on Aerospace/Defense Sensing, Simulation and Controls.*

Kaijser, T. (1975) A limit theorem for partially observed Markov chains. *Ann. Probab.*, **3**, 677–696.

Kailath, T. and Frost, P. A. (1968) An innovations approach to least-squares estimation–Part II: Linear smoothing in additive white noise. *IEEE Trans. Automat. Control*, **13**, 655–660.

Kailath, T., Sayed, A. and Hassibi, B. (2000) *Linear Estimation.* Prentice-Hall.

Kaleh, G. K. and Vallet, R. (1994) Joint parameter estimation and symbol detection for linear or nonlinear unknown channels. *IEEE Trans. Commun.*, **42**, 2406–2413.

Kalman, R. E. and Bucy, R. (1961) New results in linear filtering and prediction theory. *J. Basic Eng., Trans. ASME, Series D*, **83**, 95–108.

Kéribin, C. and Gassiat, E. (2000) The likelihood ratio test for the number of components in a mixture with Markov regime. *ESAIM Probab. Statist.*, **4**, 25–52.

Kesten, H. (1972) Limit theorems for stochastic growth models. I, II. *Adv. Appl. Probab.*, **4**, 193–232.

Kieffer, J. C. (1993) Strongly consistent code-based identification and order estimation for constrained finite-state model classes. *IEEE Trans. Inform. Theory*, **39**, 893–902.

Kim, C. and Nelson, C. (1999) *State-Space Models with Regime Switching: Classical and Gibbs-Sampling Approaches with Applications.* MIT Press.

Kim, S., Shephard, N. and Chib, S. (1998) Stochastic volatility: Likelihood inference and comparison with ARCH models. *Rev. Econom. Stud.*, **65**, 361–394.

Kitagawa, G. (1987) Non-Gaussian state space modeling of nonstationary time series. *J. Am. Statist. Assoc.*, **82**, 1023–1063.

— (1996) Monte-Carlo filter and smoother for non-Gaussian nonlinear state space models. *J. Comput. Graph. Statist.*, **1**, 1–25.

Kohn, R. and Ansley, C. F. (1989) A fast algorithm for signal extraction, influence and cross-validation in state space models. *Biometrika*, **76**, 65–79.

Kong, A., Liu, J. S. and Wong, W. (1994) Sequential imputation and Bayesian missing data problems. *J. Am. Statist. Assoc.*, **89**.

Koopman, S. J. (1993) Disturbance smoother for state space models. *Biometrika*, **80**, 117–126.

Kormylo, J. and Mendel, J. M. (1982) Maximum-likelihood detection and estimation of Bernoulli-Gaussian processes. *IEEE Trans. Inform. Theory*, **28**, 482488.

Koski, T. (2001) *Hidden Markov Models for Bioinformatics.* Kluwer.

Krishnamurthy, V. and Rydén, T. (1998) Consistent estimation of linear and non-linear autoregressive models with Markov regime. *J. Time Ser. Anal.*, **19**, 291–307.

Krishnamurthy, V. and White, L. B. (1992) Blind equalization of FIR channels with Markov inputs. In *Proc. IFAC Int. Conf. Adapt. Systems Control Signal Process.*

Krishnamurthy, V. and Yin, G. G. (2002) Recursive algorithms for estimation of hidden Markov models and autoregressive models with Markov regime. *IEEE Trans. Inform. Theory*, **48**, 458–476.

Krogh, A., Mian, I. S. and Haussler, D. (1994) A hidden Markov model that finds genes in E. coli DNA. *Nucleic Acids Res.*, **22**, 4768–4778.

Krolzig, H.-M. (1997) *Markov-switching Vector Autoregressions. Modelling, Statistical Inference, and Application to Business Cycle Analysis.* Springer.

Kuhn, E. and Lavielle, M. (2004) Coupling a stochastic approximation version of EM with an MCMC procedure. *ESAIM Probab. Statist.*, **8**, 115–131.

Kukashin, A. V. and Borodovsky, M. (1998) GeneMark.hmm: new solutions for gene finding. *Nucleic Acids Res.*, **26**, 1107–1115.

Künsch, H. R. (2000) State space and hidden Markov models. In *Complex Stochastic Systems* (eds. O. E. Barndorff-Nielsen, D. R. Cox and C. Klupپelberg). CRC Press.

— (2003) Recursive Monte-Carlo filters: algorithms and theoretical analysis. Preprint ETHZ, seminar für statistics.

Kushner, H. J. and Clark, D. S. (1978) *Stochastic Approximation Methods for Constrained and Unconstrained Systems.* Springer.

Kushner, H. J. and Yin, G. G. (2003) *Stochastic Approximation and Recursive Algorithms and Applications.* Springer, 2nd ed.

Laarhoven, P. J. V. and Arts, E. H. L. (1987) *Simulated Annealing: Theory and Applications.* Reidel Publisher.

Lange, K. (1995) A gradient algorithm locally equivalent to the EM algorithm. *J. Roy. Statist. Soc. Ser. B*, **57**, 425–437.

Lauritzen, S. L. (1996) *Graphical Models.* Oxford University Press.

Lavielle, M. (1993) Bayesian deconvolution of Bernoulli-Gaussian processes. *Signal Process.*, **33**, 67–79.

Lavielle, M. and Lebarbier, E. (2001) An application of MCMC methods to the multiple change-points problem. *Signal Process.*, **81**, 39–53.

Le Gland, F. and Mevel, L. (1997) Recursive estimation in HMMs. In *Proc. IEEE Conf. Decis. Control*, 3468–3473.

— (2000) Exponential forgetting and geometric ergodicity in hidden Markov models. *Math. Control Signals Systems*, **13**, 63–93.

Le Gland, F. and Oudjane, N. (2004) Stability and uniform approximation of nonlinear filters using the hilbert metric and application to particle filters. *Ann. Appl. Probab.*, **14**, 144–187.

Lehmann, E. L. and Casella, G. (1998) *Theory of Point Estimation.* Springer, 2nd ed.

Leroux, B. G. (1992) Maximum-likelihood estimation for hidden Markov models. *Stoch. Proc. Appl.*, **40**, 127–143.

Levine, R. A. and Casella, G. (2001) Implementations of the Monte Carlo EM algorithm. *J. Comput. Graph. Statist.*, **10**, 422–439.

Levine, R. A. and Fan, J. (2004) An automated (Markov chain) Monte Carlo EM algorithm. *J. Stat. Comput. Simul.*, **74**, 349–359.

Levinson, S. E., Rabiner, L. R. and Sondhi, M. M. (1983) An introduction to the application of the theory of probabilistic functions of a Markov process to automatic speech recognition. *Bell System Tech. J.*, **62**, 1035–1074.

Liporace, L. A. (1982) Maximum likelihood estimation of multivariate observations of Markov sources. *IEEE Trans. Inform. Theory*, **28**, 729–734.

Lipster, R. S. and Shiryaev, A. N. (2001) *Statistics of Random Processes: I. General theory.* Springer, 2nd ed.

Liu, C. and Narayan, P. (1994) Order estimation and sequential universal data compression of a hidden Markov source by the method of mixtures. *IEEE Trans. Inform. Theory*, **40**, 1167–1180.

Liu, J. and Chen, R. (1995) Blind deconvolution via sequential imputations. *J. Am. Statist. Assoc.*, **430**, 567–576.

— (1998) Sequential Monte-Carlo methods for dynamic systems. *J. Am. Statist. Assoc.*, **93**, 1032–1044.

Liu, J., Chen, R. and Logvinenko, T. (2001) A theoretical framework for sequential importance sampling and resampling. In *Sequential Monte Carlo Methods in Practice* (eds. A. Doucet, N. De Freitas and N. Gordon). Springer.

Liu, J., Wong, W. and Kong, A. (1994) Covariance structure of the Gibbs sampler with applications to the comparisons of estimators and augmentation schemes. *Biometrika*, **81**, 27–40.

Liu, J. S. (1994) The collapsed Gibbs sampler with applications to a gene regulation problem. *J. Am. Statist. Assoc.*, **89**, 958–966.

— (1996) Metropolized independent sampling with comparisons to rejection sampling and importance sampling. *Stat. Comput.*, **6**, 113–119.

— (2001) *Monte Carlo Strategies in Scientific Computing.* Springer.

Louis, T. A. (1982) Finding the observed information matrix when using the EM algorithm. *J. Roy. Statist. Soc. Ser. B*, **44**, 226–233.

Luenberger, D. G. (1984) *Linear and Nonlinear Programming.* Addison-Wesley, 2nd ed.

MacDonald, I. and Zucchini, W. (1997) *Hidden Markov and Other Models for Discrete-Valued Time Series.* Chapman.

MacEachern, S. N., Clyde, M. and Liu, J. (1999) Sequential importance sampling for nonparametric bayes models: The next generation. *Can. J. Statist.*, **27**, 251–267.

Mayne, D. Q. (1966) A solution of the smoothing problem for linear dynamic systems. *Automatica*, **4**, 73–92.

Meng, X.-L. (1994) On the rate of convergence of the ECM algorithm. *Ann. Statist.*, **22**, 326–339.

Meng, X.-L. and Rubin, D. B. (1991) Using EM to obtain asymptotic variance-covariance matrices: The SEM algorithm. *J. Am. Statist. Assoc.*, **86**, 899–909.

— (1993) Maximum likelihood estimation via the ECM algorithm: A general framework. *Biometrika*, **80**, 267–278.

Meng, X.-L. and Van Dyk, D. (1997) The EM algorithm–an old folk song sung to a fast new tune. *J. Roy. Statist. Soc. Ser. B*, **59**, 511–567.

Mengersen, K. and Tweedie, R. L. (1996) Rates of convergence of the Hastings and Metropolis algorithms. *Ann. Statist.*, **24**, 101–121.

Metropolis, N., Rosenbluth, A. W., Rosenbluth, M. N., Teller, A. H. and Teller, E. (1953) Equations of state calculations by fast computing machines. *J. Chem. Phys.*, **21**, 1087–1092.

Meyn, S. P. and Tweedie, R. L. (1993) *Markov Chains and Stochastic Stability.* Springer.

Neal, R. M. (1997) Markov chain Monte Carlo methods based on 'slicing' the density function. *Tech. Rep.*, University of Toronto.

— (2003) Slice sampling (with discussion). *Ann. Statist.*, **31**, 705–767.

Neveu, J. (1975) *Discrete-Time Martingales.* North-Holland.

Niederreiter, H. (1992) *Random Number Generation and Quasi-Monte Carlo Methods.* SIAM.

Nielsen, S. F. (2000) The stochastic EM algorithm: estimation and asymptotic results. *Bernoulli*, **6**, 457–489.

Nummelin, E. (1978) A splitting technique for Harris recurrent Markov chains. *Z. Wahrscheinlichkeitstheorie und Verw. Gebiete*, **4**, 309–318.

— (1984) *General Irreducible Markov Chains and Non-Negative Operators.* Cambridge University Press.

Orchard, T. and Woodbury, M. A. (1972) A missing information principle: Theory and applications. In *Proceedings of the 6th Berkeley Symposium on Mathematical Statistics*, vol. 1, 697–715.

Ó Ruanaidh, J. J. K. and Fitzgerald, W. J. (1996) *Numerical Bayesian Methods Applied to Signal Processing.* Springer.

Ostrowski, A. M. (1966) *Solution of Equations and Systems of Equations.* Academic Press, 2nd ed.

Pearl, J. (1988) *Probabilistic Reasoning in Intelligent Systems: Networks of Plausible Inference.* Morgan Kaufmann.

Peskun, P. H. (1973) Optimum Monte Carlo sampling using Markov chains. *Biometrika*, **60**, 607–612.

— (1981) Guidelines for chosing the transition matrix in Monte Carlo methods using Markov chains. *J. Comput. Phys.*, **40**, 327–344.

Petrie, T. (1969) Probabilistic functions of finite state Markov chains. *Ann. Math. Statist.*, **40**, 97–115.

Petris, G. and Tardella, L. (2003) A geometric approach to transdimensional Markov chain Monte Carlo. *Can. J. Statist.*, **31**, 469–482.

Petrov, V. V. (1995) *Limit Theorems of Probability Theory.* Oxford University Press.

Pierre-Loti-Viaud, D. (1995) Random perturbations of recursive sequences with an application to an epidemic model. *J. Appl. Probab.*, **32**, 559–578.

Pitt, M. K. and Shephard, N. (1999) Filtering via simulation: Auxiliary particle filters. *J. Am. Statist. Assoc.*, **94**, 590–599.

Polson, N. G., Carlin, B. P. and Stoffer, D. S. (1992) A Monte Carlo approach to nonnormal and nonlinear state-space modeling. *J. Am. Statist. Assoc.*, **87**, 493–500.

Polson, N. G., Stroud, J. R. and Müller, P. (2002) Practical filtering with sequential parameter learning. *Tech. Rep.*, University of Chicago.

Polyak, B. T. (1990) A new method of stochastic approximation type. *Autom. Remote Control*, **51**, 98–107.

Polyak, B. T. and Juditsky, A. B. (1992) Acceleration of stochastic approximation by averaging. *SIAM J. Control Optim.*, **30**, 838–855.

Poznyak, A. S. and Chikin, D. O. (1984) Asymptotic properties of procedures of stochastic approximation with dependent noise. *Autom. Remote Control*, **1**, 78–93.

Press, W., Teukolsky, S., Vetterling, W. and Flannery, B. (1992) *Numerical Recipes in C: The Art of Scientific Computing.* Cambridge University Press, 2nd ed. URL `http://www.numerical-recipes.com/`.

Proakis, J. G. (1995) *Digital Communications.* McGraw-Hill.

Punskaya, E., Doucet, A. and Fitzgerald, W. (2002) On the use and misuse of particle filtering in digital communications. In *Proc. Eur. Signal Process. Conf.*, vol. 2, 173–176.

Quintana, F. A., Liu, J. and del Pino, G. (1999) Monte-Carlo EM with importance reweighting and its applications in random effects models. *Comput. Statist. Data Anal.*, **29**, 429–444.

Rabiner, L. R. (1989) A tutorial on hidden Markov models and selected applications in speech recognition. *Proc. IEEE*, **77**, 257–285.

Rabiner, L. R. and Juang, B.-H. (1993) *Fundamentals of Speech Recognition.* Prentice-Hall.

Raj, B. (2002) Asymmetry of business cycles: the Markov-switching approach. In *Handbook of Applied Econometrics and Statistical Inference* (eds. A. Ullah, A. T. K. Wan and A. Chaturvedi), 687–710. Dekker.

Rauch, H., Tung, F. and Striebel, C. (1965) Maximum likelihood estimates of linear dynamic systems. *AIAA Journal*, **3**, 1445–1450.

Richardson, S. and Green, P. J. (1997) On Bayesian analysis of mixtures with an unknown number of components (with discussion). *J. Roy. Statist. Soc. Ser. B*, **59**, 731–792.

Ripley, B. (1987) *Stochastic Simulation.* Wiley.

Ristic, B., Arulampalam, M. and Gordon, A. (2004) *Beyond Kalman Filters: Particle Filters for Target Tracking.* Artech House.

Robbins, H. and Monro, S. (1951) A stochastic approximation method. *Ann. Math. Statist.*, **22**, 400–407.

Robert, C. P. (2001) *The Bayesian Choice.* Springer, 2nd ed.

Robert, C. P. and Casella, G. (2004) *Monte Carlo Statistical Methods.* Springer, 2nd ed.

Robert, C. P., Celeux, G. and Diebolt, J. (1993) Bayesian estimation of hidden Markov chains: A stochastic implementation. *Statist. Probab. Lett.*, **16**, 77–83.

Robert, C. P., Rydén, T. and Titterington, M. (1999) Convergence controls for MCMC algorithms, with applications to hidden Markov chains. *J. Comput. Graph. Statist.*, **64**, 327–355.

— (2000) Bayesian inference in hidden Markov models through reversible jump Markov chain Monte Carlo. *J. Roy. Statist. Soc. Ser. B*, **62**, 57–75.

Robert, C. P. and Titterington, M. (1998) Reparameterisation strategies for hidden Markov models and Bayesian approaches to maximum likelihood estimation. *Stat. Comput.*, **8**, 145–158.

Roberts, G. O. and Rosenthal, J. S. (1998) Markov chain Monte Carlo: Some practical implications of theoretical results. *Canad. J. Statist.*, **26**, 5–32.

— (2001) Optimal scaling for various Metropolis-Hastings algorithms. *Statist. Sci.*, **16**, 351–367.

— (2004) General state space Markov chains and MCMC algorithms. *Probab. Surv.*, **1**, 20–71.

Roberts, G. O. and Tweedie, R. L. (1996) Geometric convergence and central limit theorems for multidimensional Hastings and Metropolis algorithms. *Biometrika*, **83**, 95–110.

— (2005) Understanding MCMC. In preparation.

Rosenthal, J. S. (1995) Minorization conditions and convergence rates for Markov chain Monte Carlo. *J. Am. Statist. Assoc.*, **90**, 558–566.

— (2001) A review of asymptotic convergence for general state space Markov chains. *Far East J. Theor. Stat.*, **5**, 37–50.

Rubin, D. B. (1987) A noniterative sampling/importance resampling alternative to the data augmentation algorithm for creating a few imputations when the fraction of missing information is modest: the SIR algorithm (discussion of Tanner and Wong). *J. Am. Statist. Assoc.*, **82**, 543–546.

— (1988) Using the SIR algorithm to simulate posterior distribution. In *Bayesian Statistics 3* (eds. J. M. Bernardo, M. H. DeGroot, D. Lindley and A. Smith), 395–402. Clarendon Press.

Sakalauskas, L. (2000) Nonlinear stochastic optimization by the Monte-Carlo method. *Informatica (Vilnius)*, **11**, 455–468.

— (2002) Nonlinear stochastic programming by Monte-Carlo estimators. *European J. Oper. Res.*, **137**, 558–573.

Sandmann, G. and Koopman, S. J. (1998) Estimation of stochastic volatility models via Monte Carlo maximum likelihood. *J. Econometrics*, **87**, 271–301.

Schervish, M. J. (1995) *Theory of Statistics.* Springer.

Schick, I. C. and Mitter, S. K. (1994) Robust recursive estimation in the presence of heavy-tailed observation noise. *Ann. Statist.*, **22**, 1045–1080.

Scott, D. J. and Tweedie, R. L. (1996) Explicit rates of convergence of stochastically ordered Markov chains. In *Athens Conference on Applied Probability and Time Series: Applied Probability in Honor of J. M. Gani*, vol. 114 of *Lecture Notes in Statistics.* Springer.

Scott, S. L. (2002) Bayesian methods for hidden Markov models: recursive computing in the 21st century. *J. Am. Statist. Assoc.*, **97**, 337–351.

Seber, G. A. F. (1983) Capture-recapture methods. In *Encyclopedia of Statistical Science* (eds. S. Kotz and N. Johnson). Wiley.

Segal, M. and Weinstein, E. (1989) A new method for evaluating the log-likelihood gradient, the Hessian, and the Fisher information matrix for linear dynamic systems. *IEEE Trans. Inform. Theory*, **35**, 682–687.

Serfling, R. J. (1980) *Approximation Theorems of Mathematical Statistics.* Wiley.

Shephard, N. and Pitt, M. (1997) Likelihood analysis of non-Gaussian measurement time series. *Biometrika*, **84**, 653–667. Erratum in 91:249–250, 2004.

Shiryaev, A. N. (1966) On stochastic equations in the theory of conditional Markov process. *Theory Probab. Appl.*, **11**, 179–184.

— (1996) *Probability.* Springer, 2nd ed.

Shtarkov, Y. M. (1987) Universal sequential coding of messages. *Probl. Inform. Transmission*, **23**, 3–17.

Shumway, R. and Stoffer, D. (1991) Dynamic linear models with switching. *J. Am. Statist. Assoc.*, **86**, 763–769.

Stephens, M. (2000a) Bayesian analysis of mixture models with an unknown number of components - an alternative to reversible jump methods. *Ann. Statist.*, **28**, 40–74.

— (2000b) Dealing with label switching in mixture models. *J. Roy. Statist. Soc. Ser. B*, **62**, 795–809.

Stratonovich, R. L. (1960) Conditional Markov processes. *Theory Probab. Appl.*, **5**, 156–178.

Tanizaki, H. (1996) *Nonlinear Filters: Estimation and Applications.* Springer.

— (2003) Nonlinear and non-Gaussian state-space modeling with Monte-Carlo techniques: a survey and comparative study. In *Handbook of Statistics 21. Stochastic processes: Modelling and Simulation* (eds. D. N. Shanbhag and C. R. Rao), 871–929. Elsevier.

Tanizaki, H. and Mariano, R. (1998) Nonlinear and non-Gaussian state-space modeling with Monte-Carlo simulations. *J. Econometrics*, **83**, 263–290.

Tanner, M. and Wong, W. (1987) The calculation of posterior distributions by data augmentation. *J. Am. Statist. Assoc.*, **82**, 528–550.

Tanner, M. A. (1993) *Tools for Statistical Inference.* Springer, 2nd ed.

Teicher, H. (1960) On the mixture of distributions. *Ann. Math. Statist.*, **31**, 55–73.

— (1961) Identifiability of mixtures. *Ann. Math. Statist.*, **32**, 244–248.

— (1963) Identifiability of finite mixtures. *Ann. Math. Statist.*, **34**, 1265–1269.

— (1967) Identifiability of mixtures of product measures. *Ann. Math. Statist.*, **38**, 1300–1302.

Titterington, D. M., Smith, A. F. M. and Makov, U. E. (1985) *Statistical Analysis of Finite Mixture Distributions.* Wiley.

Tugnait, J. (1984) Adaptive estimation and identification for discrete systems with Markov jump parameters. *IEEE Trans. Automat. Control*, **27**, 1054–1065.

Van der Merwe, R., Doucet, A., De Freitas, N. and Wan, E. (2000) The unscented particle filter. In *Adv. Neural Inf. Process. Syst.* (eds. T. K. Leen, T. G. Dietterich and V. Tresp), vol. 13. MIT Press.

Van Overschee, P. and De Moor, B. (1993) Subspace algorithms for the stochastic identification problem. *Automatica*, **29**, 649–660.

— (1996) *Subspace Identification for Linear Systems. Theory, Implementation, Applications.* Kluwer.

Viterbi, A. J. (1967) Error bounds for convolutional codes and an asymptotically optimal decoding algorithm. *IEEE Trans. Inform. Theory*, **13**, 260–269.

Wald, A. (1949) Note on the consistency of the maximum likelihood estimate. *Ann. Math. Statist.*, **20**, 595–601.

Wei, G. C. G. and Tanner, M. A. (1991) A Monte-Carlo implementation of the EM algorithm and the poor man's Data Augmentation algorithms. *J. Am. Statist. Assoc.*, **85**, 699–704.

Weinstein, E., Oppenheim, A. V., Feder, M. and Buck, J. R. (1994) Iterative and sequential algorithms for multisensor signal enhancement. *IEEE Trans. Acoust., Speech, Signal Process.*, **42**, 846–859.

Welch, L. R. (2003) Hidden Markov models and the Baum-Welch algorithm. *IEEE Inf. Theory Soc. Newslett.*, **53**.

West, M. and Harrison, J. (1989) *Bayesian Forecasting and Dynamic Models.* Springer.

Whitley, D. (1994) A genetic algorithm tutorial. *Stat. Comput.*, **4**, 65–85.

Williams, D. (1991) *Probability with Martingales.* Cambridge University Press.

Wonham, W. M. (1965) Some applications of stochastic differential equations to optimal nonlinear filtering. *SIAM J. Control*, **2**, 347–369.

Wu, C. F. J. (1983) On the convergence properties of the EM algorithm. *Ann. Statist.*, **11**, 95–103.

Younes, L. (1988) Estimation and annealing for Gibbsian fields. *Ann. Inst. H. Poincaré Probab. Statist.*, **24**, 269–294.

— (1989) Parametric inference for imperfectly observed Gibbsian fields. *Probab. Theory Related Fields*, **82**, 625–645.

Young, S. (1996) A review of large-vocabulary continuous-speech recognition. *IEEE Signal Process. Mag.*, **13**.

Zangwill, W. I. (1969) *Nonlinear Programming: A Unified Approach.* Prentice-Hall.

Zaritskii, V., Svetnik, V. and Shimelevich, L. (1975) Monte-Carlo techniques in problems of optimal data processing. *Autom. Remote Control*, **12**, 2015–2022.

Zeitouni, O. and Dembo, A. (1988) Exact filters for the estimation of the number of transitions of finite-state continuous-time Markov processes. *IEEE Trans. Inform. Theory*, **34**.

Zeitouni, O. and Gutman, M. (1991) On universal hypothesis testing via large deviations. *IEEE Trans. Inform. Theory*, **37**, 285–290.

Zeitouni, O., Ziv, J. and Merhav, N. (1992) When is generalized likelihood ratio test optimal? *IEEE Trans. Inform. Theory*, **38**, 1597–1602.

Ziv, J. and Merhav, N. (1992) Estimating the number of states of a finite-state source. *IEEE Trans. Inform. Theory*, **38**, 61–65.

Index

Springer Series in Statistics

(continued from p. ii)

Huet/Bouvier/Poursat/Jolivet: Statistical Tools for Nonlinear Regression: A Practical Guide with S-PLUS and R Examples, 2nd edition.
Ibrahim/Chen/Sinha: Bayesian Survival Analysis.
Jolliffe: Principal Component Analysis, 2nd edition.
Knottnerus: Sample Survey Theory: Some Pythagorean Perspectives.
Kolen/Brennan: Test Equating: Methods and Practices.
Kotz/Johnson (Eds.): Breakthroughs in Statistics Volume I.
Kotz/Johnson (Eds.): Breakthroughs in Statistics Volume II.
Kotz/Johnson (Eds.): Breakthroughs in Statistics Volume III.
Küchler/Sørensen: Exponential Families of Stochastic Processes.
Kutoyants: Statistical Influence for Ergodic Diffusion Processes.
Lahiri: Resampling Methods for Dependent Data.
Le Cam: Asymptotic Methods in Statistical Decision Theory.
Le Cam/Yang: Asymptotics in Statistics: Some Basic Concepts, 2nd edition.
Liu: Monte Carlo Strategies in Scientific Computing.
Longford: Models for Uncertainty in Educational Testing.
Manski: Partial Identification of Probability Distributions.
Mielke/Berry: Permutation Methods: A Distance Function Approach.
Molenberghs/Verbeke: Models for Discrete Longitudinal Data.
Pan/Fang: Growth Curve Models and Statistical Diagnostics.
Parzen/Tanabe/Kitagawa: Selected Papers of Hirotugu Akaike.
Politis/Romano/Wolf: Subsampling.
Ramsay/Silverman: Applied Functional Data Analysis: Methods and Case Studies.
Ramsay/Silverman: Functional Data Analysis, 2nd edition.
Rao/Toutenburg: Linear Models: Least Squares and Alternatives.
Reinsel: Elements of Multivariate Time Series Analysis, 2nd edition.
Rosenbaum: Observational Studies, 2nd edition.
Rosenblatt: Gaussian and Non-Gaussian Linear Time Series and Random Fields.
Särndal/Swensson/Wretman: Model Assisted Survey Sampling.
Santner/Williams/Notz: The Design and Analysis of Computer Experiments.
Schervish: Theory of Statistics.
Shao/Tu: The Jackknife and Bootstrap.
Simonoff: Smoothing Methods in Statistics.
Singpurwalla and Wilson: Statistical Methods in Software Engineering: Reliability and Risk.
Small: The Statistical Theory of Shape.
Sprott: Statistical Inference in Science.
Stein: Interpolation of Spatial Data: Some Theory for Kriging.
Taniguchi/Kakizawa: Asymptotic Theory of Statistical Inference for Time Series.
Tanner: Tools for Statistical Inference: Methods for the Exploration of Posterior Distributions and Likelihood Functions, 3rd edition.
van der Laan: Unified Methods for Censored Longitudinal Data and Causality.
van der Vaart/Wellner: Weak Convergence and Empirical Processes: With Applications to Statistics.
Verbeke/Molenberghs: Linear Mixed Models for Longitudinal Data.
Weerahandi: Exact Statistical Methods for Data Analysis.
West/Harrison: Bayesian Forecasting and Dynamic Models, 2nd edition.